Microbes in Land Use Change Management

AF765249

Microbes in Land Use Change Management

Edited by

Jay Shankar Singh

Department of Environmental Microbiology, School for Environmental Sciences, Babasaheb Bhimrao Ambedkar University, Lucknow, India

Shashank Tiwari

Department of Environmental Microbiology, School for Environmental Sciences, Babasaheb Bhimrao Ambedkar University, Lucknow, India

Chhatarpal Singh

Agro Environmental Development Society (AEDS), Rampur, India

Anil Kumar Singh

Department of Horticulture, Institute of Agricultural Sciences, Banaras Hindu University, Varanasi, India

ELSEVIER

Elsevier
Radarweg 29, PO Box 211, 1000 AE Amsterdam, Netherlands
The Boulevard, Langford Lane, Kidlington, Oxford OX5 1GB, United Kingdom
50 Hampshire Street, 5th Floor, Cambridge, MA 02139, United States

Copyright © 2021 Elsevier Inc. All rights reserved.

No part of this publication may be reproduced or transmitted in any form or by any means, electronic or mechanical, including photocopying, recording, or any information storage and retrieval system, without permission in writing from the publisher. Details on how to seek permission, further information about the Publisher's permissions policies and our arrangements with organizations such as the Copyright Clearance Center and the Copyright Licensing Agency, can be found at our website: www.elsevier.com/permissions.

This book and the individual contributions contained in it are protected under copyright by the Publisher (other than as may be noted herein).

Notices
Knowledge and best practice in this field are constantly changing. As new research and experience broaden our understanding, changes in research methods, professional practices, or medical treatment may become necessary.

Practitioners and researchers must always rely on their own experience and knowledge in evaluating and using any information, methods, compounds, or experiments described herein. In using such information or methods they should be mindful of their own safety and the safety of others, including parties for whom they have a professional responsibility.

To the fullest extent of the law, neither the Publisher nor the authors, contributors, or editors, assume any liability for any injury and/or damage to persons or property as a matter of products liability, negligence or otherwise, or from any use or operation of any methods, products, instructions, or ideas contained in the material herein.

British Library Cataloguing-in-Publication Data
A catalogue record for this book is available from the British Library

Library of Congress Cataloging-in-Publication Data
A catalog record for this book is available from the Library of Congress

ISBN: 978-0-12-824448-7

For Information on all Elsevier publications
visit our website at https://www.elsevier.com/books-and-journals

Publisher: Stacy Masucci
Acquisitions Editor: Linda Versteeg-Buschman
Editorial Project Manager: Aleksandra Packowska
Production Project Manager: Debasish Ghosh
Cover Designer: Victoria Pearson

Typeset by MPS Limited, Chennai, India

Working together to grow libraries in developing countries
www.elsevier.com • www.bookaid.org

Contents

List of contributors

Charith Raj Adkar-Purushothama Pavilion for Applied Cancer Research, Department of Biochemistry, Faculty of Medicine and Health Sciences, Sherbrooke University, Sherbrooke, QC, Canada

Zahoor Ahmad Department of Botany, University of Central Punjab, Bahawalpur, Pakistan

VT Anju Department of Biochemistry and Molecular Biology, School of Life Sciences, Pondicherry University, Pondicherry, India

Muhammad Ashar Ayub Institute of Soil and Environmental Sciences, Faculty of Agriculture, University of Agriculture Faisalabad, Pakistan

Shashi Lata Bharati Department of Chemistry, North Eastern Regional Institute of Science and Technology, Nirjuli, India

Arpita Bhattacharya Division of Microbial Technology, CSIR-National Botanical Research Institute, Lucknow, India; Academy of Scientific and Innovative Research (AcSIR), Ghaziabad, India

Avantika Chandra Laboratory of Applied Microbiology, Department of Environmental Science and Engineering, Indian Institute of Technology (Indian School of Mines), Dhanbad, India

Ram Chandra Department of Microbiology, School of Environmental Sciences, Babasaheb Bhimrao Ambedkar University (A Central University), Lucknow, India

Jairam Chaudhary Indian Institute of Farming Systems Research, Meerut, India

Priyanka Chauhan Division of Microbial Technology, CSIR-National Botanical Research Institute, Lucknow, India; Academy of Scientific and Innovative Research (AcSIR), Ghaziabad, India

Pankaj Kumar Chaurasia PG Department of Chemistry, L.S. College
(A Constituent Unit of B.R.A. Bihar University) Muzaffarpur, India

G. Chethan Kumar Indian Institute of Farming Systems Research, Meerut, India

N. Deepa Department of Studies in Microbiology, University of Mysore, Mysore, India

Madhu Dyavaiah Department of Biochemistry and Molecular Biology, School of Life Sciences, Pondicherry University, Pondicherry, India

Ahmed Abd El Fattah El Baroudy Soil and Water Department, Faculty of Agriculture, Tanta University, Tanta, Al Gharbiyah governorate, Egypt

Rajeeva Gaur Centre of Excellence, Department of Microbiology, Dr. Rammanohar Lohia Avadh University, Ayodhya, India

Ved Prakash Giri Division of Microbial Technology, CSIR-National Botanical Research Institute, Lucknow, India; Department of Botany, Lucknow University, Lucknow, India

Kuldeep Gupta Department of Molecular Biology and Biotechnology, Tezpur University, Tezpur, India

Pratishtha Gupta Laboratory of Applied Microbiology, Department of Environmental Science and Engineering, Indian Institute of Technology (Indian School of Mines), Dhanbad, India

Anwar ul Haq Institute of Soil and Environmental Sciences, Faculty of Agriculture, University of Agriculture Faisalabad, Pakistan

Ahmed Emdadul Hoque Ecology and Environment Laboratory, Department of Botany, University of Dhaka, Dhaka, Bangladesh

Tahmina Jamali Ecology and Environment Laboratory, Department of Botany, University of Dhaka, Dhaka, Bangladesh

Monu Jariyal Regional Centre of Organic Farming Panchkula, Ministry of Agriculture & Farmers Welfare, Government of India, India

Ram Jatan CSIR-National Botanical Research Institute, Lucknow, India; Academy of Scientific and Innovative Research (AcSIR), Ghaziabad, India

Hinnan Khalid Agriculture Department, Government of Punjab, Pakistan

Irina Kravchenko Research Center of Biotechnology, Russian Academy of Sciences, Moscow, Russia

Amit Kumar Indian Institute of Farming Systems Research, Meerut, India

Vipin Kumar Laboratory of Applied Microbiology, Department of Environmental Science and Engineering, Indian Institute of Technology (Indian School of Mines), Dhanbad, India

Charu Lata Academy of Scientific and Innovative Research (AcSIR), Ghaziabad, India; CSIR-National Institute of Science Communication and Policy Research, New Delhi, India

Manabendra Mandal Department of Molecular Biology and Biotechnology, Tezpur University, Tezpur, India

Amrit Lal Meena Indian Institute of Farming Systems Research, Meerut, India

Lalit Krishan Meena Indian Institute of Farming Systems Research, Meerut, India

Aradhana Mishra Division of Microbial Technology, CSIR-National Botanical Research Institute, Lucknow, India; Academy of Scientific and Innovative Research (AcSIR), Ghaziabad, India

A.K.M. Nazrul Islam Ecology and Environment Laboratory, Department of Botany, University of Dhaka, Dhaka, Bangladesh

Ajay Neeraj School of Environment and Sustainable Development, Central University of Gujarat, Gandhinagar, India

Shipra Pandey Division of Microbial Technology, CSIR-National Botanical Research Institute, Lucknow, India; Academy of Scientific and Innovative Research (AcSIR), Ghaziabad, India; Department of Chemical Engineering, Indian Institute of Technology Bombay, Mumbai, India

Subhaswaraj Pattnaik Department of Microbiology, School of Life Sciences, Pondicherry University, Puducherry, India

Rupa Rani Laboratory of Applied Microbiology, Department of Environmental Science and Engineering, Indian Institute of Technology (Indian School of Mines), Dhanbad, India

Muzamil Ahmad Rather Department of Molecular Biology and Biotechnology, Tezpur University, Tezpur, India

Raman Kumar Ravi Department of Agriculture, Uttaranchal College of Science and Technology, Dehradun, India; School of Environment and Sustainable Development, Central University of Gujarat, Gandhinagar, India

Muhammad Zia ur Rehman Institute of Soil and Environmental Sciences, Faculty of Agriculture, University of Agriculture Faisalabad, Pakistan

Sinha Sahab Department of Environment and Sustainable Development, Institute of Environment and Sustainable Development, Banaras Hindu University, Varanasi, India

Vyacheslav M. Semenov Institute of Physicochemical and Biological Problems of Soil Science, Russian Academy of Sciences, Pushchino, Moscow region, Russia

Ayushi Sharma Department of Biotechnology and Bioinformatics, Jaypee University of Information Technology, Waknaghat, India

Rahul Shrivastava Department of Biotechnology and Bioinformatics, Jaypee University of Information Technology, Waknaghat, Solan, India

Busi Siddhardha Department of Microbiology, School of Life Sciences, Pondicherry University, Puducherry, India

Mohd A. Siddiqui School of Environment and Sustainable Development, Central University of Gujarat, Gandhinagar, India

Kshitij Singh Department of Microbiology, School of Environmental Sciences, Babasaheb Bhimrao Ambedkar University (A Central University), Lucknow, India

Rajeev Pratap Singh Department of Environment and Sustainable Development, Institute of Environment and Sustainable Development, Banaras Hindu University, Varanasi, India

Satyendra Pratap Singh Division of Microbial Technology, CSIR-National Botanical Research Institute, Lucknow, India; Division of Pharmacognosy, CSIR-National Botanical Research Institute, Lucknow, India

Shalini Singh Laboratory of Applied Microbiology, Department of Environmental Science and Engineering, Indian Institute of Technology (Indian School of Mines), Dhanbad, India

Snigdha Singh School of Environment and Sustainable Development, Central University of Gujarat, Gandhinagar, India

Ruchi Soni Regional Centre of Organic Farming, Nagpur, India, Ministry of Agriculture & Farmers Welfare, Government of India, India

M.Y. Sreenivasa Department of Studies in Microbiology, University of Mysore, Mysore, India

Ibha Suhani Department of Environment and Sustainable Development, Institute of Environment and Sustainable Development, Banaras Hindu University, Varanasi, India

Ekaterina Tikhonova Research Center of Biotechnology, Russian Academy of Sciences, Moscow, Russia

Ashutosh Tripathi Division of Microbial Technology, CSIR-National Botanical Research Institute, Lucknow, India; Academy of Scientific and Innovative Research (AcSIR), Ghaziabad, India

Manikant Tripathi Centre of Excellence, Department of Microbiology, Dr. Rammanohar Lohia Avadh University, Ayodhya, India

Sonam Tripathi Department of Microbiology, School of Environmental Sciences, Babasaheb Bhimrao Ambedkar University (A Central University), Lucknow, India

Wajid Umar Institute of Environmental Science, Hungarian University of Agricultural and Life Sciences, Godollo, Hungary

Jitendraa Vashistt Department of Biotechnology and Bioinformatics, Jaypee University of Information Technology, Waknaghat, India

Pratibha Verma Division of Microbial Technology, CSIR-National Botanical Research Institute, Lucknow, India; Academy of Scientific and Innovative Research (AcSIR), Ghaziabad, India

Hiranmai R. Yadav School of Environment and Sustainable Development, Central University of Gujarat, Gandhinagar, India

Husnain Zia Institute of Soil and Environmental Sciences, Faculty of Agriculture, University of Agriculture Faisalabad, Pakistan

Foreword

Soil, an infinitely renewable resource, supports ecosystem functioning and agriculture for human survival. Under natural conditions it can take 500-1000 years to form an inch of soil due to various abiotic and biotic activities. Unfortunately, more than half (52%) of all fertile, food-producing soils globally are now considered degraded due to land-use change (LUC) due to anthropogenic activities and climatic disturbances. The LUC-like conversion of natural ecosystems to croplands also contributes to changes in the composition and activities of soil microbial communities and biogeochemical processes affecting soil functioning. The LUC significantly reduces soil organic carbon, increases greenhouse gas emissions, and reduces beneficial soil microbial diversity. Urbanization is generally accompanied by LUC and environmental changes in chemistry and climate, which consequently impact the activities of soil microbial communities too. Soil quality has been defined as the continued capacity of the soil to function as a vital living system to sustain biological activity and the supply of ecosystem services. Soil microorganisms and their biomass regulate ecosystem processes such as nutrient cycling through decay and decomposition. There has been a global concern that such continued land degradation intensification due to LUC could change the soil quality irreversibly. Therefore there is a need to investigate microbial responses, to better understand the effects of LUC and soil disturbances on ecosystem functioning. Soil degradation due to LUC and the role of microbes in the management of ecological restoration need to be recognized, alongside soil disturbances. Given the technological advances, soil degradation is potentially reversible through planned ecosystem restoration and by beneficial microbial services and introducing sustainable agricultural practices that regenerate soil fertility and stability. It is essential that soil health should be given a central position in decisions made to LUC and that it is recognized as a vital resource for the future of humanity.

The role of microbial services in LUCs management has recently received significant attention from researchers with its potential to improve the soil health status and contribution to ecosystem services. Limited literature is available on the role of microbes in LUC management. Editors very timely invited the leading experts to contribute articles in the area of LUC management. *Microbes in Land Use Change Management* gives an overview of the current status of microbial-mediated management of soil degradation. I compliment the editors for compiling this volume and hope the book will benefit various stakeholders involved in agriculture research and environmental sustainability.

Trilochan Mohapatra
Department of Agricultural Research & Education, and Indian Council of Agricultural Research, Ministry of Agriculture and Farmers Welfare, Government of India, Krishi Bhavan, New Delhi, India
February 1, 2021

Preface

Agro-economy has a noteworthy contribution toward the development of the nation. The agriculture sector played a crucial role to nurture the economic prosperity of developed countries and also served in the growth of developing states across the globe. The rising demand to feed the population rapidly on a hike puts huge pressure on agrarian practices. As per reports, agro-production needs a leap of 50% to fulfill the requirements of the world population in a few decades, which may cross 10 billion. Such a quantum leap in agro production can be achieved either by the acquisition of more land or by enhancing the productivity of available land. The first option seems a remote dream due to the limited availability of cultivable land. However, environment-friendly management to enhance soil fertility without harming the planet may be a healthier way to achieve food security.

The agro-environmentalists have shown their concern for the irregular use of chemical fertilizers and pesticides disturbing the agro-ecosystem. Excessive input of chemical fertilizers to increase the production is deteriorating the physicochemical properties of soil, consequently reducing the organic matter, microbial diversity, and fertility leading to land degradation. In addition, these unnatural processes not only lead to the emission of greenhouse gases polluting the environment but also contaminate the groundwater through leaching, affecting human health. Globally, adequate money is being invested to restore the soil characteristics for the improvement of crop yield. Microbial inoculants or biofertilizers, a sustainable ecofriendly approach, and promising technology of biological origin may reduce the use of conventional chemical fertilizers and hence mitigate its negative impact on soil and the environment. They make useful nutrients available to agro-ecosystems such as N, P, K, and other nutrients, antibiotics, and hormones including the enrichment of rhizospheric region. The land-use changes considerably impact the indigenous microbial diversity in soils, which derails the ecosystem functioning. Therefore sustainable land-management approaches like organic farming, improvised biofertilizer input, and crop rotation may be useful in the restoration of soil and agricultural productivity.

The present manuscript aspires to provide state-of-the-art information along with vital microbial services addressing the current challenges of natural and anthropogenic interventions of land-use changes destroying the agro-environmental sustainability. The meticulous scrutiny of the manuscripts focuses on prospects and perspectives to enhance our understanding of microbial services related to the reclamation of soil, restoration of degraded, and disturbed ecosystems. The leading experts in the subject area have contributed articles making the manuscript informative. The present book *Microbes in Land Use Change Management* would be instrumental to provide up-to-date and effective information to graduate students, research scholars, agro-environmentalists, policymakers, etc., working in the area.

Editors
Jay Shankar Singh[1], Shashank Tiwari[1], Chhatarpal Singh[2] and Anil Kumar Singh[3]
[1]Department of Environmental Microbiology, School for Environmental Sciences, Babasaheb Bhimrao Ambedkar University, Lucknow, India, [2]Agro Environmental Development Society (AEDS), Rampur, India, [3]Department of Horticulture, Institute of Agricultural Sciences, Banaras Hindu University, Varanasi, India

Temperature sensitivity of litter and soil organic matter decomposition: perspective of soil microbial community structure and function

1

Irina Kravchenko[1], *Ekaterina Tikhonova*[1] *and Vyacheslav M. Semenov*[2]
[1]Research Center of Biotechnology, Russian Academy of Sciences, Moscow, Russia,
[2]Institute of Physicochemical and Biological Problems of Soil Science, Russian Academy of Sciences, Pushchino, Moscow region, Russia

1.1 Introduction

Since the industrial revolution, the average global temperature of the atmosphere has increased by 1.5°C, and it is predicted to further increase by 4°C–7°C in the next century (IPCC, 2007). The world's soils are estimated to currently contain 3000 Gt carbon, which is three times the estimated amount of atmospheric C-CO_2 (Tarnocai et al., 2009). About 40% of that carbon is stored in boreal forests, which cover a large part of the northern hemisphere (Denman et al., 2007). Thus even small changes in the soil carbon pool may severely affect the atmospheric CO_2 concentration, thereby further affecting the global air temperature. The decomposition of organic material (OM) has been shown to be more sensitive to temperature than net primary production (NPP) (Kirschbaum, 2000), further supporting the hypothesis that changes in the global air temperature could affect the rate of net C-exchange between the atmosphere and biosphere (Cox et al., 2000). Therefore reliable predictions of ecosystem responses to climate change require a thorough understanding of the factors and processes controlling the temperature sensitivity of soil organic matter's (SOM) decomposition (Conant et al., 2011; Davidson and Janssens, 2006). Some of these factors and processes are not yet well understood or under debate and, hence, need further research (Fig. 1.1).

An imbalance in the carbon cycle is a crucial challenge to human civilization in the 21st century (Ciais et al., 2013). The main sign of anthropogenic disturbances in the global carbon cycle and the dominant cause of global warming is an increase in the concentration of CO_2 in the atmosphere (Lal, 2010). The weighted average CO_2 concentration in the atmosphere in 2018 reached a new all-time high of 407.8 ± 0.1 ppm, 147% of the preindustrial level (WMO Greenhouse Gas Bulletin, 2019). Bioproductivity, biodiversity, and biogeochemical cycles of elements are considered to be the central nodes in the response of terrestrial ecosystems to an

Microbes in Land Use Change Management. DOI: https://doi.org/10.1016/B978-0-12-824448-7.00004-8
© 2021 Elsevier Inc. All rights reserved.

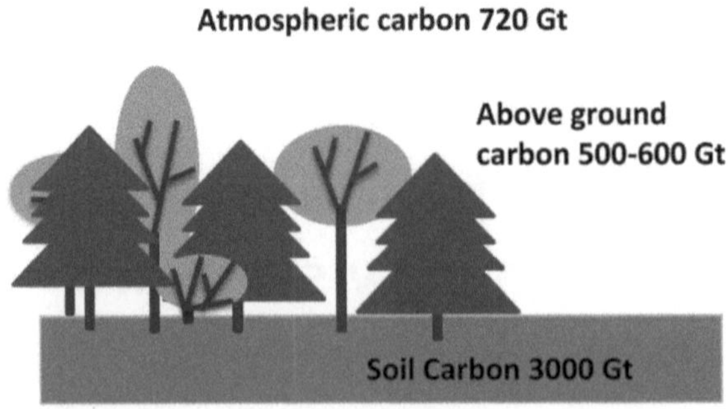

Figure 1.1 Carbon pools in the terrestrial ecosystems.
Source: Based on data from Tarnocai, C., Canadell, J.G., Schuur E.A.G., Kuhry, P., Mazhitova, G., Zimov, S., 2009. Soil organic carbon pools in the northern circumpolar permafrost region. Glob. Biogeochem. Cycles. 23, GB2023.

increase in atmospheric CO_2 concentration and global warming (Rustad, 2008). The first to be affected by global warming will be soil microorganisms and plants. The global warming will likely to lead an increase in photosynthesis, NPP, ecosystem, and soil respiration (Wu et al., 2011).

Decarbonization of industry, energy and transport terrestrial ecosystems is strategic directions in solving the problem of excess carbon in the atmosphere. Decarbonization provides for a decrease in the volume of fossil fuel combustion by switching to low-carbon or carbon-free energy sources and carriers, the use of advanced transport technologies, increasing the efficiency of units, direct capture of CO_2 from gas emissions from industrial production (Åhman et al., 2017; Bataille et al., 2016, 2018; Gambhir and Tavoni, 2019). Recarbonization of terrestrial ecosystems is achieved through refresterization of felling areas, inconveniences and degraded lands, increasing the biological productivity of crops and the progressive accumulation of C_{org} in the soil to the level of undisturbed or natural ecosystems (Lal et al., 2012; Lozano-García et al., 2020).

It is important not only to convert atmospheric CO_2 into new biomass but also to prevent the immediate return of associated carbon to the atmosphere as a result of fires or the decomposition of plant litter accumulating on the soil surface. Therefore the recarbonization of terrestrial ecosystems should provide for soil carbon sequestration, which is understood as the photosynthetic transfer of C-CO_2 from the atmosphere to plant biomass, exudates, plant residues, and other OMs into SOM (humus) (Chenu et al., 2019; Olson et al., 2014). The sequestered carbon can be predominantly used to maintain the biological activity and functionality of the soil, or, conversely, be deposited (Baveye et al., 2020; Stockmann et al., 2013). In the first case, the active decomposition of plant residues with a violent release of CO_2, rapid and durable stabilization of decomposition products by the mineral part of the soil is envisaged. In the second case, the use of slowly decomposing plant material and techniques that slow down the turnover of C_{org} in the soil provide long-term storage of C_{org} in the soil and minimal emission, but without a significant increase in soil fertility. In both cases, the decomposition of fresh organic matter is considered as the initial and critical stage in the conversion of biomass carbon into SOM.

Soils receive their carbon (C) via two major pathways: the "green" pathway via readily available plant root exudates and the "brown" pathway via slowly decomposing dead plant material in the form of litter (Moore et al., 2005). Since most of the soil C input is originating from the plant detritus production, litter fall has a considerable influence on the soil food web in terms of mineralization processes, nutrient cycling, and the formation and sequestration of organic matter (van der Wal et al., 2017).

According to modern concepts, the persistence of plant residues and SOM to biodegradation is not due to the biochemical strength of organic substances but is due to the influence of physicochemical, biological, and environmental factors (Lehmann et al., 2020; Prescott, 2010; Schmidt et al., 2011). Temperature is one of the main external determinants of the activity of decomposers and the rate of decomposition. The detected response of heterotrophic respiration of microorganisms and decomposition of organic matter to an increase (decrease) in temperature characterizes the temperature sensitivity of destructors and organic matter. The thermal sensitivity of organic matter decomposition is controlled by combinations of temperature with two, three, or multiple variables interacting with each other (Fierer et al., 2005; Frey et al., 2013; Conant et al., 2011; Feng and Simpson, 2008; Leifeld and von Lützow, 2014; Blagodatskaya et al., 2016; Davidson and Janssens, 2006; Petraglia et al., 2019). These factors most often include the quality and availability (protection) of the substrate, the structure and functional profile of the microbial community, moisture, aeration, nutrient properties, and soil pH.

According to the theory, more complex and more protected organic substances require higher activation energies for decomposition; therefore they demonstrate a high-temperature sensitivity, in comparison with simple and labile substances. However, some factors accompanying the decomposition can weaken or hide the temperature sensitivity of individual compounds. Soil microorganisms can adapt to elevated temperatures with prolonged exposure to temperature, as a result of which the thermal sensitivity of decomposition is less than expected or demonstrates different dependencies (Kögel-Knabner, 2002; Blagodatskaya et al., 2016). It is essential to understand how the substrate availability and succession of decomposers change during decomposition, and whether there are differences in temperature sensitivity between rapidly decomposing and slowly decomposing substrates. The purpose of our work was: (1) to show the general way of transforming plant residues into SOM; (2) to assess the temperature response of decomposition of plant debris and SOM; and (3) to determine the temperature sensitivity of the soil microbial community.

1.2 Soil organic matter

1.2.1 General characteristics

Living individuals and nonliving organic matter found in the soil of natural and thermally altered OMs of biological origin, regardless of the source and stage of decomposition, except for the aerial part of growing plants, form an organic soil continuum (Baldock and Broos, 2011). Recently, a novel model theory, the "soil continuum model," has been put forward challenging the most widely accepted current views on SOM transformations (Lehmann and Kleber, 2015). The organic continuum of the soil is characterized by circulation and renewal of organic matter, heterogeneity and diversity of material composition, polyphasic state, unity of dynamism and stability, influx and alienation, growth and withering away of biomass, stabilization and destabilization, simultaneous presence of living and nonliving, new and old, active and inert (Fig. 1.2).

The inanimate component of the organic continuum of the soil is SOM – a heterogeneous, but integral substance with characteristic features and properties. It performs a specific range of functions and ecosystem services, the content and quality of which can be characterized by the same indicators for all soil conditions. At the ecosystem level, SOM is a special compartment of a functioning ecosystem, a spare but moderately dynamic pool of energy and biophilic elements, a product of the transformation of dying biomass by consumers and decomposers, whose activity is controlled by environmental factors (Manlay et al., 2007). According to Stevenson (1994), SOM is all OM in soils, including residues, light fraction, microbial biomass, and water-soluble compounds, and stabilized organic matter (humus). From bio-physicochemical concepts, soil organic matter is a set of individual particles and ensembles of biomolecules of wholly and partially transformed biota residues. It differs in size, mass, chemical structure, age, and protection, has different nature and strength of internal and external chemical bonds, and is characterized by volumetric configuration and spatial irregularity of the location of mineral particles in the conglomerate (Semenov et al., 2013). In a modified form, the last definition of SOM is given as follows. SOM is s a set of different-sized particles and ensembles of biomolecules of plant, animal, and microbial origin, which are in a free, aggregated and bound state by soil minerals. This definition combines the mineral and organic parts of the solid and liquid phases, emphasizes the role of organic matter in the formation of soil aggregates, and assumes quantitative fractionation.

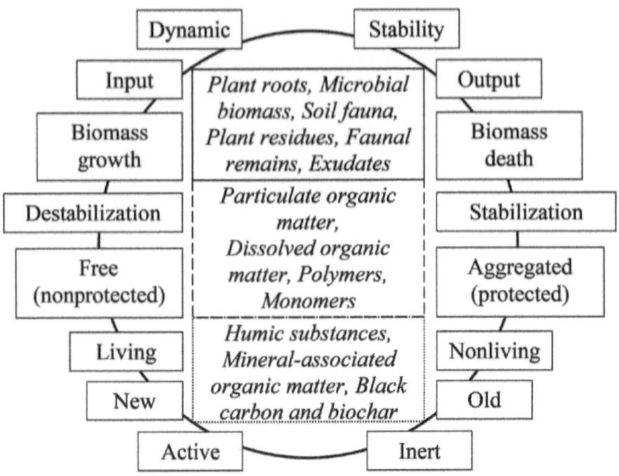

Figure 1.2 Schematic presentation of the continuum in the soil organic components.

1.2.2 Hierarchical structure of soil organic matter

The soil contains plenty of individual organic compounds and substances with unique properties and differing in chemical structure and form, properties, functions, accumulation, and other characteristics. Their list is constantly expanding due to the use of new methods and tools that allow detecting and identifying substances in small quantities (Schmidt et al., 2011; Simpson and Simpson, 2012). The SOM contains products of primary and secondary metabolism of plants, soil fungi, and bacteria: lignin, polyphenols, proteins, carbohydrates, lipids and phospholipids, waxes, resins, nucleic acids, chlorophyll, pigments, hydrocarbons, alcohols, organic acids, vitamins, and other compounds (Kögel-Knabner, 2002, 2017). Melanin, glomalin, polyketides, and hydrophobins are of particular importance in the formation of SOM properties (Schaeffer et al., 2015). The predominant part of the SOM components is considered "molecularly uncharacterized," that is, has the unknown composition and undeciphered structure (Hedges et al., 2000). The soil contains artificially synthesized substances that enter the soil with agrochemicals. According to the modern nomenclature, SOM includes: (1) plant residues; (2) particulate organic matter (POM); (3) dissolved substances; (4) humus in the form of biomolecules and humic substances (HSs) chemically bound by mineral particles; and (5) coals and charred materials (Baldock and Broos, 2011; Paul, 2016; Schaeffer et al., 2015; Stockmann et al., 2013).

1.2.3 Plant residues

Plant residues include fallen off and dead plant parts on the surface or in the root layer of the soil, which are free or part of mega-aggregates 10–2 mm in size. Larger fragments of plant residues do not belong to SOM, being an independent fraction of mortmass (detritus). Plant debris can be removed from the soil sample, depending on the procedure for preparing the soil for chemical analysis. Plant residues play a crucial role in maintaining soil biogenicity and are the primary source of CO_2 released in situ from the soil. According to the known estimates, the stocks of tree mortmass and forest waste on the surface and inside the soil are 73 ± 6 and 43 ± 3 Pg C, respectively (Pan et al., 2011). 2.83 Pg of carbon is being concentrated in forest litter and 36–72 Pg C in the deadwood (Bani et al., 2018). During logging, an average of 10%–15% of the aboveground biomass remains on the soil surface (Baruya, 2015). The mass of wood residues on the soil surface varies depending on the region, species, age, type of site, and wood assortment from 1 to 500 t/ha with a median value of 18 t/ha (Berg and McClaugherty, 2014). Underground plant waste in forests and grasslands accounts for an average of 48% and 33% of annual inputs of plant residues (Freschet et al., 2013). It is believed that during the decomposition of the world reserves of forest litter (leaves, needles, large and small branches, bark, flowers, cones, etc.) and dead wood, 0.06 and 2.1–11 Pg C-CO_2 per year are formed, respectively (Bani et al., 2018).

1.2.4 Particulate organic matter

A significant part of SOM is presented by POM, consisting of biota residues ranging from 2 to 0.053 mm in the free or occluded form in the soil and separated from soil aggregates by dispersing the soil in a solution of sodium hexamethophosphate (Cambardella and Elliott, 1992). C-POM accounts for 20%–48% of soil C_{org} (Semenov et al., 2019a). The POM fraction contains 2.8–7.6 times more potentially mineralizable carbon than SOM in whole. Many POM components are a high-energy substrate for soil organisms; therefore in the places of POM localization, the most populated microbial biomass microzones with high biological activity and increased enzyme content are formed (Lavallee et al., 2020; Wander, 2004).

1.2.5 Dissolved organic matter

Groups of hydrophobic and hydrophilic substances present dissolved organic matter (DOM), both simple (amino acids, sugars) and more complex (amino sugars, phenols, fulvic acids, humic acids, etc.) compounds <0.45 μm in size (Bolan et al., 2011; Kalbitz et al., 2000; Zsolnay, 2003). The DOM usually accounts for less than 2% of the total C_{org} (von Lützow et al., 2007). The source of DOM is plant debris, discrete solid particles, dead microbial biomass, and humus. DOM is formed during the enzymatic depolymerization of biopolymers. The organic matter dissolved in the soil solution is localized in the system of soil pores or is in the form of adsorbed by a solid mineral mass. Due to the relative simplicity of its chemical composition and mobility, DOM is rapidly utilized by microorganisms, migrates along with the soil profile, or is adsorbed by soil minerals. Therefore the low content of DOM in the soil is due to their rapid consumption in intra soil processes.

1.2.6 Humus

Humus is a well-decomposed OM of predominantly microbial source, which is <53 μm in size and bound by soil minerals (Baldock and Broos, 2011). It is corresponding by the totality of characteristics of mineral-associated organic matter (MAOM) or the continuum of organic substances related to HSs. According to one definition, humus is "a subsystem of SOM formed from OMs and compounds of the plant, animal, and microbial origin that have passed humification and nonhumification stages of stabilization, with a time of complete decomposition of the constituent components >10 years" (Ivanov et al., 2017). Humus is the most stable part of SOM, it is represented by organic compounds mainly of microbial origin chemically bound by the surface of fine particle size fractions and includes HSs (Ivanov et al., 2009, 2017; Cotrufo et al., 2013; Hu et al., 2020a,b; Kallenbach et al., 2016; Simpson et al., 2007). Humus is not a molecule or a series of molecules, but rather a form of organic matter transformed by microorganisms and stabilized. The source of MAOM is low molecular weight compounds that are washed out from plant residues (Cotrufo et al., 2013; Lavallee et al., 2020) or are formed during the enzymatic degradation of POM in vivo (Liang et al., 2017) or ex vivo

(Grandy and Neff, 2008). MAOM is a highly stable, long-lasting pool of SOMs with the oldest C (Haddix et al., 2020; Kögel-Knabner et al., 2008a,b; Paul, 2016; von Lützow et al., 2006). The MAOM fraction is fundamentally different from POM in terms of formation paths, chemical composition, stability, and functions; it is confined to particle size fractions of dust and clay (less than 63−20 μm) and usually has a density of more than 1.6−1.85 g/cm^3 (Lavallee et al., 2020). MAOM accounts for 50%−90% of C_{org}.

HS traditionally belongs to the dominant humus component. HS is understood as a biogenic, chemically active, and naturally stable heterogeneous continuum of biomolecules that form polymer-supramolecular assemblies through random chemical transformations and nonvalence interactions of various precursor molecules (Semenov et al., 2013). This term is used as a generic name for fractions of a dark-colored material that are soluble in alkalis (humic acids), in alkalis and acids (fulvic acids), and insoluble (nonextractable) components (humin) of humus (Ivanov et al., 2017). The reality of the presence of HS in the soil, the validity of the designation of an alkaline extract by a particular class of chemical substances, the truth of the mechanisms of HS formation, the authenticity of the properties attributed to HS, and other aspects are still in the focus of attention of researchers from different disciplines (Baveye and Wander, 2019; Kleber and Lehmann, 2019; Lehmann and Kleber, 2015; Olk et al., 2019). New methods of HS isolation and identification are being developed; new mechanisms of HS formation in natural environments are being proposed (Mohinuzzaman et al., 2020; Wells and Stretz, 2019). There is an understanding that if HS actually exist in soils, then it is not in the form and in the amounts that are preparatively released by chemical fractionation (Ivanov et al., 2017).

1.2.7 Black carbon, biochar

The SOM contains substances of pyrogenic (char and herbaceous coals, biochar, soot and its varieties, such as black carbon), as well as nonpyrogenic (carbon-like, highly aromatic substances) origin (Bird et al., 2015; Chen et al., 2020; Landry and Matthews, 2017; Reisser et al., 2016). Often, the entire range of products of thermochemical conversion of OMs, which can be biomass or fossil fuels, are conventionally designated black carbon. At the same time, biochar refers to black carbon, while not all black carbon is biochar (Spokas et al., 2012). Charcoal and herbaceous coals are formed by incomplete combustion of plant biomass during fires and burns. Biochar (biochar) is introduced into the soil as a reclamation agent. Particulate soot and black carbon particles enter the soil with atmospheric deposition. Carbon-like substances are a product of aerobic decomposition of plant residues at natural temperature.

According to some estimates, 56−129 Tg of pyrogenic carbon enter the soil annually (Bird et al., 2015). The reserves of pyrogenic carbon in terrestrial ecosystems are estimated at 86,000 Tg; on a global scale, soil contains an average of 13.7% of soil C_{org} and can reach 60% (Reisser et al., 2016). According to these authors, soils with a clay content of more than 50% contain significantly more pyrogenic carbon ($>$30% of C_{org}) than with a clay content of less than 5% ($<$6% of C_{org}), and alkaline soils contain at least 1.5 times more than acidic soils. The annual flux of nonpyrogenic carbon-like carbon formed from straw residues and root decomposition can be 5.4 Tg C per year (Chen et al., 2020). Due to its strong aromatic structure and pronounced hydrophobicity, coals and carbon-like substances are exceptionally durable, remaining in the soil from decades for hundreds of years (Bird et al., 2015). Biochar production is one of the promising pathways for carbon sequestration (Smith, 2016; Spokas et al., 2012).

1.3 Decomposition and stabilization of plant residues in soil

In terrestrial ecosystems, a significant portion of primary plant production enters the soil as fresh plant material, forming a detritus pool as an available substrate for soil microbiota and the primary material for the formation of soil humus (Didion et al., 2016). In the mixed forests of the Moscow region, the amount of leaf litter is 2.2–7.0 t/ha per year or 20%–80% of the value of primary production (Vedrova, 1997).

Destruction of OMs and residues into separate fragments, leaching of soluble compounds, mineralization and stabilization of decomposition products and metabolites are the main stages in the transformation of mortmass in soil. As a result of enzymatic attacks, OMs are broken down into separate fragments and particles (POM). The large structural polymers contained in POM gradually depolymerize and mineralize to CO_2 (Cotrufo et al., 2013; Haddix et al., 2020; Lehmann and Kleber, 2015). During degradation, POM is stabilized by occlusion and aggregation, and the depolymerization products chemically interact with the surface of the mineral particles to form mineral-bound organic matter (MAOM), which is often referred to as organo-mineral complexes.

According to the modified model (Fig. 1.3), POM is formed as a result of partial decomposition and physical fragmentation of the structural parts of the plant. In contrast, MAOM is formed after washing out the soluble organic substances in the fresh material and/or transformation of POM by soil microorganisms. The advantage of the modified model is that it allows soil enrichment with organic matter and the formation of MAOM also due to forest litter and plant litter on the soil surface.

The rate of decomposition of plant residues directly depends on the initial strength of plant biomolecules and the combination of external conditions that control the enzymatic activity of the soil microbiome (Ekschmitt et al., 2005). Simply, predominantly water-soluble substrates are directly utilized by heterotrophs, while complex large biomolecules of polymers and multicomponent complexes, are depolymerized by extracellular enzymes (Burns et al., 2013). For successful extracellular degradation of complex organic heteropolymers, it is necessary that the organic substrate is present in the soil and is available for microorganisms and/or extracellular enzymes; extracellular enzymes were protected and functional; there was the correct combination of enzymes and microorganisms; the chemical and physical conditions of the environment favored catalysis (Burns et al., 2013). The decomposition of organic matter in the soil is carried out only if in the same place. At the same time, there is a

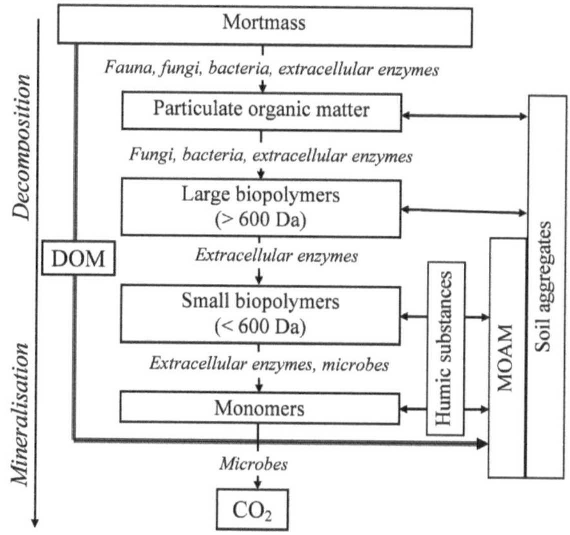

Figure 1.3 Sequential stages, decomposition, mineralization, and stabilization of plant residues in soil.
Source: Adapted from Cotrufo, M.F., Wallenstein, M.D., Boot, C.M., Denef, K., Paul, E., 2013. The microbial efficiency-matrix Sstabilization (MEMS) framework integrates plant litter decomposition with soil organic matter stabilization: Do labile plant inputs form stable soil organic matter? Glob. Change Biol. 19(4), 988−995; Cotrufo, M.F., Soong, J.L., Horton, A.J., Campbell, E.E., Haddix, M.L., Wall, D.H., et al., 2015. Formation of soil organic matter via biochemical and physical pathways of litter mass loss. Nat. Geosci. 8, 776−779; Haddix, M.L., Gregorich, E.G., Helgason, B.L., Janzen, H., Ellert, B.H., Cotrufo, M.F., 2020. Climate, carbon content, and soil texture control the independent formation and persistence of particulate and mineral-associated organic matter in soil. Geoderma. 363, 114160; Lavallee, J.M., Soong, J.L., Cotrufo, M.F., 2020. Conceptualizing soil organic matter into particulate and mineral-associated forms to address global change in the 21st century. Glob. Change Biol. 26(1), 261−273; Lehmann, J., Kleber, M., 2015. The contentious nature of soil organic matter. Nature. 528, 60−68.

water film, active enzymes, a substrate available and susceptible to enzymes, as well as favorable conditions of temperature, aeration, and pH of the environment (Ekschmitt et al., 2005; Schimel and Weintraub, 2003). A deviation from the norm of one of these factors can be the reason for the slowing down of decomposition, regardless of the favorableness of other factors. The main factors that control the rate and extent of decomposition of organic residues are geographic variables (latitude and altitude of locations), average annual temperature, average annual precipitation and yearly evapotranspiration, quality of residues (e.g., N content, C: N ratio, lignin and polyphenol content, lignin: N ratio), types of vegetation and the remains (Prescott, 2010; Zhang et al., 2008). Globally and regionally, the rate of decomposition of organic residues is controlled by the temperature and humidity of the environment, while under local conditions, it is controlled by the chemical composition of the residues, the molecular organization of structures, their resistance to enzymes, and the spatial reach of microorganisms (Bradford et al., 2016).

Discrete particles, large biomolecules, biopolymers, and small monomers, formed during the decomposition of organic residues, are subjected to humification/nonhumification stabilization in the soil. Stabilization of organic matter in the soil means an increase in its resistance to biotic and abiotic effects with an increase in the time of its turnover in the soil due to the acquisition of a protected state arising from various processes and phenomena (Semenov et al., 2013; Sollins et al., 1996; Kögel-Knabner et al., 2008a,b). By the nature of the dominant processes, physical, chemical, and biological stabilization are distinguished (Semenov et al., 2013; Kögel-Knabner et al., 2008a,b; Krull et al., 2003; Sollins et al., 1996; von Lützow et al., 2006, 2007, 2008). It is believed that the chemical strength of compounds and physical protection is responsible for the short-term and medium-term defense of organic matter from several years to decades, while the long-term preservation of SOM components for hundreds and thousands of years is provided by the stabilization of organic compounds by the mineral part of the soil (von Lützow et al., 2006; Kögel-Knabner et al., 2008b; Kögel-Knabner, 2017).

From experiments in which decomposition was assessed by weight loss in mesh bags, it was found that SOM is formed mainly due to the slowly decomposing part of litter and residues resistant to decomposition (Berg and McClaugherty, 2014). According to the new concept: (1) plant tissues are not particularly resistant to degradation, therefore the stability of SOM is controlled not due to the selective preservation of potent compounds contained in plant tissues, but as a result of the physical, chemical, and biological protection of decomposition products; (2) soluble and rapidly decomposing components of plant residues are more efficiently utilized by microorganisms and stabilized by a mineral matrix than difficult-to-decompose compounds; (3) the decomposition products of microbial biomass are stabilized more strongly than plant residues; (4) the preservation of organic matter is not due to some unique internal properties that give strength, but to the fact that external physicochemical, biological, and environmental conditions wholly or partially limit the rate of decomposition of organic matter; and (5) for a more significant sequestration of carbon in the soil, one should not slow down the deterioration of plant residues, but rather quickly and to a greater extent expose them to microbial and physicochemical reactions, creating physical and chemical protection of the transformation products from further decomposition (Pinskii et al., 2017; Semenov et al., 2001, 2004; Trofimov et al., 2012; Castellano et al., 2015; Cotrufo et al., 2013, 2015;Hu et al., 2020a,b; Janzen, 2006; Schmidt et al,. 2011; von Lützow et al., 2006).

We have found that the C_{org} content in the soil amended with plant residues after a year of incubation was the same or slightly higher than before gestation (Table 1.1). This indicates that the carbon of the plant residues almost wholly offset the mineralization losses of soil carbon.

In the control variants without plant residues, the decrease in C_{org} was accompanied by a decline in the content of carbon of discrete particles (C_{POM}) and potentially mineralizable (C_0) organic matter by 1.2–1.5 and 2.6–2.8 times, respectively. The soil incubated with tree debris contained 1.1–1.7 times more

Table 1.1 Content of total (C_{org}), particulate (C_{POM}), potentially mineralizable (C_0) organic matter, and microbial biomass (C_{mic}) in the soil after 365-day incubation with various plant residues (Semenov et al., 2019b).

Experimental variant	C_{org}, % from soil mass	C_{POM}		C_0		C_{mic}	
		mg/100 g	% from C_{org}	mg/100 g	% from C_{org}	mg/100 g	% from C_{org}
Soil under the forest (back ground I)	1.67 (1.88)	562 (680)	33.8 (36.0)	49 (128)	2.9 (6.8)	16	1.0
I + oak leaves	1.83	642	35.1	74	4.1	20	1.1
I + aspen leaves	1.89	669	35.5	64	3.4	19	1.0
I + aspen branches	1.97	889	45.0	78	4.0	16	0.8
I + tree roots	1.94	936	48.3	76	3.9	16	0.8
Arable soil (back ground II)	1.32 (1.44)	155 (230)	11.7 (15.9)	27 (77)	2.0 (5.4)	6	0.4
II + barley straw	1.49	411	27.5	51	3.4	8	0.5
II + barley roots	1.56	434	27.8	61	3.9	8	0.5
II + overground mass of clover	1.49	340	22.8	47	3.1	10	0.7
II + clover roots	1.41	309	21.9	44	3.1	11	0.8

Notes: Data in parentheses are for the original soil before incubation.

C_{POM} than the soil set without plant debris. The arable soil with barley and clover residues had 2.0–2.8 times more C_{POM} than without plant residues. In general, according to the experiment, practically all types of plant residues compensated for C_{POM} consumed in the control soil during the annual incubation. Partially decomposed plant residues and decomposition products created an increased level of C_0 in the incubated soil of forest and arable land by 1.3–1.6 and 1.6–2.2 times, respectively, as compared to the variants without plant residues. However, in no case was there a complete compensation for the potentially mineralizable organic matter contained in the original soil samples before incubation. The decomposition of plant residues contributed to an increase in the carbon of microbial biomass. At the same time, highly decomposable plant residues gave a more significant increase in microbial biomass in the soil. On the contrary, the greatest increase in the C_{POM} content in soil was found in variants with poorly decomposed plant residues. The positive effect of poorly decomposed plant residues on the soil C_0 content was longer than that of strongly decomposed ones.

Based on the obtained relationships, it can be assumed that the transformation of plant residues proceed along two paths: (1) fragmentation to POM during the usage of the substrate by microorganisms; and (2) synthesis of microbial biomass. POM and microbial biomass are the primary sources of potentially mineralizable organic matter in the soil.

Decomposition products of microbial biomass, together with biopolymers released from plant residues, interact with each other and with fractions of dust and clay and form humus, which is a regular part of the SOM. Root residues play a more significant role in the formation of SOM than aboveground biomass; carbon from roots is characterized by 2.4 times longer lifetime than from stems (Rasse et al., 2005).

1.4 Environmental factors controlling temperature sensitivity of soil organic matter decomposition

When studying the temperature sensitivity of SOM decomposition, several factors must be considered. These factors can be categorized as environmental, and others linked to the temperature sensitivity of the OM. In soils of forest ecosystems, the rate of decomposition of litter is determined by several biophysical, chemical, and biological factors, including humidity, temperature, pH, the availability of nitrogen and oxygen, the composition of the microbial population. Soil microbial communities modulate SOM dynamics by catalyzing litter decomposition. Microbial communities at various stages of litter decomposition were studied mostly with traditional microbiological methods (12–14) and more recently using molecular approach (15–20); however, the studies mainly concerned broad-leaved forest litter (17, 21, 22).

Generally, four significant environmental constraints are recognized. The first is soil moisture, one of the main factors determining the activity of soil microbial communities and, accordingly, regulating the processes and rates of destruction. Drought reduces water films in the soil, thus inhibiting the diffusion of soluble extracellular enzymes and reducing the substrate availability for microorganisms at the reaction sites. It was found that low (below 30%) or excessive (above 150%) moisture level significantly inhibits the rate of decomposition of cellulose.

Another important biophysical factor affecting the destructive activity of soil organisms is temperature. Temperature can be considered as a prime factor in determining the rates of litter decomposition, and decomposition is more sensitive to temperature than the primary production. Most researchers believe that temperature is the main abiotic factor in the decomposition of organic matter. The effect of temperature is primarily due to its influence on the growth and activity of soil organisms, as well as on the saturation of the soil with oxygen and soluble nutrients. The literature shows that the optimal temperature for cellulose decomposition in the soil is 35°C–37°C, although decomposition can occur at lower temperatures of 0°C–5°C. Some authors mentioned that temperature affects the intensity and rate of mineralization of plant residues six times stronger than humidity.

The third is physicochemical protection of the OM from microorganisms, which involves physical occlusion of OM in the interior of soil aggregates and chemical adsorption of OM to mineral surfaces through covalent or electrostatic bonds (Lutzow et al., 2006). And the fourth is flooding, which reduces oxygen diffusion and thus often leads to anaerobic decomposition, which is much slower than aerobic decomposition.

Kinetic theory predicts that the chemical composition of SOM represents a dominant influence on the temperature response of decomposition. However, empirical observations and modeling indicate that other factors constrain this relationship. Many research questions related to these factors are central to a thorough understanding of temperature sensitivity in litter decomposition under otherwise similar conditions (Davidson and Janssens, 2006). In some cases, extreme pH or high concentrations of toxins may also severely limit OM decomposition. However, such conditions could be regarded as other types of chemical protection in this context.

One of the first empirical observations consistent with modeling had demonstrated increasing temperature sensitivity for the uptake of carbon monomers over microbial cell membranes (Hopkins et al., 2014). Another important statement is the reduction of temperature response by the increased content soil of nitrogen compounds (Hu et al., 2020a,b). The suggested mechanism may be the influence of nitrogen on the chemical composition of organic matter, mediating a direct effect on temperature response. Nitrogen availability in terrestrial ecosystems has doubled relative to preindustrial levels, so it may be vital in understanding the net impact of temperature increase on CO_2 return to the atmosphere.

1.4.1 Temperature sensitivity of organic matter decomposition

An important factor influencing the temperature sensitivity of SOM decomposition is the quality of the OM. According to kinetic theory, the temperature sensitivity of OM should be inversely related to its "quality," that is, temperature sensitivity should increase with reductions in the degradability of the carbon forms (Davidson and Janssens, 2006; Knorr et al., 2005). However, the results of empirical studies do not necessarily follow this pattern. Some studies have shown that the decomposition of stable compounds (low-quality carbon) is not temperature sensitive (Giardina and Ryan, 2000), others that are both labile and stable organic compounds respond similarly to temperature increases (Conen et al., 2006; Fang et al., 2005), and some recent studies have shown that the temperature sensitivity increases with decreased quality of the OM (Wetterstedt et al., 2010; Hartley and Ineson, 2008; Fierer et al., 2005). Thus despite these research efforts, no consensus has yet been reached on likely effects of increases in atmospheric temperature on the decomposition of SOM.

According to "carbon quality theory," the temperature responses of OM are governed by the carbon quality of the OM, that is, how easily it can be degraded, which depends largely on the number of enzymatic steps required for microorganisms to decompose it (Bosatta and Ågren, 1999). The theory is based on the Arrhenius function, which describes the dependence of the rate of a given chemical reaction on temperature and the activation energy required for it to occur. The carbon quality theory predicts that the decomposition of low-quality carbon substrates (which require high activation energy for degradation) should have higher temperature sensitivity than the decomposition of higher quality carbon substrates (Davidson and Janssens, 2006).

Much of the variation in the results from studies of the temperature sensitivity of OM decomposition stems from variations in the carbon quality of the material used. Still, contrasting responses of several key processes involved in decomposition may also act antagonistically or synergistically (Ågren and Wetterstedt, 2007). Fig. 1.4 presents a conceptual illustration of the key processes and parameters involved: 1(S), the substrate release rate; 2(D), the substrate diffusion rate; 3(μ), the substrate uptake rate; and 4(K), the half-saturation constant. Substrate release rate (1, S) is directly related to carbon quality while uptake rate (3, D) and half-saturation constant (4, K) reflect microorganism characteristics and diffusion rate (2, D) reflects prevailing environmental condition.

The conceptual model in Fig. 1.4 describes the decomposition of a substrate from a sizeable organic polymersome distance from a microorganism. It begins with substrate release (1) by exoenzymes and diffusion of the substrate (2) to the microorganism at rates S and D, respectively. At the surface of the microorganism, the substrate's uptake (3) is described by the Michaelis–Menten equation, with a maximal rate μ and (4) half-saturation constant K. All of these processes are temperature-dependent. The overall temperature response is governed by the integrative changes in activation energy for all of them (Ågren and Wetterstedt, 2007).

1.4.2 Organic chemical composition of the organic material concerning the decomposition

The organic chemical composition of litter and SOM and its influence on degradation and recalcitrance have been a key focus in terrestrial biogeochemistry and soil ecology (Berg, 2000; Bosatta and Ågren, 1999). Much of the general understanding of variation in organic degradability stems from a combination of various wet chemical digestion protocols and analytical techniques (Berg and McClaugherty, 2003). Based on such an approach, the general understanding is that cellulose and hemicellulose are easily decomposed. These are followed by polymers of aromatic or alkyl carbon compounds; or polymers made up of aromatic and alkyl carbons such as lignins, cutins, suberins (Berg and McClaugherty, 2003).

However, based on investigations utilizing a range of other analytical techniques, for example, nuclear magnetic resonance (NMR) spectroscopy, Fourier transform infrared spectroscopy; pyrolysis gas chromatography mass spectroscopy this view appears too simplistic. For example, based on several NMR spectroscopy approaches it is evident that SOM composition is generally dominated by O-alkyl C and di-O-alkyl C accompanied by varying proportions of alkyl C methoxy/N-alkyl C, aromatic,

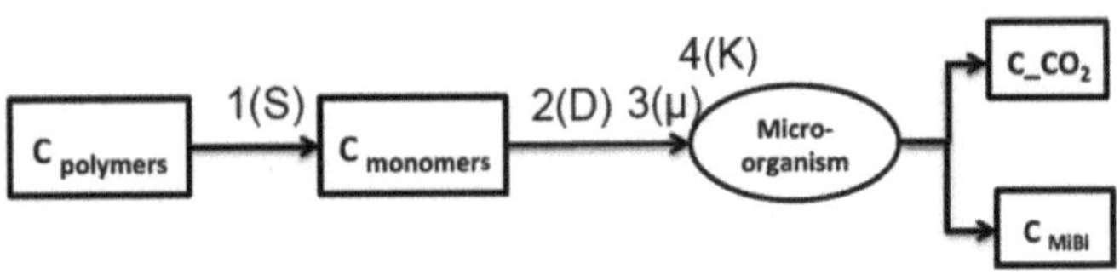

Figure 1.4 Conceptual presentation of the critical processes and parameters affecting the temperature the sensitivity of OM decomposition.

O-aromatic C, and carbonyl C (Preston et al., 2006). Alkyl carbon originates from fatty acid chains, and therefore in litter mainly comes from surface waxes and cutin (Lorenz et al., 2000). Methoxy carbon originates from lignin, and N-alkyl carbon mostly from protein compounds (Nelson and Baldock, 2005). O-alkyl carbon originates from carbohydrates such as cellulose and hemicellulose, aromatic and O-aromatic molecules originate from lignin and tannins, the carbonyl carbon is mainly a constituent of lipids or amino acids (Kögel-Knabner, 2002). O-alkyl and di-O-alkyl carbons are significant constituents of cellulose and hemicellulose and are commonly considered to be readily degradable, yet surprisingly they represent up to 20%–50% of SOM (Erhagen et al., 2013; Lorenz et al., 2000).

In contrast, polymers of aromatic carbons are considered to be recalcitrant to microbial degradation. Thus the relative contribution of aromatic polymers in SOM should be substantially higher in comparison to the plant and microbial biomass constituting the carbon source. Published data reveal that the relative proportions of aromatic carbons are approximately the same in litter and SOM (Erhagen et al., 2013; Preston et al., 2006; Kögel-Knabner, 2002; Lorenz et al., 2000). Both examples above clearly indicate that simply assigning carbohydrate polymers as easily degradable and aromatic polymers as recalcitrant is far too simplistic.

1.4.3 The role of nitrogen in decomposition

Since the late 20th century, anthropogenic inputs of nitrogen to the terrestrial ecosystem have exceeded natural inputs from nitrogen fixation (Galloway et al., 2008). This has significant implications as nitrogen is the most strongly limiting nutrient for carbon fixation in the terrestrial ecosystem (Vitousek et al., 1997), and thus a vital determinant of the net ecosystem balance (De Vries et al., 2006). Several studies have also shown that nitrogen strongly affects decomposition rates (Janssens et al., 2010; Berg, 2000), but there is little knowledge of the effects of nitrogen deposition on the temperature sensitivity of decomposition. Increases in nitrogen deposition have been shown to affect soil carbon storage positively, because they reduce saprotrophic respiration via mechanisms that shift the composition of the OM towards more chemically stable compounds (Janssens et al., 2010; Liu and Greaver, 2010). High N content also appears to suppress activities of ligninolytic fungi and their enzymes (Berg and Matzner, 1997). However, N has complex effects in decomposition processes as high N concentrations appear to enhance degradation in early stages of fresh litter decomposition but suppress it in late stages (Berg, 2000; Berg and Matzner, 1997). Thus it plays a key role in the sequestration of soil carbon. Because of the importance of carbon sequestration for the climate, the interactive effects of soil N content and temperature on the temperature sensitivity of CO_2 production from OM decomposition is a critical issue.

1.4.4 Microbial metabolism

Another critical aspect that requires better understanding is the partitioning of the resources acquired via the decomposition of organic carbon between respiration and biomass growth in microorganisms. This partitioning is often expressed in terms of the carbon use efficiency (CUE), a measure of the proportion of a utilized carbon source that is converted into microbial biomass (del Giorgio and Cole, 1998). CUE is highly sensitive to changes in environmental conditions such as temperature increases (Wetterstedt and Ågren, 2011; Allison et al., 2010).

Microorganisms obtain both energy and substrates for cell growth from the decomposition of OM, via processes often referred to as catabolic (breaking down) and anabolic (building up), respectively. Ratios of soil microorganisms catabolic and anabolic reaction rates depend on the stoichiometry of the available carbon and nutrient sources. The balance between anabolic and catabolic reaction rates is often referred to as CUE. It provides a measure of the proportion of acquired carbon that is used to synthesize new microbial biomass. How this relationship is affected by temperature is relatively poorly understood (Manzoni et al., 2012; Allison et al., 2010; Drotz et al., 2010). However, it is generally believed that the CUE should decrease with temperature increases (Steinweg et al., 2008; Hall and Cotner, 2007), because of relatively high increases in maintenance respiration rates driven by rises in energy costs of maintaining ion gradients across the cell membranes and increases in protein turnover rates (Hall and Cotner, 2007). Some more recent studies have supported this view (Frey et al., 2013; Steinweg et al., 2008), but others have found that the CUE is not temperature-sensitive (Dijkstra et al., 2011; Drotz et al., 2010).

To estimate the CUE and to monitor amounts of newly synthesized microbial biomass and $^{13}C\text{-}CO_2$ respired by the microorganisms, a ^{13}C-labeled substrate can be added. Then CUE can be calculated by dividing the amount of ^{13}C incorporated into biomass by the total amount contained and respired as $^{13}C\text{-}CO_2$. The fate of a carbon substrate follows a well-established pattern (especially for labeled ^{13}C-glucose), which can be divided into several distinct metabolic phases (Ilstedt et al., 2003): an initial response reflecting the microbial population's capacity to use it substrate-induced respiration (Anderson and Domsch, 1978) followed by a lag phase until the microorganisms start to grow, and then a specific growth rate reflected in an exponential increase in CO_2 production (Fig. 1.5).

The proportion of carbon in plant litter transformed by microorganisms into biomass (CUE) is a central factor determining global land-atmosphere CO_2 exchange. CUE was highly sensitive to whether carbon monomers or polymers were degraded, yet the temperature had no clear effect on CUE.

1.5 Microbial communities and litter decomposition under different temperature regimes

Litter decomposition is a fundamental process regulating the carbon cycle in the ecosystems. Currently, much is understood about how changes in abiotic characteristics or substrate quality affect decomposition rates of litter. At the same time, the role of microbial communities, the engines of the litter decomposition, remains obscure. This knowledge gap may be significant for predicting ecosystem responses to climate change.

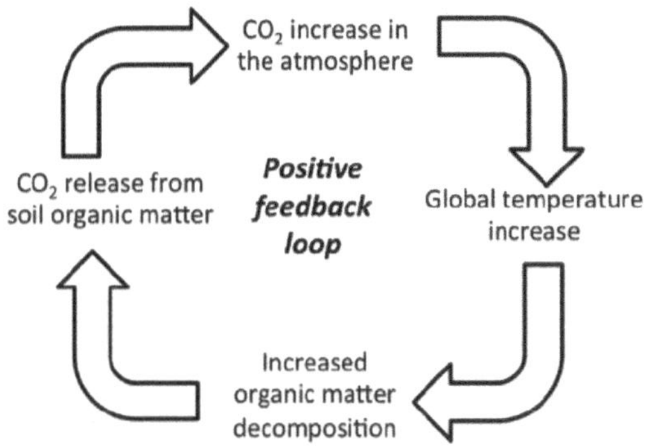

Figure 1.5 Effect of increasing organic matter as the temperature increases the effect of increasing soil organic matter decomposition with increased temperature and positive feedback.

Most of the models that describe the biogeochemical processes in the terrestrial ecosystems ignore the size and composition of the microbial communities (Schimel, 2001). However, recent work based on laboratory incubations and field experiments (Allison et al., 2013; Cleveland et al., 2014; Martiny et al., 2017) has demonstrated that decomposition rates are affected by bacterial and fungal community composition. Because the factors regulating decomposition are often context-dependent and can vary in their influence in a range of spatial and temporal scales, it may be hypothesized that responses of the litter decomposition to climatic change would be depended on.

Microbial community composition may shift as organisms that are physiologically better and have adapted to the new conditions become more abundant and vice versa. Even if relative abundances of the taxa within a community do not change, ecosystem processes could be affected by changes in the absolute quantity of the community. These mechanisms could therefore influence ecosystem responses to environmental change on generational timescales (Kardol et al., 2010).

Because physiological responses and indirect processes coincide, studying mechanisms requires independent manipulation of abiotic conditions and biological communities. Still, fewer studies have examined indirect effects on ecosystem processes driven by microorganisms. For litter decomposition changes in both plant (quantity and quality of substance intake) and microbial communities may contribute to the environmental response. Thus there is a need for studies that directly examine the importance of the microbial community in responses to environmental change. These are experiments with soil microcosms under controlled laboratory conditions.

1.5.1 Decomposing microbial communities in boreal forest soils

Bacteria and fungi are the principal decomposers in boreal forest soils. They are responsible for more than 95% of the decomposition of OM, and the microbial biomass in the boreal forest is dominated by fungi (Högberg and Högberg, 2002). Fungi play critical ecological roles in forests. The fungal community has a dual function in being either free-living saprotrophs, that is, decomposing OM or biotrophs through the mycorrhizal relationship with living plants. Due to their interactions with forest trees, fungi contribute to the soil C balance by affecting the NPP as far as carbon mineralization and sequestration (Crowther et al., 2019). Fungal mycelia, a substantial part of soil biomass, may represent a vital C sink since the necromass of many mycorrhizal species contributes to long-term C storage in boreal regions (Clemmensen et al., 2013). On the other hand, the decomposition of fungal mycelia (Brabcová et al., 2018) and the contribution of saprotrophic and mycorrhizal fungi to the decomposition of SOM contribute to ecosystem-level C mineralization (Zak et al., 2019). Much of the high relative contribution from fungi to the total microbial biomass in forest soils most likely originating from the mycorrhizal fungi (Clemmensen et al., 2015). Besides being crucial in carbon and nutrient cycling, the microorganisms also contribute most significantly to the secondary biomass production in the soil, both from saprotrophic and mycorrhizal fungi (Clemmensen et al., 2015; Štursová et al., 2020). Thus both the accumulation of soil carbon and the rate of soil carbon decomposition rely on biomass from both primary plant production and microbial heterotrophic secondary production. The vertical distribution of fungi in a soil profile reflects a distinct successional gradient with saprotrophic fungi dominating the young and less decomposed material. In contrast, the older humus material deeper down in the profile is dominated by mycorrhizal fungi (Lindahl et al., 2007).

Temperature can be considered as a prime factor in determining the rates of litter decomposition (Hobbie, 1996), and it was shown that decomposition is more sensitive to temperature than the primary production (Kirschbaum, 2000). Most researchers believe that temperature is the main abiotic factor in the decomposition of organic matter. The effect of temperature is primarily due to its influence on the growth and activity of soil organisms, as well as on the saturation of the soil with oxygen and soluble nutrients. The literature shows that the optimal temperature for cellulose decomposition in soil is 35°C–37°C, although decomposition can occur at lower temperatures of 0°C–5°C (Ball, 1997). Some authors have mentioned that temperature affects the intensity and rate of mineralization of plant residues six times stronger than humidity (Schlesinger, 1984). Microbial litter decomposition proceeds through numerous mechanisms, especially heterotrophic consumption of organic composites and CO_2 release, which can add more than 20% to soil surface CO_2 efflux, which is known as soil respiration.

Microbial decomposition of litter can be divided into phases, in each of which specific products are formed, making it possible to develop certain groups of microorganisms. The initial stages of decomposition are dominated by the release (mainly by leaching) of Ca, K, and Mg, while for N, P, and S there may be a temporary increase in their content in plant residues. Intermediate decay products are a nutrient substrate for different populations supporting the diversity of soil microbiota and its succession changes during decomposition (Berg and McClaugherty, 2008). The interaction of other representatives of the biodegradation block in soils is an essential factor determining the intensity of plant litter decomposition (Rakhleeva et al., 2011). The rate of decomposition of nutrient-rich litter is high at the initial stage, but in the later stages, the situation may change to the opposite.

1.6 Temperature influence on the microbial decomposition of forest litter: a case study in the mixed-wood forest in European Russia

1.6.1 Litter decomposition processes at the early stage of the microbial degradation: a microcosm study

Wood litter represents a mixture of organic substances contained in the annually dying plant parts. Litter fall in terrestrial ecosystems signifies a crucial pathway for nutrient return to the soil, so its biogeocenosis value is very high and can markedly modify the biotic and abiotic conditions on the forest floor and altering soil C and nutrient cycling. Despite the sufficiently high degree of study of primary productivity of forest communities, the transformation of the above-ground part of the woody vegetation is not adequately investigated and requires detailed analysis. The decomposition of leaf litter contributes 5%–45% to total soil respiration, thereby playing a pivotal role in determining atmospheric CO_2 concentration (Borken and Beese, 2005).

The total wood fall in Russia is estimated at 35–55 center C ha^{-1} in the coniferous zone and 65–90 center C ha^{-1} in deciduous forests with the value of leaf fall of 8–15 center C ha^{-1} (Basilevich, 1993). Minimal studies on litter decomposition were conducted in temperate Russian forest ecosystems. Hence comprehensive have to be undertaken to understand the turnover rate of nutrients and other elements in these sensitive ecosystems.

According to modern climate models, the 21st century will be characterized by a global increase in air temperature (Meehl et al., 2007). For regions of middle and high latitudes, as well as polar regions, the temperature is projected to increase by 3°C–5°C. Temperature rise can increase the flow of organic residues into the soil, but at the same time, increase the rate of microbial decomposition. The long-term effects of global warming on SOM are not yet apparent. According to some estimates, after a short burst of mineralization affecting a small active (labile) pool of organic matter, the process will slow down, on the other and will not change fundamentally, or, on the contrary, will increase, as the mineralization of a large stable pool due to its greater thermal sensitivity will increase (Baveye et al., 2020).

It was hypothesized that the temperature change is the primary driver of microbial activity and differentiation of microbial communities at the initial stage of transformation of tree litter. To test this hypothesis the series of microcosm experiments with gray forest soil from forest biocenosis in the Moscow region supplemented with crushed leaves and twigs at the rate of 0.5% by weight was carried out at constant temperatures of 5°C, 15°C, and 25°C for 28 days (Kravchenko et al., 2019). Evaluation of soil respiration (CO_2 flux) has demonstrated apparent differences in the dynamics and intensity of basal and substrate-induced soil respiration for different temperatures (Fig. 1.6). When incubated at 5°C, the magnitude of the basal respiration was low during the all observation period, and a significant increase after substrates addition was recorded after 8–10 days (Fig. 1.6). At 15°C and 25°C the intensity of basal respiration was significantly higher, increased

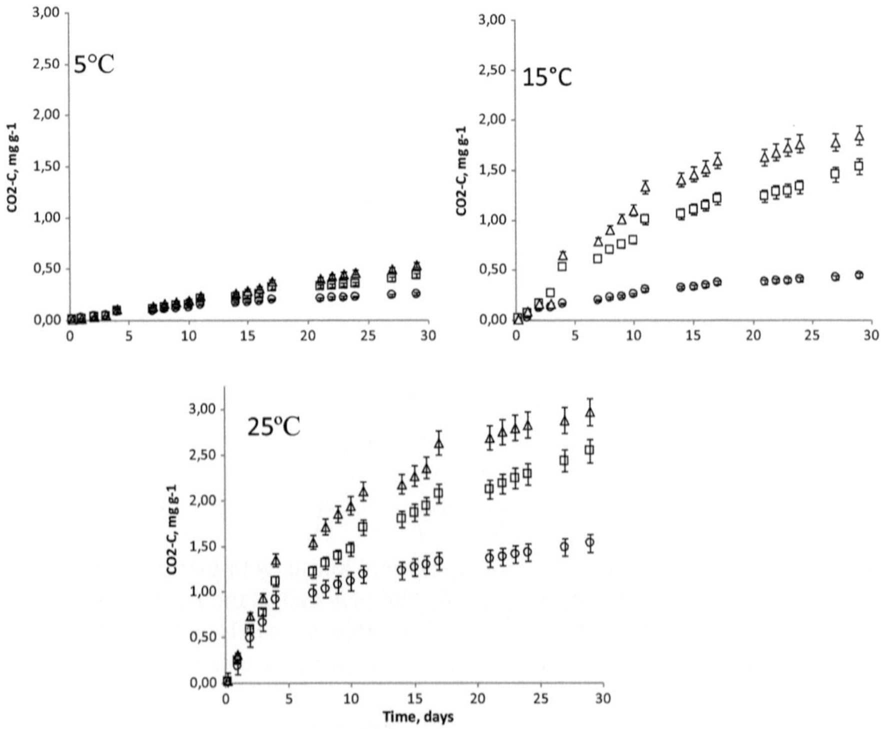

Figure 1.6 Cumulative CO_2-C production in the incubation experiments with different values of temperature: ○, Control, □, twigs, Δ, leaves.

immediately after the beginning of incubation and reached a maximum value after 8–10 days. A similar character of the dynamics of respiratory activity induced by substrates was described earlier in the works of other researchers (Thiessen et al., 2013; Malik et al., 2016).

To assess the effect of temperature on respiratory activity during the introduction of plant litter, the response coefficients for leaves (C_L) and twigs (C_T) were applied. The most significant response to substrate application was recorded for the 15°C experiment (Table 1.2). Addition of leaf fragments increased respiratory activity up to 4.15 times, and twigs addition to 3.45 times. For both other variants, the coefficients were twice as low. The dynamics of substrates decomposition is presented in Fig. 1.7. At a low temperature, a slight uniform decrease was observed throughout the observation period, and the total decrease for leaves was 0.28 mg C (6.8% of the initial content) and for twigs 0.19 (4.16%). At 15°C and 25°C substrates were decomposed most intensively from 5 to 20 days, and then the rate slowed down (Fig. 1.7). Leaves and twigs losses were 1.58 mg (37%) and 1.45 mg (33.9%) at 15°C and 1.09 mg C (23.49%) and 1.02 mg C (21.81%) at 25°C, respectively.

Even though the absolute values of cumulative respiration at 15°C and 25°C differed significantly, the amount of entirely oxidized substrate was similar. Thus an

Table 1.2 Influence of incubation temperature on the response of the microbial community to the substrates addition and decomposition rate of plant residues.

Temperature (°C)	Coefficient of the microbial community response to the substrate addition		Decomposition rate (mg C/(g day))	
	C_L	C_T	Leaves	Twigs
5	2.11	1.71	0.24	0.16
15	4.15	3.45	1.11	0.83
25	1.94	1.66	1.22	0.83

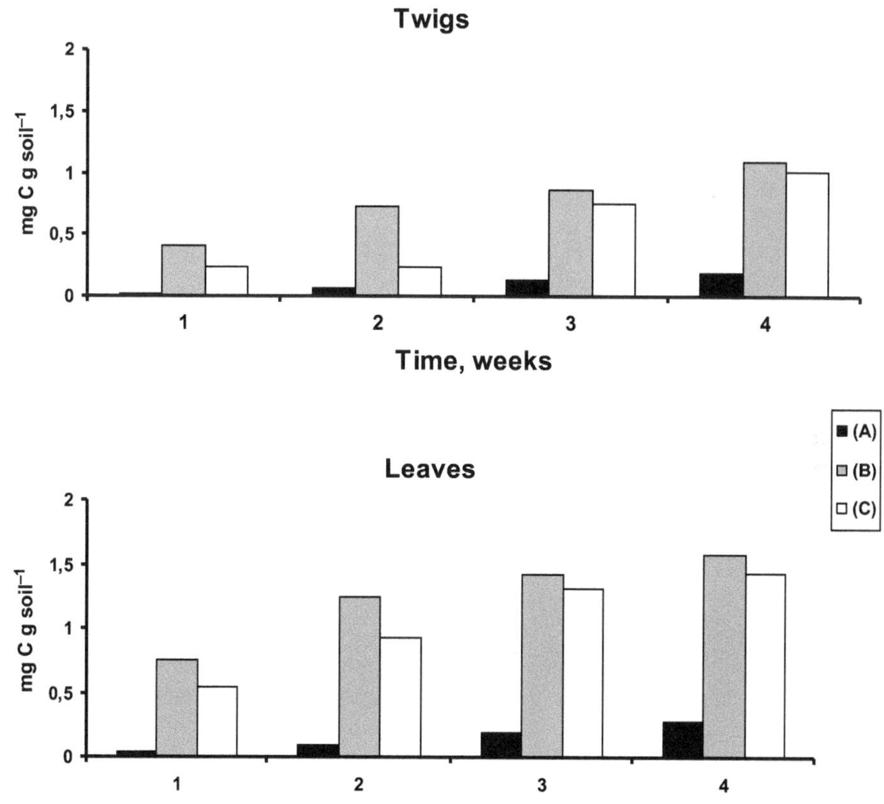

Figure 1.7 Dynamics of C loss from decomposing aspen leaves and twigs litter in the soil incubation experiments at (A) 5°C, (B) 15°C, and (C) 25°C.

increase in temperature led to a change in the ratio of carbon of fresh plant material and SOM in the CO_2 emission pool towards an increase in the proportion of soil carbon (so-called priming effect).

The principles of chemical kinetics suggest that the temperature sensitivity of the process depends on the quality of the substrate, and it can be assumed that the methods in which the share of the SOM higher than fresh organic substrates, should be more sensitive to temperature. SOM consists of complex organic molecules, and their decomposition requires high energy costs for activation. Therefore an increase in temperature will have a more significant impact on the dynamics of these components.

The temperature sensitivity of the substrate decomposition rate was characterized by a temperature coefficient (Q_{10}), which shows how the intensity of CO_2 emission has increased with the temperature. For the basal respiration, the main substrate is SOM, Q_{10} for the temperature range of 5°C–15°C was 1.75, and at 15°C–25°C increased to 3.4. At the same time in variants with leaves and twigs, the reverse pattern was observed—high value at low temperatures and decrease at high temperatures (Table 1.3). The Q_{10} in the present study was measured using the respiration of litter treated soils minus soil without litter (control). Such a technique was similar to the natural condition of litter decomposition. Still, it involved the risk of triggering a "priming effect" because of the input of labile substrates. The obtained values of Q_{10} were similar to that obtained at the initial stages of leaf litter decomposition in other model experiments with the soil of the broad-leaved conifer forest (Li et al., 2021), but were higher than those reported for the tundra and temperate forest (Christiansen et al., 2016; Tao et al., 2019), and lower than that reported for temperate grassland (Fierer et al., 2005). Therefore our findings confirmed that the temperature dependency of litter decomposition was linked to the ecosystem type and plant properties, and Q_{10} should be considered when predicting the feedback of global C cycle on climate change.

1.6.2 General characteristics of the soil microbial communities at the early stage of the microbial degradation

Soil microorganisms are integral parts of forest ecosystems and play a critical role in the decomposition of organic matter and the immobilization and mineralization of nutrients. Soil microbial biomass acts as an essential ecological indicator and is responsible for the decomposition and mineralization of plant residues in the soil (Schloter et al., 2018). Quantitative characteristics of microbial biomass are one of the main parameters, which reflect the increase in microbial activity in the decomposition of complex organic compounds. Litter treated soils did not show a significant increase in the microbial biomass at all temperature regimes and biomass

Table 1.3 Temperature coefficients (Q_{10}) for microbial processes in forest soil.

Variant **Temperature interval**	**Temperature coefficient (Q_{10})** **5°C–15°C**	 **5°C–25°C**
Basal respiration	1.75	3.44
Leaves decomposition	3.44	1.61
Wood litter decomposition	3.54	1.65

varied from 136 to 150 μg/ g for all variants. A possible explanation may be a lack of nitrogen available to microbial growth. The content of C: N in leaves material was 46.45 and in twigs 63.08. Another reason may be that the introduced substrates are used only for the transition of resting forms of soil microorganisms to the active state and further maintenance of their hydrolytic activity.

The share of microbial biomass carbon in the gross content of SOM (C_{MIC}: C_{ORG}) is the essential ecological and physiological parameter, characterizing the trophic level of microbial communities and soil biological activity (Marinari et al., 2006). This ratio is also an indicator of the availability of soil carbon for microorganisms, and the narrowing of this ratio indicates the stability of the organic substrate or the presence of conditions that prevent the development of its microorganisms (Semenov et al., 2018). Litter addition increased the content of SOM especially in experiments with 15°C and 25°C; however, the C fraction of microbial biomass remained constant and varied between 0.5% and 0.6% of the organic C content.

The method of real-time PCR (qPCR) estimation of the ribosomal gene's copies number was applied to obtain quantitative data of microorganisms in the process of primary stages of litter decomposition in the soil. The number of bacterial copies of 16s rRNA gene ranged from 4×10^8 to 9×10^9 per g of soil. Incubation with leaves and twigs led to an increase in the number of bacteria but no more than 2–7 times compared to the control. The number of fungal copies (18s rRNA) ranged from 6 to 340×10^6 per g of soil and increased in variants with substrates by 2–5 times. The number of archaea did not change significantly and was $0.3–0.7 \times 10^6$ per g of soil. The data obtained are consistent with the results of the determination of microbial biomass and indicate that in the primary stages of decomposition of fresh plant material there is an increase in the activity of hydrolytic microorganisms, but there is no significant increase in the number.

One of the approaches to understanding the functioning of microbial soil communities is their differentiation into ecologically significant groups, for example, copiotrophs and oligotrophs, autochthonous and zymogenous, r- and k-strategies, as well as eukaryotes (fungi) and prokaryotes (bacteria). The ratio of bacteria: fungi (B: F) well reflects the processes of decomposition of substrates, the transformation of nutrients and, as a consequence, the ability of soil ecosystems to self-regulation (Strickland and Rousk, 2010). The B: F ratio was in a narrow range from 1.10 to 1.44 for all variants, reflecting the approximate equilibrium in the number with a slight predominance of the bacterial component over the fungal one.

1.6.3 Succession of cultivated soil bacteria in response to temperature changes during litter decomposition

Over recent years, a growing interest in the study of soil bacteria, which is an important but poorly investigated component of the hydrolytic microbial communities of forest soils, can be observed. Bacteria are involved in the decomposition of plant residues, as well as of dead fungal mycelium; they also play a crucial role in the soil processes of the carbon, nitrogen, and phosphorus cycles.

Analysis of a model bacterial group, the members of which have specific characteristics and grow on the same nutrient medium, is a promising method for assessment of succession changes in the composition of microbial communities. It makes it possible to determine the proportion of different bacterial taxa and carry out analysis of the community using conventional synecological indicators. A complex of saprotrophic aerobic and facultative anaerobic bacteria growing on glucose-yeast-peptone medium supplemented with cycloheximide is widely used as such a model in soil microbiology (Dobrovol'skaya et al., 1997).

The annual cycle of plant residues transformation was simulated in a series of aerobic microcosms with gray forest soil (Gleyic Phaeozems) from the secondary small-leaved forest in the Moscow region. Microcosms were incubated for 360 days in thermostats at temperature simulating: (1) the average temperature in spring and autumn in the southern Moscow oblast (2°C); (2) the average summer temperature (12°C); and (3) global temperature increase (22°C). To identify the agents of transformation of plant residues, fragments of fallen leaves and thin branches of Norway maple, crushed to a size of <0.5 mm, were added to the soil (5% of the soil weight) (Tikhonova et al., 2019).

Succession studies can explain the biodynamics of soils and are therefore widely used in soil microbiology. The study indicated that litter addition did not increase the number of bacterial colony-forming unit (CFU) (Fig. 1.8) significantly, as well as in bacterial abundance and the value of microbial biomass at all tested temperatures. The succession coefficient (Cs), which shows the ratio of the bacterial quantity determined by direct counts to the number of bacteria according to the CFU number, was calculated at various stages of succession. Addition of leaves and branches resulted in a decrease in the Cs value after 1 month of incubation, which may be associated with a reduction of the diversity of microorganisms in the system and an increase in the proportion of rapidly growing bacteria (r-strategists). After 6 months, the Cs value increased, which indicated changes in the community and the succession stages at which microorganisms with a K strategy prevailed. Assessment of the Shannon index of microbial diversity showed that elevated temperature did not increase the index value, while low temperature decreased it significantly.

Taxonomic analysis of cultivated bacterial communities has revealed the predominance of the Beta- and Gammaproteobacteria, as well as Actinobacteria and Firmicutes, which agreed with the published data (Tlaskal et al., 2016). The predominant cultures were isolated and identified to a species level, and their abundance was monitored. At the early stages of succession, *Stenotrophomonas rhizophila (Xanthomonadaceae)* dominated, and its share was 30%–70% independent of the temperature. Patterns for some other predominant cultures are presented on Fig. 1.9. Temperature rise led to a significant increase (from 20% to 50%) in the abundance *Microbacterium* during succession. On the contrary, *Janthinobacterium* were minor (less than1%) at elevated temperatures, but reached 80% during low-temperature decomposition of the litter. *Bacillus* composed a significant part of the microbial community (15%–20%) at all stages of decomposition of plant litter.

Therefore the stability of the microbial community composition was shown at an elevated temperature and increased carbon inflow to the soils of boreal forests. At

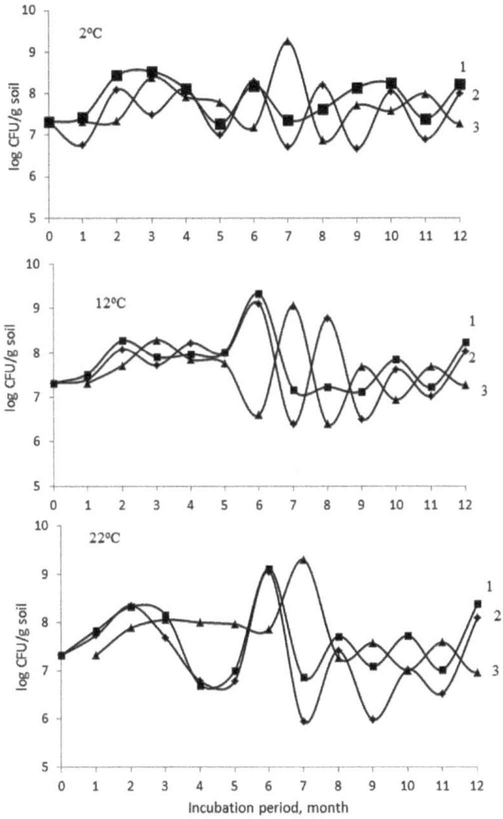

Figure 1.8 Temperature effect on the dynamics of cultivated saprotrophic bacteria number during succession initiated by addition of leaves and branch material: (1) litter addition; (2) control; (3) branch addition.

the same time, a decrease in temperature resulted in lower bacterial diversity and mono domination of psychrotolerant bacteria.

1.6.4 Succession changes of cultivated soil fungi in response to temperature changes during litter decomposition

For forest biomes the decomposition of plant substrates by soil fungi have been studied in sufficient detail: successional changes in the composition of species at different stages of destruction, properties and relationships of dominants, chemical changes during transformation (see, e.g., Krishna and Mohan, 2017; López-Mondéjar et al., 2018; Francioli et al., 2020). Recent studies have investigated the decomposition of leaf litter over an extended period, along with analyzing bacterial and fungal community composition (Purahong et al., 2016). Modern sequencing techniques have shifted from the relatively low taxonomic resolution to the

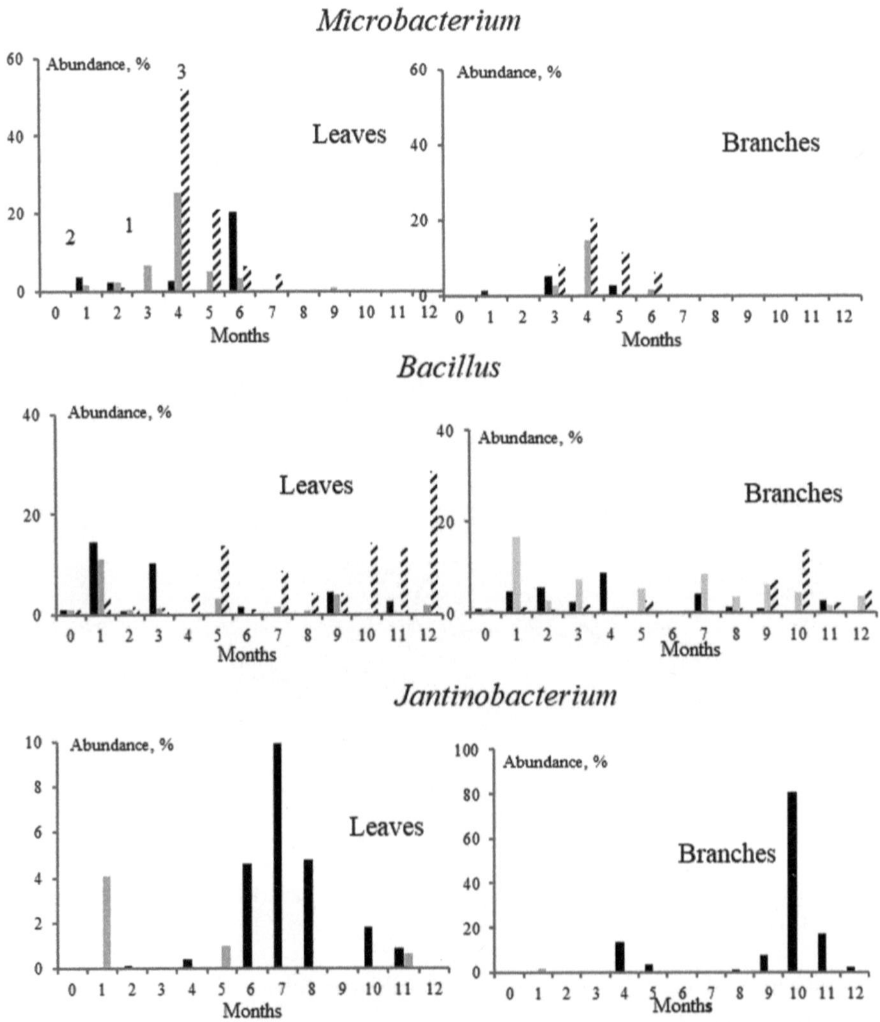

Figure 1.9 Changes in the abundance indices of some predominant species of soil bacteria in microcosms during succession, initiated by addition of leave and branch material; incubation at (1) 12°C, (2) 2°C, and (3) 22°C.

restrictions of useful and updated comparisons, which means that taxa might remain uncharacterized (Gilbert et al., 2014).

In boreal forests, up to 95% of the carbon from plant residues pass through the soil microbial biomass (Berg and Laskowski, 2006), and fungi generally dominate the microbial biomass and respiration (Joergensen and Wichern, 2008; Högberg et al., 2007). Decomposer fungi cycle up to 36 Gt carbon (C) per year through soils, releasing more CO_2 to the atmosphere than human emissions (Talbot, 2017). Nevertheless, we still have a limited understanding of how diverse fungal communities coordinate their activity to drive the decay of dead organic matter. Basidiomycetes fungi are considered especially important due to their production of ligninolytic enzymes essential for degradation of plant material (Osono and Takeda, 2002). Fungi are considered to be the main participants in the decomposition of litter in boreal forests, as they form a wide range of extracellular enzymes that allow efficient decomposition of various fractions of plant material (Krishna and Mohan, 2017). Plant litter consists of several groups of compounds, and dominant C costly components are soluble organic compounds, hemicellulose, cellulose, and lignin (Berg and McClaugherty, 2008). Soluble organic compounds consist of sugars, low molecular weight phenolics (tannins), hydrocarbons, and glycerides. These are generally easily degraded or lost from the litter through dissolution and leaching during the first stage of decomposition. Ascomycetous fungi constitute a large part of the fungal community in forest litter (Lindahl et al., 2007; O'Brien et al., 2005). However, the lignin-degrading capacity of most ascomycetes appears to be limited (Osono and Takeda, 2002).

Fungi are the significant decomposers of forest litter, but the identity and ecology of most litter fungi are still poorly understood. Despite their importance, the specific physiology of soil fungi is often disregarded in process-oriented researches. Studies have revealed correlations between fungal diversity and the local environmental conditions, including the chemical and physical soil properties (Větrovský et al., 2019). So, it is necessary to identify the diversity and successional variability of the fungal community during the decomposition of plant material as a priority area in the field of soil microbiology.

Usually, fungal succession in soil is studied by isolating the pure cultures (Frankland, 1998). The main drawback of this method is the discrimination against slow-growing and nonculturable fungal species (Lindahl and Boberg, 2008). Recently, molecular techniques have been developed, enabling DNA-based identification of the whole fungal community (O'Brien et al., 2005; Lindahl et al., 2007). Studies using the cultivation method together with molecular studies may be the most appropriate in the course of soil fungi in the soil environments and reveal the typical patterns during litter decomposition. Numerous experimental data showed a high degree of response of microscopic fungi to changes in environmental conditions and opportunities to use characteristics of fungal communities as a proxy to assess the quality of natural environments. The most appropriate and often used parameters for such analysis are taxonomic composition and indices of fungal species diversity.

Fig. 1.10 presents data of the culture-based soil fungal enumeration during the decomposition of the litter substrates under different temperatures. At the beginning of the incubation (time zero), the number of fungal CFU in soil samples was calculated as 3.9×10^3 in fresh soil equivalent to 1 g of oven-dried soil. Throughout the first 5−6 months of incubation, fungal CFU number remained practically unchanged regardless of substrate application and incubation temperature (Fig. 1.10). Then two peaks of increase in the number of viable (colony-forming) fungal propagules were observed only in the variants with the addition of the substrate. It was found that amendments with OMs increased the quantity of the fungal CFU significantly. After 6−7 months the number of CFU has increased up to 2−8

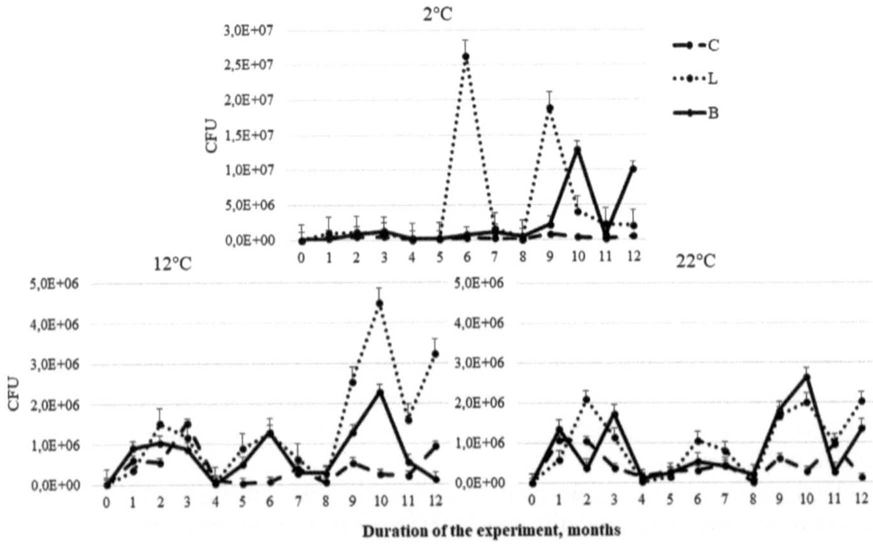

Figure 1.10 Annual dynamics of the cultivable fungal population's abundance (CFU) in soil samples of the microcosm study. *B*, Branch litter addition; *C*, control, without substrate addition; *L*, lives litter addition.

times and after 10 months by 1–2 orders of magnitude and reached $1.3–4.0 \times 10^6$ g^{-1} of soil.

The fungal colony counts showed different behavior in samples incubated at different temperatures. Control samples stored at 2°C demonstrated no increase in the number of viable colonies during all incubation period (12 months). In samples with leaves amendment, two peaks in the colonies number were found: 2.6×10^7 CFU/g was detected after 6 months and 1.8×10^7 CFU/g after 9 months (Fig. 1.10). Branches addition resulted in the drastic increase up to 1.4×10^7 CFU/g after 10 months of incubation and did not differ from the control ($P \leq .05$) in all other time.

A more complicated picture was observed at 12°C and 22°C incubations. At 12°C, there was an increase in the number of CFU in the soil samples with the introduction of substrates after 2, 6, and 10 months of incubation. After 2 and 6 months, the samples with both leaves and branches showed 1.00×10^6 CFU/g while after 10 months was 4.4×106 and 2.1×10^6 CFU/g in samples with leaves and twigs, respectively (Fig. 1.10). At an elevated temperature (22°C) leaves addition had increased CFU number up to 20×10^6 g^{-1} after 2 and 10 months and 1.0×10^6 after 6 months. Addition of branches showed 1.1, 1.8, and 2.7×10^6 g^{-1} after 1, 3, and 10 months of incubation, respectively.

It is interesting to note that the maximum fungal CFU values were found in soil samples incubated at 2°C, which indicates that the majority of the micromycete community in gray forest soil may be psychrophilic or psychrotolerant organisms.

We measured fungal hyphal length as a metric for total soil fungal abundance. Full hyphal length (32–104 m/g) increased during the first 3 months for all variants, reached up to 15,000–19,000 m/g at 12°C and 12,500–10,500 m/g at two and 22°C, and maintained at this level to the end of the incubation period. At the first stage of succession, the mycelium biomass in control soil was 0.4–0.6 mg/g. During incubation for all variants, mycelial biomass increased from the 3rd to the 6th months (intermediate succession stage) and remained approximately the same to the end (Fig. 1.11). In the control soil, it reached 5–16 mg/g of soil and increased to 10–17 and 15–30 mg in the samples amended with leaves and branches, respectively

The introduction of plant substrates led to a significant increase in the ratio of mycelium biomass of fungi and spores reflects the state of the community of soil fungi. Spore biomass was significantly lower than mycelium biomass in all variants, and varied from 0.06 to 0.30 mg/g (Fig. 1.11) and constituted about 3%–5% of the total fungal mass.

Integral indicators of the development of fungal soil complexes, the number of CFUs and the amount of biomass are widely used in research on the ecology of soil fungi. Culture-based methods constitute an excellent complement to modern DNA-based approaches. Also, it was shown that the number of CFUs is positively correlated with enzymes and respiratory activity, so may be applied to characterize the relative abundance of active microorganisms with specific functions (Blagodatskaya and Kuzyakov, 2013). Culture-dependent methods are not ideal for evaluating the actual composition of natural microbial communities but are useful for understanding development, and potential functions of soil microorganisms (Van Insberghe et al., 2013).

The diversity of soil fungi has been studied by conventional culture-based techniques, which are widely used in fungal ecological studies. However, culture-dependent approaches are typically selective because only a small fraction of microbial taxa grow under the conditions used for and slow-growing species that do not produce such reproductive structures can be missed by these approaches.

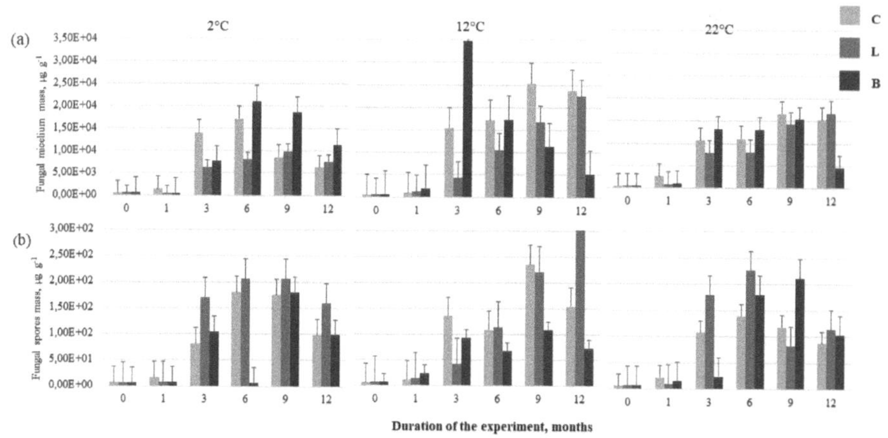

Figure 1.11 Changes of fungal mycelia (A) and spores (B) biomass in the soil at different succession stages of the microcosm study. *B*, Branch litter addition; *C*, control, without substrate addition; *L*, lives litter addition;.

During litter decomposition fungal succession was observed, wherein different fungal taxa dominate different stages of decay. Based on the culture-based observations, we have divided soil fungi into early, intermediate and late decomposers of forest litter. In most previous studies, fungi from the Ascomycota phylum were found to get the priority development during the initial stages of litter, and their abundance has decreased during the process of degradation (Osono, 2007; Voříšková and Baldrian, 2013) Our study demonstrated that *Penicillium* and *Mucor*, the diverse genera occurring worldwide and using a wide range of substrates, were the most-important fungal phyla at the early succession stages. These micromycetes are widespread and commonly isolated from the soils of northern regions (Wallenstein et al., 2007), including gray forest soils (Sorokin et al., 2008). The present investigation showed the increase of the Ascomycota diversity at all succession stages, including the late stage (Fig. 1.12). Ascomycetes are able to decompose complex substrates such as cellulose and are key agents of organic decomposition in soils (Frąc et al., 2018).

The introduction of substrates has increased the diversity of soil fungi and changed the structure of their community. The relative abundance of some native genera dominating the original soil decreased, and on the contrary, micromycetes that were minor became dominant.

Fungal community analysis showed that at the beginning of the experiment, the communities of micromycetes demonstrated the minor differences from the native

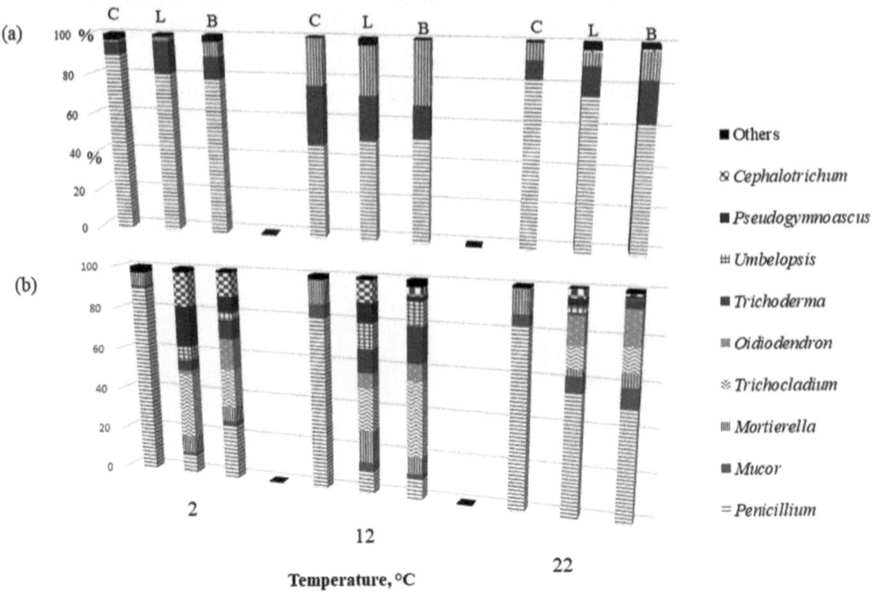

Figure 1.12 The relative abundances of the fungal genera in the (A) initial and (B) late successional stages in microcosm studies at different temperature regimes. The abundances that accounted for <1% of all of the classified sequences belong to others in the bar graph. *B*, Branch litter addition; *C*, control, without substrate addition; *L*, lives litter addition.

soil (Fig. 1.3), and the most abundant phylum was Ascomycota followed by Mucoromycota. Only two fungal dominants at the genera level were found, namely, *Penicillium* (Ascomycota) and *Mucor* (Mucoromycota).

In the initial stage of litter decomposition (1 month), in contrast to the control soil, high abundance of *Mortirella* (Mortierellomycota) (24%–31%) was found, but the diversity of fungal community remained quite low. From month three (early stage) diversity is increased, and the new dominants appeared. Along with *Penicillium* (Eurotiales order) and *Mucor* (Mucorales), the dominant group includes the primary saprotroph *Mortierella* (Mortierellales). At 2°C and 12°C, the proportion of *Mucor* decreased, and the abundance of *Trichocladium* (Sordariales) increased.

After the 12 months of the experiment (late phase of decomposition succession), the fungal communities were characterized by the increase in Ascomycota diversity and supplemented by representatives of the genera *Oidiodendron* (Euroteales), *Trichoderma* (Hypocreales), *Pseudogymnoascus* (Incertae sedis), and *Cephalotrichum* (Microascales) (Fig. 1.11). All these organisms are typical inhabitants of the gray forest soils of European Russia. In the control samples, the composition of micromycetes remained almost unchanged throughout the experiment.

The highest fungal diversity was found at the 12°C. Representatives of the genera *Trichocladium* (20%), *Mortierella* (15%), and *Umbrellaopsis* (Mortierellales) (12%), were predominant in leaves –amended samples but were absent in all other variants. The relative abundance of *Trichoderma* increased significantly (17%) in the variant with the branch's addition at 12°C (Fig. 1.11).

We designed a two-factor incubation experiment in the laboratory: one was litter quality (leaf and branch), while the other was temperature regime (2°C, 12°C, and 22°C). The results demonstrated a rapid shift in the composition of the fungal community. Although the quality of substrates is a critical factor influencing the composition of the soil fungal communities, experimental climate change modifies their succession. Soils with different temperature harbored compositionally distinct fungal communities when compared to soils with the same litter additions (Fig. 1.12). This study showed many similarities with results obtained by other researchers who studied different litter materials, highlighting that the successional process we observed seems to be universal.

In line with previous findings, the results of our study indicate that litter decomposition in forest soil is a highly complex process mediated by various fungal taxa. After 12 months of experimentally increased warming in boreal forest soil, fungal biomass values are not significantly different from those in controls. However, it has been demonstrated that warming affects fungal communities toward populations that are more adapted to the use of complex carbon sources. Members of the Ascomycota display increased diversity after simulated warming, and these shifts may affect the C cycling processes that involve soil fungi.

Our study was based on the conventional culture-based method and direct microscopic evaluation of fungal biomass. The combination of these methods is widely used in synecology study of soil environments. Ecological indicators such as frequency, the share of representatives of individual taxa, diversity, and dominance allow describing the structure of the soil fungi complexes. Comparison of the quantitative and qualitative characteristics with similar characteristics of microscopic fungi in the control soils can give an idea of the extent of the differences. More complex and comparable studies to assess the variables affected by global change in a wide range of forest ecosystems will be required to predict their future.

Although care should be taken when extrapolating results from laboratory microcosms to field conditions, this study demonstrated significant potential effects of temperature on different groups of fungi participated in forest litter decay. Global climate warming may cause shifts in vegetation and alter the litter input. We studied the decomposition of varying litter types under soil conditions to identify factors responsible for changes in the community structure and succession of fungal decomposers. This will help to predict the consequences of induced changes on the abundance and activity of microbial decomposers and recognize if the decomposition process and resulting quality and quantity of SOM will be affected at various sites.

1.7 Conclusion

Results provide an exciting insight into litter decomposition. We identified a diverse array of taxa assimilating, demonstrating the specialization of decomposer population's members that could collectively degrade more than one substrate. Bacterial diversity changes were different during the initial and other stages of decomposition. Microbial succession on decomposing aspen litter is rapid and initial communities are replaced with distinct assemblage characteristic for particular stages of decomposition. After 12 months of experimentally increased warming in boreal forest soil, fungal biomass values are not significantly different from those in controls. However, it has been demonstrated that warming affects fungal communities toward populations that are more adapted to the use of complex carbon sources. Members of the Ascomycota display increased diversity after simulated warming, and these shifts may affect the C cycling processes that involve soil fungi.

Our results indicate that not only does climate change and transplantation have differential legacy effects among bacteria and fungi, but also that bacterial communities might be less functionally redundant than fungi with regards to decomposition. We showed that decomposition responses to climate depend on the composition of microbial communities, which is not considered in terrestrial carbon models. Microbial communities varied in their effects on both mass loss and types of carbon decomposed in an interactive manner not predicted by current theory. Contrary to the traditional paradigm, bacterial communities appeared to have a more substantial impact on forest litter decomposition rates than fungi. Furthermore, bacterial communities shifted more rapidly in response to changing climates than fungi. This information is critical to improving global terrestrial carbon models and predicting ecosystem responses to climate change.

Thus it may be time to reevaluate both the role of microbial community composition in its decomposition response to climate and the relative roles of bacterial and fungal communities in decomposition. Changes in communities seem to be driven not only by physicochemical variables of litter but temperature change.

Microbial ecologists have traditionally considered fungi to be the primary agents of decomposition and for bacteria to play a minor role. Our results indicate that not only does climate change have differential effects among bacteria and fungi, but also those bacterial communities might be less functionally redundant than fungi with regards to decomposition. Thus it may be essential to reevaluate both the role of microbial community composition in its decomposition response to climate and the relative positions of bacterial and fungal communities in decomposition.

This chapter represents a contribution to our understanding of the intrinsic and external controls acting on the temperature sensitivity of decomposition, and thus to the regulation of CO_2 return to the atmosphere under a changing climate.

References

Ågren, G.I., Wetterstedt, J.A.M., 2007. What determines the temperature response of soil organic matter decomposition? Soil Biol. Biochem. 39 (7), 1794–1798.

Åhman, M., Nilsson, L.J., Johansson, B., 2017. Global climate policy and deep decarbonization of energy-intensive industries. Clim. Policy 17 (5), 634–649.

Allison, S.D., Wallenstein, M.D., Bradford, M.A., 2010. Soil-carbon response to warming dependent on microbial physiology. Nat. Geosci. 3 (5), 336–340.

Allison, S.D., Lu, Y., Weihe, C., Goulden, M.L., Martiny, A.C., Treseder, et al., 2013. Microbial abundance and composition influence litter decomposition response to environmental change. Ecology 94 (3), 714–725.

Anderson, J.P.E., Domsch, K.H., 1978. A physiological method for the quantitative measurement of microbial biomass in soils. Soil Biol. Biochem. 10, 215–221.

Baldock, J.A., Broos, K., 2011. Soil organic matter. In: Huang, P.M., Li, Y., Sumner, M.E. (Eds.), Handbook of Soil Sciences: Properties and Processes, second ed. Taylor & Francis Group, Boca Raton, FL, pp. 11.1–11.52.

Ball, A.S., 1997. Microbial decomposition at elevated CO_2 levels: effect of litter quality. Glob. Change Biol. 3, 379–386.

Bani, A., Pioli, S., Ventura, M., Panzacchi, P., Borruso, L., Tognetti, R., et al., 2018. The role of microbial community in the decomposition of leaf litter and deadwood. Appl. Soil Ecol. 126, 75–84.

Baruya, P., 2015. World Forest and Agricultural Crop Residue Resources for Cofiring. IEA Clean Coal Centre, London, p. 66.

Basilevich, N.I., 1993. Biological Productivity of Ecosystem in the North Eurasia. Nauka, Moscow, p. 293 (in Russian).

Bataille, C., Waisman, H., Colombier, M., Segafredo, L., Williams, J., Jotzo, F., 2016. The need for national deep decarbonization pathways for effective climate policy. Clim. Policy 16 (1), S7–S26.

Bataille, C., Åhman, M., Neuhoff, K., Nilsson, L.J., Fischedick, M., Lechtenböhmer, S., et al., 2018. A review of technology and policy deep decarbonization pathway options for making energy-intensive industry production consistent with the Paris Agreement. J. Clean Prod. 187, 960–973.

Baveye, P.C., Schnee, L.S., Boivin, P., Laba, M., Radulovich, R., 2020. Soil organic matter research and climate change: merely re-storing carbon versus restoring soil functions. Front. Environ. Sci.

Baveye, P.C., Wander, M., 2019. The (bio)chemistry of soil humus and humic substances: why is the "new view" still considered novel after more than 80 years? Front. Environ. Sci. 7 (27), 1–6.

Berg, B., 2000. Litter decomposition and organic matter turnover in northern forest soils. For. Ecol. Manage. 133 (1–2), 13–22.

Berg, B., Laskowski, R., 2006. Litter decomposition: a guide to carbon and nutrient turnover, Advances in Ecological Research, 38. Elsevier, Amsterdam, p. 448.

Berg, B., Matzner, E., 1997. Effect of N deposition on decomposition of plant litter and soil organic matter in forest systems. Environ. Rev. 5 (1), 1–25.

Berg, B., McClaugherty, C., 2003. Plant Litter: Decomposition, Humus Formation, Carbon Sequestration. United Kingdom. Springer.

Berg, B., McClaugherty, C., 2008. Plant Litter: Decomposition, Humus Formation, Carbon Sequestration. Springer-Verlag, Berlin, Heidelberg.

Berg, B., McClaugherty, C., 2014. Plant Litter: Decomposition, Humus Formation, Carbon Sequestration, third ed. Springer-Verlag, Berlin, Heidelberg, p. 315.

Bird, M.I., Wynn, J.G., Saiz, G., Wurster, C.M., McBeath, A., 2015. The pyrogenic carbon cycle. Annu. Rev. Earth Planet. Sci. 43, 273–298.

Blagodatskaya, E., Kuzyakov, Y., 2013. Active microorganisms in soil: critical review of estimation criteria and approaches. Soil Biol. Biochem. 67, 192–211.

Blagodatskaya, E, Blagodatsky, S., Khomyakov, N., Myachina, O., Kuzyakov, Y., 2016. Temperature sensitivity and enzymatic mechanisms of soil organic matter decomposition along an altitudinal gradient on Mount Kilimanjaro. Sci. Rep. 6, 22240.

Bolan, N.S., Adriano, D.C., Kunhikrishnan, A., James, T., McDowell, R., Senesi, N., 2011. Dissolved organic matter: biogeochemistry, dynamics, and environmental significance in soils. Adv. Agro. 110, 1–75.

Borken, W, Beese, F, 2005. Soil respiration in pure and mixed stands of European beech and Norway spruce following removal of organic horizons. Canadian Journal of Forest Research 35 (11), 2756–2764. Available from: https://doi.org/10.1139/x05-192.

Bosatta, E., Ågren, G., 1999. Soil organic matter quality interpreted thermodynamically. Soil Biol. Biochem. 31 (13), 1889–1891.

Brabcová, V., Štursová, M., Baldrian, P., 2018. Nutrient content affects the turnover of fungal biomass in forest topsoil and the composition of associated microbial communities. Soil Biol. Biochem. 118, 187–198.

Bradford, M.A., Berg, B., Maynard, D.S., Wieder, W.R., Wood, S.A., 2016. Understanding the dominant controls on litter decomposition. J. Ecol. 104, 229–238.

Burns, R.G., DeForest, J.L., Marxsen, J., Sinsabaugh, R.L., Stromberger, M.E., Wallenstein, M.D., et al., 2013. Soil enzymes in a changing environment: current knowledge and future directions. Soil Biol. Biochem. 58, 216–234.

Cambardella, C.A., Elliott, E.T., 1992. Particulate soil organic-matter changes across a grassland cultivation sequence. Soil Sci. Soc. Am. J. 56 (3), 777–783.

Castellano, M.J., Mueller, K.E., Olk, D.C., Sawyer, J.E., Six, J., 2015. Integrating plant litter quality, soil organic matter stabilization, and the carbon saturation concept. Glob. Change Biol. 21 (9), 3200–3209.

Chen, X., Ye, X., Chu, W., Olk, D.C., Cao, X., Schmidt-Rohr, K., et al., 2020. Formation of char-like, fused-ring aromatic structures from a nonpyrogenic pathway during decomposition of wheat straw. J. Agric. Food Chem. 68 (9), 2607–2614.

Chenu, C., Angers, D.A., Barré, P., Derrien, D., Arrouays, D., Balesdent, J., 2019. Increasing organic stocks in agricultural soils: knowledge gaps and potential Innovations. Soil Till. Res. 188, 41–52.

Christiansen, C.T., Haugwitz, M.S., Priemé, A., Nielsen, C.S., Elberling, B., Michelsen, A., et al., 2016. Enhanced summer warming reduces fungal decomposer diversity and litter mass loss more strongly in dry than in wet tundra. Glob. Change Biol. 23, 406–420.

Ciais, P., Sabine, C., Bala, G., Bopp, L., Brovkin, V., Canadell, J., et al., 2013. Carbon and other biogeochemical cycles. In: Stocker, T.F., et al., (Eds.), Climate Change 2013: The Physical Science Basis. Contribution of Working Group I to the Fifth Assessment Report of the Intergovernmental Panel on Climate Change. Cambridge University Press, Cambridge, United Kingdom and New York, pp. 465–570.

Clemmensen, K.E., Bahr, A., Ovaskainen, O., Dahlberg, A., Ekblad, A., Wallander, H., et al., 2013. Roots and associated fungi drive long-term carbon sequestration in boreal forest. Sci. 339, 1615–1618.

Clemmensen, K.E., Finlay, R.D., Dahlberg, A., Stenlid, J., Wardle, D.A., Lindahl, B.D., 2015. Carbon sequestration is related to mycorrhizal fungal community shifts during long-term succession in boreal forests. N. Phytol. 205, 1525–1536.

Cleveland, C.C., et al., 2014. Litter quality versus soil microbial community controls over decomposition: a quantitative analysis. Oecologia 174, 283–294.

Conant, R.T., Ryan, M.G., Ågren, G.I., Birge, H.E., Davidson, E.A., Eliasson, P.E., et al., 2011. Temperature and soil organic matter decomposition rates - synthesis of current knowledge and a way forward. Glob. Change Biol. 17, 3392–3404.

Conen, F., Leifeld, J., Seth, B., Alewell, C., 2006. Warming mineralises young and old soil carbon equally. Biogeosciences 3 (4), 515–519.

Cotrufo, M.F., Wallenstein, M.D., Boot, C.M., Denef, K., Paul, E., 2013. The microbial efficiency-matrix stabilization (MEMS) framework integrates plant litter decomposition with soil organic matter stabilization: do labile plant inputs form stable soil organic matter? Glob. Change Biol. 19 (4), 988–995.

Cotrufo, M.F., Soong, J.L., Horton, A.J., Campbell, E.E., Haddix, M.L., Wall, D.H., et al., 2015. Formation of soil organic matter via biochemical and physical pathways of litter mass loss. Nat. Geosci. 8, 776–779.

Cox, P.M., Betts, R.A., Jones, C.D., Spall, S.A., Totterdell, I.J., 2000. Acceleration of global warming due to carbon-cycle feedbacks in a coupled climate model. Nature 408 (6813), 750–750.

Crowther, T.W., van den Hoogen, J., Wan, J., Mayes, M.A., Keiser, A.D., Mo, L., et al., 2019. The global soil community and its influence on biogeochemistry. Science 365, eaav0550.

Davidson, E.A., Janssens, I.A., 2006. Temperature sensitivity of soil carbon decomposition and feedbacks to climate change. Nature 440, 165–173.

De Vries, W., Reinds, G.J., Gundersen, P., Sterba, H., 2006. The impact of nitrogen deposition on carbon sequestration in European forests and forest soils. Glob. Change Biol. 12 (7), 1151–1173.

del Giorgio, P.A., Cole, J.J., 1998. Bacterial growth efficiency in natural aquatic systems. Annu. Rev. Ecol. Syst. 29, 503–541.

Denman, K.L., Brasseur, G., Chidthaisong, A., Ciais, P., Cox, P.M., Dickinson, R.E., et al., 2007. Couplings between changes in the climate system and biogeochemistry. In: Solomon, S., Qin, D., Manning, M., Chen, Z., Marquis, M., Averyt, K.B., Tignor, M., Miller, H.L. (Eds.), Climate Change 2007: The Physical Science Basis. Contribution of Working Group I to the

Fourth Assessment Report of the Intergovernmental Panel on Climate Change. Cambridge University Press, Cambridge, United Kingdom and New York.

Didion, M., Repo, A., Liski, J., Forsius, M., Bierbaumer, M., Djukic, I., 2016. Towards harmonizing leaf litter decomposition studies using standard tea bags-a field study and model application. Forests 7 (8), 167.

Dijkstra, P., Thomas, S.C., Heinrich, P.L., Koch, G.W., Schwartz, E., Hungate, B.A., 2011. Effect of temperature on metabolic activity of intact microbial communities: evidence for altered metabolic pathway activity but not for increased maintenance respiration and reduced carbon use efficiency. Soil Biol. Biochem. 43 (10), 2023–2031.

Dobrovol'skaya, T.G., Chernov, I., Yu, Zvyagintsev, D.G., 1997. Characterizing the structure of bacterial communities. Microbiology (Mosc.) 66, 342–348.

Drotz, S.H., Sparrman, T., Nilsson, M.B., Schleucher, J., Öquist, M.G., 2010. Both catabolic and anabolic heterotrophic microbial activity proceed in frozen soils. Proc. Natl. Acad. Sci. U S A. 107 (49), 21046–21051.

Ekschmitt, K., Liu, M., Vetter, S., Fox, O., Wolters, V., 2005. Strategies used by soil biota to overcome soil organic matter stability – why is dead organic matter left over in the soil? Geoderma 128, 167–176.

Erhagen, B., Öquist, M., Sparrman, T., Haei, M., Ilstedt, U., Hedenström, M., et al., 2013. Temperature response of litter and soil organic matter decomposition is determined by chemical composition of organic material. Glob. Change Biol. 19 (11).

Fang, C., Smith, P., Moncrieff, B.J., Smith, U.J., 2005. Similar response of labile and resistant soil organic matter pools to changes in temperature. Nature 433, 57–59.

Feng, X.J., Simpson, M.J., 2008. Temperature responses of individual soil organic matter components. J. Geoph. Res. Biogeosci. 113 (G3), G03036.

Fierer, N., Craine, J.M., McLauchlan, K., Schimel, J.P., 2005. Litter quality and the temperature sensitivity of decomposition. Ecology 86 (2), 320–326.

Frąc, M., Hannula, S.E., Bełk, M., Jędryczka, M., 2018. Fungal biodiversity and their role in soil health. Front. Microbiol. 9, 707.

Francioli, D., van Rijssel, S.Q., van Ruijven, J., Termorshuizen, A.J., Cotton, T.E.A., Dumbrell, A.J., et al., 2020. Plant functional group drives the community structure of saprophytic fungi in a grassland biodiversity experiment. Plant Soil .

Frankland, J.C., 1998. Fungal succession-unravelling the unpredictable. Mycol. Res. 102 (1), 1–15.

Freschet, G.T., Cornwell, W.K., Wardle, D.A., Elumeeva, T.G., Liu, W., Jackson, B.G., et al., 2013. Linking litter decomposition of above- and below-ground organs to plant–soil feedbacks worldwide. J. Ecol. 101, 943–952.

Frey, S.D., Lee, J., Mellilo, J.M., Six, J., 2013. The temperature response of soil microbial efficiency and its feedback to climate. Nat. Clim. Change 3, 395–398.

Galloway, J.N., Townsend, A.R., Erisman, J.W., Bekunda, M., Cai, Z., Freney, J.R., et al., 2008. Transformation of the nitrogen cycle: recent trends, questions, and potential solutions. Science 320 (5878), 889–892.

Gambhir, A., Tavoni, M., 2019. Direct air carbon capture and sequestration: how it works and how it could contribute to climate-change mitigation. One Earth 1 (4), 405–409.

Giardina, C.P., Ryan, M., 2000. Evidence that decomposition rates of organic carbon in mineral soil do not vary with temperature. Nature 404 (6780), 858–861.

Gilbert, J.A., Jansson, J.K., Knight, R., 2014. The earth microbiome project: successes and aspirations. BMC Biol. 12, 69.

Grandy, A.S., Neff, J.C., 2008. Molecular C dynamics downstream: the biochemical decomposition sequence and its impact on soil organic matter structure and function. Sci. Total Environ. 404 (2–3), 297–307.

Haddix, M.L., Gregorich, E.G., Helgason, B.L., Janzen, H., Ellert, B.H., Cotrufo, M.F., 2020. Climate, carbon content, and soil texture control the independent formation and persistence of particulate and mineral-associated organic matter in soil. Geoderma 363, 114160.

Hall, E.K., Cotner, J.B., 2007. Interactive effect of temperature and resources on carbon cycling by freshwater bacterioplankton communities. Aquat. Microb. Ecol. 49, 35–45.

Hartley, I.P., Ineson, P., 2008. Substrate quality and the temperature sensitivity of soil organic matter decomposition. Soil Biol. Biochem. 40 (7), 1567–1574.

Hedges, J.I., Eglinton, G., Hatcher, P.G., Kirchman, D.L., Arnosti, C., Derenne, S., et al., 2000. The molecularly-uncharacterized component of nonliving organic matter in natural environments. Org. Geochem. 31 (10), 945–958.

Hobbie, S.E., 1996. Temperature and plant species control over litter decomposition in Alaskan tundra. Ecol. Monogr. 66, 503–522.

Hopkins, F.M., Filley, T.R., Gleixner, G., Lange, M., Top, S.M., Trumbore, S.E., 2014. Increased belowground carbon inputs and warming promote loss of soil organic carbon through complementary microbial responses. Soil Biol. Biochem. 76, 57–69.

Högberg, M.N., Högberg, P., 2002. Extrametrical ectomycorrhizal mycelium contributes one-third of microbial biomass and produces, together with associated roots, half the dissolved organic carbon in a forest soil. N. Phytol. 154 (3), 791–795.

Högberg, M.N., Högberg, P., Myrold, D.D., 2007. Is microbial community composition in boreal forest soils determined by pH, C-to-N ratio, the trees, or all three? Oecologia 150 (4), 590–601.

Hu, R., Wang, X., Xu, J., Zhang, Y., Su, Y.P.X., 2020a. The mechanism of soil nitrogen transformation under different biocrusts to warming and reduced precipitation: from microbial functional genes to enzyme activity. Sci. Total Environ. 722, 137849.

Hu, Y., Zheng, Q., Noll, L., Zhang, S., Wanek, W., 2020b. Direct measurement of the in situ decomposition of microbial-derived soil organic matter. Soil Biol. Biochem. 141, 107660.

Ilstedt, U., Giesler, R., Nordgren, A., Malmer, A., 2003. Changes in soil chemical and microbial properties after a wildfire in a tropical rainforest in Sabah, Malaysia. Soil Biol. Biochem. 35 (8), 1071–1078.

IPCC, 2007. Working Group I Contribution to the IPCC Fourth Assessment Report. Climate Changes the Physical Science Basis. Cambridge University Press, Cambridge, United Kingdom.

Ivanov, I.V., Pesochina, L.S., Semenov, V.M., 2009. Biological mineralization of organic matter in the modern virgin and plowed chernozems, buried chernozems, and fossil chernozems. Eur. Soil Sci. 42 (10), 1109–1119.

Ivanov, A.L., Kogut, B.M., Semenov, V.M., Turina Oberlander, M.I., Waksman Schanbacher, N., 2017. The development of theory on humus and soil organic matter: from Turin and Waksman to present days. Byulleten Pochvennogo instituta im. V.V. Dokuchaeva 90, 3–38.

Janssens, I.A., Dieleman, W., Luyssaert, S., Subke, J.A., Reichstein, M., Ceulemans, R., et al., 2010. Reduction of forest soil respiration in response to nitrogen deposition. Nat. Geosci. 3 (5), 315–322.

Janzen, H.H., 2006. The soil carbon dilemma: shall we hoard it or use it? Soil Biol. Biochem. 38, 419–424.

Joergensen, R.G., Wichern, F., 2008. Quantitative assessment of the fungal contribution to microbial tissue in soil. Soil Biol. Biochem. 40 (12), 2977–2991.

Kalbitz, K., Solinger, S., Park, J.-H., Michalzik, B., Matzner, E., 2000. Controls on the dynamics of dissolved organic matter in soils: a review. Soil Sci. 165 (4), 277–304.

Kallenbach, C.M., Frey, S.D., Grandy, A.S., 2016. Direct evidence for microbial-derived soil organic matter formation and its ecophysiological controls. Nat. Commun. 7, 13630.

Kardol, P., Cregger, M.A., Campany, C.E., Classen, A.T., 2010. Soil ecosystem functioning under climate change: plant species and community effects. Ecology 91, 767–781.

Kirschbaum, M.U.F., 2000. Will changes in soil organic carbon act as a positive or negative feedback on global warming? Biogeochemistry 48 (1), 21–51.

Kleber, M., Lehmann, J., 2019. Humic substances extracted by alkali are invalid proxies for the dynamics and functions of organic matter in terrestrial and aquatic ecosystems. J. Environ. Qual. 48 (2), 207–216.

Knorr, W., Prentice, L.C., House, J.I., Holland, A.E., 2005. Long-term sensitivity of soil carbon turnover to warming. Nature 433 (7023), 298–301.

Kögel-Knabner, I., 2002. The macromolecular organic composition of plant and microbial residues as inputs to soil organic matter. Soil Biol. Biochem. 34 (2), 139–162.

Kögel-Knabner, I., 2017. The macromolecular organic composition of plant and microbial residues as inputs to soil organic matter: fourteen years on. Soil Biol. Biochem. 105, A3–A8.

Kögel-Knabner, I., Ekschmitt, K., Flessa, H., Guggenberger, G., Matzner, E., Marschner, B., et al., 2008a. An integrative approach of organic matter stabilization in temperate soils: linking chemistry, physics, and biology. J. Plant Nutr. Soil Sci. 171 (1), 5–13.

Kögel-Knabner, I., Guggenberger, G., Kleber, M., Kandeler, E., Kalbitz, K., Scheu, S., et al., 2008b. Organo-mineral associations in temperate soils: integrating biology, mineralogy, and organic matter chemistry. J. Plant Nutr. Soil Sci. 171 (1), 61–82.

Kravchenko, I.K., Tikhonova, E.N., Ulanova, R.V., Menko, E.V., Sukhacheva, M.V., 2019. Effect of temperature on litter decomposition, soil microbial community structure and biomass in a mixed-wood forest in European Russia. Curr. Sci. 116 (5), 765–772.

Krishna, M.P., Mohan, M., 2017. Litter decomposition in forest ecosystems: a review. Energy Ecol. Environ. 2 (4), 236–249.

Krull, E., Baldock, J.A., Skjemstad, J.O., 2003. Importance of mechanisms and processes of the stabilisation of soil organic matter for modelling carbon turnover. Funct. Plant Biol. 30 (2), 207–222.

Lal, R., 2010. Managing soils and ecosystems for mitigating anthropogenic carbon emissions and advancing global food security. BioScience 60, 708–721.

Lal, R., Lorenz, K., Hüttl, R.F., Schneider, B.U., von Braun, J., 2012. Research and development priorities towards recarbonization of the biosphere. In: Lal, R., Lorenz, K., Hüttl, R., Schneider, B., von Braun, J. (Eds.), Recarbonization of the Biosphere. Springer, Dordrecht, pp. 533–544.

Landry, J.S., Matthews, H.D., 2017. The global pyrogenic carbon cycle and its impact on the level of atmospheric CO_2 over past and future centuries. Glob. Change Biol. 23 (8), 3205–3218.

Lavallee, J.M., Soong, J.L., Cotrufo, M.F., 2020. Conceptualizing soil organic matter into particulate and mineral-associated forms to address global change in the 21st century. Glob. Change Biol. 26 (1), 261–273.

Lehmann, J., Kleber, M., 2015. The contentious nature of soil organic matter. Nature 528, 60–68.

Lehmann, J., Hansel, C.M., Kaiser, C., Kleber, M., Maher, K., Manzoni, S., et al., 2020. Persistence of soil organic carbon caused by functional complexity. Nat. Geosci. 13, 529–534.

Leifeld, J., von Lützow, M., 2014. Chemical and microbial activation energies of soil organic matter decomposition. Biol. Fertil. Soils 50, 147–153.

Li, R., Zhang, Y., Yu, D., Wang, Y., Zhao, X., Zhang, R., et al., 2021. The decomposition of green leaf litter is less temperature sensitive than that of senescent leaf litter: an incubation study. Geoderma 381, 114691.
Liang, C., Schimel, J.P., Jastrow, J.D., 2017. The importance of anabolism in microbial control over soil carbon storage. Nat. Microbiol. 2, 17105.
Lindahl, B., Boberg, J., 2008. Distribution and function of litter Basidiomycetes in coniferous forests. In: Boddy, L., Frankland, J., van West, P. (Eds.), Ecology of Saprotrophic Basidiomycetes. Elsevier Academic Press, London, pp. 183–196. (British Mycological Society Symposium Series, Volume 28).
Lindahl, B.D., Ihrmark, K., Boberg, J., Trumbore, S.E., Högberg, P., Stenlid, J., et al., 2007. Spatial separation of litter decomposition and mycorrhizal nitrogen uptake in a boreal forest. N. Phytol. 173 (3), 611–620.
Liu, L.L., Greaver, T.L., 2010. A global perspective on belowground carbon dynamics under nitrogen enrichment. Ecol. Lett. 13 (7), 819–828.
López-Mondéjar, R., Brabcová, V., Štursová, M., Davidová, A., Jansa, J., Cajthaml, T., et al., 2018. Decomposer food web in a deciduous forest shows high share of generalist microorganisms and importance of microbial biomass recycling. ISME J. 12 (7), 1768–1778.
Lorenz, K., Preston, C.M., Raspe, S., Morrison, I.K., Feger, K.H., 2000. Litter decomposition and humus characteristics in Canadian and German spruce ecosystems: information from tannin analysis and C-13 CPMAS NMR. Soil Biol. Biochem. 32 (6), 779–792.
Lozano-García, B., Francaviglia, R., Renzi, G., Doro, L., Ledda, L., Benítez, C., et al., 2020. Land use change effects on soil organic carbon store. An opportunity to soils regeneration in Mediterranean areas: implications in the 4p1000 notion. Ecol. Indic. 119, 106831.
Lutzow, M.V., Kogel-Knabner, I., Ekschmitt, K., Matzner, E., Guggenberger, G., Marschner, B., et al., 2006. Stabilization of organic matter in temperate soils: mechanisms and their relevance under different soil conditions - a review. Eur. J. Soil Sci. 57 (4), 426–445.
Malik, Z.A., Pandey, R., Bhatt, A.B., 2016. Anthropogenic disturbances and their impact on vegetation in Western Himalaya, India. J. Mt. Sci. 13, 69–82.
Manlay, R.J., Feller, C., Swift, M.J., 2007. Historical evolution of soil organic matter concepts and their relationships with the fertility and sustainability of cropping systems. Agric. Ecosyst. Environ. 119, 217–233.
Manzoni, S., Taylor, P., Richter, A., Porporato, A., Agren, G.I., 2012. Environmental and stoichiometric controls on microbial carbon-use efficiency in soils. N. Phytol. 196 (1), 79–91.
Marinari, S., Mancinelli, R., Campiglia, E., Grego, S., 2006. Chemical and biological indicators of soil quality in organic and conventional farming systems in Central Italy. Ecol. Indic. 6, 701–711.
Martiny, J.B.H., et al., 2017. Microbial legacies alter decomposition in response to simulated global change. ISME J. 11, 490–499.
Meehl, G.A., Stocker, T.F., Collins, W.D., Friedlingstein, P., Gaye, A.T., Gregory, J.M., et al., 2007. Climate Change 2007: The Physical Science Basis. Contribution of Working Group I to the Fourth Assessment Report of the Intergovernmental Panel on Climate Change. Cambridge University Press, Cambridge, United Kingdom and New York.
Mohinuzzaman, M., Yuan, J., Yang, X., Senesi, N., Li, S.-L., Ellam, R.M., et al., 2020. Insights into solubility of soil humic substances and their fluorescence characterisation in three characteristic soils. Sci. Total Environ. 720, 137395.
Moore C., J., McCann, K., de Ruiter C., P., 2005. Modeling trophic pathways, nutrient cycling, and dynamic stability in soils. Pedobiologia 49 (6), 499–510. Available from: https://doi.org/10.1016/j.pedobi.2005.05.008.

Nelson, P.N., Baldock, J.A., 2005. Estimating the molecular composition of a diverse range of natural organic materials from solid-state C-13 NMR and elemental analyses. Biogeochemistry 72 (1), 1–34.

O'Brien, H.E., Parrent, J.L., Jackson, J.A., Moncalvo, J.M., Vilgalys, R., 2005. Fungal community analysis by large-scale sequencing of environmental samples. Appl. Environ. Microbiol. 71 (9), 5544–5550.

Olk, D.C., Bloom, P.R., Perdue, E.M., McKnight, D.M., Chen, Y., Farenhorst, A., et al., 2019. Environmental and agricultural relevance of humic fractions extracted by alkali from soils and natural waters. J. Environ. Qual. 48 (2), 217–232.

Olson, K.R., Al-Kaisi, M.M., Lal, R., Lowery, B., 2014. Experimental consideration, treatments, and methods in determining soil organic carbon sequestration rates. Soil Sci. Soc. Am. J. 78 (2), 348–360.

Osono, T., 2007. Ecology of ligninolytic fungi associated with leaf litter decomposition. Ecological Research 22, 955–974. Available from: https://doi.org/10.1007/s11284-007-0390-z.

Osono, T., Takeda, H., 2002. Comparison of litter decomposing ability among diverse fungi in a cool temperate deciduous forest in Japan. Mycologia 94 (3), 421–427.

Pan, Y., Birdsey, R.A., Fang, J., Houghton, R., Kauppi, P.E., Kurz, W.A., et al., 2011. A large and persistent carbon sink in the world's forests. Science 333 (6045), 988–993.

Paul, E.A., 2016. The nature and dynamics of soil organic matter: plant inputs, microbial transformations, and organic matter stabilization. Soil Biol. Biochem. 98, 109–126.

Petraglia, A., Cacciatori, C., Chelli, S., Fenu, G., Calderisi, G., Gargano, D., et al., 2019. Litter decomposition: effects of temperature driven by soil moisture and vegetation type. Plant Soil 435, 187–200.

Pinskii, D.L., Maltseva, A.N., Zolotareva, B.N., Dmitrieva, E.D., 2017. Transformation kinetics of corn and clover residues in mineral substrates of different composition. Eur. Soil Sci. 50, 681–687.

Prescott, C.E., 2010. Litter decomposition: what controls it and how can we alter it to sequester more carbon in forest soils? Biogeochemistry 101, 133–149.

Preston, C.M., Trofymow, J.A., Flanagan, L.B., 2006. Decomposition, $\delta^{13}C$, and the "lignin paradox.". Can. J. Soil Sci. 86 (2), 235–245.

Purahong, W., Wubet, T., Lentendu, G., Schloter, M., Pecyna, M.J., Kapturska, D., et al., 2016. Life in leaf litter: novel insights into community dynamics of bacteria and fungi during litter decomposition. Mol. Ecol. 25 (16), 4059–4074.

Rakhleeva, A.A., Semenova, T.A., Striganova, B.R., Terekhova, V.A., 2011. Dynamics of zoomicrobial complexes upon decomposition of plant litter in spruce forests of the southern taiga. Eur. Soil Sci. 44, 38–48.

Rasse, D.P., Rumpel, C., Dignac, M.-F., 2005. Is soil carbon mostly root carbon? Mechanisms for a specific stabilization. Plant Soil 269, 341–356.

Reisser, M., Purves, R.S., Schmidt, M.W.I., Abiven, S., 2016. Pyrogenic carbon in soils: a literature-based inventory and a global estimation of its content in soil organic carbon and stocks. Front. Earth Sci. 4 (80), 1–14.

Rustad, L.E., 2008. The response of terrestrial ecosystems to global climate change: towards an integrated approach. Sci. Total Environ. 404 (2–3), 222–235.

Schaeffer, A., Nannipieri, P., Kästner, M., Schmidt, B., Botterweck, J., 2015. From humic substances to soil organic matter – microbial contributions. In honour of Konrad Haider and James P. Martin for their outstanding research contribution to soil science. J. Soils Sediments 5 (9), 1865–1881.

Schimel, J., 2001. Biogeochemical models: implicit versus explicit microbiology. In: Schulze, E.D., et al., (Eds.), Global Biogeochemical Cycles in the Climate System. Academic, New York:, pp. 177–183.

Schimel, J.P., Weintraub, M.N., 2003. The implications of exoenzyme activity on microbial carbon and nitrogen limitation in soil: a theoretical model. Soil Biol. Biochem. 35, 549–563.

Schlesinger, W.H., 1984. Soil organic matter: a source of atmospheric CO_2. In: The Role of Terrestrial Vegetation in the Global Carbon Cycle: Measurement by Remote Sensing (Scope 23) ed. by G.M. Woodwell, London, pp. 111–127.

Schloter, M., Nannipieri, P., Sørensen, S.J., van Elsas, J.D., 2018. Microbial indicators for soil quality. Biol. Fertil. Soils 54, 1–10.

Schmidt, M.W.I., Torn, M.S., Abiven, S., Dittmar, T., Guggenberger, G., Janssens, I.A., et al., 2011. Persistence of soil organic matter as an ecosystem property. Nature 478, 49–56.

Semenov, V.M., Ivannikova, L.A., Kuznetsova, T.V., Tulina, A.S., Kudeyarov, V.N., 2001. Kinetic analysis of the decomposition and mineralization of plant residues in gray forest soil. Eur. Soil Sci. 34, 503–511.

Semenov, V.M., Ivannikova, L.A., Kuznetsova, T.V., Semenova, N.A., 2004. The role of plant biomass in the formation of the active pool of soil organic matter. Eur. Soil Sci. 37, 1196–1204.

Semenov, V.M., Tulina, A.S., Semenova, N.A., Ivannikova, L.A., 2013. Humification and nonhumification pathways of the organic matter stabilization in soil: a review. Eur. Soil Sci. 46, 355–368.

Semenov, V.M., Kogut, B.M., Zinyakova, N.B., Masyutenko, N.P., Malyukov, S., Lebedeva, T.N., et al., 2018. Biologically active organic matter in soils of European Russia. Eur. Soil Sci. 51, 434–447.

Semenov, V.M., Lebedeva, T.N., Pautova, N.B., 2019a. Particulate organic matter in noncultivated and arable soils. Eur. Soil Sci. 52 (4), 396–404.

Semenov, V.M., Pautova, N.B., Lebedeva, T.N., Khromychkina, D.P., Semenova, N.A., Lopes de Gerenyu, V.O., 2019b. Plant residues decomposition and formation of active organic matter in the soil of the incubation experiments. Eur. Soil Sci. 52 (10), 1183–1194.

Simpson, M.J., Simpson, A.J., 2012. The chemical ecology of soil organic matter molecular constituents. J. Chem. Ecol. 38, 768–784.

Simpson, A.J., Simpson, M.J., Smith, E., Kelleher, B.P., 2007. Microbially derived inputs to soil organic matter: are current estimates too low? Environ. Sci. Technol. 41, 8070–8076.

Smith, P., 2016. Soil carbon sequestration and biochar as negative emission technologies. Glob. Change Biol. 22 (3), 1315–1324.

Sollins, P., Homann, P., Caldwell, B.A., 1996. Stabilization and destabilization of soil organic matter: mechanisms and controls. Geoderma 74 (1–2), 65–105.

Sorokin, N.D., Evgrafova, S.Yu, Grodnitskaya, I.D., Bogorodskaya, A.V., 2008. Ecological characteristics of microflora in cryogenic forest soils of the North of Central Siberia. Contemp. Probl. Ecol. 6, 859–865.

Spokas, K.A., Cantrell, K.B., Novak, J.M., Archer, D.W., Ippolito, J.A., Collins, H.P., et al., 2012. Biochar: a synthesis of its agronomic impact beyond carbon sequestration. J. Environ. Qual. 41 (4), 973–989.

Steinweg, J.M., Plante, A.F., Conant, R.T., Paul, E.A., Tanaka, D.L., 2008. Patterns of substrate utilization during long-term incubations at different temperatures. Soil Biol. Biochem. 40 (11), 2722–2728.

Stevenson, F.J., 1994. Humus Chemistry: Genesis, Composition, Reactions. John Wiley and Sons, New York, p. 496.

Stockmann, U., Adams, M.A., Crawford, J.W., Field, D.J., Henakaarchchi, N., Jenkins, M., et al., 2013. The knowns, known unknowns and unknowns of sequestration of soil organic carbon. Agric. Ecosyst. Environ. 164 (1), 80–99.

Strickland, M.S., Rousk, J., 2010. Considering fungal:bacterial dominance in soils - methods, controls, and ecosystem implications. Soil Biol. Biochem. 42, 1385–1395.

Štursová, M., Kohout, P., Human, R.Z., Baldrian, P., 2020. Production of fungal mycelia in a temperate coniferous forest shows distinct seasonal patterns. J. Fungi 6, 190.

Talbot, J.M., 2017. Fungal communities and climate change. In: Dighton, J., White, J.G. (Eds.), The Fungal Community: Its Organization and Role in the Ecosystem, fourth ed. CRC Press, Taylor & Francis Group, Boca Raton, FL.

Tao, B.X., Zhang, B.H., Dong, J., Liu, C.Y., Cui, Q., 2019. Antagonistic effect of nitrogen additions and warming on litter decomposition in the coastal wetland of the Yellow River Delta, China. Ecol. Eng. 131, 1–8.

Tarnocai, C., Canadell, J.G., Schuur, E.A.G., Kuhry, P., Mazhitova, G., Zimov, S., 2009. Soil organic carbon pools in the northern circumpolar permafrost region. Glob. Biogeochem. Cycles 23, GB2023.

Thiessen, S., Gleixner, G., Wutzler, T., Reichstein, M., 2013. Both priming and temperature sensitivity of soil organic matter decomposition depend on microbial biomass – an incubation study. Soil Biol. Biochem. 57, 739–748.

Tikhonova, E.N., Men'ko, E.V., Ulanova, R.V., Li, H., Kravchenko, I.K., 2019. Effect of temperature on the taxonomic structure of soil bacterial communities during litter decomposition. Microbiology 88 (6), 781–785.

Tlaskal, V., Voriskova, J., Baldrian, P., 2016. Bacterial succession on decomposing leaf litter exhibits a specific occurrence pattern of cellulolytic taxa and potential decomposers of fungal mycelia. FEMS Microbiol. Ecol. 92 (11), 1–10.

Trofimov, S.Y., Lazarev, A.S., Fokin, A.D., 2012. Mineralization of organic-matter labile fragments in the humus-accumulative horizon of soddy-podzolic soil. Eur. Soil Sci. 45, 1110–1119.

Van Insberghe, D., Hartmann, M., Stewart, G.R., Mohn, W.W., 2013. Isolation of a substantial proportion of forest soil bacterial communities detected via pyrotag sequencing. Appl. Environ. Microbiol. 79, 2096–2098.

van der Wal, A., Gunnewiek, P., Boer, W., 2017. Soil-wood interactions: influence of decaying coniferous and broadleaf logs on composition of soil fungal communities. Fungal Ecol. 132–134.

Vedrova, E.F., 1997. Organic matter decomposition in forest litters. Eur. Soil Sci. 30 (2), 181–188.

Větrovský, T., Kohout, P., Kopecký, M., Machac, A., Man, M., Bahnmann, B.D., et al., 2019. A meta-analysis of global fungal distribution reveals climate-driven patterns. Nat. Commun. 10, 5142.

Vitousek, P.M., Mooney, H.A., Lubchenco, J., Melillo, J.M., 1997. Human domination of Earth's ecosystems. Science 277 (5325), 494–499.

von Lützow, M., Kögel-Knabner, I., Ekschmitt, K., Matzner, E., Guggenberger, G., Marschner, B., et al., 2006. Stabilization of organic matter in temperate soils: mechanisms and their relevance under different soil conditions – a review. Eur. J. Soil Sci. 57, 426–445.

von Lützow, M., Kögel-Knabner, I., Ekschmitt, K., Flessa, H., Guggenberger, G., Matzner, E., et al., 2007. SOM fractionation methods: relevance to functional pools and to stabilization mechanisms. Soil Biol. Biochem. 39, 2183–2207.

von Lützow, M., Kögel-Knabner, I., Ludwig, B., Matzner, E., Flessa, H., Ekschmitt, K., et al., 2008. Stabilization mechanisms of organic matter in four temperate soils: development and application of a conceptual model. J. Plant Nutr. Soil Sci. 171, 111–124.

Voříšková, J., Baldrian, P., 2013. Fungal community on decomposing leaf litter undergoes rapid successional changes. ISME Journal 7 (3), 477–486. Available from: https://doi.org/10.1038/ismej.2012.116.

Wallenstein, M.D., McMahon, S., Schimel, J., 2007. Bacterial and fungal community structure in arctic tundra tussock and shrub soils. FEMS Microbiol. Ecol. 59 (2), 428–435.

Wander, M., 2004. Soil organic matter fractions and their relevance to soil function. In: Magdoff, F., Weil, R.R. (Eds.), Soil Organic Matter in Sustainable Agriculture. CRC Press, Boca Raton, FL, pp. 67–102.

Wells, M.J.M., Stretz, H.A., 2019. Supramolecular architectures of natural organic matter. Sci. Total Environ. 671, 1125–1133.

Wetterstedt, J.A.M., Ågren, G.I., 2011. Quality or decomposer efficiency - which is most important in the temperature response of litter decomposition? A modelling study using the GLUE methodology. Biogeosciences 8, 477–487.

Wetterstedt, J.A.M., Persson, T., Agren, G.I., 2010. Temperature sensitivity and substrate quality in soil organic matter decomposition: results of an incubation study with three substrates. Glob. Change Biol. 16 (6), 1806–1819.

WMO Greenhouse Gas Bulletin. The State of Greenhouse Gases in the Atmosphere Based on Global Observations Through 2018, 2019, WMO. No 15. 8 p. https://library.wmo.int/doc_num.php?explnum_id = 10100.

Wu, Z., Dijkstra, P., Koch, G.W., Peñuelas, J., Hungate, B.A., 2011. Responses of terrestrial ecosystems to temperature and precipitation change: a meta-analysis of experimental manipulation. Glob. Change Biol. 17, 927–942.

Zak, D.R., Pellitier, P.T., Argiro, W.A., Castillo, B., James, T.Y., Nave, L.E., et al., 2019. Exploring the role of ectomycorrhizal fungi in soil carbon dynamics. N. Phytol. 223, 33–39.

Zhang, D., Hui, D., Luo, Y., Zhou, G., 2008. Rates of litter decomposition in terrestrial ecosystems: global patterns and controlling factors. J. Plant Ecol. 1 (2), 85–93.

Zsolnay, Á., 2003. Dissolved organic matter: artefacts, definitions and functions. Geoderma 113 (3–4), 187–209.

Further reading

De Nobili, M., Bravoa, C., Chen, Y., 2020. The spontaneous secondary synthesis of soil organic matter components: a critical examination of the soil continuum model theory. Appl. Soil Ecol. 154, 103655.

Glassmana, S.I., Weihe, C., Li, J., Albright, M.B.N., Looby, C.I., Martinya, A.C., et al., 2018. Decomposition responses to climate depend on microbial community composition. PNAS 115 (47), 11994–11999.

Makkonen, M., et al., 2012. Highly consistent effects of plant litter identity and functional traits on decomposition across a latitudinal gradient. Ecol. Lett. 15, 1033–1041.

Semenov, V.M., Ivannikova, L.A., Kuznetsova, T.V., Semenova, N.A., Tulina, A.S., 2008. Mineralization of organic matter and the carbon sequestration capacity of zonal soils. Eur. Soil Sci. 41 (7), 717–730.

von Lützow, M., Kögel-Knabner, I., 2009. Temperature sensitivity of soil organic matter decomposition — what do we know? Biol. Fert. Soils 46, 1–15.

Wang, S., Ruan, H., Han, Y., 2010. Effects of microclimate, litter type, and mesh size on leaf litter decomposition along an elevation gradient in the Wuyi Mountains, China. Ecol. Res. 25, 1113–1120.

Adaptation of bacterial communities and plant strategies for amelioration and eco-restoration of an organometallic industrial waste polluted site

2

Sonam Tripathi, Kshitij Singh and Ram Chandra
Department of Microbiology, School of Environmental Sciences, Babasaheb Bhimrao Ambedkar University (A Central University), Lucknow, India

2.1 Introduction

Adaptation is the evolutionary process whereby an organism adjusts to living in its habitat. The literary meaning of adaptation (from the Latin adaptation, "making suitable to" or "making consistent with") is understood as a complex of physiological, morphological, and behavioral responses of an organism, aimed at changing its rate of growth/metabolism and viability (survivability), which is genetically programmed and inherent in the organism. The ultimate goal of adaptation is to ensure the survival of a given population of organisms and the species to which the condition they belong and where stress is understood to mean pressure, strain, and force, or, in other words, a considerable change in the intensity of a certain factor (e.g., temperature). The word "shock" denotes a blow, a collision, or an impact, that is, a drastic effect on an organism, which acts over shorter periods than the length of the cell cycle or the time required to develop an adaptive response; the effect of a shock is also considerable in intensity. Hence, bacteria have evolved to exploit the temporal and spatial structure of their native environments, mapping commonly occurring patterns of stimuli to high-fitness responses (Freddolino and Tavazoie, 2012; Tagkopoulos et al., 2008). Thus adaptation occurs through both the acquisition of requisite biochemical and biophysical functions, such as enzymatic capabilities and membrane properties, and the evolution of a regulatory network that responds to the environment by deploying the organism's phenotypic capacities in a context-appropriate fashion as required (Hottes et al., 2013). In principle, bacteria may grow very slowly in a new environment because they lack necessary biochemical capabilities and biophysical properties or because they express these existing capacities inappropriately at the initial (Tripathi et al., 2021). In a new environment, however, the same bacteria may grow poorly due to regulatory constraints or biochemical deficiencies. Adaptation to such conditions can proceed through the

Microbes in Land Use Change Management. DOI: https://doi.org/10.1016/B978-0-12-824448-7.00005-X
© 2021 Elsevier Inc. All rights reserved.

acquisition of new cellular functions due to the gain of functional mutations (horizontal gene transfer) or via modulation of cellular networks. When bacteria encounter a new challenge in their environment, such as treatment with an antibiotic or a poor nutrient source, their population faces tremendous selective pressure to evolve to grow better under the new conditions. We typically think of bacterial evolution in terms of what is gained: a bacterium might, for example, acquire an antibiotic resistance gene, or modify an existing enzyme to make better use of a nutrient source (Eitam et al., 2012). Therefore adaptation is not always a simple matter where the ideal phenotype evolves for a given environment. An organism must be viable at all stages of its development and all stages of its evolution. This places constraints on the evolution of development, behavior, and structure of organisms (Nikolaev, 2004). Thus all adaptations help organisms survive in their ecological niches. The adaptive traits may be structural, behavioral, or physiological. Structural adaptations are physical features of an organism, such as shape, body covering, armament, and internal organization. Behavioral adaptations are inherited systems of behavior, whether inherited in detail as instincts, or as a neuropsychological capacity for learning. Adaptation differs from flexibility, acclimatization, and learning, all of these are changes during life that are not inherited. The flexibility deals with the relative capacity of an organism to maintain itself in different habitats: its degree of specialization. Acclimatization describes automatic physiological adjustments during a lifetime (Matin, 1992), learning means improvement in behavioral performance during the life cycle (Welch and Bartlett, 1996). Extremophilic microorganisms thrive in harsh environments where other organisms cannot even survive due to adverse environments. Extremophiles are taxonomically widely distributed and have a functionally diverse group that includes thermophiles (high temperatures in volcanic springs), psychrophiles (low temperatures in the cold polar regions), acidophiles (low pH conditions), alkalophiles (high pH conditions), halophiles (very high salt concentrations), and barophiles/piezophiles (high pressure in the deep sea) (Cowan et al., 2015). Extremophiles have the potential to produce various biomolecules of high relevance for white, gray, and red biotechnological sectors. These microorganisms produce extremophilic enzymes as protective organic biomolecules (extremolytes), which convey survival characteristics under extreme environmental conditions. However, the presence of highly sophisticated mechanisms of adaptation together with the availability of a sweet of novel biochemical pathways sustaining peculiar physiological or metabolic capabilities converts extremophiles into a current focus of applied research to exploit their biotechnological potential for industrial application (Rampelotto, 2013). Biocatalysts are whole microbial cells or enzymes that can be used in biochemical reactions of modern biotechnology. Some bacteria can grow and survive in a very adverse condition either due to their special physiological adaptation or a broad range of their genetic frame. These groups of bacteria are known as an extremophile. Extremophiles have various strategies for their growth and survival in diverse extreme conditions. Recently, interest in biocatalysts has increased with biotechnology growing and developing as a strategy to achieve a bio-based economy. White biotechnology (or industrial) requires tackling the environmental and economic issues associated with increasing demands for energy and fuel and resulting petroleum-based product prices. It uses biocatalysts to convert renewable resources into biopolymers, biomaterials, fine chemicals, and biofuels such as waste and by-products. Gray biotechnology (or environmental) applies biocatalysts to polluted areas for bioremediation, while red biotechnology (or pharmaceutical/medical) utilizes microorganisms for pharmaceutical development. To date, the majority of enzymes on the market are of bacterial or fungal origin, although few are archaea derived, most of which were produced by mesophilic microorganisms that are often inhibited under the extreme conditions of many industrial processes. Therefore to define robust biocatalysts, the search for new sources of separation, experimental procedures, and analytical methods is expanding recently. Specifically, extremophiles received attention and several have been obtained in pure culture, their genomes have been also analyzed, and their enzymes have been identified by either academic or industrial applications (Tripathi et al., 2021a, 2021b).

In an environment, important extremophilic enzymes have been reported as a ligninolytic enzyme that constitutes laccase, lignin peroxidase (LiP), and manganese peroxidase (MnP). Such enzymes function in normal to diverse conditions on a wide range of their substrates. Microorganisms with thermostable enzyme systems reduce the possibility of microbial contamination in long-lasting, large-scale industrial reactions. The mechanisms for many thermotolerant enzymes have been reported due to their structural properties (i.e., copper, iron, manganese, and calcium ions) specificity (hydrophobic interaction, salt bridge, disulfide bridge, and hydrogen bond), saturated fatty acid, and α-helical structure it retains extremophilic property and stability (Rampelotto, 2013; Welch and Bartlett, 1996). The ever-increasing demand in industrial sectors for these enzymes necessitates the production of enzymes from cheap raw materials. It is a well-known fact that most agro-industrial residues in nature are lignocellulosic waste. The production of enzymes from agro-industrial residues is of great significance owing to two reasons: (1) the process helps in the reuse of industrial wastes and thereby reduces the problems arising out of their disposal; and (2) the enzymes produced from the residues can be utilized for various industrial applications. In general, a plant strategy includes the mechanism and responses of the plant in their habitat for growth, survival, and reproduction. The phytoremediation process in the polluted environment for the growth and survival of plants is a common strategy for the eco-restoration of any polluted site. The main strategies include phytostabilization, phytovolatilization, phytosequestration, rhizofiltration, phytoextraction, phytotransformation, phytostimulation. Phytoremediation potential is also assisted by a unique microbial community known as rhizospheric bacterial communities for adaptation of plant growing at the polluted site for eco-restoration.

2.2 Environmental adaptation of bacterial communities

Extremophilic microorganisms have established a diversity of molecular strategies to survive in an extreme environment. During the past few decades, it has been revealed that microbial communities can grow in the most extreme conditions, including extremes of temperature, pH, salinity, and pressure. These extremophilic microorganisms can produce various enzymes and proteins to regulate their physiological functions under extreme conditions. Extremophilic enzymes are considered superior to normal enzymes, due to broad range adaptation and various industrial applications even under extreme conditions in completely denatured. Extremophilic microorganisms thrive in even harsh environments where other organisms cannot even survive. Therefore extremophiles are taxonomically widely distributed and have a functionally diverse group which includes thermophiles (high temperatures in volcanic springs), psychrophiles (low temperatures in the cold polar regions), acidophiles (low pH conditions), alkalophilic (high pH conditions), halophiles (very high salt concentrations), and barophiles/piezophiles (high pressure in the deep sea) (Cowan et al., 2015). Extremophiles have the potential to produce biomolecules of high relevance for white, gray, and red biotechnological sectors. These microorganisms produce extremophilic enzymes and protective organic biomolecules (extremolytes), which convey survival characteristics under extreme environmental conditions. The various thermophile mechanisms are shown in Fig. 2.1.

2.2.1 Thermophiles

Thermophilic microorganisms are one of the most observed extremophiles over the last centuries. They have the potential to live at higher temperatures between 41°C and 122°C (Dumorné et al., 2017). A wide number of enzymes from thermophilic microorganisms have been characterized, such as laccase, LiP, and MnP. Enzymes from thermophilic microorganisms are capable of accepting proteolysis and extreme situations like the presence of denaturing agents and organic solvents and high salinity (Dumorné et al., 2017). The use of these enzymes includes the possibility to reduce the risk of contamination, keeping a low adhesiveness, and greater solubility of substrates. The thermozymes possess the physical property and electrostatic interactions to keep activity at higher temperatures. The main mechanism in thermophiles is the role of heat shock proteins including the chaperones GroEL, GroES, and DnaK, to assist protein folding and DNA-repair systems are also active to respond to DNA damage as shown in Fig. 2.1 (Orellana et al., 2018). The stability of thermophilic proteins was attributed to greater hydrophobicity, deletion or shortening of loops, better atom packing (Russell, 1998), smaller and fewer cavities, and increased surface area buried upon oligomerization, residue substitution within and outside the secondary structures (Zuber, 1988), increased occurrence of proline residues in loops (Haney et al., 1997), decreased occurrence of thermolabile residues (Russell, 1998; Kumar et al., 2000), increased helical content, increased polar surface area, better hydrogen bonding, and better salt bridge formation (Kumar and Nussinov, 2001). Amino acids like glycine, lysine, tyrosine, and isoleucine are preferred in mesophiles. In which, glutamic acid was substituted with two positive (lysine and arginine) and one hydrophobic (alanine) residues (Chandra and Chowdhary, 2015). The percentage of proline residues from *Bacillus pumilus* W3 CotA is 9.0% (46 pro residues and 513 total residues), and those of other laccases of known structure are 8% for TvLa, 8.2% for CcLa, and 6.2% for CueO (Enguita

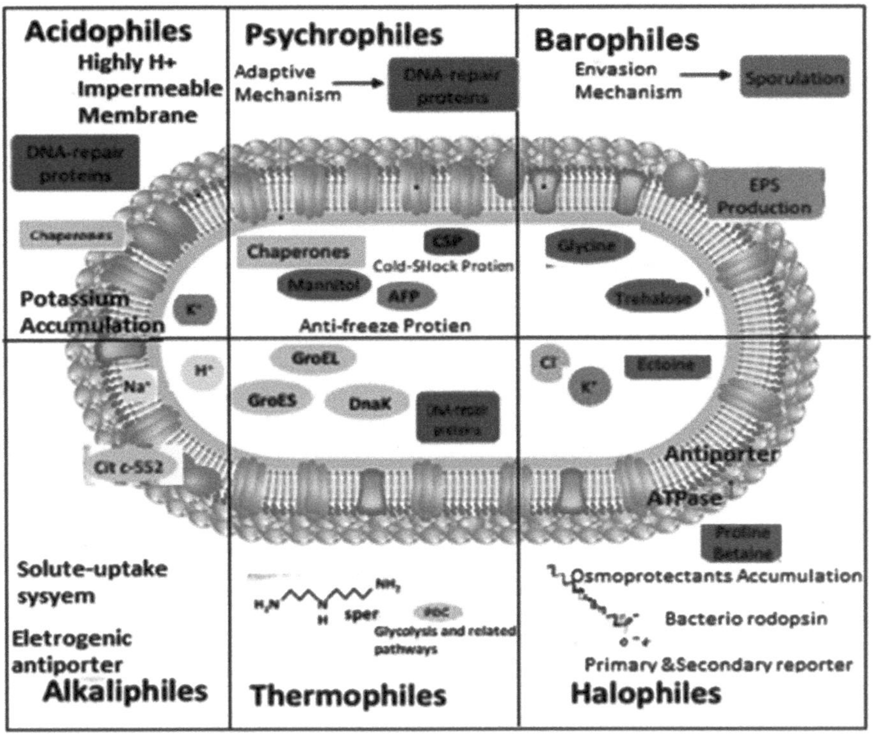

Figure 2.1 Bacterial cell membrane mechanisms of various extremophiles and their adaptation.
Source: Modified from Orellana, R., Macaya, C., Bravo, G., Dorochesi, F., Cumsille, A., Valencia, R., et al., 2018. Living at the frontiers of life: extremophiles in Chile and their potential for bioremediation. Front. Microbiol. 9, 2309.

et al., 2003). Besides, laccase Cot A from *Bacillus subtilis* at 75°C showed the maximum stability of enzymes. Copper ions also play key roles in the stability of MCOs. However, the other Ca^{2+} on an α-helix also contained residues that were proposed to be involved in binding H_2O_2 and Mn^{2+}. The removal of this Ca^{2+} may result in a conformational change in this α-helix which renders the enzyme the inactivation of the enzyme.

2.2.2 Psychrophiles

Psychrophiles are extremophiles that are adapted to an extremely cold environment. Psychrophiles are true extremophiles because they are not only adapted to low temperature, but also certain environmental constraints. Many psychrophiles live in biotopes having more than one stress factor. In the ocean depths and sediments, they are faced with extremely high pressures or high salt concentration and therefore they are termed as piezo-psychrophiles or holo-psychrophiles (Feller and Gerday,

2003). A diverse range of psychrophilic microorganisms, belonging to gram-positive bacteria (*Arthrobacter, Bacillus*, and *Micrococcus* sp.), and gram-negative bacteria (*Pseudoalteromonas, Pseudomonas* sp., *Psychrobacter, Polaribacter, Polaromonas, Moritella, Moraxella*, and *Vibrio*), archaea (*Methanococcoides, Halorubrum* sp., and *Methanogenium*) and several fungi (*Cladosporium* and *Penicillium*) can be found in these environments and display striking cold-adaptive characteristics (Feller and Gerday, 2003). The cellular adaptations at low temperatures and their molecular mechanisms are not fully understood so far. Research has included on membrane fluidity regulation, protein synthesis maintenance, and cold-acclimation protein (CAP) production, and the mechanisms of freeze tolerance or avoidance (Feller and Gerday, 2003). The microbial membranes are lipid bilayers that must have the proper fluidity to support the permeability properties and motions of essential membrane proteins. Another important and general feature of cold-adapted microorganism appears to be CAPs as shown in Fig. 2.2 (Hébraud and Potier, 1999). This set of ~20 proteins is synthesized at low temperatures, but not at milder temperatures, during steady-state growth (Hebraud et al., 1994; Berger et al., 1996). Interestingly, some of the few CAPs identified in cold-adapted bacteria, such as the RNA chaperone CspA, are cold-shock proteins in mesophiles (Berger et al., 1997). These CAPs have been proposed to be essential for maintaining both growth and the cell cycle at low temperatures, but their function is still poorly understood. The ability of psychrophilic enzymes to catalyze reactions at low temperatures offers great industrial and biotechnological potential. While incorporating unique features in their proteins and membranes, psychrophiles have developed adaptive mechanisms to perform their metabolic functions at low temperatures (Feller and Gerday, 2003).

2.2.3 Acidophiles

Acidophiles are organisms that grow at an optimum pH below 3–4. These are a diverse group of organisms included in archaea, bacteria, fungi, algae, and protozoa growing in acidic conditions, reported from natural environments like solfataric fields, sulfuric pools, and geysers, and artificial environments like areas associated with human activities like mining of coal and metal ores (Sharma et al., 2012). While, most of the extremely acidophiles belong to the archaeal group that includes *Acidianus, Desulfurococcus, Metallosphaera, Stygiolobus, Sulfolobus, Sulfurisphaera, Pyrococcus, Thermoplasma*, and *Picrophilus* (Bertoldo et al., 2004). Enzymes from microorganisms that can survive under extreme pH could be particularly useful for application under highly acidic conditions. However, one of the striking properties of acidophilic microorganisms is their ability to maintain a neutral pH internally and so the intracellular enzymes from these microorganisms do not need to be adapted to extreme growth conditions as shown in Fig. 2.3A and B (Kumar et al., 2011). However, this does not account for extracellular proteins that have to function in low pH environments in the case of acidophiles. One of the most striking properties of acidophilic microorganisms is their use of proton pumps to maintain a neutral pH internally and so the intracellular enzymes from these

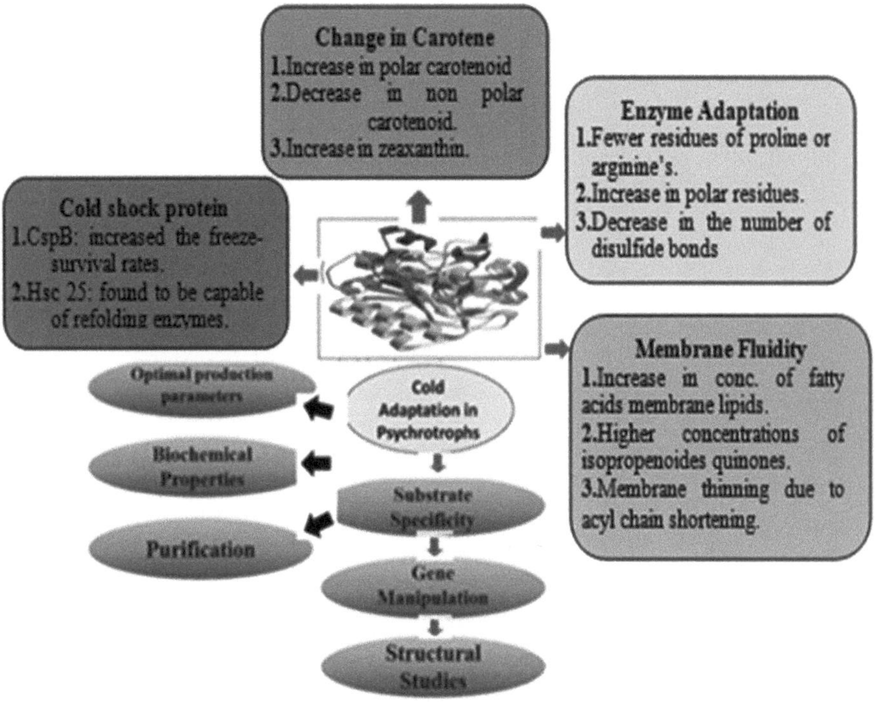

Figure 2.2 Protein involvement in the survival of psychrophiles and their adaptation strategies.
Source: Modified from Furhan, J., 2020. Adaptation, production, and biotechnological potential of cold-adapted proteases from psychrophiles and psychrotrophs: recent overview. J. Genet. Eng. Biotechnol. 18 (1), 1–13.

microorganisms do not need to be adapted to extreme growth conditions (Kumar et al., 2011). However, the extracellular enzyme proteins of acidophiles have to function at low pH. How these extracellular proteins operate at low pH values is yet poorly understood. For the cells to survive in the aggressive condition of pH, acidophiles utilize several strategies. To withstand low pH, acidophiles employ a range of mechanisms such as a positively charged membrane surface a high internal buffer capacity, over expression of H^+ exporting enzymes, and unique transport systems (Wiegel and Kevbrin, 2004). Besides, it has been predicted that the organisms develop specific metabolic properties, genetic features, structural and functional characteristics of their macromolecules that help maintain the pH and distinguish them from their neutrophilic counterparts (Baker-Austin and Dopson, 2007). Other stability features of acidophilic organisms such as DNA-repair proteins, ATP synthase, chaperones, membrane highly impermeable to protons, and potassium antiporter releases protons towards the extracellular medium (Orellana et al., 2018). Many acidophilic proteins have a surface that is negatively charged and low pH as shown in Fig. 2.3A and B (Fushinobu et al., 1998). In this figure:

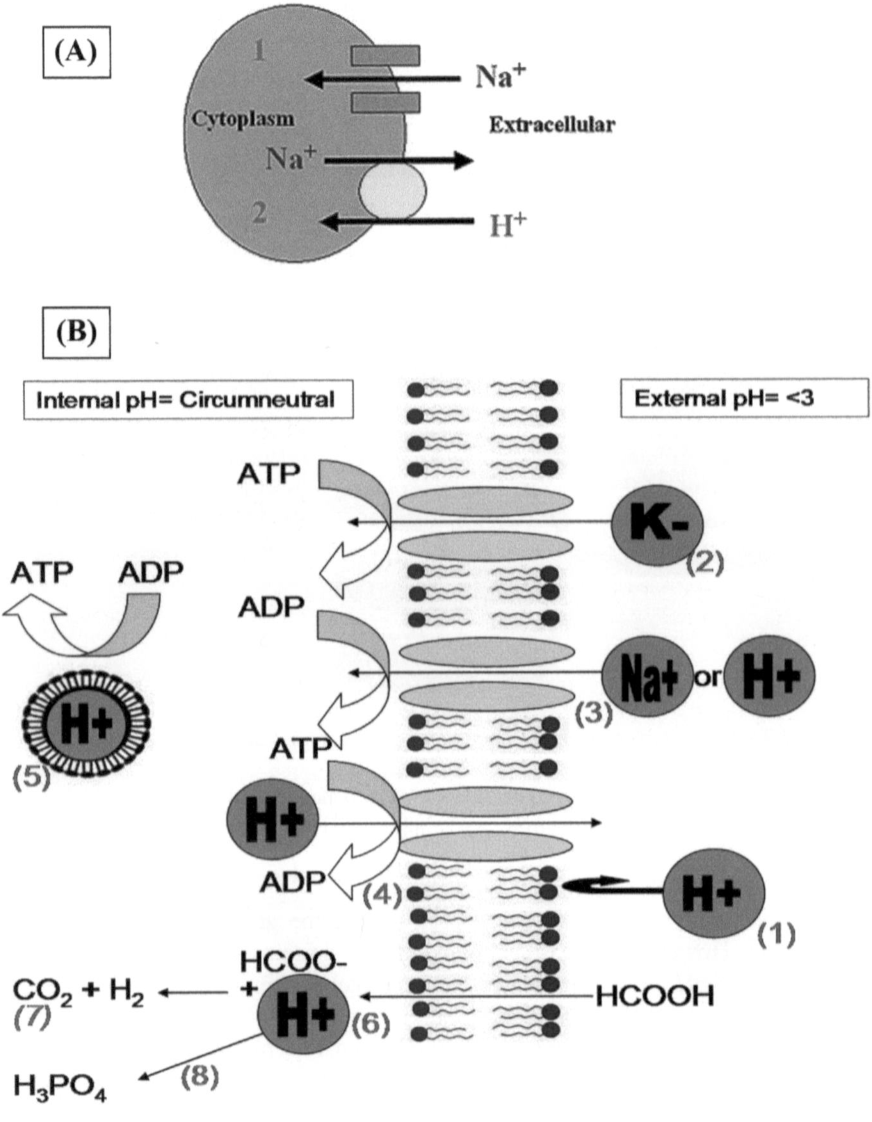

Figure 2.3 (A and B) Methods of pH homeostasis and energy generation in acidophiles. *Source*: Based on Baker-Austin, C., Dopson, M., 2007. Life in acid: pH homeostasis in acidophiles. Trends Microbiol. 15 (4), 165–171.

(1) direction of the transmembrane electrochemical gradient (pH) and blocking of H^+ by the cell membrane; (2) reversed membrane potential through potassium transport, a modification towards maintaining a stable Donnan potential; (3) secondary transporter protein; the H^+ and Na^+ gradient is harnessed to drive the transport of nutrients and solutes; (4) proton pump actively removes H^+, balancing the

energy gained from the H^+ entry to the cytoplasm; (5) vesicles containing protons avoid acidification of the cytoplasm, but still generate ATP from the electrochemical gradient (in *Acidithiobacillus ferrooxidans*); (6) uncouplers (uncharged compounds), such as organic acids, permeate the membrane and release their H^+, leading to acidification of the cytoplasm; (7) to avoid this, heterotrophic acidophiles may degrade the uncouplers; and (8) alternatively, cytoplasmic enzymes or chemicals may bind or sequester the protons. Because of their activity and stability at extremely low pH values, the ligninolytic enzymes may be suitable for biotechnological applications.

2.2.4 Alkalophiles

Alkalophiles are high pH (>8.5 pH) growing bacteria which is adapted to survive and reproduce in the alkaline environment. Many different kinds of alkaliphilic microorganisms, including bacteria belonging to the genera *Bacillus, Micrococcus, Pseudomonas*, and *Streptomyces* and eukaryotes such as yeasts and filamentous fungi, have been reported from a variety of environments as shown in Table 2.1 (Duckworth et al., 1996; Groth et al., 1997; Horikoshi, 1999). Alkaliphilic *Bacillus* sp. is the most common alkaliphilic bacteria that have been reported from various environments including not only alkaline environments such as soda lakes and indigo balls but also neutral soils. For the adaptation in an alkaline environment, sodium carbonate is generally used to adjust the pH to around 10 because alkaliphiles usually require at least some sodium ions (Horikoshi, 1999). This sodium requirement has been found to depend on a Na^+ cycle consisting of sodium efflux and influx systems. As sodium efflux systems, Na^+/H^+ antiporters of alkaliphilic *Bacillus* sp. are essential components of the Na^+ cycle (Horikoshi, 1999; Padan et al., 2005). Na^+/H^+ antiporters extrude Na^+ against H^+ uptake, thus supporting acidification of the cytoplasm directly. A Na^+/H^+ activity defective mutant of *Bacillus halodurans* C-125 could not grow at a pH value above 9.0 (Krulwich, 1995). The Mrp system which consists of seven hydrophobic subunits (MrpABCDEFG) is the dominant Na^+/H^+ antiporter of the alkaliphilic *Bacillus* sp. Some other Na^+/H^+ antiporters, for example, NhaC, have supporting roles in pH homeostasis at low Na^+ concentration, especially at near-neutral pH as shown in Fig. 2.3A and B (Horikoshi, 1999; Krulwich et al., 2009). Na^+ reentry routes such as Na^+/solute symporters and Na^+ channels also play important roles in the Na^+ cycle (Horikoshi, 1999; Padan et al., 2005).

In extremely alkaline conditions, utilization of a sodium-motive force (smf) is advantageous rather than a proton motive force (pmf) (Horikoshi, 1999; Padan et al., 2005). Alkaliphilic *Bacillus* sp., then, uses an smf to take up nutrients and to rotate flagellar instead of a pmf which is generally used by most neutralophilic bacteria (Krulwich et al., 2001). The typical alkaliphilic *Bacillus* species have a voltage-gated Na^+ channel, which has a role in pH homeostasis and chemotaxis (Fujinami et al., 2009; Fujinami et al., 2007). Alkaliphilic *Bacillus* sp. uses the pmf to make ATP from ADP and inorganic phosphate in oxidative phosphorylation even in alkaline pH conditions where the pmf is about one-third compared to a

Table 2.1 The detail of bacteria showing extremophilic properties.

Extremophiles	Microorganisms	Optimum temperature (°C)	Optimum pH	mW^b (kDa)	References
Thermophiles	*Enterobacter* sp.	45	5.0	NR	
	Bacillus HR03	55	5.5	20	
	Cladosporium cladosporioides	40	5.0	50	
	Pseudomonas putida	40	8.0	39.5	
	Aquisalibacillus elongatus	40	8.0	69	
Psychrophiles	*Streptomyces psammetichus*	32	8.0	NR	
	Bacillus sp. MSK-1	30	7.0	NR	
	Aspergillus flavus PUF5	25	4.0	NR	
	Trichoderma muroiana IS 1037	30	4.5	NR	
	Pleurotus sp.	28	5.5	NR	
	Tricholoma giganteum AGHP	30	5.0	66	
	Agaricus blazer	20	7.0	66	
Acidophiles	*Fomitopsis pinicola*	35	2.5	54.3	
	Piptoporus betulinus	30	2.3	52.2	
	Gymnopilus sapineus	30	4.2	33.3	
	Bacillus sp. WT	37	5.0	180	
	Staphylococcus saprophyticus	32	3.0	NR	
	Alteromonas MMB-1	37	5.0	46	
	Acidiphilium organovorum	37	3.0	64	
Alkalophiles	*Bacillus subtilis* MTCC 2414	35	9.0	37	
	P. putida F6	30	8.0	59	
	B. subtilis WPI	25	8.0	55	
	Penicillium pinohilum TERI DB1	25	8.0	NR	
	Haloferax volcanii strain SB01	45–50	6.0	75	
Halophiles	*Chromohalobacter* sp. (3%)	45	6.5	60	Abomohra et al. (2017)
	Aquisalibacillus elongates (2%)	40	6.0	75	
	Bacillus safensis sp. strain S31 (3%)	30	5.5	NR	Ali et al. (2015)
Barophiles	*Shewanella sp. strains PT99*	70 MPa	9.0	NR	
	Moritella strain	50 MPa	8.5	NR	

neutral pH. The respiratory chain and the F_1F_o-ATP synthase of *Bacillus pseudofirmus* OF4 are all proton-coupled (Krulwich et al., 2001; Wang and Ng, 2004). Alkaliphiles-specific amino acid residues or motifs were found in the membrane-embedded a- and c-subunits of the F_1F_o-ATP synthase (Wang and Ng, 2004). Moreover, it is thought that the high content of cytochrome c is useful for keeping a high $\Delta\Psi$ (Goto et al., 2005). Therefore one of the most striking properties of alkaliphilic microorganisms is their use of proton pumps to maintain a neutral pH internally and so the intracellular enzymes from these microorganisms do not need to be adapted to extreme growth conditions. However, the extracellular enzyme proteins of alkaliphiles function at alkaline pH. How these extracellular proteins operate at high pH values is yet poorly understood. For the cells to survive in the aggressive conditions of pH, alkaliphiles utilize several strategies. Alkaliphiles have negatively charged cell wall polymers in addition to peptidoglycan which may reduce the charge density at the cell surface and help to stabilize the cell membrane (Wiegel and Kevbrin, 2004; Horikoshi, 1999).

2.2.5 *Halophiles*

Halophilic microorganisms are capable of living at a high salt concentration (at least 1 M NaCl) and have evolved various chemical, structural, and physiological modifications that allow proteins with physicochemical properties to be selective and stable (Delgado-García et al., 2012; Jackson et al., 2001). The halophilic microorganism can be categorized into three groups according to optimum salt concentration: (1) extreme halophilic capable of developing at 2500–5200 mM NaCl; (2) moderate halophilic developing at 500–2500 mM NaCl; and (3) slightly halophilic capable of developing at 200–500 mM NaCl. The production of halophilic enzymes, such as laccase, LiP, and MnP has been reported for some halophiles *Chromohalobacter* sp., *Aquisalibacillus elongates, Bacillus safensis, Aspergillus sclerotiorum, and Mucor racemosus* shown in Table 2.1 (Abomohra et al., 2017; Ali et al., 2015; Bonugli-Santos et al., 2010). High salt concentrations critically affect the solubility, binding, stability, and crystallization of proteins as shown in Fig. 2.4 (Irimia et al., 2003). The interactions between proteins and protein subunits in solution are also altered by salts. The electrostatic interactions between charged amino acids are also disrupted, with significant effects on the structure and function of the protein (Dennis and Shimmin, 1997). Water is essential for the proper functioning of the native structure and to prevent protein aggregation (Persson and Halle, 2008; Spitzer, 2011). As salt ions increase within the cell, water is removed from protein surface hydrophobic regions until proteins are no longer hydrated sufficiently. In general, non-halophilic proteins are less able to compete with salt and at relatively lower ionic concentrations lose their structure and function. Nonetheless, in the presence of high ionic concentration, halophilic proteins can compete effectively with salt ions for hydration and preserve their functional conformation (Dym et al., 1995; Britton et al., 2006). This applies in particular to proteins from halophilic microorganisms that use the osmotic stabilization mechanisms of salt (i.e., halophilic archaea and some halophilic bacteria). Such halophilic

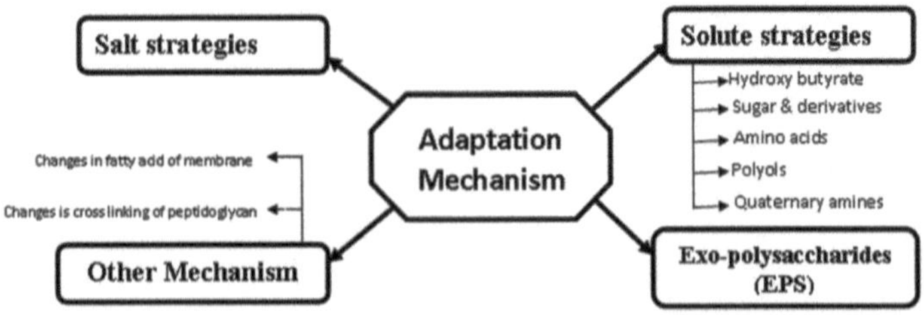

Figure 2.4 Mechanisms of adaptation of halophiles in a saline environment.
Source: Modified from Gontia-Mishra, I., Sapre, S., Tiwari, S.,2017. Diversity of halophilic bacteria and actinobacteria from India and their biotechnological applications. NISCAIR-CSIR, India 46 (8), 1575–1587.

proteins have developed a specific set of molecular characteristics that help them compete for water with ions and maintain a stable shell of hydration (DasSarma et al., 2010; Bolhuis et al., 2008; Karan et al., 2012). Halophilus enzymes use various adaptation mechanisms and are very stable in the presence of organic solvents and low water activity (Jackson et al., 2001). Halophilic enzymes show a large number of negatively-charged amino acid residues such as serine and threonine compared to non-halophilic microorganisms. Moreover, extremely halophilic microorganisms accumulate compatible solutes (i.e., glycine, betaine, sugars, polyols, amino acids, and ectoines) which help them to maintain an environment isotonic with the growth medium (Kumar et al., 2011). These substances also help to protect cells against stresses like temperature and freezing. Besides, halophilic proteins have also been found to contain a low content of bulky hydrophobic side chains on their surface, compared to non-halophilic proteins. The number of larger hydrophobic amino acid residues (phenylalanine, isoleucine, and leucine) is reduced compared to small (glycine and alanine) and borderline hydrophobic (serine and threonine) amino acid residues (Pire et al., 2004; Britton et al., 2006; Tadeo et al., 2009; Karan et al., 2012). These findings are also consistent with increased flexibility, increased surface hydration, and reduced surface hydrophobicity of halophilic proteins (Karan et al., 2012). In halophilic proteins, oppositely charged neighboring residues often interact to form salt bridges and these salt bridges play significant roles in protein folding, structure, and oligomerization as shown in Fig. 2.4. While the underlying molecular mechanisms of halophilic protein function are not yet fully understood, available studies have started to shed considerable light on their adaptation strategies for high salinity and relatively low water activity.

2.2.6 Barophiles

A barophile is an organism that survives in a high-pressure environment. Barophiles are a type of extremophile. An example of a high-pressure habitat is the deep-sea environment, such as ocean floors and deep lakes where the pressure can exceed 380 atm. Another is the subsurface rocks with high lithostatic pressures.

Barophiles that cannot survive outside their high-pressure habitats are referred to as obligate barophiles. Those that can live at high pressures and in a less extreme environment are referred to as *barotolerants*. *Halomonas salaria*, a gram-negative proteobacteria, is an example of an obligate barophile. It needs a pressure of 1000 atm. Many barophiles are sensitive to ultraviolet rays and are susceptible to UV radiation. They lack the essential mechanisms of DNA repair to counter the effects of UV radiation. Thus many of them grow in darkness. Because of this, they also tend to be psychrophilic. That means they live under cold temperatures, for example, about 2°C–3°C below 100 m. Another example of barophile is the *xenophyophores*, which are single-celled eukaryotes found in the deepest ocean trench (i.e., about 6.6 miles below the surface). The deep-sea is regarded as an extreme environment with high hydrostatic pressures (up to 110 MPa), predominantly low temperatures (1°C–2°C) but with occasional regions of extremely high temperature (up to 375°C) at hydrothermal vents, darkness, and low nutrient availability. The molecular basis of piezophiles is now being investigated extensively focusing on aspects of gene regulation and the function of certain proteins in deep-sea isolates as shown in Fig. 2.5. Little attention has been paid, however, to the potential biotechnological applications of piezophiles compared with other extremophiles. In contrast to non-piezophiles, it is necessary to set up a specialized apparatus for the high-pressure cultivation of piezophiles. The simplest and most convenient system for high-pressure cultivation is a pressure syringe, generally made of stainless steel or titanium, which can be used at 100–200 MPa. In this case, bacteria in a suitable culture medium are put into a sterilized plastic bag or tube, and hydrostatic pressure is applied using a hand pump. For aerobic microorganisms, oxygenated fluor inert is added to the culture. This system is used for endpoint measurement of high-pressure growth until decompression. Some laboratories have built specialized reactors for continuous high-pressure cultivation of piezophiles.

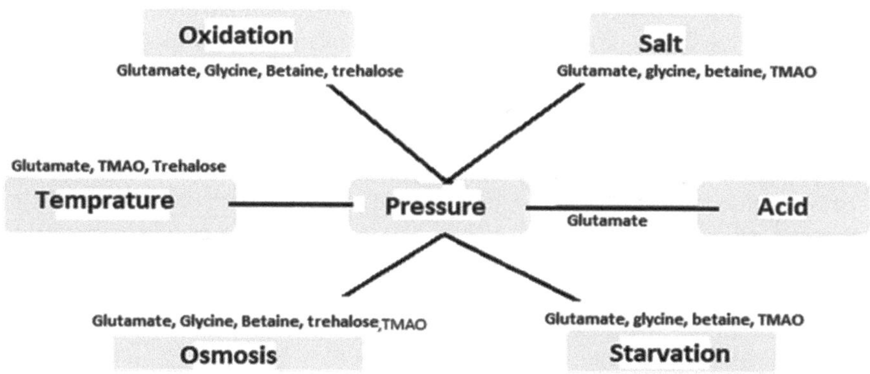

Figure 2.5 Relationship between adaptation strategies at high pressure of barophiles. *Source*: Modified from Zhang, Y., Li, X., Bartlett, D.H., Xiao, X.,2015. Current developments in marine microbiology: high-pressure biotechnology and the genetic engineering of piezophiles. Curr. Opin. Biotechnol. 33, 157–164.

2.3 Plant strategies in amelioration and eco-restoration of a polluted site

Plant strategies include the mechanism and response of plants for growth survival and reproduction in the specific habitat. The plant strategies may be adaptive strategies through changes in their genotype second may be reproductive strategies and third may be resource allocation strategies or ecological strategies. In all the situation plant make trade-offs when responding to their environment. The plant strategies for adaptation C-S-R triangle theory has been postulated by *JP Grime* in 1977 based on *universal adaptive strategies theory*. The third categorizes planta according to how they respond under verifying levels of stress and completion. According to this theory, plants develop strategies that demonstrate the resource trade-off between growth, reproduction, and maintenance. The C-S-R triangle theory tells about three components for plant adoption in the environment: C-Competitors, S-stress tolerance, R- ruderal. This theory remained dominant for plant strategy for several decades. But in early *1980s, David Tillman* introduced the L-H-S strategy which focuses on leaf and seed mass. This trait can be measured and compared between species. But recent studies have revealed a biochemical mechanism for plant strategies for organic-metallic compounds. The heavy metals can bring about ultrastructural biochemical and molecular changes in the plant tissue which can provide the basis for growth, maintenance, and reproduction of the plant. It is accomplished by stimulation of ROS production, direct interaction with cellular macromolecules such as protein and amino acids, and the third process is a displacement of essential metal ions from specific binding sites, which can disrupt important cellular function (Sharma and Dietz, 2009). Thus the plants have developed several strategies to overcome the unfavorable impacts of heavy metal stress as shown in Fig. 2.6. This includes direct scavenging of cytotoxic species formation of metal-ligand complexes and their sequestration into vacuoles, immobilization of heavy metal to reduce their uptake by plants. There are various biochemical strategies, that is, glutathione production, detoxification of methylglyoxal are two important strategies in phytoextraction of heavy metals in their tissue. While under physiological strategies there is synthesis of various cellular biomolecules, for example, Phytochelating (PC) production which is short-chained, thiol-rich repetitions. The peptide of low molecular weight compounds are synthesized in response to heavy metals stress. The PC are found in the cytosol, where they form complex with metal. PC complexes are then transported to vacuoles (Song and Zhao, 2014). This transport is believed to be mediated by an mg-ATP-dependent carrier or ATP binding cassette transporter (Sytar et al., 2013). The sulfhydryl and carboxyl group of PC attach to the heavy metal such as Cd, Cu, Ag, Zn, Pb, Ni, and Ar that are present in cationic or anionic form Cd^{2+} ion has been reported the most effective stimulation of PCs synthesis, where they are 4–6 fold stronger in inducing PCs synthesis than Cu^{2+} or Zn^{2+} in cell structure reported in *Rauwolfia serpentina* (Gupta et al., 2013). Another PC is reported as Metallothionein or prolines which are five carbon multifunctional amino acids in the plant that act as metabolic

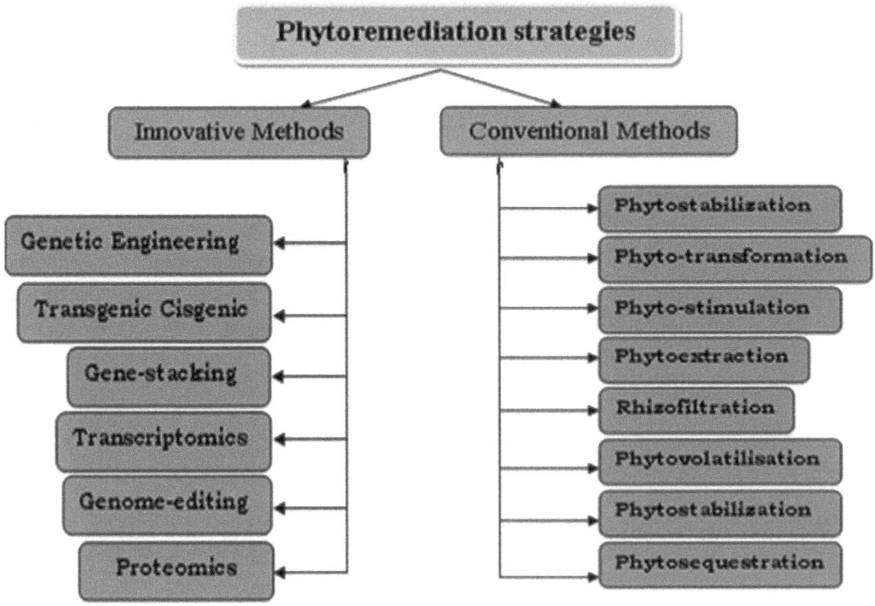

Figure 2.6 Various approaches and plant strategies of phytoremediation.
Source: Modified from Sarwar, N., Imran, M., Shaheen, M.R., Ishaque, W., Kamran, M.A., Matloob, A., et al., 2017. Phytoremediation strategies for soils contaminated with heavy metals: modifications and future perspectives. Chemosphere 171, 710–721.

Osmolite, a constituent of the cell wall, a free radical scavenger, and an antioxidant. Besides, there are various organic acids and amino acids, that is, malate, oxalate, etc., present in the cell also act as an efficient heavy metal-binding molecule. Similarly, there are also morphological strategies for symbiotic association with bacteria, fungi, and mycorrhiza. There are various phytoremediation strategies for amelioration and eco-restoration of the various organometallic industrial waste polluted site as shown in Table 2.4. Phytoremediation covers aspects for cleaning and revegetation of polluted areas using potential plant species (Adriano et al., 2004; Pulford and Watson, 2003). The concept of phytoremediation includes multiple methods and applications. They vary in the mechanism by which toxins may be eliminated or deteriorated by plants. Phytoextraction or rhizofiltration is the procedure in which plants are used to eradicate organic or inorganic substances from the polluted site and store them for the utilizable product. Likewise, phytovolatilization is the technique used by plants to extract pollutants via volatilization (Moreno et al., 2005). Phytostabilization is the technology used to immobilize inorganic pollutants, including heavy metals and metalloids, and to restrict movement in water or dust. The mechanism of this technology will increase the oxidation through microbial behavior of organic pollutants, including pesticides and hydrocarbons, associated with the root crop and speed up the processing of those pollutants through nontoxicity. (Berti and Cunningham, 2000). Usually, metals from soil solution that act as a bioactive reservoir of heavy metals and plant nutrients are taken up by plants. The availability of heavy metals in soil is influenced by factors such as soil pH, organic matter, root exudates, microbial biomass, and strategic metal oxides (Sarwar et al., 2017). Once taken up by plant roots, a particular heavy metal can either concentrate in root tissues (phytoimmobilization) or translocate through symplastic and/or apoplastic pathways to the aerial parts of the plant through xylem vessels and in the shoot, metals are normally accumulated in vacuoles (cellular organelles with low metabolic activity). Hyperaccumulator plants may be an important resistance mechanism to keep deleterious metals away from important cellular metabolic processes. The cell wall binding, successful transportation of metal ions into vacuoles, chelation of the metal ions with proteins and peptides, and complex growth rely on the metal tolerance ability of a plant (Memon and Schröder, 2009).

2.3.1 Phytostabilization

Phytostabilization is the mechanism in which heavy metals are immobilized by metal-tolerant plant species by root penetration and aggregation, permeation on roots, or deposition within the root system. This method substantially decreases metal mobility and metal bioavailability for entry into the ecosystem (Wong, 2003). Plants can immobilize heavy metal in soils by root sorption, precipitation, complexation, or reduction of valence in the rhizosphere, requiring the development of a safe and solid plant layer before the reconstruction of human activity on the land to be handled. (Bech et al., 2012; Ghosh and Singh, 2005; Yoon et al., 2006; Wuana and Okieimen, 2011). Phytostabilization is intended to eliminate pollutants within the vadose region by roots aggregation or rhizospheric accumulation. This prevents off-site contamination through their migration via wind and water erosion, leaching, and soil dispersion. Phytostabilization reduces the mobility, and therefore the risk, of inorganic contaminants without necessarily removing them from the site. This technology does not generate contaminated secondary waste that needs further treatment. It also enhances soil fertility, thereby achieving ecosystem restoration. However, since the contaminants are left in place, the site requires regular monitoring to ensure that the optimal stabilizing conditions are maintained. If soil amendments are used to enhance immobilization, they may need to be periodically reapplied to maintain their effectiveness (Bolan et al., 2013). While some organic contaminants undergo microbial or chemical degradation, inorganic contaminants such as metal(loid)s are immutable. Therefore, containment of metal (loid) s through Phytostabilization is critical in managing contaminated sites. Phytostabilization also refers to establishing a plant cover on the surface of the contaminated soils, which reduces their exposure to wind, water, and direct contact with humans or animals. Phytostabilization can be enhanced by using soil amendments that are effective in the immobilization of metal (loid)s and plant species that are tolerant of high levels of contaminants (Kumpiene et al., 2007; Vangronsveld et al., 1995). Phytostabilization is a good option for large land areas where the removal of substantial amounts of polluted soil is not a viable solution. Besides, wildlife can safely eat the plants, due to the low accumulation of pollutants in their aerial parts.

2.3.2 Phytovolatilization

Phytovolatilization is commonly observed as fluid sources of some inorganic compounds from plants that may be volatilized including Se, As, and Hg. (Terry et al., 2000; Sakakibara et al., 2010). Volatilization is the separation of the air spaces of the farm, and its resulting dissemination into ambient air, assuming the ambient air is less polluted. For volatile and semivolatile organic compounds (VOCs), volatilization from the plant, or phytovolatilization, can represent a major loss mechanism (Dushenkov, 2003). Phytovolatilization has largely focused on the common groundwater contaminants, trichloroethylene (TCE) and tetrachloroethylene (PCE), using traditional phytoremediation plants such as willow (*Salix* sp.) and hybrid poplar (*Populus* sp. × P. sp.) (Andraski et al., 2003). Phytovolatilization of other VOCs has been explored to a lesser degree. In an experiment with alfalfa, most of the applied benzene volatilized directly from the sandy soil, preventing estimates of direct phytovolatilization. Phytovolatilization of other VOCs has been explored to a lesser degree. In an experiment with alfalfa, most of the applied benzene volatilized directly from the sandy soil, preventing estimates of direct phytovolatilization. phytovolatilization is often considered beneficial, as phytovolatilization of the contaminant generally results in substantial dilution and photochemical decay in the atmosphere (Andraski et al., 2005). Conversely, phytovolatilization may be viewed as a risk in urban areas, where exposure potential exists and air quality is degraded, such as nonattainment areas, although minimal evidence related to this concern is present in current literature. Phytovolatilization is another approach that involves the conversion of metal into volatile form and its release into the atmosphere through stomata (Ghosh and Singh, 2005). This technique is primarily useful for Hg as the mercuric ion is transformed into a relatively less toxic elemental form. As the volatile form of Hg released into the atmosphere may be recycled back to the soil by precipitation, so this technique presents a temporary solution to the problem in the phytoremediation techniques shown in Fig. 2.7.

2.3.3 Phytosequestration

Phytosequestration is the ability of plants to sequester certain contaminants in the root zone. This is accomplished through several of the plant's physiological mechanisms. Phytochemicals extruded in the root zone can immobilize or precipitation of the target contaminant. The transport proteins associated with the root also can irreversibly bind and stabilize target contaminants. Contaminants can also be taken up by the root and sequestered in the vacuoles in the root system. Phytosequestration is one of the environmentally friendly technologies that use plants to clean up soil from trace element contamination. The uptake and accumulation of pollutants vary from plant to plant and also from species to species within a

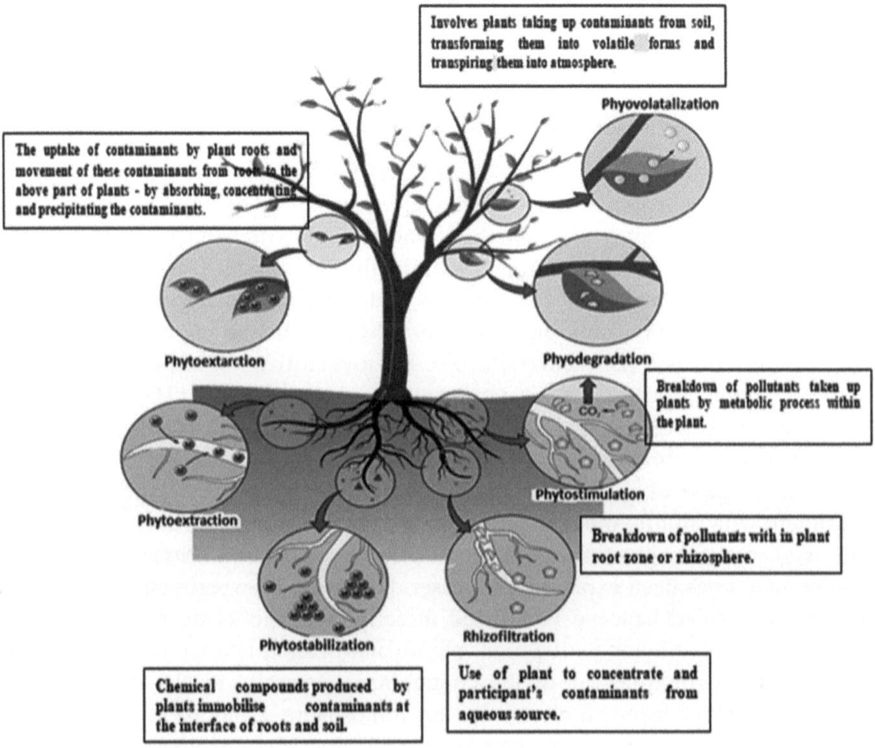

Figure 2.7 Mechanism and plant strategies of phytoremediation.
Source: Modified from Favas, P.J., Pratas, J., Varun, M., D'Souza, R., Paul, M.S., 2014. Phytoremediation of soils contaminated with metals and metalloids at mining areas: potential of native flora. Environ. Risk Assess. Soil Contamination 3, 485–516.

genus. Phytosequestration is mainly performed by several photosynthetic mechanisms such as C3, C4, and crassulacean acid metabolism pathways of plants, carboxysomes of cyanobacteria, and pyrenoids of microalgae. For effective mitigation of global climate change, it is required to stabilize the CO_2 concentration to viable levels. It requires various permutations and combinations of naturally existing and engineering strategies.

2.3.4 Rhizofiltration

Rhizofiltration is a type of phytoremediation, which refers to the approach of using hydroponically cultivated plant roots to remediate contaminated water through absorption, concentration, and precipitation of pollutants. It also filters through water and dirt. Rhizofiltration is the adsorption onto plant roots or absorption into plant roots associated bacteria that are in the solution surrounding the root zone (rhizosphere). Rhizofiltration is used to decontaminate groundwater. The technique of rhizofiltration is used for the remediation of wastewater by aquatic or land plants (Henry, 2000). Pb, Cd, Cu, Ni, Zn, and Cr can be extracted using rhizofiltration. Terrestrial plants are widely preferred for rhizofiltration as they have fibrous root systems with fast growth. The rhizofiltration technique can be constructed either as floating rafts on ponds or as tank systems. One of the main disadvantages of rhizofiltration involves growing plants in a greenhouse first and then transferring them to the remediation site. Great care must be taken to maintain an optimum pH in the effluent solution. Surface water can be treated using rhizofiltration (phytoextraction) in ponds, engineered tanks, natural wetlands, or constructed wetlands (Huang et al., 1997; Thayaparan et al., 2013). The constructed wetland treatment system has been applied the world over for biofilters of various organic pollutants from industrial wastewater and sewage as a green sustainable technology. The constructed wetland (CW) treatment system has been reported from various part of the world for removal of lead (Pb), cadmium (Cd), iron (Fe), nickel (Ni), chromium (Cr), and copper (Cu), using continuous free surface flow wetland. Besides, the rhizofiltration techniques have been widely reported for the treatment of municipal wastewater but at presently constructed wetland are successfully applied for various industrial wastewater treatments. The most recent application includes brewery and tannery wastewater treatment (Billore et al., 2001; Kucuk et al., 2004; Kadlec and Wallace, 2009). The surface and subsurface flow constructed wetland treatment have been used for industrial wastewater treatment. In subsurface flow constructed wetland, there may be either vertical flow or horizontal flow system. The detailed application of the rhizofiltration technique used for various industrial wastewater treatments is reported in Table 2.2.

Some hydrophytes are also used in a constructed wetland treatment system for the accumulation of various heavy metals. The *Typha latifolia, Eichhornia crassipes, Lemna gibba*, and *Pistia stratiotes* are also reported for the Cd, Cu, and Pb removal showed in Table 2.3. The rhizofiltration potential by *Arundo donax L.* has been reported from France in the constructed wetland for Bordeaux mixture containing effluent treatment. The potential rhizospheric bacterial communities from *Phragmites australis* and *Phragmites communis* has been reported for the decolorization and degradation of distillery pulp paper mill wastewater respectively from the constructed wetland in the pot and natural drainage system. In some cases, the contaminated water can be used as irrigation water in which the contaminants then undergo rhizodegradation and phytodegradation. Rhizofiltration that involves the use of plants (both terrestrial and aquatic) to absorb, concentrate and precipitate contaminants from polluted aqueous sources with low contaminant concentration in their roots finds its application in dealing with effluents discharged from industries and agricultural runoff as well as acid mine drainage shown in Fig. 2.8. It is more commonly used for lead (Pb), cadmium (Cd), copper (Cu), nickel (Ni), zinc (Zn), and chromium (Cr), which are primarily retained within the roots (Galal et al., 2018). As a cost-effective technology, it has been used for the treatment of surface or groundwater containing significant proportions of heavy metals such as Cr, Pb, and Zn.

Table 2.2 Detailed application of rhizofiltration in constructed wetland systems used for various industrial wastewater treatments.

Type of wastewater	Type of CW	Location	Treatment efficiency (%)	References
Pulp and paper	HF	USA	NH_4-N 31–88, TSS 55–68, TP 14–31	Thut (1990)
	HF	USA	BOD 80–90	Thut (1990)
	HF	USA	BOD 27–58	Hammer et al. (1993)
	HF	USA	BOD 67–84	Boyd et al. (1993)
	FWS	USA	BOD 6–7, TSS 33–81, TP 32–53, NO_3-N 64–80	Tettleton et al. (1993)
	FWS	China	BOD 18–33, TSS 51	Xianfa and Chuncai (1995)
	HF	India	Resin and fatty acids 92–96	Chowdhury et al. (2010)
Tannery	FWS	Turkey	COD 30, NH_4-N 95, Cr 45–55	Kucuk et al. (2004)
	HF	Portugal	BOD 48, COD 59, TSS 72, TKN 25	Calheiros et al. (2007, 2009)
	HF	Portugal	BOD 48, COD 59, TSS 72, TKN 25	Calheiros et al. (2012)
Textile	VF	UK	Acid blue 113, reactive blue 171, both 98	Pervez et al. (1999)
	HF	Tanzania	COD 68–72, color 72–77, sulfate 53–59	Mbuligwe (2005)
	VF	Portugal	COD 64, TOC 71, AO7 74	Davies et al. (2005)
Distillery	HF	India	BOD 84, COD 64, TKN 59, TP 79	Billore et al. (2001)
	HF	Mexico	BOD 85, COD 80, TKN 75, NO_3-N 57, NH_4-N 2–10, $SO4^{2-}$69	Olguín et al. (2008)
	HF	UK	Cu 53–85	Murphy et al. (2009)
Brewery	FWS	USA	NO_3-N 86, PO_4-P 33	Kadlec and Wallace (2009)
	HF	South Africa	COD 20, NH_4-N 80, NO_3-N 86, PO^4-P 33	Crous and Britz (2010)

CW, Constructed wetlands; *FWS*, free water surface; *HF*, horizontal subsurface flow; *VF*, vertical flow.

Table 2.3 Various plant species reported for plant strategies of phytoremediation.

Subprocesses	Plant species	Collection chamber	Pollutants	Toxicity from these Pollutants	References
Phytostabilization	• Rose plant • *Brassica juncea* • *Iris sibirica* • *Typha domingensis* • *Vossia cuspidate* • *Hordeum vulgare* • *Typha domingensis*	Reduction in rhizosphere	Pb, Cd, Zn, As, Cu, Cr, Se, U, hydrophobic organics—PAHs, PCBs, dioxins, furans, pentachlorophenol, DDT, dieldrin	Elevated blood pressure, lung damage, and fragile bones	Andreazza et al. (2015)
Phytovolatilization	• *Brassica* sp. (wild type) • Wetland plants • Phreatrophytic trees • Legumes • *Helianthus annuus* • *Pennisetum annuus*	Release to the atmosphere	Chlorinated solvents such as carbon tetrachloride, TCE, MC PCE, CT, Hg (mercuric ion) Se	Exposure may cause skin and other body tissues to turn gray or blue-gray	Bañuelos et al. (2005)
Phytosequestration	• *Brassica* sp. • *Pteris vittate* • *Puccinellia frigida* • *Zea mays* • *Triticum* sp.	Root to stem to leaves	Pb, Zn, Cd, As, Cu, Cr, Se, U, hydrophobic organics, etc.	Dysfunction, muscle twitching, and elevated blood pressure	Zhang et al. (2016)
Rhizofiltration	• *Phragmites australis* • *Warnstorfia fluitans* • *Micranthemum umbrosum* • *Phalari arundinacea* • *Noccaea caerulescens*	Aerial parts/ root	Heavy metals such as Pb, Cd, Zn, Ni, Cu, Radionuclides—Cs, Sr, U, hydrophobic organics, and radionuclides	Carcinogenic, mutagenic, endocrine disruptor, lung damage	Islam et al. (2015)

(*Continued*)

Table 2.3 (Continued)

Subprocesses	Plant species	Collection chamber	Pollutants	Toxicity from these Pollutants	References
Phytoextraction	• *Ricinus communis* • Brassica napus • *Pteris vittate* • *Puccinellia frigida* • *Helianthus annuus* • *Pennisetum annuus* • *Brachiaria decumbens* • *Pisum sativum* • *Jatropha curcas* • *Brachiaria decumbens*	Shoots	Pb, Cd, Zn, Ni, Cu with EDTA addition for Pb, radionuclides, pentachlorophenol, aliphatic compounds (short chained)	Brain and kidney damage, elevated levels result in liver cirrhosis, and chronic anemia	Yang et al. (2017)
Phytotransformation	• Phreatophytic trees grasses (rye, fescue, Bermuda grass, sorghum, switchgrass, reed canary grass); • Legumes (clover, alfalfa, cowpeas)	Root	Herbicides - atrazine, alachlor, mixtures BTEX, chlorinated aliphatic compounds TCE, nutrients $-NO_3^-$, NH_4^+, PO_4^{3-}, ammunition wastes, RDX	Hair loss, carcinogenic, mutagenic	Gao et al. (2000)
Phytostimulation	• *Brassica campestris* • *Citrullus varifistulosus* • *Lagenaria siceraira* • *Sesamum indicum* • *Zea mays* • Triticum sp.	Root to shoot	Organic matter, phosphates, alkalizing agents, and biosolids, Pb, Zn, Cd, As, etc.	Dizziness, fatigue, etc.	Schnoor (1997)

BTEX, Benzene toluene ethylbenzene and xylene; *CT*, carbon tetrachloride; *MC*, methylene chloride; *PAHs*, polycyclic aromatic hydrocarbons; *PCBs*, polychlorinated biphenyl; *PCE*, tetrachloroethylene; *RDX*, rapid detonating explosive; *TCE*, trichloroethylene; *DDT*, dichlorodiphenyltrichloroethane.

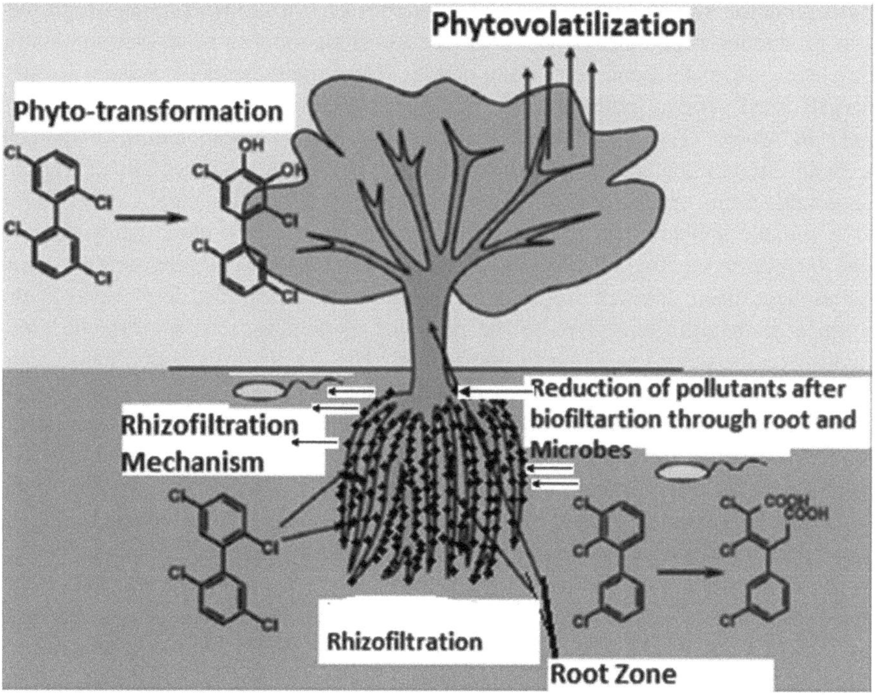

Figure 2.8 Schematic representation of rhizofiltration through root zone due to bacterial biodegradation and phytovolatilization of pollutants through transformation.
Source: Modified from Favas, P.J., Pratas, J., Varun, M., D'Souza, R., Paul, M.S., 2014. Phytoremediation of soils contaminated with metals and metalloids at mining areas: potential of native flora. Environ. Risk Assess. Soil Contamination 3, 485–516.

2.3.5 Phytoextraction

Phytoextraction is a process of phytoremediation in which plants remove dangerous elements or compounds from soil or water, usually heavy metals and metalloids that have a high density and may be toxic to organisms even at relatively low concentrations (Chandra et al., 2017, 2018). The heavy metals that plants extract are toxic to the plants as well, and the plants used for phytoextraction are known hyperaccumulators that sequester extremely large amounts of heavy metals in their tissues. Phytoextraction can also be performed by plants that uptake lower levels of pollutants, but due to their high growth rate and biomass production, may remove a considerable amount of other contaminants from the soil and polluted site (Sharma et al., 2020). Phytoextraction is the most important phytoremediation approach for the removal of metals and metalloids from contaminated soils water, biosolids, and sediments (Seth et al., 2012; Ali et al., 2013). It is more suitable for a commercial application as compared to other phytoremediation techniques. A number of factors affect the efficiency of phytoextraction including soil properties, metal availability to plants, metal speciation, and concerned plant's characteristics. Plants selected for phytoextraction should possess the characteristics of fast growth rate and high biomass production capable for metal translocation from shoot to root. Organometallic complexes are metal containing compounds. These organometallic wastes are discharged from various industries, for example, distilleries, tanneries, and pulp and paper industries. The degradation and detoxification of organometallic waste is threat in environment due to their complexity. Therefore the detection of organic compounds from organometallic sludge is possible only after their extraction by using suitable solvent. The analysis and characterization of organic compounds is done by gas-chromatography or GC-MS analysis. Various complex organic compounds have been detected from sugarcane molasses-based distillery waste in the presence of metallic ions. Due to the presence of metallic ions the organic compounds become more venerable for its toxicity. The details of various organic compounds and their toxicity have been reported in Table 2.3. Generally, plant shoot metal concentration and shoot biomass mainly determine a suitability plant species for phytoextraction of metals. There are variable metal accumulation patterns in different plants for different heavy metals (Chandra and Kumar, 2017; Chandra et al., 2017, 2018). Depending upon these parameters, two different phytoextraction approaches have been used, that is, use of hyperaccumulator plants with relatively low biomass production and use of plants with relatively higher aboveground biomass production but lesser metal accumulation such as *Brassica juncea* as shown in Fig. 2.9 (Chandra et al., 2017, 2018; Ali et al., 2013). While some researchers preferred hyperaccumulation and metal tolerance as characters of vital importance to pollutant removal (Sharma et al., 2020). The plants with low biomass and high metal accumulation are easy to dispose after harvesting an economical concentration of heavy metals is reported (Ahmad et al., 2015). To evaluate the phytoextraction potential of growing native plants the bioconcentration factor is calculated as the following formula which has also been reported earlier (Yoon et al., 2006). Similarly, translocation factor (TF) is important to evaluate the metabolic activities and health of individual plants growing on polluted sites. TF of any metal may be calculated by the ratio of metal concentration in plants aerial parts(shoot) and metal concentration in plants root growing on sludge (Chandra et al., 2017) (Fig. 2.10; Table 2.4).

$$\mathrm{BCF} = \frac{\text{Concentration of metal in plant root}}{\text{Concentration of metal at contaminated site(sludge)}}$$

$$\mathrm{TF} = \frac{\text{Concentration of metal in corresponding plant roots}}{\text{Concentration of metal in plant shoots}}$$

2.3.6 Phytotransformation

Phytotransformation also referred to as phytodegradation, is the breakdown of organic contaminants sequestered by plants via metabolic processes within the

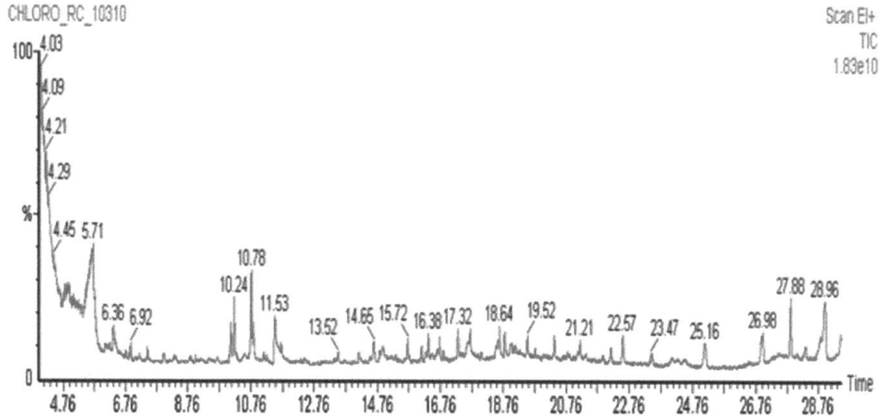

Figure 2.9 Total ion chromatogram of trimethylsilyl derivatives of organic compounds extracted from pulp and paper effluent's sludge.
Source: Reproduce with permission from Chandra, R., Yadav, S., Yadav, S., 2017. Phytoextraction potential of heavy metals by native wetland plants growing on chlorolignin containing sludge of pulp and paper industry. Ecol. Eng. 98, 134–145.

plant; or with enzymatic action. The organic contaminants are degraded into simpler compounds that are integrated with plant tissue, which in turn, foster plant growth. Remediation of a site by Phytotransformation is dependent on direct uptake of contaminants from the media and accumulation in the vegetation. Direct uptake of chemicals into plant tissue via the root system is dependent on uptake efficiency, transpiration rate, and concentration of the chemical in soil water. Uptake efficiency depends on chemical speciation, physical/chemical properties, and plant characteristics, whereas transpiration rate depends on plant type, leaf area, nutrients, soil moisture, temperature, wind conditions, and relative humidity. The release of volatile contaminants to the atmosphere via plant transpiration, called phytovolatilization, is a form of phytotransformation. Although the transfer of contaminants to the atmosphere may not achieve the goal of a complete remediation, phytovolatilization may be desirable in that prolonged soil exposure and the risk of groundwater contamination are reduced. Phytotransformation can be employed to remediate sites contaminated with organic compounds. Certain enzymes produced by plants can breakdown and convert chlorinated solvents (e.g., TCE), ammunition wastes, and herbicides. This technology also can be used to remove contaminants from petrochemical sites and storage areas, fuel spills, landfill leachates, and agricultural chemicals. Successful implementation of this technology requires that the transformed compounds that accumulate within the plant be nontoxic or significantly less toxic than the parent compounds. In some applications, phytotransformation may be used in concert with other remediation technologies or as a polishing treatment. The study revealed that the pesticide atrazine was transformed to ammeline with no mineralization to CO_2. The toxicity and fate of contaminants need to be evaluated and understood properly. The bound residue or metabolism of the tree or grass used also must be considered as a critical success factor for this technology.

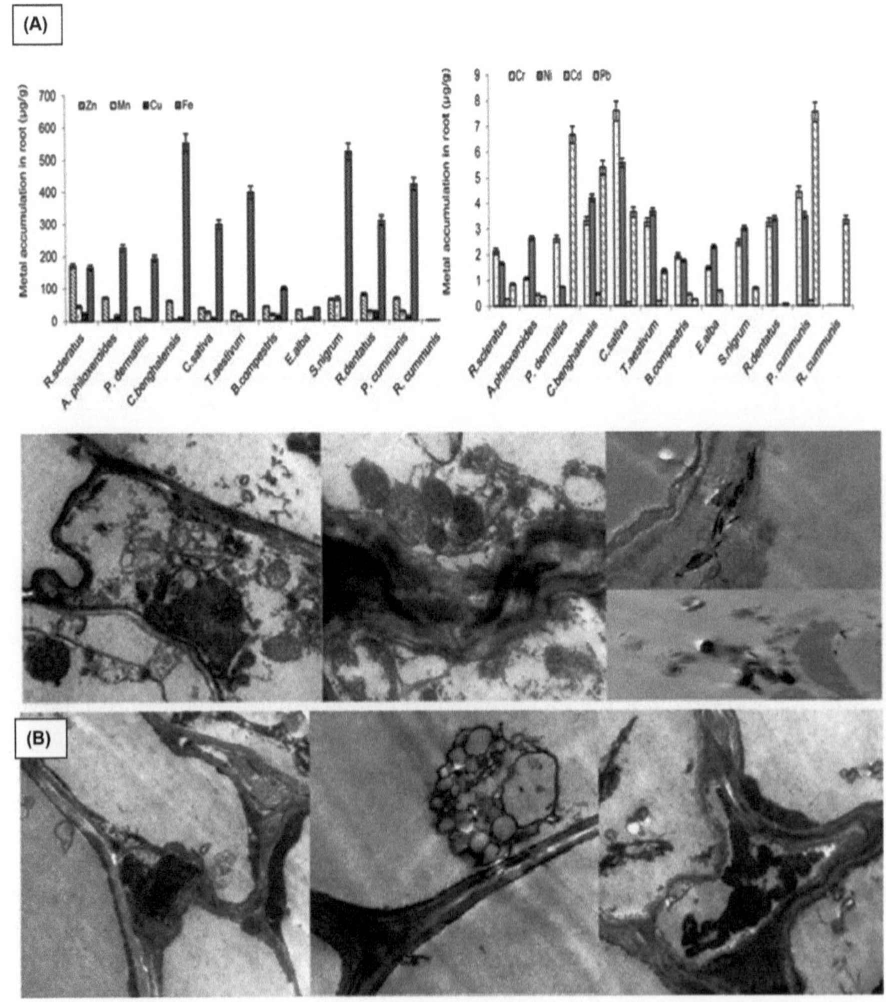

Figure 2.10 (A) Comparative analysis of heavy metal accumulation pattern in the root of native wetland plants growing at the polluted site. (B) Transsion electron microscopy (TEM) images of native plants root after phytoextraction of heavy metals.
Source: Courtesy Chandra, R., Kumar, V., 2017. Phytoextraction of heavy metals by potential native plants and their microscopic observation of root growing on stabilised distillery sludge as a prospective tool for in situ phytoremediation of industrial waste. Environ. Sci. Pollut. Res. 24 (3), 2605–2619.

2.3.7 Phytostimulation

The breakdown of organic pollutants in the soil through enhanced microbial activity in the plant root zone or rhizosphere is phytostimulation, also referred to as

Table 2.4 Various endocrine distrupting chemicals (EDCs) and other organic pollutants and their toxicity present in distillery sludge identified by gas charomatography-mass spectroscopy (GC-MS).

SNo	RT	Compound	Toxicity
1.	7.46	Propanoic acid	Stress, premenstrual syndrome
2.	10.12	2-2-Bi-1,3-dioxolane	Carcinogenic, allergy reaction
3.	10.78	Trimethylenebis-1,3-dioxolane	Bone disease, lung cancer
4.	11.53	1,3-Dioxolane,2-(phenylmethyl)	Nausea, headaches, delirium
5.	11.72	1,3-Dioxane,2-methyl	Inhalation, ingestion
6.	13.51	Decane, 1 bromo-2-methyl	Vision loss, headache
7.	14.18	2-Pyridinenmethanol	Irritates the skin
8.	14.64	Heptane,2,2,3,3,5,6,6-heptamethyl	Skin or eye irritation
9.	15.71	Pentadecanone	Oral, chest or abdominal pain
10.	16.16	2-Pentadecanone	Vision loss, vomiting
11.	16.37	Benzene dicarboxilic acid	Epigastric pain
12.	17.32	Dibutyl phthalate	Possible kidney damage
13.	17.70	Heptacosane	Nausea, vomiting
14.	18.63	1-Decanol,2-hexyl	Central nervous system damage
15.	18.79	Phytol	Diarrhoea, fever
16.	19.52	Tritetracontane	Blood forming
17.	20.38	Hentricontane	Inhalation, ingestion
18.	21.21	Undecane, 3,5-dimethyl	Acute lethal toxicity
19.	22.18	Tetracontane	Diarrhoea
20.	22.56	Phthalic acid	Strong eye irritant, vomiting
21.	23.46	Sulfurous acid	Diarrhoea, fever, vomiting
22.	25.16	Nonadecane, 2-methyl	Parageusia, pink disease
23.	26.97	Geranylgeraniol	Pain, acute oral toxicity
24.	27.87	Heptacosane	Skin redness, eye redness
25.	28.95	Trans, trans-3,5-heptadien-2-one	Nausea, headaches

RT, Retention time (Min).
Source: Courtesy Chandra, R., Yadav, S., Yadav, S.,2017. Phytoextraction potential of heavy metals by native wetland plants growing on chlorolignin containing sludge of pulp and paper industry. Ecol. Eng. 98, 134–145.

enhanced rhizosphere biodegradation, rhizodegradation, or plant-assisted bioremediation/degradation.

In many ways, microbial development is promoted in the rhizosphere:

- Indigenous microbe communities are enriched by compounds such as sugars, carbohydrates, amino acids, acetates, and enzymes that are exudated from the roots;
- Root networks provide oxygen to the rhizosphere, which ensures aerobic transformations;
- Increases usable organic carbon by fine-root biomass;
- Mycorrhizae fungi, which develop inside the rhizosphere, can degrade organic pollutants that, due to special enzymatic pathways, can not be converted solely by bacteria; and
- Plants boost the ecosystem for increased microbial species and activity.

This approach is effective in the elimination from soil and sediments of organic compounds such as contaminants, aromatics, and aromatic polynuclear carbohydrates. In deeper environments, low levels of emission can be used at sites where phytostimulation is being applied. High levels of contaminants can be toxic to plants. For a good design, microbial degradation and dense root systems are needed. Until using this technology, the toxicity and fate of pollutants must be determined and recognized. Plants can consist of trees, herbs, and legumes. Phytostimulation could be cheaper and easier than many other technologies. The advantages and disadvantages of all these technologies have been shown in Table 2.5.

2.4 Characteristics features of rhizospheric bacteria in an organometallic industrial waste polluted site

Plant-microbe interaction is known as the morphological and physiological strategy of a plant for its growth, survival, and reproduction in polluted environmental conditions. For beneficial interaction and stability, rhizobacteria produce various inorganic and organic compounds which play a vital role in assisting phytoremediation of complex organometallic industrial waste. Such compounds are mutually released by plant and microbes both, like siderophore, indoleacetic acid (IAA), hydrogen cyanide (HCN), ammonia, 1-aminocyclopropane-1-carboxylic acid (ACC) deaminase, antibiotic resistance. Some other activities are also performed by rhizobacteria like phosphate, potassium, and zinc solubilization.

Siderophore mainly helps in a chelating variety of metals in addition to iron, and they have a wide range of chemical structures and specific properties. For instance, siderophores function as biocontrols, biosensors, and have the potential for bioremediation of metal and organic compounds in polluted areas (Ahmed and Holmström, 2014). The other important parameter is IAA production which stimulates and facilitates plant growth by production of longer roots with an increased number of root hairs and root laterals that are involved in nutrient uptake. Moreover, it promotes inhibition or delay abscission of leaves, induces flowering, and fruiting (Mohite, 2013). Furthermore, some rhizobacteria produce HCN which inhibits the electron transport and disrupts the energy supply to the cell, and leads to the death of living organisms. Cyanide forms stable complexes with the essential elements (Cu^{2+}, Fe^{2+}, and Mn^{2+}) for the protein function and therefore is considered a toxic substance to most living organisms (Abd El-Rahman et al., 2019). The ability of some bacterial strains to control nematodes was attributed to their ability to produce HCN. Besides HCN, ammonia production by rhizobacteria accumulates and supplies nitrogen to their host plant and promotes root and shoot elongation and their biomass (Backer et al., 2018). Several soil microorganisms are reported to produce ACC deaminase which promotes plant growth by lowering plant ethylene levels, for example, *Pseudomonas putida, Streptomyces* sp. (Jaemsaeng et al., 2018). Moreover, it has been reported that some ACC deaminase-producing bacteria promote plant growth under a variety of stressful conditions, such as flooding,

Table 2.5 Detail mechanism, advantage, and disadvantage of various plant strategies of phytoremediation.

Subprocesses	Mechanism concepts	Advantage	Disadvantage	References
Phytostabilization	Stabilizing of pollutants by binding, holding or decreased leaching	• It circumvents the removal of soil, low cost and is less disruptive and enhances ecosystem restoration • No disposal of hazardous material/biomass is required • Very effective when rapid immobilization is needed to preserve ground and surface water	•Contaminant remain in soil•Application of extensive fertilization/soil amendments•Mandatory monitoring required	USEPA (2000) Kumpiene et al. (2007)
Phytovolatilization	Plants takes up volatile organic pollutants, that is, selenium, mercury	• The contaminant, mercuric ion, may be transformed into a less toxic substance (i.e., elemental Hg)	• The mercury released into the atmosphere is likely to be recycled by precipitation and then redeposited back into lakes and oceans, repeating the production of methylmercury by anaerobic bacteria	USEPA (2000)
Phytosequestration	Reduce the mobility of the contaminant and prevent migration to soil, water and air by immobilization of target contaminants in the root zone	• Reduction in dispersal of dust and contaminants by wind • Plant process more easily controlled than those of microorganisms	• Toxicity and bioavailability of degradation products remain largely unknown	Licht and Isebrands (2005)
Rhizofiltration	Uptake of metal in plant root	• Ability to use both terrestrial and aquatic plants for either in situ or ex situ applications • Contaminants do not have to be translocated into shoots	• Constant need to adjust pH • Plants may first need to be grown in greenhouse/nursery • There is periodic harvesting and plant disposal • Tank design should be well engineered	Sharma et al. (2015)

(*Continued*)

Table 2.5 (Continued)

Subprocesses	Mechanism concepts	Advantage	Disadvantage	References
Phytoextraction	Uptake and concentration of metals via direct uptake into plant tissue with subsequent removal of the plants	• Cost is fairly inexpensive compared to conventional methods • Contaminant permanently removed from soil • Amount of waste material that must be disposed of is decreased up to 95%	• Metal bioavailability within the rhizosphere • Rate of metal uptake by roots • Proportion of metal "fixed" within the roots • Cellular tolerance to toxic metals	Meyerholt (2013)
Phytotransformation	Plants uptake and degradation of pollutants	• Pollutants will be transformed in to less-toxic forms • Both economically and environmentally friendly	• Requires more than one growing season to be efficient • Soil must be less than 3 ft. in depth and groundwater within 10 ft. of the surface • Contaminants may still reenter the food chain through animals or insects that eat plant material	Singh and Prasad (2011)
Phytostimulation	• Carried out by bacteria or other microorganisms flourishing in the rhizosphere	• In situ practice resulting in no disturbance • No removal of contaminated materials • Complete mineralization of the contaminant can occur • Low installation and maintenance cost	• Development of extensive root zone required takes time • Root depth limited due to physical structure of soil • Organic matter from plant may be used as a C source instead of contaminant decrease amount of contaminant biodegradation	Rungwa et al. (2013)

saline conditions, and drought, for example, *B urkholderia tropica, B urkholderia caribensis, Bombus terricola* (Onofre-Lemus et al., 2009). Some bacteria like *Pseudomonas sp., Bacillus cereus, Bacillus megaterium, Pseudochrobactrum glaciale*, etc. also show ligninolytic enzyme activity which plays a key role in degrading and detoxifying lignocellulosic waste from the polluted site. The major ligninolytic enzymes are laccase, LiP, MnP, and versatile peroxidase (Kumar and Chandra, 2020). Rhizobacteria have also been reported for solubilizing various nutrients and an organic compound containing phosphate (P), zinc (Zn), and potassium (K) (Kalayu, 2019; Etesami et al., 2017)

The application of plant growth promoting (PGPR) has been extended to remediate contaminated industrial sites in association with plants. Environmental pollutants containing the complex organometallic industrial waste have drawn the attention of researchers worldwide for sustainable development. Due to the complexity of organic pollutants, they do not degrade and persist in the environment for a longer time. Moreover, they also have a binding tendency with another pollutant during transformation. They mainly include total petroleum hydrocarbons and polycyclic aromatic hydrocarbons coming from the exploration and consumption of fossil fuel, polychlorinated biphenyls (PCBs) are widely used in the industrial process and are most degradation-resistant, and other chlorinated aromatics used as PCB replacement such as polychlorinated terphenyls, halogenated compounds like PCE and TCE and pesticides like atrazine and bentazon. Heavy metals are the primary inorganic contaminants, which include cadmium, chromium, copper, lead, mercury, nickel, and zinc (Zhuang et al., 2007). Some examples of PGPR-assisted the phytoremediation of heavy metal and organic pollutants and have been shown in Table 2.6. Various characteristics features of rhizospheric bacteria play a vital role at the polluted site of industrial waste, such as:

2.4.1 Siderophore

Siderophores are small organic molecules produced by microorganisms under iron-limiting conditions which enhance the uptake of iron to the microorganisms. In the environment, the ferric form of iron is insoluble and inaccessible at physiological pH (7.35−7.40). Under this condition, microorganisms synthesize siderophores which have a high affinity for ferric iron. Heavy-metal contaminated samples can be detoxified by applying siderophores, which explicate their role in bioremediation. Siderophores are divided into three main families depending on the characteristic functional group, that is, hydroxamates, catecholate, and carboxylates. Most of the bacterial siderophores are catecholates (i.e., enterobactin), and some are carboxylates (i.e., rhizobactin) and hydroxamates (i.e., ferrioxamine B). Mugineic acid is the most common and the first identified phytosiderophore (Fig. 2.11) (Ahmed and Holmström, 2014).

2.4.2 IAA production

In plants, IAA production plays a key role in both roots and shoots development. IAA is produced by some beneficial bacteria in the rhizosphere, where it acts as a

Table 2.6 Some examples of PGPR-assisted phytoremediation of heavy metals and organic pollutants.

PGPR	Host plant	Heavy metal	References
Pseudomonas aeruginosa	Indian mustard (*Brassica juncea* L.) and pumpkin (*Cucurbita maxima* L.)	Cd	Sinha and Mukherjee (2008)
Pseudomonas putida	Canola (*Brassica napus* L.)	Ni	Farwell et al. (2006)
P. putida	Arugula (*Eruca sativa* L).	Cd	Kamran et al. (2015)
Pseudomonas asplenii	Canola (*B. napus* L.) and reed (*Phragmites australis*)	Cu	Reed and Glick (2005), Reed et al. (2005)
Pseudomonas fluorescens Avm and *Rhizobium leguminosarum*	Alfalfa (*Medicago sativa* L.)	Cu	
Kluyvera ascorbata	Canola (*B. napus* L.)	Ni, Pb, Zn, and Cr	Burd et al. (1998)
Azotobacter chroococcum HKN-5	Indian mustard (*B. juncea*)	Pb, Zn	Wu et al. (2006)
Mesorhizobium huakuii subsp. rengei B3	Milk-vetch (*Astragalus sinicus*)	Cd	Sriprang et al. (2003)
K. ascorbata SUD165	Indian mustard (*B. juncea*)	Pb, Zn, Ni	
B. subtilis SJ-101	Indian mustard (*B. juncea*)	Ni	Zaidi et al. (2006)
		Organic pollutants	
Azospirillum lipoferum	Wheat (*Triticum aestivum*)	Crude oil	Muratova et al. (2005)
Azospirillum brasilense	Tall fescue (*Festuca arundinacea*)	PAHs	Huang et al. (2004)
Enterobactor cloacae	Tall fescue (*F. arundinacea*)	TPHs	Huang et al. (2005)
P. fluorescens 2–79	Wheat (*T. aestivum*)	TCE	Yee et al. (1998)
P. fluorescens F113	Alfalfa (*M. sativa*)	PCBs	Villacieros et al. (2005)
P. putida Flav1-1	Arobidopsis (*Arabidopsis thaliana*)	PCBs	Narasimhan et al. (2003)
Azospirillum lipoferum	Wheat (*T. aestivum*)	Crude oil	
Enterobactor cloacae	Tall fescue (*F. arundinacea*)	TPHs	Ebadi et al. (2018)
P. fluorescens	Alfalfa (*M. sativa*)	PCBs	Toussaint et al. (2012)

PAHs, Polycyclic aromatic hydrocarbons; *PCBs*, polychlorinated biphenyls; *TCE*, trichloroethylene; *TPHs*, total petroleum hydrocarbons.

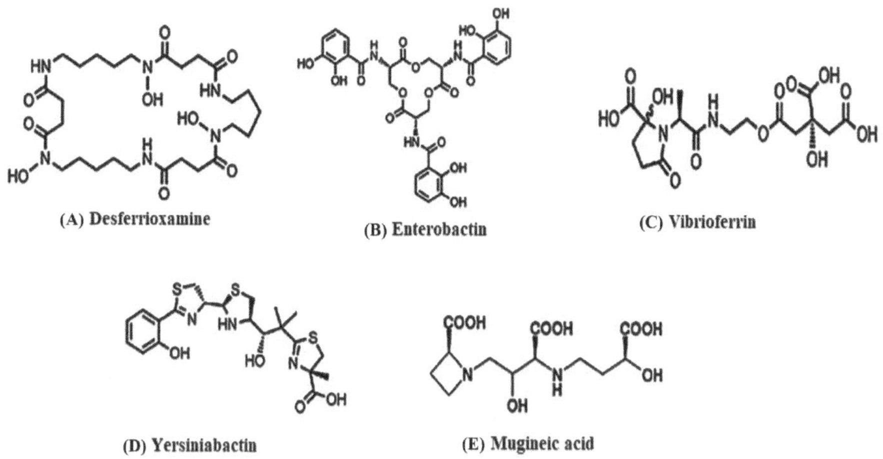

Figure 2.11 Various forms of siderophores. (A) Hydroxamate siderophore: desferrioxamine. (B) Catecholate siderophore: enterobactin. (C) Carboxylate siderophore: vibrioferrin. (D) Mixed siderophore: yersiniabactin. (E) Phytosiderophore: mugineic acid.

signaling molecule that has significant effects on the communication between plants and microorganisms and promotes plant growth. The property of synthesizing IAA is considered an effective tool for screening beneficial microorganisms suggesting that IAA-producing bacteria have a profound effect on plant growth. Tryptophan is the main precursor for IAA biosynthesis in microbes (Mohite, 2013). A positive test of IAA production is shown in Fig. 2.12(A).

2.4.3 Phosphate solubilization

The phosphate solubilizing ability of bacteria can convert insoluble forms of phosphorus (P) to access forms. Hence, rhizospheric bacteria with phosphate solubilizing activity play a key role at an industrial waste polluted site in the bioremediation of phosphate-containing pollutants (Walia et al., 2017). A positive test of phosphate solubilization production is shown in Fig. 2.12D.

2.4.4 Exopolysaccharide production

Rhizobacteria exopolysaccharide (EPS) is also an important feature of rhizospheric bacteria which can also help to mitigate stress conditions at polluted sites by reducing the content of Na^{+} or other cationic ions available for plant uptake. Important role exhibited by EPSs includes protection from desiccation, microbial aggregation, plant-microbe interaction, surface attachment, bioremediation and its use by many industries for stabilizing, thickening, coagulating, gelling, suspending for film-forming (Li et al., 2001).

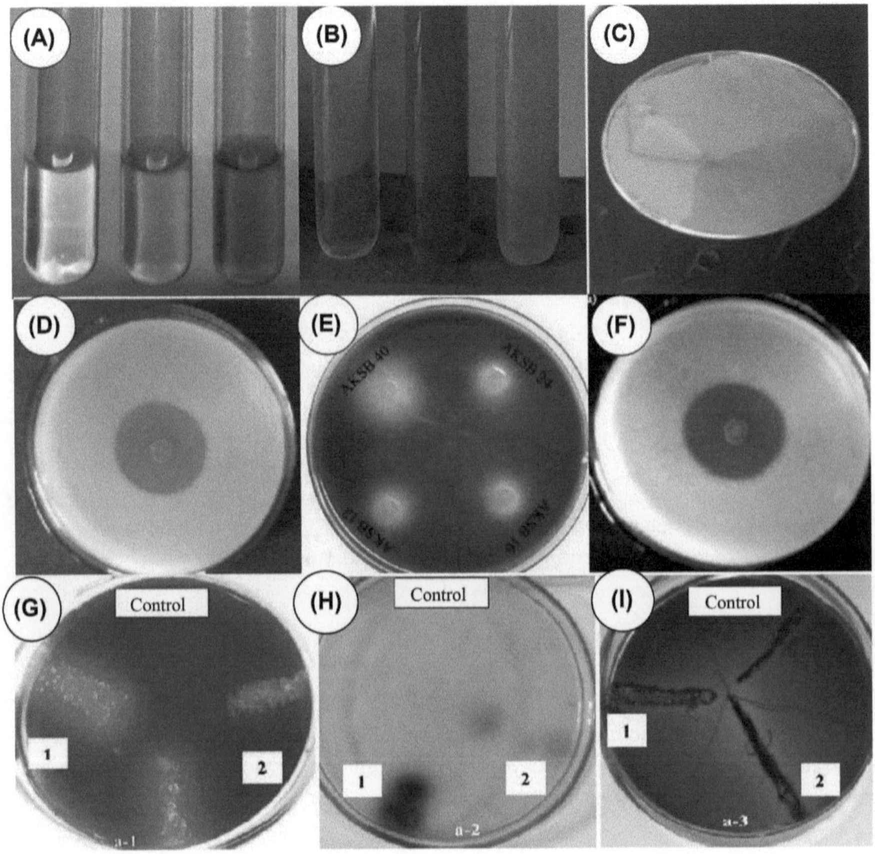

Figure 2.12 PGPR test. (A) Indole acetic acid test. (B) Ammonia production. (C) Hydrogen cyanide production. (D) Phosphate solubilization. (E) Potassium solubilization. (F) Zinc solubilization. (G) MnP (mangnese peroxidase). (H) Laccasse. (I) Lip (lignine peroxidase). *Source*: Based on Akintokun, A.K., Taiwo, M.O., 2016. Comparison of single culture and the consortium of growth-promoting rhizobacteria from three tomato (*Lycopersicon esculentum Mill*) varieties. Adv. Plants Agric. Res. 5 (1), 00167; Sharma, P., Tripathi, S., Chandra, R., 2020. Highly efficient phytoremediation potential of metal and metalloids from the pulp paper industry waste employing *Eclipta alba* (L) and *Alternanthera philoxeroide* (L): biosorption and pollution reduction. Bioresour. Technol. 319, 124–147.

2.4.5 Ammonia and HCN production

The production of HCN and ammonia (NH_3) is the important rhizobacterial activity of plant growth-promoting strains. HCN is also used in the chelating of metal ions as well as indirectly involved in making the availability of phosphate. Besides HCN, ammonia production by rhizobacteria accumulates and supplies nitrogen to their host plant and promotes root and shoot elongation and their biomass (Singh et al., 2019). To check the HCN production in the laboratory, filter paper soaked in reagent solution (2.5 g picric acid, 0.125 g, and sodium carbonate, 1000 mL of distilled water) can be used. A positive test for HCN production will turn filter paper yellow to orange/brown. A positive test of ammonia and HCN production is shown in Fig. 2.12B and C.

2.4.6 ACC deaminase production

ACC deaminase regulates ethylene levels in plants by metabolizing its precursor ACC into aketobutyric acid and ammonia. ACC deaminase bacteria could be used for the development of plant-inoculant systems for phytoremediation of the contaminated soil environment. Rhizobacterium having ACC deaminase activity can help in reducing phytotoxicity of metal and enhanced its accumulation in plant roots (Arshad et al., 2007).

2.4.7 Antibiotic resistance production

The production of one or more antibiotics is the mechanism most commonly associated with the ability of rhizobacteria to act as antagonistic agents against phytopathogens. Antibiotics encompass a heterogeneous group of organic, low-molecular-weight compounds that are deleterious to the growth or metabolic activities of other microorganisms (Beneduzi et al., 2012). The multidrug resistance pattern of rhizospheric bacteria has been also reported from the rhizosphere of *P. australis*, growing in constructed wetland treatment of distillery effluent treatment process.

2.4.8 Carbohydrate utilization profile production

Plant roots exude a huge diversity of organic compounds with carbon, hydrogen, and oxygen (CHO) atoms that attract microbial populations, especially those able to metabolize plant-exuded compounds and proliferate in this microbial habitat. Most of the pollutants also consist of CHO atoms and are therefore metabolized by rhizospheric bacteria (Beneduzi et al., 2012).

2.4.9 Ligninolytic enzyme activity

Some potential rhizobacteria show ligninolytic enzyme activity which consists of three oxidative enzymes: LiP, MnP, and laccase. These enzymes have their mechanisms to mitigate the pollution load through the biological remediation of industrial recalcitrant chemicals present in wastewaters. Ligninolytic enzymes play an important role in the degradation of lignocellulosic materials of soil and other waste. These enzymes degrade not only lignocellulose, but also recalcitrant environmental pollutants such as crude oil wastes, textile wastes, distillery, and tannery wastes, pulp and paper mill wastes and wastewaters, and organochlorine agrochemicals, and also bioremediate polluted soils as shown in Fig. 2.12G–I (Sharma et al., 2020; Zainith et al., 2020).

An example of PGPR isolates for positive test showing appreciable P, K, and Zn solubilization, IAA, HCN, ammonia, and siderophore production has been shown in Table 2.7.

2.5 Challenges and future prospects

The microbial communities that survive at the extremes of pH (> pH 8.5 < pH 5.0) temp (> 45°C– < 15°C) pressure (0.500 atm) salinity (1.0 m NaCl) in high concentration of recalcitrant organic matter or heavy metals represent one of the last frontiers for biotechnological and industrial discovery. The application for most extremophiles is based on their biomolecules, primarily enzymes, proteins, lipids, or other small molecules. The well-known success application of an extremophiles derived product is *Taq* DNA polymerase. Similarly, the cold adaptation in bacteria is regulated through intrinsic properties of nucleic acid, lipid, and enzyme which allow the microbes to grow in adverse conditions. Besides plant strategies for phytoextraction of heavy metals and amelioration of organic pollutants, it also opens broad prospects for their research and biotechnological application for

Table 2.7 Characteristic features of rhizospheric bacteria responsible for assisting phytoremediation strategy of the plant.

PGPR trait observed	B1	B2	B3	B4	B5	B6
Plant hormone and nutrient bioavailability						
IAA production	+++	+++	+++	+++	+++	+++
Ammonia production	+++	+++	+++	+++	+++	+++
Siderophore	+++	+	++	++	++	+
Biocontrol						
HCN production	+++	++	+++	+++	++	+++
Ligninolytic enzyme						
Laccase	++	+	++	++	++	++
Lip	+	++	+	+	+	+++
MnP	+++	+	+	+++	+	+
Mineral solubilization						
Zn solubilization (zone in mm)	11.4	7.3	8.6	10.6	8.4	8.2
K solubilization (zone in mm)	5.2	6.6	7.7	8.2	6.8	9.3
P solubilization (zone in mm)	12	16	16	14	13	14

eco-restoration of a polluted site for sustainable development. Therefore to ensure additional gains further research in the following area is warranted:

1. Application of constructed wetland treatment system after bacterial degradation of organic pollutants for industrial wastewater treatment as a tertiary treatment technology.
2. Application of phytoremediation technology for sewage wastewater treatment and recycling of treated water for pollution prevention of aquatic resources.
3. Manipulation of metal transporter and their cellular targeting to specific cell types such as vacuoles to allow for safe compartmentation of heavy metal from the polluted site.
4. Application of genetically engineered plant for the production of specific proteins and phytoremediation process at a large scale.
5. To reveal the rhizospheric bacterial metagenomes from the polluted site to understand the mechanism of phytoremediation of a complex industrial waste polluted site.
6. Screening of medical food-producing plant capable to accumulate a high amount of heavy metal from the polluted site, so that they can be maintained for the production of quality of food and drug policy formulation for decontamination of polluted soil and aquatic sources may be taken into consideration.
7. Development of metal recovery technology from plant capable of phytoextraction and hyperaccumulation of heavy metal from the polluted site.
8. To elucidate the mechanism of plant growth-promoting activity of rhizospheric bacterial community growing at the polluted site.

The above challenges and prospects will establish the most sustainable ecosystem with good agricultural products.

Acknowledgment

This work is financially supported by the Department of Biotechnology, Government of India. Vide Letter No. BT/PR13922/BCE/8/1129/2015, to Prof. R. Chandra, Department of Environmental Microbiology, BBAU, Lucknow.

References

Abd El-Rahman, A.F., Shaheen, H.A., Abd El-Aziz, R.M., Ibrahim, D.S., 2019. Influence of hydrogen cyanide-producing rhizobacteria in controlling the crown gall and root-knot nematode, *Meloidogyne incognita*. Egypt. J. Biol. Pest. Co. 29 (1), 41.

Abomohra, A.E.F., El-Sheekh, M., Hanelt, D., 2017. Screening of marine microalgae isolated from the hypersaline Bardawil lagoon for biodiesel feedstock. Renew. Energy 101, 1266–1272.

Adriano, D.C., Wenzel, W.W., Vangronsveld, J., Bolan, N.S., 2004. Role of assisted natural remediation in environmental cleanup. Geoderma 122 (2–4), 121–142.

Ahmed, E., Holmström, S.J., 2014. Siderophores in environmental research: roles and applications. Microb. Biotechnol. 7 (3), 196–208.

Ahmad, A., Hadi, F., Ali, N., 2015. Effective phytoextraction of cadmium (Cd) with increasing concentration of total phenolics and free proline in *Cannabis sativa* (L) plant under various treatments of fertilizers, plant growth regulators and sodium salt. Int. J. Phytoremediation 17 (1), 56–65.

Akintokun, A.K., Taiwo, M.O., 2016. Comparison of single culture and the consortium of growth-promoting rhizobacteria from three tomato (*Lycopersicon esculentum Mill*) varieties. Adv. Plants Agric. Res. 5 (1), 00167.

Ali, H., Khan, E., Sajad, M.A., 2013. Phytoremediation of heavy metals—concepts and applications. Chemosphere 91 (7), 869–881.

Ali, I., Akbar, A., Anwar, M., Prasongsuk, S., Lotrakul, P., Punnapayak, H., 2015. Purification and characterization of a polyextremophilic α-amylase from an obligate halophilic *Aspergillus penicillioides* isolate and its potential for souse with detergents. BioMed. Res. Int. 2015, Article ID 245649.

Andraski, B.J., Sandstrom, M.W., Michel, R.L., Radyk, J.C., Stonestrom, D.A., Johnson, M. J., et al., 2003. Simplified method for detecting tritium contamination in plants and soil. J. Environ. Qual. 32 (3), 988–995.

Andraski, B.J., Stonestrom, D.A., Michel, R.L., Halford, K.J., Radyk, J.C., 2005. Plant-based plume-scale mapping of tritium contamination in desert soils. Vadose Zone J. 4 (3), 819–827.

Andreazza, R., Bortolon, L., Pieniz, S., Bento, F.M., Camargo, F.A.D.O., 2015. Evaluation of two Brazilian indigenous plants for phytostabilization and phytoremediation of copper-contaminated soils. Braz. J. Biol. Sci. 75 (4), 868–877.

Arshad, M., Saleem, M., Hussain, S., 2007. Perspectives of bacterial ACC deaminase in phytoremediation. Trends Biotechnol. 25 (8), 356–362.

Backer, R., Rokem, J.S., Ilangumaran, G., Lamont, J., Praslickova, D., Ricci, E., et al., 2018. Plant growth-promoting rhizobacteria: context, mechanisms of action, and roadmap to commercialization of biostimulants for sustainable agriculture. Front. Plant Sci. 9, 1473.

Baker-Austin, C., Dopson, M., 2007. Life in acid: pH homeostasis in acidophiles. Trends Microbiol. 15 (4), 165–171.

Bañuelos, G., Terry, N., LeDuc, D.L., Pilon-Smits, E.A., Mackey, B., 2005. Field trial of transgenic Indian mustard plants shows enhanced phytoremediation of selenium-contaminated sediment. Environ. Sci. Technol. 39 (6), 1771–1777.

Bech, J., Duran, P., Roca, N., Poma, W., Sánchez, I., Roca-Pérez, L., et al., 2012. Accumulation of Pb and Zn in *Bidens triplinervia* and *Senecio* sp. spontaneous species from mine spoils in Peru and their potential use in phytoremediation. J. Geochem. Explor. 123, 109–113.

Beneduzi, A., Ambrosini, A., Passaglia, L.M., 2012. Plant growth-promoting rhizobacteria (PGPR): their potential as antagonists and biocontrol agents. Genet. Mol. Biol. 35 (4), 1044–1051.

Berger, F., Morellet, N., Menu, F., Potier, P., 1996. Cold shock and cold acclimation proteins in the psychrotrophic bacterium *Arthrobacter globiformis* SI55. J. Bacteriol. 178 (11), 2999–3007.

Berger, F., Normand, P., Potier, P., 1997. capA, a cspA-like gene that encodes a cold acclimation protein in the psychrotrophic bacterium *Arthrobacter globiformis* SI55. J. Bacteriol. 179 (18), 5670–5676.

Berti, W.R., Cunningham, S.D., 2000. Phytoremediation of Toxic Metals: Using Plants to Clean-up the Environment. pp. 71–88.

Bertoldo, C., Dock, C., Antranikian, G., 2004. Thermoacidophilic microorganisms and their novel biocatalysts. Eng. Life Sci. 4 (6), 521–532.

Billore, S.K., Singh, N., Ram, H.K., Sharma, J.K., Singh, V.P., Nelson, R.M., et al., 2001. Treatment of a molasses based distillery effluent in a constructed wetland in central India. Water Sci. Technol. 44 (11–12), 441–448.

Bolan, N., Mahimairaja, S., Kunhikrishnan, A., Choppala, G., 2013. Phosphorus–arsenic interactions in variable-charge soils in relation to arsenic mobility and bioavailability. Sci. Total Environ. 463, 1154–1162.

Bolhuis, A., Kwan, D., Thomas, J.R., 2008. Halophilic adaptations of proteins. In: Siddiqui, K.S., Thomas, T. (Eds) (Eds.), Protein Adaptation in Extremophiles. Nova Science Publishers, Inc., New York, pp. 71–104.

Bonugli-Santos, R.C., Durrant, L.R., Da Silva, M., Sette, L.D., 2010. Production of laccase, manganese peroxidase and lignin peroxidase by Brazilian marine-derived fungi. Enzyme Microb. Technol. 46 (1), 32–37.

Boyd, J., McDonald, O., Hatcher, D., Portier, R.J., Conway, R.B., 1993. Interfacing constructed wetlands with traditional wastewater biotreatment systems. Constructed Wetlands for Water Quality Improvement. Lewis Publishers, Boca Raton, FL, pp. 453–459.

Britton, K.L., Baker, P.J., Fisher, M., Ruzheinikov, S., Gilmour, D.J., Bonete, M.J., et al., 2006. Analysis of protein solvent interactions in glucose dehydrogenase from the extreme halophile *Haloferax mediterranei*. PNAS. 103 (13), 4846–4851.

Burd, G.I., Dixon, D.G., Glick, B.R., 1998. A plant growth-promoting bacterium that decreases nickel toxicity in seedlings. Appl. Environ. Microbiol. 64 (10), 3663–3668.

Calheiros, C.S., Rangel, A.O., Castro, P.M., 2007. Constructed wetland systems vegetated with different plants applied to the treatment of tannery wastewater. Water Res. 41 (8), 1790–1798.

Calheiros, C.S., Rangel, A.O., Castro, P.M., 2009. Treatment of industrial wastewater with two-stage constructed wetlands planted with *Typha latifolia and Phragmites australis*. Bioresour. Technol. 100 (13), 3205–3213.

Calheiros, C.S., Quitério, P.V., Silva, G., Crispim, L.F., Brix, H., Moura, S.C., et al., 2012. Use of constructed wetland systems with *Arundo* and *Sarcocornia* for polishing high salinity tannery wastewater. J. Environ. Manage. 95 (1), 66–71.

Chandra, R., Chowdhary, P., 2015. Properties of bacterial laccases and their application in bioremediation of industrial wastes. Environ. Sci.: Process. Impacts. 17 (2), 326–342.

Chandra, R., Kumar, V., 2017. Phytoextraction of heavy metals by potential native plants and their microscopic observation of root growing on stabilised distillery sludge as a prospective tool for in situ phytoremediation of industrial waste. Environ. Sci. Pollut. Res. 24 (3), 2605–2619.

Chandra, R., Yadav, S., Yadav, S., 2017. Phytoextraction potential of heavy metals by native wetland plants growing on chlorolignin containing sludge of pulp and paper industry. Ecol. Eng. 98, 134–145.

Chandra, R., Kumar, V., Tripathi, S., Sharma, P., 2018. Heavy metal phytoextraction potential of native weeds and grasses from endocrine-disrupting chemicals rich complex distillery sludge and their histological observations during in-situ phytoremediation. Ecol. Eng. 111, 143–156.

Chowdhury, P., Viraraghavan, T., Srinivasan, A., 2010. Biological treatment processes for fish processing wastewater – a review. Bioresour. Technol. 101 (2), 439–449.

Cowan, D.A., Ramond, J.B., Makhalanyane, T.P., De Maayer, P., 2015. Metagenomics of extreme environments. Curr. Opin. Microbiol. 25, 97–102.

Crous, L., Britz, P., 2010. The use of constructed wetland technology in the treatment and beneficiation of brewery effluent for aquaculture. In: the IWA International Conference on Wetland Systems for Water Pollution Control, Venice, Italy, pp. 1254–1259.

DasSarma, P., Coker, J.A., Huse, V., DasSarma, S., 2010. Halophiles, biotechnology. In: Flickinger, M.C., Hoboken, N.J. (Eds) (Eds.), Encyclopedia of Industrial

Biotechnology, Bioprocess, Bioseparation, and Cell Technology. John Wiley and Sons, pp. 2769–2777.

Davies, L.C., Carias, C.C., Novais, J.M., Martins-Dias, S., 2005. Phytoremediation of textile effluents containing azo dye by using *Phragmites australis* in a vertical flow intermittent feeding constructed wetland. Ecol. Eng. 25 (5), 594–605.

Delgado-García, M., Valdivia-Urdiales, B., Aguilar-González, C.N., Contreras-Esquivel, J.C., Rodríguez-Herrera, R., 2012. Halophilic hydrolases as a new tool for the biotechnological industries. J. Sci. Food Agric. 92 (13), 2575–2580.

Dennis, P.P., Shimmin, L.C., 1997. Evolutionary divergence and salinity-mediated selection in halophilic archaea. Microbiol. Mol. Biol. Rev. 61 (1), 90–104.

Duckworth, A.W., Grant, W.D., Jones, B.E., Van Steenbergen, R., 1996. Phylogenetic diversity of soda lake alkaliphiles. FEMS Microbiol. Ecol. 19 (3), 181–191.

Dumorné, K., Córdova, D.C., Astorga-Eló, M., Renganathan, P., 2017. Extremozymes: a potential source for industrial applications. J. Microbiol. Biotechnol. 27 (4), 649–659.

Dushenkov, S., 2003. Trends in phytoremediation of radionuclides. Plant Soil 249 (1), 167–175.

Dym, O., Mevarech, M., Sussman, J.L., 1995. Structural features that stabilize halophilic malate dehydrogenase from an archaebacterium. Science 267 (5202), 1344–1346.

Ebadi, A., Sima, N.A.K., Olamaee, M., Hashemi, M., Nasrabadi, R.G., 2018. Remediation of saline soils contaminated with crude oil using the halophyte Salicornia persica in conjunction with hydrocarbon-degrading bacteria. J. Environ. Manage. 219, 260–268.

Eitam, H., Agmon, R., Asher, A., Brosh, A., Orlov, A., Izhaki, I., et al., 2012. Protein deprivation attenuates Hsp expression in fat tissue. Cell Stress. Chaperones. 17 (3), 339–347.

Enguita, F.J., Martins, L.O., Henriques, A.O., Carrondo, M.A., 2003. Crystal structure of a bacterial endospore coat component a laccase with enhanced thermostability properties. J. Biol. Chem. 278 (21), 19416–19425.

Etesami, H., Emami, S., Alikhani, H.A., 2017. Potassium solubilizing bacteria (KSB): mechanisms, promotion of plant growth, and future prospects a review. J. Soil Sci. Plant Nutr. 17 (4), 897–911.

Farwell, A.J., Vesely, S., Nero, V., Rodriguez, H., Shah, S., Dixon, D.G., et al., 2006. The use of transgenic canola (*Brassica napus*) and plant growth-promoting bacteria to enhance plant biomass at a nickel-contaminated field site. Plant Soil 288 (1–2), 309–318.

Favas, P.J., Pratas, J., Varun, M., D'Souza, R., Paul, M.S., 2014. Phytoremediation of soils contaminated with metals and metalloids at mining areas: potential of native flora. Environ. Risk Assess. Soil Contamination 3, 485–516.

Feller, G., Gerday, C., 2003. Psychrophilic enzymes: hot topics in cold adaptation. Nat. Rev. Microbiol. 1 (3), 200–208.

Freddolino, P.L., Tavazoie, S., 2012. Beyond homeostasis: a predictive-dynamic framework for understanding cellular behavior. Annu. Rev. Cell Dev. Biol. 28, 363–384.

Fujinami, S., Sato, T., Trimmer, J.S., Spiller, B.W., Clapham, D.E., Krulwich, T.A., et al., 2007. The voltage-gated Na^+ channel NaVBP co-localizes with methyl-accepting chemotaxis protein at cell poles of alkaliphilic *Bacillus pseudofirmus* OF_4. Microbiology (Reading, Engl.) 153 (Pt. 12), 4027.

Fujinami, S., Terahara, N., Krulwich, T.A., Ito, M., 2009. Motility and chemotaxis in alkaliphilic Bacillus species. Future Microbiol. 4 (9), 1137–1149.

Furhan, J., 2020. Adaptation, production, and biotechnological potential of cold-adapted proteases from psychrophiles and psychrotrophs: recent overview. J. Genet. Eng. Biotechnol. 18 (1), 1–13.

Fushinobu, S., Ito, K., Konno, M., Wakagi, T., Matsuzawa, H., 1998. Crystallographic and mutational analyses of an extremely acidophilic and acid-stable xylanase: biased distribution of acidic residues and importance of Asp37 for catalysis at low pH. Protein Eng. 11 (12), 1121–1128.

Galal, T.M., Eid, E.M., Dakhil, M.A., Hassan, L.M., 2018. Bioaccumulation and rhizofiltration potential of *Pistia stratiotes L.* for mitigating water pollution in the Egyptian wetlands. Int. J. Phytoremediation. 20 (5), 440–447.

Gao, J., Garrison, A.W., Hoehamer, C., Mazur, C.S., Wolfe, N.L., 2000. Uptake and phytotransformation of organophosphorus pesticides by axenically cultivated aquatic plants. J. Agric. Food Chem. 48 (12), 6114–6120.

Ghosh, M., Singh, S.P., 2005. A comparative study of cadmium phytoextraction by accumulator and weed species. Environ. Pollut. 133 (2), 365–371.

Gontia-Mishra, I., Sapre, S., Tiwari, S., 2017. Diversity of halophilic bacteria and actinobacteria from India and their biotechnological applications. NISCAIR-CSIR, India 46 (8), 1575–1587.

Goto, T., Matsuno, T., Hishinuma-Narisawa, M., Yamazaki, K., Matsuyama, H., Inoue, N., et al., 2005. Cytochrome c and bioenergetic hypothetical model for alkaliphilic *Bacillus* spp. J. Biosci. Bioeng. 100 (4), 365–379.

Groth, I., Schumann, P., Rainey, F.A., Martin, K., Schuetze, B., Augsten, K., 1997. *Bogoriella caseilytica* gen. nov., sp. nov., a new alkaliphilic actinomycete from a soda lake in Africa. Int. J. Syst. Evol. 47 (3), 788–794.

Gupta, V.K., Agarwal, S., Singh, P., Pathania, D., 2013. Acrylic acid grafted cellulosic *Luffa cylindrical* fiber for the removal of dye and metal ions. Carbohydr. Polym. 98 (1), 1214–1221.

Hammer, D.A., Pullin, B.P., McMurry, D.K., Lee, J.W., 1993. Testing color removal from pulp mill wastewaters with constructed wetlands. Constructed Wetlands for Water Pollution Improvement. CRC Press/Lewis Publishers, Boca Raton, FL, pp. 449–452.

Haney, P., Konisky, J., Koretke, K.K., Luthey-Schulten, Z., Wolynes, P.G., 1997. Structural basis for thermostability and identification of potential active site residues for adenylate kinases from the archaeal genus *Methanococcus*. Proteins 28 (1), 117–130.

Hébraud, M., Potier, P., 1999. Cold acclimation and cold shock response in psychrotrophic bacteria. J. Mol. Microbiol. Biotechnol. 1 (2), 211–219.

Hebraud, M., Dubois, E., Potier, P., Labadie, J., 1994. Effect of growth temperatures on the protein levels in a psychrotrophic bacterium, *Pseudomonas fragi*. J. Bacteriol. 176 (13), 4017–4024.

Henry, J.R., 2000. Prepared By Jeanna R. Henry. National Network of Environmental Management Studies (NNEMS). Fellow. May–August 2000.

Horikoshi, K., 1999. Alkaliphiles: some applications of their products for biotechnology. Microbiol. Mol. Biol. Rev. 63 (4), 735–750.

Hottes, A.K., Freddolino, P.L., Khare, A., Donnell, Z.N., Liu, J.C., Tavazoie, S., 2013. Bacterial adaptation through loss of function. PLoS Genet. 9 (7), e1003617.

Huang, J.W., Chen, J., Berti, W.R., Cunningham, S.D., 1997. Phytoremediation of lead-contaminated soils: role of synthetic chelates in lead phytoextraction. Environ. Sci. Technol. 31 (3), 800–805.

Huang, X.D., El-Alawi, Y., Penrose, D.M., Glick, B.R., Greenberg, B.M., 2004. A multi-process phytoremediation system for removal of polycyclic aromatic hydrocarbons from contaminated soils. Environ. Pollut. 130 (3), 465–476.

Huang, X.D., El-Alawi, Y., Gurska, J., Glick, B.R., Greenberg, B.M., 2005. A multi-process phytoremediation system for decontamination of persistent total petroleum hydrocarbons (TPHs) from soils. Microchem. J. 81 (1), 139–147.

Irimia, A., Ebel, C., Madern, D., Richard, S.B., Cosenza, L.W., Zaccaï, G., et al., 2003. The oligomeric states of *Haloarcula marismortui* malate dehydrogenase are modulated by solvent components as shown by crystallographic and biochemical studies. J. Mol. Biol. 326 (3), 859–873.

Islam, M.N., Jung, H.Y., Park, J.H., 2015. Subcritical water treatment of explosive and heavy metals co-contaminated soil: removal of the explosive, and immobilization and risk assessment of heavy metals. J. Environ. Manage 163, 262–269.

Jackson, C.R., Langner, H.W., Donahoe-Christiansen, J., Inskeep, W.P., McDermott, T.R., 2001. Molecular analysis of microbial community structure in an arsenite-oxidizing acidic thermal spring. Environ. Microbiol. 3 (8), 532–542.

Jaemsaeng, R., Jantasuriyarat, C., Thamchaipenet, A., 2018. Positive role of 1-aminocyclopropane-1-carboxylate deaminase-producing endophytic *Streptomyces* sp. GMKU 336 on flooding resistance of mung bean. Agric. Nat. Resour. 52 (4), 330–334.

Kadlec, R.H., Wallace, S.D., 2009. Treatment Wetlands, second ed. CRC Press.

Kalayu, G., 2019. Phosphate solubilizing microorganisms: promising approach as biofertilizers. Int. J. Agron. 2019, Article ID 4917256.

Kamran, M.A., Syed, J.H., Eqani, S.A.M.A.S., Munis, M.F.H., Chaudhary, H.J., 2015. Effect of plant growth-promoting rhizobacteria inoculation on cadmium (Cd) uptake by *Eruca sativa*. Environ. Sci. Pollut. Res. 22 (12), 9275–9283.

Karan, R., Capes, M.D., DasSarma, S., 2012. Function and biotechnology of extremophilic enzymes in low water activity. Aquat. Biosyst. 8 (1), 4.

Krulwich, T.A., 1995. Alkaliphiles: 'basic' molecular problems of pH tolerance and bioenergetics. Mol. Microbiol. 15 (3), 403–410.

Krulwich, T.A., Hicks, D.B., Ito, M., 2009. Cation/proton antiporter complements of bacteria: why so large and diverse? Mol. Microbiol. 74 (2), 257–260.

Krulwich, T.A., Ito, M., Guffanti, A.A., 2001. The Na^+-dependence of alkaliphily in *Bacillus*. Biochim. Biophys. Acta. 1505 (1), 158–168.

Kucuk, O.S., Sengul, F., Kapdan, I.K., 2004. Removal of ammonia from tannery effluents in a reed bed constructed wetland. Water Sci. Technol. 48 (11–12), 179–186.

Kumar, A., Chandra, R., 2020. Ligninolytic enzymes and its mechanisms for degradation of lignocellulosic waste in environment. Heliyon 6 (2), e03170.

Kumar, S., Nussinov, R., 2001. How do thermophilic proteins deal with heat? Cell Mol. Life Sci. 58 (9), 1216–1233.

Kumar, S., Ma, B., Tsai, C.J., Nussinov, R., 2000. Electrostatic strengths of salt bridges in thermophilic and mesophilic glutamate dehydrogenase monomers. Proteins 38 (4), 368–383.

Kumar, L., Awasthi, G., Singh, B., 2011. Extremophiles: a novel source of industrially important enzymes. Biotechnology 10 (2), 121–135.

Kumpiene, J., Lagerkvist, A., Maurice, C., 2007. Stabilization of Pb-and Cu-contaminated soil using coal fly ash and peat. Environ. Pollut. 145 (1), 365–373.

Li, P., Harding, S.E., Liu, Z., 2001. Cyanobacterial exopolysaccharides: their nature and potential biotechnological applications. Biotechnol. Genet. Eng. Rev. 18 (1), 375–404.

Licht, L.A., Isebrands, J.G., 2005. Linking phytoremediated pollutant removal to biomass economic opportunities. Biomass Bioenergy 28 (2), 203–218.

Matin, A., 1992. Physiology, molecular biology and applications of the bacterial starvation response. J. Appl. Microbiol. 73, 49–57.

Mbuligwe, S.E., 2005. Comparative treatment of dye-rich wastewater in engineered wetland systems (EWSs) vegetated with different plants. Water Res. 39 (2–3), 271–280.

Memon, A.R., Schröder, P., 2009. Implications of metal accumulation mechanisms to phytoremediation. Environ. Sci. Pollut. Res. 16 (2), 162–175.
Meyerholt, B., 2013. Phytoextraction of Cadmium: the advantages, limitations, cost and potential improvements of this remediation technique. UF/IFAS University of Florida, pp. 1–22.
Mohite, B., 2013. Isolation and characterization of indole acetic acid (IAA) producing bacteria from rhizospheric soil and its effect on plant growth. J. Soil Sci. Plant Nutr. 13 (3), 638–649.
Moreno, F.N., Anderson, C.W., Stewart, R.B., Robinson, B.H., 2005. Mercury volatilisation and phytoextraction from base-metal mine tailings. Environ. Pollut. 136 (2), 341–352.
Muratova, A.Y., Turkovskaya, O.V., Antonyuk, L.P., Makarov, O.E., Pozdnyakova, L.I., Ignatov, V.V., 2005. Oil-oxidizing potential of associative rhizobacteria of the genus Azospirillum. Microbiology 74 (2), 210–215.
Murphy, C., Hawes, P., Cooper, D.J., 2009. The application of wetland technology for copper removal from distillery wastewater: a case study. Water Sci. Technol. 60 (11), 2759–2766.
Narasimhan, K., Basheer, C., Bajic, V.B., Swarup, S., 2003. Enhancement of plant-microbe interactions using a rhizosphere metabolomics-driven approach and its application in the removal of polychlorinated biphenyls. Plant Physiol. 132 (1), 146–153.
Nikolaev, Y.A., 2004. Extracellular factors of bacterial adaptation to unfavorable environmental conditions. Appl. Biochem. Microbiol. 40 (4), 327–336.
Olguín, E.J., Sánchez-Galván, G., González-Portela, R.E., López-Vela, M., 2008. Constructed wetland mesocosms for the treatment of diluted sugarcane molasses stillage from ethanol production using *Pontederia sagittata*. Water Res. 42 (14), 3659–3666.
Onofre-Lemus, J., Hernández-Lucas, I., Girard, L., Caballero-Mellado, J., 2009. ACC (1-aminocyclopropane-1-carboxylate) deaminase activity, a widespread trait in Burkholderia species, and its growth-promoting effect on tomato plants. Appl. Environ. Microbiol. 75 (20), 6581–6590.
Orellana, R., Macaya, C., Bravo, G., Dorochesi, F., Cumsille, A., Valencia, R., et al., 2018. Living at the frontiers of life: extremophiles in Chile and their potential for bioremediation. Front. Microbiol. 9, 2309.
Padan, E., Bibi, E., Ito, M., Krulwich, T.A., 2005. Alkaline pH homeostasis in bacteria: new insights. Biochim. Biophys. Acta. 1717 (2), 67–88.
Persson, E., Halle, B., 2008. Cell water dynamics on multiple time scales. PNAS. 105 (17), 6266–6271.
Pervez, A., Headley, A.D., Terzis, E., 1999. The treatment of azo dyes using reedbed treatment systems. In: Wetlands Remediation: An International Conference, pp. 187–194.
Pire, C., Marhuenda-egea, F.C., Esclapez, J., Alcaraz, L., Ferrer, J., José Bonete, M., 2004. Stability and enzymatic studies of glucose dehydrogenase from the Archaeon *Haloferax mediterranei* in reverse micelles. Biocatal. Biotransformation 22 (1), 17–23.
Pulford, I.D., Watson, C., 2003. Phytoremediation of heavy metal-contaminated land by trees —a review. Environ. Int. 29 (4), 529–540.
Rampelotto, P.H., 2013. Extremophiles and extreme environments. Life 3 (3), 482–485.
Reed, M.L., Glick, B.R., 2005. Growth of canola (*Brassica napus*) in the presence of plant growth-promoting bacteria and either copper or polycyclic aromatic hydrocarbons. Can. J. Microbiol. 51 (12), 1061–1069.
Reed, M.L.E., Warner, B.G., Glick, B.R., 2005. Plant growth–promoting bacteria facilitate the growth of the common reed *Phragmites australis* in the presence of copper or polycyclic aromatic hydrocarbons. Curr. Microbiol. 51 (6), 425–429.

Rungwa, S., Arpa, G., Sakulas, H., Harakuwe, A., Timi, D., 2013. Phytoremediation—an eco-friendly and sustainable method of heavy metal removal from closed mine environments in Papua New Guinea. Procedia Environ. Sci. 6, 269—277.

Russell, N.J., 1998. Molecular adaptations in psychrophilic bacteria: potential for biotechnological applications. Biotechnology of Extremophiles. Springer, Berlin, Heidelberg, pp. 1—21.

Sakakibara, M., Watanabe, A., Inoue, M., Sano, S., Kaise, T., 2010. Phytoextraction and phytovolatilization of arsenic from As-contaminated soils by Pteris vittata. In:Proceedings of the Annual International Conference on Soils, Sediments, Water and Energy. vol. 12, No. 1, p. 26.

Sarwar, N., Imran, M., Shaheen, M.R., Ishaque, W., Kamran, M.A., Matloob, A., et al., 2017. Phytoremediation strategies for soils contaminated with heavy metals: modifications and future perspectives. Chemosphere 171, 710—721.

Schnoor, J.L., 1997. Phytoremediation: Ground-Water Remediation. Technologies Analysis Center Technology Evaluation Report, TE-98-01, p. 37.

Seth, C.S., Remans, T., Keunen, E., Jozefczak, M., Gielen, H., Opdenakker, K., et al., 2012. Phytoextraction of toxic metals: a central role for glutathione. Plant Cell Environ. 35 (2), 334—346.

Sharma, S.S., Dietz, K.J., 2009. The relationship between metal toxicity and cellular redox imbalance. Trends Plant Sci. 14 (1), 43—50.

Sharma, A., Kawarabayasi, Y., Satyanarayana, T., 2012. Acidophilic bacteria and archaea: acid stable biocatalysts and their potential applications. Extremophiles 16 (1), 1—19.

Sharma, S., Singh, B., Manchanda, V.K., 2015. Phytoremediation: role of terrestrial plants and aquatic macrophytes in the remediation of radionuclides and heavy metal contaminated soil and water. Environ. Sci. Pollut. Res. 22 (2), 946—962.

Sharma, P., Tripathi, S., Chandra, R., 2020. Highly efficient phytoremediation potential of metal and metalloids from the pulp paper industry waste employing *Eclipta alba* (L) and *Alternanthera philoxeroide* (L): biosorption and pollution reduction. Bioresour. Technol. 319, 124—147.

Singh, A., Prasad, S.M., 2011. Reduction of heavy metal load in food chain: technology assessment. Rev. Environ. Sci. Biotechnol. 10 (3), 199.

Singh, M., Singh, D., Gupta, A., Pandey, K.D., Singh, P.K., Kumar, A., 2019. Plant growth promoting rhizobacteria: application in biofertilizers and biocontrol of phytopathogens. PGPR Amelioration in Sustainable Agriculture. Woodhead Publishing, pp. 41—66.

Sinha, S., Mukherjee, S.K., 2008. Cadmium—induced siderophore production by a high Cd-resistant bacterial strain relieved Cd toxicity in plants through root colonization. Curr. Microbiol. 56 (1), 55—60.

Song, C.L., Zhao, X.H., 2014. Structure and property modification of an oligochitosan-glycosylated and crosslinked soybean protein generated by microbial transglutaminase. Food Chem. 163, 114—119.

Spitzer, J., 2011. From water and ions to crowded biomacromolecules: in vivo structuring of a prokaryotic cell. Microbiol. Mol. Biol. Rev. 75 (3), 491—506.

Sriprang, R., Hayashi, M., Ono, H., Takagi, M., Hirata, K., Murooka, Y., 2003. Enhanced accumulation of Cd^{2+} by a *Mesorhizobium* sp. transformed with a gene from Arabidopsis thaliana coding for phytochelatin synthase. Appl. Environ. Microbiol. 69 (3), 1791—1796.

Sytar, O., Kumar, A., Latowski, D., Kuczynska, P., Strzałka, K., Prasad, M.N.V., 2013. Heavy metal-induced oxidative damage, defense reactions, and detoxification mechanisms in plants. Acta Physiol. Plant 35 (4), 985—999.

Tadeo, X., López-Méndez, B., Trigueros, T., Laín, A., Castaño, D., Millet, O., 2009. Structural basis for the aminoacid composition of proteins from halophilic archea. PLoS Biol. 7 (12), e1000257.

Tagkopoulos, I., Liu, Y.C., Tavazoie, S., 2008. Predictive behavior within microbial genetic networks. Science 320 (5881), 1313–1317.

Terry, N., Zayed, A.M., De Souza, M.P., Tarun, A.S., 2000. Selenium in higher plants. Annu. Rev. Plant Biol. 51 (1), 401–432.

Tettleton, R.P., Howell, F.G., Reaves, R.P., 1993. Performance of a Constructed Marsh in the Tertiary Treatment of Bleach Kraft Pulp Mill Effluent: Results of a 2-year Pilot Project. CRC Press/Lewis Publishers, Boca Raton, FL, pp. 437–440.

Thayaparan, M., Iqbal, S.S., Chathuranga, P.K.D., Iqbal, M.C.M., 2013. Rhizofiltration of Pb by *Azolla pinnata*. Int. J. Environ. Sci. 3 (6), 1811–1821.

Thut, R.N., 1990. Treatment of pulp mill effluent by an artificial marsh-large-scale pilot study. In: Proceeding of the TAPPI Environmental Conference, TAPPI, Washington, DC, p. 121–127.

Toussaint, J.P., Pham, T.T.M., Barriault, D., Sylvestre, M., 2012. Plant exudates promote PCB degradation by a rhodococcal rhizobacteria. Appl. Microbiol. Biotechnol. 95 (6), 1589–1603.

Tripathi, S., Sharma, P., Singh, K., Purchase, D., & Chandra, R., 2021a. Translocation of heavy metals in medicinally important herbal plants growing on complex organometallic sludge of sugarcane molasses-based distillery waste. Environmental Technology & Innovation 22, 101434. Available from: https://doi.org/10.1016/j.eti.2021.101434.

Tripathi, S., Singh, K., Mishra, A., Chandra, R., 2021b. Organo-metallic pollutants of distillery effluent and their toxicity on freshwater fish and germinating Zea mays seeds. International Journal of Environmental Science and Technology 1–14. Available from: https://doi.org/10.1007/s13762-021-03233-1.

USEPA, 2000. National Primary Drinking Water Regulations.

Vangronsveld, J., Van Assche, F., Clijsters, H., 1995. Reclamation of a bare industrial area contaminated by non-ferrous metals: in situ metal immobilization and revegetation. Environ. Pol. 87 (1), 51–59.

Villacieros, M., Whelan, C., Mackova, M., Molgaard, J., Sánchez-Contreras, M., Lloret, J., et al., 2005. Polychlorinated biphenyl rhizoremediation by *Pseudomonas fluorescens* F113 derivatives, using a *Sinorhizobium meliloti* nod system to drive bph gene expression. Appl. Environ. Microbiol. 71 (5), 2687–2694.

Walia, A., Guleria, S., Chauhan, A., Mehta, P., 2017. Endophytic bacteria: role in phosphate solubilization. Endophytes: Crop Productivity and Protection. Springer, Cham:, pp. 61–93.

Wang, H.X., Ng, T.B., 2004. A novel laccase with fair thermostability from the edible wild mushroom (*Albatrella dispansus*). Biochem. Biophys. Res. Commun. 319 (2), 381–385.

Welch, T.J., Bartlett, D.H., 1996. Isolation and characterization of the structural gene for OmpL, a pressure-regulated porin-like protein from the deep-sea bacterium Photobacterium species strain SS9. J. Bacteriol. 178 (16), 5027–5031.

Wiegel, J., Kevbrin, V.V., 2004. Alkalithermophiles. pp. 193–198.

Wong, M.H., 2003. Ecological restoration of mine degraded soils, with emphasis on metal contaminated soils. Chemosphere 50 (6), 775–780.

Wu, S.C., Cheung, K.C., Luo, Y.M., Wong, M.H., 2006. Effects of inoculation of plant growth-promoting rhizobacteria on metal uptake by *Brassica juncea*. Environ. Pollut. 140 (1), 124–135.

Wuana, R.A., Okieimen, F.E., 2011. Heavy metals in contaminated soils: a review of sources, chemistry, risks and best available strategies for remediation. ISRN Ecol. Article ID 402647.

Xianfa, L., Chuncai, J., 1995. Constructed wetland systems for water pollution control in North China. Water Sci. Technol. 32 (3), 349–356.

Yang, Y., Ge, Y., Zeng, H., Zhou, X., Peng, L., Zeng, Q., 2017. Phytoextraction of cadmium-contaminated soil and potential of regenerated tobacco biomass for recovery of cadmium. Sci. Rep. 7 (1), 1–10.

Yee, D.C., Maynard, J.A., Wood, T.K., 1998. Rhizoremediation of trichloroethylene by a recombinant, root-colonizing *Pseudomonas fluorescens*strain expressing toluene *ortho*-Monooxygenase constitutively. Appl. Environ. Microbiol. 64 (1), 112–118.

Yoon, J.M., Aken, B.V., Schnoor, J.L., 2006. Leaching of contaminated leaves following uptake and phytoremediation of RDX, HMX, and TNT by poplar. Int. J. Phytoremediation 8 (1), 81–94.

Zaidi, S., Usmani, S., Singh, B.R., Musarrat, J., 2006. Significance of *Bacillus subtilis* strain SJ-101 as a bioinoculant for concurrent plant growth promotion and nickel accumulation in *Brassica juncea*. Chemosphere 64 (6), 991–997.

Zainith, S., Chowdhary, P., Mani, S., Mishra, S., 2020. Microbial ligninolytic enzymes and their role in bioremediation. Microorg. Sustain. Environ. Health 179.

Zhang, Y., Li, X., Bartlett, D.H., Xiao, X., 2015. Current developments in marine microbiology: high-pressure biotechnology and the genetic engineering of piezophiles. Curr. Opin. Biotechnol. 33, 157–164.

Zhang, H., Guo, Q., Yang, J., Ma, J., Chen, G., Chen, T., et al., 2016. Comparison of chelates for enhancing *Ricinus communis* L. phytoremediation of Cd and Pb contaminated soil. Ecotoxicol. Environ. Saf. 133, 57–62.

Zhuang, X., Chen, J., Shim, H., Bai, Z., 2007. New advances in plant growth-promoting rhizobacteria for bioremediation. Environ. Int. 33 (3), 406–413.

Zuber, H., 1988. Temperature adaptation of lactate dehydrogenase structural, functional and genetic aspects. Biophys. Chem. 29 (1–2), 171–179.

Effect of soil biofilms on ecological function and impact on soil properties

3

VT Anju[1], *Busi Siddhardha*[2] *and Madhu Dyavaiah*[1]
[1]Department of Biochemistry and Molecular Biology, School of Life Sciences, Pondicherry University, Pondicherry, India, [2]Department of Microbiology, School of Life Sciences, Pondicherry University, Pondicherry, India

3.1 Introduction

Microorganisms thrive in the soil participate in complex and important interactions such that they improve the health of soil and plants. Soil microbial interactions can be grouped into beneficial as well as pathogenic. Beneficial microorganisms develop biocommunication among them through symbiosis or mutualism. They also found to be in association with plants exhibiting favorable roles (Witzany, 2011). Stream bed biofilms are often regarded as the positive indicators of overall ecosystem health. In ecological perspectives, the application of biofilms in soil is infinite. They are involved in the biogeochemical cycling, decomposition of organic matter, nutrient dynamics, crop protection, and so on (Balcázar et al., 2015). Formation of biofilms is regarded as one such soil bacterial interactions which impacts plant and soil health (Sun et al., 2020). In general, biofilms are known to be the population of microbial cells that are protected from harsh conditions by exopolysaccharide (EPS) matrix. Microbial biofilms are formed as a result of attachment, growth, and aggregation of individual cells on natural or artificial surfaces (Rana et al., 2020). Biofilms may be comprised of single species but most frequently multiple species communities are encased in a matrix of exudates, polysaccharides, and detritus. One of the main advantages of microorganisms found in nature is that they can change their lifestyle from free-floating to adhered based on their surrounding physiological or physicochemical properties (Balcázar et al., 2015).

In nature more bacteria exist in their sessile mode than planktonic cells. In addition, soil harbors innumerable microorganisms in their planktonic and sessile mode of growth. These soil biofilms provide improved fitness advantage of protection from desiccation and more chances of horizontal gene transfer. Also, soil microorganisms benefit the physicochemical parameters of soil. They contribute to the enhanced water permeability through bioclogging in soil pores, soil microaggregation, which helps in surviving environmental stress factors and outlines the soil micro environments (Wu et al., 2019). In the case of plant health, biofilms associated with the rhizosphere region of plants safeguard them from plant pathogens and

Microbes in Land Use Change Management. DOI: https://doi.org/10.1016/B978-0-12-824448-7.00006-1
© 2021 Elsevier Inc. All rights reserved.

act as biocontrol agents. Globally, rhizosphere-associated bacteria and biofilms have a prominent role in agriculture. They strongly impact the growth of crop plants by enhanced nutrient availability and uptake, improved immunity of plants toward stress conditions and plant growth (Harjai and Sabharwal, 2017). In this chapter, beneficial soil-microbe interactions, formation of soil biofilms, their impact on soil and plant health and ecosystem will be discussed.

3.2 Soil biofilms and characteristics

Soil microbiome is the collective term for all microorganisms found in soil. Soil microbiome is the key driven factor for many of the biogeochemical cycling, responsible for storage of carbon and maintaining fertility of soil (Liang and Balser, 2011). Soil microbiome in turn results in the direct or indirect improvement of animal and plant health (Ahmad et al., 2017). The abundance of microorganisms in soil is influenced by several biotic and abiotic factors. These factors are pH, quality and quantity of organic carbon, soil oxygen and redox status, moisture of soil, phosphorus and nitrogen content, texture and structure of soil, temperature, and plant species. There is always a variation in the abundance of microbial taxa in different soils. The dominant group of soil microbiome is contributed by bacteria and fungi. It is present as approximately 10^2–10^3 times more than the number of other microorganisms present in soil (Fierer, 2017).

Due to the availability of large surface in soil systems, microorganisms either attach or interact to the surrounding surfaces, colonize and stimulate their survival and growth. These microorganisms absorb the nutrients and aggregate their small communities to escape from the varying environmental circumstances and develop into highly resistant biofilm structures (Rodríguez and Bishop, 2007). Biofilm formation is characterized with the initiation of irreversible adhesion to phenotype alteration, enhanced growth rate, and changed gene expression pattern. According to the literature available, biofilms are the structural bacterial communities attached to a live or inert surface and are enclosed in self-produced EPS matrix. EPS matrix has several counter parts such as lipids, proteins, EPSs, and extracellular DNA. Biofilm formation is initiated with the reversible binding of individual cells to the surface, followed by irreversible attachment and secretion of EPS matrix. This results in the development of microbial monolayers on substrate and subsequent formation of micro colonies. In the final step, biofilm structure is matured with the formation of special biofilm structures and finally free-floating cells are dispersed to a new target. Every stage of biofilm growth is unique and differs in their morphology and regulation process (Ahmad et al., 2017).

Formation of biofilms in soil is impacted by several physical, chemical, and biological factors. These soil biofilms can be diverse in nature. There are bacterial biofilms present in the soil, associated with plants or found in water bodies. Other than the bacteria, there are biofilms formed by fungi and protozoa (Burmølle et al., 2007; Nannipieri et al., 2002).

The functions of these soil biofilms are infinite. The production of extracellular enzymes by biofilms and other free living microorganisms helps in the degradation of organic material present in the soil. In compared to planktonic cells, biofilms attached or near to the roots and litters have more nutrient availability (Ahmad and Husain, 2017). There are heterotrophic populations of bacteria and fungi that are able to produce lignolytic and cellulolytic enzymes for the degradation of organic matter. Thus these activities are influenced by the physiological ability and improved growth rate of biofilms. In general, several processes such as mineralization, turnover of nutrients, and soil fertility are directly controlled by soil biofilms. Also they have the ability to perform bioremediation of organic and metal pollutants (von Canstein et al., 2002). Soil biofilms also provide extended protection to plants from the attack of pathogenic microorganisms and also enhance their survival (Burmølle et al., 2005).

3.3 Mechanism of biofilm formation

The formation of biofilms is a complex and highly regulated process. In nature, biofilm architecture is developed to rescue the planktonic cells from unfavorable environmental factors. The important cell components and molecules, which are integral parts of bacteria such as flagella, lipopolysaccharides, and EPSs along with functional signals generated by bacteria, are central elements in the establishment of biofilms. Biofilm structure is defined as the communities of bacteria encased in EPS matrix and self-protected from external stimuli. They are initially adhered to a biotic or inert surface in reversible manner followed by the irreversible attachment. Once the attachment is proper, multiplication of bacteria occurs and a three-dimensional architecture of bacterial communities develops which may be infused by channels. These channels stand for the circulation or exchange of water, enzymes, signals, and nutrients and excludes toxins and also boosts metabolic cooperativity (Rinaudi and Giordano, 2010).

The dynamic process of biofilm formation includes five stages including initial attachment, specific attachment, microcolony formation, biofilm growth, and detachment of biofilm cells. The conditioning of surface for the adequate attachment of cells and formation of biofilm is required which facilitated by the production of some vital biomolecules (proteins, carbohydrates, and other molecules) (Burmølle et al., 2011). The first reversible attachment is mediated by flagella and type IV pili. In this, flagella help in the attachment to the surface and pili provide twitching motility for the aggregation and microcolony formation (Sutherland, 2001). Irreversible attachment is modulated by the production of matrix especially EPSs (Fig. 3.1).

EPS matrix consists of EPSs, extracellular DNA, alginate, and extracellular proteins. These EPSs serve as important components of biofilm architecture, scaffolds for other biofilm components and help in the attachment to the surface. EPSs vary in their structure and properties from one biofilm to another (Rabin et al., 2015).

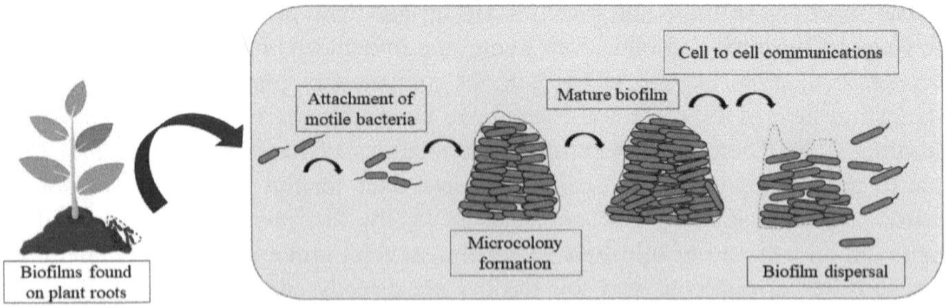

Figure 3.1 The mechanism of biofilm formation in soil. The different stages of biofilm formation are represented.

Biofilm structure changes from a thin layer to a mushroom shape during the initial stages. Cells within the mature biofilms perform specialized functions and communicate each other through the production of various signals. The final stage is dispersal of biofilm structure which is vital for the initiation of new biofilm formation at a different target site (Rabin et al., 2015). In soil, biofilms are found generally in carbon rich soil surfaces and clay minerals and transform into aggregates through the production of EPS (Ma et al., 2017; Nunan, 2017). Most of microorganisms associated with plants in the rhizosphere region are in their biofilm mode rather than the planktonic growth stage. There are biofilms of plant growth-promoting bacteria (PGPB) associated with plant roots which improves the growth of plants, yield, and quality of crops (Seneviratne et al., 2010).

3.4 Soil biofilm interactions

Individual cells of biofilm communities communicate with each other and execute several functions due to their close vicinity. These interactions are collectively termed as sociomicrobiology. It has been observed that several inter and intraspecies communications or interactions occur in the diverse and complex biofilm communities. These interactions are important in the formation of biofilm also (Parsek and Greenberg, 2005). There are some bacteria and fungi that are able to produce cell signaling molecules or signals which leads to a highly strong cell-to-cell signaling or quorum-sensing (QS) mechanisms (Garnett and Matthews, 2012). These interactions help in the production of agents which can be pathogenic as well as beneficial.

Biofilms which are in cell-to-cell contact exhibit cooperative interactions. This cooperative behavior results in the formation of dual or multispecies biofilms. Multispecies biofilms are made of more than one bacterial species and cooperate among themselves against the environmental stress (Ren et al., 2015; Faust and Raes, 2012). Cross feeding is a phenomenon observed in most of the biofilms due to the ability to share or exchange metabolites produced by one bacterium to another member. As most of biofilm microorganisms are auxotrophic, cross feeding is very necessary for the survival of organisms that lack essential metabolic pathways (Zengler and Zaramela, 2018; Pande et al., 2016). For example, cross feeding is observed in interdomain interactions of biofilms of *Streptococcus mutans* and *Candida albicans*. In these interactions, *S. mutans* produces lactic acid and it benefits *C. albicans* (Sztajer et al., 2014). In another case of metabolic cooperation, ammonia oxidizers are furthered from the nitrite oxidizing bacteria (e.g., *Nitrospira moscoviensis*) which produces ammonia and provides to ammonia oxidizers (Flemming et al., 2016). Co-metabolism is a common phenomenon observed in biofilms especially in soil. The resistance or recalcitrance nature of biofilm is highly influenced by co-metabolism. It is well studied that the highly resistant nature of biofilms toward antibiotics, surfactants, and pollutants is owed to their cell-to-cell interactions. Co-metabolism is very crucial and helpful in the biodegradation of pollutants or contaminants, turnover of organic matter, and nutrient cycling (Olsen, 2015; Flemming, 2010).

Another interesting interaction of soil microorganisms with their eukaryotic host is called inter/intra cell signaling pathways which helps to co-ordinate among their populations. Bacteria produce various signal molecules which co-ordinate the gene expression when a threshold population density reaches and this process is called as QS. There are different types of QS signal produced by diverse range of microorganisms which structure their communities. Among these, acylated homoserine lactone (AHL) is the widely produced signals by Gram negative bacteria. AHL signals are produced mainly by soil and plant related rhizobia, *Pseudomonas* sp., *Pantoeae, Chromobacteria, Serratia*, and *Pectobacteria* (Teplitski et al., 2011). These AHL QS decreased the virulence in pathogens of plants and animals indicating their importance in their natural environment through intraspecies interactions. AHL signaling is found common in natural aquatic biofilms and soil microcosms (Gantner et al., 2006). A few examples of bacteria associated with plants (tomato, wheat, and some legumes) and that produce AHL signaling are *Pseudomonas putida* (Gantner et al., 2006) and *Sinorhizobium meliloti*. These bacteria are colonized in the rhizosphere region and followed by the formation of micro colonies on roots and later produce signals in their natural environments (Gao and Teplitski, 2008).

Bacteria and biofilms associated with plants perform different functions. Some bacteria are found in the nodules of roots and enhance plant growth. A soil bacterium, *Rhizobium leguminosarum bv. Viciae* associated with nodules of pea can produce four types of QS signals and regulate overall gene expressions. These QS systems are found to be involved in the rhizobial physiology and transfer of symbiont plasmids (Lithgow et al., 2000; Boyer and Wisniewski-Dyé, 2009). There are interspecific interactions between soil bacterial communities which improve soil properties and plant health. A study showed that competitive interactions among carbon limited soil bacteria exhibited enhanced expression of several genes which are involved in the production of secondary metabolites and chemotaxis. The secondary metabolites produced by them inhibited the growth of soil borne plant pathogenic fungi. Two bacterial strains, *Pseudomonas* sp. A21 and *Pedobacter* sp. V48 isolated from dune soil, inhibited the growth of a bioindicator fungus in laboratory conditions due to their competitive interspecies interactions in mixed cultures (Garbeva and de Boer, 2009) (Fig. 3.2).

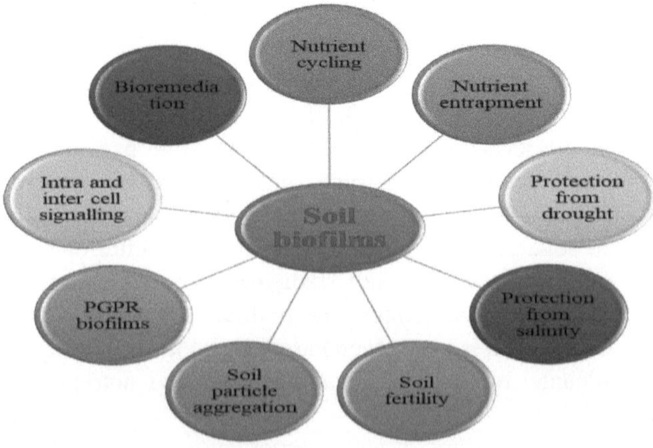

Figure 3.2 Functional roles of biofilms formed in soil. Functions of soil biofilm in soil and plant health are listed.

3.5 Role of biofilms on soil health and plant health

Soil health is determined by the availability and maintenance of carbon content, nutrient cycling, structure of soil, and control of pests and pathogens (Ahmad and Husain, 2017). Biofilm biodiversity influences the ecosystem function and thus soil microbial ecology is of great importance (Besemer, 2015). Biofilms in soil forms biocarriers which effectively control the subsurface contamination plumes. Biofilms perform this by reducing the permeability of subsurface and prevent contamination. Biobarriers has enhanced lifetime when compared to the physical and chemical containment methods (Pintelon et al., 2012). In a study by Komlos et al. (2004), they showed that biobarrier containing *Klebsiella oxytoca* and *Burkholderia cepacia* effectivity controlled the contaminant, trichloroethylene, and also degraded it (Komlos et al., 2004).

As the biofilm formation progresses, EPS is produced and protects the bacterial population form the environment. EPS has several roles in biofilm architecture and also improves soil properties. EPS is primarily defined as the mass fraction of stable organic matter. EPS may bind to mineral surface once released by biofilms and influence several biogeochemical process. EPS can change the charge and hydrophilicity of mineral surfaces, cause mineral dissolution, biomineralization, and bioleaching. Generally, EPS-N of proteins and EPS-P of nucleic acid moieties bind to montmorillonite and kaolinite and goethite and Al (hydro) oxide layers, respectively (Omoike and Chorover, 2006; Liu et al., 2013). Lin et al. (2016) studied the binding of EPS moieties from *P. putida* to montmorillonite, kaolinite, and goethite surfaces of clay minerals (Lin et al., 2016). This will enhance the internal adhesion of formed soil aggregates.

Microbial biofilms can cause bioclogging of the soil which is observed in several engineered systems. It is observed in sand filters, irrigated agriculture, petroleum engineering systems, bioremediation plants, infiltration ponds, and constructed wetlands. Sealed or fully clogged soil is witnessed in water saturated systems. It is also reported in partially saturated systems. Soil biofilms are able to reduce hydraulic properties of saturated water systems. In a study conducted by group of researchers, it was reported that biofilms of *P. putida* reduced the unsaturated hydraulic properties of sandy soil. They reported the water retention ability of soil contributed by biofilms (Volk et al., 2016). Also, precipitation of calcite by microbes helps to strengthen, plug, and improve the properties of soil. This can improve the shear strength of soil and reduce the permeability of tropical soil and sand by preventing the water flow (Umar et al., 2016). Soon et al. (2014) showed that *Bacillus megaterium* reduced the permeability of soil by decreasing the hydraulic conductivity by 90% which was due to its calcite precipitation property (Soon et al., 2014).

It has been reported that extensive use of chemical fertilizers reduced the properties of agricultural soils, soil fertility, beneficial microbial populations, and crop productivity. Alternative methods which reduce the extensive use of chemical fertilizers without reducing the fertility of soils and with crop improvement are popular demands (Hvistendahl, 2010). There are several biofilms capable of nitrogen fixation which enhance the quality of soil and improve crop production. These nitrogen fixers are incorporated into fungal–bacterial biofilms (FBBs). FBBs are bacterial biofilms formed on biotic surface consisting of fungal communities. FBB lives in metabolic cooperation benefiting each other. The incorporation of rhizobial strains into FBBs results in the formation of fungal-rhizobial biofilms. Thus fungal-rhizobial biofilms can act as a biofilmed inoculum for the production of biocontrol agent and biofertilizer in agriculture especially in nitrogen deficient soils (Seneviratne et al., 2008). There are studies that show that FBBs are more effective than their monocultures in soil.

In a study, the ability of FBB to restore the soil properties that deteriorated by traditional practices in tea plantation was studied. The study concluded the role of FBB in deteriorated conventional agricultural soils. They found that soil properties were improved with less pest infestation on plants owing to the less application of chemical fertilizers. Biological nitrogen fixation and growth of crop plants were improved much more when coupled with the biofilmed biofertilizers. A decreased leaf transpiration (c. 40%) and enhanced soil organic carbon content (c. 20%) were experienced after the combined application of fertilizers. As the soil biofilm and microbial communities were profound, an improved soil performance was also observed. Therefore biofilmed biofertilizers are found to be beneficial for nutrient depleted soils and agroecosystems with enhanced crop production (Seneviratne et al., 2011).

There are bipartite interactions exists between mycorrhizal fungi and bacteria. These bipartite interactions can be modified to tripartite symbiotic interactions along with plant roots. In general, bacteria exist as a sheet of cells encased in EPS matrix in specialized arbuscular mycorrhizal fungi (AMF) or ectomycorrhizal fungi structures. These biofilms associated with fungi exhibit several functional roles in plants. They improve the soil quality, plant growth, and fitness of plants in all ecosystems (Torres-Cortés et al., 2015; Chen et al., 2018; Guennoc et al., 2018). The advantages of such symbiosis are enhanced root development, sporulation, recognition of plant roots and AMF associations, establishment of mycelial growth, antagonistic activities, production of antifungal and antimicrobial substances and enhanced nitrogen fixation and phosphate solubilization (Wagner et al., 2019). These interactions benefit all members as they live in a symbiotic relationship. Thus the bacteria contribute to the bacteria-AMF interactions and signaling through the production of carbon sources for fungi such as citric acid and maleic acid and also modify fungal structures (Kannan et al., 2011).

In another study, the interaction of periphytic biofilms and dicyandiamide (DCD) (nitrification inhibitor) in paddy fields was investigated. Peripytic biofilms are common in paddy lands consists of bacteria, algae, protozoa, epiphytes, metazoans, and detritus. They are observed on the substrates of shallow water. The major roles of these periphytic biofilms in paddy systems are nutrient cycling, nitrogen assimilation, ammonia volatilization, denitrification, nitrification, and production of EPS. The study showed that DCD application indirectly favored periphytic biofilms and subsequently resulted in increased denitrification process. The growth of periphytic biofilms improved the nutrient cycling in paddy lands by augmenting DCD degradation and nitrogen utilization (Zhao et al., 2020).

Periphytic biofilms have important roles in nutrient cycling and energy flow in aquatic ecosystems. Among these biofilms, cyanophytes are of great importance owing to their ability to fix atmospheric nitrogen and enhance the crop yield by reducing the use of chemical fertilizers. Thus these biofilms can enhance the paddy field fertility and improve the yield. Biomass of periphyton biofilms itself acts as a nutrient source which can be diverted into the paddy fields. Therefore paddy field periphyton biofilms can be utilized to enhance the efficiency of fertilizers and rice yield (Su et al., 2017).

One of the soil bacteria, which is omnipresent in nature is *Bacillus* sp. This genus is often present as endophytes of plants and also in association with rhizosphere region. Thus the role of *Bacillus* sp in its plant growth-promoting and biocontrol properties is well documented. They are already being utilized by current agriculture systems due to their ability to form highly stress resistant endospores, strong cell wall morphology, produce antimicrobial compounds, and develop into resistant biofilm structures (Timmusk and Nevo, 2011). In a study, *Bacillus subtilis* had the ability to form biofilm on root surface and eventually suppress the wilt of tomato plants (Chen et al., 2013). Another strain, *B. subtilis* UMAF6614 secreted surfactin which helped the formation of biofilms and eventually produced several suppressive factors such as bacillomycins, lipopeptides, and fengycins which suppressed phytopathogens (Zeriouh et al., 2014). A group of researchers identified a compound, bacillomycin D from *Bacillus amyloliquefaciens* SQR9 which served as a signal to form biofilms and also inhibited a common fungal plant pathogen, *Fusarium oxysporum* (Xu et al., 2013). Xu et al. (2014) studied a wheat endophyte, *Bacillus cereus* 0–9 which showed significant biocontrol against wheat sharp eyespot owing to their ability to colonize and form biofilm on roots (Xu et al., 2014). The importance of soil biofilms in contributing the soil and plant health is reviewed and studied by several researchers. However, the role of multispecies biofilms in soil is less explored.

3.6 Plant growth promoting rhizobacteria biofilms

The region around plants which is termed as a pleasant area surrounding the plant roots is known as rhizosphere. Several important interactions occur in the rhizosphere region among plants, microorganisms, soil particles, and soil flora. This region is known as the first line in between the roots of plant and soil pests owing to the interactions with microorganisms and exudates produced by them. Thus the microbes present in the rhizosphere region are the epitome for sustainable agriculture (Gupta et al., 2017). There are plant growth-promoting rhizobacteria (PGPR) which are defined as useful biofertilizers and enhance the growth of plants by several nutrient cycling mechanisms. PGPRs are proven as an environmental practice for enhanced agriculture yield (Lugtenberg and Kamilova, 2009). PGPRs associated with plant roots often exist in their biofilm stage. Biofilm growth protects these beneficial bacteria from environmental stress. Initially bacteria colonize in and around the roots which is influenced by the signals from root exudates. These PGPRs acquire their biofilm mode once they interact with plants in order to escape from biotic and abiotic conditions. Biofilms can persist in the rhizosphere region and escape from plant defense strategies especially from free radical attack. Also they are protected from UV light, antimicrobial agents, and desiccation owing to their ability to produce EPS and antioxidant molecules. Biofilms in the rhizosphere region enable the production of antimicrobial molecules against plant pathogens and phytohormones for plant growth and development and solubilize mineral nutrients, mineralize organic compounds, and execute nutrient exchange between the plant and soil (Amaya-Gómez et al., 2020).

Some of the important strains of PGPRs for enhanced crop production are *Bacillus, Agrobacterium, Enterobacter, Caulobacter, Burkholderia, Chromobacterium, Acinetobacter, Micrococcous, Alcaligenes, Arthrobacter, Azospirillum, Azotobacter, Actinoplanes, Beijerinckia, Streptomyces, Erwinia, Flavobacterium, Spirillospora, Rhizobium, Azorhizobium, Allorhizobium*, and *Serratia* sp. There are both extracellular and intracellular PGPRs classified based on their position in rhizosphere region (Gray and Smith, 2005; Franco et al., 2007; Bharti et al., 2013). Table 3.1 represents a few PGPR strains and their functional roles.

PGPRs can alter the properties of different types of soil. For instance, saline soils are always a major concern in limiting the productivity. Saline soils are not suitable for crops as it prevents photosynthesis, and interferes with lipid metabolism and translation (Paul and Lade, 2014). PGPRs are reported to enhance the growth of crop plants and yield in saline soils. They increase the root and shoot length necessary for nutrient acquisition and growth. This eventually results in enhanced hydration and chlorophyll production and also provides resistance to pathogens (Qurashi and Sabri, 2012). Kang et al. (2014) reviewed the effect PGPRs on growth and productivity of wheat, mustard, tobacco, tomatoes, cucumbers, bell peppers, and barley (Kang et al., 2014).

Kasim et al. (2016) reported that biofilms of PGPB provide protection to crop plants from different types of stress and enhance the yield even in saline soils. The stress tolerance ability of biofilms efficiently protects crop plants from these

Table 3.1 Microbial biofilms associated with plants and their functions.

Plant associated biofilms	Colonization	Natural habitat	Functional roles	References
Pseudomonas putida T6SS	Root of crop plants	Rhizosphere and phyllosphere	Plant warden against phytopathogens	Bernal et al. (2017)
Pseudomonas aeruginosa L10	Roots of reed	Rhizosphere and bacterial endophyte	Hydrocarbon-degrading, biosurfactant producing PGPR	Wu et al. (2018)
Azospirillum brasilense	Roots of a variety of cereals	Rhizosphere	Promote the growth of host plants and root proliferation	Burdman et al. (2000)
Bacillus cereus	Plant roots	Rhizosphere	Enhanced phytoremediation against cadmium heavy metals	Jan et al. (2019)
Pseudomonas fluorescens	Plant roots	Rhizosphere	Biocontrol the pests of many crops	David et al. (2018)
P. putida KT2440	Seeds	Rhizosphere	Enhanced bioremediation	Espinosa-Urgel et al. (2000)
Paenibacillus polymyxa	Peanut roots	Rhizosphere	Biocontrol against crown rot disease	Haggag and Timmusk (2008)
P. polymyxa	Wheat plant roots	Rhizosphere	Protect plants from pathogen attack and abiotic stress	Timmusk et al. (2005)
Microbacterium sp.	Sugarcane roots	Rhizosphere	Enhanced biological nitrogen fixation	Lin et al. (2012)
Pseudomonas entomophila FAP1	Wheat plant roots	Rhizosphere	Tolerance to external stress, improved plant growth and crop yield	Ansari and Ahmad (2018)

extreme conditions (Kasim et al., 2016). In the above study, *Bacillus amyloliquifaciens*, a PGPR strain is able to form strong biofilm which alleviated the saline stress on barley grains. Further, these PGPRs could be effective in promoting the growth of barley grains under saline conditions (Kasim et al., 2016). In laboratory conditions, lettuce seeds in saline atmosphere showed significant germination and growth when inoculated with *Azospirillum* (Barassi et al., 2006).

The PGPR biofilms are also associated with drought tolerance for crop plants. It is also known that various signaling molecules produced by them provide the ability to escape from drought stress. Plants have the ability to sense and respond to these signals and provide protection from desiccation. PGPRs also stimulate induced systemic tolerance in plants to escape from drought stress (Wang et al., 2019). Wang et al. (2019) reported that biofilms of *B. amyloliquefaciens* 54 enhanced the survival of tomato plants by contributing to its drought tolerance. The strains forming hyper robust biofilm boosted the tolerance of plants against drought conditions by improved survival, water content, and root vigor (Wang et al., 2019). *Pseudomonas fluorescens* DR7 isolated from the rhizosphere region of foxtail millet showed drought tolerance. The drought tolerance was influenced positively by the production of EPS and ACC deaminase by the PGPR. This PGPR strain protected the plants from drought stress and enhanced the seed germination and seedling production in vitro. Also, the strain improved the soil texture, its moisture, and root adhered to soil/root tissue ratio and efficiently colonized to plant roots facilitating drought tolerance (Niu et al., 2018).

The EPS produced by biofilms is the major factor influencing the drought tolerance to plants and subsequently their growth. EPS is able to retain the soil water content and growth even in the dry sandy soils. EPS produced by biofilms colonized on roots can protect the plant roots from drought stress for a long time. EPS also provides the proper attachment of bacteria to the roots, bioremediation, drought tolerance, microcolony formation, and plant-bacteria interactions. It has been reported that EPS producing biofilms and PGPR generate high amounts of sugars and free amino acids under desiccation. Thus EPS production by PGPRs and their biofilms is a better method to prevent the desiccation (Naseem et al., 2018).

Another application of PGPR biofilms is in bioremediation. The extensive ability of bacterial biofilms and to produce EPS protects them from adverse conditions. Thus they can survive and degrade the highly recalcitrant heavy metals. Biofilms and EPS can entrap and biomineralize the precipitated metals which are difficult to remove by natural degradation. In a study, the ability of PGPR bacteria *P. fluorescens*, in the removal of heavy metals such as zinc and lead was studied. The ability of biofilm formation enhanced the removal of heavy metals from the metal contaminated soil (Meliani and Bensoltane, 2016). Cadmium is one of the heavy metals associated with high pollution and affects the plant growth. Thus the ability of phytoremediation of heavy metals can be enhanced by biofilm forming PGPRs. In a comparative study, biofilm forming *Aeromonas* sp augmented to the rhizosphere region of *Vetiveria zizanioides* and enhanced the uptake of cadmium metals than the non-biofilm formers. PGPR biofilms also improved the root and shoot length of plants which in turn enhanced the uptake of metals by plants (Itusha et al., 2019).

In another study, biofilm formation and EPS production by *Pseudomonas aeruginosa* directed the entrapment or bioaccumulation of heavy metals. *P. aeruginosa* is one of the widely suitable biofilm forming PGPR in cadmium bioremediation (Chien et al., 2013; Chellaiah, 2018). *P. aeruginosa* JP-11 developed resistance to cadmium to a higher concentrations (1000 ppm) and the cadmium resistance was associated with *czcABC* genes (Chakraborty and Das, 2014).

The PGPR and biofilms are imperative for several microbial-plant interactions because they enhance soil fertility and improve the growth and yield of crops. PGPR is a good substitute for the extensive use of chemical fertilizers and pesticides in agriculture. They also are important as they recycle the essential nutrients and minerals, degradation of organic and inorganic pollutants, assimilation and mineralization of nutrients, stimulation of plant growth, and biocontrol of pests and pathogens. Thus the application of PGPR biofilms and EPS is crucial for sustainable agriculture (Gupta et al., 2017).

3.7 Conclusion

Soil is one of the natural habitats of a myriad of microorganisms. Among these, more than 99% occur in their biofilm stage. Biofilms consist of single or multispecies microbial communities protected from external stimuli in their EPS matrix. This biofilm matrix helps the microorganisms to escape from adverse conditions. Soil biofilms interact through intra or interspecies communications such as cooperation, competition, and mutualism. These interactions contribute to the degradation of soil organic matter and other nondegradable materials. They are involved in the modification of soil physicochemical properties. Soil biofilms and secreted EPS play a pivotal role in the bioclogging of soil pores, influencing permeability of water in soil. Also, it enhances the soil quality by forming strong, internally adhered soil microaggregates. Overall, soil biofilms are involved in soil micro niches configuration by augmenting the soil microbial community and metabolic activity. Thus soil fertility and ecosystem health are directly related to the diverse soil microbiome due to more resources. One of the essential soil biofilm is PGPBs or PGPRs, generally found in the rhizosphere region. These PGPR biofilms have several roles, including improved soil health, plant health, and biogeochemical cycling involvement. Owing to their highly recalcitrant nature contributed by the EPS matrix, biofilms are resistant to desiccation and heavy metals. Thus they donate a great opportunity in the degradation of recalcitrant heavy metals.

References

Ahmad, I., Husain, F.M. (Eds.), 2017. Biofilms in Plant and Soil Health. John Wiley & Sons, Ltd., Chichester, UK.

Ahmad, I., Khan, M.S., Altaf, M.M., Qais, F.A., Ansari, F.A., Rumbaugh, K.P., 2017. Biofilms: an overview of their significance in plant and soil health. Biofilms in Plant and Soil Health. John Wiley & Sons, Ltd., Chichester, UK, pp. 1–25.

Amaya-Gómez, C.V., Porcel, M., Mesa-Garrig, L., Gómez-Álvarez, M.I., 2020. A framework for the selection of plant growth-promoting rhizobacteria based on bacterial competence mechanisms. Appl. Environ. Microbiol. 86, e00760–20.

Ansari, F.A., Ahmad, I., 2018. Biofilm development, plant growth promoting traits and rhizosphere colonization by *Pseudomonas entomophila* FAP1: a promising PGPR. Adv. Microbiol 08, 235–251.

Balcázar, J.L., Subirats, J., Borrego, C.M., 2015. The role of biofilms as environmental reservoirs of antibiotic resistance. Front. Microbiol. 6, 2016.

Barassi, C.A., Ayrault, G., Creus, C.M., Sueldo, R.J., Sobrero, M.T., 2006. Seed inoculation with *Azospirillum* mitigates NaCl effects on lettuce. Sci. Hortic. (Amsterdam) 109, 8–14.

Bernal, P., Allsopp, L.P., Filloux, A., Llamas, M.A., 2017. The *Pseudomonas putida* T6SS is a plant warden against phytopathogens. ISME J 11, 972–987.

Besemer, K., 2015. Biodiversity, community structure and function of biofilms in stream ecosystems. Res. Microbiol. 166, 774–781.

Bharti, N., Yadav, D., Barnawal, D., Maji, D., Kalra, A., 2013. *Exiguobacterium oxidotolerans*, a halotolerant plant growth promoting rhizobacteria, improves yield and content of secondary metabolites in *Bacopa monnieri* (L.) Pennell under primary and secondary salt stress. World J. Microbiol. Biotechnol. 29, 379–387.

Boyer, M., Wisniewski-Dyé, F., 2009. Cell-cell signalling in bacteria: not simply a matter of quorum. FEMS Microbiol. Ecol. 90, 1–19.

Burdman, S., Okon, Y., Jurkevitch, E., 2000. Surface characteristics of *Azospirillum brasilense* in relation to cell aggregation and attachment to plant roots. Crit. Rev. Microbiol. 26, 91–110.

Burmølle, M., Hansen, L.H., Sørensen, S.J., 2005. Use of a whole-cell biosensor and flow cytometry to detect AHL production by an indigenous soil community during decomposition of litter. Microb. Ecol. 50, 221–229.

Burmølle, M., Hansen, L.H., Sørensen, S.J., 2007. Establishment and early succession of a multispecies biofilm composed of soil bacteria. Microb. Ecol. 54, 352–362.

Burmølle, M., Kjøller, A., Sørensen, S.J., 2011. Biofilms in soil. In: Gliński, J., Horabik, J., Lipiec, J. (Eds.), Encyclopedia of Earth Sciences Series. Springer, Dordrecht, Netherlands, pp. 70–75.

Chakraborty, J., Das, S., 2014. Characterization and cadmium-resistant gene expression of biofilm-forming marine bacterium *Pseudomonas aeruginosa* JP-11. Environ. Sci. Pollut. Res. 21, 14188–14201.

Chellaiah, E.R., 2018. Cadmium (heavy metals) bioremediation by *Pseudomonas aeruginosa*: a minireview. Appl. Water Sci. 8, 154.

Chen, Y., Yan, F., Chai, Y., Liu, H., Kolter, R., Losick, R., et al., 2013. Biocontrol of tomato wilt disease by *Bacillus subtilis* isolates from natural environments depends on conserved genes mediating biofilm formation. Environ. Microbiol. 15, 848–864.

Chen, M., Arato, M., Borghi, L., Nouri, E., Reinhardt, D., 2018. Beneficial services of arbuscular mycorrhizal fungi – from ecology to application. Front. Plant Sci. 9, 1270.

Chien, C.-C., Lin, B.-C., Wu, C.-H., 2013. Biofilm formation and heavy metal resistance by an environmental *Pseudomonas* sp. Biochem. Eng. J 78, 132–137.

David, B.V., Chandrasehar, G., Selvam, P.N., 2018. *Pseudomonas fluorescens*: A plant-growth-promoting rhizobacterium (PGPR) with potential role in biocontrol of pests of crops. In: Crop Improvement through Microbial Biotechnology. Elsevier, Amsterdam, The Netherlands, pp. 221–243.

Espinosa-Urgel, M., Salido, A., Ramos, J.L., 2000. Genetic analysis of functions involved in adhesion of *Pseudomonas putida* to seeds. J. Bacteriol 182, 2363–2369.

Faust, K., Raes, J., 2012. Microbial interactions: from networks to models. Nat. Rev. Microbiol. 10, 538–550.

Fierer, N., 2017. Embracing the unknown: disentangling the complexities of the soil microbiome. Nat. Rev. Microbiol. 15, 579–590.

Flemming, H.-C., 2010. Biodeterioration of synthetic materials - a brief review. Mater. Corros. 61, 986–992.

Flemming, H.-C., Wingender, J., Szewzyk, U., Steinberg, P., Rice, S.A., Kjelleberg, S., 2016. Biofilms: an emergent form of bacterial life. Nat. Rev. Microbiol. 14, 563–575.

Franco, C., Michelsen, P., Percy, N., Conn, V., Listiana, E., Moll, S., et al., 2007. Actinobacterial endophytes for improved crop performance. Australas. Plant Pathol. 36, 524.

Gantner, S., Schmid, M., Dürr, C., Schuhegger, R., Steidle, A., Hutzler, P., et al., 2006. In situ quantitation of the spatial scale of calling distances and population density-independent N-acylhomoserine lactone-mediated communication by rhizobacteria colonized on plant roots. FEMS Microbiol. Ecol. 56, 188–194.

Gao, M., Teplitski, M., 2008. RIVET - a tool for in vivo analysis of symbiotically relevant gene expression in *Sinorhizobium meliloti*. Mol. Plant Microbe Interact 21, 162–170.

Garbeva, P., de Boer, W., 2009. Inter-specific interactions between carbon-limited soil bacteria affect behavior and gene expression. Microb. Ecol. 58, 36–46.

Garnett, J., Matthews, S., 2012. Interactions in bacterial biofilm development: a structural perspective. Curr. Protein Pept. Sci. 13, 739–755.

Gray, E.J., Smith, D.L., 2005. Intracellular and extracellular PGPR: commonalities and distinctions in the plant-bacterium signaling processes. Soil Biol. Biochem. 37, 395–412.

Guennoc, C.M., Rose, C., Labbé, J., Deveau, A., 2018. Bacterial biofilm formation on the hyphae of ectomycorrhizal fungi: a widespread ability under controls? FEMS Microbiol. Ecol. 94, fiy093.

Gupta, G., Snehi, S.K., Singh, V., 2017. Role of PGPR in biofilm formations and its importance in plant health. Biofilms in Plant and Soil Health. John Wiley & Sons, Ltd., Chichester, UK, pp. 27–42.

Haggag, W.M., Timmusk, S., 2008. Colonization of peanut roots by biofilm-forming *Paenibacillus polymyxa* initiates biocontrol against crown rot disease. J. Appl. Microbiol. 104, 961–969.

Harjai, K., Sabharwal, N., 2017. Biofilm formation and quorum sensing in rhizosphere. Biofilms in Plant and Soil Health. John Wiley & Sons, Ltd., Chichester, UK, pp. 111–130.

Hvistendahl, M., 2010. China's push to add by subtracting fertilizer. Science 327, 801.

Itusha, A., Osborne, W.J., Vaithilingam, M., 2019. Enhanced uptake of Cd by biofilm forming Cd resistant plant growth promoting bacteria bioaugmented to the rhizosphere of *Vetiveria zizanioides*. Int. J. Phytoremediation 21, 487–495.

Jan, M., Shah, G., Masood, S., Iqbal Shinwari, K., Hameed, R., Rha, E.S., et al., 2019. *Bacillus cereus* enhanced phytoremediation ability of rice seedlings under cadmium toxicity. Biomed Res. Int. 8134651.

Kang, S.M., Khan, A.L., Waqas, M., You, Y.H., Kim, J.H., Kim, J.G., et al., 2014. Plant growth-promoting rhizobacteria reduce adverse effects of salinity and osmotic stress by regulating phytohormones and antioxidants in *Cucumis sativus*. J. Plant Interact 9, 673–682.

Kannan, V.R., Suganya, S., Solomon, E.K., Balasubramanian, V., Ramesh, N., Rajesh, P., 2011. Analysis of interaction between arbuscular mycorrhizal fungi and their helper bacteria by MILPA model. Res. Plant Biol 1, 48–62.

Kasim, W.A., Gaafar, R.M., Abou-Ali, R.M., Omar, M.N., Hewait, H.M., 2016. Effect of biofilm forming plant growth promoting rhizobacteria on salinity tolerance in barley. Ann. Agric. Sci. 61, 217–227.

Komlos, J., Cunningham, A.B., Kamper, A.K., Sharp, R.R., 2004. Biofilm barriers to contain and degrade dissolved trichloroethylene. Environ. Prog. 23, 69–77.

Liang, C., Balser, T.C., 2011. Microbial production of recalcitrant organic matter in global soils: implications for productivity and climate policy. Nat. Rev. Microbiol. 9, 75–75.

Lin, L., Guo, W., Xing, Y., Zhang, X., Li, Z., Hu, C., et al., 2012. The actinobacterium *Microbacterium* sp. 16SH accepts pBBR1-based pPROBE vectors, forms biofilms, invades roots, and fixes N 2 associated with micropropagated sugarcane plants. Appl. Microbiol. Biotechnol. 93, 1185–1195.

Lin, D., Ma, W., Jin, Z., Wang, Y., Huang, Q., Cai, P., 2016. Interactions of EPS with soil minerals: a combination study by ITC and CLSM. Colloids Surf. B: Biointerfaces 138, 10–16.

Lithgow, J.K., Wilkinson, A., Hardman, A., Rodelas, B., Wisniewski-Dye, F., Williams, P., et al., 2000. The regulatory locus cinRI in *Rhizobium leguminosarum* controls a network of quorum-sensing loci. Mol. Microbiol. 37, 81–97.

Liu, X., Eusterhues, K., Thieme, J., Ciobota, V., Höschen, C., Mueller, C.W., et al., 2013. STXM and NanoSIMS investigations on EPS fractions before and after adsorption to goethite. Environ. Sci. Technol. 47, 3158–3166.

Lugtenberg, B., Kamilova, F., 2009. Plant-growth-promoting rhizobacteria. Annu. Rev. Microbiol. 63, 541–556.

Ma, W., Peng, D., Walker, S.L., Cao, B., Gao, C.H., Huang, Q., et al., 2017. *Bacillus subtilis* biofilm development in the presence of soil clay minerals and iron oxides. NPJ Biofilms Microbiomes 3, 0–1.

Meliani, A., Bensoltane, A., 2016. Biofilm-mediated heavy metals bioremediation in PGPR *Pseudomonas*. J. Bioremediation Biodegrad. 7, 370.

Nannipieri, P., Kandeler, E., Ruggiero, P., 2002. Enzyme activities and microbiological and biochemical processes in soil. In: Burns, R.G., Dick, R.P. (Eds.), Enzymes in the Environment: Activity, Ecology and Applications. Marcel Dekker, New York, pp. 1–33.

Naseem, H., Ahsan, M., Shahid, M.A., Khan, N., 2018. Exopolysaccharides producing rhizobacteria and their role in plant growth and drought tolerance. J. Basic Microbiol. 58, 1009–1022.

Niu, X., Song, L., Xiao, Y., Ge, W., 2018. Drought-tolerant plant growth-promoting rhizobacteria associated with foxtail millet in a semi-arid agroecosystem and their potential in alleviating drought stress. Front. Microbiol. 8, 2580.

Nunan, N., 2017. The microbial habitat in soil: scale, heterogeneity and functional consequences. J. Plant Nutr. Soil Sci. 180, 425–429.

Olsen, I., 2015. Biofilm-specific antibiotic tolerance and resistance. Eur. J. Clin. Microbiol. Infect. Dis. 34, 877–886.

Omoike, A., Chorover, J., 2006. Adsorption to goethite of extracellular polymeric substances from *Bacillus subtilis*. Geochim. Cosmochim. Acta 70, 827–838.

Pande, S., Kaftan, F., Lang, S., Svato, A., Germerodt, S., Kost, C., 2016. Privatization of cooperative benefits stabilizes mutualistic cross-feeding interactions in spatially structured environments. ISME J 10, 1413–1423.

Parsek, M.R., Greenberg, E.P., 2005. Sociomicrobiology: the connections between quorum sensing and biofilms. Trends Microbiol 13, 27–33.

Paul, D., Lade, H., 2014. Plant-growth-promoting rhizobacteria to improve crop growth in saline soils: a review. Agron. Sustain. Dev. 34, 737–752.

Pintelon, T.R.R., Picioreanu, C., van Loosdrecht, M.C.M., Johns, M.L., 2012. The effect of biofilm permeability on bio-clogging of porous media. Biotechnol. Bioeng. 109, 1031–1042.

Qurashi, A.W., Sabri, A.N., 2012. Bacterial exopolysaccharide and biofilm formation stimulate chickpea growth and soil aggregation under salt stress. Brazilian J. Microbiol. 43, 1183–1191.

Rabin, N., Zheng, Y., Opoku-Temeng, C., Du, Y., Bonsu, E., Sintim, H.O., 2015. Biofilm formation mechanisms and targets for developing antibiofilm agents. Future Med. Chem. 7, 493–512.

Rana, K.L., Kour, D., Yadav, A.N., Yadav, N., Saxena, A.K., 2020. Agriculturally important microbial biofilms: biodiversity, ecological significances, and biotechnological applications. New and Future Developments in Microbial Biotechnology and Bioengineering: Microbial Biofilms. Elsevier, pp. 221–265.

Ren, D., Madsen, J.S., Sørensen, S.J., Burmølle, M., 2015. High prevalence of biofilm synergy among bacterial soil isolates in cocultures indicates bacterial interspecific cooperation. ISME J 9, 81–89.

Rinaudi, L.V., Giordano, W., 2010. An integrated view of biofilm formation in rhizobia. FEMS Microbiol. Lett. 304, 1–11.

Rodríguez, S.J., Bishop, P.L., 2007. Three-dimensional quantification of soil biofilms using image analysis. Environ. Eng. Sci. 24, 96–103.

Seneviratne, G., Zavahir, J.S., Bandara, W.M.M.S., Weerasekara, M.L.M.A.W., 2008. Fungal-bacterial biofilms: their development for novel biotechnological applications. World J. Microbiol. Biotechnol. 24, 739–743.

Seneviratne, G., Weerasekara, M.L.M.A.W., Seneviratne, K.A.C.N., Zavahir, J.S., Kecskés, M.L., Kennedy, I.R., 2010. Importance of biofilm formation in plant growth promoting rhizobacterial action. pp. 81–95.

Seneviratne, G., Jayasekara, A.P.D.A., De Silva, M.S.D.L., Abeysekera, U.P., 2011. Developed microbial biofilms can restore deteriorated conventional agricultural soils. Soil Biol. Biochem. 43, 1059–1062.

Soon, N.W., Lee, L.M., Khun, T.C., Ling, H.S., 2014. Factors affecting improvement in engineering properties of residual soil through microbial-induced calcite precipitation. J. Geotech. Geoenviron. Eng 140, 04014006.

Su, J., Kang, D., Xiang, W., Wu, C., 2017. Periphyton biofilm development and its role in nutrient cycling in paddy microcosms. J. Soil Sediment. 17, 810–819.

Sun, X., Cai, P., Sørensen, S.J., Mortimer, M., Gao, C., Huang, Q., et al., 2020. Interspecific interactions in dual-species biofilms of soil bacteria: effects of fertilization practices. J. Soil Sediment. 20, 1494–1501.

Sutherland, I.W., 2001. Biofilm exopolysaccharides: a strong and sticky framework. Microbiology 147, 3–9.

Sztajer, H., Szafranski, S.P., Tomasch, J., Reck, M., Nimtz, M., Rohde, M., et al., 2014. Cross-feeding and interkingdom communication in dual-species biofilms of *Streptococcus mutans* and *Candida albicans*. ISME J 8, 2256–2271.

Teplitski, M., Merighi, M., Gao, M., Robinson, J., 2011. Integration of cell-to-cell signals in soil bacterial communities. In: Witzany, G. (Ed.), Biocommunication in Soil Microorganisms. Soil Biology, 23. Springer, Berlin, Heidelberg, pp. 369–401.

Timmusk, S., Nevo, E., 2011. Plant root associated biofilms: perspectives for natural product mining. Bacteria in Agrobiology: Plant Nutrient Management. Springer, Berlin, pp. 285–300.
Timmusk, S., Grantcharova, N., Wagner, E.G.H., 2005. *Paenibacillus polymyxa* invades plant roots and forms biofilms. Appl. Environ. Microbiol. 71, 7292–7300.
Torres-Cortés, G., Ghignone, S., Bonfante, P., Schüßler, A., 2015. Mosaic genome of endobacteria in arbuscular mycorrhizal fungi: transkingdom gene transfer in an ancient mycoplasma-fungus association. Proc. Natl. Acad. Sci. U. S. A 112, 7785–7790.
Umar, M., Kassim, K.A., Ping Chiet, K.T., 2016. Biological process of soil improvement in civil engineering: a review. J. Rock Mech. Geotech. Eng. 8, 767–774.
Volk, E., Iden, S.C., Furman, A., Durner, W., Rosenzweig, R., 2016. Biofilm effect on soil hydraulic properties: experimental investigation using soil-grown real biofilm. Water Resour. Res. 52, 5813–5828.
von Canstein, H., Li, Y., Leonhäuser, J., Haase, E., Felske, A., Deckwer, W.-D., et al., 2002. Spatially oscillating activity and microbial succession of mercury-reducing biofilms in a technical-scale bioremediation system. Appl. Environ. Microbiol. 68, 1938–1946.
Wagner, K., Krause, K., Gallegos-Monterrosa, R., Sammer, D., Kovács, Á.T., Kothe, E., 2019. The ectomycorrhizospheric habitat of Norway spruce and *Tricholoma vaccinum*: promotion of plant growth and fitness by a rich microorganismic community. Front. Microbiol. 10, 307.
Wang, D.C., Jiang, C.H., Zhang, L.N., Chen, L., Zhang, X.Y., Guo, J.H., 2019. Biofilms positively contribute to *Bacillus amyloliquefaciens* 54-induced drought tolerance in tomato plants. Int. J. Mol. Sci. 20, 6271.
Witzany, G., 2011. Introduction: key levels of biocommunication of bacteria. In: Witzany, G. (Ed.), Biocommunication in Soil Microorganisms. Springer, pp. 1–34.
Wu, T., Xu, J., Xie, W., Yao, Z., Yang, H., Sun, C., et al., 2018. *Pseudomonas aeruginosa* L10: a hydrocarbon-degrading, biosurfactant-producing, and plant-growth-promoting endophytic bacterium isolated from a reed (*Phragmites australis*). Front. Microbiol. 9, 1087.
Wu, Y., Cai, P., Jing, X., Niu, X., Ji, D., Ashry, N.M., et al., 2019. Soil biofilm formation enhances microbial community diversity and metabolic activity. Environ. Int. 132, 105116.
Xu, Z., Shao, J., Li, B., Yan, X., Shen, Q., Zhang, R., 2013. Contribution of bacillomycin D in *Bacillus amyloliquefaciens* SQR9 to antifungal activity and biofilm formation. Appl. Environ. Microbiol. 79, 808.
Xu, Y.B., Chen, M., Zhang, Y., Wang, M., Wang, Y., Huang, Q.B., et al., 2014. The phosphotransferase system gene *ptsI* in the endophytic bacterium *Bacillus cereus* is required for biofilm formation, colonization, and biocontrol against wheat sharp eyespot. FEMS Microbiol. Lett. 354, 142–152.
Zengler, K., Zaramela, L.S., 2018. The social network of microorganisms - how auxotrophies shape complex communities. Nat. Rev. Microbiol. 16, 383–390.
Zeriouh, H., de Vicente, A., Pérez-García, A., Romero, D., 2014. Surfactin triggers biofilm formation of *Bacillus subtilis* in melon phylloplane and contributes to the biocontrol activity. Environ. Microbiol. 16, 2196–2211.
Zhao, Y., Liu, H., Wang, R., Wu, C., 2020. Interactions between dicyandiamide and periphytic biofilms in paddy soils and subsequent effects on nitrogen cycling. Sci. Total Environ. 718, 137417.

Alteration in microbial population density composition in different land use systems

4

Ajay Neeraj[1], *R. Hiranmai Yadav*[1] *and Raman Kumar Ravi*[2]
[1]School of Environment and Sustainable Development, Central University of Gujarat, Gandhinagar, India, [2]Department of Agriculture, Uttaranchal College of Science and Technology, Dehradun, India

4.1 Introduction

Land use pattern throughout the continents has gone through drastic changes over the last few decades, as the enormous population growth leading to the stress of food, shelters, and other resources. Developmental activities including industrialization, urbanization, and the modernization are the basic causes of these changes. The land use pattern change has a great impact on the associated ecosystems. Changes in the climatic conditions and patterns of land use systems are subjected to disturb the biogeochemical cycling in the environment which is carried through endogenous microorganisms of the soil in these ecosystems shows in Fig. 4.1. It is now widely accepted that the change in land use can drastically alter the function of an ecosystem. Changes in nutrient cycles, structure, and soil carbon sequestration (C) which occur when natural ecosystems are converted for agricultural use are particularly well documented. Land use change use influences the economic growth, quality of life, environmental resources management, and national food supply. The microorganisms of any soil ecosystems expeditiously respond to soil changes about the chemical or physical properties of soil (Jackson et al., 2007) (Fig. 4.1).

Therefore the understanding of soil microbial community and their responses toward differential land use pattern might be extremely significant for the mitigation of land degradation and desertification processes. Hence the present literature will provide an insight into the changes in the land use pattern and their probable adverse effects on the community of the soil microorganisms.

4.2 Impact of land use on microbial community change in a different land-use system

Land-use change and cover a significant, rapid and comprehensive process. Changes in land use and land cover are human-driven and, in many cases, also

Microbes in Land Use Change Management. DOI: https://doi.org/10.1016/B978-0-12-824448-7.00007-3
© 2021 Elsevier Inc. All rights reserved.

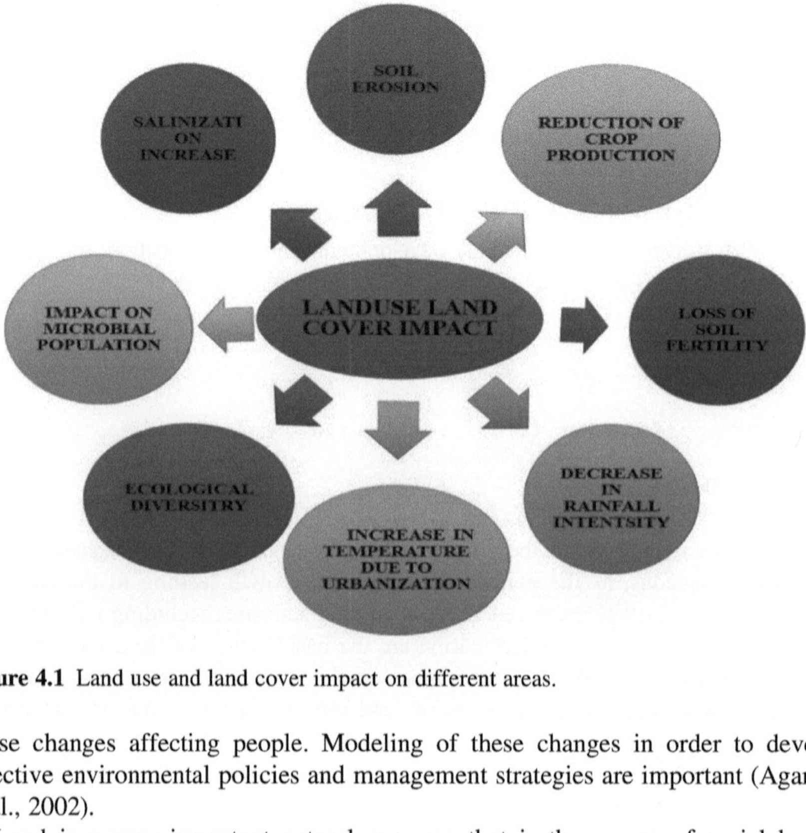

Figure 4.1 Land use and land cover impact on different areas.

cause changes affecting people. Modeling of these changes in order to develop effective environmental policies and management strategies are important (Agarwal et al., 2002).

Land is a very important natural resource that is the source of social health. Because of the land resources, change of use is very important for the management of the country's economic development, quality, and environmental resources. supply. The rate of land use is increasing due to increasing urbanization expectations. Land-use change is an important factor of global change and has a wide impact on the structures, processes, and operations of ecosystems (Rathee, 2014). Although agricultural production systems are considered the main cause of rapid carbon loss (C) (Spohn et al., 2016). forest regeneration or reforestation (i.e, afforestation) carried out on previously cultured or uncultured land can capture C in above-ground biomass and soil organic matter (SOM). Restoring deforestation or redistribution forests is an approach that minimizes the effects of climate change. Terrestrial oil microorganism are decomposites of plant waste and soil organic carbon (SOC) in the ecosystem, which can soil C and regulate many inputs and losses routes of nitrogen (N). Change in the microbial community structure and function transform the ecosystem processes in, such as the decomposition of plant litter, and nutrient availability. This assumption may be influenced by microbial degradation of debris and SOM due to land use change, which regulates the exchange of carbon and nitrogen in the soil, the balance of terrestrial ecosystems. Thus estimating the effects of land use change on soil microbial group composition is significant for a better understanding of anthropological impacts on the global carbon cycle. Agricultural land microbial changes in the structure of the use of plant species during change community organization and biomass changed mainly SOC and nitrogen input can be affected for example, Reforestation has been found in previous studies usually soil C and N input enhance and stimulate microorganisms because they are a source of nutrients and energy for microorganisms. Moreover, it has a great effect on microbial activity in various chemical compositions of plant waste (litter) and soil organic matter. The land remains changes and microbial community structure using the quantity of different microbial groups typically differ regarding sources and the amount of carbon. For example, Urbanová et al. (2015) compared the effects of soil properties of tree species in the forestry ecosystem and reported on a vast proportion of difference in microbial group structure, especially in the fungus. Meanwhile, G-bacteria have recently been established to prefer plant originated carbon, and G + bacteria have been known to prefer older SOM-derived carbon (Kramer and Gleixner, 2008). Relatives to fungal abundance of bacteria (i.e., F:B ratio) are usually sensitive to disturbance of soils associated proportion with high frequencies of plowing (Strickland and Rousk, 2010).

Some studies have suggested that the affect in agricultural land use on the composition of soil microbial community and biomass remains a widely discussed topic (Zhang et al., 2016). Traditional tillage used in agricultural methods can result in a bacteria rather than fungal-dominated system likened to no-tillage methods. Some studies have suggested that the impact of changes in agricultural land use on the structure of soil microbial community and biomass remains a widely debated topic, since traditional tillage is used in agricultural methods rather than in a fungal-dominated system. It may lead to more bacteria in the system. Even agricultural methods alter the quality of the substrate and can also change the fungal and bacteria ratio, because a substrate with a low carbon and nitrogen ration helps bacteria as well as with high carbon, nitrogen ratio usually favors fungi (Bossuyt et al., 2001; Kihara et al., 2012). Thus the expected changes in the use of agricultural land may affect the structure of the microbial community and biomass, but more specific data have been warned that structure of the microbial community and biomass can be influenced by changes in land use, such as pH, soil depth, moisture and temperature, and soil properties (Rousk et al., 2010). Especially, the pH of the soil F: B ratios may affect the relative concentration and variety of bacteria (Firer and Jackson, 2008). Some studies have shown that fungal bacteria acid-tolerant, result in the dominance of fungal in acid soil, (Joergensen and Wichern, 2008). Meanwhile, Stevenson et al. (2014) has reported that soil moisture is very important for all microbial communities. In general, fungi will be less susceptible to changes in moisture than bacteria because their chitinous cell walls make them more resistant to changes in moisture and temperature. Furthremore, the wealth of microbial communities in the soil structure and biomass are strongly dependent on increasing the depth of the soil due to different carbon amount availability (Hansel et al., 2008) Studies showed that a rise in input litter of deforestation region, may increase soil carbon sequestration.Agricultural landshift toward the shrubland and woodland can also reduce soil nutrients and soil erosion (Cheng et al., 2013; Deng et al.,

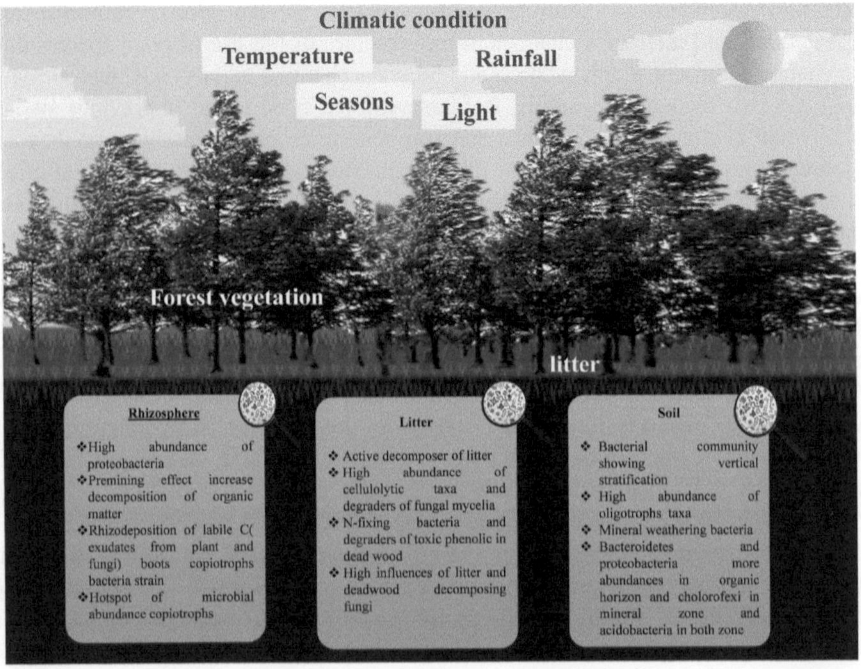

Figure 4.2 Role of microorganism and the function in forest vegetation.
Source: Salvador Lladó et al. Microbiol. Mol. Biol. Rev. 2017.

2014). The information regarding the microbial response to land use change is lacking in literature (Fig. 4.2).

Zhang et al. (2019) investigated these hypothesis, through analyzing composition of soil microbail community along with the biomass which were deratives of phospholipid fatty acids (PLFAs) and respiration of basal microbes for the period of 30 days incubation from the soil profile of 0–10 and 10–30 cm from the associated land with two different fractions viz. applying forestry land and planted land in contrast with nearby cropland and uncultured lands. Furthurmore, the physicochemical properties include soil pH, moisture content, temperature range, and soil organic carbon percentage and nitrogen (in the form of SOC and SON), carbon and nitrogen (LC and LN) and recalcitrant carbon and recalcitrant nitrogen (RC and RN) (Zhang et al., 2016).

4.2.1 Forest land

Microorganisms play an important role in soil through degradation of organic matters, carbon, and soil nutrient rotation in forest ecosystems functions (Koranda et al., 2013). Minor changes in microbial biomass or microbes community composition can affect the soil organic matter turnover and nutrient cycling (Huang et al., 2013) shows in Fig. 4.2. Therefore the soil is considered a key indicator reflecting the effects of the land use of forest on soil microbial properties (Li et al., 2014). Forest land use changes often show little or no effect but mineral soil alters the microbial community profiles in the organic layer. In the forest land use change the soil microbial community but there is a slight or no effect on the nutrient of soil. Another important cause of forest management and land use disturbances affect soil microbes. They can change the characteristics of the soil, and then affect the diversity and composition of the microbial community (Hartmann et al., 2014). An additional analysis across that microbial biomass, PLFA content, and soil enzyme activity are higher in summertime than in cold season (Chang et al., 2016).

4.2.2 Agriculture land

Deforestation from the conversion of forests to cultivated land use is an essential issue around the world. This phenomenon affects soil organic matter (SOM) and due to the change in environmental condition and with soil loss soil health nad size degradation affects the soil microbial community (Raiesi and Beheshti, 2015). Different type of fertilizer and their chamical properties affect the soil nutruents and soil profile (Wright et al., 2007). Specific change in the microbial community structure and function due to the consequences of change in land use intensity or with history of agriculture practies (Grayston et al., 2004). Storage of soil carbon and nitrogen material change due to land use change in the range of habite. (Monkiedje et al., 2006), Which is usually seen as the major factors causing changes in microbial community structure (Cookson et al., 2007). Management of agriculture soil from forest and natural grassland variesit receiving large inputs of nutrients, such as chemical and organic fertilizers that can often cultivated and irrigated agricultural soils. These agricultural practices can have an impact on specific functional groups as the structure of the microbial community and Naitrifiars and Denifritifrs which are important to the cycle of the nitrogen in the soil (Pernes-Debuyser and Tessier, 2004).

4.2.3 Grassland

Soil type is more important than the type of plant to affect the structure of the microbial community in the grassland (Singh et al., 2007). Plants studied that influenced microbial activity around their roots, is reflected in different grass species on micro-community structure of soil as shows in Fig. 4.3. Low pH leads to nutrient depletion, slow nitrification, and mineralization of acidic grass soils. (Williams et al., 1999). The role of microbes in the soil do not get inputs of nutrients from fertilizers, which is particularly important, such as large-scale pastures and rugged land; where phosphorus is 75% and can be provided by Mychorrizal fungal, 20% nitrogen respectively and nitrogen fixing bacteria (Van der Heijden et al., 2008). It has been widely studied that microbial community structure in soil grass, tend to investigate those factors which control their variety and functionality (Rooney et al., 2006). Hedlund et al. (2003) studied that with field sites, there is no regular effect of grassland diversity on soil microbial communities. Bezemer et al. (2006) studied that soil microbial community structure is influenced by manipulating plant species diversity (Fig. 4.3).

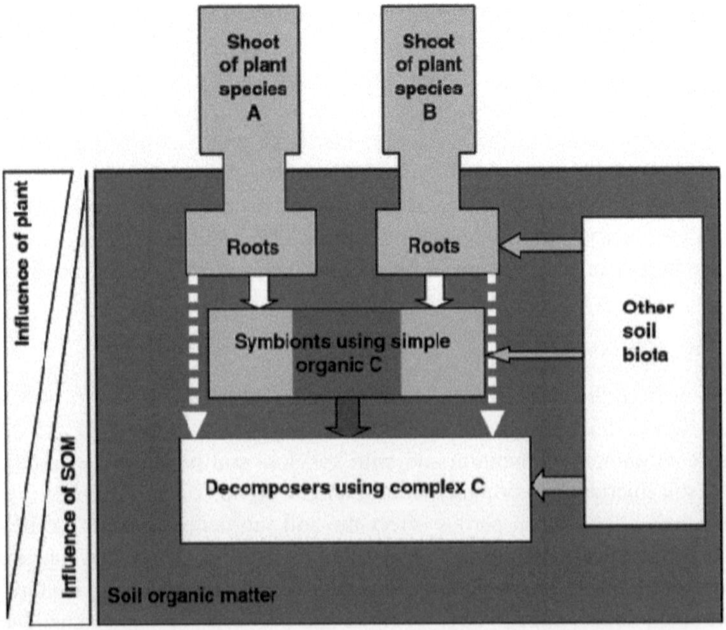

Figure 4.3 A conceptual model of how grassland plants affect soil microbial diversity via carbon inputs to the soil (Millard and Singh, 2010).

4.2.4 Barren land

It was shown that, in most terrestrial ecosystems, the structure of the soil microbial community is strongly influenced by the soil. Temperature, soil moisture, soil pH, and another abiotic factor change the microbes polulation which may have influenced the rate of decomposition and soil mineral cycling (Clark et al., 2009). Using vegetation land and other natural forest land benefited soil chemical properties of bare land. The pH value of the alkaline soil in bare land is from 8.48 to 8.18. The bare land has low soil nutrient, high pH, rather than vegetation type land (Tian et al., 2017).

4.3 Microbial population in selected land use system in Indo-Gangetic plains: a case study

4.3.1 Microbial population and soil stress

Heavy metal contamination, saltinity, and loss of moisture can slow down plant growth and can affect their performance. The AM (arbuscular mycorrhizae) can increase the development and efficiency of plants under stress conditions due to the formation of extensive networks of hyphae and production of biochemicals such as glomalin. These capacities can lead to better absorption of water and nutrients and better soil structure. However, it should be noted that elevated levels of stress can turn a symbiotic relationship between the two associates in a parasitic association as unfavorable circumstances can adversely affect the performance of AM (Rillig, 2004; Hildebrandt et al., 2007; Miransari et al., 2008, 2009a, 2009b). Adverse effects of AM (arbuscular mycorrhizae) on the development of plants under stress may be due to the adverse effects of stress on the functioning and development of the AM. These effects include reduction of the colonization rate and spore maturation and hyphal fungi growthdecline (Jahromi et al., 2008; Evelin et al., 2009). Consequently, the significant role of AM in the ecosystem is now considered as an important detailof such arrangements.

4.3.2 Forest land

Kumar and Ghoshal (2017) reported that the impact of land use change in Uttar Pradesh forest across all land use type PLFAs in bacteria community higher in natural forest and lowest in agroecosystem. Total PLFA, fungal: bacteria ratio (F/B) and SOC trend were tracked similarly to total bacterial PLFA also reported that the fungus was found to be more sensitive to land use change compared to bacterial PLFA. In all types of land observed a strong positive relationship amongst the contents of COS and G^{-}, the total number of bacteria, fungi, and total number of PLFA ratio F/B. The microbial community of soil more readily responds to alterations in the soil in any ecosystem, depending on the chemical or physical properties of the soil (Jackson et al., 2007). Uttar Pradesh, a major part of this state is IGP largest food grain -producing regions in the globle and the foundation of the food security of India. Rapid population growth, urbanization, and the demand for food increase the pressure on land resources, which leads to the fact that forests and woodlands are regularly transformed into agricultural land and, in turn, and lead to the loss of biodiversity unavailability of plant and soil microbial diversity.

4.3.2.1 Agriculture land

In the region of the Indo-Gangetic plains studied long history of cultivation and the impact of changes and undisturbed soil mediated tillage on glomalin-related soil protein and SOC and its sub fractions (Singh et al., 2016). The IGP in terms of the structural composition and diversity of the microbial population of the soil is mainly limited to the surface horizons, as most studies focus only on the rhizosphere soil. Modifications in environmental conditions with soil depth give rise to variation between surface and subsurface microorganism populations, which are also poorly understood in IGP soils. Significant differences were observed in microbial populations belonging to different bioclimate IGP. In the surface layer a large population of bacteria is recorded in a humid subhumid bioclimate, then in semiarid; however bacterial population more or less the same in both wet and dry subhumid bioclimate. The fungal population is high in the subhumid bioclimatic zone and lowest in the humid bioclimate.

Reducing the microbial population can be explained by a combination of biological and environmental factors. Reducing the number of labile carbon and sustainable increase in the number of substrates in the organic horizon could have a negative impact on the microbial population. Most microbial community observed in the culture system legume-based IGP is attributed to higher nutrient content of the SOC along with microbial activity stimulated. Several root exudates and root residues of the rotary. Decomposed crops in the soil can affect the availability of nutrients for microbes in the soil (Srivastava et al., 2014).

4.4 Salinity and microbial population

Soil microbes are considered indicators of changes in sensitivity to the environment due to their large ratio of surface area to volume ratio, cell membrane permeability, and a high refresh rate, which leads to quick response to changes in nutrient availability (Troxler et al., 2012). Organisms with Small groups that exhibit changes in community structure in response to the gradients of the adjacent medium, may carry more specific functions such as fermentation and methanogenesis, while more general groups may have a high efficient redundancy (Morrissey et al., 2014). The heterogeneity of the soil microbial community is expected, which depends on the type and biomass of the plants, as well as on the processing conditions. It is believed that any change in litteryield, variety or quality leads to a change in the function and structure of the microbic population (Anderson and Domsch, 1989; Haynes, 1999). Highre ratio C_{mic}/C_{org} in high salinity traeatment, as the ratio frequently increase due to the incresre availability of susbtrate as wll as microbibal popualtin incresase dute to nutrient supply (Anderson and Domsch, 1989; Haynes, 1999). soil salinity feccet in the microbial community and also decrease the soil respiration, biomass, minearalization processes. The fraction increases with a high salt content, which indicates either a shift towards fungi, or a change in the morphology of fungal cells, or the accumulation of ergosterol in the soil (Wichern et al., 2006).

4.5 Conclusion

Present situation industrialization and urbanization increase because of a rapid increase of population requires more land for the infrastructure due to these enthroprogenic activities increase the land degradation which the effect can be seen in the different land use pattern (forest, grassland, and agriculture land) there major role of microorganism play a important role in in the balancing of ecosystem density is higher in forest and agriculture land as comparison to barren and grassland. Different types of plants, due to its quality and quantity entering the litter, a strong influence on the community of soil microbes/biomass under different land uses and the depths of the earth. Land use also increase the soil salinity problem due to soil layer erosion. So the proper management of land use may influence the soil attributes and microbial population in Indo-Gangetic plain.

References

Agarwal, C., Green, G.M., Grove, J.M., Evans, T.P., Schweik, C.M., 2002. A review and assessment of land-use change models: dynamics of space, time, and human choice. Gen. Tech. Rep. NE-297. US Department of Agriculture, Forest Service, Northeastern Research Station, Newtown Square, PA, pp. 1954-1984.

Anderson, T.H., Domsch, K.H., 1989. Ratios of microbial biomass carbon to total organic carbon in arable soils. Soil Biol. Biochem. 21, 471–479.

Bezemer, M.T., Lawson, C.S., Hedlund, K., Edwards, A.R., Brook, A.J., Igual, J.M., et al., 2006. Plant species and functional group effects on abiotic and microbial soil properties and plant–soil feedback responses in two grasslands. J. Ecol. 94, 839–904.

Bossuyt, H., Denef, K., Six, J., Frey, S.D., Merckx, R., Paustian, K., 2001. Influence of microbial populations and residue quality on aggregate stability. Appl. Soil Ecol. 16 (3), 195–208.

Chang, E.H., Chen, T.H., Tian, G.L., Hsu, C.K., Chiu, C.Y., 2016. Effect of 40 and 80 years of conifer regrowth on soil microbial activities and community structure in subtropical low mountain forests. Forests 7, 244.

Cheng, X., Yang, Y., Li, M., Dou, X., Zhang, Q., 2013. The impact of agricultural land use changes on soil organic carbon dynamics in the Danjiangkou Reservoir area of China. Plant Soil 366 (1-2), 415–424.

Clark, J.S., Campbell, J.H., Grizzle, H., Acosta-Martìnez, V., Zak, J.C., 2009. Soil microbial community response to drought and precipitation variability in the Chihuahuan Desert. Microb. Ecol. 57 (2), 248–260.

Cookson, W.R., Osman, M., Marschner, P., Abaye, D.A., Clark, I., Murphy, D.V., et al., 2007. Controls on soil nitrogen cycling and microbial community composition across land use and incubation temperature. Soil Biol. Biochem. 39, 744–756.

Deng, Q., Cheng, X., Yang, Y., Zhang, Q., Luo, Y., 2014. Carbon–nitrogen interactions during afforestation in central China. Soil Biol. Biochem. 69, 119–122.

Evelin, H., Kapoor, R., Giri, B., 2009. Arbuscular mycorrhizal fungi in alleviation of salt stress: a review. Ann. Bot. 104, 1263–1280.

Fierer, N., Strickland, M.S., Liptzin, D., Bradford, M.A., Cleveland, C.C., 2009. Global patterns in belowground communities. Ecology letters 12 (11), 1238–1249.

Grayston, S.J., Campbell, C.D., Bardgett, R.D., Mawdsley, J.L., Clegg, C.D., Ritz, K., et al., 2004. Assessing shifts in microbial community structure across a range of grasslands of different management intensity using CLPP, PLFA, and community DNA techniques. Appl. Soil Ecol. 25, 63–84.

Hansel, C.M., Fendorf, S., Jardine, P.M., Francis, C.A., 2008. Changes in bacterial and archaeal community structure and functional diversity along a geochemically variable soil profile. Appl. Environ. Microbiol 74 (5), 1620–1633.

Hartmann, M., Niklaus, P.A., Zimmermann, S., Schmutz, S., Kremer, J., Abarenkov, K., Frey, B., 2014. Resistance and resilience of the forest soil microbiome to logging-associated compaction. The ISME journal 8 (1), 226–244.

Haynes, R.J., 1999. Size and activity of the soil microbial biomass under grass and arable management. Biol. Fertil. Soils 30, 210–216.

Hedlund, K., Santa Regina, I., Van der Putten, W.H., Lepš, J., Díaz, T., Korthals, G.W., et al., 2003. Plant species diversity, plant biomass and responses of the soil community on abandoned land across Europe: idiosyncracy or above-belowground time lags. Oikos 103, 45–58.

Hildebrandt, U., Regvar, M., Bothe, H., 2007. Arbuscular mycorrhiza and heavy metal tolerance. Phytochemistry 68, 139–146.

Huang, Z., Wan, X., He, Z., Yu, Z., Wang, M., Hu, Z., et al., 2013. Soil microbial biomass, community composition and soil nitrogen cycling in relation to tree species in subtropical China. Soil Biol. Biochem. 62, 68–75.

Jackson, R.B., Fierer, N., Schimel, J.P., 2007. New directions in microbial ecology. Ecology 88, 1343. 1344.

Jahromi, F., Aroca, R., Porcel, R., Ruiz-Lozano, J.M., 2008. Influence of salinity on the in vitro development of *Glomus* intraradices and on the in vivo physiological and molecular responses of mycorrhizal lettuce plants. Microb. Ecol. 55, 45–53.

Joergensen, R.G., Wichern, F., 2008. Quantitative assessment of the fungal contribution to microbial tissue in soil. Soil Biol. Biochem. 40 (12), 2977–2991.

Kihara, J., Martius, C., Bationo, A., Thuita, M., Lesueur, D., Herrmann, L., et al., 2012. Soil aggregation and total diversity of bacteria and fungi in various tillage systems of sub-humid and semi-arid Kenya. Appl. Soil Ecol. 58, 12–20.

Koranda, M., Kaiser, C., Fuchslueger, L., Kitzler, B., Sessitsch, A., Zechmeister-Boltenstern, S., et al., 2013. Seasonal variation in functional properties of microbial communities in beech forest soil. Soil Biol. Biochem. 60, 95–104.

Kramer, C., Gleixner, G., 2008. Soil organic matter in soil depth profiles: distinct carbon preferences of microbial groups during carbon transformation. Soil Biol. Biochem. 40 (2), 425–433.

Kumar, C.M., Ghoshal, N., 2017. Impact of land-use change on soil microbial community composition and organic carbon content in the dry tropics. Pedosphere 27 (5), 974–977.

Li, H., Ye, D., Wang, X., Settles, M.L., Wang, J., Hao, Z., et al., 2014. Soil bacterial communities of different natural forest types in Northeast China. Plant Soil 383 (12), 203–216.

Millard, P., Singh, B.K., 2010. Does grassland vegetation drive soil microbial diversity? Nutrient Cycling in Agroecosystems 88 (2), 147–158.

Miransari, M., Bahrami, H.A., Rejali, F., Malakouti, M.J., 2008. Using arbuscular mycorrhiza to reduce the stressful effects of soil compaction on wheat (Triticum aestivum L.) growth. Soil Biology and Biochemistry 40, 1197–1206.

Miransari, M., Rejali, F., Bahrami, H.A., Malakouti, M.J., 2009a. Effects of soil compaction and arbuscular mycorrhiza on corn (Zea mays L.) nutrient uptake. Soil and Tillage Research 103, 282–290.

Miransari, M., Rejali, F., Bahrami, H.A., Malakouti, M.J., 2009b. Effects of arbuscular mycorrhiza, soil sterilization, and soil compaction on wheat (Triticum aestivum L.) nutrient uptake. Soil and Tillage Research 104, 48–55.

Monkiedje, A., Spiteller, M., Fotio, D., 2006. The effect of land use on soil health indicator in peri-urban agriculture in the humid forest zone of southern Cameroon. Journal of Environmental Quality 35, 2402–2409.

Morrissey, E.M., Gillespie, J.L., Morina, J.C., Franklin, R.B., 2014. Salinity affects microbial activity and soil organic matter content in tidal wetlands. Global change biology 20 (4), 1351–1362.

Pernes-Debuyser, A., Tessier, D., 2004. Soil physical properties affected by long-term fertilization. Eur. J. Soil Sci. 55 (3), 505–512.

Raiesi, F., Beheshti, A., 2015. Microbiological indicators of soil quality and degradation following conversion of native forests to continuous croplands. Ecol. Indic. 50, 173–185.
Rathee, G., 2014. Trends of land-use change in India. Urbanization in Asia. Springer, New Delhi, pp. 215–238.
Rillig, M.C., 2004. Arbuscular mycorrhizae and terrestrial ecosystem processes. Ecol. Lett. 7, 740–754.
Rooney, D., Kennedy, N., Deering, L., Gleeson, D., Clipson, N., 2006. Effect of sheep urine deposition on the bacterial community structure in an acidic upland grassland soil. Appl. Environ. Microbiol. 72 (11), 7231–7237.
Rousk, J., Brookes, P.C., Bååth, E., 2010. Investigating the mechanisms for the opposing pH relationships of fungal and bacterial growth in soil. Soil Biology and Biochemistry 42 (6), 926–934.
Singh, B.K., Munro, S., Potts, J.M., Millard, P., 2007. Influence of grass species and soil type on rhizosphere microbial community structure in grassland soils. Appl. Soil Ecol. 36 (2-3), 147–155.
Singh, A.K., Rai, A., Singh, N., 2016. Effect of long term land use systems on fractions of glomalin and soil organic carbon in the Indo-Gangetic plain. Geoderma 277, 41–50.
Spohn, M., Novák, T.J., Incze, J., Giani, L., 2016. Dynamics of soil carbon, nitrogen, and phosphorus in calcareous soils after land-use abandonment—a chronosequence study. Plant Soil 401 (1-2), 185–196.
Srivastava, A.K., Velmourougane, K., Bhattacharyya, T., Sarkar, D., Pal, D.K., Prasad, J., Thakre, S., 2014. Impacts of agro-climates and land use systems on culturable microbial population in soils of the Indo-Gangetic Plains, India. Current Science 1464–1469.
Stevenson, B.A., Hunter, D.W., Rhodes, P.L., 2014. Temporal and seasonal change in microbial community structure of an undisturbed, disturbed, and carbon-amended pasture soil. Soil Biol. Biochem. 75, 175–185.
Strickland, M.S., Rousk, J., 2010. Considering fungal: bacterial dominance in soils—methods, controls, and ecosystem implications. Soil Biol. Biochem. 42 (9), 1385–1395.
Tian, Q., Taniguchi, T., Shi, W.Y., Li, G., Yamanaka, N., Du, S., 2017. Land-use types and soil chemical properties influence soil microbial communities in the semiarid Loess Plateau region in China. Scientific reports 7 (1), 1–9.
Troxler, T.G., Ikenaga, M., Scinto, L., et al., 2012. Patterns of soil bacteria and canopy community structure related to tropical peatland development. Wetlands 32, 769–782.
Urbanová, M., Šnajdr, J., Baldrian, P., 2015. Composition of fungal and bacterial communities in forest litter and soil is largely determined by dominant trees. Soil Biol. Biochem. 84, 53–64.
Van der Heijden, M.G., Bardgett, R.D., Van Straalen, N.M., 2008. The unseen majority: soil microbes as drivers of plant diversity and productivity in terrestrial ecosystems. Ecol. Lett. 11 (3), 296–310.
Wichern, J., Wichern, F., Joergensen, R.G., 2006. Impact of salinity on soil microbial communities and the decomposition of maize in acidic soils. Geoderma 137 (1-2), 100–108.
Williams, B.L., Shand, C.A., Sellers, S., Young, M.E., 1999. Impact of synthetic sheep urine on N and P in two pastures in the Scottish uplands. Plant Soil 214 (1-2), 93.
Wright, A.L., Dou, F., Hons, F.M., 2007. Crop species and tillage effects on carbon sequestration in subsurface soil. Soil Sci 172 (2), 124–131.
Zhang, Q., Wu, J., Yang, F., Lei, Y., Zhang, O., Cheng, X., 2016. Alterations in soil microbial community composition and biomass following agricultural land use change. Sci. Rep. 6, 36587.

Molecular characterization of bacterial community succession and analysis of physiochemical properties in a compost of solid organic waste from Gandhinagar, Gujarat, India

5

Mohd A. Siddiqui and R. Hiranmai Yadav
School of Environment and Sustainable Development, Central University of Gujarat, Gandhinagar, India

5.1 Introduction

The increasing population and urbanization are forcing the world to face the problem of solid waste management which is considered as one of the biggest environmental challenges. A sustainable approach is needed to overcome the pollution caused by the solid waste by converting it into useful products by composting and vermicomposting. The organic solid waste treatment by composting is a natural process that is widely practiced for waste management. Vermicomposting involves earthworms for the process of degradation of biodegradable materials. The resultant product is rich in nutrients, enzymes and plant growth promoting substances (Bajsa et al., 2003).

The approaches by different municipalities for managing this solid waste through monitoring, collection, transport, processing, recycling and disposal highly varies. Among the available technologies of solid waste management, composting and vermicomposting are economic and environmentally friendly and also reduce the waste and help to improve soil quality through supplementing organic rich nutrients. As practiced presently by selected organizations, there is a need to perform composting at home, in the neighborhood, parks, schools, etc. It can be considered as the most important short-term issue in solid waste management in absence of a proper landfill site. A number of compostable materials are collected by municipalities and provide to the farmers, who use it to increase soil fertility (Verma, 2010).

The present-day anthropogenic activities are adding increased waste material to the environment resulting in air, soil, and water pollution and adding to global warming (Peter, 2010). The increased rate of waste production at local and global levels due to increased population and industrialization urges to opt for waste

Microbes in Land Use Change Management. DOI: https://doi.org/10.1016/B978-0-12-824448-7.00008-5
© 2021 Elsevier Inc. All rights reserved.

management options that are economic and ecofriendly. Reducing the waste generation is the best practice to reduce the accumulation. The organic wastes and wastes generated from industries in developing countries are traditionally directly incinerated or deposited in landfills. This leads to production of greenhouse gases in addition to utilizing the land resources. These wastes if not treated properly lead to environmental and health problems (Zhang et al., 2013).

The term composting denotes the process of controlled decomposition and stabilization of organic wastes by biological means under aerobic conditions (Dees and Ghiorse, 2001). Co-composting however is simultaneous composting of two or more types of waste materials with the potential benefit of enhancing the quality of the final compost product (Das et al., 2011; Petric et al., 2012). Compost reduces the bulk of material, weight, moisture content, potential odor, and pathogens and improves the nutrient value of the product. It is an economic and ecofriendly process (Scheutz et al., 2011). The end product of composting can be utilized as a fertilizer by amending the soil and acting as a soil conditioner. It is also used as a pellet for fuel, microbial control agents are derived from it, and it benefitted as a potential heat recovery process (Zajonc et al., 2014; Larkin and Tavantzis, 2013). Compost prepared from agroindustrial wastes could be utilized in crop systems as a biofertilizer or soil conditioner as the waste is considered a clean biowaste (Ben et al., 2007).

5.1.1 Solid organic waste

The combined residential, industrial, and commercial activities in a given area produce large quantities of unwanted or useless solid materials and are dumped into different places that promote environmental pollution. These solid wastes can be characterized on the basis of its origin or its contents. A proper management of this solid waste reduces or eliminates its adverse impacts on the environment and human health and supports economic development and improved life quality. Municipal solid waste (MSW) forming a large source of organic solid waste constitutes varying organic waste from industrial, institutional and residential sources. The waste from the cafeteria, wood plates, office and classroom papers, disposable tableware, boxes, food scrabs, clothing appliances, and newspapers are the main categories of MSW. It is very difficult to reach an exact rate of MSW generation because of its components and variability of MSW production (Singh et al., 2014) (Table 5.1).

5.1.2 Generation of solid organic waste

Generation of solid organic waste is a natural habit of human life. As a result of improved lifestyle and social status of mankind the generation of MSW is directly proportional to the economic growth of people in terms of kg/capita/day (Kumar and Kaushal, 2015). The rapid population growth, fast industrialization, and urbanization directly affects the MSW generation (Singh and Sharma, 2002; Minghua et al., 2009). As compared to any other environmental pollutant generation rate

Table 5.1 Composition of organic waste.

S. no.	Classification	Composition	General
1	Organic waste	Food waste	Beef, chicken, vegetables
		Garden waste	Dried leaves, twigs
		Textile/ rubber waste	Leather, clothes, carpet
		Paper waste	Newspaper, box, paper cups
2	Nonorganic waste	Plastic	Polyethylene terephthalate, high density polyethylene, polyvinyl chloride, low-density polyethylene, polypropylene, polystyrene, multilayer plastic
		Glass	Glass plates, glass windows, light bulbs
		Metal	Ferrus, alluminum, zinc, chromium

Source: Alias, A.R., Mohd Mokhlis N.A., Zainun, N.Y., 2017. Baseline for food waste generation—A case study in Universiti Tun Hussein Onn Malaysia cafeterias. IOP Conference Series: Materials Science and Engineering 271.

including greenhouse gases also, the rate of MSW generation is high at present time. The percent of urbanization of India increased from 27.81% in 2001 to 31.16% in 2011 being the second most populated country in the world. The regular expansion of human population applies a huge pressure on natural resources for shelter, food, and economic growth (Hoornweg et al., 2013) (Table 5.2).

The generation of management of solid organic waste of a country may differ from another country, of a state to another state, of a city to other city even it can be different area to area within the city. In Indian cities the range of MSW generation is between 0.3 and 0.6 kg/capita/day and it is estimated that by volume annual increase in MSW generation is approximately 1.33% per capita (Pattnaik and reddy, 2010). The municipalities are facing a number of problems in providing effective and dynamic system to the society for managing generated MSW properly in developing countries like India. The unavailability of appropriate collection system, lack of technical expertise and insufficient financial resources are the main reasons of failure of municipalities (Sujauddin et al., 2008; Guerrero et al., 2013).

5.1.3 Management of solid organic waste

The process of waste management is of great concern for environmentalists and composting is one of the methods that allows the conversion of organic waste into a nutrient rich end product that could be applied to agricultural fields. Marketing of the quality compost is of concern to reduce the waste accumulation and obtain quality products. There is increasing interest in composting of solid wastes due to the environmental protection strategies and problems arising from the landfills. The

Table 5.2 Generation of solid organic waste.

S. no.	Category	Source of generation	Type of solid waste
1	Municipal solid wastes	Residential- family dwellings, low, medium and high-rise buildings, apartments, etc. Commercial- stores, restaurants, markets, hotels, shopping complexes, repair workshops, etc. Open area- streets, alleys, parks, vacant plots, play grounds, beaches, highways, recreational areas, etc. Treatment plant sites- water, waste water and industrial treatment processes, etc.	Food wastes, rubbish, ashes, special wastes Food wastes, rubbish, ashes, construction wastes, special wastes, occasional hazardous wastes Special wastes, rubbish Treatment wastes/residential sludges
2	Industrial wastes	Textiles, paper and allied products, chemicals and related products, rubber, food and kindred products, furniture, printing, leather and leather products, glass, petroleum and refineries, etc.	Chemicals, metals, scrap products, gypsum, asbestos resins, glass, organic dyes, glues, etc.
3	Hazardous wastes	Industries- chemicals and oil refineries, ordinance factories, fireworks, etc. Hospitals- biomedical wastes	Volatile organic chemicals, inflammable substances, toxic gases and liquids, etc. Chemicals, pathological wastes, infectious wastes, sharp objects, pharmaceutical wastes, laboratory wastes, pressurized cans, etc.

Source: Upadhyay, V.P., 2005. Eco tools for urban waste management in India. J. Hum. Ecol. 18, 4, 253-269.

nitrogen content and organic matter in a compost plays a significant role during the process of composting and thereby on the end product (Grigatti et al., 2004). There is variation in the quality of compost prepared from different substrates as the constituents varies (Tognetti et al., 2005). Co-composting is preferred for its interactive qualities and improved product quality (Cooperband, 2000). Temperature and nitrogen are significant parameters during the composting process.

5.1.3.1 Composting

Composting of this MSW is the best option to manage this waste properly. There are five main groups including bacteria, actinomycetes, fungi, algae, and protozoa which make microbial population of soils. For the decomposition of organic waste bacteria are the most important microbes among all these categories of microorganisms. Bacteria utilize organic waste for their own metabolism and as a result they form some simple and useful compounds which are beneficial to soil health, soil fertility, growth of plants, and over all to keep the balance of the natural ecosystem. Generally, under the aerobic environment, composting is the controlled biodegradation or biotransformation of organic matter. As a result of this organic materials are transformed into an end product which is stable and a material-like soil called compost (Dees and Ghiorse, 2001) (Fig. 5.1).

5.1.3.2 Vermicomposting

Earthworms convert the organic wastes into organic rich materials and help to improve the worm biomass. They maintain aerobic conditions in the organic wastes by proper mixing and microbial decomposition of the substrate in the intestine of earthworms during the process of vermicomposting to help in improving the quality of vermicompost (Bhat et al., 2018) (Fig. 5.2).

The common use of compost and vermicompost in agriculture, forestry, landscaping, and environmental restoration, it is essential that it is perfectly compatible with plant growth and it increases production and the quality of plant biomass. It is therefore necessary to evaluate the absence of phytotoxic substances in the compost and vermicompost, which informs not only about compost's quality but also on the proper handling of the composting process. The procedure for assessing the phytotoxicity as an indicator of compost maturity can be grouped into two types: (1) seed germination and seedling elongation tests with extracts of compost; and (2) direct seeding tests on substrate made wholly or partly by compost. For the compost to be considered mature, the results of germination and elongation, or plant biomass should be at least 80%–90% of those obtained in a control. These tests have proven

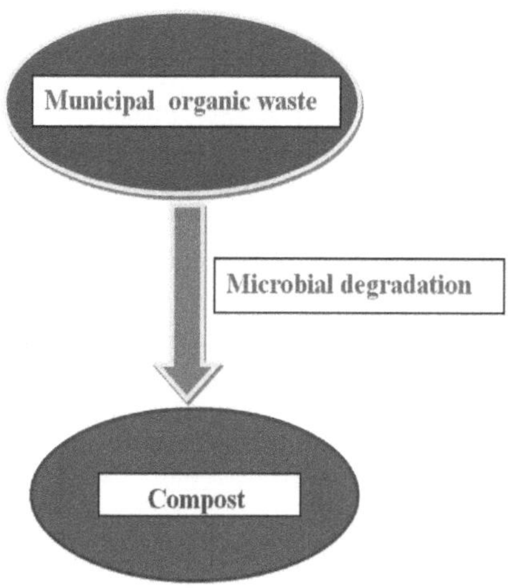

Figure 5.1 Composting process.

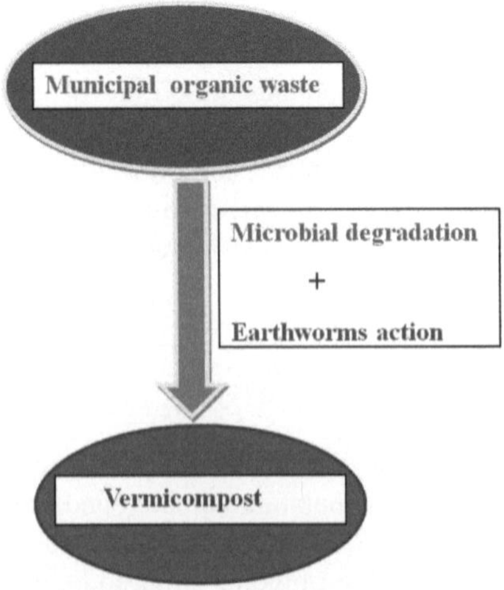

Figure 5.2 Vermicomposting process.

to be sensitive to various toxins such as salts, organic acids, ammonia, or metals and can be used to evaluate the process conditions and the degree of stabilization of organic matter and the suitability of the raw materials. However, the great variability in testing methods makes the comparison and interpretation of results difficult, so it is necessary to advance in the standardization of procedures and determine phytotoxicity thresholds for potential phytotoxic compounds in compost (Barral and Paradelo, 2011).

5.2 Microbial succession during composting

In solid waste degradation a number of microorganisms play a vital role along with the rodents and insects. Bacteria play the most important role among them and therefore, the effective bacteria can be used for planned decomposition of solid organic waste. Recognizing enormous importance of garbage decomposition,the objective of present study was to isolate and characterize the more efficient bacteria from the prepared compost from different organic waste. This study was attempted in order to develop a reproducible protocol and to search for more active decomposer bacteria, which might produce useful components for plant nutrition through successful and effective degradation of municipal organic waste (Zaved et al., 2008) (Fig. 5.3).

5.2.1 Bacterial community

With the most significant role in municipal organic waste decomposition, bacteria are the most abundant group among all other major groups of microbial

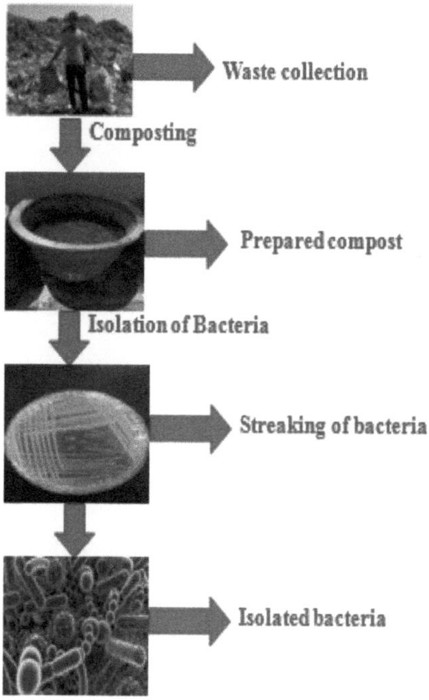

Figure 5.3 Bacteria succession during composting.

population of soil such as fungi, actinomycetes, algae, and protozoa. Due to the secretion of different important compounds these bacteria are very important for soil health and fertility, plant growth, and for maintaining balance of the natural ecosystem. By utilizing the available material for their own metabolism, microbes reduce the accumulating waste in soil. Mainly under aerobic condition, composting is the controlled bio decomposition or transformation of organic compounds, where it is turned into a stable and soil-like end product called compost. There are many other microbes besides bacteria which are also responsible for the decomposition of solid organic waste. The process of decomposition of organic waste is started by the secretion of biological substances called enzymes by the microbes (Saha, 2012).

5.3 Variation in physicochemical properties during composting

The five different samples were collected from five different treatments and the composite of these were assessed for physicochemical parameters. Different parameters and their values are shown in the following Table 5.3.

Table 5.3 Physicochemical characterization of composite.

S. no.	Parameters	Value	Comparative references
1	pH	7.54	Suthar, 2008; Yadav and Garg, 2011
2	Electrical conductivity (μs)	310.80	Yadav and Garg, 2011; Kalemelawa et al., 2012
3	Moisture content (%)	90.89	Zhang et al., 2007; Siddiqqui and Yadav, 2017
4	Pore space volume (mL)	1.74	Jain et al., 2019; Siddiqqui and Yadav, 2017
5	Percentage pore space (%)	2.99	Jain et al., 2019; Siddiqqui and Yadav, 2017
6	Bulk density (mg/ m^3)	345.55	Jain et al., 2019; Siddiqqui and Yadav, 2017
7	Particle density 9 mg/m^3)	355.19	Jain et al., 2019; Siddiqqui and Yadav, 2017
8	Organic carbon (%)	12.20	Suthar, 2008; Kalemelawa et al., 2012
9	Organic matter (%)	21.03	Suthar, 2008; Kalemelawa et al., 2012
10	Total nitrogen (mg/ kg)	468.60	Suthar, 2008; Yadav and Garg, 2011
11	Phosphorous (%)	0.0058	Suthar, 2008; Rashad et al., 2010; Yadav and Garg, 2011
12	Potassium (%)	0.64	Suthar, 2008; Rashad et al., 2010; Yadav and Garg, 2011
13	Zinc (ppm)	10.39	Yadav and Garg, 2011; Rameshwar and Argaw, 2016
14	Ferrous (ppm)	10.48	Yadav and Garg, 2011; Rameshwar and Argaw, 2016
15	Manganese (ppm)	0.86	Siddiqqui and Yadav, 2017; Rameshwar and Argaw, 2016
16	Copper (ppm)	12.29	Yadav and Garg, 2011; Rameshwar and Argaw, 2016

Table 5.4 Morphological characters of bacteria colony isolated from different treatments.

S.no.	Strain	Color	Form	Margin	Elevation
1	C_1	Yellow	Circular	Entire	Convex
2	C_4	Light yellow	Circular	Entire	Raised
3	C_6	Creamy	Circular	Entire	Convex
4	C_7	Creamy	Irregular	Lobate	Flat
5	C_8	White	Irregular	Undulate	Flat
6	C_{10}	White	Circular	Undulate	Convex

Table 5.5 16S rRNA sequencing of different bacteria.

S. no.	Marking	Bacteria	Accession no.	Sequence
1	C_1	*Staphylococcus hominis*	MH671870	AGCGTTATCCGGAATTATTGGGCGTAAAGCGCGCGTAGGCGGTTTTTAAGTCTGA TGTGAAAGCCCACGGCTCAACCGTGGAGGGTCATTGGAAACTGGAAAACTTGA GTGCAGAAGAGGAAAGTGGAATTCCATGTGTAGCGGTGAAATGCGCAGAGATA TGGAGGAACACCAGTGGCGAAGGCGACTTTCTGGTCTGTAACTGACGCTGATG TGCGAAAGCGTGGGGATCAAACAGGATTAGATACCCTGGTAGTCCACGCCGTA AACGATGAGTGCTAAGTGTTAGGGGGTTTCCGCCCCTTAGTGCTGCAGCTAACG CATTAAGCACTCCGCCTGGGGAGTACGACCGCAAGGTTGAAACTCAAAGGAAT TGACGGGGACCCGCACAAGCGGTGGAGCATGTGGTTTAATTCGAAGCAACGCG AAGAACCTTACCAAATCTTGACATCCTTTGACCCTTCTAGAGATAGAAGTTTCC CCTTCGGGGGACAAAGTGACAGGTGGTGCATGGTTGTCGTCAGCTCGTGTCGT GAGATGTTGGGTTAAGTCCCGCAACGAGCGCAACCCTTAAGCTTAGTTGCCAT CATTAAGTTGGGCACTCTAAGTTGACTGCCGGTGACAAACCGGAGGAAGGTG GGGATGACGTCAAATCATCATGCCCCTTATGATTTGGGCTACACACGTGCTAC AATGGACAATACAAAG
2	C_4	*Streptomyces werraensis*	MH671871	GTCGAACGATGAACCTCCTTCGGGAGGGGATTAGTGGCGAACGGGTGAGTAAC ACGTGGGCAATCTGCCCTGCACTCTGGGACAAGCCCTGGAAACGGGGTCTAA TACCGGATACTGATCATCTTGGGCATCCTTGGTGATCGAAAGCTCCGGCGGTG CAGGATGAGCCCGCGGCCTATCAGCTTGTTGGTGAGGTAATGGCTCACCAAG GCGACGACGGGTAGCCGGCCTGAGAGGGCGACCGGCCACACTGGGACTGAG ACACGGCCCAGACTCCTACGGGAGGCAGCAGTGGGGAATATTGCACAATGG GCGAAAGCCTGATGCAGCGACGCCGCGTGAGGGATGACGGCCTTCGGGTTG TAAACCTCTTTCAGCAGGGAAGAAGCGAAAGTGACGGTACCTGCAGAAGAA GCGCCGGCTAACTACGTGCCAGCAGCCGCGGTAATACGTAGGGCGCGAGCG TTGTCCGGAATTATTGGGCGTAAAGAGCTCGTAGGCGGCTTGTCGCGTCGGT TGTGAAAGCCCGGGGCTTAACTCCGGGTCTGCAGTCGATACGGGCAGGCTA GAGTTCGGTAGGGGAGATCGGAATTCCTGGTGTAGCGGTGAAATGCGCAGA TATCAGGAGGAACACCGGTGGCGAAGGCGGATCTCTGGGCCGATACTGACG CTGAGGAGCGAAAGCGTGGGGAGCGAACAG

(Continued)

Table 5.5 (Continued)

S. no.	Marking	Bacteria	Accession no.	Sequence
3	C_6	*Streptomyces matensis*	MH671873	CGTCGTCCGTGCCGCAGCTAACGCATTAAGTGCCCCGCCTGGGGAGTACGGCCG CAAGGCTAAAACTCAAAGGAATTGACGGGGGCCCGCACAAGCGGCGGAGCAT GTGGCTTAATTCGACGCAACGCGAAGAACCTTACCAAGGCTTGACATACACCG GAAAGCATCAGAGATGGTGCCCCCCTTGTGGTCGGTGTACAGGTGGTGCATGG CTGTCGTCAGCTCGTGTCGTGAGATGTTGGGTTAAGTCCCGCAACGAGCGCAA CCCTTGTCCCGTGTTGCCAGCAGGCCCTTGTGGTGCTGGGGACTCACTGGTTTA CCGCCGGGGTCAACTCGGAGGAAGGTGGGGACGACGTCAAGTCATCATGCCC CTTATGTCTTGGGCTGCACACGTGCTACAATGGCCGGTACAATGAGCTGCGAT ACCGCGAGGTGGAGCGAATCTCAAAAAGCCGGTCTCAGTTCGGATTGGGGTC TGCAACTCGACCCCATGAAGTCGGAGTCGCTAGTAATCGCAGATCAGCATTG CTGCGGTGAATACGTTCCCGGGCCTTGTACACACCGCCCGTCACGTCACGAA AGTCGGTAACACCCGAAGCCGGTGGCCCAACCCCTTGTGGGAGGGAGCTG
4	C_7	*Bacillus licheniformis*	MH671869	GAGGGTGATCGGCCACACTGGGACTGAGACACGGCCCAGACTCCTACGGGAGG CAGCAGTAGGGAATCTTCCGCAATGGACGAAAGTCTGACGGAGCAACGCCGCG TGAGTGATGAAGGTTTTCGGATCGTAAAACTCTGTTGTTAGGGAAGAACAAGTA CCGTTCGAATAGGGCGGTACCTTGACGGTACCTAACCAGAAAGCCACGGCTAA CTACGTGCCAGCAGCCGCGGTAATACGTAGGTGGCAAGCGTTGTCCGGAATTAT TGGGCGTAAAGCGCGCGCAGGCGGTTTCTTAAGTCTGATGTGAAAGCCCCCGG CTCAACCGGGGAGGGTCATTGGAAACTGGGGAACTTGAGTGCAGAAGAGGAGA GTGGAATTCCACGTGTAGCGGTGAAATGCGTAGAGATGTGGAGGAACACCAGT GGCGAAGGCGACTCTCTGGTCTGTAACTGACGCTGAGGCGCGAAAGCGTGGG GAGCGAACAGGATTAGATACCCTGGTAGTCCACGCCGTAAACGATGAGTGCTA AGTGTTAGAGGGTTTCCGCCCTTTAGTGCTGCAGCAAACGCATTAAGCACTCC GCCTGGGGAGTACGGTCGCAAGACTGAAACTCAAAGGAATTGACGGGGGCCC GCACAAGCGGTGGAGCATGTGGTTTAATTCGAAGCAACGCGAAGAACCTTAC CAGGTCTTGACATCCTCTGACAACCCTAGAGATAGGGCTTCCCCTTCGGGGGCA

				GAGTGACAGGTGGTGCATGGTTGTCGTCAGCTCGTGTCGTGAGATGTTGGGTTAA GTCCCGCAACGAGCGCAACCCTTGATCTTAGTTGCCAGCATTCAGTTGGGCACTC TAAGGTGACTGCCGGTGACAAACCGGAGGAAGGTGGGGATGACGTCAAATCATC ATGCCCCTTATGACCTGGGCTACACACGTGCTACAATGGGCAGAACAAAGGGCA GCGAAGCCGCGAGGCTAAGCCAATCCCACAAATCTGTTCTCAGTTCGGATCGCA GTCTGCAACTCGACTGCGTGAAGCTGGAATCGCTAGTAATCGCGGATCAGCATG CCGCGGTGAATACGTTCCCGGGCCTTGTACACACCGCCCGTCACACCACGAG
5	C_8	*Bacillus cereus*	MH671874	GTTGGTGAGGTAACGGCTCACCAAGGCAACGATGCGTAGCCGACCTGAGAGGGTG ATCGGCCACACTGGGACTGAGACACGGCCCAGACTCCTACGGGAGGCAGCAGTA GGGAATCTTCCGCAATGGACGAAAGTCTGACGGAGCAACGCCGCGTGAGTGATG AAGGCTTTCGGGTCGTAAAACTCTGTTGTTAGGGAAGAACAAGTGCTAGTTGAA TAAGCTGGCACCTTGACGGTACCTAACCAGAAAGCCACGGCTAACTACGTGCCA GCAGCCGCGGTAATACGTAGGTGGCAAGCGTTATCCGGAATTATTGGGCGTAAA GCGCGCGCAGGTGGTTTCTTAAGTCTGATGTGAAAGCCCACGGCTCAACCGTGG AGGGTCATTGGAAACTGGGAGACTTGAGTGCAGAAGAGGAAAGTGGAATTCCA TGTGTAGCGGTGAAATGCGTAGAGATATGGAGGAACACCAGTGGCGAAGGCGA CTTTCTGGTCTGTAACTGACACTGAGGCGCGAAAGCGTGGGGAGCAAACAGGA TTAGATACCCTGGTAGTCCACGCCGTAAACGATGAGTGCTAAGTGTTAGAGGG TTTCCGCCCTTTAGTGCTGAAGTTAACGCATTAAGCACTCCGCCTGGGGAGTA CGGCCGCAAGGCTGAAACTCAAAGGAATTGACGGGGGCCCGCACAAGCGGTG GAGCATGTGGTTTAATTCGAAGCAACGCGAAGAACCTTACCAGGTCTTGACAT CCTCTGAAAACCCTAGAGATAGGGCTTCTCCTTCGGGAGCAGAGTGACAGGT GGTGCATGGTTGTCGTCAGCTCGTGTCGTGAGATGTTGGGTTAAGTCCCGCAA CGAGCGCAACCCTTGATCTTAGTTGCCATCATTAAGTTGGGCACTCTAAGGTG ACTGCCGGTGACAAACCGGAGGAAGGTGGGGATGACGTCAAATCATCATGCC CCTTATGACCTGGGCTACACACGTGCTACAATGGACGGTACAAAGAGCTGCA AGACCGCGAGGTGGAGCTAATCTCATAAAACCGTTCTCAGTTCGGATTGTAG GCTGCAACTCGCCTACATGAAGCTGGAATCGCTAGTAATCGCGGATCAGCATG CCGC

(*Continued*)

Table 5.5 (Continued)

S. no.	Marking	Bacteria	Accession no.	Sequence
6	C_{10}	*Bacillus lonarensis*	MH671872	GGAATTATTGGGCGTAAAGCGCGCAGGCGGCTTCTTAAGTCTGATGTGAAATC TTGCGGCTCAACCGCAAGCGGCCATTGGAAACTGGGAAGCTTGAGTACAGAAG AGGAGAGTGGAATTCCACGTGTAGCGGTGAAATGCGTAGATATGTGGAGGAAC ACCAGTGGCGAAGGCGACTCTCTGGTCTGTAACTGACGCTGAGGCGCGAAAGC GTGGGGAGCAAACAGGATTAGATACCCTGGTAGTCCACGCCGTAAACGATGAG TGCTAGGTGTTAGGGGTTTCGATGCCCTTAGTGCCGAAGTAAACACATTAAGCA CTCCGCCTGGGGAGTACGACCGCAAGGTTGAAACTCAAAGGAATTGACGGGGA CCCGCACAAGCAGTGGAGCATGTGGTTTAATTCGAAGCAACGCGAAGAACCTT ACCAGGTCTTGACATCCTTTGACCACCCTAGAGATAGGGCTTTCCCTTCGGGGA CAAAGTGACAGGTGGTGCATGGTTGTCGTCAGCTCGTGTCGTGAGATGTTGGGT TAAGTCCCGCAACGAGCGCAACCCTTGATCTTAGTTGCCAGCATTCAGTTGGGC ACTCTAAGGTGACTGCCGGTGACAAACCGGAGGAAGGTGGGGACGACGTCAAA TCATCATGCCCCTTATGACCTGGGCTACACACGTGCTACAATGGATGGTACAAA GGGAAGCGAAACCGCGAGGTGGAGCCAATCCCATAAAGCCATTCTCAGTTCGG ATTGTAGGCTGCAACTCGCCTACATGAAGCCGGAATTGCTAGTAATCGCGGATC AGCATGCCGCGGTGAATACGTTCCCGGGTCTTGTACA

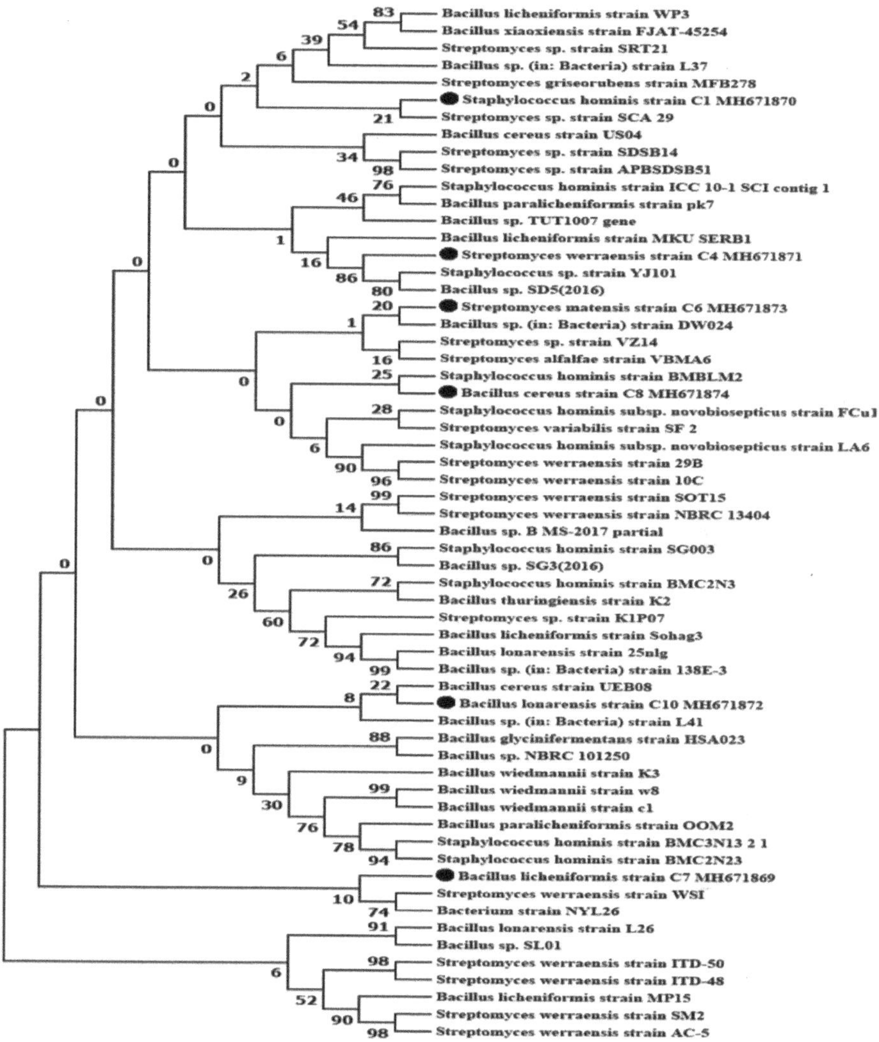

Figure 5.4 Phylogenetic tree of isolated bacteria.

5.4 Molecular characterization of bacterial community

5.4.1 Morphological characters of isolated bacterial colony

Compost collected from the different treatments was subjected to bacterial isolation and various morphological characters of bacterial colony like color, shape, elevation, margin, and surface were observed (Table 5.4).

5.4.2 Molecular characterization of isolated bacterial strains

Determination of microbial presence and diversity using molecular techniques have made the discovery of previously unknown microbes possible. In the present study the isolated bacteria were subjected to DNA analysis and sequencing (Table 5.3) Based on the molecular analysis, the bacterial species identified were *Staphylococcus hominis*, *Streptomyces werraensis*, *Streptomyces matensis*, *Bacillus licheniformis*, *Bacillus cereus*, and *Bacillus lonarensis* (Table 5.5).

5.4.3 Phylogenetic tree of isolated bacteria

The phylogenetic tree showing evolutionary relationship among the isolated bacteria municipal organic waste decomposition (Fig. 5.4).

5.5 Conclusion

Finally, it may be concluded that the composts prepared from the different organic wastes are rich in microorganisms specially bacteria which have potential to degrade MSW for the betterment of proper disposal of solid waste and mitigation of pollution. The present study is very helpful in isolation and characterization of bacteria which may have more and more potential to biodegrade MSW.

References

Alias, A.R., Mohd Mokhlis, N.A., Zainun, N.Y., 2017. Baseline for food waste generation – A case study in Universiti Tun Hussein Onn Malaysia cafeterias. IOP Conference Series: Materials Science and Engineering 271.

Bajsa, O., Nair, J., Mathew, K., Ho, G.E., 2003. Vermiculture as a tool for domestic waste water management. Water Sci. Technol. 48, 25–132.

Barral, M.T., Paradelo, R., 2011. A review on the use of phytotoxicity as a compost quality indicator. Dyna. Soil Dyna. Plant 5 (2), 36–44.

Ben, R.F., Prevost, D., Yezza, A., Tyagi, R.D., 2007. Agro industrial waste materials and waste water sludge for rhizobial inoculant production: a review. Bioresour. Technol. 98, 3535–3546.

Bhat, S.A., Singh, J., Vig, A.P., 2018. Earthworms as organic waste managers and biofertilizer producers. Waste Biom. 9, 1073–1086.

Cooperband, L., 2000. Sustainable use of by-products in landmanagement. In: Bartels, J.M., Dick, W.A. (Eds.), Land Application of Agricultural, Industrial, and Municipal By-Products. SSSA Book, Madison, WI, pp. 215–235. Series Nr 6.

Das, M., Uppal, H.S., Singh, R., Beri, S., Mohan, K.S., Gupta, V.C., et al., 2011. Co-composting of physic nut (*Jatropacurcas*) deoiled cake with rice straw and different animal dung. Bioresour. Technol. 102, 6541–6546.

Dees, P.M., Ghiorse, W.C., 2001. Microbial diversity in hot synthetic compost as revealed by PCR amplified rRNA sequences from cultivated isolates and extracted DNA. Microb. Ecol. 35, 207–2016.

Grigatti, M., Ciavatta, C., Gessa, C., 2004. Evolution of organic matterfrom sewage sludge and garden trimming during composting. Bioresour. Technol. 91, 163–169.

Guerrero, L.A., Maas, G., Hogland, W., 2013. Solid waste management challenges for cities in developing countries. Waste Manag 33 (1), 220–232.

Hoornweg, D., Bhada-Tata, P., Kennedy, C., 2013. Waste production must peak this century. Nature 502, 615–617.

Jain, M.S., Daga, M., Kalamdhad, A.S., 2019. Variation in the key indicators during composting of municipal solid organic wastes. Sustain. Environ. Res. 29 (9), 1–8.

Kalemelawa, F., Nishihara, E., Endo, T., Ahmad, Z., Yeasmin, R., Tenywa, M.M., et al., 2012. An evaluation of aerobic and anaerobic composting of banana peels treated with different inoculums for soil nutrient replenishment. Bioresour. Technol. 126, 375–382.

Kumar, P., Kaushal, R.K., 2015. Avenues of collection and disposal of municipal solid wastes management in India—a review. Int. J. Eng. Sci. Invent. Res. Dev. 1, 458–466.

Larkin, R.P., Tavantzis, S., 2013. Use of biocontrol organisms and compost amendments for improved control of soil borne diseases and increased potato production. Am. J. Potato Res. 90, 261–270.

Minghua, Z.F., Xiumin, A., Rovetta, H., Qichang, F., Vicentini, L., Bingkai, A., et al., 2009. Municipal solid waste management in Pudong New Area, China. Waste Manag 29, 1227–1233.

Peter, H., 2010. Population and environment. Population, waste and chemicals. Am. Assoc. Advancement Sci. Available from: http://atlas.aaas.org/index.php?part = 2&sec = waste > .

Petric, I., Helic, A., Avdic, E.A., 2012. Evolution of process parameters and determination of kinetics for co-composting of fraction of municipal solid waste with poultry manure. Bioresour. Technol. 117, 106–116.

Rameshwar, H.Y., Argaw, A., 2016. Manurial value of khat waste vermicompost from Awday, Harar town, Ethiopia. Int. J. Recycl. Org. Waste Agric. 5, 105–111.

Rashad, F.M., Saleh, W.D., Moselhy, M.A., 2010. Bioconversion of rice straw and certain agro-industrial wastes to amendments for organic farming systems: 1. Composting, quality, stability and maturity indices. Bioresour. Technol. 101, 5952–5960.

Saha, A., 2012. Biochemical and molecular characterization of bacterial consortium of municipal solid wastes, 9.

Scheutz, C., Pedicone, A., Pedersen, G.B., Kjeldsen, P., 2011. Evaluation of respiration in compost landfill biocovers intended for methane oxidation. Waste Manag. 31, 895–902.

Siddiqqui, M.A., Yadav, R.H., 2017. Physicochemical changes during the bioconversion of solid organic wastes using vermitechnology. Asian Pac. J. Reprod. I (Lviii), 65–71.

Singh, A., Sharma, S., 2002. Composting of a crop residue through treatment with microorganisms and subsequent vermicomposting. Bioresour. Technol. 85 (2), 107–111.

Singh, G.K., Gupta, K., Chaudhary, S., 2014. Solid waste management: its sources, collection, transportation and recycling. Int. J. Environ. Sci. Dev. 5 (4), 347–351.

Sujauddin, M., Huda, M.S., Hoque, R., 2008. Household solid waste characteristics and management in Chittagong, Bangladesh. Waste Manag 28, 1688–1695.

Suthar, S., 2008. Bioconversion of post harvest crop residues and cattle shed manure into value-added products using earthworm *Eudrilus eugeniae* Kinberg. Ecol. Eng 3 (2), 206–214.

Tognetti, C., Laos, F., Mazzarino, M.J., Hernandez, M.T., 2005. Composting vs. vermicomposting: a comparison of end product quality. Compost. Sci. Util. 1, 6–13.

Upadhyay, V.P., 2005. Eco tools for urban waste management in India. J. Hum. Ecol. 18 (4), 253–269.

Verma, R., 2010. Composting and vermicomposting, a simple way. Wetlands, Biodivers. Clim. Change 1–14.

Yadav, A., Garg, V.K., 2011. Recycling of organic wastes by employing *Eisenia fetida*. Bioresour. Technol. 102, 2874–2880.

Zajonc, O., Frydrych, J., Jezerska, L., 2014. Pelletization of compost for energy utilization. In: 2014 International Conference on Agricultural and Biosystems Engineering, IERI Procedia 8:2-10.

Zaved, H.K., Rahman, M., Rahman, M.M., Rahman, A., Arafat, S.M.Y., Rahman, M.S., 2008. Isolation and characterization of effective bacteria for solid waste degradation for organic manure. Sci. Technol. J. 8 (2), 44–55.

Zhang, R., El-Mashad, H.M., Hartman, K., Wang, F., Liu, G., Choate, C., et al., 2007. Characterization of food waste as feedstock for anaerobic digestion. Bioresour. Technol. 98, 929–935.

Zhang, L., Sun, X.Y., Tian, Y., Gong, X.X.Q., 2013. Effects of brown sugar and calcium superphosphate on the secondary fermentation of green waste. Bioresour. Technol. 131, 68–75.

Molecular technologies for the early detection of fungal phytopathogens associated with cereal crops

6

N. Deepa[1], *Charith Raj Adkar-Purushothama*[2] *and M.Y. Sreenivasa*[1]
[1]Department of Studies in Microbiology, University of Mysore, Mysore, India, [2]Pavilion for Applied Cancer Research, Department of Biochemistry, Faculty of Medicine and Health Sciences, Sherbrooke University, Sherbrooke, QC, Canada

6.1 Introduction

Phytopathogens are present in all the cereal growing regions of the world, and it not only affects the growth and seedling survival but also reduces the quality and production of cereals (Brown et al., 2011; Weiss et al., 1987). It is estimated that the effect of phytopathogens on cereals leads to great economic loss up to hundred billion dollars every year. Nearly 70%–80% of crop diseases are by fungal phytopathogens causing significant reduction in yield and quality of cereal crops, fruits and vegetables. Association of fungal phytopathogens with cereals and due to secretion of variety of metabolites and toxins they are harmful to humans and animals posing threat to agricultural products as become unfit for human consumption (Christensen and Kaufmann, 1969; Goyal and Prasal, 2010). Cereals contaminating fungi are broadly classified into field fungi and storage fungi, where *Fusarium, Curvularia, Alternaria,* and *Cladosporium* being field fungi whereas *Penicillium* and *Aspergillus* form to be storage fungi (Deepa et al., 2016a) Among a total of 130 maize and sorghum samples ten diversified species of *Fusarium* and *Aspergillus* were isolated and similarly among 80 maize samples collected throughout Karnataka screened high frequency of (96.5%) *Fusarium* and (41.7%) *Aspergillus* species including species of *Penicillium, Drechslera, Nigrospora, Curvularia, Alternaria, Chaetomium,* and *Phomopsis* were isolated (Sreenivasa et al., 2011a,b) (Table 6.1).

Rice being the primary crop for world's half of the world's population has been widely associated with *Pyricularia oryzae* causing a disease rice blast which leads to 10%–30% of loss yearly (Talbot, 2003). Postharvest phytopathogenic fungi study was conducted on 44 sorghum samples recording 93.2% of *Fusarium* species and 88.6% of *Aspergillus* species (Sreenivasa et al., 2010). Wheat being another important cereal crop grows more than 240 million hectares globally and yields 729 million tons approximately by 30 million farmers and causes 12%–20% of loss due to occurrence of fungal phytopathogens (FAO of United nations, 2016). Association of *Fusarium*

Microbes in Land Use Change Management. DOI: https://doi.org/10.1016/B978-0-12-824448-7.00009-7
© 2021 Elsevier Inc. All rights reserved.

Table 6.1 Phytopathogenic fungi and the commodities they affect.

Cereal crops	Fungal phytopathogens
Maize (*Zea mays*)	*Aspergillus flavus, Fusarium moniliforme, Fusarium semitectum, Nigrospora, Cuvularia, Lasiodiplodia, Bipolaris, Arthrinium, Rhizopus, Phoma, Rhizoctonia, Chaetomium, Penicillium citrinum, Penicillium funiculosum, Aspergillus wentii*
Paddy (*Oryza sativa*)	*Bipolaris maydis, F. semitectum, Cladosporium cladosporioides, Nigrospora oryzae, Curvularia lunata, Curvularia genticulatus, Caulophilus oryzae, Curvularia eragrostidis, Curvularia pallescens, Phoma* sp., *Colletotrichum* species, *Aspergillus* species, *Penicillium* species, *Altenaria padwickii, Altenaria alternata, Altenaria longissimi, Fusarium verticillioides*
Sorghum (*Sorghum bicolor*)	*B. maydis, F. semitectum, C. cladosporioides, N. oryzae, C. lunata, C. pallescens, Phoma* species, *Setosphaeria rostrata, Aspergillus niger, Eurotium chevalieri, Eurotium rubrum, Chaetomium* species, *A. flavus, F. moniliforme, P. citrinum, A. longissimi, A. alternata*
Peanuts	*C. cladosporioides, Lasiodiplodia theobromae, Pestalotiopsis guepinii, A. niger, Penicillium pinophilum, Chaetomium* species, *A. flavus*
Soyabean	*Arthrinium phaeospermum, L. theobromae, F. semitectum, C. cladosporioides, N. oryzae, C. lunata, C. pallescens, Phoma* species, *Epicoccum nigrum, P. guepinii, A. niger, A. wentii, Aspergillus restrictus, Aspergillus penicillioides, E. rubrum, Eupen cinnamopurpureum, Chaetomium* species, *A. flavus, F. moniliforme, P. citrinum, A. alternata*
Barley (*Hordeum vulgare*)	*Penicilium species, Rhizopus species, Stemphylium species, Trichoderma viride, Epococcum purpurascens, Oedocephalum species*
Oats (*Avena sativa*)	*Septoria avenae, F. verticillioides, Rhizopus species, Penicillium species*
Pearl millet (*Pennisetum glauccum*)	*Aspergillus species, Penicillium species, Eurotium species, Alternaria species, F. verticillioides, Cladosporium species, Curvularia species, Drechslera species, Rhizopus species, Stachybotrys chartarum*
Rye (*Secale cereale*)	*Claviceps purpurea, F. verticillioides, Rhizopus species*
Wheat (*Triticum* spp.)	*Alternaria triticina, Glomerella graminicola, Cladosporium species, Cochliobolus sativus, Gibberela avenacea, C. purpurea*

verticillioides pathogen in sorghum and maize leads to worldwide economic losses (Deepa and Sreenivasa, 2017).

Fungal identification has become the cornerstone of plant pathology (Borman et al., 2008). Conventional methods like morphological features, microscopic

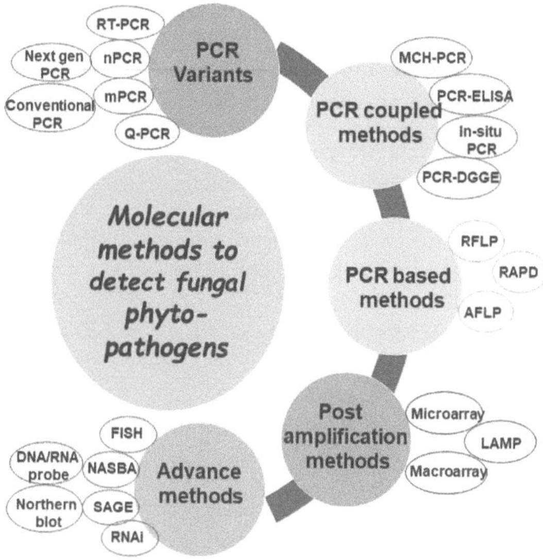

Figure 6.1 Chart representing various methods for detection of fungal phytopathogens.

features, and cross fertility are not accurate and also noted to be difficult on the following traditional analytical methods in identifying several other species of organisms as it requires the physiology and taxonomy expatriation which is of labor intensive and time consuming (Leslie and Summerell, 2006). Certain advanced fungal disease detection and diagnosis techniques are available such as polymerase chain reaction (PCR), nested PCR, real-time PCR, quantitative PCR (qPCR), immunocapture PCR, magnetic capture-hybridization PCR (MCH-PCR), PCR restriction fragment length polymorphism (PCR-RFLP), in situ PCR, PCR-DGGE, cooperational PCR, droplet digital PCR, loop-mediated isothermal amplification (LAMP), multiplex PCR, fluorescence in situ hybridization (FISH), reverse transcriptase PCR (RT-PCR), and DNA microarrays (Tsui et al., 2011; Kumar et al., 2016). Recently, a study was conducted on advanced molecular techniques such as RFLP and rapid amplified polymorphic DNA (RAPD) for strain level variations in *F. verticillioides* association with cereals and LCMS for production of fumonisin confirmation among toxigenic strains (Deepa et al., 2018). The present chapter is a study on molecular technologies that combat the early detection of fungal phytopathogens that are in association with cereal crops (Fig. 6.1).

6.2 Molecular techniques for the detection of fungal phytopathogens

A number of techniques are in routine use for the detection of fungal phytopathogens to encounter the industry's need for accurate and quick results but don't meet the conventional methods (Deepa and Sreenivasa, 2019) (Table 6.2).

Table 6.2 Precedence and hindrance of molecular methods for the early detection of phytopathogenic fungi.

Methods	Precedence	Hindrance
Nested PCR	Accurate, specificity, sensitivity, amplification with less concentration	Risk of contamination is more
Multiplex PCR	Accurate, cost effective, time saver, more results with less sample	Reduce sensitivity
Real-time PCR	Time efficient, cost effective, sensitivity, specificity	Expertization is required, instrument is costly
Reverse transcriptase PCR	Obtains quantitative and qualitative data, simple, rapid, cost effective, agarose gel is not required, specificity, sensitivity	Higher expertise is required, contamination is more, extremely sensitive
Conventional PCR	Rapid and precise results with species-specific primers	Costly labor required
Next generation PCR	Quick, informative data, cost effective, simple, high yield	Low yield of high-quality sequences, expensive
MCH-PCR	Purifies and concentrates DNA	Laborious
PCR-ELISA	Simpler by combining two methods, electrophoresis not required	Time consuming
In situ hybridization	In demand due to light microscopy, short tissues can be maximized use	Time consuming, difficult with low number of copies of DNA/RNA
PCR-DGGE	Studies on diversity	Time consuming, poorly reproducible
RFLP	Quick, simple, accurate in identification of population, stable, reproducible	High labor requirements, costly, requires radio isotope
RAPD	Quick, simple, efficient, probes not required	Reproducibility is not feasible, mismatch occurs more
AFLP	High reproducibility, different part of genome can be analyzed	Complex and lengthy protocol
Microarray	Bulk analysis, does not require large amount of DNA sample	Requires more amount of mRNA
Macroarray	Quick, easy, data for thousands of genes	Expensive, time consuming
LAMP	Rapid, sensitive, highly specific	Great risk of contamination

(*Continued*)

Table 6.2 (Continued)

Methods	Precedence	Hindrance
Northern blotting	Detects the size	Applied to small samples
SAGE	Precise knowledge is not required	Requires specificity
FISH	Quick, rapid characterization	Difficult
NASBA	Sensitive, isothermal	Temperature should be below 42°C
RNA interference	Efficient for high tissue samples	Specificity of reagents is required

AFLP, Amplified fragment length polymorphism; *FISH*, fluorescence in situ hybridization; *LAMP*, loop-mediated isothermal amplification; *PCR*, polymerase chain reaction; *RFLP*, restriction fragment length polymorphism.

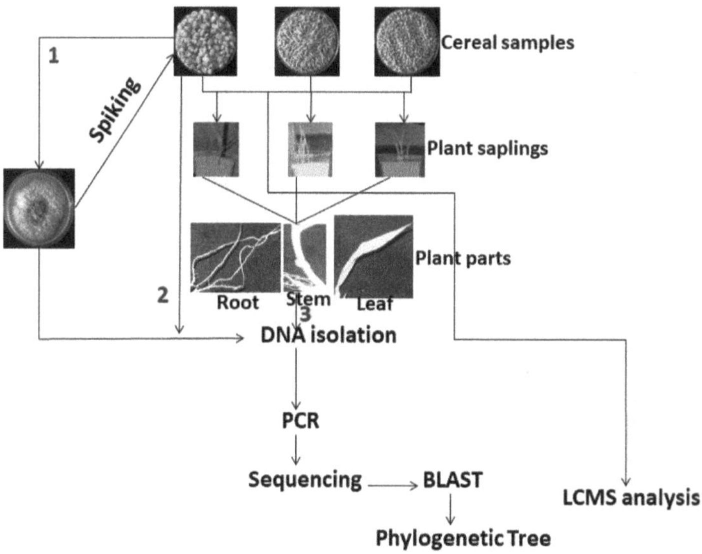

Figure 6.2 Representation to check the specificity and sensitivity of the fungal phytopathogen by molecular method in pure culture, cereal sample, and plant parts.

The PCR has been used in many epidemiological analysis-like fungal taxonomical studies and also been applied for the fungal pathogens' detection in plant tissues (Jurado et al., 2006; Baird et al., 2008; Chandra et al., 2010). For the first time nested PCR and multiplex PCR kit were developed for the early detection of fumonisin producing *F. verticillioides* in pure cultures and its sensitivity was tested among cereal samples and its plant parts directly from root, stem, and leaves (Deepa et al., 2016b,c). Recently molecular methods with variants of PCR are in use to check the specificity and sensitivity of the fungal phytopathogen in culture, directly from cereal samples and plant parts (Fig. 6.2).

6.3 Variants of polymerase chain reaction

6.3.1 *Conventional polymerase chain reaction*

Plant pathology field has combated great revolution by inviting PCR and its methods which synthesizes million copies of specific DNA using specific primers. Earlier PCR was the method mainly used for detecting the disease-causing bacteria and viruses but in recent days it is broadly used for the detection of plant pathogens. Many methods like use of cetyltrimethyl ammonium bromide, phenol – chloroform, silica matrix purification and many more are used for phytopathogens (Fang and Ramasamy, 2015; Mancini et al., 2016). Nearly 135 cereal samples from different districts of Karnataka were screened with conventional PCR among which 69 samples were found to be associated with phytopathogenic *Fusarium* infection and 51 samples were associated with *F. verticillioides* (Deepa et al., 2016a) In 1993 leaf spot disease was recorded among plantain and banana caused by *Mycosphaerella fijiensis* and *Mycosphaerella musicola* was detected by PCR technology (Henson and French, 1993). Cereal samples collected from different districts of Karnataka was screened with PCR method which reported the association of 103 *Fusarium* isolates from maize, 64 phytopathogenic *Fusarium* species were found to be associated with sorghum samples and 64 isolates were infected with fumonisin producing *F. verticillioides* (Sreenivasa et al., 2007, 2008). This method requires specific primers for DNA replication process but nowadays for accurate and specific results it requires DNA probe and nested primers. PCR method enables phytopathogens *Zymoseptoria tritici* and *Blumeria graminis* detection and identification even in a latent period without symptoms (Fones and Gurr, 2015; Zeng et al., 2010). The method gives rapid and precise results with use of specific primers but requires much labor and is cost effective.

6.3.2 *Nested polymerase chain reaction*

nPCR involves two consecutive cycles in which single primer is used to DNA amplification, then the same amplified DNA sequence acts as target DNA for second round PCR using two internal primers. Investigations reported *Fusarium solani* in soyabean roots with single or combination of FSg1/FSg2 and FSgEF1/FSgEF2 primers (Li and Haitman, 2003) in ocular samples using Fusofor/Fusorev primers (Emma et al., 2000), *Fusarium oxysporum* in artificial and natural soil using WiltNF-1/WiltNR-1 primers (Garcia et al., 1999), in basil (Chiocchetti et al., 2001) and *Fusarium niveum* in infected plant tissues and soil using Fn-1/Fn-2 primers (Zhenggang et al., 2005). One forward and two reverse primers VERTF-1/VERTR and VERTF-2 respectively was applied in detection of fumonisin producing *F. verticillioides* associated with cereal samples using nested PCR (Deepa et al., 2016b). This method uses two sets of primers which increase the yield and specificity of target DNA amplification but risk of contamination is more due to two cycles of amplification (Rahman et al., 2013).

6.3.3 Multiplex polymerase chain reaction

The method is widely acceptable and widely been use in plant pathology when plant is been infected with more than one pathogen. The mPCR involves more than one species-specific primer in single PCR tube, further amplified product is gel electrophoresed. Fumonisin producing *F. verticillioides* was specifically detected by mPCR in pure cultures, cereal samples and plant materials using one forward and two reverse primers (Deepa et al., 2016c). Four phytopathogens *Septoria tritici, Stagonospora nodorum, Puccinia striiformis, Puccinia recondite* was detected by mPCR in wheat (Fraaije et al., 2001). Chen et al. (2014) developed single step mPCR reaction for detection of three fungal phytopathogens *Puccinia gramini, Puccinia triticia, B. graminis* causing stem rust, leaf rust and powdery mildew respectively in wheat. mPCR was developed to detect trichothecene (tri5, tri6) and fumonisin (FUM1, FUM13) from *Fusarium* species associated with rice and finger millet from southern parts of India (Ramana et al., 2011). *F. verticillioides* and *Fusarium subglutinans* was simultaneously detected by mPCR in Brazil by developing a gaoB gene coding for galactose oxidase in infected maize samples (Faria et al., 2012). This method can be time and money saver using several pair of primers in single reaction followed but interference of primer and probes reduce sensitivity.

6.3.4 Real-time polymerase chain reaction (Q-PCR)

Real-time PCR detects amplicons through specific fluorescent signal eliminating postamplification steps. It is a versatile technique for accurate, sensitive and quantifying of template DNA in detection of phytopathogenic fungi. The method is more sensitive and specific than normal PCR. It identifies symptomless infections with higher sensitivity than conventional PCR (Ginzinger, 2002; Schaad and Frederick, 2002). Real-time PCR detects as less as 300 fg of the DNA. Q-PCR method applied in detecting the fungal phytopathogens was grouped into sequence specific and nonspecific method (Schena et al., 2004). This method is mainly used for quantification and also identification of disease-causing fungi like *Alternaria verscicoli, Cladosporium cladosporoiodes, Stachybotys chartarum,* and *Alternaria alternata* (Black, 2009). This method is automated and no need of post amplification analysis but it is cost effective and complexity due to both thermal cycling and fluorescence detection at a time.

6.3.5 Reverse transcriptase polymerase chain reaction

The RT-PCR is a diagnostic technique that detects target mRNA. The method is primarily used in detecting phytopathogens like *Mycosphaeralla graminicola* and *Oidium neolycopersici* in wheat and tomato respectively (Guo et al., 2005; Matsuda et al., 2005). RT-PCR used to quantify *Fusarium graminearum* causing the disease *Fusarium* ear blight in cereals like wheat, rye, barley, oat and maize (Brown et al., 2011). In the method, RNA is reverse transcribed with reverse transcriptase enzyme in production of cDNA, later cDNA is amplified either by conventional or other method. The method gives quantitative data than conventional PCR and it is the most recurrently claimed technique applied in phytopathology for analyzing fungal gene expression during disease development but is time consuming and requires expensive equipment and reagents and the other major disadvantage is that it is unable to distinguish living or dead fungi.

6.3.6 Next generation polymerase chain reaction

Next generation RNA sequencing was advanced in *F. verticillioides* to study the transcriptional changes associated with resistant C0441 and susceptible C0354 maize genotypes. They found 6951 differentially expressed genes (Lanubile et al., 2014). Ju et al. (2017) reported eight quantitative trait loci and 43 genes associated with 57 SNP correlated with *F. verticillioides* stalk rot resistance through linkage mapping and genome wide association analysis.

6.4 Polymerase chain reaction coupled methods

6.4.1 Magnetic capture-hybridization polymerase chain reaction

MCH-PCR is a novel technology combining with both extraction and purification of DNA involving hybridization with ssDNA probe on magnetic beads. *Nectria galligena* was detected from fruits and cereal plants (Langrell and Barbara, 2001). Botrytis root rot disease-causing fungi *Botrytis aclada* in onion was detected by this method (Walcott et al., 2004). MCH-PCR purifies and concentrates DNA simultaneously by removing nontarget DNA that inhibits amplification reaction.

6.4.2 Polymerase chain reaction enzyme-linked immune sorbent assay

The method combines both PCR and enzyme-linked immune sorbent assay (ELISA) technique into a single method in which method is similar to ELISA which letsthe recognition of nuclic acids instead of proteins (Shamloul and Hadidi, 1999). It does not require electrophoretic separation or dot hybridization but can be easily automated. Phytopathogens like *Phytophthera* and *Pythium* species was detected by this method (Bailey et al., 2002). mPCR assay was developed to discriminate *Fusarium* species between toxigenic and nontoxigenic strains with FUM1, FUM8 and FUM21 gene (Divakara et al., 2014). Simultaneously, fum producing *F. verticillioides* isolates from mPCR assay were also positive using competitive direct ELISA.

6.4.3 In situ polymerase chain reaction

This method is a combination of PCR and in situ hybridization to amplify specific genes of cells/tissues and further amplified tissues will be stained (Long, 1998). In situ PCR is mainly used to identify spores and mycelia of *B. graminis* on Barley (Bindslev et al., 2002). It is a time consuming method due to hybridization step and also in demand due to light microscopy. Short tissues can be used for maximum but it is difficult in identifying target DNA/RNA with low number of copies.

6.4.4 Polymerase chain reaction-DGGE (denaturing gradient gel electrophoresis)

The PCR-Denaturing Gradient Gel Electrophoresis is the molecular technique in which phytopathogens target DNA will be amplified initially using PCR and was furter subjected to denaturing electrophoresis. PCR-DGGE is a technique recently applied to detect phytophthora from plants and environment samples (Rytkönen et al., 2011). Spores of *Trichoderma harzianum* and *Arthrobotrys oligospora* was detected from plant samples applying PCR-DGGE method for 14 days and 2 months respectively. The method is time consuming, very poorly reproducible and fragments with varied sequences but having similar melting behavior were not correctly separated. Hence, it is mainly used in diversity study and detection of pathogenic fungi.

6.5 Polymerase chain reaction-based methods

6.5.1 Polymerase chain reaction restriction fragment length polymorphism

Combination of amplification of target region and later restriction enzyme digestion of the PCR products. The method differentiates pathogenic and nonpathogenic strains of *Pythium myriotylum* (Gómez-Alpizar et al., 2011). RFLP was conducted to study the genetic variation with six different enzymes (*Hae* III, *Hinf* I, *Kpn* I, *Hind* III, *EcoR* I, and *Xho* I) for the 10 strains of *F. verticillioides* isolated from cereal samples recorded identical pattern with two *Hae* III and *Hinf* I endonucleases whereas other four enzymes reported strain level differentiation (Deepa et al., 2018). PCR-RFLP was conducted to detect and discriminate variations among 33 fumonisin producing *F. verticillioides* isolates irrespective of their origin and host (Patino et al., 2006). Visentin et al. (2012) described the ITS-RFLP discrimination between *F. verticillioides* and *Fusarium proliferatum*. *Fusarium* species DNA with mating populations and the telomorphs of *F. Verticillioide* (A), *F. proliferatum* (D) and *F. Subglutinans* (E), respectively was analyzed by RFLP (Schlacht et al., 1997).

6.5.2 Polymerase chain reaction amplified fragment length polymorphism

Amplified fragment length polymorphism (AFLP) analysis indicated the presence of genetic diversity not only between Italian strains and the American reference but also among the Italian isolates (Covarelli et al., 2012). A wide range of pathogenicity in terms of effects on seed germination, seedling development and symptoms produced on detached leaves were not correlated with the different fumonisin production. High genetic variability was investigated using AFLP from the 3840 maize kernels tested in southern Brazil and they reported that essential genes FUM1 and FUM8 for fumonisin biosynthesis showed that 98% were *F. verticillioides* species (Silva et al., 2017).

6.5.3 Polymerase chain reaction random amplified polymorphic DNA

RAPD reports the intense method to identify the genotypic differences among the fungi using short oligodeoxynucleotide primers (Williams et al., 1990). RAPD was mainly used to correlate taxa-based differences using their morphology, physiology and toxin production to distinguish fungal species and the pathogenicity variations within the species (Zhang et al., 1997; Bayman and Cotty, 1993; Jones and Dunkle, 1993). RAPD method was conducted to study the genetic variation among the ten strains of *F. verticillioides* isolated from different cereal samples using three RAPD primers (R1, R14, R16). The method reported *F. verticillioides* strains FVM86, FVP200, FVP19 and FVP31, FVS3, FVS36 representing same banding pattern for R1 primer, *F. verticillioides* strains FVP19, FVP31 and FVP31, FVS3resembled banding pattern similar to R14, and *F. verticillioides* strains FVM86, FVM146, FVP19 recorded similar banding pattern as R16 and remaining isolates differed from each other (Deepa et al., 2018). RAPD was not sensitive enough to detect fumonisin producing fungi, hence specific primers from known sequences of genes producing fumonisin were investigated and were used to compare the DNA-based strategies for identification of fumonisin producing and nonproducing strains (Waalwijk et al., 2004; Gonzalez-Jaen et al., 2004). Moeller et al. (1999) reported *F. verticillioides* and *F. subglutinans* identification based on RAPD sequences of fragments that were consecutively used for the infected maize kernelsanalysis. Although RAPD has been used as a diagnostic tool in many laboratories for the identification of *Fusarium* species (Blackwell et al., 1999). RAPD is not considered reliable because of the lack of reproducibility of the technique.

6.6 Postamplification techniques

6.6.1 Microarray

Microarray is a DNA chip technology which allows the analysis of thousands of mRNAs simultaneously. It is also used to observe gene expression. Thousands of DNA probes are arrayed with a small surface area onto a support matrix that includes glass chip or nylon filters. It is mainly used for the identification of *Aspergillus* and *Candida* species (Singh and Kumar, 2013). The method is easy to use because it does not require large scale DNA sequences but it requires large amount of mRNA. *F. verticillioides* maize photosystem reported the resistant maize compared to susceptible line with higher enzymatic antioxidant levels before infection with basic defense against the fungus (Lanubile et al., 2012).

6.6.2 Macroarray

It is also known as DNA hybridization technique or reverse dot plot method (Singh and Kumar, 2013). This method is a molecular tool for diagnostic and epidemiological studies. It is primarily used for the identification of *A. alternata, Aspergillus fumigatus, F. solani, Candida albicans, Cladosporium herbarum* (Sato et al., 2010).

6.6.3 Loop-mediated isothermal amplification

The method is novel nucleic acid amplification mainly considered as an alternative to the PCR. It amplifies the target nucleic acid under isothermal conditions with high specificity. The method does not require thermal cycler instead it requires single temperature for DNA amplification for production of temperature changes (Tsui et al., 2011). It mainly requires DNA polymerase and four set of primers with two inner and two outer primers recognizing six unique sequences on target DNA (Fakruddin, 2011). Assochyta blight disease caused by *Ascochyta rabiei* in chicken peas, *Penicillium marneffei* and recently *Ophiostoma clavatum* blue stain fungus was diagnosed by LAMP method respectively (Das et al., 2012; Sun et al., 2010; Villari et al., 2013). The method is also used to identify the pathogenic fungus such as *Ochroconis gallopava* (Tsui et al., 2011). It is a rapid, sensitive and highly specific but designing of primer is complex and recognizes one specific pathogen at a time. The method is also with a great risk of contamination

6.7 Advanced molecular methods

6.7.1 DNA/RNA probe-based methods

The method uses probe for diagnosis of plant disease caused by microbes with great speed and sensitivity. It is pains taking as backbone to the current knowledge where probe is used for analyzing nucleic acid without its amplification. In earlier days probes were applied as substitute to PCR for fungal identification. In recent days they are used in conjunction with PCR (McCartney et al., 2003). Sandhu et al. (1995) developed 21 specific nucleic acid probes targeting large subunit rRNA genes from organisms to detect 100 fungi among which nearly 50 fungal phytopathogens like *Aspergillus* species, *Candida* species, and *Helminthosporium* species.

6.7.2 Northern blotting

Northern blot technique is also known as RNA blot, mainly used to transfer the RNA onto a carrier to identify the pathogenic fungi. RNA from each tissue is purified and examined later the same is loaded on agarose gel that migrates RNA towards bottom of gel where RNA moves faster compared to larger RNA (Kim et al., 2010). *Magnaporthe grisea* is identified in rice using combination of PCR and northern blotting (Qi and Yang, 2002). It mainly used to detect size of RNA but can be applied only on small sample of the genes.

6.7.3 Serial analysis of gene expression

Serial analysis of gene expression (SAGE) is a sequence-based and comprehensive method for the quantitative gene expression profiling that allows the identification of multiple transcripts. It is a method applied to detect the fungal phytopathogen *B. graminis* on barley. In this method acquaintance of the particular genome is not required but requires specificity of tag sequence is its limitation.

6.7.4 Fluorescence in situ hybridization

FISH is a molecular technique mainly applied for quick characterization of microorganism such as fungal phytopathogens. It is with greater speed, resolution and safety for the development of multiple targets detection and analysis (Tsui et al., 2011). In this method, Fluorescent probe is used that binds to the part of chromosome with which it exhibits greater degree of homology (Baschien et al., 2001). Many numbers of probes such as locus-specific probe, centromeric repeat probe, and whole chromosome probe is in use (Volpi and Bridger, 2008). The method is mainly used in detection of *Aureobaisdium pullulans* fungi, *Euotiomycetes*, dotheideomycetes, aquatic hyphomycetes fungus (Baschien et al., 2008; Baker et al., 2004; Tsui et al., 2011). It is mainly used for nondividing cells, even though the probe preparation method is very difficult, it is necessary to identify the particular DNA sequence.

6.7.5 NASBA (nucleic acid sequence-based amplification)

Nucleic acid sequence-based amplification is a very sensitive, isothermal and transcription-based amplification designed for detection of RNA. This method is not applied only in basic research but is also applied in application-oriented fields like clinical medicine development and infectious disease diagnostic diagnosis (Sergentet-Thevenot et al., 2008). Many numbers of methods like PCR, LAMP, RCA (rolling circle replication) does not amplify RNA directly with high sensitivity (Saiki et al., 1992; Lizardi et al., 1998; Lee et al., 2009). Pan-*Candida* and Pan-*Aspergillus* was detected using 16S and 28S rRNA targets respectively with 100% specificity (Zhao et al., 2009).

6.7.6 RNA interference

RNAi is a latest technology that identifies and controls the phytopathogenic fungi. It is also an effective technique highly applied for tissue−specific to boutthe mycotoxigenic fungi causing crop plants infection (Panwar et al., 2012). The method is used in *Zea mays* for identifying *Aspergillus flavus* and *F. graminearum* in *Hordeum vulgare* (Masanga et al., 2015; Chen et al., 2006).

6.8 Conclusion

Phytopathogens growth may encourage due to poor agricultural practices which increases the risk of toxin production in cereal crops that enter food chain. Even though many detection techniques are user friendly with specificity, sensitivity, robustness and cost effective is in practise, phytopathogens attack is not minimized. Hence there is need for the accurate and rapid detection and identification of the pathogen. Presently many numbers of diagnostic laboratories have been using many molecular techniques for the identification and detection of the phytopathogenic disease caused by them. Current molecular technologies like DNA barcoding, RNA-sequence-based next generation sequence, SAGE, FISH, helping to overcome the challenges by developing novel disease control strategies. The chapter explains the recent developments for the early detection of fungal phytopathogens in association with cereals.

References

Bailey, A.M., Mitchell, D.J., Manjunath, K.L., Nolasco, G., Niblett, C.L., 2002. Identification to the species level of the plant pathogens *Phytophthora* and *Pythium* by using unique sequences of the ITS1 region of ribosomal DNA as capture probes for PCR Elisa. FEMS Microbiol. Lett. 207, 153–158.

Baird, R., Abbas, H.K., Windham, G., Williams, P., Baird, S., Ma, P., et al., 2008. Identification of select fumonisin forming *Fusarium* species using PCR applications of the polyketide synthase gene and its relationship to fumonisin production *in vitro*. Int. J. Mol. Sci. 9, 554–570.

Baker, B.J., Lutz, M.A., Dawson, S.C., Bond, P.L., Banfield, J.F., 2004. Metabolically active eukaryotic communities in extremely acidic mine drainage. Appl. Environ. Microbiol 70, 6264–6271.

Baschien, C., Manz, W., Neu, T.R., Szewzyk, U., 2001. Fluorescence in situ hybridization of freshwater fungi. Int. Rev. Hydrobiol. 86, 371–381.

Baschien, C., Manz, W., Neu, T.R., Marvanova, L., Szewzyk, U., 2008. In situ detection of freshwater fungi in an alpine stream by new taxon-specific fluorescence in situ hybridization probes. Appl. Environ. Microbiol 74, 6427–6436.

Bayman, P., Cotty, P.J., 1993. Genetic diversity in *Aspergillus flavus*: association with aflatoxin production and morphology. Can. J. Bot 71, 23–34.

Bindslev, L., Oliver, R.P., Johansen, B., 2002. *In situ* PCR for detection and identification of fungal species. Mycol. Res. 106, 277–279.

Black, J., 2009. Quantitative Real-Time Polymerase Chain Reaction (qPCR) of Filamentous Fungi in Carpet. RTI Press, Research Triangle Park, NC, pp. 1–12.

Blackwell, B.A., Gilliam, J.T., Savardm, M.E., Miller, D., Duvick, J.P., 1999. Oxidative deamination of hydrolysed fumonisin B1 (AP1) by cultures of *Exophiala spinifera*. Nat. Toxins 7, 31–38.

Borman, A.M., Linton, C.J., Miles, S.J., Johnson, E.M., 2008. Molecular identification of pathogenic fungi. J. Antimicrob. Chemother 61, 7–12.

Brown, N.A., Bass, C., Baldwin, T.K., Chen, H., Massot, F., Carion, P.W., et al., 2011. Characterisation of the *Fusarium graminearum* in wheat and floral inter-action. J. Pathog.

Chandra, N.S., Udaya Shankar, A.C., Niranjana, S.R., Wulff, E.G., Mortensen, C.N., Prakash, H.S., 2010. Detection and quantification of fumonisins from *Fusarium verticillioides* in maize grown in southern India. World J. Microbiol. Biotechnol. 26, 71–78.

Chen, X., Steed, A., Harden, C., Nicholson, P., 2006. Characterisation of *Arabidopsis thaliana–Fusarium graminearum* interactions and identification of variation in resistance among ecotypes. Mol. Plant Pathol. 7, 391–403.

Chapter 4: Food and Agriculture systems in climate change mitigation. 2016. The state of Food and Agriculture. 2016. 2016. Food and Agricultural Organization of the United Nations. Rome.

Chen, S., Cao, Y., Li, T., Wu, X., 2014. Simultaneous detection of three wheat pathogenic fungal species by multiplex PCR. Phytoparasitica 43, 449–460.

Chiocchetti, A., Sciaudone, L., Durando, F., Garibaldi, A., Migheli, Q., 2001. PCR detection of *Fusarium oxysporum* f. sp. basilici on basil. Plant Dis. 85, 607–611.

Christensen, C.M., Kaufmann, H.H., 1969. Grain Storage: The Role of Fungi in Quality Loss. University of Minnesota Press, Minneapolis, MN, p. 172.

Covarelli, L., Stifano, S., Beccari, G., Raggi, L., Lattanzio, V.M.T., Albertini, E., 2012. Characterization of *Fusarium verticillioides* strains isolated from maize in Italy: Fumonisin production, pathogenecity and genetic variability. Food Microbiol 31, 17–24.

Das, A., Babiuk, S., McIntosh, M.T., 2012. Development of a loop-mediated isothermal amplification assay for rapid detection of capripoxviruses. J. Clin. Microbiol 50, 1613–1620.

Deepa, N., Sreenivasa, M.Y., 2017. *Fusarium verticillioides*, a globally important pathogen of agriculture and livestock: a review. J. Vet. Med. Res. 4, 1084–1092.

Deepa, N., Sreenivasa, M.Y., 2019. Molecular methods and key genes targeted for the detection of fumonisin producing *Fusarium verticillioides* – an updated review. Food Biosci. 32, 1–7.

Deepa, N., Nagaraja, H., Sreenivasa, M.Y., 2016a. Prevalence of fumonisin producing *Fusarium verticillioides* associated with cereals grown in Karnataka (India). Food Sci. Hum. Well. 5 (3), 156–158.

Deepa, N., Adkar-Purushothama, C.R., Sreenivasa, M.Y., 2016b. Nested PCR method for early detection of fumonisin producing *Fusarium verticillioides* in pure cultures, cereal samples and plant parts. Food Biotechnol. 30, 18–29.

Deepa, N., Adkar-Purushothama, C.R., Sreenivasa, M.Y., 2016c. Multiplex PCR for the early detection of fumonisin producing *Fusarium verticillioides*. Food Biosci. 13, 84–88.

Deepa, N., Rakesh, S., Sreenivasa, M.Y., 2018. Morphological, pathological and mycotoxicological variations among *Fusarium verticillioides* isolated from cereals. 3 Biotech. 8, 105–114.

Divakara, S.T., Santosh, P., Aiyaz, M., Ramana, M.V., Hariprasad, P., Nayaka, S.C., et al., 2014. Molecular identification and characterization of *Fusarium* species associated with sorghum seeds. J. Sci. Food Agric 91, 1666–1673.

Emma, E.M.J., Nora, M.C., Sarah, C., Anthony, A.S.D., Hamish, M.A.T., Melville, M.M., et al., 2000. Rapid detection and identification of *Candida*, *Aspergillus* and *Fusarium* species in ocular samples using nested PCR. J. Clin. Microbiol 38, 2902–2908.

Fakruddin, M., 2011. Loop mediated isothermal amplification (LAMP) – an alternative to polymerase chain reaction (PCR). Bangladesh Res. Publ. J. 5, 425–439.

Fang, Y., Ramasamy, R.P., 2015. Current and prospective methods for plant disease detection. Biosensors 5, 537–561.

Faria, C.B., Abe, C.A.L., Novais da-silva, C., Tessmann, D.J., Barbosa-Tesmann, L.P., 2012. New PCR assays for the identification of *Fusarium verticillioides*, *F. subglutinans*, and other species of the *Gibberella fujikuroi* complex. Int. J. Mol. Sci. 13, 115–132.

Fones, H., Gurr, S., 2015. The impact of *Septoria tritici* blotch disease on wheat: an EU perspective. Fungal Genet. Biol. 79, 3–7.

Fraaije, B.A., Lovell, D.J., Coelho, J.M., Baldwin, S., Hollomon, D.W., 2001. PCR-based assays to assess wheat varietal resistance to blotch (*Septoria tritici* and *Stagonospora nodorum*) and rust (*Puccinia striiformis* and *Puccinia recondite*) diseases. Eur. J. Plant Pathol. 107, 905–917.

Garcia, P., Bainbridge, B.W., Heala, J.B., Perez, A.E., Jimenez, D.R.M., 1999. A simple PCR based method for the detection of the chickpea-wilt pathogen *Fusarium oxysporum* f.sp. *ciceris* in artificial and natural soils. Eur. J. Plant Pathol. 105, 251–259.

Ginzinger, D.G., 2002. Gene quantification using real-time quantitative PCR: an emerging technology hits the mainstream. Exp. Hematol. 30, 503–512.

Gómez-Alpizar, L., Saalau, E., Picado, I., Tambong, J.T., Saborio, F., 2011. A PCR-RFLP assay for identification and detection of *Pythium myriotylum*, causal agent of the cocoyam root rot disease. Lett. Appl. Microbiol 52, 185–192.

Gonzalez-Jaen, M.T., Mirete, S., Patino, B., Lopez-Errasquin, E., Vazquez, C., 2004. Genetic markers for the analysis of variability and for production of specific diagnostic sequences in fumonisin-producing strains of *Fusarium verticillioides*. Eur. J. Plant Pathol. 110, 525–532.

Goyal, A., Prasal, R., 2010. Some important diseases and their impact on wheat production. In: Arya, A., Perelló, A.E. (Eds.), Management of Fungal Plant Pathogens. Center for Agriculture and Bioscience International, Oxford, pp. 362–372.

Guo, J.R., Schnieder, F., Beyer, M., Verreet, J.A., 2005. Rapid detection of *Mycosphaerella graminicola*in wheat using reverse transcription-PCR assay. J. phytopathol 153, 674–679.

Henson, J.M., French, R., 1993. The polymerase chain reaction and plant disease diagnosis. Annu. Rev. Phytopathol 31, 81–109.

Jones, M.J., Dunkle, L.D., 1993. Analysis of *Cochliobulus carbonum* races by PCR amplification with arbitrary and gene-specific primers. Phytopathology 83, 366–370.

Ju, M., Zhou, Z., Mu, C., Zhang, X., Gao, J., Liang, Y., et al., 2017. Dissecting the genetic architecture of *Fusarium verticillioides* seed rot resistance in maize by combining QTL mapping and genome-wide association analysis. Sci. Rep. 7, 46446–46457.

Jurado, M., Vazquez, C., Marin, S., Sanchis, V., Gonzalez-Jaen, M.T., 2006. PCR-based strategy to detect contamination with mycotoxigenic *Fusarium* species in maize. Syst. Appl. Microbiol. 29, 681–689.

Kim, S.W., Li, Z., Moore, P.S., Monaghan, A.P., Chang, Y., Nichols, M., et al., 2010. A sensitive non-radioactive northern blot method to detect small RNAs. Nucleic Acids Res. 38, 98–110.

Kumar, A, Shaik, F, Kumar, DS, 2016. DNA Microarray analysis. Signal and image processing in Medical Applications. 39–47.

Langrell, S.R.H., Barbara, D.J., 2001. Magnetic capture hybridization for improved PCR detection of *Nectria galligena* from lignified apple extracts. Plant Mol. Biol. Report 19, 5–11.

Lanubile, A., Bernardi, J., Marocco, A., Logrieco, A., Paciolla, C., 2012. Differential activation of defense genes and enzymes in maize genotypes with contrasting levels of resistance to *Fusarium verticillioides*. Environ. Exp. Bot. 78, 39–46.

Lanubile, A., Ferrarini, A., Maschietto, V., Delledonne, M., Marocco, A., Bellin, D., 2014. Functional genomic analysis of constitutive and inducible defense responses to *Fusarium verticillioides* infection in maize genotypes with contrasting ear rot resistance. BMC Genomics 15, 710–722.

Lee, D., LaMura, M., Allnutt, T.R., Powell, W., 2009. Detection of genetically modified organisms (GMOs) using isothermal amplification of target DNA sequences. BMC Biotechnol. 9, 7–18.

Leslie, J.F., Summerell, B.A., 2006. The *Fusarium* Laboratory Manual. Blackwell Publishing, Ames, IA.

Li, S., Haitman, G.L., 2003. Molecular detection of *F. solani* f. sp. glycines in soyabean roots and soil. Plant Pathol. 52, 74–83.

Lizardi, P.M., Huang, X., Zhu, Z., Bray-Ward, P., Thomas, D.C., Ward, D.C., 1998. Mutation detection and single-molecule counting using isothermal rolling-circle amplification. Nat. Genet. 19, 225–232.

Long, A.A., 1998. *In-situ* polymerase chain reaction: foundation of the technology and today's options. Eur. J. Histochem 42, 101–109.

Mancini, V., Murolo, S., Romanazzi, G., 2016. Diagnostic methods for detecting fungal pathogens on vegetable seeds. Plant Pathol. 65, 691–703.

Masanga, J.O., Matheka, J.M., Omer, R.A., Ommeh, S.C., Monda, E.O., Alakonya, A.E., 2015. Downregulation of transcription factor aflR in *Aspergillus flavus* confers reduction to aflatoxin accumulation in transgenic maize with alteration of host plant architecture. Plant Cell Rep. 34, 1379–1387.

Matsuda, Y., Sameshima, T., Moriura, N., Inoue, K., Nonomura, T., Kakutani, K., et al., 2005. Identification of individual powdery mildew fungi infecting leaves and direct detection of gene expression by single conidium polymerase chain reaction. Phytopathology 95, 1137–1143.

McCartney, H.A., Foster, S.J., Fraaije, B.A., Ward, E., 2003. Molecular diagnostics for fungal plant pathogens. Pest. Manag. Sci. 59, 129–142.

Moeller, E.M., Chelkowski, J., Geiger, H.H., 1999. Species-specific PCR assays for the fungal pathogens *Fusarium moniliforme* and *Fusarium subglutinans* and their application to diagnose maize ear rot. J. Phytopathol 147, 497–508.

Panwar, V., McCallum, B., Bakkeren, G., 2012. Endogenous silencing of *Puccinia triticina* pathogenicity genes through in planta-expressed sequences leads to the suppression of rust diseases on wheat. Plant J. 73, 521–532.

Patino, B., Mirete, S., Vazquez, C., Jimenez, M., Rodriguez, M.T., Gonzalez-Jaen, M.T., 2006. Characterization of *F. verticillioides* strains by PCR-RFLP analysis of the intergenic spacer region of the rDNA. J. Sci. Food Agric. 86, 429–435.

Qi, M., Yang, Y., 2002. Quantification of *Magnaporthe grisea* during infection of rice plants using real-time polymerase chain reaction and northern blot/phosphoimaging analyses. Phytopathology 92, 870–876.

Rahman, M.T., Uddin, M.S., Sultana, R., Moue, A., Setu, M., 2013. Polymerase chain reaction (PCR): a short review. Anwer Khan Mod. Med. Coll. J. 4, 30–36.

Ramana, M.V., Balakrishna, K., Murali, H.C.S., Batra, H.V., 2011. Multiplex PCR based strategy to detect contamination with mycotoxigenic *Fusarium* species in rice and finger millet collected from southern India. J. Sci. Food Agric. 91, 1666–1673.

Rytkönen, K.T., Williams, T.A., Renshaw, G.M., Primmer, C.R., Nikinmaa, M., 2011. Molecularevolution of the metazoan PHD-HIF oxygen-sensing system. Mol. Biol. Evol. 28, 1913–1926.

Saiki, R.K., Scharf, S., Faloona, F., Mullis, K., Horn, G., Erlich, H., et al., 1992. Enzymatic amplification of beta-globin genomic sequences and restriction site analysis for diagnosis of sickle cell anemia. Biotechnology 24, 476.

Sandhu, G.S., Kline, B.C., Stockman, L., Roberts, G.D., 1995. Molecular probes for diagnosis of fungal infections. J. Clin. Microbiol. 33, 2913–2919.

Sato, T., Takayanagi, A., Nagao, K., Tomatsu, N., Fukui, T., Kawaguchi, M., et al., 2010. Simple PCR- based DNA microarray system to identify human pathogenic fungi in skin. J. Clin. Microbiol. 48, 2357–2364.

Schaad, N.W., Frederick, R.D., 2002. Real-time PCR and its application for rapid plant disease diagnostics. Can. J. Plant Pathol. 24, 250–258.

Schena, L., Nigro, F., Ippolito, A., Gallitelli, D., 2004. Real-time quantitative PCR: a new technology to detect and study phytopathogenic and antagonistic fungi. Eur. J. Plant Pathol. 110, 893–908.

Schlacht U.F., Moller E.M., Geiger H.G. 1997. Genetic diversity of *Gibberella fujikuroi* isolates from different geographic origins. Proceedings of the Fifth European *Fusarium* Seminar. Cereal Research and Communication. 25: 557–600.

Sergentet-Thevenot, D., Montet, M.P., Vernozy-Rozand, C., 2008. Challenges to developing nucleic acid sequence-based amplification technology for the detection of microbial pathogens in food. Rev. Med. Vet. 159, 514–527.

Shamloul, A.M., Hadidi, A., 1999. Sensitive detection of potato spindle tuber and temperate fruit tree viroids by reverse transcription-polymerase chain reaction-probe capture hybridization. J. Virol. Methods 80, 145–155.

Silva, J.J., Viaro, H.P., Ferranti, L.S., Oliveira, A.L.M., Ferreira, J.M., Ruas, C.F., et al., 2017. Genetic structure of *Fusarium verticillioides* populations and occurrence of fumonisins in maize grown in southern Brazil. Crop Prot. 99, 160–167.

Singh, A., Kumar, N., 2013. A review on DNA microarray technology. Int. J. Curr. Res. Rev. 5, 01–05.

Sreenivasa, M.Y., Dass, R.S., Charithraj, A.P., Janardhana, G.R., 2007. PCR method for the detection of genus *Fusarium* and fumonisin producing isolates from freshly harvested sorghum grains grown in Karnataka, India. J. Food Saf. 28, 236–247.

Sreenivasa, M.Y., Gonzalez Jaen, M.T., Dass, R.S., Charithraj, A.P., Janrdhana, G.R., 2008. A PCR based assay for the detetction and differentiation of potential fumonisin producing *Fusarium verticillioides* isolated from maize kernels. Food Biotechnol. 22, 160–170.

Sreenivasa, M.Y., Dass, R.S., Janardhana, G.R., 2010. Survey of post-harvest fungi associated with sorghum grain produced in Karnatka (India). J. Plant Prot. Res. 50, 335–339.

Sreenivasa, M.Y., Charithraj, A.P., Janardhana, G.R., 2011a. Diversity of *Fusarium* species associated with maize and sorghum grains in Karnataka, India. Fungal Sci. 26, 111–123.

Sreenivasa, M.Y., Dass, R.S., Charithraj, A.P., Janardhana, G.R., 2011b. Mycological evaluation of maize grains produced in Karnataka (India) for the post-harvest fungal contamination. World Appl. Sci. J. 13, 688–692.

Sun, J., Li, X., Zeng, H., Xie, Z., Lu, C., Xi, L., et al., 2010. Development and evaluation of loop-mediated isothermal amplification (LAMP) for the rapid diagnosis of *Penicillium marneffei* in archived tissue samples. FEMS Immunol. Med. Microbiol. 58, 381–388.

Talbot, N.J., 2003. On the trail of a cerial killer: exploring the biologyof *Magnaporthe grisea*. Annu. Rev. Microbiol. 57, 177–202.

Tsui, C.K., Woodhall, J., Chen, W., Lévesque, C.A., Lau, A., Schoen, C.D., et al., 2011. Moleculartechniques for pathogen identification and fungus detection in the environment. IMA Fungus 2, 177–189.

Villari, C., Tomlinson, J.A., Battisti, A., Boonham, N., Capretti, P., Faccoli, M., 2013. Use of loop-mediated isothermal amplification for detection of *Ophiostoma clavatum*, the primary blue stain fungus associated with *Ips acuminatus*. Appl. Environ. Microbiol. 79, 2527–2533.

Visentin, I., Montis, V., Doll, K., Alabouvette, C., Tamietti, G., Karlovsky, P., et al., 2012. Transcription of genes in biosynthetic pathway for fumonisin mycotoxins is epigenetically and differentially regulated in the fungal maize pathogen *Fusarium verticillioides*. Eukaryot. Cell 11, 252–259.

Volpi, E.V., Bridger, J.M., 2008. Fish glossary: an overview of the fluorescence in situ hybridization technique. Biotechniques 45, 385–386.

Waalwijk, C., van der Lee, T., de Vries, I., Hesselink, T., Arts, J., Kema, G.H.J., 2004. Synteny in toxigenic *Fusarium* species: the fumonisin gene cluster and the mating type region as examples. Eur. J. Plant Pathol. 110, 533–544.

Walcott, R.R., Gitaitis, R.D., Langston, D.B., 2004. Detection of *Botrytis aclada* in onion seed using magnetic capture hybridization and the polymerase chain reaction. Seed Sci. Technol. 32, 425–438.

WEISS, MV, 1987. Compendium of wheat diseases. Second edition. In: Compendium of wheat diseases. Second edition., merican Phytopathological Society St. Paul, Minnesota USA.

Williams, J.G.K., Kubelik, A.R., Rafalski, J.A., Tingey, S.V., 1990. DNA polymorphisms amplified by arbitrary primers are useful as genetic markers. Nucleic Acids Res. 18, 615–643.

Zeng, X., Luo, Y., Zheng, Y., Duan, X., Zhou, Y., 2010. Detection of latent infection of wheat leaves caused by *Blumeria graminis* f.sp.*tritici* using nested PCR. J. Phytopathol. 158, 227–235.

Zhang, A.W., Riccioni, G.L., Chen, L., Ma, W.D., Petersen, W.L., 1997. Using PCR to distinguish *Diaporthe phaseolorum* and *Phompsis longicolla* from other soybean fungal pathogens and to detect them in soybean tissues. Plant Dis. 81, 1143–1149.

Zhao, Y., Park, S., Kreiswirth, B.N., Ginocchio, C.C., Veyret, R., Laayoun, A., et al., 2009. Rapid real-time nucleic acid sequence-based amplification-molecular beacon platform to detect fungal and bacterial bloodstream infections. J. Clin. Microbiol. 47, 2067–2078.

Zhenggang, Z., Jingyu, Z., Yucoo, W., Xiaobo, Z., 2005. Molecular detection of *F. oxysporum, F. niveum* and *Mycospaerella meloxis* in infected plant tissues and soil. FEMS Microbiol. Lett. 249, 39–47.

Applicability of fungi in agriculture and environmental sustainability

7

Pankaj Kumar Chaurasia[1] *and Shashi Lata Bharati*[2]
[1]PG Department of Chemistry, L.S. College (A Constituent Unit of B.R.A. Bihar University) Muzaffarpur, India, [2]Department of Chemistry, North Eastern Regional Institute of Science and Technology, Nirjuli, India

7.1 Introduction

Fungi or funguses are any member of the group of eukaryotic organisms which include yeasts, molds, and much familiar mushrooms like microorganisms and they have been classified as a separate kingdom from other life kingdoms of plants and animals (https://en.wikipedia.org/wiki/Fungus). They are abundantly present worldwide. Due to their small structural sizes and cryptic lifestyles in soil or on dead materials, most fungi are inconspicuous. They are well-known principal ecological decomposers. The presence of chitin in their cell walls is the cause of their place in separate kingdom from bacteria, plants, and some protists. It is well known that fungi are heterotroph like animals and do not perform the function of photosynthesis for the preparation of food which means they cannot form their food and depend on other sources for it. They receive their food by absorbing the dissolved molecules by secreting their digestive enzymes. These differences along with some other differences place fungi in a single group of similar related microorganisms, namely eumycota, that is, true fungi or eumycetes. It should be clear that slime molds or myxomycetes and water molds or oomycetes are different from the aforementioned fungal group (https://en.wikipedia.org/wiki/Fungus).

Most of the fungi grow as a cylindrical, thread-like structure known as hyphae which grow at their tips. Tips emergences give new hyphae by the process of branching. Fungi have some unique features like some species grow as unicellular yeasts in which reproduction occurs by budding or fission. Dimorphic fungi can switch between a yeast phase and a hyphal phase in response to environmental conditions. Glucans and chitin both are present in the cell wall of fungi and thus these are only organisms that have these structural components in the cell wall. Glucans are also present in plants while chitin in arthropod's exoskeleton (Bowman and Free, 2006).

For the detailed study of fungi, their origin, sources, culture, production, growth, developments, and wide varieties of their biotechnological roles; readers can visit different works already completed in this field (Chaurasia and Bharati, 2020a,b).

Microbes in Land Use Change Management. DOI: https://doi.org/10.1016/B978-0-12-824448-7.00010-3
© 2021 Elsevier Inc. All rights reserved.

Dwivedi (2020) have shown and discussed the indigenous vesicular arbuscular mycorrhizal fungi isolated from wheat cultivars of Pratapgarh UP, India and their species diversity and distribution. Singh and Bhatt (2020) discussed the diversity, distribution, and ethnomycology of wild edible mushrooms of Rudraprayag district. Devi and Tayung (2020) discussed the present and future prospects of lichen diversity and its associated fungi. Sridharan and Krishnaswamy (2020) discussed the different methods of fungus characterization. Zhou et al. (2020) discussed theophylline degrading *Aspergillus* fungi from isolation, characterization, and application points of view and demonstrated their application in production of methylxanthines with theophylline. They collected and identified seven fungi that were tea-derived. They found for the first time that *Aspergillus ustus* and *Aspergillus tamarii* were capable of degrading theophylline through *N*-demethylation and oxidation. Through a new microbial synthesis platform with *A. ustus* and *A. tamarii* as appropriate starter strains, they were also capable of biologically developing 3-methylxanthine and xanthine efficiently from theophylline (Zhou et al., 2020). Chand et al. (2020) isolated and characterized endophytic fungi from *Vanda cristata*, an epiphytic orchid. They used roots, stems, and leaves of *V. cristata* for this purpose, and morphological and molecular study were used for identification of endophytic fungi. Further, in order to understand qualitatively and quantitatively the growth-promoting effect, endophytic fungi were subjected to auxin synthesis, phosphate solubilization, ammonia synthesis, and elicitor growth test (Chand et al., 2020).

7.2 Applications of fungi

Fungi are highly valuable microorganisms for humans. They have wide varieties of applications and are known for their significant role in various fields like in medicinal industries, food industries, synthetic chemistry, environmental sciences, fungicides, biotechnology researches, genetic engineering, cosmetics, and so on. An important review article by Hyde et al. (2019) shows different valuable potential of fungi in the field of biotechnology. In this article they reviewed fifty ways through which fungi may be useful in the field of biotechnology. Fungi have been effectively used in the field of antibacterial antibiotics. In industries, the term antibiotics are used for antibacterial agents but according to Waksman (1947) it may defined as (Hyde et al., 2019) "a chemical substance, produced by microorganisms (including fungi), which has the capacity to inhibit the growth of and even to destroy bacteria and other microorganisms." Penicillin V, cephalosporin C, pleuromutilin, fusidic acid, and retapamulin are some fungal metabolites that are developed to antibacterial drugs (Hyde et al., 2019). Besides, they have discussed other various aspects of fungi in the field of human-based medicinal biotechnology under topics like antimycotics, biofilm inhibitors, anticancer agents, antidiabetes, traditional medications, in control of cardiovascular diseases, antiviral agents, immune-suppressive agents, immune-modulatory agents, etc. They further discussed strategies against different plant diseases and showed the significances of fungi for solving the problems of plants as use of endophytes in bio-control of plant diseases, fungal uses in bio-control of insects, nematodes and weeds, fungal antagonists, antagonistic fungi in bio-control of rusts and smuts. Further, their significances in field of crop and forestry as bio-fertilizers (arbuscular mycorrhizae), ectomycorrhizal fungi in forestry, orchid mycorrhizae and endophyte's uses in the field of biotechnology, fungal-based growth promoting hormones etc. have also been discussed by Hyde et al. (2019). Various other biotechnological aspects of fungi in the field of food and beverages dealing with mushrooms biotechnology, food coloring, food flavoring, functional food and nutraceuticals, the field of wine, beer, and spirits and many more and biotechnology of fungi in saving the planet considering the topic under mycoremediation, mycofumigation, biomass to biofuel, packed-bed bioreactor, plastic-bioremediation, polycyclic-aromatic hydrocarbons, and so on have also been dealt under this review article (Hyde et al., 2019).

Sant'Anna da Silva et al. (2020) discussed the use of secondary metabolites from endophytic fungi in the development of natural and ecofriendly agrochemicals. Due to the presence of several chemical groups; secondary metabolite production from fungi are complex processes. Many secondary metabolites like nectnapyrone, 4-hydroxyphenylethanol, a novel norsesquiterpene ϒ-lactone having insecticidal activity is produced from *Diaporthe eres*. Secondary metabolites like duroquinone, naphthalene, adamantine derivatives from *Aspergillus flavus* and *Nigrospora oryzae* have larvicidal activity against *Hyblaea purea*, *Atteva fabriciella*, and *Eligma narcissus* (Sant'Anna da Silva et al., 2020). Uddin Pk et al. (2020) discussed the role of edible mushrooms in the treatment of cancer. Various mycotoxins and their different impacts on human health were discussed by Anita (2020). Similarly, various types of applications and uses of fungi have also been covered inside this book edited by the author (Chaurasia and Bharati, 2020a,b).

Yuvaraj and Ramasamy (2020) discussed the uses of fungi in the field of agriculture. It may show their role in various physiological processes, water and mineral uptake, stomatal movement, chemical change, and biosynthesis of compounds termed bio-stimulants, auxins, lignan, and ethylene to boost the flexibility of plants to ascertain and cope environmental stresses like drought, salinity, heat, cold, and significant metals (Yuvaraj and Ramasamy, 2020). Meyer et al. (2020) discussed fungal biotechnology on an interesting topic "growing a circular economy with fungal biotechnology-a white paper." Fungi can transform the various organic matters in different useful products and provide opportunities to tackle the different challenges. Fungal biotechnology can advance the transition from our petroleum-based economy into a bio-based circular economy. They have a strong ability to produce sources of food, chemicals, textiles, fuels, materials for construction, automotive and transportation industries, for furniture and beyond. In this paper, they interestingly demonstrated the opportunity and challenges of fungal biotechnology (Meyer et al., 2020). Chaurasia and Bharati (2019) have discussed the role of fungi in the field of bioremediation of dyes based on recent studies. They showed that fungus based bio-treatment of various toxic dyes pollutants are green techniques in order to clean the aqueous environment and may play a significant role in making lives safe living in dye polluted environments. Deshmukh et al. (2016) discussed diverse metabolic activities of fungi in bioremediation and showed that they may play major role in the field of remediation biotechnology and discussed the various groups of fungi from different habitats having a role in biological remediation of toxic components, textile dyes, persistent organic pollutants, textile effluents, polyaromatic hydrocarbons, leather industries and many more. Fungal-based bioremediation of toxic pollutants are almost cost effective green technology. They also discussed fungal enzymes like laccases, peroxidases, catalases, and cyrochrome P450 monooxygeneses (Deshmukh et al., 2016). They also showed the role of fungi bioremediation of heavy metals in this review article (Deshmukh et al., 2016). Chaurasia et al. (2019) discussed the important applications of fungi for the biological treatment of contaminated soil polluted with organic and inorganic matters. Li et al. (2020) documented the descriptive role of fungi in the soil bioremediation contaminated with petroleum hydrocarbons and poisonous harmful metals. A few works have been completed on the role of fungi in response to complex pollutant stress. In this article they also discussed the ability of new methodologies like gradient diffusion film technology and metabolomics in order to overcome present bottlenecks in understanding fungi-based bioremediation (Li et al., 2020).

7.3 Fungal enzymes and their biotechnological applications

Most of the applications of fungi are due to their enzymatic activities after its secretion. Fungal enzymes are known for their various important biotechnological applications in pharmaceutical applications, industrial applications, food industries, textile industries, paper and pulp industries, bioengineering technology, bioremediation, environmental safety, bio-sensoring, dairy industries, waste water managements, bio-transformations, organic synthesis, etc. (Chaurasia and Bharati, 2020a,b; Kango et al., 2019; Chaurasia et al., 2013, 2014a,b,c,d, 2015a,b, 2016, 2018; Agarwal et al., 2018; Kushwaha et al., 2017, 2018; Sharma et al., 2016). There are a lot of applications known based on the activities of fungal enzymes like laccase, cellulase, lactase, peroxidase, lipase, xylanase, pectinase, fungal amylase etc. (https://en.wikipedia.org/wiki/Fungal_extracellular_enzyme_activity) (Singh and Gupta 2020; Thapa et al., 2020; Chaurasia and Bharati, 2017, 2020a,b; Khan and Selamoglu, 2020; Chowdhary et al., 2020; Chandra et al., 2020; Basit et al., 2020; Sudeep et al., 2020; Elyasi Far et al., 2020; Jayasekara and Ratnayake, 2019; Bharati and Chaurasia, 2018; Ravalason et al., 2008; Kissi et al., 2001).

7.4 Fungi and their roles in the field of agriculture

Fungi along with above discussed valuable applications also show advantageous roles in the field of agriculture. Filamentous fungi greatly help in soil aggregation and degrade organic matters in soil (Yuvaraj and Ramasamy, 2020). This property makes fungi valuable from the agriculture point of view. Except it, different bound species of fungi play an important role for the production of valuable organic molecules and necessary for maintaining soil health by providing necessary organic components inside soil. In the modern time of fast growing world where a lot of chemicals like plant growth regulators and chemical fertilizers (Islam, 2008; Metting, 1993) are being used in order to enhance the production in the field of agriculture and in order to accomplish the requirements of continuously growing heavy populations. It is a well-known fact that such chemicals used in the field of agriculture for enhancing the crop production may have various harmful effects on the health of human in the future. In this direction the use of microbes may be helpful in promoting green technologies.

7.4.1 Mycorrhizal association and agriculture

Fungi have the ability to grow in all environments but they prefer to grow abundantly in acidic soil environment. This is because at neutral pH medium of the soil they have competition with bacteria. Fungi have tendency to form mycorrhizal association (symbiosis) with plants *via* hyphae network. This symbiotic association in the soil between plants and fungi play significant role in flow of nutrients from soil to plants. Phosphate, zinc, copper, and other minerals from soil are smoothly provided to plants by this symbiotic interaction (mycorrhizal association). Fungi help plants in mineral uptake by receiving nutrients like sugar from the roots of plants (Yuvaraj and Ramasamy, 2020). Such relationships are called mycorrhizae network (Smith and Read, 1997). Also, fungi are known for their strong secretion of enzymes which help in degrading the organic molecules, substances, and strongly help in maintaining soil health by balancing the carbon and other necessary nutrients in soil. The fungal enzymes secreted in soil also help in the decomposition of dead materials present inside soil (Yuvaraj and Ramasamy, 2020).

7.4.2 Functioning of mycorrhizae and agriculture

Mycorrhiza is composed of two terms *viz.* "myco" refers to "fungi" while "rhiza" refers to the roots system of the plants. It is a symbiotic association between fungi and roots of plants in which both are mutually benefited from each other. Fungi colonize the root tissues of plants acting as host and provide soil mineral nutrition to plants and in reverse fungi find food from plants. In particular cases, mycorrhizae may have parasitic association with host plants (Johnson et al., 1997). These mycorrhizae play a significant role in the field of soil biology, plants' nutrition, and soil chemistry.

Mycorrhizae are mainly of two types (https://en.wikipedia.org/wiki/Mycorrhiza). One is endomycorrhizae and the other is ectomycorrhizae. Endomycorrhize can be further divided into arbuscular mycorrhiza vesicular-arbuscular mycorrhizae, ericoid mycorrhizae, orchid mycorrhizae, and monotropoid mycorrhizae. In arbuscular mycorrhyzae, hyphae penetrate the cells of plant and produce balloon like vesicles or dichotomously branching invaginations (arbuscules) in order to exchange nutrition. Surface area of contact between hyphae and cell cytoplasm is greatly increased by these arbuscules which make easy nutritional facilitation between them. This type of mycorrhizal association is found in 85% of all plant families and occurs in several crop species (Wang and Qiu, 2006). Glycoprotein glomalin, may be one of the major stores of carbon in the soil, is produced by the hyphae of arbuscular mycorrhizal fungi (International Institute for Applied Systems Analysis, 2019). Ericoid mycorrhizae have intra-radical phase consisting of dense coils of hyphae in the outermost layer of root cells. They have also been shown to keep substantial saprophytic competences which would facilitate plants to obtain nutrients from not-yet-decomposed materials via decomposing actions of their ericoid partners. In orchid mycorrhizal association; hyphae penetrate the cells of root and form coils in order to exchange nutrition. Plants of monotropoid mycorrhiza are mainly heterotrophic or mixotrophic derive their carbon from the fungus partner. This mycorrhizal symbiosis is nonmutualistic parasite types.

Ectomycorrhizae are symbiotic associations between the roots of around 10% of plant families, mostly woody plants including the birch, dipterocarp, eucalyptus, oak, pine, and rose (Wang and Qiu, 2006) families, orchids, ("Orchids and fungi: An unexpected case of symbiosis," 2011), and fungi belonging to the Basidiomycota, Ascomycota, and Zygomycota (https://en.wikipedia.org/wiki/Mycorrhiza). Nutrients can be shown to move between different plants through the fungal network. Carbon has been shown to move from paper birch trees into Douglas-fir trees thereby promoting succession in ecosystems (Simard et al., 1997).

Mycorrhizal association makes the plants safer from microbial soil borne pathogens and found to be helpful in plant defense either above the ground or below the ground. Enzymes excreted by mycorrhizae are poisonous for the soil born organism-like nematodes (Azcón-Aguilar and Barea, 1997). Mycorrhizae may help in making the plants safe from insects, drought, toxicity, and diseases (https://en.wikipedia.org/wiki/Mycorrhiza).

7.4.3 Fungi in food and nutrition

Fungi are also known for their important applications in the field of food industries. Fungal fermentation plays a role in the production of the majority of the citric acid of the world. Fungal bioprocessing is used to produce the majority of the flavors used in food production. Many enzymes used in the food industries during food processing come from fungi. In this regard, the review article written by Copetti (2019) described the food ingredients obtained from yeasts and molds species and their role in field of food industries. Fungi and fungal-based enzymes are extremely valuable in the field of food industries. Enzymes may be used to enhance the quality, flavor, and stability of food materials. High specificity of enzymes and generation of lesser by products during the use of enzymatic processes in the food processing makes them valuable and very useful in the field of food industries. Various enzymes like laccases, amylases, proteases, lipases, lactases and pectinases have their proved roles in this field. Fungal enzymes have now proved their importance in the field of dairy, bakery, brewing, fruit processing, wine industries, beverages, etc. Saxena et al. (2001) significantly described the various roles of fungal enzymes in the field of food processing. Sharma et al. (2019) showed the uses of various enzymes in the field of food industries. Table 7.1 adapted from the review article of Sharma et al. (2019) showed the significant uses of various enzymes in the food sector. Chaurasia and Bharati (2017) comprehensively described the various possible uses of laccases in the field of different food industries.

7.4.4 Mushroom farming

In the present time of highly demanding era of nutrition; mushroom biotechnology and mushroom farming is a hot area in order to provide better nutrition and nutrients to big populations. Mushrooms are actually spore bearing fruiting bodies of fungi fleshy in structure and may grow above moister-containing grounds, soil and dead organic sources. The cultivated white button mushroom, *Agaricus bisporus* is the standard name for "mushroom." Fungi (Basidiomycota, Agaricomycetes) having stem (stipe), a cap (pileus), and gills (lamellae, sing. lamella) on the underside of the cap are generally referred to as mushroom. A variety of other gilled fungi with

Table 7.1 Various biotechnologically valuable enzymes and their areas of applications.

S.no.	Enzymes	Applications/industries
1	Glucose oxidase, pentosanase, protease, amylase (Novamyl), amyloglycosidases alpha-amylases	Baking industries
2	Amyloglucosidase	Alcohol production
3	Pectinase, glucose oxidase, laccase	Wine and fruit juice
4	Protease	Meat industries
5	Catalases, protease, lactase	Dairy industries
6	Beta-glucanases, alpha-amylases, proteases, maltogenic amylases	Brewing industries
Use of marketed enzymes in food industries		
7	Protease	Baking, brewing, diary, fish, meat, starch, and vegetable
8	Lipases	Fats and oil
9	Alpha-amylase	Baking, brewing, distilling, and starch
10	Chymosin	Cheese
11	Beta-glucanase	Brewing
12	Alpha-glucan transferase	Starch
13	Xylanase	Baking and starch
14	Pullulanase	Brewing and starch
15	Alpha-acetolactate decarboxylase	Brewing
16	Phytase	Starch
17	Maltogenic amylase	Baking and starch

Source: Adapted from Sharma, K., Vaidya, D., Sharma, N., 2019. Recent trends of enzymes in food industry: a review. Int. J. Chem. Stud. 7 (2), 1525–1529.

or without with stem are also described under "mushroom." By extension, the term "mushroom" can also refer to either the entire fungus when in culture, the thallus (called a mycelium) of species forming the fruiting bodies called mushrooms, or the species itself (https://en.wikipedia.org/wiki/Mushroom). Edible mushrooms are full of nutritional values because they are full of carbohydrates, minerals (P, K, Zn, Ca, Fe, Mg, Mn,), vitamins (B1-thiamine, B2-riboflavin, B3-niacin, B5-pantothenic acid, B6 and D, folate), proteins, selenium, copper etc. 92% water, 4% carbohydrates, 2% protein, and less than 1% fat are present in raw brown mushrooms. Per 100 g raw mushroomcontains94 kJ energy, vitamins B (20% or more of the daily value) like riboflavin, niacin and pantothenic acid, selenium (37% DV) and copper (25% DV) and a moderate source (10%−19% DV) of phosphorus, zinc, and potassium. Presence of vitamin D depends upon the postharvest handling (sunlight exposure). Seeing these great nutritional equations inside mushrooms, mushroom farming is a hot sector in the present time. Mushrooms used extensively in cooking are being commercially produced in mushrooms farms. *A. bisporus* (button mushrooms when small or Portobello mushrooms when larger) is widely used mushrooms in various food dishes. Several other cultivated species are straw mushrooms (*Volvariella volvacea*), oyster mushrooms (*Pleurotus ostreatus*), shiitakes (*Lentinula edodes*), and enokitake (*Flammulina* spp.) (Stamets, 2000). Many other mushrooms are harvested from wild for personal consumption or for commercial uses. Milk mushrooms, morels, chanterelles, truffles, black trumpets, and porcini mushrooms (*Boletus edulis*) (also known as king boletes) demand a high price on the market. All mushrooms are not safe because many of them are poisonous (https://en.wikipedia.org/wiki/Fungus; https://en.wikipedia.org/wiki/Mushroom). Kumla et al. (2020) describes about the mushrooms cultivation and production of lignocellulolytic enzymes. There are many more works that have been performed by various researchers worldwide on mushroom biotechnology (Cunha Zied et al., 2020; Spoorthi et al., 2020; Ho et al., 2020).

7.4.5 Biological pest control by the use of fungi

Fungi are also capable of controlling pests and helping to save the crop and agricultural environment from the attack of such pests. They may be used as fungal-based pesticides also called as myco-pesticides like myco-herbicides, myco-insecticides and myco-fungicides. Such types of fungal-based pesticides may be ecologically safe and more specific alternative in comparison to chemical pesticides (Moore et al., 2011). Some commonly used species are *Lecanicillium lecanii*, *Beauveria bassiana*, and *Trichoderma* spp (http://kumu.brocku.ca/BIOL_2P96_Jan_2013-group07/Fungi_Uses_in_Pest_Management#_note-textbook_fungi).
Entomopathogenic fungi are known for their role in pest control. Such fungi are parasite in nature on pests or insects and either kill them or strongly deactivate them (https://en.wikipedia.org/wiki/Entomopathogenic_fungus).

An old but relevant paper by Gillespie and Claydon (1989) demonstrates the use of entomogenous fungi in pest control. In this chapter, biology of entomogenous fungi has been examined and studies on the entomophthoraceae, *B. bassiana*,

Beauveria brongniartii, Metarhizium anisopliae, and *Verticillium lecanii* have been reviewed (Gillespie and Claydon, 1989). Robles-Acosta et al. (2019) performed a work under greenhouse conditions on entomopathogenic fungi as biological control agents of *Phyllocoptruta oleivora. P. oleivora*, a citrus rust mite, is a major pest of citrus affecting quality and yields both. Results of their study showed that under the green-house conditions, entomopathogenic fungi like *B. bassiana and M. anisopliae* were effective in control of *P. oleivora* populations. A review article on endophytic entomopathogenic fungi written by Mantzoukas and Eliopoulos (2020) shows the capability of such fungi in biologically control the plants pests. Rizwan et al. (2019) performed work on the evaluation of the entomopathogenic fungi as a non-traditional control of the rice leaf roller, *Cnaphalocrocis medinalis*. Branine et al. (2019) discussed about endophytic insect-pathogenic fungi and their biology and applications. They provided an overview on endophytic insect-pathogenic fungi by reviewing what is currently known about their evolution, ecology, and mechanisms of insect and plant colonization. They also address real-world applications of endophytic insect-pathogenic fungi (Branine et al., 2019). Similarly, there are several articles which present the advantageous roles of fungi for controlling plant pests (Baron et al., 2019; Maina et al., 2018; Singh et al., 2017).

7.4.6 Mechanism behind the action of fungi on insects/pests

Microscopic spores of such fungi become attached with the body surface of insects and germinate under suitable conditions of temperature and humidity. Then, grow as hyphae, colonize the cuticles of insects and then, bore by enzymatic hydrolysis and reach to the cavity of insect's body (Fernandes et al., 2012) where fungal cells proliferate inside the bodies of insects and become a reason behind the death of insects (https://en.wikipedia.org/wiki/Entomopathogenic_fungus).

7.5 Fungi in environmental sustainability

Along with various aforementioned biotechnological roles of fungi; they are also proven to show a promising role in making soil environments clean and safe. Various fungi and their associated enzymes are significantly known for their promising applications in the field bioremediation or detoxification of various organic and inorganic toxic pollutants. In organic pollutants, they have role in detoxification of dyes, dye's byproducts, petroleum wastes, xenobiotcis, polyaromatic hydrocarbons, pesticides, industrial wastes containing pharmaceutical and other organic components and other while under the branch of inorganic matter's detoxification, they have promising roles in heavy metal stabilization and detoxification of their influences inside soil. In this way fungi may be very much helpful in keeping the soil healthy by detoxifying the toxicity of various pollutants and heavy metals present inside soil due to different human activities (Chaurasia et al., 2019).

7.5.1 Heavy metal remediation

There are a number of works that have been performed by researchers worldwide on the study of the ability of fungi in bioremediation of heavy metals and their affirmative roles in soil and environment sustainability. Zhang et al. (2020) show the application of an earthworm gut fungus *Trichoderma brevicompactum* QYCD-6 in the removal of heavy metals and tolerance. They have evaluated the tolerance ability of this fungus as well as its ability to remove multimetals. In this study they isolated fungus from *Pheretima tschiliensis* and performed tolerance study using heavy metals like copper (II), chromium (II), zinc (II), cadmium (II) and lead (II) and found that this fungus was able to have maximum tolerance for metal Pb(II) with minimum inhibitory concentration value of 1600 mg/L on potato dextrose medium. In their experiment, they found a moderate result of removal (45.9%) for multimetals while the highest result of removal (97.5%) for Pb(II) by the use of fungus and based on their study showed that fungus *T. brevicompactum* QYCD-6 may be utilized for the bioremediation of heavy metals from the metals-polluted sites (Zhang et al., 2020). Lin et al. (2020) performed a case study in eastern China in order to know the fungal adaptability to heavy metal contamination and found which fungal species has strong adaptability to heavy metals. Their study showed that some fungi can have ability to reduce metals toxicity inside soil and have ability to generate a good environment inside soil helpful in proper growth of crops (Lin et al., 2020). Dhalaria et al. (2020) reviewed arbuscular mycorrhizal fungi as potential agents in ameliorating the stress of heavy metals in plants. This review article focuses on heavy metals uptake, accumulation, and arbuscular mycorrhizal effect in ameliorating heavy metal stress in plants (Dhalaria et al., 2020). Oladipo et al. (2018) isolated and identified filamentous fungal species from the soils of gold and gemstone mine site and performed assessments for their heavy metals (cadmium, copper, lead, arsenic, and iron) tolerance. Their study showed notable tolerance in heavy metal-rich media by indigenous filamentous fungal species from gold and gemstone mine sites (Oladipo et al., 2018).

7.5.2 Dyes remediation

White rot fungi and other fungi as well as their associated lignolytic enzymes are found to be interestingly valuable in the detoxification and decolorization of various dyes that are proved on the basis of various researches in this field. Singh et al. (2020) performed investigation on decolorization of azophloxine dyes using cocultures of lignolytic enzymes producing white rot fungi *Trametes versicolor* and *Phanerochaete chrysosporium*. Improved dye decolorization was observed by the use of coculturing of fungal strains *T. versicolor* and *P. chrysosporium*. Coculuturing of these two strains extended the laccase activity 8.2 times in comparison mono-cultivation of *T. versicolor*. *P. chrysosporium* is a non-laccase producing fungus but due to synergistic interaction with *T. versicolor* and hyphal modification, it produced the enzyme (Singh et al., 2020). The secretome of the ascomycetous fungus *Myrothecium roridum* IM6482 was studied by Jasińska et al. (2019) by using a bottom-up proteomic approach for identifying the enzymes potentially useful for the decolorization of azo dyes. Laccase was more effective in decolorization of dyes like acid blue 113 (66% decolorization), acid red 27 (91%), direct blue 14 (79%), and acid orange 7 (80%) over other proteins (Jasińska et al., 2019). Chaurasia et al. (2019) reviewed the recent works on roles of fungi and associated enzymes in the decolorization of various dyes.

Yesilada et al. (2018) discussed the applicability of white rot fungi in bioremediation of various textile dyes. White rot fungi and their associated enzymes like laccases, peroxidases, can be successfully used for the remediation processes of polluted wastewater and other pollutants. In this paper, they have reviewed the roles of such fungi in decolorization, detoxification, and bioremediation of textile dyes. They have also evaluated the effects of immobilization, enzyme mediators, and various conditions along with advantages and disadvantages of treatment methods (Yesilada et al., 2018). Przystaś et al. (2018) performed a study of decolorization using fungal biomass immobilized on different solid supports. Decolorization of two dyes were studied by immobilized biomasses of fungal strains *P. ostreatus*—BWPH, *Gleophyllum odoratum*—DCa, and *Polyporus picipes*—RWP17 in order to check their efficacy. Fungal strain *P. picipes* (RWP17) showed the best result of decolorization and toxicity reduction (Przystaś et al., 2018). There are many more works that describe the significances of fungi in the field of dye decolorization (Al-Tohamy et al., 2020; Gao et al., 2020).

7.5.3 Petroleum remediation

Fungi are also very useful in the bioremediation of petroleum wastes and polyaromatic hydrocarbons. A mini-review written by Li et al. (2020) describes the role of fungi in bioremediation of petroleum hydrocarbons and metal-contaminated soils. They also describe the mechanism involved in the biotransformation of such substances (Li et al., 2020). Ogbonna et al. (2020) identified five hydrocarbon utilizing fungi namely, *Penicillium citrinum, Penicillium brocae, Fusarium solani, Kodamaea ohmeri*, and *Lentinus squarrosulus*. They performed characterization of crude oil contaminated soil associated with fungi and bacteria. According to them these organisms utilize crude oil as a carbon source, thus, they can be used for bioremediation of crude oil polluted environments (Ogbonna et al., 2020). Al-Hawash et al. (2019) used *Aspergillus* sp. RFC-1, a filamentous fungus, for the biodegradation and removal of different petroleum hydrocarbons. Removal efficiencies of crude oil, naphthalene, phenanthrene, and pyrene were 60.3%, 97.4%, 84.9%, and 90.7%, respectively while degradation efficiencies of crude oil, naphthalene, phenanthrene, and pyrene were 51.8%, 84.6%, 50.3%, and 55.1%, respectively. In conclusion, *Aspergillus* sp. RFC-1 can be used to degrade crude oil and the three types of polyaromatic hydrocarbons (Al-Hawash et al., 2019). Barnes et al. (2018) did experiments on bioremediation ability of hydrocarbon utilizing fungi. They isolated 10 fungal isolates with the ability to degrade crude oil. All the isolates were isolated from select marine substrates, such as mangrove sediments, Arabian Sea sediments, and tarballs. These fungal isolates were able to degrade long-chain *n*-alkanes. In the degradation of crude oil, Mangrove fungus #NIOSN-M126 (*P. citrinum*) was very effective. It was able to reduce total crude oil by 77% and the individual *n*-alkane fraction by an average of 95.37%. Their experiment proves the efficiency of fungus to be an effective bioremediating agent (Barnes et al., 2018). Olukunle and Oyegoke (2016) did a study on the role of fungi isolated from cow dung contaminated soils in crude oil biodegradation. Readers may also be interested in some other works (Dawoodi et al., 2015; Steliga, 2012; Das and Chandran, 2011).

7.6 Conclusion

This chapter deals with information on the fungal role in agriculture and environmental sustainability based on some selected and recent available data and researches in these fields. Above discussions clearly indicate about the valuable applications of fungi in various areas of biotechnology, agriculture, and environmental managements. Fungi have promising roles in the field of agriculture and may be useful in increasing crop production, pest control, heavy metal bioremediation, in bioremediation of dyes and petroleum wastes contaminated soil and thus cause sustainable developments in the field of agricultural environment. Fungal-based researches in these areas need special attention in the present scenario of the growing world in which everywhere is a great demand of the development of green technology and fungal applicability and their advantageous roles may be better suited in accomplishment of such demands.

Acknowledgment

Dr. P.K. Chaurasia is thankful to the PG Department of Chemistry, LS College, Muzaffarpur (India) and Dr. S.L. Bharati is thankful to the Department of Chemistry, NERIST, Arunachal Pradesh (India) for providing facilities during the course of work.

References

Agarwal, S., Gupta, K.K., Chaturvedi, V.K., Kushwaha, A., Chaurasia, P.K., Singh, M.P., 2018. The potential application of peroxidase enzyme for the treatment of industry wastes. In: Bharati, S.L., Chaurasia, P.K. (Eds.), Research Advancements in Pharmaceutical, Nutritional, and Industrial Enzymology. IGI Global, United States, ISBN 9781522552376.

Al-Hawash, A.B., Zhang, X., Ma, F., 2019. Removal and biodegradation of different petroleum hydrocarbons using the filamentous fungus *Aspergillus* sp. RFC-1. Microbiologyopen 8 (1), e00619.

Al-Tohamy, R., Sun, J., Fareed, M.F., et al., 2020. Ecofriendly biodegradation of reactive black 5 by newly isolated *Sterigmatomyces halophilus* SSA1575, valued for textile azo dye wastewater processing and detoxification. Sci. Rep. 10, 12370.

Anita, K., 2020. Mycotoxins and their impacts on human health. In: Chaurasia, P.K., Bharati, S.L. (Eds.), Research Advances in the Fungal World: Culture, Isolation, Identification, Classification, Characterization, Properties and Kinetics. Nova Science Publishers, Inc., New York, pp. 257–270. ISBN: 978-1-53617-197-6.

Azcón-Aguilar, C., Barea, J.M., 1997. Arbuscular mycorrhizas and biological control of soil-borne plant pathogens – an overview of the mechanisms involved. Mycorrhiza 6 (6), 457–464.

Barnes, N.M., Khodse, V.B., Lotlikar, N.P., Meena, R.M., Damare, S.R., 2018. Bioremediation potential of hydrocarbon-utilizing fungi from select marine niches of India. 3 Biotech. 8 (1), 21.

Baron, N.C., Rigobelo, E.C., Zied, D.C., 2019. Filamentous fungi in biological control: current status and future perspectives. Chil. J. Agric. Res. 79 (2).

Basit, A., Jiang, W., Rahim, K., 2020. Xylanase and Its Industrial Applications [Online First]. IntechOpen., DOI: 10.5772/intechopen.92156. Available from: https://www.intechopen.com/online-first/xylanase-and-its-industrial-applications.

Bharati, S.L., Chaurasia, P.K. (Eds.), 2018. Research Advancements in Pharmaceutical, Nutritional, and Industrial Enzymology. IGI Global, United States, ISBN: 978-1522552376 (hardcover), ISBN: 9781522552383 (ebook), pp. 1–549. DOI: 10.4018/978-1-5225-5237-6.

Bowman, S.M., Free, S.J., 2006. The structure and synthesis of the fungal cell wall. BioEssays 28 (8), 799–808.

Branine, M., Bazzicalupo, A., Branco, S., 2019. Biology and applications of endophytic insect-pathogenic fungi. PLoS Pathog. 15 (7), e1007831.

Chandra, P., Enespa, Singh, R., et al., 2020. Microbial lipases and their industrial applications: a comprehensive review. Microb. Cell Fact. 19, 169.

Chand, K., Shah, S., Sharma, J., Paudel, M.R., Pant, B., 2020. Isolation, characterization, and plant growth-promoting activities of endophytic fungi from a wild orchid *Vanda cristata*. Plant Signal. Behav. 15 (5). Article: 1744294.

Chaurasia, P.K., Bharati, S.L., 2017. Significance of laccases in food chemistry and related bioremediation. In. In: Grumezescu, A.M., Holban, A.M. (Eds.), Multi Volume Handbook of Food Bioengineering. Elsevier, pp. 299–335. (vol. 3: Soft Chemistry and Food Fermentation, ISBN: 978-0-12-811412-4).

Chaurasia, P.K., Bharati, S.L., 2019. Recent myco-dye decolorization studies (mini-review). J. Biotechnol. Bioeng. 3 (4), 27–31.

Chaurasia, P.K., Bharati, S.L., 2020a. Fungi and their associated enzymes. In: Chaurasia, P.K., Bharati, S.L. (Eds.), Research Advances in the Fungal World: Culture, Isolation, Identification, Classification, Characterization, Properties and Kinetics. Nova Science Publishers, Inc, New York, pp. 257–270. ISBN: 978-1-53617-197-6.

Chaurasia, P.K., Bharati, S.L. (Eds.), 2020b. Research Advances in the Fungal World: Culture, Isolation, Identification, Classification, Characterization, Properties and Kinetics. Nova Science Publishers, Inc, New York, ISBN: 978-1-53617-197-6.

Chaurasia, P.K., Yadav, A., Yadav, R.S.S., Yadava, S., 2013. Purification and characterization of laccase from *Coriolopsis floccosa* MTCC-1177 and its use in the selective oxidation of aromatic methyl group to aldehyde without mediators. J. Chem. Sci. 125 (6), 1395–1403.

Chaurasia, P.K., Yadav, R.S.S., Yadava, S., 2014a. Purification and characterization of yellow laccase from *Trametes hirsuta* MTCC-1171 and its application in synthesis of aromatic aldehydes. Process. Biochem. 49, 1647–1655.

Chaurasia, P.K., Yadava, S., Bharati, S.L., Singh, S.K., 2014b. Syntheses of aromatic aldehydes by laccase of *Pleurotus ostreatus* MTCC-1801. Synth. Commun. 44 (17), 2535–2544.

Chaurasia, P.K., Yadava, S., Bharati, S.L., Singh, S.K., 2014c. Selective oxidation and N-coupling of amines by purified laccase of *Xylaria polymorpha* MTCC-1100. Russ. J. Bioorg. Chem. 40 (4), 455–460.

Chaurasia, P.K., Yadava, S., Bharati, S.L., Singh, S.K., 2014d. Syntheses of aromatic aldehydes by laccase without the help of mediators. Green Chem. Lett. Rev. 7 (1), 100–104.

Chaurasia, P.K., Bharati, S.L., Sharma, M., Singh, S.K., Yadav, R.S.S., Yadava, S., 2015a. Fungal laccases and their biotechnological significances in the current perspective: a review. Curr. Org. Chem. 19 (19), 1916–1934.

Chaurasia, P.K., Bharati, S.L., Yadava, S., Yadav, R.S.S., 2015b. Purification, characterization and synthetic application of a thermally stable laccase from *Hexagonia tenuis* MTCC-1119. Indian. J. Biochem. Biophys. 52 (1), 60–67.

Chaurasia, P.K., Bharati, S.L., Sarma, C., 2016. Laccases in pharmaceutical chemistry: a comprehensive appraisal. Mini-rev. Org. Chem. 13, 430–451.

Chaurasia, P.K., Bharati, S.L., Mani, A., 2018. Enzymatic treatment of petroleum based hydrocarbons (Chapter 19). In: Pathak, V.M., Navneet (Eds.), Microbial Tools and Techniques for Environmental Waste Management. IGI Global, United States, pp. 396–408. ISBN 9781522535409.

Chaurasia, P.K., Bharati, S.L., Mani, A., 2019. Significances of fungi in bioremediation of contaminated soil. In: Singh, J.S., Singh, D.P. (Eds.), New and Future Developments in Microbial Biotechnology and Bioengineering. Elsevier, pp. 281–294. ISBN: 978-0-444-64191-5.

Chowdhary, P., Hare, V., Mani, S., Singh, A.K., Zainith, S., Raj, A., et al., 2020. Recent advancement in the biotechnological application of lignin peroxidase and its future prospects. Microorganisms for Sustainable Environment and Health, pp. 1–16.

Copetti, M.V., 2019. Fungi as industrial producers of food ingredients. Curr. Opin. Food Sci. 25, 52–56.

Cunha Zied, D., Sánchez, J.E., Noble, R., Pardo-Giménez, A., 2020. Use of spent mushroom substrate in new mushroom crops to promote the transition towards a circular economy. Agronomy 10 (9), 1239.

Das, N., Chandran, P., 2011. Microbial degradation of petroleum hydrocarbon contaminants: an overview. Biotechnol. Res. Int. 2011, Article ID 941810, 13 pages.

Dawoodi, V., Madani, M., Tahmourespour, A., Golshani, Z., 2015. The study of heterotrophic and crude oil-utilizing soil fungi in crude oil contaminated regions. J. Bioremed. Biodeg. 6, 270.

Deshmukh, R., Khardenavis, A.A., Purohit, H.J., 2016. Diverse metabolic capacities of fungi for bioremediation. Indian. J. Microbiol. 56 (3), 247–264.

Devi, D., Tayung, K., 2020. A review on lichen diversity and its associated fungi: present and future prospects. In: Chaurasia, Bharati (Eds.), Research Advances in the Fungal World: Culture, Isolation, Identification, Classification, Characterization, Properties and Kinetics. Nova Science Publishers, Inc., New York, pp. 89–114. ISBN: 978-1-53617-197-6.

Dhalaria, R., Kumar, D., Kumar, H., Nepovimova, E., Kuca, K., Islam, M.T., et al., 2020. Arbuscular mycorrhizal fungi as potential agents in ameliorating heavy metal stress in plants. Agronomy 10, 815.

Dwivedi, O.P., 2020. Species diversity and distribution of indigenous vesicular arbuscular mycorrhizal fungi isolated from different wheat cultivars of Pratapgarh (Uttar Pradesh) India. In: Chaurasia, Bharati (Eds.), Research Advances in the Fungal World: Culture, Isolation, Identification, Classification, Characterization, Properties and Kinetics. Nova Science Publishers, Inc., New York, pp. 21–34. ISBN: 978-1-53617-197-6.

Elyasi Far, B., Ahmadi, Y., Yari Khosroshahi, A., Dilmaghani, A., 2020. Microbial alpha-amylase production: progress, challenges and perspectives. Adv. Pharm. Bull. 10 (3), 350–358.

Fernandes, E.G., Valério, H.M., Feltrin, T., Van Der Sand, S.T., 2012. Variability in the production of extracellular enzymes by entomopathogenic fungi grown on different substrates. Braz. J. Microbiol. 43 (2), 827–833.

Gao, T., Qin, D., Zuo, S., et al., 2020. Decolorization and detoxification of triphenylmethane dyes by isolated endophytic fungus, *Bjerkandera adusta* SWUSI4 under non-nutritive conditions. Bioresour. Bioprocess. 7, 53.

Gillespie, A.T., Claydon, N., 1989. The use of entomogenous fungi for pest control and the role of toxins in pathogenesis. Pestic. Sci. 27 (2), 203–2015.

Ho, L.-H., Zulkifli, N.A., Tan, T.-C., 2020. In: Passari, A.K., Sánchez, S. (Eds.), Edible Mushroom: Nutritional Properties, Potential Nutraceutical Values, and Its Utilisation in Food Product Development. In: An Introduction to Mushroom. IntechOpen., DOI: 10.5772/intechopen.91827. Available from: https://www.intechopen.com/books/an-introduction-to-mushroom/edible-mushroom-nutritional-properties-potential-nutraceutical-values-and-its-utilisation-in-food-pr.

Hyde, K.D., Xu, J., Rapior, S., et al., 2019. The amazing potential of fungi: 50 ways we can exploit fungi industrially. Fungal Divers. 97, 1–136.

International Institute for Applied Systems Analysis, 2019. Plants and fungi together could slow climate change. Available from: https://phys.org/news/2019-11-fungi-climate.html.

Islam, K.R., 2008. Lecture on Soil Physics, Personal Collection of R. Islam. The Ohio State University School of Environment and Natural Resources, Columbus, OH.

Jasińska, A., Soboń, A., Góralczyk-Bińkowska, A., et al., 2019. Analysis of decolorization potential of *Myrothecium roridum* in the light of its secretome and toxicological studies. Environ. Sci. Pollut. Res. 26, 26313–26323.

Jayasekara, S., Ratnayake, R., 2019. In: Pascual, A.R., Martín, M.E.E. (Eds.), Microbial Cellulases: An Overview and Applications. Cellulose. IntechOpen., DOI: 10.5772/intechopen.84531. Available from: https://www.intechopen.com/books/cellulose/microbial-cellulases-an-overview-and-applications.

Johnson, N.C., Graham, J.H., Smith, F.A., 1997. Functioning of mycorrhizal associations along the mutualism–parasitism continuum. New Phytol. 135 (4), 575–585.

Kango, N., Jana, U.K., Choukade, R., 2019. Fungal enzymes: sources and biotechnological applications. In: Satyanarayana, T., Deshmukh, S., Deshpande, M. (Eds.), Advancing Frontiers in Mycology & Mycotechnology. Springer, Singapore.

Khan, U., Selamoglu, Z., 2020. Use of enzymes in dairy industry: a review of current progress. Arch. Razi Inst. 75 (1), 131–136.

Kissi, M., Mountadar, M., Assobhei, O., Gargiulo, E., Palmieri, G., Giardina, P., et al., 2001. Roles of two white-rot basidiomycete fungi in decolorisation and detoxification of olive mill waste water. Appl. Microbiol. Biotechnol. 57, 1–2.

Kumla, J., Suwannarach, N., Sujarit, K., Penkhrue, W., Kakumyan, P., Jatuwong, K., et al., 2020. Cultivation of mushrooms and their lignocellulolytic enzyme production through the utilization of agro-industrial waste. Molecules (Basel, Switzerland) 25 (12), 2811.

Kushwaha, A., Agarwal, S., Gupta, K.K., Maurya, S., Chaurasia, P.K., et al., 2017. Laccase enzyme from white rot fungi: an overview and its' application. In: Singh, M.P., Verma, V., Singh, A.K. (Eds.), Incredible World of Biotechnology. Nova Science Publishers, Inc, New York, pp. 25–41. ISBN: 978-1-53611-097-5.

Kushwaha, A., Maurya, S., Pathak, R.K., Agarwal, S., Chaurasia, P.K., Singh, M.P., 2018. Laccase from white rot fungi having significant role in food, pharma, and other industries. In: Bharati, S.L., Chaurasia, P.K. (Eds.), Research Advancements in Pharmaceutical, Nutritional and Industrial Enzymology. IGI Global, United States, ISBN 9781522552376.

Li, Q., Liu, J., Gadd, G.M., 2020. Fungal bioremediation of soil co-contaminated with petroleum hydrocarbons and toxic metals. Appl. Microbiol. Biotechnol. 104, 8999–9008.

Lin, Y., Xiao, W., Ye, Y., et al., 2020. Adaptation of soil fungi to heavy metal contamination in paddy fields—a case study in eastern China. Environ. Sci. Pollut. Res. 27, 27819–27830.

Maina, U.M., Galadima, I.B., Gambo, F.M., Zakaria, D., 2018. A review on the use of entomopathogenic fungi in the management of insect pests of field crops. J. Entomol. Zool. Stud. 6 (1), 27–32.

Mantzoukas, S., Eliopoulos, P.A., 2020. Endophytic entomopathogenic fungi: a valuable biological control tool against plant pests. Appl. Sci. 10, 360.

Metting, F.B., 1993. Structure and physiological ecology of soil microbial communities. In: Metting Jr., N.F.B. (Ed.), Soil Microbial Ecology. Marcel Dekker, Inc., New York, pp. 3–25.

Meyer, V., Basenko, E.Y., Benz, J.P., Braus, G.H., Caddick, M.X., Csukai, M., et al., 2020. Growing a circular economy with fungal biotechnology: a white paper. Fungal Biol. Biotechnol. 7, 5.

Moore, D., Robson, G.D., Trinci, A.P.J., 2011. 21st Century Guidebook to Fungi, first ed. Cambridge, New York.

Ogbonna, D.N., Douglas, S.I., Awari, V.G., 2020. Characterization of hydrocarbon utilizing bacteria and fungi associated with crude oil contaminated soil. Microbiol. Res. J. Int. 30 (5), 54–69.

Oladipo, O.G., Awotoye, O.O., Olayinka, A., Bezuidenhout, C.C., Maboeta, M.S., 2018. Heavy metal tolerance traits of filamentous fungi isolated from gold and gemstone mining sites. Braz. J. Microbiol. 49 (1), 29–37.

Olukunle, O.F., Oyegoke, T.S., 2016. Biodegradation of crude-oil by fungi isolated from cow dung contaminated soils. Niger. J. Biotechnol. 31, 46–58.

Orchids and fungi: An unexpected case of symbiosis. Am. J. Bot. July 12, 2011.

Przystaś, W., Zabłocka-Godlewska, E., Grabińska-Sota, E., 2018. Efficiency of decolorization of different dyes using fungal biomass immobilized on different solid supports. Braz. J. Microbiol. 49 (2), 285–295.

Ravalason, H., Jan, G., Mollé, D., et al., 2008. Secretome analysis of *Phanerochaete chrysosporium* strain CIRM-BRFM41 grown on softwood. Appl. Microbiol. Biotechnol. 80 (4), 719–733.

Rizwan, M., Atta, B., Sabir, A.M., et al., 2019. Evaluation of the entomopathogenic fungi as a non-traditional control of the rice leaf roller, *Cnaphalocrocis medinalis* (Guenee) (Lepidoptera: Pyralidae) under controlled conditions. Egypt. J. Biol. Pest Control 29, 10.

Robles-Acosta, I.N., Chacón-Hernández, J.C., Torres-Acosta, R.I., Landeros-Flores, J., Vanoye-Eligio, V., Arredondo-Valdés, R., 2019. Entomopathogenic fungi as biological control agents of *Phyllocoptruta oleivora* (Prostigmata: Eriophyidae) under greenhouse conditions. Fla. Entomol. 102 (2), 303–308.

Sant'Anna da Silva, A.P., Xavier da Kunha, R., et al., 2020. Secondary metabolites from endophytic fungi: potential for bio-based products developments. In: Chaurasia, Bharati (Eds.), Research Advances in the Fungal World: Culture, Isolation, Identification, Classification, Characterization, Properties and Kinetics. Nova Science Publishers, Inc., New York, pp. 115–138. ISBN: 978-1-53617-197-6.

Saxena, R.K., Gupta, R., Saxena, S., Gulati, R., 2001. Role of fungal enzymes in food processing. Appl. Mycol. Biotechnol. 1, 353–386.

Sharma, M., Chaurasia, P.K., Yadav, A., Yadav, R.S.S., Yadava, S., Yadav, K.D.S., 2016. Purification and characterization of a thermally stable yellow laccase from *Daedalea flavida* MTCC-145 with higher catalytic performance towards selective synthesis of substituted benzaldehydes. Russ. J. Bioorg. Chem. 42 (1), 59–68.

Sharma, K., Vaidya, D., Sharma, N., 2019. Recent trends of enzymes in food industry: a review. Int. J. Chem. Stud. 7 (2), 1525–1529.

Simard, S.W., Perry, D.A., Jones, M.D., Myrold, D.D., Durall, D.M., Molina, R., 1997. Net transfer of carbon between ectomycorrhizal tree species in the field. Nature 388 (6642), 579–582.

Singh, D., Gupta, N., 2020. Microbial laccase: a robust enzyme and its industrial applications. Biologia 75, 1183–1193. Available from: https://doi.org/10.2478/s11756-019-00414-9.

Singh, D., Raina, T.K., Singh, J., 2017. Entomopathogenic fungi: an effective biocontrol agent for management of insect populations naturally. J. Pharm. Sci. Res. 9 (6), 830–839.

Singh, J., Das, A., Yogalakshmi, K.N., 2020. Enhanced laccase expression and azo dye decolourization during co-interaction of *Trametes versicolor* and *Phanerochaete chrysosporium*. SN Appl. Sci. 2, 1095.

Singh, U., Bhatt, R.P., 2020. Wild edible mushrooms of Rudraprayag district, Uttarakhand Himalaya: diversity, distribution and ethnomycology. In: Chaurasia, Bharati (Eds.), Research Advances in the Fungal World: Culture, Isolation, Identification, Classification, Characterization, Properties and Kinetics. Nova Science Publishers, Inc., New York, pp. 61–68. ISBN: 978-1-53617-197-6.

Smith, S.E., Read, D.J., 1997. Mycorrhizal Symbiosis, second ed. Academic Press, London, United Kingdom, p. 605.

Spoorthi, N.S., Sandhyashree, B.M., Sangeetha, K.S., Yashaswini, Sushma, S.J., 2020. Energy management in smart mushroom factory. Int. J. Eng. Res. Technol. NCCDS – 2020 8 (13).

Sridharan, R., Krishnaswamy, V., 2020. Ancient and recent methods of characterization of fungi and its breakthrough in their identification. In: Chaurasia, Bharati (Eds.), Research Advances in the Fungal World: Culture, Isolation, Identification, Classification, Characterization, Properties and Kinetics. Nova Science Publishers, Inc., New York, pp. 185–212. ISBN: 978-1-53617-197-6.

Stamets, P., 2000. Growing Gourmet and Medicinal Mushrooms. Ten Speed Press, Berkeley, CA, pp. 233–248, ISBN 978-1-58008-175-7.

Steliga, T., 2012. Role of fungi in biodegradation of petroleum hydrocarbons in drill waste. Pol. J. Environ. Stud. 21 (2), 471–479.

Sudeep, K.C., Upadhyaya, J., Joshi, D.R., Lekhak, B., Chaudhary, D.K., Pant, B.R., et al., 2020. Production, characterization, and industrial application of pectinase enzyme isolated from fungal strains. Fermentation 6 (59). Available from: https://doi.org/10.3390/fermentation6020059.

Thapa, S., Mishra, J., Arora, N., et al., 2020. Microbial cellulolytic enzymes: diversity and biotechnology with reference to lignocellulosic biomass degradation. Rev. Environ. Sci. Biotechnol. 19, 621–648.

Uddin Pk, M.M., Pervin, R., Rahman, M.M., 2020. Role of edible mushrooms in cancer treatment. In: Chaurasia, Bharati (Eds.), Research Advances in the Fungal World: Culture, Isolation, Identification, Classification, Characterization, Properties and Kinetics. Nova Science Publishers, Inc., New York, pp. 213–220. ISBN: 978-1-53617-197-6.

Wang, B., Qiu, Y.L., 2006. Phylogenetic distribution and evolution of mycorrhizas in land plants. Mycorrhiza 16 (5), 299–363.

Waksman, S.A., 1947. What is an antibiotic or an antibiotic substance?. Mycologia 39, 565–569.

Yesilada, O., Birhanli, E., Geckil, H., 2018. Bioremediation and decolorization of textile dyes by white rot fungi and laccase enzymes. In: Prasad, R. (Ed.), Mycoremediation and Environmental Sustainability. Fungal Biology. Springer, Cham.

Yuvaraj, M., Ramasamy, M., 2020. In: Mirmajlessi, S.M., Radhakrishnan, R. (Eds.), Role of Fungi in Agriculture, Biostimulants in Plant Science. IntechOpen, DOI: 10.5772/intechopen.89718. Available from: https://www.intechopen.com/books/biostimulants-in-plant-science/role-of-fungi-in-agriculture.

Zhang, D., Yin, C., Abbas, N., et al., 2020. Multiple heavy metal tolerance and removal by an earthworm gut fungus *Trichoderma brevicompactum* QYCD-6. Sci. Rep. 10, 6940.

Zhou, B., Ma, C., Xia, T. et al., 2020. Isolation, characterization and application of theophylline-degrading *Aspergillus* fungi. Microb. Cell Fact 19, 72. Available from: https://doi.org/10.1186/s12934-020-01333-0.

Further reading

From Wikimedia Commons, the free media repository, 2020. File: Fungi Collage.jpg, https://commons.wikimedia.org/wiki/File:Fungi_collage.jpg.

From Wikimedia Commons, the free media repository, 2020. File: Penicillium labeled cropped.jpg, https://commons.wikimedia.org/wiki/File:Penicillium_labeled_cropped.jpg.

From Wikimedia Commons, the free media repository, 2020. File:Mycorrhizal network.svg, https://commons.wikimedia.org/wiki/File:Mycorrhizal_network.svg.

From Wikipedia, the free encyclopedia, 2020. Entomopathogenic fungus, https://en.wikipedia.org/wiki/Entomopathogenic_fungus.

From Wikipedia, the free encyclopedia, 2020. Fungal extracellular enzyme activity. https://en.wikipedia.org/wiki/Fungal_extracellular_enzyme_activity.

From Wikipedia, the free encyclopedia, 2020. Fungus. https://en.wikipedia.org/wiki/Fungus.

From Wikipedia, the free encyclopedia, 2020. Mushroom. https://en.wikipedia.org/wiki/Mushroom

From Wikipedia, the free encyclopedia, 2020. Mycorrhiza. https://en.wikipedia.org/wiki/Mycorrhiza.

Fungi Uses in Pest Management, 2020. http://kumu.brocku.ca/BIOL_2P96_Jan_2013-group07/Fungi_Uses_in_Pest_Management#_note-textbook_fungi.

Litterfall decomposition of selected plant species and nutrient cycling in Madhupur Sal (*Shorea robusta Roth*) forest of Bangladesh

8

A.K.M. Nazrul Islam, Tahmina Jamali and Ahmed Emdadul Hoque
Ecology and Environment Laboratory, Department of Botany, University of Dhaka, Dhaka, Bangladesh

8.1 Introduction

Forest plants accumulate nutrients in their biomass for long periods until they are harvested, but a major portion of nutrients absorbed found their way back to the surface through shedding of the dead parts litter. It is decomposed by soil organisms, mainly microbes slowly or rapidly, with the release of nutrients and are utilized by the plants for their further growth (Pandey and Singh, 1984). This cycling of nutrients between soil and plants performs ecological, physiological, and biochemical functions repeatedly. In undisturbed natural forests, litter production and decomposition are thought to stay in an equilibrium situation whereby the production equals decomposition and the litter turnover rate remains the same over periods of time (Evans, 1981). Decomposition of litter forms the most important link in the recycling of nutrients in the forest ecosystem (Maclean and Wein, 1978). If the decomposition is fast, organic matter status of soil may deplete, and the mobile organic nutrient elements may be lost by leaching or run off. If it is slow, nutrients accumulate in litter, a net immobilization and nutrient unavailability may result.

About 60%–90% of fallen litters are leaves (Zavitkovski and Newton, 1971; Radwann et al., 1984). Therefore the decomposition rate of leaf litters ultimately determines the amount of nutrients to be released. The decomposability of leaf litter, therefore, influences primarily the efficiency of nutrient cycling. In addition, climatic factors, the litter of different plant species decompose at different extents and rates and these differences are attributed to the chemical composition of the leaf litter (Fogel and Cromack, 1977). The soluble and insoluble as well as organic and inorganic constituents are the regulatory factors of decomposition (Berg et al., 1982). The forest resources of Bangladesh are not sufficient, and the entire forest cover is less than 0.9% of the land area (Rashid et al., 1995). Out of 2.46 million ha forest land in Bangladesh, a vast area is topographically plain and is termed as

Microbes in Land Use Change Management. DOI: https://doi.org/10.1016/B978-0-12-824448-7.00011-5
© 2021 Elsevier Inc. All rights reserved.

the plain land forest covering an area of 120,000 ha. The plain land forest is also known as "Sal forest" due to the existence of dominant tree "Sal" (*Shorea robusta*) and it covers about 32% (Khan, 1990) of the total forest land. Sal forests are the only tropical moist deciduous forest of Bangladesh (Champion, 1933).

Several workers attempted to quantify the site litterfall and decomposition as an important pathway for the transfer of litter mass and mineral to the soil surface in forest ecosystems. Loria-Naranjo et al. (2019) investigated mangrove leaf litter decomposition in a seasonal tropical environment. They analyzed the influence of weather conditions on the decay rate, nitrogen, and phosphorus content as a decomposition indicator of leaf litter of *Rhizophora racemosa*. Intensive studies have been carried out in many parts of the world (Bray and Gorham, 1964; Hopkins, 1966; Birk and Simpson, 1980); but most of these studies involved comparison between litter decomposition of single species at different sites (Sharma and Ambasht, 1987). The aim of the present investigation was to observe decomposition of leaf litter of 10 tree species for a period of 360 days and nutrient cycling in the Sal forest of Bangladesh: changes in nutrient concentration of N, P, K, Ca, Mg, Fe, Mn, and Zn and also relative mass loss rates of leaf litter during decomposition by microbes and soil organisms.

8.2 Overview of the Madhupur Sal forest

8.2.1 Environmental conditions

The Madhupur Sal forest is in the north central part of Bangladesh, Latitude 24°30′ to 24°36′N; Longitude 90°05′ to 90°15′E (Agroecological region 28; FAO, 1988; Fig. 8.1) and the forest map is shown in Fig. 8.2. The forest is dominated by *S. robusta* locally known as "Sal or Garjan" and in the high land is locally called "Challas" intercepted by numerous depressions of valleys locally known as: "Baid" lands which are cultivated. Forest and cultivation are thus intricately mixed up. The high lands containing the forest look like numerous islands surrounded by narrow depression. The canopy layer is continuous at about 50–60 ft. from the ground (Narul Islam et al., 2010). Sal itself supplies 85%–90% trees. The plants of the Madhupur forest are deciduous in nature (Plates 8.1–8.4). In late winter (February to middle of March) the leaves; color change from green to yellow before defoliation. The forest floor is thus covered with a thick layer of leaves at the end of February (Plates 8.1–8.4).

8.2.2 Temperature and rainfall

Mean monthly temperature of 1985–94 (maximum and minimum) of the nearest two districts (Mymensingh and Tangail) of the forest are given in Table 8.1. The mean maximum and minimum monthly temperature was 32.49°C (April) and 11.95°C (January). The average annual highest and lowest rainfall was 3411 and 1584 mm and 2592 and 1570 mm in Mymensingh and Tangail respectively. The

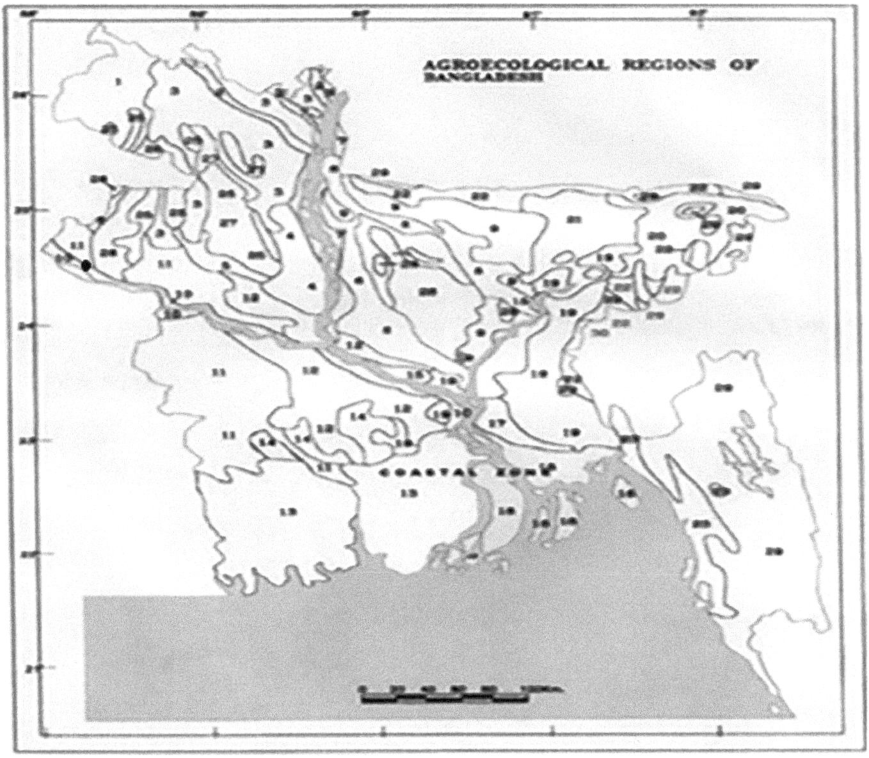

Figure 8.1 Agro-ecological regions showing location of Madhupur Sal forest shown with "•" sign. (1) old Himalayan piedmont plain, (2) active Tista floodplain, (3) Tista Mander floodplain, (4) Karatoya-Bangali floodplain, (5) lower Atrai Basin, (6) lower Purnabhaba floodplain, (7) active Brahmaputra-Jumuna floodplain, (8) Young Brahamaputra and Jamuna floodplains, (9) old Brahmaputra floodplain, (10) active Ganges floodplain, (11) high Ganges river floodplain, (12) low Ganges river floodplain, (13) Ganges tidal floodplain, (14) Gopalganj-Khulna Bils, (15) Arial Bil, (16) middle Meghna river floodplain, (17) lower Meghna river floodplain, (18) Young Meghna Estuarine floodplain, (19) old Meghna Estuarine floodplain, (20) Eastern Surma-Kusiyara floodplain, (21) Sylhet basin, (22) Northern and Eastern piedmont plains, (23) Chittagong coastal plain, (24) St. Martin's Coral Island, (25) level Barind Tract, (26) high Barind Tract, (27) North-Eastern Barind Tract, (28) Madhupur Tract, (29) Northern and Eastern Hills, (30) Akhaura Terrace.

lowest rainfall was in the December (00–14 mm), January (00–08 mm), and February (00–59 mm) and the highest rainfall was in June (603 mm), July (789 mm), and August (512 mm). The region lies in the monsoon belt and receives nearly 75% of the annual precipitation during the above months.

The dominant tree species of this forest is *S. robusta* (Sal) which forms extensive stands often mixed with numerous broad-leaved species including *Dillenia pentagyna*, *Semecarpus anacardium*, *Adina cordifolia*, *Terminalia bellirica*, *Mallotus philippinensis*, *Grewia microcos*, *Garuga pinnata*, *Lagerstroemia* sp.

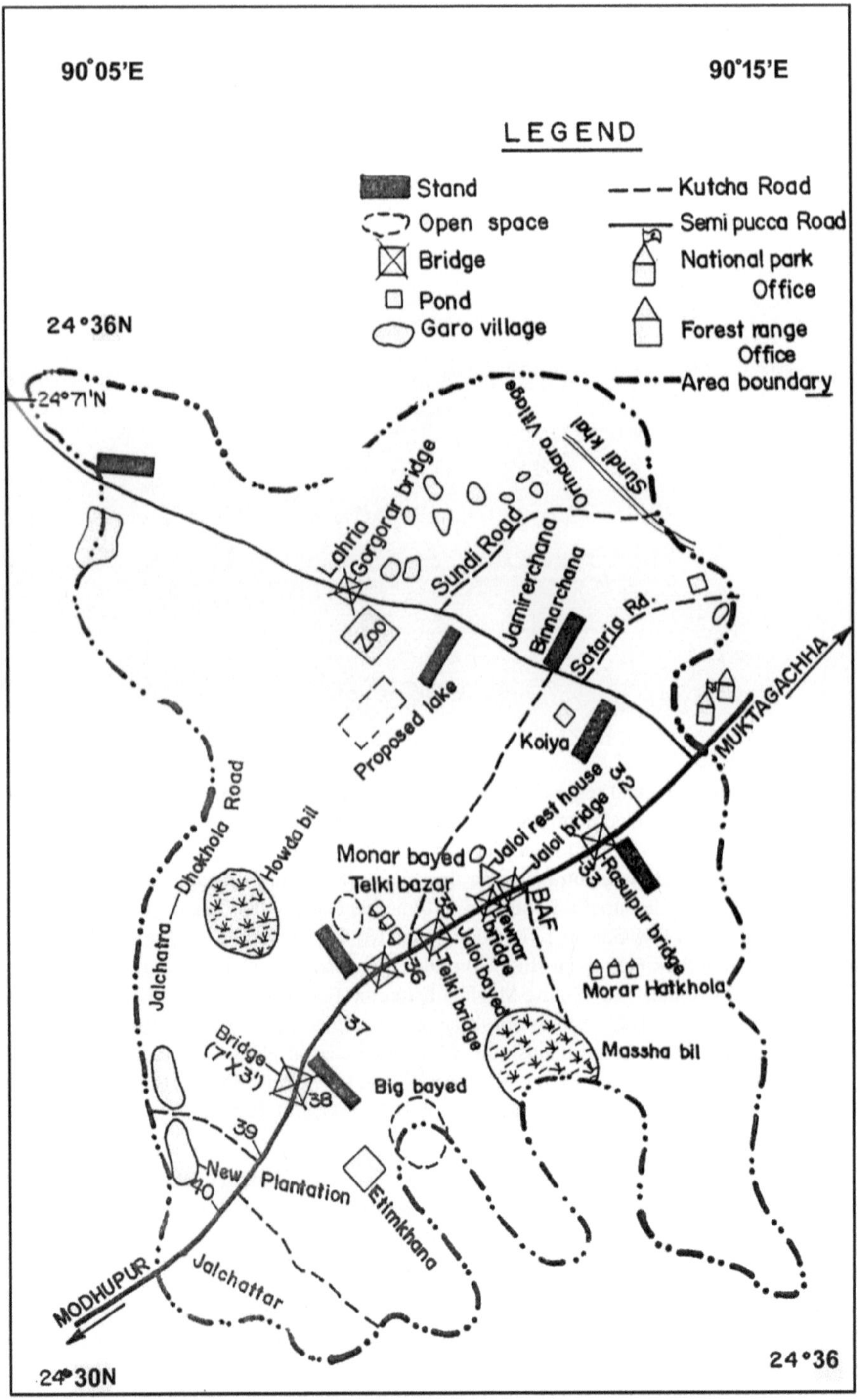

Figure 8.2 Ecological regions showing location of Madhupur Sal forest shown with “x” sign.

Plate 8.1 Leaves of several species on the forest floor.

Plate 8.2 *Shorea robusta.*

Plate 8.3 *Dillenia pentagyna.*

Plate 8.4 *Phyllanthus embelica* after defoliation.

8.3 Geology

8.3.1 Origin of the Madhupur forest

There are many controversies about the origin of the Madhupur tract. Some information about the origin is given. Morgan and McIntire (1959) recognized the tract as a Pleistocene terrace. They observed much similarity between the Gulf coast terraces and the Madhupur tract. They opined that sediments examined in the Pleistocene areas of Bangladesh were deposited as floodplains of earlier "Ganges-Brahmaputra river system." According to them it has been uplifted, and to justify it they have mentioned the present altitude and sharp escarpment of the tract.

Islam (1976) in his study of Quaternary Geology of the Dhaka region observed that the Madhupur clay was deposited in a shallow freshwater basin. Then with gradual infilling of the basin with sediments, the condition changed into a flood plain condition. There is also the opinion that Madhupur tract sediments were deposited not in a flood plain but in a deltaic environment. The Madhupur tract should therefore be called an uplifted delta.

8.4 Materials and methods

Soil Analysis: Soil samples were collected from 0 to 10 cm depth. Soils were extracted immediately after sampling (Gupta and Rorison, 1974; Nazrul Islam and Rorison, 1978). 10 g fresh soil samples were weighed and extracted with 50 mL neutral ammonium acetate on an end shaker for 16 hours at c. 20°C for exchangeable elements. Samples were centrifuged at 3000 RPM for 10 minutes. The tubes were washed with twice with 25 mL solution and centrifugation was repeated. The supernatant was filtered with Whatman No.1 filter paper. Calcium, magnesium,

Table 8.1 Mean monthly maximum and minimum temperature (°C) at different months and rainfall (mm. average annual) at different years of Mymensingh and Tangail Districts (Data of 1985–94).

A. Mymensingh												
Temperature	January	February	March	April	May	June	July	August	September	October	November December	December
Maximum	24.98	27.10	30.51	32.49	31.34	31.26	31.30	31.60	31.13	31.31	29.65	26.36
Minimum	11.95	14.64	18.49	22.16	23.34	25.49	25.89	26.11	25.24	22.90	18.04	13.11
B. Tangail												
Maximum	24.77	27.63	31.30	34.18	32.86	32.29	31.65	31.79	31.64	31.30	29.16	25.88
Minimum	11.76	14.44	18.79	22.50	23.60	25.59	26.05	26.31	25.79	23.61	18.58	13.61
Rainfall in different years	1985	1986	1987	1988	1989	1990	1991	1992	1993	1994		
A. Mymensingh	2008	3053	2239	3134	2457	2364	3411	1584	3215	1604		
B. Tangail	1624	2089	1573	1645	2567	1570	2592	2592	2950	1805		

manganese, iron, and Zn were determined by Atomic Absorption Spectrometry, and potassium was analyzed by the flame photometer.

Plates (8.1–8.3) show the leaves of various species on the forest floor and (8.4) defoliated tree.

Ten plant species including *A. cordifolia* Hook, *Buhinia purpurea* L., *Burserra serrata* Wall. ex. Colebar, *D. pentagyna* Roxb. *G. microcos* L., *Miliusa velutina* Hook. f., *Schleichera trijuga* Willd., *S. anacardium* L. f., *S. robusta* Roth., *T. bellirica* (Gaertn.) Roxb. were selected for the investigation. *S. Robusta* is the dominant species and the remaining nine are associated species. Litter bag technique was used for the determination of the decomposition rate of leaf litter. Freshly fallen leaves were collected from the ground surface; litter bags 20 cm × 20 cm of 1 mm mesh fine non-decomposing nylon nets were used for the eight species while for *D. pentagyna* and *S. anacardium* 40 cm × 30 cm size bags of the same mesh was used for collection in February, 1998. For each species 24 bags were placed on surface soil on March 1998. After 60 days four bags of each species were collected. Soils and other adhering materials were removed by using brush and then the bags were opened. The litters were taken out from the bags and the particles attached with the litters were removed by brush, air dried, and then weighed. The weight loss was calculated. From the weight of material remaining after decomposition and concentration of each of the elements, the amounts of nutrients retained, and the amount of nutrient released or immobilized were calculated. Air dried materials were made powder by grinder and were then analyzed for C, N, P, K, Ca, Mg, Fe, Mn, and Zn. Plant samples were made to ash at 450°C in a muffle furnace for 16 hours and the ash was dissolved in N/10 nitric acid. The solution was then analyzed for Ca, Mg, Mn, and Fe by atomic absorption spectrophotometry, and K by flame photometer.

Soil samples were collected from litter free and litter added areas of study sites, dried in the sun, grounded and were analyzed for physicochemical properties. Organic carbon was determined by wet oxidation method of Walkley and Black (1934). Total nitrogen was determined by Jackson (1961).

8.5 Results

8.5.1 Edaphic features

Physicochemical properties of soil of Madhupur Sal forest are given in Table 8.2. Deforested areas are cultivated for rice and other crops and have high organic matter. Soil pH is acidic, and the value ranged from 4.8 to 6.3. In general, exchangeable calcium was higher than magnesium. In lower layer calcium content showed an increasing trend under the effect of evaporation. The exchangeable Mn and Fe contents showed seasonal variation. Maximum value of Mn was found in June–July during the rainy season when the forest floor was flooded. Manganese content is very high in forest soil whereas iron content is very high in deforested soil. Deforested soil (i.e., cultivated soil) high dose of fertilizer is used for high yielding rice cultivation at least two times in a year.

Table 8.2 Physicochemical properties of soils (milliequivalent/100 g) of Madhupur Sal forest (values are in range); Data in parenthesis are the values of deforested areas.

pH	Organic carbon %	Total N %	C/N ratio	P	Ca	Mg	Mn	Fe
4.8–6.3 (4.9–6.4)	0.49–0.53 (0.70–0.80)	0.8–1.0 (0.09–0.13)	5.2–6.1 (6.2–7.8)	2.20–2.30 (3.60–3.3.65)	1.02–2.51 (1.15–2.91)	1.00–1.24 (0.82–1.15)	184.39–308.70 (1.20–1.35)	33.30–33.3 (90.78–130.70)

8.5.2 Decomposition of leaf litter

Decomposition and accompanied release of nutrients from leaf litter of 10 species were investigated and showed significant variations in the rate of decomposition among the species over the periods. Decomposition of leaf litter of the species was evaluated from the amount of weight loss from litter bags at 60 days intervals up to 360 days of 10 species, that is, *A. cordifolia, B. purpurea, B. serrata, D. pentagyna G. microcos, M. velutina, S. trijuga, S. anacardium, S. robusta, T. bellirica.* is presented in Table 8.3. The results show that the weight loss rate varied with species and time. The highest initial decomposition was noted in *M. velutina* and *T. bellirica* (0.86 g/bag) followed by *S. robusta* (0.62 g/bag). There was very slow decomposition in the first 60 days in *B. purpurea* (0.05 g/bag) and *A. cordifolia* (99%) 06 g/bag. After 360 days the proportion of leaf litter decomposition was highest in *A. cordifolia* (99%) and the least in *S. trijuga* (42%). Initially *A. cordifolia* leaf litter showed slow decomposition. After 120 days there was a sharp rise in the decomposition rate of *A. cordifolia* leaf litter. The amount of litter remaining after decomposition for different periods among the species is shown in Fig. 8.3.

8.5.3 Concentration of ash and nutrients

The initial concentrations of ash and nutrients in a leaf subjected to decomposition are given in Table 8.3. Ash was lowest in *A. cordifolia* (2.52%) *M. velutina* (2.99%) and the highest value was noted in *B. Serrata* (15.14%) followed by *S. anacardium* (14.95%). The percentage of carbon varied from 51.12 (*B. purpurea*) to 57.93 (*T. bellirica*) and the nitrogen concentration was 0.35%, 1.28%, 0.69%, 0.53%, 0.31%, 0.38%, 1.28%, 0.66%, 1.33%, and 0.34% in *A. cordifolia, B. purpurea, B. serrata, D. pentagyna, G. microcos, M. velutina, S. trijuga, S. anacardium, S. robusta*, and *T. bellirica* respectively giving corresponding C/N ratio of 162, 40, 78, 100, 185, 145, 41, 83, 41, and 170. Concentrations of P, K, Ca, and Mg were within the range of 0.08 (*G. microcos*) to 0.25 (*B. purpurea*); 0.17 (*G. microcos*) to 0.90 (*B. purpurea*); 0.26 (*M. velutina*) to 1.80 (*S. robusta*); and 0.09 (*S. trijuga*) to 0.56 (*B. purpurea*) respectively. *G. microcos* had the lowest concentration of macronutrients except calcium and magnesium. This species had the highest concentration of iron (250 ppm) and Zn (125 ppm).

8.5.4 Changes due to decomposition

Changes in concentration of ash, nutrients, and C/N ratio due to decomposition at different periods are described below.

8.5.4.1 Ash

After decomposition, the ash content of litters increased, and after 60 days in decomposed litters, ash content was highest in *S. anacardium* (29.01) followed by

Table 8.3 Initial ash and nutrients of 10 tree species litter subjected to decomposition.

Ash, C/N and nutrients (%)	***Adina cordifolia***	***Bauhinia purpurea***	***Bursera serrata***	***Dillenia pentagyna***	***Grewia microcos***	***Miliusa velutina***	***Schleichera trijuga***	***Semecarpus anacardium***	***Shorea robusta***	***T. bellirica***
Ash	2.52	14.40	15.14	14.80	3.62	2.99	7.94	14.95	8.00	3.63
Carbon	56.82	51.12	54.12	53.12	57.28	54.92	52.32	55.02	54.21	57.93
Nitrogen	0.35	1.28	0.69	0.53	0.31	0.38	1.28	0.66	1.33	0.34
Phosphorus	0.09	0.25	0.19	0.15	0.08	0.10	0.09	0.20	0.11	0.12
Potassium	0.19	0.90	0.35	0.30	0.17	0.20	0.50	0.33	0.55	0.18
Calcium	0.27	0.79	0.30	0.28	0.27	0.26	1.69	0.29	1.80	0.29
Magnesium	0.36	0.56	0.16	0.10	0.34	0.33	0.09	0.35	0.11	0.35
C/N ratio	162	40	78	100	185	145	41	83	41	170
Iron	240	123	140	145	250	245	75	147	70	247
Mn	202	206	428	458	208	207	86	450	80	210
Zn	120	64	66	63	125	119	80	55	79	119

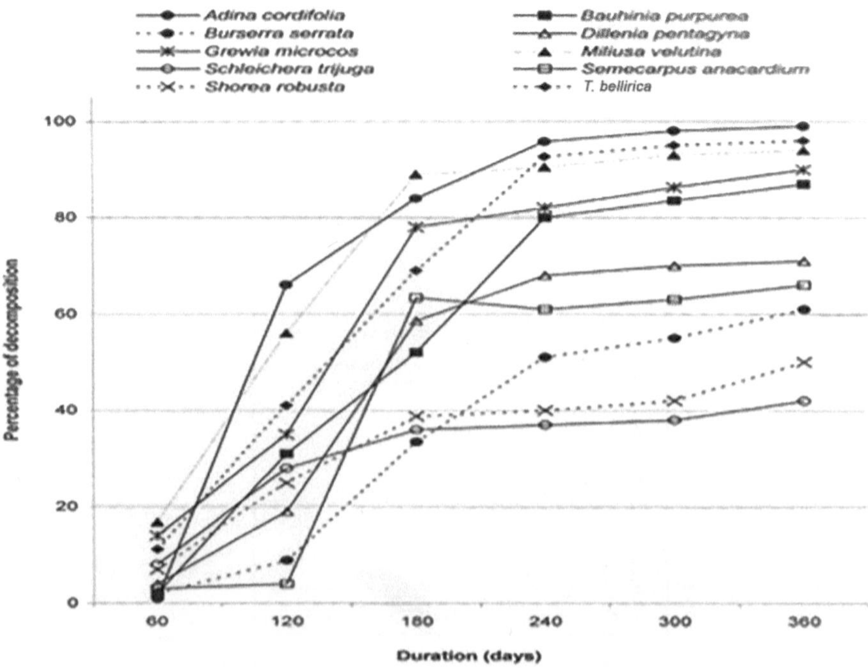

Figure 8.3 Weight loss due to decomposition of leaf litters of 10 tree species.

B. serrata (28.47) and *D. pentagyna* (27.97) and *B. purpurea* (23.25). The values of the remaining species were in the range of 11.06–12.15 (Table 8.3).

8.5.4.2 Carbon

After 60 days of decomposition the highest percentage of carbon contained in *A. cordifolia* (52.66%) followed by *S. trijuga* and *S. robusta*, *T. bellirica* (52.1%, 51.62%, and 50.69%, respectively). The lowest value was noted in *S. anacardium* (40.98%). Finally, after 360 days decomposed carbon was highest in *D. pentagyna* (41.05%) followed by *S. anacardium* and *B. serrata* 40.66% and 40.13%, respectively. The lowest carbon value after 360 days was found in *A. cordifolia* (23.49%).

8.5.4.3 Nitrogen

Changes of nitrogen concentration are given in Fig. 8.4. The lowest percentage of N was noted in *G. microcos* (20%) and the highest percentage was found in *S. trijuga* (1.41%) followed by *S. robusta* (1.38%) after 360 days of decomposition. The highest concentration of N in decomposing leaf litter was 1.68 in *B. purpurea* (60 days); 1.58 in *S. trijuga* (180 days), and 1.57 for *S. robusta* (180 days).

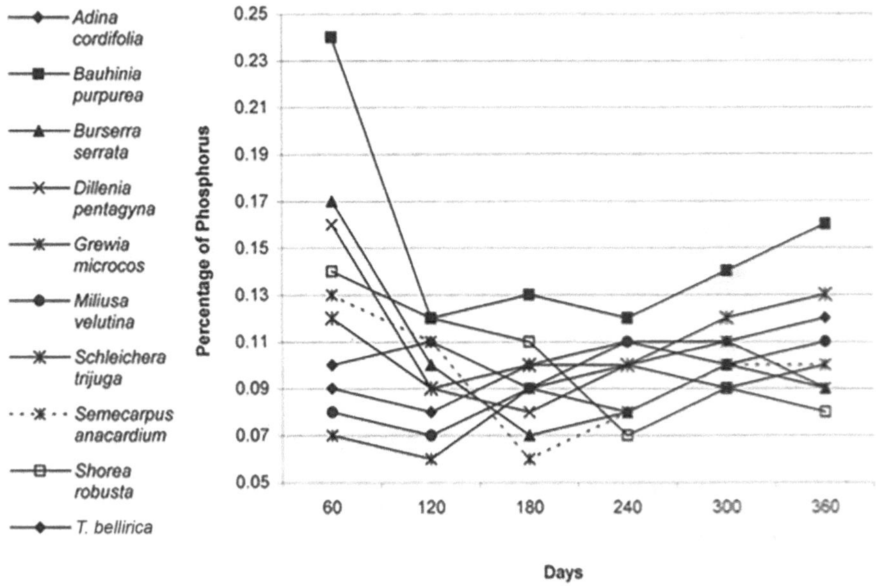

Figure 8.4 Percentage of nitrogen in decomposing leaf litters of 10 species.

8.5.4.4 C/N ratio

This ratio was variable for all the species (Fig. 8.5). There were occasional increases and decreases in C/N ratio after some periods of decomposition. Finally, after 360 days decomposed litter, this ratio was highest in *G. microcos* (127%) followed by *T. bellirica* (116%) and *M. velutina* (112%). The lowest ratio was 20% noted in *S. robusta* and *B. purpurea* (24%).

8.5.4.5 Phosphorus

Phosphorus concentration varied from 0.11% (120 days) to 0.24% (360 days) in *A. cordifolia*; 0.20% (120 days) to 0.39% (300 days) in *B. purpurea*; 0.09% (120 days) to 0.23% (360 days) in *G. microcos*; 0.10% (120 days) to 0.25% (360 days) in *M. velutina*; 0.18% (120 days) to 0.38% (360 days) in *S. trijuga*; 0.09% (120 days) to 0.19% (60 days) in *S. anacardium*; 0.19% (120 days) to 0.37% (180 and 300 days) in *S. robusta*; 0.10% (120 days) to 0.24% (360 days) in *T. bellirica* (Fig. 8.6).

8.5.4.6 Calcium

In all 10 species calcium concentration increased over the initial concentration after 60 days of decomposition and it decreased abruptly after 60 days (Table 8.5). The highest concentration of calcium was observed in *S. robusta* (1.96%) and in *S triju-ga* (1.95%). Decomposition after 360 days resulted in Ca concentration of 0.07%, 0.11%, 0.10%, 0.11%, 0.05%, 0.06%, 0.49%, 0.12%, 0.48%, and 0.05% in leaf

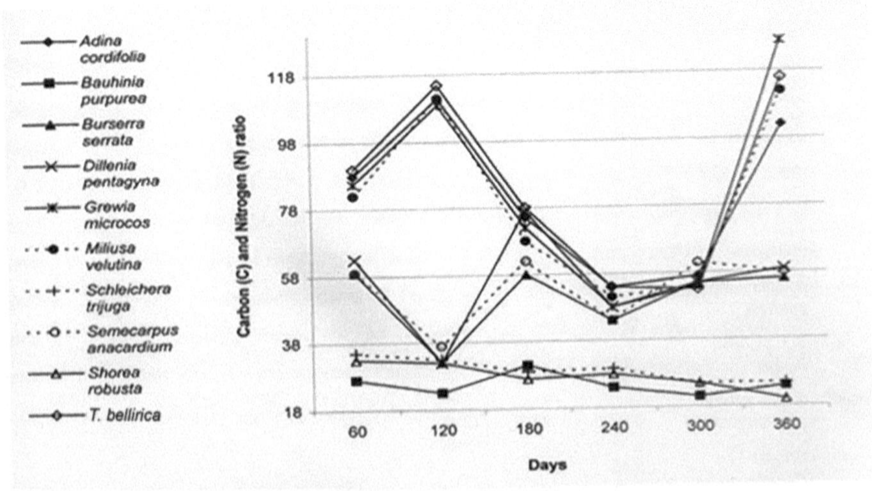

Figure 8.5 Variation of C/N ratio of decomposing leaf litters in 10 species.

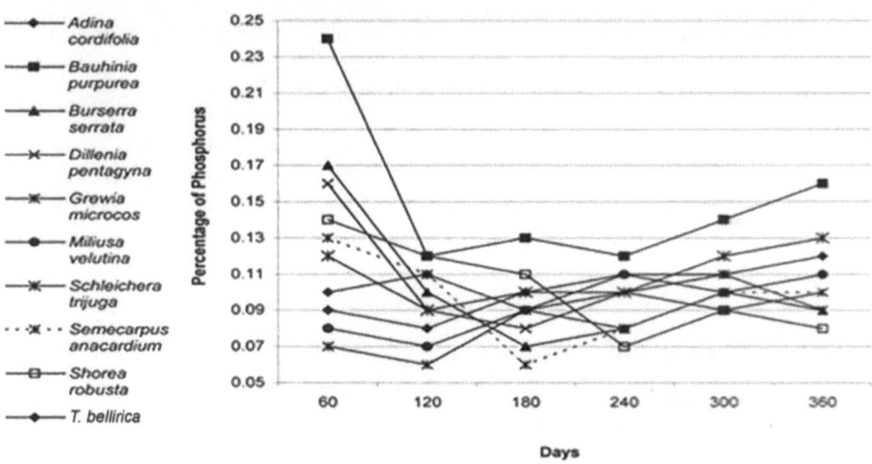

Figure 8.6 Percentage of phosphorus in decomposing leaf litters of 10 species.

litters of *A. cordifolia, B. purpurea, B. serrata, D. pentagyna, G. microcos, M. velutina S. anacardium, S. robusta, and T. bellirica*, respectively.

8.5.4.7 Magnesium

Decomposed litter of *S. trijuga* contained less magnesium than undecomposed litter (Tables 8.4 and 8.5). However, in other species the values were higher in the decomposed period. After 360 days of decomposition *B. purpurea* contained highest Mg (0.37%) and the lowest value was found in *G. microcos* (0.12%).

Table 8.4 Percentage of ash and carbon (values in parenthesis) in decomposing leaf litters at different days of 10 tree species.

Days	*Adina cordifolia*	*Bauhinia purpurea*	*Bursera serrata*	*Dillenia pentagyna*	*Grewia microcos*	*Miliusa velutina*	*Schleichera trijuga*	*Semecarpus anacardium*	*Shorea robusta*	*T. bellirica*
60	11.94 (52.66)	23.25 (46.09)	28.47 (42.18)	27.97 (42.68)	12.10 (49.96)	11.02 (50.16)	12.15 (52.10)	29.01 (40.98)	11.06 (51.62)	12.20 (50.66)
120	15.50 (50.07)	41.80 (34.38)	22.05 (45.37)	21.50 (44.96)	14.96 (47.07)	16.96 (48.66)	15.10 (48.39)	21.65 (47.73)	17.01 (47.79)	15.40 (48.70)
180	26.21 (45.70)	60.97 (23.98)	36.95 (37.86)	35.05 (48.68)	29.09 (42.70)	27.71 (42.98)	28.01 (43.10)	35.75 (38.68)	26.97 (43.50)	28.21 (43.07)
240	29.98 (40.82)	58.70 (24.81)	42.05 (33.50)	41.90 (33.95)	29.95 (36.28)	50;90 (38.02)	30.02 (29.09)	40.97 (34.05)	49.85 (30.00)	30.20 (37.82)
300	39.29 (45.83)	52.11 (27.37)	28.72 (41.43)	27.80 (41.01)	39.75 (34.83)	38.68 (35.50)	40.17 (36.08)	29.01 (42.34)	37.98 (35.96)	40.30 (35.38)
360	58.70 (23.49)	55.45 (25.37)	32.65 (40.13)	31.56 (41.05)	58.95 (25.40)	40.05 (24.68)	60.00 (35.89)	31.15 (40.66)	39.96 (35.19)	59.02 (24.40)

Table 8.5 Percentage of calcium and magnesium (values in parenthesis) in decomposing leaf litters of 10 tree species.

Days	*Adina cordifolia*	*Bauhinia purpurea*	*Bursera serrata*	*Dillenia pentagyna*	*Grewia microcos*	*Miliusa velutina*	*Schleichera trijuga*	*Semecarpus anacardium*	*Shorea robusta*	*T. bellirica*
60	0.51 (0.15)	1.08 (0.64)	0.34 (0.14)	0.33 (0.15)	0.48 (0.12)	050 (0.13)	1.95 (0.11)	0.33 (0.15)	1.96 (0.14)	0.50 (0.16)
120	0.40 (0.23)	0.23 (0.33)	0.31 (0.33)	0.30 (0.34)	0.38 (0.20)	0.39 (0.21)	1.10 (0.19)	0.29 (0.33)	1.11 (0.22)	0.38 (0.20)
180	0.17 (0.24)	0.07 (0.30)	0.07 (0.11)	0.06 (0.12)	015 (0.21)	0.16 (0.22)	0.84 (0.06)	0.08 (0.11)	0.85 (0.08)	0.14 (0.23)
240	0.08 (0.17)	0.05 (0.31)	0.03 (0.15)	0.04 (0.16)	0.08 (0.14)	0.07 (0.15)	0.12 (0.08)	0.05 (0.14)	0.13 (0.10)	0.08 (0.19)
300	0.14 (0.19	0.11 (0.35)	0.09 (0.16)	0.10 (0.17)	0.11 (0.16)	0.13 (0.17)	0.18 (0.16)	0.07 (0.16)	0.19 0.18)	0.12 (0.18)
360	0.07 (0.15)	0.11 (0.37)	0.10 (0.14)	0.11 (0.15)	0.05 (0.12)	0.06 (0.13)	0.49 (0.09)	0.12 (015)	0.48 (0.11)	0.05 (0.14)

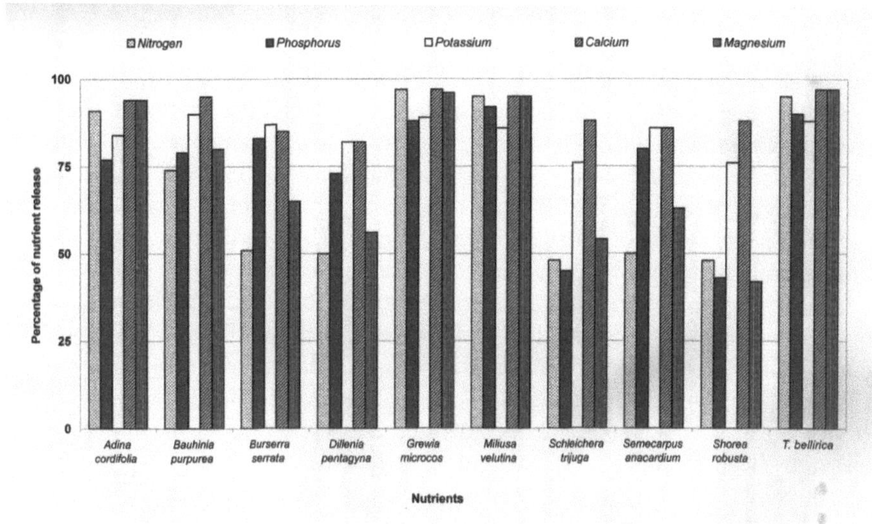

Figure 8.7 Release of nutrients from leaf litter over 360 days of decomposition.

8.5.5 Release of nutrients

The amount of nutrients viz. N, P, K, Ca, Mg, Fe, Mn, and Zn released as a result of decomposition of leaf litter were calculated by subtracting the amount present in decomposed litter from the initial content. At some decomposition periods the amount increased over the initial amount and then, the amount of increase was taken to be immobilization of nutrients by decomposing leaf litter from the available nutrient pool of the soil. The total release of N, P, K, Ca and Mg from litter bags over a period of 360 days of decomposition are given in Fig. 8.7. The relative proportion of release was the highest in case of Ca, Mg, and Mn. Among the species, *T. bellirica*, showed the maximum nutrient release and *S. robusta* showed the lowest rate; 97% of N in *G. microcos* was released and 48% release was noted in *S. trijuga and S. robusta*. There was little variation in the relative release of N, Ca, and Mg in *A. cordifolia, G. microcos, M. velutina* and *T. bellirica*. The pattern of release of micronutrients Fe, Mn, and Zn was different from the macronutrients. In most cases there was immobilization and, in some cases (nutrients or species) a net release was observed over a period of 360 days. Attention must be paid to the variety of leaves of plant species because the leaves of different species do decompose at different rates.

8.5.6 Role of microbes in land management

The decay of organic material-litter decomposition is a critical process for life on Earth and is an essential part of the global carbon cycle. The simple rates of decay of plant material, that is, the loss of weight over time, would vary in any one environmental condition according to the anatomical structure of leaf and litter involved. Analysis of the C and N content of decaying leaf tissue and the C:N ratio may prove to be a useful measurement to describe the microbial colonization. As litterfall and decomposition are influenced by physical and biological factors hence to provide valuable information about diversity of leaf litter microbes and fungi involved in this process will be useful. There is evidence that species belonging to class Zygomycota and Ascomycota and their anamorphs are involved in decomposition of leaf litter (Jatav et al., 2020). It is necessary to determine how the microbial functional pathways involved in nitrogen cycling and vary according to the changes in the microbial community composition induced by mixing of litter.

Investigating the effects of individual tree species on fungal species in leaf litter allows a mechanistic understanding of how tree diversity affects the diversity and composition of fungal species at the community level. The role of microorganisms in the input and output of nutrients is not considered in many cases. Microbial activity is particularly important when considering N and rates of fixation should be measured because blue green algae and bacteria associated with litters can be significant contributors to the nitrogen budget to any forest ecosystem. Litter decomposition and nitrogen metabolism are the key determinants of nutrient cycling in forest ecosystems. However, how processes and functional groups are affected by mixed litter decomposition is poorly understood. The difference of litter pH and nitrate content between mixed litter and Larix litter are determinants of changes in microbial functional potentials of nitrogen metabolism.

Knowledge of the environmental factors affecting leaf element stoichiometric patterns can reveal nutrient fluxes of ecosystems and their responses to global change. However, understanding of continental patterns of leaf nitrogen (N) and phosphorus (P) stoichiometry in relation to solar radiation is still insufficient. when predicting the effects of global dimming on ecosystem nutrient fluxes, one must distinguish the roles of direct, diffuse, and spectral components of solar radiation. All this information will of some use for land management in future.

8.6 Discussion

Soil pH gives some measure of general level of fertility (Wilde, 1954). Grubb (1964) noted low pH (4.2) with poor exchangeable in Montane Forest soil. Ovington and Madgwick (1957) have shown a positive correlation between pH of soils at different depths in woodlands and also those of tree leaves and litter layers, despite the very wide range of ecological conditions.

Higher exchangeable calcium than magnesium and increasing tendency of calcium is possible under the effect of evaporation. Lutz and Chandler (1955) noted that in temperate climates calcium usually increases in lower horizon. Ponnamperuma (1972) explained that deep rooted plants frequently build up the concentration of lime in surface soil as a result of contribution made by the leaf fall annually. The increase of calcium in the 0–10 cm soil in the present investigation may be due to decomposition of litter; subsequent decrease may be due to leaching. Seasonal variation of manganese and iron in June–July (rainy season) when the forest floor is flooded is interesting. Waterlogging of soil showed a high concentration of Mn (Gill, 1970; Nazrul Islam and Rorison, 1981; Nazrul Islam and Saha, 1980).

Litterfall helps maintain nutrient return in forest and the dynamics between litter fall and its decay is part of the proper functioning of ecosystems. The decomposition of leaf litter is a complex microbial phenomenon involving several interacting processes of breakdown and synthesis. Litter fall and the associated nutrient return are a significant supply of resources in forest ecosystems. In mixed stands both litter quantity and quality can change under different species compositions.

The rates of pathways and products largely depend on the environmental conditions and inherent litter quality characteristics such as soluble, insoluble, organic, and inorganic chemical constituents (Zinn et al., 2002; Bernhard, 1998; McClaugherty et al., 1985). Patterns of nutrient release and immobilization (incorporation into decomposing litter material from available nutrient pool of the soil) also differ in circumstances and substrate quality. Litter plays a central role in the nutrient budgets of forests by supplying inflow to nutrient turnover through decomposition. However, studies of litter decomposition and nutrient dynamics in forest ecosystems are limited.

In the present investigation, decomposition of leaf litters of *A. cordifolia, B. purpurea, B. serrata, D. pentagyna, G. microcos, M. velutina, S. trijuga, S. anacardium, S. robusta, T. bellirica* was noted for a continuous period of 360 days with periodic observation of 60 days intervals. In all these litters the greater proportion of mass loss occurred within 60–120 days. Further weight loss was gradual with time of decomposition. Only in three species, that is, *B. serrata, D. pentagyna*, and *S. anacardium*, the first 60–120 days showed a slow decomposition. Then 120–180 days, further weight loss was gradual with time of decomposition.

However, this two-phase decomposition—an early rapid phase followed by a later slow one occurs during decomposition of almost all litters as mentioned by McClaugherty et al. (1984), but the rate varies with species (Shanks and Olson, 1962). In the present study *A. cordifolia* litter showed the highest amount (99%) of mass loss after 360 days followed by *T. bellirica* (96%), *M. velutina* (94%), *G. microcos* (90%). Decomposition of 90%–99% litter within 1 year appears to be high. The disappearance of 90% litter in *Pinus caribaea* within 1 year was reported by Osman et al. (1997). Leaf litters of a particular species may vary at different geographical conditions owing to variations in climate (Berg and Argen, 1984; McGrath et al., 2000; Torreta and Takada, 1999; Khanna, 1997) and population of soil microorganisms (Singh and Gupta, 1977; Swift et al., 1979). The importance of soil fauna and microbes in the regulation of plant litter decomposition in the tropics was reported by Warren and Zou (2002). In the present study fast decomposition of *A. cordifolia, T. bellirica, M. velutina*, and *G. microcos* litter which form the major proportion of total litter, would result in little accumulation of organic matter in soil surface. These species demand little nutrients in soil for even high biomass accumulation (Osman et al., 1992) and thus can be grown in poorly fertile soils.

In the early rapid stage of litter decomposition high levels of nutrients and low C/N ratio accelerate the rate of mass loss (Berg and Staaf, 1980; Bosatta and Staaf,

1982). Slow decomposition of *A. cordifolia, B. purpurea, B. serrata, D. pentagyna and S. anacardium* litter during 0–60 days may be for its poorest contents of macronutrients N, P, K, and Ca and wide C/N ratio. However, an immobilization of large quantities of N and Ca during this stage may have compensated for this poor initial content and added momentum to the rate of further decomposition. *S trijuga* and *S. robusta* contained higher nutrients and lower C/N ratio. This result indicated that the effect of nutrient levels and C/N ratio are more pronounced in the early decomposition stage but not in the later stage. Berg and Staaf (1980) suggested that nutrient levels affected rate of weight loss only at initial stage whereas refractory substances such as lignin and tannins overtook in the final stage. Climatic condition (season) had an effect on the leaf litter nitrogen (high in dry season) and phosphorus content (high in rainy season). Seasonal variation in nitrogen and phosphorus is possibly a result of changes in microbial activity. This study highlights the need for an integral approach to research on nutrient cycling in deciduous forest of calcifuge habitat because of its complex decay patterns. The present result shows similarity with the work of Loria-Naranjo et al. (2019). Succession of micro fungi on leaf litter of *Acacia catechu* from Ratangarh forest, Datia, Madhya Pradesh, India was evaluated by Jatav et al. (2020) to find out the fungal diversity on leaf litter.

Concentration of ash increased, and carbon generally decreased with time to decomposition suggesting higher loss of organic matter than inorganic constituents. Although there is evidence that considerable leaching of inorganic substances might occur during decomposition (Yavitt and Fahey, 1986); the accumulation of ash indicated that leaching was not probably acute in the present study. However, N, Ca, Mg, Fe, Mn, and Zn concentrations of decomposed litters were generally higher; and C/N ratio, P and K concentrations decreased. when the amount of an element is decomposed litter decreased over the initial content a net release of that element was obtained; but when it increased a net immobilization was affected. Release and immobilization showed different patterns for different species and nutrients. Nitrogen was released throughout the entire duration from *B. purpurea, S. trijuga*, and *S. robusta* leaf litter but an initial immobilization was recorded in other species.

Roots of forest trees draw nutrients from relatively greater soil depth and accumulate in their biomass, a larger proportion of which is returned on surface soil in litters (Pandey and Singh, 1984). Due to decomposition, the released nutrients should be expected to be incorporated in the soils. In sloping lands, receiving high rainfall in the monsoon and sandy loam soils of good internal drainage like the present one, some soluble nutrients are leached and lost too.

Litter fall and the associated nutrient return are a significant supply of resources in forest ecosystems and both litter quantity and quality can change under different species compositions. Further investigation on the structure of leaf anatomy of various plant species and microbes and fungi involved in leaf litter decomposition of deciduous forest may provide information about the mechanism of in nutrient cycling in forest ecosystem. Information on the chemical composition of leaf and litter is also useful in studies of food chain, biogeochemical cycling, and physiological ecology.

Knowledge of the environmental factors affecting leaf element stoichiometry patterns can reveal nutrient fluxes of ecosystems and their responses to global change. However, understanding continental patterns of leaf nitrogen (N) and phosphorus (P) stoichiometry in relation to solar radiation is still insufficient. When predicting the effects of global dimming on ecosystem nutrient fluxes, one must distinguish the roles of direct, diffuse, and spectral components of solar radiation. The present study has implications for all future macroecological research involving the effects of solar radiation on ecosystems.

References

Berg, B., Argen, G., 1984. Decomposition of needle litter and its organic chemical components: theory and field experiments. Long-term decomposition in a Scots pine (*Pinus sylvestris*) forest. III. Can. J. Bot. 62, 2880–2888.

Berg, B., Staaf, H., 1980. Decomposition rate and chemical changes of Scots pine needle litter. II. Influence of chemical composition. Ecol. Bull. 32, 373–390.

Berg, B., Hannas, K., Popoff, T., Theander, O., 1982. Changes in organic chemical composition of needle litter during decomposition. Long-term decomposition in a Scots pine forest. I. Can. J. Bot. 60, 1310–1319.

Bernhard, R.F., 1998. Changes in relationships between initial litter quality and CO_2 release during early laboratory decomposition of tropical leaf litters. Eur. J. Soil Biol. 343, 117–122.

Birk, E.M., Simpson, R.W., 1980. Steady state and continuous input model of litter accumulation and decomposition in Australia Eucalypt forests. Ecology 61, 481–485.

Bosatta, E., Staaf, H., 1982. The control of nitrogen turnover in forest litter. Oikos 39 (2), 143–151.

Bray, R.J., Gorham, E., 1964. Litter production in forests of the world. Adv. Ecol. Res. 2, 101–157.

Champion, H.G., 1933. A preliminary survey of the forest types of India and Burma. Indian For. Rec. (N.S.) 1, 1–6.

Evans, J., 1981. Plantation Forestry in the Tropics. Oxford Science Publications, United Kingdom, p. 471.

FAO, 1988. Agroecological Regions. Vol.2. UNDP. Rome.

Fogel, R., Cromack Jr., K., 1977. The effect of habitat and substrate quality on Douglas fir litter decomposition in Western Oregon. Can. J. Bot. 55, 1632–1640.

Gill, C.J., 1970. The flooding tolerance of woody species – a review. Forestry Abstr. 31 (4), Published by the Commonwealth Agricultural Bureau. Farnham Royal, England.

Grubb, P.J., 1964. A comparison of Montane and lowland rainforest forest in Equador. J. Ecol. 51, 567–601.

Gupta, P.L., Rorison, I.H., 1974. Effects of storage in the soluble phosphorus and potassium content in Derbyshire soils. J. Appl. Ecol. 11, 1197–1204.

Hopkins, B., 1966. Vegetation of Olokemiji forest reserve, Nigeria. IV. The litter and soil with special reference to their seasonal changes. J. Ecol. 54, 687–704.

Islam, M.A., 1976. Quaternary geology of the Dhaka region. Rajshahi Univ. Stud. 7, 155–171.

Jackson, M.L., 1961. Soil Chemical Analysis. Printice Hall, Englewood, NJ.

Jatav, B.K., Sharma, T.K., Dassani, S., 2020. Succession of micro fungi on leaf litter of *Acacia catechu* in Datia, Madhya Pradesh, India. J. Pure Appl. Microbiol. 14, 581–590.

Khan, M.S., 1990. Towards Sustainable Development of Genetic Resources of Bangladesh. IUCN. Ministry of Environment and Forest and NCS. BARC, Dhaka, Bangladesh.

Khanna, P.K., 1997. Nutrient cycling under mixed species tree system in Southeast Asia. Agrofor. Syst. 38 (1), 99–120.

Loria-Naranjo, M., Sibaia-Cordero, J.A., Cortes, J., 2019. Mangrove leaf litter decomposition in a seasonal tropical environment. J. Coast. Res. 35, 122–129.

Lutz, H.L., Chandler, R.F., 1955. Forest Soils. John Wiley and Sons, New York.

Maclean, D.A., Wein, R.W., 1978. Weight loss and nutrient changes in decomposing litter and forest floor material in New Brunswick forest stands. Can. J. Bot. 56, 2730–2749.

McClaugherty, C.A., Pastor, J., Aber, J.D., Melillo, J.M., 1984. Decomposition dynamics of fine roots in forested ecosystem. Oikos 42 (3), 378–386.

McClaugherty, C.A., Pastor, J., Aber, J.D., Melillo, J.M., 1985. Forest litter decomposition in relation to soil nitrogen dynamics and litter quality. Ecology 6 (1), 266–275.

McGrath, D.N., Comerford, N.B., Dureya, M.L., 2000. Litter dynamics and monthly fluctuations in soil phosphorus availability in an Amazonian agroforest. For. Ecol. Manag. 131, 167–181.

Morgan, J.P., McIntire, W.G., 1959. Quaternary geology of the Bengal Basin, East Pakistan and India. Bull. Geol. Soc. Am. 70, 31–342.

Nazrul Islam, A.K.M., Rorison, I.H., 1978. Field investigation of seasonal oxidation-reduction conditions in soil and of changes in the mineral contents of shoots of associated species. Dhaka Univ. Studies, B. XXVI, 57–65.

Nazrul Islam, A.K.M., Rorison, I.H., 1981. Effects of calcium and manganese concentrations varied separately on the growth and chemical concentration of *Rumex acetosa*, *Molinia caerulea* and *Eriophorum angustifolium*. Z. pflanzenphysiol 104, 467–474.

Nazrul Islam, A.K.M., Saha, U.K., 1980. Some aspects of physiology and ecology of soybean under water-logged and non-waterlogged conditions. Bangladesh J. Bot. 12, 54–59.

Nazrul Islam, M., Rezaur Rahman, M., Mominul Haque, M., Monimul Huque, M., 2010. Physico-mechanical properties of chemically treated coir reinforced polypropylene composites. Composites Part A: Appl. Sci. Manuf. 41 (2), 192–198.

Osman, K.T., Gafur, M.A., Jamali, T., Kamaluddin, S., 1997. Growth and nutrient dynamics of a young *Pinus caribaea* Morelet plantation on a denuded hill soil of Bagladesh. Indian For. 123 (2), 127–135.

Osman, K.T., Haque, S.M., Forkan, M., 1992. Above ground biomass and nutrient distribution in *Acacia auriculiformis*, *Dipterocarpus turbinatus* and *Pinus caribaea* plantation in Chittagong University hills. Indian J. For. 15 (3), 193–197.

Ovington, J.D., Madgwick, H.A.I., 1957. Afforestation and soil reaction. J. Soil Sci. 8, 141–149.

Pandey, U., Singh, J.S., 1984. Nutrient changes and release during decomposition of leaf litter in a Himalayan (India) Oak-conifer forest. Can. J. Bot. 62, 1824–1831.

Ponnamperuma, F.H., 1972. The chemistry of submerged soils. Advan. Agron. 24, 96.

Radwann, M.A., Constance, A., Harrington, A., Kraft, J.M., 1984. Litterfall and nutrient returns in red alder (*Alnus rubra*) stands in Western Washington, Bellingham, WA. Plant Soil 79, 343–352.

Rashid, S.H., Rahman, M.M., Hossain, A.B.M.E., 1995. Some soil properties of Chandra Sal forest at Gazipur, Bangladesh. Jahangirnagar Univ. J. Sci. 20, 87–96.

Shanks, R.E., Olson, J.S., 1962. First year breakdown of leaf litter in Southern Appalachian forests. Science (N.Y) 134, 194–195.

Sharma, E., Ambasht, R.S., 1987. Litterfall, decomposition and nutrient release in an age sequence of *Plnus nepalensis* plantation stands in the Eastern Himalaya. J. Ecol. 75, 997–1010.

Singh, J.S., Gupta, S.R., 1977. Plant decomposition and soil respiration in terrestrial ecosystems. Bot. Rev. 43, 449–528.

Swift, M.J., Heal, O.W., Anderson, J.M., 1979. Decomposition in terrestrial ecosystems. Ecology 5, 1–37.

Torreta, N.K., Takada, H., 1999. Carbon and nitrogen dynamics of decomposing leaf litter in a tropical hill evergreen forest. Eur. J. Soil Biol. 35 (2), 57–63.

Walkley, A., Black, C.A., 1934. An examination determent method for determining soils organic matter and a proposed modification of chromic acid titration method. Soil Sci. 37, 29–38.

Warren, M.W., Zou, X.M., 2002. Soil macrofauna and litter nutrients in three tropical tree plantations on a disturbed site in Puerto Rico. For. Ecol. Manag. 170 (1), 161–171.

Wilde, S.A., 1954. Reaction of soils: facts and fallacies. Ecology 35, 89–99.

Yavitt, J.B., Fahey, T.J., 1986. Litter decay and leaching from the forest floor in *Pinus contorta* (Lodgepole pine) ecosystems. J. Ecol. 74 (2), 525–546.

Zavitkovski, J., Newton, M., 1971. Litterfall and litter accumulation in red alder stands in Western Oregon. Plant Soil 35, 257–268.

Zinn, Y.L., Resek, D.V.S., Silva, J.E., 2002. Soil organic carbon as affected by afforestation with *Eucalyptus* and *Pinus* in the Cerrado region of Brazil. For. Ecol. Manag. 166, 285–294.

Bioactivity of soil microorganisms for agriculture development

9

Manikant Tripathi and Rajeeva Gaur
Centre of Excellence, Department of Microbiology, Dr. Rammanohar Lohia Avadh University, Ayodhya, India

9.1 Introduction

Soil system is one of the most appropriate reservoirs of almost all types of microorganisms with different nutritional classes, because it provides a variety of plant diversity from bryophytes to pteridophytes including the existence of animal life of chordates and non chordates. Biological activities make the soil rich in a variety of organic and inorganic compounds. In terrestrial life, the primary producers and the organic carbon are the most abundant on soil system. The organic carbon and nitrogen exist in various forms maintained by biogeochemical cycles supporting the activities of soil microorganisms. A variety of bacteria, fungi, yeast, algae, protozoa, nematodes help in organic matter decomposition and mineral transformation. These activities are rich in different levels of physicochemical and nutritional conditions depending on the presence of specific microbial types. Such activities are generally supported by microbial shifting according to the environmental conditions for their growth and metabolites production which is required for specific reactions and transformations of organic matter decomposition, plant growth promotion, antibiosis, production of antimicrobial specific compounds, organic acids, mineral acids, and several other types of metabolites. These activities directly or indirectly help competent and specific group of microorganisms, but several mechanisms may also exist for microorganisms survival like biofilm formation or modification of metabolites or the ability to resist the intake of metabolites inside the cells. These aspects are seen simultaneously with a variety of microorganisms, interactions, their final survival and long-term existence.

The soil's physical, chemical, and microbiological status is maintained by microorganisms. Soil biodynamic nature is important for sustainable agriculture development (Paustian et al., 2016; Tscharntke et al., 2012). The sustainable agriculture provides long-term persistent production by crops management technologies such as crop diversification, recycling of plant nutrients, pest control, etc. (Tripathi et al., 2020). Both biotic and abiotic factors affect the agriculture system which directly changes the crop productivity, soil health, and fertility (Chodak et al., 2015). Plants associated with soil microbial community, such as mycorrhizal fungi and plant growth promoting microorganisms (PGPM), help in crops' growth and development under abiotic and biotic stresses (Yadav et al., 2017a,b; Begum et al., 2019).

Microbes in Land Use Change Management. DOI: https://doi.org/10.1016/B978-0-12-824448-7.00012-7
© 2021 Elsevier Inc. All rights reserved.

Microbial survival at culturable and nonculturable forms is important for ecosystem development. The culturable microbial potentials are being used for commercial production of microbial metabolites like ethanol, organic acids, amino acids, polysaccharides, enzymes, antibiotics, hormones, vitamins, and several others, while several groups help in microbial transformation of chemicals, organic matter decomposition, soil mineralization, biogeochemical cycles of carbon, nitrogen, sulfur, phosphorous, biocontrol of phytopathogens, plant growth promotion activities like nitrogen fixation, phosphate solubilization, metal leaching from ores, and several others showing useful as well as harmful effects which indicate the bioactive potentials of the soil ecosystem (Fig. 9.1). The ecology of microbial system is so complex that we could know only 1%–2% of the microorganisms existing within a culturable one. Their interactions, survival, shifting and their impact of soil fertility, plant growth promotion, and their management for better cause are still not well studied.

Bioactive compound and survival of plants and others microbes are well documented. It is also well established by Kloepper, Martin, Garret, Lynch, Schorth, and several other scientists who established many facts about plant growth promoting rhizobacteria (PGPR) activities. These beneficial microbes play a direct and indirect role suppressing several phytopathogens. In spite of that, pathogens also live friendly without harming microbial community but they do harm plants under favored circumstances. It is clear from several research findings that plant and animal pathogenic microbes exist in soil ecosystem without harming microbial community for a longer period and survive with common nutrients and seeking favorable animal or microbial host, regeneration, and spreading with favorable hosts.

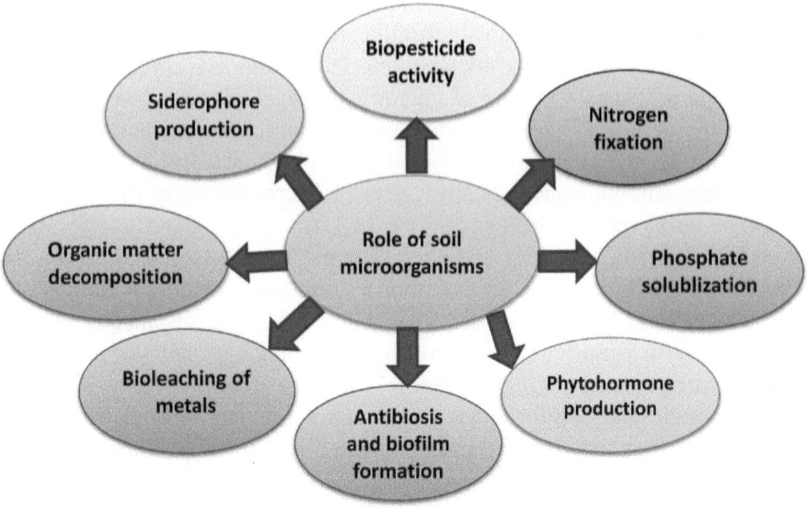

Figure 9.1 Applications of soil microorganisms for sustainable agriculture development.

Microbial community development with newer strain or organisms may be explained through the process of recombination of bacteria through transformation, conjugation, and transduction, while process of protoplast fusion, mutation, and recombinant DNA technology may also exist for developing new strains, genes, and species. Natural system has almost all functions because microbial genome and specific enzymes are required for recombinant DNA technology. Metagenomic studies have proved that genetic pool of microorganisms may be amplified and used for microbial identification and development of newer microorganism. The role of antibiotic producing microbial system, endophytes, and free living N_2 fixing microbial community is important for agriculture developments (Franche et al., 2009; Singh and Gaur, 2016a,b; Yadav et al., 2016; Tripathi et al., 2020). Further, the microbes produce hormones for plants which improve metabolisms and plant growth. Moreover, microbes associated to plants as endophytes are another interesting area for survival and protection as symbiotic association. Several friendly microbial associations helping other beneficial microbes as a probiotic can also be established as an effective bioactivity for plant and soil health. The application of nanotechnology can also be useful for sustainable agriculture development (Tripathi et al., 2018). This chapter covers different components which take part in organic matter decomposition, soil nutrients recycling, xenobiotic compounds status, and its microbial degradation.

9.2 Soil microbes in organic matter decomposition

The experimental facts established by most researchers show that in soil starch, cellulose, hemicelluloses, lignin, pectin, and chitin are the main carbon source comes from dead plants, animals, and microorganisms. They are a slightly or more complex form of carbon along with nitrogen and other minerals. The organic and inorganic forms of carbon and nitrogen are the most essential components which are recycled by microbial activities. These can be explained by carbon and nitrogen cycles. Similarly the other organic and inorganic phosphorus and sulfur are transformed through microbial system. In these processes, microbial enzymes, that is, amylases, cellulases, xylanases, ligninase, peroxidases, phenoxidases, peroxidases, and oxidoreductases play a role in degradation and mineralization of organics. There are some common bacteria like *Bacillus, Arthrobacter, Pseudomonas, Xanthomonas, Clostridium, Proteus, Achromobacter, Acinetobacter*, etc. and several fungi like *Aspergillus, Penicillium, Humicola, Paecilomyces, Fusarium*, and yeast like *Candida, Torula, Endomycosis*, etc. that help in decomposing organic matter (Shah et al., 2016; Singh and Gaur, 2016a,b; Yadav et al., 2017a,b; Nicolás et al., 2019; Tripathi et al., 2020).

All the microbes exist in several status with a variety of functions, which are generally referred to as strains. These microbes exist from psychrophilic, psychrotrophic, mesophilic, thermotolerant to thermophilic range which can also survive at different pH, water activity, and nutritional conditions in culturable form. There are some bacteria, fungi and yeast which have the ability to utilize a variety of carbon sources if simple carbon source is not available. For example, *Pseudomonas, Bacillus, Escherichia coli*, and *Clostridium* can utilize more than 50 different carbon and nitrogen sources and even *Pseudomonas* can utilize more than a hundred different types of carbon sources. Therefore, it is well established that these bacteria are ancient and have almost all short of such genomic status to survive in any environment with a variety of carbon and nitrogen sources. Gaur et al. (2010) also showed that *Bacillus* sp. tolerated extreme conditions of chemicals without affecting cellulase and amylase production along with alkali tolerance activity. Such microbial systems are acquainted with their genomic expression differentially in different environments from mild to extreme depending on the genetic makeup of the microbial strain, thus adaptation appears to be a prerequisite for high tolerance.

Microbes exist in colony, therefore, one group degraded complex form of carbon and nitrogen sources by their specific enzyme system and tolerate huge amount of simple glucose or fructose or several disaccharides like maltose, arabinose, cellobiose, and simple form of peptides, peptone, amino acids for those who do not have machinery to do such functions. The other groups grow well with such microbes and may help directly or indirectly to several groups. For example, free living N_2 fixing bacteria fix nitrogen which may be consume by the organisms which degrade complex organic carbon. This approach is a new dimension where microbial associates must be explored in order to increase higher decomposition rate in soil system because this can be important factor for soil fertility. Similarly, phosphorus, sulfur and other micronutrients are recycled from organic to inorganic forms by certain enzymes produced by variety of microorganisms. Research updates confirm that in soil, variety of bacteria, fungi, yeasts and algae have been reported to produce amylases, proteases, cellulases, xylanases and others enzymes which are required for the transformations of organic and inorganics at different pH, temperature and aeration levels (Fig. 9.2). This clearly indicate that microbes of all ranges of physicochemical and nutritional levels coexist in the soil and even at any season or extreme conditions, they shift their population at particular environment without losing the microbial population/biomass status. Only, qualitative levels may deviate with specific microbial units. Several studies have shown that if soil is treated with solarization, that is, mulching of soil by polyethylene sheet, increased the temperature by 8°C from non-mulched soil, but the population of bacteria, fungi and actinomycetes remain same only the shift of the microbes from mesophilic to thermotolerant/thermophilic during summers when atmospheric temperature was 36°C–39°C on an average (Gaur, 1992). Cellulase and amylase producing microorganisms were also found with the similar trends. This confirm that the microbial shifting always take place in soil where organic matter is abundant as this is directly proportional to the microbial population of different nutritional and physiological class.

The role of organic matter decomposers is very important for better soil productivity. When one crop is removed, the straw and roots provide lignocellulosic component to soil, helping the soil fertility in this way to increase the microbial biomass and supporting several beneficial microbial counts. In an experiment, it has

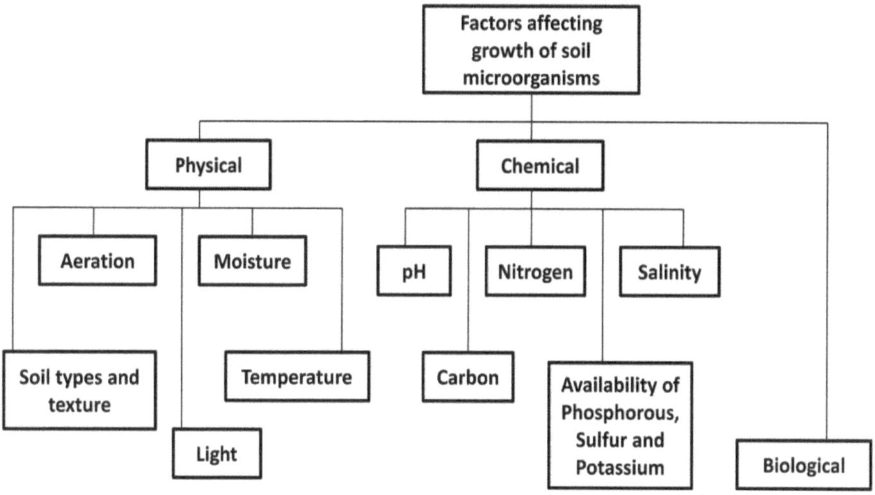

Figure 9.2 Factors influencing growth and activity of soil microorganisms.

been found that the pot inoculated by 1% bagasse along with *Trichoderma* and cellulolytic *Bacillus* inoculums increased the plant growth (wheat, maize grain and mung bean). Further, the nodulation in pea and gram plants increased by 10%–15% resulting in better plant growth (Gaur unpublished data). Gaur (1992) reported that the addition of same contents increased the population of *Azotobacter* by 25% leading to a better wheat plant yield. Several researchers also reported the bacteria, viz., *Bacillus* sp., *Pseudomonas* sp., *Xanthomonas* sp., *Arthrobacter* sp., and fungi, viz., *Aspergillus* sp., *Penicillium* sp., *Humicola* sp. were found to produce *exo*-gluconases, *endo*-gluconases, β-glucosidases, etc. responsible for cellulose decomposition, while exoxylananses, endoxylanases, and xylanase for hemicelluloses degradation, and mono and dioxygenases, peroxidases, laccases for lignin degradation. Likewise, pectinases, chitinases, and proteases producing microorganisms have been found in higher numbers where such carbon and nitrogen contents are available. It clearly indicates that microbial activity is required for soil health. Direct detection of such microbes which have induced the nature of such enzymes production can be an indicator of soil health as they release a huge amount of simple carbohydrates for the growth of N_2 fixers, phosphate solublizers, phytohormones, and microbial metabolites required for plant growth.

Soil of different types from loam to clay or sandy loam, from alkaline, saline to acidic or neutral levels vary the microbial contents along with its bioactivity. Such variation may be discussed in the light of available organic compounds. Microbes also have such variations that they may grow at high salt or alkaline to acidic environment and the ability to produce all short of enzymes responsible for organic matter decomposition and chemical transformations. But, the activity may be slow, therefore, microbial status is affected by the availability of organic carbon and nitrogen. Microbes mainly free living N_2 fixation by *Azotobacter* could be able to fix nitrogen from acidic and alkaline soil with same potency. When tested in laboratory (Gaur unpublished), the status of other enzymes like cellulases and amylases are also produced by bacteria and fungi isolated from alkaline soils. The quantitative estimation of enzymes of several microbes was reported to be unaffected in the presence of several changed conditions (Tiwari and Gaur, 2019). These facts confirm the bioactivity of microbes in different soil conditions with available organic carbon and nitrogen sources because such availability is essential for increasing the population of a particular microorganism.

9.3 Soil microbes in plant growth promotion

The microbes which directly and indirectly by their actions enhancing plant growth are collectively known as PGPM. Not only plant rhizospheric microbes fall under this category but also other microbes which are directly not associated with rhizosphere may also perform several useful functions which help plant growth and yield. Therefore the bacteria, fungi, and even certain algae play an important role in such fields. Microorganisms such as *Rhizobium*, *Azotobacter*, *Azospirillum*, *Streptomyces*, *Pseudomonas*, *Enterobacter*, *Bacillus*, *Variovorax*, *Klebsiella*, *Burkholderia*, *Serratia*, and mycorrhizal fungi are termed as PGPM which promote plant growth and development under both normal and stress conditions (Yadav et al., 2016, 2017a,b; Singh and Gaur, 2016a,b; Vimal et al., 2017; Tripathi et al., 2018, 2020). It has been worked out that the lignocellulosic materials provide sufficient carbon and nitrogen sources to generate higher microbial biomass by the initial microbes producing cellulases, xylanases, and degradation of lower content of lignin. These microbial groups are very important in the ecosystem as they produce huge glucose and other simple mono and disaccharides available to other groups which directly play a role in nitrogen fixation as free living N_2 fixing bacteria and fungi consume simple carbohydrates for their growth and metabolite production for plant growth promotion.

Another important microbial community is N_2 fixing (symbiotic as well as asymbiotic) bacteria. The effect of *Rhizobium* in several crop plants have been attributed in several research work for its presence in soil, nodulation process, number, and activity of nodules for plant growth and yield (Table 9.1). In spite of several outcomes of research, still its proper application and better results are limited. This point is explained here. The main cause is its application in field conditions require more friendly microbial associates and a consortium should be added with some supplements like cellulose powder or baggase powder along with cellulolytic and xylanolytic microbial community for a better result. The colonization on roots and bacteria along with nodulation process require energy flux supporting *Rhizobium*. Further, other free living N_2 fixers like *Azotobacter*, *Azospirillum* (associated to root surface), and *Clostridium* require such conditions favoring N_2 fixation activity. Inoculation of PGPR on crop fields revitalizes growth promoting activity, disease management, and maintains the nitrogen level in agricultural soil (Yu et al., 2011; Wan et al., 2020).

Table 9.1 Bioactivity of agriculturally important soil microorganisms for plant growth promotion.

S no.	Microorganism	Biological activity	References
Soil bacteria			
1.	*Bacillus* sp., *Pseudomonas* sp., *Rhizobium* sp.	Plant growth promotion by biological nitrogen fixation and phosphate solublization	Yadav et al. (2016, 2017a,b), Tripathi et al. (2020)
2	*Rhizobia*	Promoted growth of rice shoots by targeting cell signaling, division, and expansion	Wu et al. (2018)
3	*Rhizobium* *Azotobacter*	Nitrogen fixation in pulses	Yadav et al. (2017a,b)
4	*Azospirillum* spp. (Az19)	Inoculation with *Azospirillum brasilense* Ab-V5 cells enriched with exopolysaccharides and polyhydroxybutyrate results in high productivity in maize under low N fertilizer input	Oliveira et al. (2017)
5	*Bacillus pumilus*	ACC-deaminase activity in potato	Mahmood et al. (2016)
6	*Pseudomonas* sp.	Induces salinity tolerance in cotton (*Gossypium hirsutum*) and resistance to *Fusarium* root rot through the modulation of indole-3-acetic acid	Egamberdieva et al. (2015)
7	*Enterobacter* and *Burkholderia*	Siderophore production in rice roots	Souza et al. (2014)
8	*Pseudomonas putida*	Siderophore production, phosphate solublization	Ahmed and Khan (2012)
9	*Burkholderia, Pantoea, Bacillus*, and *Klebsiella*	Biological nitrogen fixation and endophytic colonization in maize	Ikeda et al. (2013)
10	*Pseudomonas aeruginosa*	Synthesis of exopolysaccharides and phytohormones	Ahmed and Khan (2012)
Soil fungi			
12.	*Trichoderma* sp.	Biocontrol activity on various plant pathogens (such as *Rhizoctonia, Alternaria, Fusarium, Curvularia, Colletotrichum, Xanthomonas*)	Harman and Uphoff (2019)
13.	Arbuscular mycorrhizal fungi	Abiotic stress tolerance in crops	Begum et al. (2019)

(*Continued*)

Table 9.1 (Continued)

S no.	Microorganism	Biological activity	References
14.	*Funneliformis mosseae*, *Paraglomus occultum*	Increased water potential and absoption rate in *Poncircus trifoliata*	Zhang et al. (2018)
15.	Arbuscular mycorrhizal fungi	Enhanced *Glycine max* L. growth of by increasing leaf proline and photosynthesis under drought condition	Pavithra and Yapa (2018)
16.	*Piriformospora indica*	Potential and significance in plant stress tolerance	Gill et al. (2016)
17.	*Rhizophagus irregularis* *Variovorax paradoxus* 5C-2	Enhanced photosynthetic rate, and increased root water conductivity and oxidative phosphorylation in tomato	Calvo-Polanco et al. (2016)

The role of soil types, available air, temperature, pH, and other components also affect simultaneously the cultivated and noncultivated systems. The rhizospheric and nonrhizospheric soils have different factors, and the qualitative and quantitative microbial population in rhizospheric may vary as compared to nonrhizospheric soil. It is well documented that rhizosphere provides several nutrients like simple carbohydrates, lipids, and proteins which attract microbes for better plant growth. It has also been observed that healthy plants have more resistance to phytopathogens. Plant growth promoting substances like hormone, vitamins, and minerals can be increased by microorganisms. The term PGPR was first coined in 1978 by Kloepper and Schroth, to describe a subset of rhizobacteria which induce increased plant growth after inoculation to seed. There are a number of bacterial genera, viz., *Pseudomonas*, *Azospirillum*, and *Bacillus* which have been included under PGPR with a wide host range such as potato, radish, sugar beet, beans, canola cotton, peanut, wheat, and vegetables. These bacteria produce a variety of plant growth promoting substances. Furthermore, iron chelating siderophores, antibiotics, and HCN are also produced by some PGPR and have been implicated for reduction of plant pathogens and deleterious microorganisms which finally leads to increased plant growth response and yield (Saha et al., 2015). Increased plant growth response (IGR) by using these bacteria is usually an indicator of the successful nonhazardous and beneficial application of the treatment for controlling plant diseases as well as various crop improvement and soil health. Such increase growth response denotes the improvement in plant growth either by chemical, that is, release of minerals or growth hormones, other growth substances, multiplication of toxic components, or biological factors (increase of beneficial microorganisms such as mycorrhiza and other PGPR), or both (Chen and Katan, 1980; Stapleton and Devay, 1986).

Increase in growth response is an important economic implication and must be taken into account when considering the use of specific microorganisms for control of soil borne diseases. Combination of selective microorganisms when applied in field condition increased the plant growth (Linderman, 1988). Biological competence of PGPM strain is another important aspect, therefore, a competent strain of different bacteria in different combinations according to the crop and level of inoculums should be optimized prior to field application. In a study, it has been found that the use of some biological control agents, 10 days before sowing, was effective rather than the application at the time of sowing through seed bacterization. The role of fluorescent Pseudomonads has been frequently isolated and confirmed for their role for plant growth (Schroth and Hancock, 1982). Most researchers now agreed that root colonization is an active process rather than passive between a soil bacterium and a growing plant root. Therefore root colonization is the process where bacteria survive in seeds or in soil, multiply in the spermosphere (seed zone) in response to seed exudates rich in carbohydrates and amino acids associate with the root surface when used as bacterization agents. Therefore rhizobacteria may exert one or three types of effects on the host plant either deleterious, neutral, or beneficial.

The theory of microbial equilibrium held that the rhizophere microflora may exist in a dynamic equilibrium means the population of any introduced microorganisms would decline rapidly following inoculation into soil (Baker and Cook, 1974). Such findings were very basic. Since then a number of researchers reported about soil and agriculture microbiology regarding the application of beneficial effects of rhizospheric microorganisms with several crops under different soil conditions (Yadav et al., 2017a,b; Tripathi et al., 2020). There are several mechanisms by which PGPR promote plant growth or inhibit soil borne plant pathogens including the production of extracellular growth promoting substances, iron chelating siderophores (Kloepper et al., 1980), antibiotics (Weller, 1988) and HCN (Voisard et al., 1989) which reduced the population of major root pathogens. PGPR have also shown promise as potential biocontrol agents for many soil borne diseases (Kloepper, 1992) and prefer rhizosphere as a habitat (Baker, 1991).

9.4 Microorganisms of soil as biocontrol agent

The general effects of PGPR on crops promotion fall into two categories plant growth and biological control (Leong, 1986; Chet et al., 1990; Sahu and Sindhu, 2011). Initial investigations on the mode of action of PGPR indicate that many bacterial strains fail to promote plant growth under xenobiotic conditions (Kloepper et al., 1980): rather, they promote plant growth indirectly by reducing the activity of native rhizoplane fungi and bacteria (Kloepper and Schroth, 1981a). On the other hand some PGPR strains promote plant growth directly by producing metabolites which stimulate plant growth independent of native soil and microflora when tested in a disease assay with *Pythium* in soil through biological control activity (Kloepper et al., 1980).

It is presumed that such biological control activity of direct action of PGPR resulted from disease escaping by roots with accelerated growth rates, but there are a number of factors which may exert on pathogens as well as a biological control agent, which may vary soil to soil as well as the rhizospheric PGPR potentials (Table 9.1). Hence, even PGPR strains which promote plant growth in xenobiotic conditions and show no antagonism toward the target pathogen may be useful as a biological control agent against a particular pathogen. Many researchers emphasize the importance of microorganisms in relation to sustainable agriculture and biocontrol of phytopathogens (Mohammed et al., 2008; Franche et al., 2009). Thus an eco-friendly strategy using beneficial microbes is developed for better yield and protection of crops.

Weller (1988) provided a comprehensive review of the use of general bacteria in the biological control of soil borne plant pathogen, although there are several fungal agents showing potentials biocontrol agents like *Trichoderma harzianum*, *Verticellium*, *Penicillium*, *Beauvaria bassiana*, and several others. Many soil fungi show insecticidal and nematicidal activities (Anke and Sterner, 2002). The mechanisms may be either through their metabolites or physical injury or even both may be attributed by several researchers. Schippers et al. (1988) reviewed rhizobacteria especially as biological control agents. Moreover, the use of certain plant promoting bacterial agents in different soil types and crops is yet to be studied along with their mathematical modeling. There are several important mold and yeast which play important role in such activity can be explored for plant growth promotion.

Kloepper and Schroth (1981b) also explained the role of pathogenic or adverse function in the rhizosphere and termed "deleterious rhizobacteria (DRB)" and Schippers et al. (1987) proposed the term deleterious microorganisms (DRMO) which were minor pathogens of plants without parasitizing plant tissues. DRB are apparently widespread in nature where they restrict the attainment of plant growth potential. Antibiotic resistant DRB strain survived in the winter on barley straw at population greater than log 6 cfu/g straw. DRB are reported on canola, pea, and lettuce. Selective PGPR strain prevented the yield reduction associated with short rotation of potato presumably by providing biological control of DRB. Although such studies validated in plot and field conditions, but still there are deviation in crop variety, that is, genetic makeup of the crops, soil types and even microbial diversity (such as the role of bacteria, filamentous fastidious and slow growing fungi and yeast). There are pathogenic agents like bacteria, fungi and protozoans or nematodes which are also controlled by such mechanisms using similar types of microbes. Therefore the deleterious microorganisms of the common crop rhizospheric studies through statistical basis analyzed for the occurrence as well as the role of nonrhizospheric microbial contents. They have certain specific enzymes like mono dioxygenases, peroxidases or others responsible for degradation of pesticides or other xenobiotic compounds of soil. Therefore detoxification of soil may indirectly affect the plant growth.

There are several other bioactivities of rhizospheric bacteria which control parasitic plant pathogens (Kloepper and Schroth, 1981a). Many bacteria such as *Bacillus* and *Arthrobacter* combination were also found effective for plant growth of several crops (Gaur unpublished). Therefore soil suppression and conducive behavior are explained on the occurrence or absence of such biological control agents mainly fluorescent Pseudomonads, *Bacillus*, *Arthrobacter* consider as indicator of soil and plant health. Inoculation of such strains of bacteria converted a conducive soil into suppression soil which is suitable for crop production. It has been observed that when these soil having such bacterial strains added in pot soils, the seed germination and increased plant growth response observed for several crops. These findings indicate that such microbial strains are efficient to survive in many types of soils and cause common beneficial effects. In general, it supports the beneficial rhizospheric microorganisms directly or they may get benefit through their own actions. Such facts are to be established with other microorganisms. Several notorious pathogens have been successfully controlled by PGPR strains. Further, the role of bacterial and fungal agents for controlling soil borne phytopathogens emerged as an area of biological control. Several soil microorganisms have been identified against the specific phytopathogen or a wide range of plant pathogens. Biocontrol of phytopathogens ultimately leads to increased agriculture crops productivity (Singh and Gaur, 2016a,b; Tripathi et al., 2020). Many rhizosperic bacteria such as *Bacillus* spp. and *Pseudomonas* spp. are used to promote plant growth (Yadav et al., 2016, 2017a,b).

9.5 Endophytes in plant growth promotion

Endophytes are the first real plant associated beneficial group of microorganisms in which bacteria and fungi are the dominating group which survive either in main roots or branching of roots for permanent existence for the support of plant. Therefore plant-microbes' relationship existence and their beneficial aspects are the main area of research identifying their direct and indirect roles in plant metabolism, mineral support system, sequestration of chelating agents, accumulation of nanoparticles accelerating hormone production, activating enzyme system, mineral uptake, controlling pathogens and also promoting the phytoallexin activity. The role of endophytes is more important than free living microorganisms. It is friendly microbes to various plant systems. The role of endophytes whether it is *Rhizobium* or any other, they play symbiotic relationship. Endophytes play very important role in improvement of the plant growth as well as to reduce infestation of phytopathogens by metabolites production, alteration in plant physiology, and the stimulation of systemic acquired resistance in host plants (Govindasamy et al., 2014; Zhou et al., 2014; Golinska et al., 2015). Singh and Gaur (2016a,b) isolated plant growth promoting endophytic actinomycetes from various medicinal plants having antagonistic and biocontrol activities for collar rot disease on chickpea and could be useful in integrated control against diverse soil borne plant pathogens. Similarly, Eljounaidi et al. (2016) reported bacterial endophytes as potential biocontrol agents of vascular wilt diseases. Whereas, Hasan (2017) isolated seven bacterial and fungal endophytes, which produced phytohormones and ammonia, with ability of phosphate solubilization.

The microbial association of endophytes during and after initiation may be a matter of research for development of sustainable agriculture and horticulture.

Endophytes molecular characterization and diversity of the plant may be studied for exploring the science. The modern molecular tools may open the secrets of endophytic microbial diversity for plants relations. The aromatic and medicinal plants diversity with crops and horticultural plants in relation to phytopathogens can reveal the relationship and endosymbionts in this aspect. Thus microbial endophytes play a vital role to improve plant growth and could be used as inoculants for sustainable agriculture development.

9.6 Soil microbes for xenobiotic compound degradation

Among a wide range of microbial diversity, a limited number of bacteria, fungi and yeast have entirely different type of metabolism, that is, utilization of aromatic hydrocarbon which enter through lignin as well as from xenobiotic compounds like pesticides and other industrial effluents in the form color compounds like melanoidins of distillery effluent (Tiwari and Gaur, 2014, 2019), pentachlorophenol in tannery effluent (Tripathi et al., 2011; Tripathi and Garg, 2013), colored compounds from textile industries (Garg et al., 2012a; Tripathi et al., 2018) and lignin from paper and pulp effluent (Garg et al., 2012b). These contents are degraded by specific group of microorganisms which produce mono/dioxygenases, peroxidases orlaccases. In natural ecosystem, bacteria, viz., *Arthrobacter*, *Pseudomonas*, *Bacillus* sp., *Xanthomonas* and fungi of basidiomycets like *Aspergillus*, *Phanerochaete*, *Ganoderma*, *Polysporus*, and others produce such enzymes which breakdown aromatic hydrocarbons via caprolate, mandalate and β-keto adipic pathways to metabolize such compounds. Most of the degradation of such compounds is run through cometabolism in which a compound is converted in another compound by the microbial enzyme without the expense of energy and carbon source while other group of microorganism metabolize that and lead to completely metabolize form up to pyruvate. Such conditions occur in the natural ecosystem mainly in terrestrial system.

Xenobiotic and recalcitrant compounds contaminants infested in the form of pesticides and lignin from plant materials, therefore, lignin degrading microbes are always present in most of the fresh soil as well as agriculture soil. The role of microorganisms in xenobiotic compounds degradation in soil is very important as they always detoxify the soil contaminants, otherwise, soil beneficial microorganisms and plant roots will be drastically affected leading to decrease in plant growth and crop yield. Soil beneficial microorganisms utilize aromatic hydrocarbons as source of carbon, therefore, increase soil microbial biomass help in the lignocellulosic degradation. Lignin is bound with cellulose and hemicelluloses tightly. The lignin degrading microbes attack on lignin, and allow cellulose and hemicelluose free to degrade by celluloytic and xylanolytic microbial communities which are abundant is soil, as they release huge glucose flux for helping to those which can only utilize simple sugar.

Fertile soil is always rich in different groups of microorganisms with all nutritional levels. To run the nutrient cycle, microbial shifting and regeneration of their population for longer period is necessary. The presence of xenobiotic degrading microbes can be an indicator of better soil health. Further the degradation pattern of specific microbe and status of certain enzymes can be estimated as an indicator for such levels. Moreover, the status of free living N_2 fixers with the xenobiotic degrader, because free living N_2 fixer always provide N_2 for such microbes for degradation of xenobiotic compounds. In a study, working with melanoidins of distillery effluent it has been confirmed that *Pseudomonas, Bacillus, Candida, Pediococcus* were able to degrade melanoidins when N_2 fixing *Azotobacter* was used in the soil at the level of 50×10^4 cfu/g soil (Gaur unpublished). It indicates that *Azotobacter* and xenobiotic degrader population may be a correct indicator system for assessing soil health. Although the aromatic hydrocarbons are always metabolizing through cometabolism, therefore, a consortium of microbial system is always recommended as soil inoculants before 10 days of sowing seeds. Such effects have been experimentally proved that if soil has high concentration of aromatic hydrocarbons. Such treatment is always beneficial for better plant growth.

The aerobic and anaerobic ratio of microbes especially for xenobiotic degradation is important. The anaerobic bacterium *Clostridium* sp. is very versatile known to produce variety of enzymes and do almost several activities proteolytic, lipolytic, cellulolytic, xylanolytic to lignolytic, therefore, there role in degradation of pesticides and other xenobiotics like lignin which is natural recalcitrant is degraded in deeper horizon of soil, when moisture is in high level. Therefore role of sulfur, phosphorus, carbon and nitrogen by aerobic to anerobic microbial system do occur simultaneously depending on the climatic conditions.

9.7 Bioleaching potentials of metals in soil with phosphate solublization

Microbial leaching is a process by which metals are dissolved from ore bearing rocks or soil where metals are in the bound form of calcium, magnesium, and iron salts by the microorganisms. The microbial action release metals such as phosphorus, copper, zinc, cobalt, molybdenum, cadmium, nickel, etc. This process releases such metals which are bound form of their salts as these are important micronutrients of plants. Variety of bacteria and fungi are identified for such process. The most commonly used bacteria which play role in leaching are *Thiobacillus thiooxidans* and *Thiobacillus ferrooxidans*. A number of other bacteria are *Bacillus* spp., *Pseudomonas* spp., *Thiothrix thiosporus*, and *Sulfolobus*. These bacteria are generally produce sulfuric acid or any other aliphatic acid. Most of them are chemolithotropic and some of chemoorganotroph as well as heterotrophic in nature. The *Thiobacillus* is Gram-negative rod shaped bacteria. It is autotrophic aerobes and obtains its energy from the oxidation of Fe^{2+} or Fe^{3+} or from the oxidation of elemental sulfur and reduced sulfur to sulfate. The chemolithrophs use these reactions for synthesis of H_2SO_4.

$$2S^\circ + 3O_2 + 2H_2O \rightarrow 2H_2SO_4$$

$$4FeSO_4 + 2H_2SO_4 + O_2 \rightarrow 2Fe_2(SO_4)_3 + 2H_2O$$

$$2FeS_2 + 7O_2 + 2H_2O \rightarrow 2FeSO_4 + 2H_2SO_4$$

$$FeS_2 + Fe_2(SO4)_3 \rightarrow 3FeSO_4 + 2S^o$$

These four reactions simultaneously run to this process by chemolithotrophic *Thiobacillus* in which other organic acids are produced by heterotrophic by utilizing sugar and mixed acid production through phosphate solublizing bacteria (PSB) following this process. Therefore in soil, all the process run to release various essential elements like phosphorous, calcium, magnesium, sulfur and other micronutrients through such mechanism. The role of specific group perform functions for plant and microbial nutrients resulting in soil fertility and better plant growth response. Since this is an entirely different group, that is, chemolithotrophs which work in both aerobic and anaerobic systems of soil depend on the nature of bacterium. Similarly, fungi like *Aspergillus* and *Penicillium* which produce organic acid by utilizing starch and other carbohydrate present in the soil ecosystem abundantly almost follow similar functions.

Phosphorous is essential mineral macronutrient for plant growth. However, the element is only available to plants in the form of the phosphate anion which absorbs directly by soil solution to plant. The process is slow in the natural ecosystem. Plants absorb phosphate only as monobasic ($H_2PO_4^-$) and dibasic (HPO_4^{-2}) ions. The solublization and mineralization of phosphorus by PSB is an important trait that can be achieved by PGPR. Many soil bacterial synthesize low molecular weight organic acids to solubilize phosphate (Alori et al., 2017). Phosphate solubilizing PGPR belonging to the genera *Arthrobacter, Bacillus, Beijerinckia, Burkholderia, Enterobacter, Microbacterium, Pseudomonas, Erwinia, Rhizobium, Mesorhizobium, Flavobacterium, Rhodococcus*, and *Serratia* have attracted researchers' attention as soil inoculants improve crops productivity (Oteino et al., 2015). Therefore these bacteria may speed up the process and meet out the required demand up to satisfactory level. The ability of PGPR to solubilize potassium rock by producing and secreting organic acids has being widely studied. Thus these bacteria ultimately improve crops' productivity in soil.

9.8 Biofilm potentials of plant growth promoting microorganisms

Microbial ecology and its correct evaluation is a very important aspect. Since then we could know only culturable microorganisms of different nutritional class and their role in either ecosystem development, soil fertility, and their relationships with rhizosphere of various crop plants as well as horticulture soil or forest soil is very strong and important. Soil microbial community can be categorized into two; useful as well as harmful. Both the groups live together and complete their function with a competitive manner. Soil microorganisms live in different associations between each other as well as their relationship with plants. Microbial ecology deals with the existence, survival, as well as their role for different activities like soil mineralization, production of microbial metabolites, xenobiotic degradation, chemolithotrophic role in leaching of metals, N_2 fixation: symbiotic and asymbiotic, phosphate solublization and mobilization, biological control of pathogens, antimicrobial compounds like antifungal or antibacterial antibiotics or any other metabolites like enzymes, hormones, polysaccharides, amino acids, organic acids and several others like H_2, CH_4, and NH_3 etc. Such a complex group and their colonies in association either live without extra shelter or under biofilm are the matter of study. It is well documented that microbes of certain genetic characteristic live in biofilm, whether they are of medical, food, agricultural, industrial oriented or related microorganisms. The basic concept of biofilm is simple, that a transparent membranous structure produced by microorganisms which is mainly made up of polysaccharide, lipids, proteins or any other type of polymers which has high tensile, permeable and some time impermeable or both along with flexible film where different types of microorganisms of different nutritional classes exist together for longer period. The film has gaseous exchange in some pockets, variation of pH, temperature and several other physicochemical levels.

The role of biofilm in agriculture may be attributed near the rhizosphere or away from the rhizosphere under the soil horizons. Some facts are hypothetical which is based on the experimental findings while some have been proved and studied under microscope as well as by isolating several groups of microbes from soil. There are number of microorganisms which produce polysaccharides (such as alginate, dextran, pullulan, xanthan, etc.) in terrestrial ecosystem in which bacteria, fungi, yeast and algae equally participate. These polysaccharides may modify or polymerized in the presence of Ca, Na or other metal ions with other macromolecules of proteins and lipids. The role of such microorganisms like *Azotobacter, Xanthomonas, Pseudomonas, Bacillus, Pediococcus, Leuconostoc*, etc., while fungi like *Aureobasidium pullulans* along with several yeasts like *Candida, Endomycoses, Torula*, etc. They produce homo and heteropolysaccharides such as pullulan and dextran are homopolysaccharides, while xanthan is heteropolysaccharide produced by *Xanthomonas campestris*. A polymorphic fungus, *A. pullulans* grow in soil as well as aquatic system and have strong genetic makeup to produce pullulan, repeating units of maltotriose (Gaur et al., 2010). It is versatile and used for biodegradable plastic. Its tensile strength, thin transparent film permeable for oxygen characteristics indicate the biofilm of agriculture orientation near the rhizosphere or any part of the soil. It is not easily soluble in water if polymerized with certain other polysaccharide. It is also produced in carbon and nitrogen limiting conditions. Natural ecosystems have such housing or encapsulating system for long term survival of microbial community where they reside with variety of microbes having wide diversity of nutritional category. The biofilm formation for the protection of microbial community in the extreme environment is existing since the inception of microbial system.

The free living N_2 fixers, *Azotobacter* also produce alginate and plant growth promoting substance may be helpful for film formation and create a sheath where other microbes live friendly with variety of metabolic diversity. The film chemical characteristics and structural diversity may be a matter of research to explore friendly associates live in biofilm which may establish in plant rhizosphere for constant production of various bioactive molecules required for plant and soil health. Biofilm may be prepared on membrane kept on nutrient plate by inoculating *Azotoacter, Arthrobacter, Pseudomonas* (plant growth promoting fluorescent strains) and others like phosphate solubilizing *Bacillus* and *A. pullulans* which produce high content of polysaccharides and such combinations has already tested showing increased plant growth response.

Certain biological control agents and iron chelating agents along with microbes producing plant growth hormone indole acetic acid and cytokinin which are under trial to develop a biofilm community of friendly microbes for an efficient approach of rhizospheric inoculants by using specialized techniques. This area of current research in agriculture will open a new avenue in agricultural microbiology. Soil is reservoir of almost all kind of microbes producing various types of organic and inorganic carbon and nitrogen sources. Almost all minerals and gases are present in this system. Therefore the basic principle of biofilm formation as well as mechanisms for protection of various nutritional groups under a common shelter is also same, as they are not much affected by physiochemical and nutritional conditions, as biofilm fulfils or provide almost all favorable conditions for their long term survival.

Biofilm microbial system in rhizosphere has not studied well, therefore, a proper methodology for the study of aerobic, anaerobic, facultative anaerobic; autotrophic to chemoautotrophic, heterotrophy to photoautotrophs status should be evaluated along with the status of fungi, yeast, bacterial counts in plant rhizospheric in order to understand the microbial life of biofilm. It is well documented that wherever microbes are in higher number with several groups and classes, biofilm support multicomplex favorable house for the existing microbes which must be carefully explore for its application in agriculture.

9.9 Recent advancements in biopotentials for sustainable agriculture

Much work in different dimensions have done by several pioneer workers, and identified many bacteria, fungi and actinomycetes for their role in plant growth directly or indirectly. Soil and plant relations along with microbial community, its interaction with different nutritional classes are associated with a chain of sequence interrupted by several biotic and abiotic factors, but the competence adaptation, microbial shifting, nutritional sink, saprophytic carbon, nitrogen, sulfur and phosphorus cycles along with the metabolic potentials and genetic status may simultaneously work together. Despite this, microbes exist in same biomass and their population for a very long period with slight fluctuation in microbial biomass or cells population. Gaur (1992) has evaluated soil microorganisms in different seasons as well as different situations like soil solarization, where soil temperature increased by 8°C−10°C to kill phytopathogens in the field conditions. It has been recorded that microbial numbers, that is, population of bacteria, fungi and actinomycetes were not reduced significantly, as thermotolerant to thermophilic populations of all groups were evaluated and even the status of cellulose, hemicelluloses and starch degrading microbial population were doing all their activities effectively in nonsolarized soils. These findings indicate that the microbial equilibrium in numbers as well as their physiological status could be maintained. Likewise, all the microbial activities for proteolytic, pectinolytic, xylanolytic, lignolytic, amylolytic status are always maintained through microbial shifting of specific microbial groups. Such findings indicate that microbes may grow and shift their status at any extreme conditions with some more incubations and alterations.

9.10 Future challenges

In soil, microbial ecology and its evaluation is one of the major objectives for beneficial microbial interactions for better plant growth. In spite of several researches from 1985 to till date not much effective information and friendly microbial consortium could be developed which may work for plant growth by several ways like biological control, hormone production, organic matter decomposition along with xenobiotic degradation, effectively. Farmers are also not getting maximum benefit of this. Therefore a consortium of such combination under a microbial film can be important in rhizosphere where they are also able to develop all the important plant growth promoting compounds. Another technical challenge is to get a complete understanding of the ecology of each beneficial microorganisms and their functions along with their limitations. The competency of beneficial bacteria should have their friendly microbial associates during seed bacterization or direct application to soil to achieve combined effects on plant growth, and they may be evaluated in such a way that their friendly association live longer period so that soil remain fertile for longer period. These associations will be evaluated up to three crops continuously.

Another challenge is the development of new assays for selecting PGPR efficiency along with their natural survival status, antagonistic nature, metabolic efficacy and rapid testing of strain. The proper carrier of microbes along with their other agents in a suitable for long term active survival, so that they may regenerate rapidly without losing its activity. Therefore, the formulation of living agent into product and development of delivery system for applying products in a manner consistent with conventional agriculture practices for the target crops. Formulation of biological control agents and other PGPR have not recorded the level of development for long term viability of Gram-negative bacteria. Gram-positive spore forming bacteria can be preserved for longer period. Other nonspore forming bacteria must be investigated. Therefore development of consortia of friendly microbial associates which may work in a wide range of crops should be developed in single system with an easily delivery process for their sustainability.

Microbes of various origin and status exist in soil which is evident from the recent findings of the Gaur et al. (unpublished data) They identified several bacterial strains of the common soil resident, which could tolerate alkaline, saline as well as other extreme conditions. Maintenance of soil fertility as well as other microbial transformation of various organic and inorganic compounds is important. Fertility of the soil may be slow to fast depend on the activity of microorganisms at extreme conditions along with available carbon, nitrogen and sulfur.

It is important to evaluate that what are the main genera which incorporate maximum percentage of occurrence in variety of crops, lower to higher plants, aromatic plants or other plant diversity. In a study before past two decades conducted by Gaur (1992) reported the concept of friendly microbial associates which are competent enough to exist in nature and do very important work with PGPR. Since past 20 years of continuous work on almost all crops mainly wheat, gram, pea, mustard, chili as well as most of vegetables plants having *Bacillus, Azotobacter*, fluorescent and nonfluorescent *Pseudomonas* sp. These microbial associates are common friends of each. The above three bacteria were commonly colonized various lower to higher plants, crops; cereals, legumes and vegetables. These friendly associates were also inoculated with *Azotobacter* and *Rhizobium* in different legumes crop showed significant increase in plant growth and yield (Gaur et al. unpublished data). These findings initiated to work with several microbial system and crops. Plant growth promotion and friendly formulation of microorganisms is significant in plant growth promotion. Therefore, the concept of microbial associate is a new dimension of research in the area of PGPR. It is required for several crops to produce in drought conditions where the microbial activity is very less.

The increasing concentration of pesticides and other xenobiotic compounds and chemical fertilizers have raised several questions for the use of any recommended bioinoculants which have oriented without such conditions and recommended for its application, but its sustainability in such soils where deleterious microorganisms and such chemical toxicity may hinder its proper activity, therefore, the biomolecules must be resistant to all such levels. Further, any bioinoculant competency must be replaced by another strain, time to time, otherwise, there is no use of such biosystem. In order to achieve such conditions, different strains in combinations along with friendly microbial associates should be designed for field applications. On the basis of such facts some recommendations have been made based on pot and field experiment results by the authors of this chapter. In this recommendation, several crops and microbial combinations are mentioned in the chapter which is important for farmers and researchers who are interested to continue their research in the area of biopotentials in agriculture. Some facts are generally based of the research findings as well as field experiment, results observed by the Indian farmers. The conclusions were drawn that the microbial system always requires better strains identification from natural resource. The competency of the strain must be checked in the presence of several soil types where different groups of microorganisms reside and interact with the strain deliberately introducing in the soil. However, such effects have proved that there are some friendly bacteria which support such bioinoculants which have been identified through pot experiments through small plots as field condition. The soil suppressive and conducive behavior towards the suppression of several phytopathogenic fungi is important.

These properties are directly proportional to the presence of some beneficial microorganisms residing in the soil. Such microbial indicators for recognizing soil health is a new dimension of future and fruitful research in this area (Gaur, 1992). Based on past research work of author in this area, he reported friendly microbial associates application in field conditions is recommended following combination of microbes for the specific crops are listed as:

1. Crop: wheat, maize, pulses

Bacterial system: *Rhizobium, Azotobacter*, phosphate solublizing *Bacillus, Pseudomonas, Arthrobacter* (Ratio 1:1:1:1)

Biocontrol agents: *T. harzianum*

Application method: through bacterization at the time of sowing, in which *T. harzianum* through wheat bran culture + 5 g soil mix per hectare, 5–10 days before sowing in any season, while after 10 days of sowing again its application in same amount is recommended.

a. Vegetables: for any vegetables except leguminous crop *Azotobacter, Azospirillum, Pseudomonas*, PSB-*Bacillus*, any suitable PSB strain *Arthrobacter* combination in 1:1:1:1 ratio during sowing through seed bacterization, while *Arthrobacter* is applied 5 days before with bagasse mixture 10 kg/ha along with cellulase producing bacteria and fungi or actinomycetes having capability of C1, Cx, and β glucosidase.

Biocontrol agent may be used in combination with *T. harzianum* and *B. bassiana* or any other suitable biocontrol agent along with *Bacillus thuringienesis* in 1:1:1 combination 5–10 days before sowing as well as 5 days after sowing with same dose. A friendly bacteria which is commonly used is *Arthrobacter* sp. This bacterium was thoroughly studied and found that it utilizes xenobiotic compounds for their growth and survival. Furthermore, the strains of *Arthrobacter* spp. frequently used for several crop plants, grasses, aromatic medicinal plants as well as in horticulture. In a study, it was observed that when *Rhizobium, Azotobacter* and *Arthrobacter* combinations was applied, the growth and yield was increased. Singh and Gaur (2016a,b) concluded reported that lignin or any other xenobiotic compound could be degraded and root growth was favored 10%–20% level without presence of any fungal pathogen. It was again tested in vitro regarding *Arthrobacter* and *Rhizoctonia, Sclerotium rolfsisi* sclerotia and chlamydospores of *Fusarium* spp., and found 80%–95% fast colonization of sclerotia and chlamydospores leading to inactive its survival. It is therefore concluded that this bacterium may be applied with any biocontrol agents 5–10 days before the sowing. Finally, more work in this direction with direct and indirect effect on several phytopathogenic fungi and disease incidence may be worked out.

9.11 Conclusion

Agriculture development is a major concern for all countries. Soil health is adversely affected by the use of chemical fertilizers and pesticides. Whereas, the use of microbial-based fertilizers and biopesticides provides an ecofriendly and healthy approach for crop productivity. The specific plant growth promotion requires some friendly microbial associates. This may be fungi, bacteria, and specific actinomycetes. There must be a culture bank of PGPR of such isolates which must be updated for its effectiveness through research and development laboratories. The effective PGPR strains should be produced in bulk and be supplied regularly to the farmers. Further, a remarkable factor is the study of microbial community, an interest of the soil where it is being applied with specific crops. Some microbial factors such as a wide range of water activity, salinity, pH, and their colonization factors should be monitored time to time. If the microbial strains are sustained in drought, it means it will work better in field conditions showing more survival in soil. These factors are also the interest of this area to find out the microbial treasure of agriculture production with quality food production. Thus, for sustainable agriculture of any country, a mobile soil health lab is essential to work on such aspects where farmers are not aware of biopotentials in agriculture.

References

Ahmed, M., Khan, M.S., 2012. Evaluation of plant growth promoting activities of rhizobacterium *Pseudomonas putida* under herbicide stress. Ann. Microbiol 62, 1531−1540.

Alori, E.T., Glick, B.R., Babalola, O.O., 2017. Microbial phosphorus solublization and its potential for use in sustainable agriculture. Front. Microbiol. 8, 971.

Anke, H., Sterner, O., 2002. Insecticidal and nematicidal metabolites from fungi. In: Osiewacz, H.D. (Ed.), Industrial Applications. The Mycota (A Comprehensive Treatise on Fungi as Experimental Systems for Basic and Applied Research), vol. 10. Springer, Berlin, Heidelberg, https://doi.org/10.1007/978−3−662−10378-4_6.

Baker, R., 1991. Molecular biology in control of pathogens. In: Kundsen, R. (Ed.), Handbook. Marcel Dekker Inc, New York, pp. 259−271.

Baker, K.F., Cook, J.R., 1974. Biological Control of Plant Pathogens. WH Freeman, San Francisco, CA, p. 6.

Begum, N., Qin, C., Ahanger, M.A., Raza, S., Khan, M.I., Ashraf, M., et al., 2019. Role of arbuscular mycorrhizal fungi in plant growth regulation: implications in abiotic stress tolerance. Front. Plant Sci. 10, 1068. Available from: https://doi.org/10.3389/fpls.2019.01068.

Calvo-Polanco, M., Sanchez-Romera, B., Aroca, R., Asins, M.J., Declerck, S., Dodd, I.C., et al., 2016. Exploring the use of recombinant inbred lines in combination with beneficial microbial inoculants (AM fungus and PGPR) to improve drought stress tolerance in tomato. Environ. Exp. Bot. 131, 47−57.

Chen, Y., Katan, J., 1980. Effect of solar heating of soil by transparent polythelene mulching on their chemical properties. Soil Sci. 130, 271−277.

Chet, I., Ordentlich, A., Shapira, R., Oppenheim, A., 1990. Mechanisms of biocontrol of soil-borne plant pathogens by rhizobacteria. Plant Soil 129 (1), 85−92.

Chodak, M., Golebiewsi, M., Morawska-Ploskona, J., Kuduk, K., Niklinska, M., 2015. Soil chemical properties affect the reaction of forest soil bacteria to drought and rewetting stress. Ann. Microbiol. 65, 1627−1637.

Egamberdieva, D., Jabborova, D., Hashem, A., 2015. *Pseudomonas* induces salinity tolerance in cotton (*Gossypium hirsutum*) and resistance to *Fusarium* root rot through the modulation of indole-3-acetic acid. Saudi J. Biol. Sci. 22, 773–779.

Eljounaidi, K., Lee, S.K., Bae, H., 2016. Bacterial endophytes as potential biocontrol agents of vascular wilt diseases–review and future prospects. Biol. Control 103, 62–68.

Franche, C., Lindstrom, K., Elmerich, C., 2009. Nitrogen-fixing bacteria associated with leguminous and non-leguminous plants. Plant Soil 321, 35–59.

Garg, S.K., Tripathi, M., Singh, S.K., Tewari, J., 2012a. Biodecolorization of textile dye effluent by *Pseudomonas putida* SKG-1 (MTCC 10510) under the conditions optimized for monoazo dye orange II color removal in simulated minimal salt medium. Int. Biodet. Biodeg. 74, 24–35.

Garg, S.K., Tripathi, M., Kumar, S., Singh, S.K., Singh, S., 2012b. Microbial dechlorination of chloroorganics and simultaneous decolorization of pulp-paper mill effluent by *Pseudomonas putida* MTCC 10510 augmentation. Environ. Monit. Assess. 184, 5533–5544.

Gaur, R., 1992. Effect of Soil Solarization on Microorganisms (PhD thesis). Lab of Mycology and Microbiology, CAS in Botany, BHU, Varanasi, India.

Gaur, R., Singh, R., Gupta, M., Gaur, M.K., 2010. *Aureobasidium pullulans*, an economically important polymorphic yeast with special reference to pullulan. Afr. J. Biotechnol. 9 (47), 7989–7997.

Gill, S.S., Gill, R., Trivedi, D.K., et al., 2016. *Piriformospora indica*: potential and significance in plant stress tolerance. Front. Microbiol. 7, 332.

Golinska, P., Wypij, M., Agarkar, G., Rathod, D., Dahm, H., Rai, M., 2015. Endophytic actinobacteria of medicinal plants: diversity and bioactivity. Antonie Van Leeuwenhoek 108, 267–289.

Govindasamy, V., Franco, C.M., Gupta, V.V., 2014. Endophytic actinobacteria: diversity and ecology. In: Verma, V.C., Gange, A.C. (Eds.), Advances in Endophytic Research. Springer India, New Delhi, India, pp. 27–59.

Harman, G.E., Uphoff, N., 2019. Symbiotic Root-Endophytic Soil Microbes Improve Crop Productivity and Provide Environmental Benefits. https://doi.org/10.1155/2019/9106395.

Hasan, S.E.-D., 2017. Plant growth-promoting activities for bacterial and fungal endophytes isolated from medicinal plant of *Teucrium polium* L. J. Adv. Res. 8 (6), 687–695.

Ikeda, A.C., Bassani, L.L., Adamoski, D., Stringari, D., Cordeiro, V.K., Glienke, C., et al., 2013. Morphological and genetic characterization of endophytic bacteria isolated from roots of different maize genotypes. Microb. Ecol. 65 (1), 154–160.

Kloepper, J.W., Schroth, M.N., 1981a. Relationship of in vitro antibiosis of plant growth promoting rhizobacteria on potato plant development and yield. Phytopathology 70, 1078–1082.

Kloepper, J.W., Schroth, M.N., 1981b. Plant growth promoting rhizobacteria and plant growth under xenobiotic, plant parasitic condition. Phytopathology 71, 642–644.

Kloepper, J.W., 1992. Plant growth promoting rhizobacteria as biological control agents. In: Mattnig, B. (Ed.), Soil Microbiology Technologies. pp. 142–152.

Kloepper, J.W., Leong, J., Teintze, M., Schroth, M.N., 1980. Enhanced plant growth by siderophores produced by plant growth promoting rhizobacteria. Nature 286, 885–886.

Leong, J., 1986. Siderophores: their biochemistry and possible role in the biocontrol of plant pathogens. Annu. Rev. Phytopathol. 24, 187–209.

Linderman, R.G., 1988. Mycorrhizal interaction with the rhizosphere microflora: the mycorhizosphere effect. Phytopathology 78, 488–505.

Mahmood, A., Turgay, O.C., Farooq, M., Hayat, R., 2016. Seed biopriming with plant growth promoting rhizobacteria: a review. FEMS Microbiol. Ecol. 92 (8). Available from: https://doi.org/10.1093/femsec/fiw112.

Mohammed, S.H., Seady, M.A., Enan, M.R., Ibrahim, N.E., Ghareeb, A., Moustafa, S.A., 2008. Biocontrol efficiency of *Bacillus thuringiensis* toxins against root-knot nematode, *Meloidogyne incognita*. J. Cell Mol. Biol. 7, 57–66.

Nicolás, C., Martin-Bertelsen, T., Floudas, D., Bentzer, J., Smits, M., Johansson, T., et al., 2019. The soil organic matter decomposition mechanisms in ectomycorrhizal fungi are tuned for liberating soil organic nitrogen. ISME J. 13, 977–988.

Oliveira, A.L.M., Santos, O.J.A.P., Marcelino, P.R.F., 2017. Maize inoculation with *Azospirillum brasilense* Ab-V5 cells enriched with exopolysaccharides and polyhydroxybutyrate results in high productivity under low N fertilizer input. Front. Microbiol. Available from: https://doi.org/10.3389/fmicb.2017.01873.

Oteino, N., Lally, R.D., Kiwanuka, S., Lloyd, A., Ryan, D., Germaine, K.J., et al., 2015. Plant growth promotion induced by phosphate solubilizing endophytic *Pseudomonas* isolates. Front. Microbiol. 6, 745.

Paustian, K., Lehmann, J., Ogle, S., Reay, D., Robertson, P.G., Smith, P., 2016. Climate-smart soils. Nature 532, 49–57.

Pavithra, D., Yapa, N., 2018. Arbuscular mycorrhizal fungi inoculation enhances drought stress tolerance of plants. Ground Water Sust. Dev. 7, 490–494. Available from: https://doi.org/10.1016/j.gsd.2018.03.005.

Saha, M., Sarkar, S., Sarkar, B., Sharma, B.K., Bhattacharjee, S., Tribedi, P., 2015. Microbial siderophores and their potential applications: a review. Environ. Sci. Pollut. Res. 10, 1–16.

Sahu, G.K., Sindhu, S.S., 2011. Disease control and plant growth promotion of green gram by siderophore producing *Pseudomonas* sp. Res. J. Microbiol. 6, 735–749.

Schippers, B., Baker, A.W., Baker, P., 1987. Interactions with deleterious and beneficial rhizosphere microorganisms and the effect of cropping practices. Annu. Rev. Phytopathol. 25, 339–358.

Schippers, B., Bakker, P., Bakker, A., van Peer, R., 1988. Crop losses due to deleterious rhizobacteria and their prevention by bacterization. In: Proceedings Brighton Crop Protection Conference Pest and Diseases. vol. 2. British Crop Protection Council, London, pp. 611–614.

Schroth, M.N., Hancock, J.G., 1982. Disease-suppressive soil and root-colonizing bacteria. Science (Washington, DC) 216, 1376–1381.

Shah, F., Nicolás, C., Bentzer, J., Ellström, M., Smits, M., Rineau, F., et al., 2016. Ectomycorrhizal fungi decompose soil organic matter using oxidative mechanisms adapted from saprotrophic ancestors. New Phytol. 209, 1705–1719.

Singh, S.P., Gaur, R., 2016a. Evaluation of antagonistic and plant growth promoting activities of chitinolytic endophytic actinomycetes associated with medical plants against *Sclerotium rolfsii* in chick pea. J. Appl. Microbiol. 121, 506–518.

Singh, S.P., Gaur, R., 2016b. Endophytic *streptomyces* spp. underscore induction of defence regulation genes and confers resistance against *Sclerotium rolfsii* in chicken pea. Biol. Control 104, 44–56.

Souza, R., Meyer, J., Schoenfeld, R., Costa, P.B., Passaglia, L.M.P., 2014. Characterization of plant growth-promoting bacteria associated with rice cropped in iron-stressed soils. Ann. Microbiol. 65, 951–964.

Stapleton, J.J., Devay, J.E., 1986. Soil solarization: a new chemical approach for management of plant pathogens and pests. Crop Prot. 5, 190–198.

Tiwari, S., Gaur, R., 2014. Decolorization of distillery spentwash (melanoidin) by immobilized consortium (bacterium and yeast) cell: entrapped into sodium alginate bead. J. Environ. Sci. Technol. 7 (3), 137–153.

Tiwari, S., Gaur, R., 2019. The application of microbial enzymes in distillery spent wash decolorization. In: Tripathi, M. (Ed.), Microbial Treatment Strategies for Waste Management. OMICS Int., Heathrow, United Kingdom, ISBN No. 978–1–63278–079-9.

Tripathi, M., Garg, S.K., 2013. Co-remediation of pentachlorophenol and Cr^{6+} by free and immobilized cells of native *Bacillus cereus* isolate: spectrometric characterization of PCP dechlorination products, bioreactor trial and chromate reductase activity. Process. Biochem. 48, 496–509.

Tripathi, M., Vikram, S., Jain, R.K., Garg, S.K., 2011. Isolation and growth characteristics of chromium (VI) and pentachlorophenol tolerant bacterial isolate from treated tannery effluent for its possible use in simultaneous bioremediation. Indian J. Microbiol. 51 (1), 61–69.

Tripathi, M., Kumar, S., Kumar, A., Tripathi, P., Kumar, S., 2018. Agro-nanotechnology: a future technology for sustainable agriculture. Int. J. Curr. Microbiol. Appl. Sci. (Special Issue 7), 196–200.

Tripathi, M., Kumar, S., Verma, T., Singh, S., Shukla, N., Pandey, R., et al., 2020. Microbial and nutritional approaches towards sustainable agriculture development. In: Verma, T., Prasad, S. (Eds.), Importance of Biotechnology in Food Production and Food Security. Shree Publishers and Distrubutors, New Delhi, India, pp. 113–134. ISBN: 978–81–8329–997-8.

Tscharntke, T., Clough, Y., Wanger, T.C., Jackson, L., Motzke, I., Perfecto, I., et al., 2012. Global food security: biodiversity conservation and the future of agricultural intensification. Biol. Conserv. 151, 53–59.

Vimal, S.R., Singh, J.S., Arora, N.K., Singh, S., 2017. Soil-plant-microbe interactions in stressed agriculture management: a review. Pedosphere 27 (2), 177–192.

Voisard, C., Keel, C., Hass, D., Defago, G., 1989. Cyanide production of *Pseudomonas fluorescens* helps suppress black rot of tobacco under xenobiotic conditions. EMBO J. 8, 351–358.

Wan, W., Qin, Y., Wu, H., Zuo, W., He, H., Tan, J., et al., 2020. Isolation and characterization of phosphorous solublizing bacteria with multiple phosphorous sources utilizing capability and their potential for lead immobilization in soil. Front. Microbiol. 11, 752.

Weller, D.M., 1988. Biological control of soil borne plant pathogens in the rhizosphere with bacteria. Annu. Rev. Phytopathol. 26, 379–407.

Wu, Q., Peng, Z., Yang, M., et al., 2018. Rhizobia promote the growth of rice shoots by targeting cell signaling, division, and expansion. Plant Mol. Biol. 97, 507–523.

Yadav, S.K., Tripathi, M., Nayak, R., Tomar, S.K., 2016. Plant growth promoting *Rhizobacteria* in agriculture and its uses. In: Singh, R.P., Singh, R.K., Singh, M. (Eds.), Recent Technological Interventions in Agriculture. pp. 213–217. ISBN No. 978–93–84215–56-9.

Yadav, S.K., Tomar, S.K., Tripathi, M., Singh, S.N., 2017a. Front line demonstration technology for pulses production using *Rhizobium* sp., *Trichoderma* sp. and phosphate solublising bacteria: a comparative study. Bull. Environ. Pharmacol. Life Sci. 6 (Special Issue 3), 576–579.

Yadav, S.K., Tripathi, M., Kumar, S., 2017b. Microbial augmentation for enhancing lentil and field pea production in agriculture fields. J. Biotechnol. Bioeng. 1, 28–30.

Yu, X., Liu, X., Zhu, T.H., Liu, G.H., Mao, C., 2011. Isolation and characterization of phosphate solubilizing bacteria walnut and their effect on growth and phosphorus mobilization. Biol. Fertil. Soils 47, 437–446.

Zhang, F., Jia-Dong, H.E., Qiu-Dan, N.I., Qiang-Sheng, W.U., Zou, Y.N., 2018. Enhancement of drought tolerance in trifoliate orange by mycorrhiza: changes in root sucrose and proline metabolisms. Not. Bot. Horti. Agrobot. Cluj-Napoc. 46 270. Available from: https://doi.org/10.15835/nbha46110983.

Zhou, H., Yang, Y., Peng, T., Li, W., Zhao, L., Xu, L., et al., 2014. Metabolites of *Streptomyces* sp., an endophytic actinomycete from *Alpinia oxyphylla*. Nat. Prod. Res. 28, 265–267.

Next generation OMICS: a tool to understand the diversity of soil microbiota and improvement of agricultural sustainability

10

Subhaswaraj Pattnaik and Busi Siddhardha
Department of Microbiology, School of Life Sciences, Pondicherry University, Puducherry, India

10.1 Introduction

10.1.1 *Microbial diversity*

Microorganisms, being the most primitive form of cellular life and representative of biosphere on Earth, have witnessed drastic environmental changes over the ages and evolved as the most diversified living system on Earth harboring a multitude of extreme microhabitats and ecological niche. Over the eras, microbial populations are proved to be involved in the evolutionary mechanisms pertaining to their pivotal role in ecological, biogeochemical, and physiological cycles. Hence, for the life forms to sustain on Earth, the microbial diversity is highly essential and could contribute immensely starting from different biological processes to highly advanced biotechnological applications (Dunlap, 2001).

10.1.2 *Microbial diversity in soil*

Soil denotes a highly complex heterogeneous environment constituting one of the largest reservoirs for inhabiting a wide spectrum of microhabitats which is dwelled by highly diverse microbial communities (Blaud et al., 2014; Armalyte et al., 2019). Soil microbiota represents one of the highly diversified and dynamic ecosystems in the soil system and constitutes an integral component of soil habitat. The spatial heterogeneous distribution of soil microhabitats' physicochemical properties corresponds to the heterogeneous diversity of microbial populations in the soil microhabitat (Ding et al., 2015). Being one of the most dynamic and diverse ecosystems on Earth, the soil ecosystem comprises of highly diversified assemblages of microbiomes consisting of eubacteria, archaeabacteria, and eukaryotic taxa including algae and fungi (Attwood et al., 2019). Soil ecosystem functions as the supporting role in maintenance of essential nutrient availability, growth, and development of soil microbiota and degradation of

Microbes in Land Use Change Management. DOI: https://doi.org/10.1016/B978-0-12-824448-7.00013-9
© 2021 Elsevier Inc. All rights reserved.

contaminants (Biswas and Sarkar, 2018). Soil aggregates function as the most dynamic ecological niche for harboring diversity of microbial communities. Based upon the structural and physicochemical characteristics of the soil aggregates, the inhabiting soil microbiota also showed heterogeneous distribution (Li et al., 2018). For example, large macroaggregates largely harbor filamentous fungi whereas water-stable macroaggregates harbor microbial populations corresponding to *Actinobacterial* sp. and α-proteobacteria. The ratio corresponding to fungal communities to bacterial communities in microaggregates is comparatively lower as compared to macroaggregates. Hence, microaggregates tend to harbor highly diversified bacterial communities belonging to taxonomical order of Actinomycetales, Rhodobacterales, Sphingobacteriales, Burkholderiales, and many more (Bach et al., 2018).

As witnessed from different studies, the diversity in microbial communities in the soil ecosystem exhibited both taxonomic and phylogenetic species richness and comparatively more heterogeneous microbial communities coexist when the soil ecosystem is maintained under the influence of organic farming as compared to conventional farming system (Lupatini et al., 2017). The structure and functional attributes of soil microbiota are inherently affected by various environmental factors including both biotic as well as abiotic factors along with irrational anthropogenic events. Though, soil microbiota face the wrath of extreme environmental stress conditions including nutrient limitation, shallow soil horizon, low water retention, and frequent fluctuations in environmental factors in the high altitude regions; the mountainous soil harbors a large diversity of microbial communities (Praeg et al., 2019). Soil microbiota have an essential role to play in maintaining the soil health by accumulation of soil micronutrients which essentially are involved in conserving soil fertility (Kumar et al., 2017).

10.1.3 Soil microbiota and maintenance of soil ecosystem

As mentioned earlier, the soil microbial communities are proven to be instrumental in maintaining the ecosystem functioning and stability by providing the resilience to environmental disturbances. In the last few decades, a considerable attention was shifted toward the role of microbial communities harboring soil ecosystem in ecosystem sustainability. In order to achieve the sustainability in soil ecosystem, not only agroeconomic practices should be rationally controlled but also microbial diversity within the soil ecosystem should also be taken into consideration (Garbeva et al., 2004). As per the recent trends, the more diverse microbial communities harboring the soil ecosystem are, the more is the functional stability of the soil ecosystem, suggesting the influential role of microbial diversity. Thus microbial diversity in soil ecosystem and ecosystem functioning with sustainability are coherently associated with each other (Maron et al., 2018). To understand and maintain the sustainability of ecosystem functioning, it is essential to define the ecological role of soil microbiota in mediating the nutrient cycling processes and other important ecosystem services (Jiao et al., 2019).

10.1.4 Microbial diversity in soil and sustainable agriculture

The physiological and metabolic activities of soil microbial communities are essential for the maintenance of soil carbon and the soil carbon sequestration proved to be influential in conservation of soil ecosystem, management of agroecosystem sustainability, and preservation of food security values (Ding et al., 2015). Since the soil microbial communities significantly contribute to soil ecosystem functioning by providing essential nutrients by organic matter mineralization, maintaining the carbon sequestration process, upholding the structural and functional attributes of soil aggregates, and protecting from soil-borne plant pathogens; soil microbiota and their interactions with soil microenvironment gained considerable attention in the last few decades (Bonanomi et al., 2016). The cooperative interactions between the plants and associated soil microbiota are found to be influential in suppressing the soil-borne plant pathogens and parasitic activities (Topalovic et al., 2020).

To achieve the agricultural sustainability with improved agricultural productivity, it is essential ensure the sustained quality maintenance of soil quality which not only essential for preserving soil ecosystem but also critically minimizing the extreme weather conditions (Kiani et al., 2017). As it is evident from earlier instances, that soil microhabitat harbors a highly diverse group of metabolically active soil microbiotas responsible for facilitating plant-microbes interactions, maintaining the biogeoenergy fluxes including carbon and nutrient cycling. These effects in turn are responsible for the maintenance of soil health and productivity which ultimately support the improved growth and development of crops, increased yield, and enhanced productivity (Zorner et al., 2018). The sustainability in the soil equality also corresponds to the conserving the structural and functional attributes of soil microbial communities. Since the soil microbial communities are actively engaged in the management of soil quality and fertility, an insight into the diversity and characteristics of soil microbiota could possibly enable the sustainability in agriculture in terms of improved quality and yield of agriculture-based products (Wang et al., 2020). The irrational and long-term application of fertilizers not only characteristically modulates the soil properties including composition (nutrient content, soil pH, and organic matter content) and fertility but also critically modulates the composition of microbial communities thriving in soil ecosystem (Hu, 2017; Wang, 2017; Zhang et al., 2015).

10.2 OMICS technologies

10.2.1 Emergence of OMICS

From historical perspective, the term "OMICS" is derived from a Greek word "Ome" which literally corresponds to "whole"/"or"/"complete." When this concept is affiliated to cellular molecules such as genes, transcripts, proteins, metabolites then the derivatization of concept in the form of genome, transcriptome, proteome, and metabolome. The technological interventions studying these aspects are termed as genomics, transcriptomics, proteomics, and metabolomics respectively (Van Emon, 2016; Nalbantoglu and Karadag, 2019). Since the completion of Human Genome Project, it was realized that complex biological processes are regulated at various level of organization starting from gene level to protein and metabolic level. The various level of organization include the different fields in molecular biology collectively termed as "OMICS." To define, the customized quantification and/or characterization of a set of biological molecules associated with the structure, function, physiology and the molecular mechanisms of a set of biological organisms is termed as "OMICS" (Biswas and Sarkar, 2018). The OMICS-based concept is described at various levels of organization starting from "genomics" (interest revolves around the genome) to proteomics (interest revolves around the large sets of proteins, particularly the proteome) and metabolomics (interest revolves around on large sets of metabolic products, collectively termed as the metabolome) (Fig. 10.1). The genomics field is further categorized into genotyping (sequence-based approach), transcriptomics (expression-based approach), and epigenomics (revolves around the epigenetic regulation of genome expression) (Vlaanderen et al., 2010). The OMICS technologies also include the concept of metagenomics and metatranscriptomics.

The OMICS-based approaches have not only revolutionized the conventional approaches to identify microbial communities harboring extreme environmental niches but also provided next generation platforms to gain an insight into differential responses at micro-level to macro-level (Brul, 2012; Pan and Barrangou, 2020).

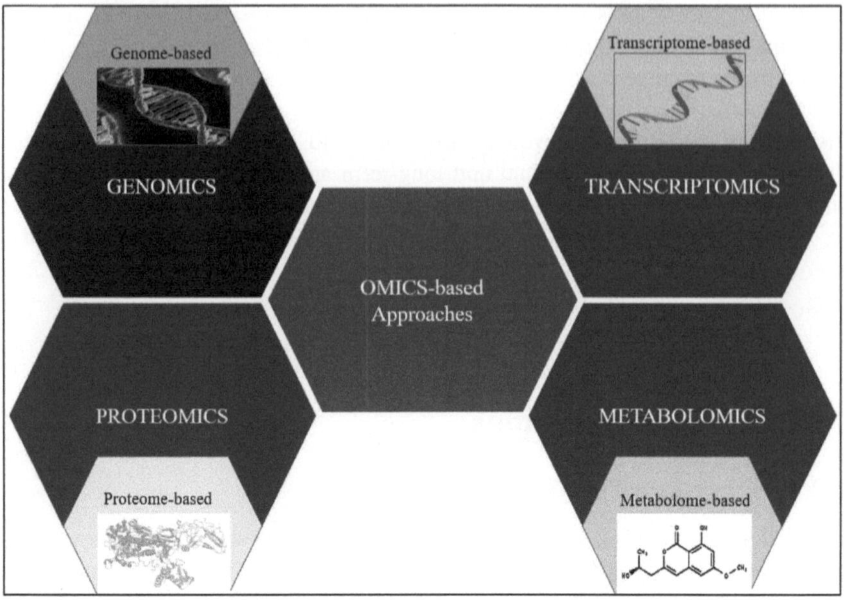

Figure 10.1 Schematic overview of different OMICS-based approaches and their putative targets.

The metagenomic analyses infer the information on the whole genome of the microbial communities isolated from different environment, their interactions profile with various biotic and abiotic factors, and their biogeochemical profiling. Meanwhile, metatranscriptomics correspond to decipher the expression profile of specific mRNA transcripts of microbial communities and their customized regulations for identifying taxonomically and functionally relevant microbiota. Similarly, understanding the expression profile of proteomes of microbial communities on a large scale basis corresponds to metaproteomics studies. The combinatorial studies of metagenomics and metaproteomics termed as "proteogenomics" further provided an insight understanding of phylogenetic association between different microbial communities (Biswas and Sarkar, 2018).

In addition to these OMICS-based approaches, a comparatively relative term "Fluxomics" gained recent recognition as it infers the study of complete set of different metabolic fluxes (collectively called as "fluxome") with specific reference to decipher an insight into the rate of metabolic activities, differential rate of metabolic reactions within biological organism. The analytical data thus generated provides vital information on cellular metabolic processes and its association with the metabolome (Chandran et al., 2020). One of the interesting aspects of OMICS-based technologies and their intervention in the biological fields is the ability to provide an insight into different biological processes through global and high throughput analytical methodologies efficiently as compared to traditional methods, which are basically time consuming and comparatively less efficient (Shi-Kai et al., 2015).

10.2.2 Applications of next generation OMICS

The recent trends in exploration of OMICS-based technological interventions has opened up new avenues to understand the biology and genetic makeup of plants and animals, the ecosystem where the biological organisms thrive and the microbial communities harboring the ecosystem. Thus the OMICS-based approaches found to be influential in maintaining a string of interactions between the microbiotas, host systems, and the concerned environment. The exploration of next generation OMICS technologies has gained considerable attention in the last few decades owing to their promising role in a wide spectrum of applications including biotechnological, pharmaceutical, agricultural fields, and also significantly contributed in next generation bioremediation approaches. Nevertheless, microorganisms are considered for production of a wide spectrum of bioactive secondary metabolites and pharmaceutically relevant enzymes; the limitations associated with conventional culture-dependent approaches in exploring microbial sources from extreme environment resulted in logistic failure in exploiting the untapped microbial products. The advent of culture-independent OMICS approached has provided an aided advantage to decipher the potential of microbiota residing in extreme environments in producing bioactive compounds and relevant enzymes of pharmaceutical importance. For example, metagenomics approaches (sequence-based as well as activity-based approach) have received widespread recognition in identifying the novel enzymes produced by microbial sources and could be implemented toward remediation of heavy metals and toxic pollutants from the environment (Ufarte et al., 2015). The OMICS-based approaches have enabled the research community an aided arsenal to explore the untapped potential of uncultivable microbial communities and their genetic potential towards the production of novel enzymes to remediate the pollutants from the environment. No doubt, the advancement in metagenomics-based technological interventions has proved to be comparatively rapid and inexpensive breakthrough in order to determine the microbial community and their untapped potential from different environmental samples. In addition, the functional attributes of "transcriptomics," "proteomics," and "metabolomics" also provided a paradigm shift in determining the bioremediation efficacy of microbial communities as compared to conventional approaches (Desai et al., 2010) (Fig. 10.2).

The advanced metagenomics approaches, as a part of microbiome engineering strategies in soil microhabitat, have provided a platform for an insight understanding of microbial composition and functional attributes. This concept could result in the development of multispecies "designer" biofertilizers with complementing strains which allow improved tolerance toward different stress conditions (Arif et al., 2020). The genomics and high throughput phenomics approaches also formed the basis for facilitating the plant breeding programs to adapt to different climatic and nutritional stress conditions. Thus the OMICS-based approaches are also considered to be influential in improving crop yield and productivity by the development of stress-resilient crop varieties thereby suggesting their agroeconomic importance in the current scenario (Dwivedi et al., 2017).

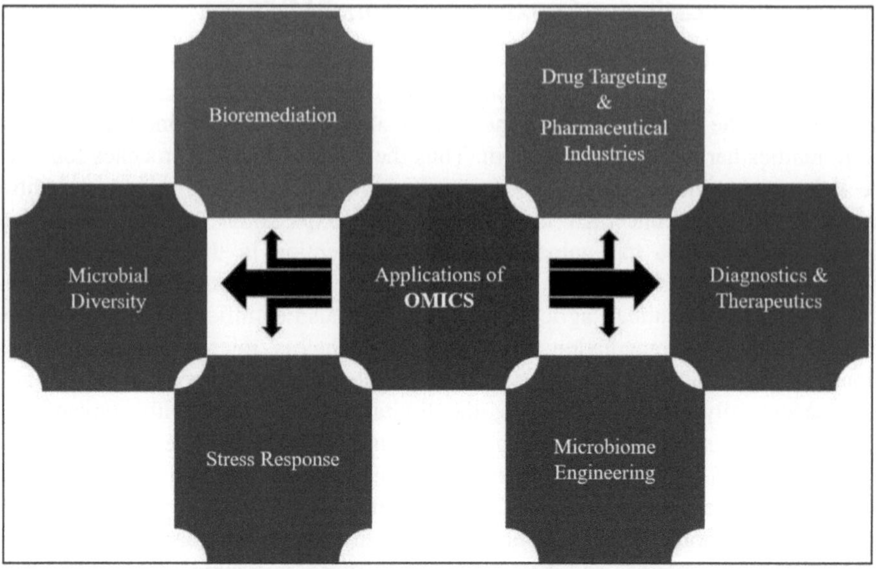

Figure 10.2 Schematic representation of multifaceted applications of different OMICS-based technologies.

The advancement in the highly sophisticated large-scale analytical OMICS-based approaches such as transcriptomics, proteomics, lipidomics, and metabolomics also proved to be highly defined unbiased approach for early diagnosis and therapeutics of several diseases and disorders. The OMICS-based strategies unravel the insight understanding of the pathways involved in disease progression by specifically identifying disease biomarkers which not only improved the current diagnostic approaches but also provided novel platform to design and develop novel therapeutic strategies (Wheelock et al., 2013). The OMICS-based technologies also characteristically provided an insight into the pathway elucidation of microbial pathogenesis which will facilitate the targeted drug development and effective therapeutics (Goff et al., 2020). The high throughput exploration of genomics, transcriptomics, proteomics, and metabolomics studies as OMICS-based approaches has provided a new dimension of ultrasensitive and sophisticated data analysis which ultimately could be explored towards targeted drug discovery, validation of drug toxicity and safety assessment, and pharmacological profiling. These implications have given an aided advantage to OMICS-based approaches in pharmaceutical applications with special reference to personalized medicines for personalized healthcare settings (Shi-Kai et al., 2015; Scelfo et al., 2018). The scientific and analytical data received from the intervention of high-throughput OMICS-based technologies gained considerable attention as the sophisticated methodologies provided an insight into the quantitative assessment of wide spectrum of putative molecular targets (e.g., DNA, RNA, protein, metabolite). As a result, the OMICS-based approaches efficiently provided analytical evidences for potential drug target for different healthcare services and targeted drug discovery programs (Paananen and Fortino, 2020). For example, the OMICS-based approaches in combination with a breakthrough advancement in next generation sequencing and bioinformatics tools are considered to be influential in determining the antimicrobial activity of bioactive natural products by specifically drug targeting approach (dos Santos et al., 2016). Apart from these wide spectrum of applications, OMICS-based approaches also thrive to unlock the mystery associated with the microbiomes in aquaculture, fisheries, food processing, and biotechnology. In addition, OMICS-based tools also create a dynamic atmosphere to identify, target, and conservation of biodiversity hotspots in an oceanic environment.

10.3 Next generation OMICS and agriculture

The massive industrialization and the population explosion have a serious impact on climate change, scarcity in groundwater level, environmental deterioration, depletion of soil fertility, and massive urbanization. All these factors again collectively pose an uphill challenge to agricultural sectors in terms of nutritional and food security and crop productivity (Pathak et al., 2018). The development of OMICS approaches particularly, metabolomics and proteomics have given a new dimension to understand the plant-microbe interactions by specifically decoding the chemical language articulated between the root exudates and the attracted/repelled microbial taxa upon the release of exudates which ultimately are helpful in enunciating the detrimental and/or beneficial effect on plant growth and promotion and developmental strategies. The OMICS-based customized approach thus extensively could be implemented to improve the agricultural yield and productivity through exploiting the beneficial plant-microbe interactions (Arif et al., 2020). Since, a rich diversity of microbial species harbor the soil ecosystem, the soil microbiota are considered to contain a dynamic and diverse source of structurally and functionally relevant genetic elements with specific functions. In order to have an insight into the functional attributes of such diverse genetic elements at ecosystem level, high throughput technological interventions in the form OMICS-based approaches in combination with computational networking and statistically relevant molecular tools are required. The OMICS-based technological interventions also facilitate the understanding of complex interactions between the diverse genetic elements, their role in ecosystem functioning and services and differential effect in response to various biotic and abiotic factors. Thus both the diversity as well the complexity of microbial communities in the soil ecosystem and their prospective role in ecosystem functioning could be understood with the advanced OMICS-based approaches across spatial and temporal scales (Attwood et al., 2019) (Fig. 10.3).

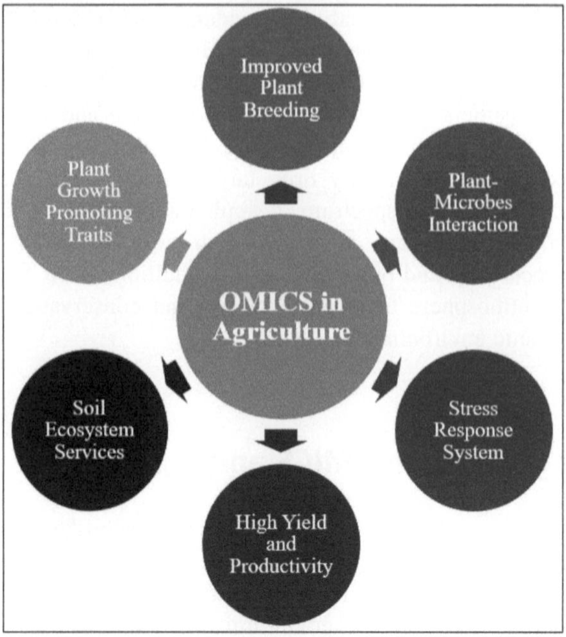

Figure 10.3 An overview of the implications of next generation, high throughput OMICS technologies on sustainable agriculture.

10.3.1 Next generation OMICS to understand microbial diversity

In earth, no doubt, microorganisms represent as the most diversified and abundant living system harboring wide spectrum of ecological niche corresponding to normal environments to extreme environmental conditions (Fierer and Jackson, 2006; Shamim et al., 2019). However, based on the conventional microbial approaches only 0.1%–1.0% of the microbial communities were successfully cultivated in laboratory conditions. In this context, the identification of microbial phyla from diverse ecological niche could be possible by implementing the high throughput, next generation metagenomic approaches which could complement the traditional culture-dependent approaches. The identification of microbial phyla through metagenomic approaches has created new avenues to predict the microbial diversity in different ecological environments. Based on the identification, these molecular tools could further facilitate in deciphering the biomedical and pharmaceutical applications of identified microbial communities (Nema, 2019; Shamim et al., 2019). The advent of metagenomic approaches followed by more advanced metatranscriptomics and single-cell genomics approaches has revolutionized the current culture-dependent approach-based understanding of the diversity of microbial communities harboring extreme environmental niche with an insight into the structural as well as functional attributes of microbial communities (Gupta et al., 2019). The recent advancement in OMICS-based approaches is considered to be state of the art tools to define the problems associated with unculturable microorganisms isolated from extreme environmental niche and their characteristic biotechnological applications (Chandran et al., 2020). It is evident that, plant biosystems provide diverse ecological niches for growth and development of highly diverse group of microbial communities starting from eubacteria and archaebacteria to eukaryotic microorganisms. In return, the microbiotas harboring plant biosystems facilitate several plant growth-promoting (PGP) traits including considerable attention for biofertilizers and biocontrol agents. In this context, the plant-associated microbiotas provide enormous potential to solve the problems associated with agricultural practices in an economical and sustainable method. Though the PGP microbiotas possess widespread applications in sustainable agriculture, their transformation in the field trial with commercial success remain obscured and urgently required technological interventions for translating the success in laboratories to the commercial field and market. In this regard, high throughput technological interventions in the form of OMICS-based approaches have provided a ray of hope in understanding diversity and functional attributes of plant-associated microbiotas (specifically termed as "phytomicrobiomes") in growth promotion, development, disease control, improved productivity, improved stress tolerance capacity and attaining sustainability in agricultural ecosystems. Understanding the complex association of phytomicrobiomes with hosts as well as environment through the metaomics-based studies also require the synthetic biology-based microbiome engineering principles which will facilitate the phytomicrobiomes in translating the agricultural sustainability to the field level (Glick, 2012; Mitter et al., 2016; Bloch et al., 2020; Ke et al., 2020).

10.3.2 Next generation OMICS for sustainable agriculture

In agricultural sectors, the advent of OMICS-based approaches have revolutionized our understanding the complexity of important and complex traits associated with crop productivity including nitrogen use efficiency, photosynthetic efficiency, water use efficiency, plant architecture, stress tolerance mechanism against various biotic and abiotic stress and nutritional quality (Pathak et al., 2018). Since the successful completion of Human Microbiome Project, a significant interest is laid upon the coordinated efforts towards the development plant microbiomes which will provide a new dimension to integrate the OMICS-based technological interventions towards identification of beneficial microbiotas associated with soil ecosystem and host plants. This will provide an impetus to characterize the microbiotas of interest for their diversity and functional attributes, their interactions with the host and environment and most importantly their influence on modern agricultural practices. This will also create an atmosphere to accelerate the agricultural productivity through sustainable agricultural practices to fulfill the growing demand on the verge of constant interference from anthropogenic events including population explosion, industrialization and global climate change (Busby et al., 2017). The OMICS-based technological interventions have revolutionized the crop productivity and yield efficiency through an integrative and comprehensive mining of several correlated genetic and epigenetic processes associated with plant growth, development and stress response mechanisms. The OMICS-based technologies also observed to be highly influential in the management of advanced crop breeding and cultivar developmental process through epigenetic reprogramming (Moshelion and Altman, 2015). The combinatorial approach considering the multifaceted OMICS-based tools with knowledge-driven integrated computational network vows to understand the complex interactions associated with the agroecosystem thereby managing several factors associated with the interactions among soil microbiotas, host plants, and soil environment. These interactions are considered to be instrumental component in enhancing crop productivity with efficient yield values (Ichihashi et al., 2020).

One of the interesting aspects of attaining sustainability in agriculture biosystems is to address the issues associated with stress tolerance mechanisms. The OMICS-based tools have successfully driven the stress tolerant genes and preparing a library of such tolerant genes as potential arsenal to fight against stress conditions. Functional genomics analysis has provided an advantage over conventional approaches to identify putative stress tolerant genes which are generally associated with the management of stress response to stress conditions such as extreme cold conditions, drought and heavy metal stress (Zhang et al., 2017; Mehta et al., 2019). The advancement in functional genomics also considered to be knowledge-driven technological interventions for improvement in crop productivity through precise and controlled molecular breeding program and marker-assisted genetic makeup for development of crop varieties with high yield and improved stress tolerance properties (Van Emon, 2016). The genomics interventions proved to be highly accurate and rapidly driven approach in inducing a suitable phenotypic variability by addition of specific genes and/or knocking down specific gene with RNAi thereby considered to be more reliable and efficient approach as compared to conventional plant breeding program (Van Emon, 2016).

In addition to genomics studies, the studies of highly sensitive transcriptome could also be considered for stress response mechanisms. The relative expression level of transcriptome is accomplished through high throughput molecular tools such as next generation RNA sequencing, digital gene expression profiling, serial analysis of gene expression and microarray analysis. This transcriptomics approach facilitates the understanding of putative stress tolerant genes and their relative expression in response to stress conditions (Le et al., 2012; Kreszies et al., 2018; Mehta et al., 2019). For example, large-scale microarray platforms were considered to be influential in identifying the differential expression level of putative stress tolerant genes. The differential expressed transcripts could be considered for stress tolerance management as these are associated with plants' defense mechanisms (Rasmussen et al., 2013; Mehta et al., 2019). Similar to that of transcriptome profile, the proteome profile is also observed to be highly dynamic in nature and corresponds to change with age, developmental stages, nutrient availability and change in the environmental conditions. Thus the proteome profile could be instrumental in defining the stress response mechanism using highly advanced proteome-based tools such as phosphoproteome, proteogenome, organellar proteome, nuclear proteome, and cell wall proteome. These proteomics-based approaches are comprehensively used in combination with mass spectrometric analysis to decipher deep insights into the differential expression of proteome fractions in response to stress conditions (Mehta et al., 2019). The interference of Proteomics-based approach also critically allows to understand the implementation of ecofriendly, biodegradable, and effective pesticides for improvement of yield and productivity of economical crop varieties. Thus the technological applications have aided an advantage over the traditional practices in attaining sustainability in agriculture (Van Emon, 2016). Similar to that of genomics and proteomics approaches, the advent of metabolomics also provide vital information on differential expression level of putative metabolomes responsible for growth dynamics, nutritional acquisition, stress response mechanism, and maintenance of plant health settings. As a result, the metadata obtained from metabolomics analysis, provide novel avenues to determine differential level of plant phenotypes and genetic diversity thereby suggesting its implications in sustainable agriculture. Hence, an in depth analysis of metadata associated with metabolomics profile provide novel avenues to identify putative target biomarkers responsible for plant response to microbes inhabiting the plants as well as the response made over different stress conditions (Nadella et al., 2012).

The OMICS-based tools, particularly high throughput next-generation sequencing technology have the inherent potential to decipher an insight into phytopathogenic fungal genome variability, pathogenic responses, diversity of pathogenesis, range of hosts and most importantly the evolutionary trend in terms of interactions when present within the host tissue. This OMICS-based data has provided a new paradigm shift in our understanding of infection strategies followed by phytopathogenic fungi thereby predicting the suitable responses from host plants to fight against the phytopathogenic fungi and their ability to create diseased conditions in hosts (AbuQamar et al., 2016). In a recently conducted experiment, the combinatorial formulae utilizing the concept of both proteomics and genomics approaches (popularly known as "proteogenomics") to decipher the pathophysiological response of phytopathogenic fungus, *Fusarium graminearum* infecting the different varieties of oats. The proteogenomics tools also facilitated in understanding the response made in oats in response to the pathophysiological infections thereby promoting the differential responses to infection by hosts and identifies the putative proteins responsible for such response (Willforss et al., 2020).

One of the interesting aspects of sustainable agriculture is the maintenance of seed nutritional values and quality of seed for future cultivars. In this context, high throughput biomarker assisted molecular tools such as marker-assisted selection, Genomic selection (GS) using genome-wide association study, phenome-wide association study, genome phenome-wide association study, transcriptome-wide association studies, proteome-wide association study, and metabolite-based genome-wide association studies could be implemented to validate the nutritional and qualitative values associated with seeds through genome mining approach and thus could be considered to accelerate the breeding and trait improvement programs in economic crop, pigeonpea (Le Signor et al., 2017; Fang and Luo, 2019; Kremling et al., 2019; Pendergrass et al., 2019; Singh et al., 2020). Similar to that of personalized medicines for human healthcare settings, agricultural sectors also need a transformation in terms of innovations and integrative approach to fulfill the budding demand for nutritional food. Similar to that of personalized medicines, the concept of "precision agriculture" also comes to existence with the interventions in the form of OMICS-based approaches to answer all the complexities associated with agricultural ecosystems, challenges associated with agricultural practices in the form of climatic variability and anthropogenic interference. The revolutionary OMICS-based approaches have given an advantage to fulfill the precision agriculture through an integrated strategies combining genomics, transcriptomics, proteomics and metabolomics data for crop improvement and agricultural sustainability (Mastronardi et al., 2018).

10.4 Current trends and future avenues

Owing to the growing incidence of massive industrialization and population explosion, there is an increasing demand for improved crop varieties with nutritional values and improved productivity. With the limited sources available, maintaining the quality of food products accomplishing the growing demand is a serious task and implementation of high throughput OMICS-based approaches have apparently reduced the relative burden from conventional strategies to improve crop productivity without compromising the nutritional quality (Pathak et al., 2018). The developmental prospects in multidimensional OMICS-based approaches allow us to understand the insight into the mechanism of biomolecular organization of complex biological systems starting from a single cellular platform to even the highly complex and diverse ecosystems (Esposito et al., 2016). The OMICS-based approaches when exploited in combination with bioinformatics tools such as computational network modeling and system biology, the combinatorial methodology critically outpaced the traditional practices of biotechnological interventions in agricultural sectors. The powerful bioinformatics tools have the inherent potential to define an insight into the occurrence of specific phenotypic traits and their relative responses towards different stress conditions irrespective of biotic and/or abiotic stress thereby critically influencing the agricultural based ecosystems in terms of yield and productivity (Esposito et al., 2016). The advancement in the multifaceted OMICS-based approaches in defining unculturable microbial diversity and their functional attributes significantly contribute to attain sustainability in agricultural ecosystem. In addition to OMICS-based technologies, the rapid development of high throughput molecular and biotechnological tools to understand plant-microbe interactions, insight mechanism of biomolecular characterization associated with plant biology, functional aspects of novel genetic resources, and genome modification tools also provide novel avenues for nutrition and food security with sustainability in agriculture under constantly changing environmental conditions (Moshelion and Altman, 2015). The biggest advancement of the evolution of high throughput OMICS-based tools corresponds to the investigation of biomolecular characterization and biomolecular pathways driven through mining the large data and analytical knowledge as compared to "hypothesis-driven" strategies in conventional approaches. Thus the OMICS tools have dramatically attracted considerable attention to define the untapped biomolecular mechanisms associated with living systems (Mehta et al., 2019). The high throughput molecular tools such as DNA barcoding in combination with OMICS-based approaches and customized bioinformatics tools to understand the community profiling of soil microbiotas with respect to the soil environment. Microbial system biology also gained considerable attention in recent times to decipher the differential metabolic pathways, genes and proteins involved in those pathways associated with soil microbiotas and thus minimizing the gap associated with earlier setup. These molecular approaches thus possess withstanding potential to be used to define soil microbiotas and their wide spectrum biomolecular interactions to achieve sustainability in agriculture (Bhatt and Barh, 2018). It is evident from various studies that parallel and customized strategies are being developed to improve the crop yield as well to maintain the agricultural sustainability. In this context, the advent of nanotechnological platforms also gained considerable attention in recent years owing to their widespread applications in PGP, tolerance towards phytopathogens, food security, and sustainability in agricultural biosystems. However, the safety issues associated with the engineered nanomaterials critically limit their translational effect from laboratory to field level. In this context, advance and high throughput OMICS-based tools could be implemented in determining the detrimental effects associated with the engineered nanomaterials and provide customized solutions for augmenting the toxicity and safety issues linked with engineered nanomaterials. In this context, highly defined multi-OMICS approaches could be instrumental in defining the responses in plant developmental pathways in response to administration of engineered nanomaterials and successfully predict the detrimental and/or beneficial effect of nanomaterials on plant physiology and thus instrumental in maintaining the sustainability in crop productivity and safe use of engineered nanomaterials (Majumdar and Keller, 2020). In the context of constant interference from anthropogenic events and climatic conditions on agricultural practices, it is important to consider an integrated and customized arsenal utilizing the high throughput OMICS-based molecular tools in associated with the advanced computational tools with network orientation for attaining the sustainability in agriculture. This combinatorial approach not only allows to integrate the improved agricultural yield and productivity with enhanced stress tolerance mechanism but also accomplishes the increasing demand for nutritional food items with improved quality and security.

10.5 Conclusion

Since the advancement of next generation sequencing tools along with single cell-based multi-OMICS approaches not only provides information of microbial diversity in various ecosystems it also provides an insight into the ecosystem functioning and services through mining the metadata associated with microbiomes in soil and their interactions with hosts and environment. The metadata obtained from OMICS-based tools could provide novel avenues to standardize the current understanding of crop improvement strategies to increase yield and sustainability. The OMICS-tools also corresponds to differential role of various biomolecular factors and their relative abundancy and expression during plant growth, development, stress response and soil health maintenance. The integrative approach combining the OMICS-based approaches with technologically-driven network mining bioinformatics tools have provided a new dimension to understand the physiological response of plants to different stress conditions including the infections associated with phytopathogens and climatic variations. The OMICS-based tools also illustrate the critical role of soil microbiotas in defining the crop productivity through growth and development as well as stress tolerance mechanism. Thus these technological interventions not only improve crop yield but also maintain the health of ecosystem functioning and services. The OMICS-based approaches in next generation plant breeding programs also facilitate in complementing the conventional approaches to improve the crop quality through traditional genetic breeding programs. Though, a significant development has been made in implementing the OMICS-based tools in sustainable agriculture, more integrative and highly customized improvement is required to attain the agricultural sustainability in the near future. The combinatorial concept of amalgamation of different OMICS approaches into a single platform and contemplating the advanced version of molecular tools to improve quality and quantity of economic crops, ultimately complements the sustainability in the agricultural ecosystem.

References

AbuQamar, S.F., Moustafa, K., Tran, L.P., 2016. 'Omics' and plant responses to *Botrytis cinerea*. Front. Plant Sci. 7, 1658.

Arif, I., Batool, M., Schenk, P.M., 2020. Plant microbiome engineering: expected benefits for improved crop growth and resilience. Trends. Biotechnol. 38 (12), 1385–1396.

Armalyte, J., et al., 2019. Microbial diversity and antimicrobial resistance profile in microbiota from soils of conventional and organic farming systems. Front. Microbiol. 10, 892.

Attwood, G.T., et al., 2019. Applications of the soil, plant and rumen microbiomes in pastoral agriculture. Front. Nutr. 6, 107.

Bach, E.M., Williams, R.J., Hargreaves, S.K., Yang, F., Hofmockel, K.S., 2018. Greatest soil microbial diversity found in micro-habitats. Soil. Biol. Biochem. 118, 217–226.

Bhatt, P., Barh, A., 2018. Bioinformatic tools to study the soil microorganisms: an in silico approach for sustainable agriculture. In: Choudhary, D.K., et al., (Eds.), In Silico Approach for Sustainable Agriculture. Springer, pp. 169–182.

Biswas, R., Sarkar, A., 2018. 'Omics' tools in soil microbiology: the state of the art. In: Adhya, T.K., et al., (Eds.), Advances in Soil Microbiology: Recent Trends and Future Prospects, Microorganisms for Sustainability 3. Springer Nature, pp. 35–64.

Blaud, A., Chevallier, T., Virto, I., Pablo, A.L., Chenu, C., Brauman, A., 2014. Bacterial community structure in soil microaggregates and on particulate organic matter fractions located outside or inside soil macroaggregates. Pedobiologia 57, 191–194.

Bloch, S.E., et al., 2020. Biological nitrogen fixation in maize: optimizing nitrogenase expression in a root-associated diazotroph. J. Exp. Bot. 71 (15), 4591–4603.

Bonanomi, G., De Filippis, F., Cesarano, G., La Storia, A., Ercolini, D., Scala, F., 2016. Organic farming induces changes in soil microbiota that affect agroecosystem functions. Soil Biol. Biochem. 103, 327–336.

Brul, S., et al., 2012. 'Omics' technologies in quantitative microbial risk assessment. Trends. Food. Sci. Technol. 27, 12–24.

Busby, P.E., et al., 2017. Research priorities for harnessing plant microbiomes in sustainable agriculture. PLoS Biol. 15 (3), e2001793.

Chandran, H., Meena, M., Sharma, K., 2020. Microbial biodiversity and bioremediation assessment through omics approaches. Front. Environ. Chem. 1, 570326.

Desai, C., Pathak, H., Madamwar, D., 2010. Advances in molecular and "-omics" technologies to gauge microbial communities and bioremediation at xenobiotic/anthropogen contaminated sites. Bioresour. Technol. 101, 1558–1569.

Ding, X., Liang, C., Zhang, B., Yuan, Y., Han, X., 2015. Higher rates of manure application lead to greater accumulation of both fungal and bacterial residues in macroaggregates of a clay soil. Soil Biol. Biochem. 84, 17–146.

dos Santos, B.S., et al., 2016. Application of omics technologies for evaluation of antibacterial mechanisms of action of plant-derived products. Front. Microbiol. 7, 1466.

Dunlap, P.V., 2001. Microbial diversity, Encyclopedia of Biodiversity, second ed.pp. 280–291.

Dwivedi, S.L., van Bueren, E.T.L., Ceccarelli, S., Grando, S., Upadhyaya, H.D., Ortiz, R., 2017. Diversifying food systems in the pursuit of sustainable food production and healthy diets. Trends. Plant. Sci. 22 (10), 842–856.

Esposito, A., Colantuono, C., Ruggieri, V., Chiusano, M.L., 2016. Bioinformatics for agriculture in the next-generation sequencing era. Chem. Biol. Technol. Agric. 3, 9.

Fang, C., Luo, J., 2019. Metabolic GWAS-based dissection of genetic bases underlying the diversity of plant metabolism. Plant J. 97, 91–100.

Fierer, N., Jackson, R.B., 2006. The diversity and biogeography of soil bacterial communities. Proc. Natl. Acad. Sci. U. S. A. 103 (3), 626–631.

Garbeva, P., van Veen, J.A., van Elsas, J.D., 2004. Microbial diversity in soil: selection of microbial populations by plant and soil type and implications for disease suppressiveness. Annu. Rev. Phytopathol. 42, 243–270.

Glick, B.R., 2012. Plant growth promoting bacteria: mechanisms and applications. Scientific 2012 963401.

Goff, A., Cantillon, D., Wildner, L.M., Waddell, S.J., 2020. Multi-omics technologies applied to tuberculosis drug discovery. Appl. Sci. 10, 4629.

Gupta, A., Sarkar, J., Sar, P., 2019. Understanding the structure and function of extreme microbiome through genomics: scope and challenges. In: Das, S., Dash, H.R. (Eds.), Microbial Diversity in the Genomic Era. Elsevier, pp. 581–610.

Hu, X., et al., 2017. Effects of over 30-year of different fertilization regimes on fungal community compositions in the black soils of northeast China. Agric. Ecosyst. Environ. 248, 113–122.

Ichihashi, Y., et al., 2020. Multi-omics analysis on an agroecosystem reveals the significant role of organic nitrogen to increase agricultural crop yield. Proc. Natl. Acad. Sci. 117 (25), 14552–14560.

Jiao, S., Xu, Y., Zhang, J., Hao, X., Lu, Y., 2019. Core microbiota in agricultural soils and their potential associations with nutrient cycling. mSystems 4, e00313–e00318.

Ke, J., Wang, B., Yoshikuni, Y., 2020. Microbiome engineering: synthetic biology of plant-associated microbiomes in sustainable agriculture. Trends. Biotechnol. 39 (3), 244–261.

Kiani, M., et al., 2017. Quantifying sensitive soil quality indicators across contrasting long-term land management systems: crop rotations and nutrient regimes. Agric. Ecosyst. Environ. 248, 123–135.

Kremling, K.A.G., Diepenbrock, C.H., Gore, M.A., Buckler, E.S., Bandillo, N.B., 2019. Transcriptome-wide association supplements genome-wide association in *Zea mays*. Gene. Genome. Genet. 9, 3023–3033.

Kreszies, T., et al., 2018. Osmotic stress enhances suberization of apoplastic barriers in barley seminal roots: analysis of chemical, transcriptomic and physiological responses. New Phytol. 221, 180–194.

Kumar, U., et al., 2017. Variation of functional diversity of soil microbial community in sub-humid tropical rice-rice cropping system under long-termorganic and inorganic fertilization. Ecol. Indic. 73, 536–543.

Le, D.T., et al., 2012. Differential gene expression in soybean leaf tissues at late developmental stages under drought stress revealed by genome-wide transcriptome analysis. PLoS One 7 (11), e49522.

Le Signor, C., et al., 2017. Genome-wide association studies with proteomics data reveal genes important for synthesis, transport and packaging of globulins in legume seeds. New Phytol. 214, 1597–1613.

Li, F., et al., 2018. Soil aggregate size mediates the responses of microbial communities to crop rotation. Eur. J. Soil Biol. 88, 48–56.

Lupatini, M., Korthals, G.W., de Hollander, M., Janssens, T.K.S., Kuramae, E.E., 2017. Soil microbiome is more heterogeneous in organic than in conventional farming system. Front. Microbiol. 7, 206.

Majumdar, S., Keller, A.A., 2020. Omics to address the opportunities and challenges of nanotechnology in agriculture. Crit. Rev. Environ. Sci. Technol. Available from: https://doi.org/10.1080/10643389.2020.1785264.

Maron, P.A., et al., 2018. High microbial diversity promotes soil ecosystem functioning. Appl. Environ. Microbiol. 84 (9), e02738–17.

Mastronardi, E., Monreal, C., DeRosa, M.C., 2018. Personalized medicine for crops? Opportunities for the application of molecular recognition in agriculture. J. Agric. Food. Chem. 66, 6457–6461.

Mehta, S., James, D., Reddy, M.K., 2019. Omics technologies for abiotic stress tolerance in plants: current status and prospects. In: Wani, S.H. (Ed.), Recent Approaches in Omics for Plant Resilience to Climate Change. Springer, pp. 1–34.
Mitter, B., Pfaffenbichler, N., Sessitsch, A., 2016. Plant–microbe partnerships in 2020. Microb. Biotechnol. 9 (5), 635–640.
Moshelion, M., Altman, A., 2015. Current challenges and future perspectives of plant and agricultural biotechnology. Trends Biotechnol. 33 (6), 337–342.
Nadella, K.D., Morla, S.S., Kumar, P.A., 2012. Metabolomics in agriculture. OMICS J. Integrat. Biol. 16 (4), 149–159.
Nalbantoglu, S., Karadag, A., 2019. Insight into the OMICS technologies and molecular medicine. In: Nalbantoglu, S., Amri, H. (Eds.), Molecular Medicine. IntechOpen, pp. 1–10.
Nema, V., 2019. The role and future possibilities of next-generation sequencing in studying microbial diversity. In: Das, S., Dash, H.R. (Eds.), Microbial Diversity in the Genomic Era. Elsevier, pp. 611–630.
Paananen, J., Fortino, V., 2020. An omics perspective on drug target discovery platforms. Brief. Bioinform. 21 (6), 1937–1953.
Pan, M., Barrangou, R., 2020. Combining omics technologies with CRISPR-based genome editing to study food microbes. Curr. Opin. Biotechnol. 61, 198–208.
Pathak, R.K., Baunthiyal, M., Pandey, D., Kumar, A., 2018. Augmentation of crop productivity through interventions of omics technologies in India: challenges and opportunities. 3 Biotech 8, 454.
Pendergrass, S.A., et al., 2019. A phenome-wide association study (PheWAS) in the population architecture using genomics and epidemiology (PAGE) study reveals potential pleiotropy in African Americans. PLoS One 14 (12), e0226771.
Praeg, N., Pauli, H., Illmer, P., 2019. Microbial diversity in bulk and rhizosphere soil of *Ranunculus glacialis* along a high-alpine altitudinal gradient. Front. Microbiol. 10, 1429.
Rasmussen, S., et al., 2013. Transcriptome responses to combinations of stresses in *Arabidopsis*. Plant Physiol. 161, 1783–1794.
Scelfo, C., et al., 2018. Towards precision medicine: the application of omics technologies in asthma management. F1000 Res. 7, 423.
Shamim, K., Mujawar, S.Y., Mutnale, M., 2019. Metagenomics a modern approach to reveal the secrets of unculturable microbes. In: Meena, S.N., Naik, M.M. (Eds.), Advances in Biological Science Research—A Practical Approach. Academic Press, pp. 177–195.
Shi-Kai, Y., et al., 2015. "Omics" in pharmaceutical research: overview, applications, challenges, and future perspectives. Chin. J. Nat. Med. 13 (1), 0003–0021.
Singh, N., Rai, V., Singh, N.K., 2020. Multi-omics strategies and prospects to enhance seed quality and nutritional traits in pigeon pea. Nucleus 63, 249–256.
Topalovic, O., Hussain, M., Heuer, H., 2020. Plants and associated soil microbiota cooperatively suppress plant-parasitic nematodes. Front. Microbiol. 11, 313.
Ufarte, L., Laville, E., Duquesne, S., Potocki-Veronese, G., 2015. Metagenomics for the discovery of pollutant degrading enzymes. Biotechnol. Adv. 33, 1845–1854.
Van Emon, J.M., 2016. The omics evolution in agricultural research. J. Agric. Food Chem. 64, 36–44.
Vlaanderen, J., et al., 2010. Application of OMICS technologies in occupational and environmental health research; current status and projections. Occup. Environ. Med. 67, 36–143.
Wang, J., et al., 2017. Impacts of inorganic and organic fertilization treatments on bacterial and fungal communities in a paddy soil. Appl. Soil. Ecol. 112, 42–50.

Wang, C.H., et al., 2020. Characterizing changes in soil microbiome abundance and diversity due to different cover crop techniques. PLoS One 15 (5), e0232453.

Wheelock, C.E., et al., 2013. Application of'omics technologies to biomarker discovery in inflammatory lung diseases. Eur. Respir. J. 42, 802–825.

Willforss, J., et al., 2020. Interactive proteogenomic exploration of response to *Fusarium* head blight in oat varieties with different resistance. J. Proteomics 218, 103688.

Zhang, Q., Zhou, W., Liang, G., Sun, J., Wang, X., He, P., 2015. Distribution of soil nutrients, extracellular enzyme activities and microbial communities across particle-size fractions in a long-term fertilizer experiment. Appl. Soil Ecol. 94, 59–71.

Zhang, L., et al., 2017. The tartary buckwheat genome provides insights into rutin biosynthesis and abiotic stress tolerance. Mol. Plant 10 (9), 1224–1237.

Zorner, P., Farmer, S., Alibek, K., 2018. Quantifying crop rhizosphere microbiome ecology: the next frontier in enhancing the commercial utility of agricultural microbes. Ind. Biotechnol. 14 (3), 116–119.

Microbial associations in ecological reclamation and restoration of marginal lands

11

Husnain Zia[1], Muhammad Ashar Ayub[1], Ahmed Abd El Fattah El Baroudy[2], Muhammad Zia ur Rehman[1], Hinnan Khalid[3], Anwar ul Haq[1], Wajid Umar[4] and Zahoor Ahmad[5]
[1]Institute of Soil and Environmental Sciences, Faculty of Agriculture, University of Agriculture Faisalabad, Pakistan, [2]Soil and Water Department, Faculty of Agriculture, Tanta University, Tanta, Al Gharbiyah governorate, Egypt, [3]Agriculture Department, Government of Punjab, Pakistan, [4]Institute of Environmental Science, Hungarian University of Agricultural and Life Sciences, Godollo, Hungary, [5]Department of Botany, University of Central Punjab, Bahawalpur, Pakistan

11.1 Introduction

In the last few years, population growth worldwide has rapidly escalated which has put an immense pressure on ecosystem (Slemp et al., 2012; Liu et al., 2015a,b). Escalated pressure enforces the population to move from rural to urban areas, which is known as urbanization. Urbanization is also accompanied by expanded urban areas and changes in the use of land (Finlayson et al., 2005; Liu et al., 2015a,b). Urban-Rural frindges is a hot topic of discussion today as it is a hub of constant land use change disturbing biodiversity, agricultural dynamics and environment (Grimm et al., 2008).

Agriculture is being negatively influenced by land-use changes, that is the major outcome of urbanization and it is estimated that by 2030 about 60% of the total world's population will move towards the urban areas (United Nations, 2007). More urban areas are being developed for the accommodation of the increased population which led to the utilization of the agricultural lands that deteriorate the land known as land degradation by human activities (Yeboah et al., 2017). Industrialization is also enhanced to meet the daily needs of the urban population; these industries are accompanied by wastewater and solid waste that is deteriorating agricultural lands indirectly (Grover, 2016).

To feed this augmented population, intensive agriculture is the only possible way to get higher production from scarce agricultural lands, which is further degrading our lands (Singh et al., 2019). Enrichment and utilization of indigenous microbial associations could be a long-lasting and sustainable approach for reclamation and management of degraded soils as well as attaining (Rashid et al., 2016).

Microbes in Land Use Change Management. DOI: https://doi.org/10.1016/B978-0-12-824448-7.00014-0
© 2021 Elsevier Inc. All rights reserved.

Achievement of sustainable agriculture can also be possible by the restoration of the marginal lands. Proper agricultural practices, nutrient management, crop rotations, application of organic as well as inorganic amendments, and several biological and chemical techniques support the land restoration (Azhar et al., 2019; Hussain, 2018). We are not only concerned with the restoration of the lands but also with sustainable agriculture that will benefit us for longer periods. Green agriculture can be accomplished by the biological remediation of the degraded lands (Thomashow et al., 2019).

Microbes are the most promising bio-agents for sustainable agriculture. Various microbes of different associations used for the restoration of the marginal soils (Simarmata et al., 2019). Screening, purification, and enrichment of microbes for any specific function can be a leading approach to tackle soil degradation and boost yield and product quality (Gouda et al., 2018). These microbes work on the nutrient enrich media that require plenty of nutrients in the soil for their growth and maintenance and sometime they can release some chemicals to solublize non-bioavailable soil nutrients. The nutrient solubilizing and oxidizing bacterial community can help plants acquire more of these nutrients in nutrient-poor, eroded or waterlogged soils (Puri et al., 2020). Microbes positively influence the soil ecosystem and enhance the soil productive biomass to a remarkable extent, improving the soil physico-chemical properties and soil production capacity (Clocchiatti et al., 2020). Soil microbial consortium (bacteria and fungi) play a vital role in maintaining soil-plant salt and mineral balance (Verma et al., 2017).

11.2 Land classification and their role for agriculture in changing environment

Food and agriculture organization (FAO) reported that the rapid growth of the world population requires high agricultural outputs to meet food demand that ultimately pressurizing our diminishing agricultural lands for healthy production. Recently they concluded based on their surveys that, to meet the food requirements of the fast-growing world we need to raise our cereal production up to 3 billion tons by 2050 (FAO, 2006) which is almost 2.6 billion tons right now (FAO, 2017).

To get higher agricultural produce, higher fertilizer application is a common practice and has become a norm which is neither economical nor sustainable approach (Hazell et al., 2010). Besides land degradation, abiotic and biotic stresses are another factor controling agricultural produce and jeopardizing agriculture sustainability (Tripathi et al., 2016). Production of staple food crops have been influenced negatively (up to 70%) by the abiotic stresses of the environment, like colder temperatures, high levels of soluble salts, high temperatures, and drought conditions (Thakur and Ojha, 2010; Ahmad et al., 2012). In term of general production potential and agriculture suitability, land has been divided to following land use classes (Table 11.1).

Table 11.1 Represent the land use capability classes following the USDA, 2008 classification systems (Akıncı et al., 2013).

Class	Description
Class I	Deep and fertile soils Plains Good for agriculture
Class II	Mild erosion Few cultivation practices should be given Moderate thick soil
Class III	Proper management required Plant alternation is mandatory
Class IV	Land that is appropriate for the pasture Can also be cultivated
Class V	Unsuitable for crops/plants Used for long-living plants Best suitable for pasture and forest
Class VI	Lands that are moderately practiced for pasture and forest use
Class VII	Lands with • High slope • High erosion • Stony, shallow, and dry soils
Class VIII	Lands that are not capable of even pasture or forestation

11.3 Causes of degraded and marginal soils

Land can be degraded by several hazards that may be natural, anthropogenic, direct or indirect participating in land deprivation. Soil degradation is the loss of soil to perform its capabilities, supporting the plant growth, hydrological regulations, filtration of the environment and many more (Lal and Anouncia, 2015).

11.3.1 Erosion

In the regions of arid and semiarid environments, the soil formation process is very slow, it took thousands of years to form a layer of soil (Blanco et al., 2017). Erosion removes that topsoil layer within months because of the shallow soils and high wind and water flow in these areas. In arid and semiarid regions degradation of the soil through topsoil layer removal by wind and water erosion have a prominent effect on soil quality (Kong et al., 2021) and erosion also is a significant component of desertification in these areas (Ravi et al., 2010). Soil erosion can happen naturally or by the mismanagement or mishandling of the soil. Two prominent factors are the main drivers of soil erosion are wind, and water. In temperate regions, due to high temperature, soils have low organic matter contents which result in the degradation of soil structure and lead to wind and water erosion (Meng et al., 2018). Among the natural hazards severe deterioration occurs because of the physical changes within the environment, for example, water erosion at the steep slopes causes severe damages to the soil (Wu et al., 2018). In this regard, human activities contribute more as compared to the natural factor. Deforestation or cutting of the trees planted alongside the fields exposes the soils directly to the sunlight which affects the soil temperature, moisture content, and soil structure (Karamage et al., 2016). One more human activity is the use of lands for grazing, and it was reported that the excessive grazing barrens the land and expose the topsoil layer to the environment that leads to erosion by wind or water (Du et al., 2019). Excessive tillage practices were also reported to be responsible for soil erosion from one place to another with both the high-speed winds and flooded water or flood irrigation (Prosdocimi et al., 2016). Water erosion mostly occurs by the flood like disasters or excessive rainfalls, especially at the times when the soils are manipulated by the tillage practice (Rodrigo-Comino et al., 2019).

11.3.2 Salinity

Salinity is the amount of salts dissolved in the water body, but in case of soil, salinity is the amount of salts present in the soil. Salts responsible for salinization consist of ions like Na^+, K^+, Ca^{2+}, Mg^{2+}, and Cl^- but mostly sodium overdominates among all cations causing salinity or sodicity (Rath et al., 2019). Salinity exists in three forms primary (by natural weathering process), secondary (dryland salinity), and territory (salinity due to irrigation) (Aslam et al., 2013). Soil degradation from the salts that comes naturally through the weathering process is very less in Pakistan (Shaygan et al., 2019). Pakistan faces high temperatures and low rainfall in most of the regions that play a significant role in secondary salinity development. In temperate regions higher evapotranspiration results in the accumulation of salts in the soil (Ijaz et al., 2020; Xie et al., 2017). The poor-quality irrigation water is also a reason behind the salinity development (Plaut et al., 2013). Higher sodium based salinity breakdown the soil aggregates and deflocculates the soil structure, movement of the water, and air within the soil is also affected. Salinity degraded the soil by losing the soil aggregates that make the soil impervious (Ali et al., 2017). Poor irrigation management increases the salinity issues in the soils of Pakistan. Earthen canals that are not lined result in the seepage of the canal water that brings the salts to the upper surface and causes salinity on 25% of Indus lands (Khan et al., 2005). Intense farming systems create hardpan in the soil that results in the perched water table and increases the evapotranspiration. Bigger loss of water from the soil surface due to the accumulation of salts and the ultimate destruction of the soil structure (Zhang et al., 2018).

11.3.3 Heavy metals

Heavy metal is a physical defination of elements based upon their higher specific gravity and include few essential plant nutrients (Fe, Mn, Zn, Ni, Cu etc) but in very minute quantities which can become toxic in higher amounts. Toxic metals not only degrade the soil but also affect the productivity and the health of the individuals that feed on the food produced from contaminated soils (Rehman et al., 2017). Iron, zinc, copper, and manganese are the essential heavy metals that are required in minute quantities while cadmium, lead, arsenic, chromium, and few others are toxic heavy metals (Ashaiekh et al., 2019). Leather tanning, alloy formation, metal cleaning, chromate plating, and textile dyeing are degrading the soil by contributing the Chromium as a major heavy metal pollutant in the soil (Tripathi et al., 2016). United States-EPA listed it as the carcinogenic material of class-A for humans (Aslam et al., 2014). Similarly, lead is a potential and highly toxic heavy metal that can exisit in soil above permissible limits due to anthropogenic activities. It comes from the mining and the industrial wastes that utilize the lead in their products. Sewage sludge application is also a source of lead in the soil (Kang et al., 2016). Cadmium is another leading heavy metal towards the deterioration of the soil and community health due to its carcinogenic effects (Naeem et al., 2019; Rehman et al., 2019, 2020).

11.3.4 Anthropogenic activities

Human intrusions in any natural system are considered anthropogenic activities, such activities can be helpful as well as harmful for the system (Akinyele et al., 2019). In agriculture, human activities are increasing rapidly because of intensive farming and increased food demands (Ayub et al., 2020, 2021a, 2021b; Wu et al., 2017). There is a vast list of human activities starting from land preparation to harvesting that increased the soil degradation (Ayub et al., 2020). Any action that exposes the soil to the wind/environment will result in the loss of soil. Manipulation (tillage practice) of the soil unnecessarily without considering the soil condition, excessive tillage practices degrade the soil structure (Nath and Rattan, 2017). Intensive agriculture results in the flush out of the soil nutrients that will affect soil fertility. It also requires the control of diseases and insect attacks on the crops that are carried out by the application of chemicals like fungicides, insecticides, pesticides, etc. These chemicals pollute the soil with different impurities present in them and also harmful to the microbial community of the soil (Prashar and Shah, 2016). Working on the farms with heavy machinery for longer periods also creates hardpans in the soil that decreases the soil porosity. Soils that are degraded by anthropogenic activities are difficult to restore as compared to the naturally degraded soils (Qu et al., 2018).

11.3.5 Soil fertility exhaustion

The green revolution led to the augmentation of the agricultural lands by the increased use of pesticides and chemicals that are the results of intense technological farming and also increased the yearly utilization of the soil (Umar et al., 2020; Hameed et al., 2021). However, environmental pollution is also the ultimate effect of technological and intensive farming including the loss of plant biodiversity. Resultantly, the extensive degradation of the soil will lead to food insecurity and declined farmer's income (Ahmad et al., 2018, 2020, Ayub et al., 2021a, 2021b;

Farooqi et al., 2021). Public and ecological health is also affected by the intensive agriculture because it requires excessive use of agrochemicals that ultimately contaminate the soil and also have detrimental effects on the consumer of the product (Ayub et al., 2020; Malik et al., 2017). Lower inputs and high yield- results in the loss of a major part of soil nutrients required to maintain the fertility status of the soil. Fertility status is declining because of the intensive farming and impure inputs that do not meet the soil requirements (Alonso-Mora et al., 2017).

11.4 Effects of marginal and degraded soils on the agriculture economy

Continuing the intensive agriculture would require 20 and 40 million metric tons of the P and N respectively by 2040 (Gregory et al., 2010). Such an alarming increase in application of chemicals fertilizer will lead to the development of marginal lands, prohibiting food production, and destroying the economic, environmental and health status of the society (Tilman et al., 2001). These situations clarify that green crop production systems are urgently needed (Khan and Khan, 2017). Crop protection and nutrient cycling are the ecological outputs of the soil that are affected by the disturbance of soil microbial activities in response to the agricultural management practices. This may also harm the functions of microorganisms in the soil (Xue et al., 2013). In Southern Chile, pasture areas are threatened because of the increased aluminum toxicity at 2000 mg/kg of dry weight which results in the lower forage quality and reduced yield of the forage (Mora et al., 2006). Water and nutrient uptake of the plants is limited because of the inhibited root growth resultantly occurred in response to the toxicity effects of Al^{3+} (Kochian et al., 2004). Therefore efficient production of the crops is threatened by the coexistence of the toxicity of Al^{3+} and the deficiency of phosphorus in the Chilean and isols (Alonso-Mora et al., 2017). A review on few types of land degradation in given in following (Table 11.2).

11.5 Role of microbial association in restoration and management of marginal soils

Nutrient management of the soil is mainly concerned with the microorganisms, being a source of nutrients itself for other organisms of soil and plants. Disease suppression, nutrient availability and many other important mechanisms involved in the management of the nutrient status of soil and plant growth can be controlled by the microorganisms (Lozano-Durán et al., 2014). There are several microbial associations that are the backbones of all the microbial associated mechanisms. Symbiotic associations and nonsymbiotic associations are playing a major role in the soil plant and microbe system.

Table 11.2 Degradation and its types in different regions of the world.

Region	Type of degradation	Effect on arable land	References
Kuwait	Salinity	12% of the entire state of Kuwait is affected by varying salinity ranges from 4.1 to 25 dS/m	Qadir et al. (2014)
Abu Dhabi Emirate	Salinity	Degradation of the land especially where the groundwater table that creating large areas of aquisalids (United States classification)	King et al. (2013), Shahid et al. (2014)
Australia	Salinity Sodicity	Most affected land of the world because of salinity and sodicity almost 17 million hectares is affected by saline soils while 340 million hectares is affected by salinity	Shahid et al. (2018)
China	Heavy metal	19.4% of the agricultural area is polluted with heavy metals	Delang (2017)
Pakistan	Erosion	50% of the rangelands are affected by water and wind erosion	Irshad et al. (2007)
	Drought Salinity and waterlogging	Balochistan is facing a severe water shortage that leads to drought conditions Almost 4.5 million hectares area of Pakistan is degraded due to salinity that occurs because of the low-quality water and water seepage through the earthen canals	Qureshi and Bhatti (2008), Qasim et al. (2017a,b)
Balochistan (Pakistan)	Overgrazing	Rangeland that constitutes 90% of Balochistan's area is eroded because of overgrazing	Anjum et al. (2010), Qasim et al. (2017a,b)

11.5.1 Microbial associations for nutrient management

11.5.1.1 Symbiotic plant-microbial associations

The biological agents, products and microorganisms have started to get more attention from world due to immense negative effects of chemical fertilizers (Solgi et al., 2018). Therefore it is necessary to explore the relations among the rhizospheric microbial community and the plants that will ultimately reduce the risk of fertilizers side effects. Among the microbial community, plant growth-promoting rhizobacteria (PGPR) are playing an important role in the establishment of environment-friendly and sustainable agricultural system (Mehmood et al., 2018).

High levels of phosphorus content are present in the agricultural farmlands but only a minute percentage (0.1%) is available in the soluble form for the plant uptake (Simpson et al., 2011). Khan and Khan (2017) isolated the microbial strains *Burkholderia* sp. (BRRh-1) that solubilizes the tricalcium phosphate (TCP) in the broth culture rather than the agar plate. While other five strains were reported, that constantly solubilize the TCP in both broth and agar media (Sarkar and Reinhold-

Hurek, 2014). Among the five isolates Khan and Khan (2017) reported that *Burkholderia* sp. (BRRh-4) and *Pseudomonas aeruginosa* (BRRh-5) gave the best results in the solubilization of the TCP. The *Burkholderia* sp. and *P. aeruginosa* are the rice associated bacterial species that gave the higher yields and plant growth in the pot experiment by producing indole-3-acetic acid (IAA), with half of the recommended fertilizer application (Oliveira-Longatti et al., 2014) (Fig. 11.1).

As an integral part of intensive agriculture, nitrogen-based fertilizers are used frequently. These fertilizers are associated with destruction of soil as well as environment that calls for the use of environment-friendly techniques to increase the nitrogen use efficiency to thus can help decreasing our dependence on N-chemical fertilizers (Igiehon and Babalola, 2018). The bacterial strain of *Azoarcus* species (wild type *Azoarcus* sp. BH72) increased the dry weight of the kallar grass developed in a nitrogen-deficient environment (Sarkar and Reinhold-Hurek, 2014). Degradation of various organic pollutants is highest by the bacterial strains of *Bacillus* spp. While strains isolated from the tissues of *Alopecurus aequalis* Sobol. and *Oxalis corniculata* L. are capable of increasing the plant tolerance to the polycyclic aromatic hydrocarbons (PAHs) (Peng et al., 2013).

11.5.1.2 Nonsymbiotic associations

This is not necessary that every microorganism will give the beneficiary or complementary results to the soil or to the crop. However, the effect of PGPR on the hormonal system of plants would be detrimental or beneficial, depending on the growing environment (Achard et al., 2006). Literatures indicate that plant growth inhibition also protects the plants from stress conditions (Chaves et al., 2003; Bechtold and Field, 2018). Thomas and Sekhar (2016) stated that *P. aeruginosa* didn't survive in the soil, it also creates a disturbance in the microbial community of the soil. So, there is a need to assess the inoculants in the field soil prior to recommendations for commercial use. *Basidiomycete, Laccaria*, and *Burkholderia* sp. strain 205 and *Laccaria bicolor* can help either alone or in association to control soil bioavailable phosphorus pool (Fontaine et al., 2016).

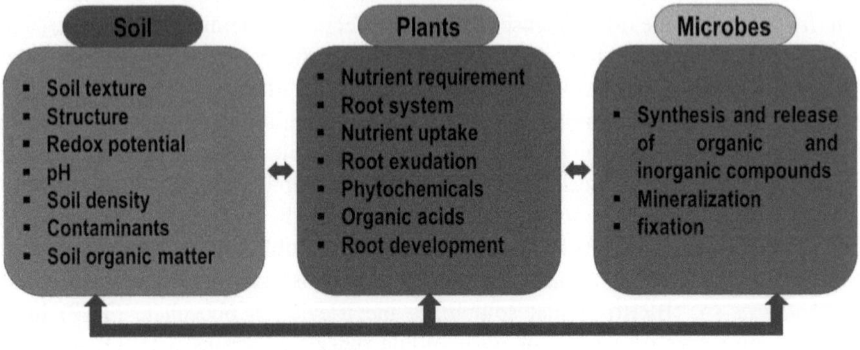

Figure 11.1 Soil-plant-microbe interaction.

11.5.1.3 Scope of nutrient and microbial consortia

Combined application of nutrients and microbial consortia can have a synergistic effect on each other. Shahane et al. (2019) conducted an experiment by comparing the application of 100% and 75% doses of required N and P fertilizers alone and with *Anabaena-Pseudomonas* biofilm formulations. They concluded that the application of *Anabaena-Pseudomonas* with 75% dose of N and P gave the best results in the rice crop as compared to the 100% dose. Gore and Navale (2017) use the three bacterial (*Pseudomonas* sp. KSB 49, 43, and 47) and three fungal (*Aspergillus* sp. KSF 3, 13, and 31) strains along with K_2O 50%, 75%, and 100% of the recommended dose. Results of the experiment revealed that the fungal consortium with 75% of K_2O showed higher uptake of potassium (K) 1198.17 kg/ha followed by the bacterial consortium that showed the maximum K uptake of 1177.21 kg/ha along with 75% of K_2O. Zinc deficiency in soil is also an important problem which is responsible for low crop production and malnutrition in humans and livestock, this can be handled in an environment-friendly way with the help of bacterial consortium. So, one strain of *Burkholderia* and two strains of *Acinetobacter* were isolated from the rhizosphere of the rice and were inoculated with the seeds of wheat. Results revealed that the residual effect of the zinc (applied in the very low dosage, 2.5 kg/ha) was highly appreciable that increased the zinc uptake and concentration in the grains of wheat. That practice also increased the bioavailability of zinc by lowering down the phytic acid concentration (Vaid et al., 2018). Certain more nutrient consortiums are given in the following Table 11.3.

11.5.2 Microbial association for the management of degraded soils

Microbial remediation can be simply defined by employing microorganisms to lower the bioavailability of pollutants (especially organic contaminants). These microorganisms are capable of breaking down (or metabolize) contaminants by using them as a food source

Table 11.3 Certain microbial strains and their role in nutrient consortium.

Microbial strain	Crop\plant	References
Comamonas testosteroni *Citrobacter freundii*	Solubilize the phosphorus and promote growth in hard red winter wheat	Baas et al. (2016)
Enterobacter cloacae	Solubilization of phosphorus and enhance the flowering in *Chrysanthemum*	Savov et al. (2012)
Pseudomonas putida	Enhanced blooming in sunflower by solubilizing the phosphate	Savov et al. (2011)
Burkholderia cepacia	Enhanced wheat growth by solubilizing the phosphorus	Song et al. (2008)

11.5.2.1 Saline agriculture

An increased amount of dissolved salts in the soil profile leads to soil salinization that ultimately affects agricultural production. Saline soils are increasing at the rate of 7% in arid and semiarid regions of the world (Orhan, 2016). Soil microbial community itself is affected by the salinity and sodicity stress that can be controlled by the isolation of the salt-tolerant microbes or introducing the genetically modified microorganisms that remediate the saline soils or support the plants that are grown on saline soils (Singh et al., 2016). Mung bean cultivation along with the PGPR improves the yield of the mung bean by the reduction of salinity stress on the crop (Aslam et al., 2013).

Rhizobial isolates were used in the salt-stressed maize to enhance the yield and growth of the maize crop (Plaut et al., 2013). The working efficiency of the chickpea isolates was more than the lentil rhizobial isolates, based on their evaluation in the fields as compared to the pot experiments (Ali et al., 2017). The *P. aeruginosa* consortium in the salt-affected soils proved to be the best consortium for the alleviation of salt stress and improve plant growth (Ebadi and Hisoriev, 2017). These bacterial counts will depend upon the root biomass as well as the root exudates produced by the roots into the rhizosphere. Salt affected soils are bio-remediated by the bacteria that are salt-tolerant and have the potential to promote the growth of plants. Enhancement of crop productivity, as well as improvement of degraded soils and alleviation of the salts stress, was attained by the inoculation of salt-tolerant bacteria with seeds (Arora et al., 2016).

11.5.2.2 Organic pollutant contaminated soils

Man-made organic compounds are mostly used in the industries as fuels and solvents as well as intermediates. Organic compounds that are harmful to humans and plants are mostly present in industrial products like paint, plastics, gasoline, and adhesives (Collins and Montgomery, 2002). Among some of these organic pollutants chlorinated hydrocarbons are the only compounds that are widely spread, toxic at extreme levels, and are common and long-lasting in the environment (Kang et al., 2016). Chen et al. (2019) reported that the immobilized-microorganisms technique can considerably enhance the removal of PAHs. An indigenous bacterium that is not yet identified and *Pseudomonas putida* both degrade PAHs.

To stimulate the effect of amendments on soil degradation, in an investigation manure was added in the soil for the bioremediation that actually showed results within 1 year of the application. The treated plot showed a 58.2% decrease in the total petroleum hydrocarbon (TPH) as compared to the untreated plot in which 15.6% decrease in TPH was observed (Ebadi and Hisoriev, 2018). Moreover, enhanced water holding capacity of the soil was a supplementary effect of the manure application that declares the environment-friendly statement correct. Physical and chemical properties of the soil, microbial activity within the soil were markedly improved while the TPH content of the soil was decreased but not as much as they need to be, so phytoremediation will be required for the impressive washout of the TPH from the soil. The high fertilizer requirements in intensive agriculture also increase the organic pollutants of the soil. Mineralization and detoxification of the harmful components of the soil by cytochrome P450 that is a monooxygenase enzymes' superfamily may originate in microbes, protists, molds, and algae (Mondal et al., 2019). Bioaccumulation and transforming of the pesticides were also done by the several microalgae that are mentioned in (Table 11.4).

11.5.2.3 Inorganic pollutant contaminated soils

Inorganic pollutants are mostly the outcome of anthropogenic activities; most abundantly heavy metals are the most toxic pollutants that are added by the human activities in the soil and water bodies (Farooqi et al., 2021; Hussain, 2021). Particular strain Z-90 was isolated from the *Burkholderia* sp. that was used with a biosurfactant for the remediation of the heavy metal contaminated soils. Zinc, manganese, and cadmium were efficiently removed due to their presence in high acid-soluble fraction while the high residual fraction of copper, lead, and arsenic showed less efficiency as compared to the acid-soluble fraction. Heavy metals were adhered to the Z-90 strain and a metal complex was formed with biosurfactant (Yang et al., 2016) (Fig. 11.2).

In the district Kasur (Punjab, Pakistan) tanneries are highly polluting the water and soil reservoirs by their chromium contaminated water that can be detoxified by the use of bacterial isolates. Word conducted by Abid and coworkers tested 20 bacterial isolates and concluded that the *Bacillus* sp. K12 is the most efficient in the

Table 11.4 Several microbes associated with the degradation of organic pollutants.

Categories	Microbes	Specificity	References
Fertilizer and pesticides	*Scenedesmus bijugatus*	Removal of pollutants associated with nitrogen and phosphorus fertilizer	Mondal et al. (2019)
	Scenedesmus sp. and *Chlorococcum* sp.	Convert α-endosulfan into endosulfan sulfate	Sethunathan et al. (2004)
	Anabaena azotica	Degrade the lindane	Zhang et al. (2012)
	Isochrysis galbana and *Dunaliella tertiolecta*	Accumulation of atrazine	Weiner et al. (2014)
	Chlamydomonas reinhardtii	Bioremediation of prometryne	Jin and Feng (2013)
Dimethyl Phthalate	*Closterium lunula*	Degrade dimethyl Phthalate	Yan et al. (2004)
PAH	*Pseudomonas putida*	PAH-degrading bacteria	Mondal et al. (2019)
Tributyltin	*Chlorella vulgaris*	Accumulation of triazine	Weiner et al. (2014)

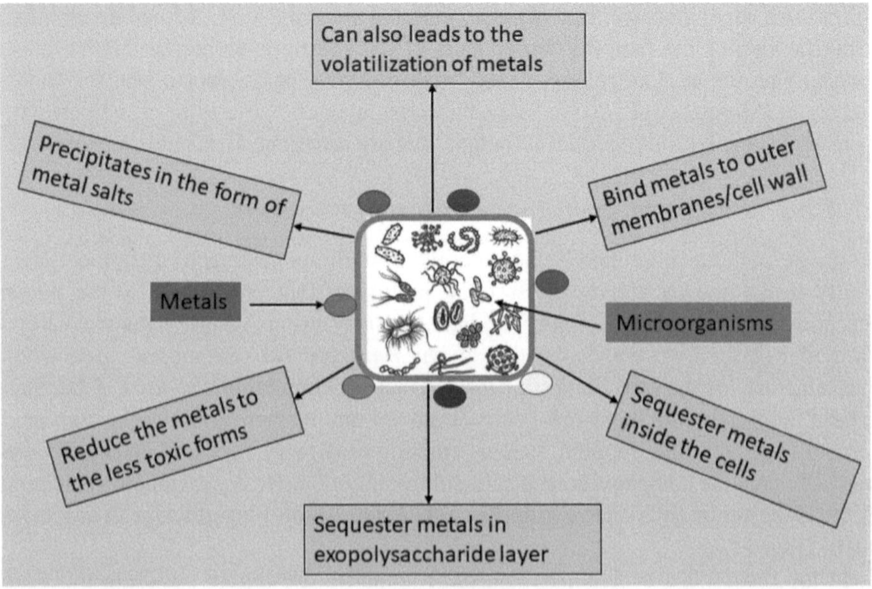

Figure 11.2 Metal-microbe interaction in bioremediation.

removal of the chromium from the soil that is named after the genus it belongs to, *Bacillus* (Abid et al., 2014). *Rhodobacter sphaeroides* isolated and cultured at 7 pH and 30°C–35°C, was firstly used for the remediation of the lead-contaminated soils and this bacterial strain changed the speciation of the lead in the soil (Li et al., 2016). *Desulfovibrio desulfuricans* a sulfate-reducing bacterium efficiently remove cadmium, nickel, and chromium from the contaminated or polluted marine water. This technique is efficiently used for green chemistry and engineering (Joo et al., 2015). *Mucor circinelloides* a bacterium isolate was used along with *Solanum nigrum* L. for the efficient removal of Lead. These microbes when used alone showed 58.6% removal efficiency, while they showed 40.2% removal efficiency when used in combination with phytoremediation. Results indicated that microbial/ phytoremediation was the best combination for the efficient removal of heavy metals from the soil (Sun et al., 2017). Strains of the *Enterobacter cloacae* KJ-46 and KJ47 were isolated from the heavy metal contaminated sites. These two strains are efficient lead removal at 68.1% and 54.2% respectively (Kang, 2015).

A bacterial strain *P. aeruginosa* was isolated as coded by (Lin et al., 2016) that is noted to be highly tolerable for the cadmium at 2200 mg/L. He also concluded that *P. aeruginosa* showed high tolerability for the lead and zinc which present it as a multi-metal resistant strain (Table 11.5).

11.5.2.4 Diseases and pest management

Soil health is the key for effective diseases and pest management up to a large extent in plants. We can control diseases and pests attack if we manage our soil

Table 11.5 Microbes that are associated with the degradation of inorganic contaminants.

Heavy metals	Microorganisms	Removal mechanisms	References
Lead	*Rhodobacter sphaeroides* *Mucor circinelloides* with *Solanum nigrum L.* *Enterobacter cloacae* strain KJ-46 Z-90 isolated from *Burkholderia* sp. *Aspergillus niger*	Changes the speciation of the lead in the soil Enhance the phytoremediation process by mobilizing the cadmium Precipitation with bacteria Complexation of the cadmium with the carboxyl groups Filamentous fungi ameliorated self-aggregation and immobilize the lead	(Li et al., 2016) Sun et al. (2017) Kang (2015) Yang et al. (2016) Yang et al. (2016)
Cadmium	Z-90 isolated from *Burkholderia* sp. *Pseudomonas aeruginosa* *Exiguobacterium undae* *A. niger*	Complexation of the cadmium with the carboxyl groups Bioremediation Precipitated as $CdCO^3$ Filamentous fungi ameliorated self-aggregation and immobilize the cadmium	Yang et al. (2016) Lin et al. (2016) Pan et al. (2018) Yang et al. (2018)
Arsenic	Z-90 isolated from *Burkholderia* sp. *Acidithiobacillus ferrooxidans*	Complexation of the cadmium with the carboxyl groups Green remediation strategy that immobilizes the arsenic	Yang et al. (2016)Yang et al. (2017)
Chromium	*Bacillus* sp. K12 *P. aeruginosa*, *Bacillus* sp., *Streptomyces* sp., and *Pseudomonas fluorescens*	Neutralization of the chromium Immobilize the chromium	Aslam et al. (2014) Akhtar et al. (2013)
Zinc	Z-90 isolated from *Burkholderia* sp.	Complexation of the cadmium with the carboxyl groups	Yang et al. (2016)

well or by improving our soil health. In the current era of chemical agriculture soil health maintenance is the biggest challenge for the researchers. *Ralstonia solanacearum*, a bacterium that causes the wilt diseases is controlled by the *Pseudomonas* spp., avirulent strains of *R. solanacearum*, *Streptomyces* spp., and *Bacillus* spp. (Yuliar et al., 2015). Enhanced/modified soil microbial communities are also playing an important role in the suppression of soil-borne diseases (Table 11.6).

Table 11.6 Biological control agents.

Microorganisms	Specific roles	References
Chryseobacterium indologenes	Pathogen to the American bullfrog	Hu et al. (2017a,b)
Xenorhabdus nematophila	Act as virulent for the insect host	Eleftherianos et al. (2018)
Streptomyces rochei	Antifungal	Yu et al. (2019)
Sphingomonas paucimobilis	Metabolize halohydrocarbons	Ostrem Loss and Yu (2018)
Burkholderia pyrrocinia	The potential biocontrol agent	Huang (2016)
Bacteriophages	Attacks the harmful bacteria	Domingo-Calap and Delgado-Martínez (2018)
Flavobacterium johnsoniae	Resistance to degradation	Alvarado-Gutiérrez et al. (2020)
Acinetobacter sp.	Contribute to the mineralization	Yan et al. (2018)
Stenotrophomonas maltophilia	Can be used as an organic amendment	Deredjian et al. (2016)
Ralstonia pickettii	The persistent gram-negative nosocomial infectious organism	Lai et al. (2016)
Pseudomonas brassicacearum	Dual role act as pathogen and plant growth promoter in tomato plants	Saleem et al. (2007)
Pseudomonas polymyxa	Nitrogen fixer	Khan and Joshi (2019)
Burkholderia sacchari	Polyhydroxyalkanoate-accumulating bacteria	Raposo et al. (2017)
Enterobacter sp.	Several species act as potential biocontrol agents	Thakur et al. (2019)
Burkholderia nodosa	Nodulation	Kaur et al. (2017)
Myroides odoratimimus	Multidrug resistance	Hu et al. (2017a,b)
Delftia acidovorans	Dissolve the gold into nanoparticles	Yusoff et al. (2017)
Burkholderia terricola	Controls bacterial wilt of tomato	Sarkar et al. (2018)
Bacillus thuringiensis	Resist the plant against pathogens	Bierne and Nielsen-LeRoux (2017)

11.6 General comments on microbial action in soil

Increasing population worldwide demands increased food production to overcome the food requirements. Illiterate farmers are using a high amount of inorganic fertilizers and chemicals to improve their production that ultimately results in soil and environmental pollution. Overpopulation is not only increasing the food demands but also increasing the Urbanization that is also polluting our water reservoirs and environment and ultimately polluting the soil. Irrigation with polluted wastewater results in the increased heavy metal contents in the soil. These heavy metals should have to be removed from the soil by the environment-friendly techniques (Rehman et al., 2017). Since the microbial revolution scientists are using their resources to overcome the soil degradation problems by microbial assistance. Bacterial strains of the genus *Brucella* have a lot of contribution in the remediation of different toxic elements including the Cr (VI) from the soil that also has previously reported by (Lakshmi et al., 2017). Recently *Brucella* sp. K12 (bacterial isolate) was introduced in the Cr (VI) contaminated soil and known to be the most efficient in increasing the Cr (VI) tolerability and lowering the concentration in the soil. This bacterium was isolated from the effluents of the tanning industry. Abid et al. (2014) concluded that reduction of the Cr concentration or detoxification by living organisms or microbial community is considered the most effective and useful remediation technique. Moreover, among all the toxic pollutants cadmium is considered the most mobile nonessential metal in the soil, which can easily move in the soil-plant system (Rehman et al., 2017).

Certain substances are produced by microorganisms that are involved in the promotion of plant growth mediated by PGPR. Currently, we are using excessive chemical fertilizers that are destroying the soil and water environment, degrading the land resources, and increasing the health risks of the community. Being an economically agricultural entity, we need to reclaim as much soil as we can instead of using chemical fertilizers and developing the marginal lands that can be possible by biologically green agriculture. We need to focus on the microbial community of the rhizosphere that is highly affected by excessive chemical applications (Gupta et al., 2019).

11.7 Future perspectives to enhance microbial assisted remediation of marginal soils

Thomas and Sekhar (2016) declared that bio-inoculant will give its best when survive for a longer period, to attain maximum benefits from the microbiomes in the future we need to focus on the resistance of microorganisms in the field soils.

11.7.1 Strategies to ensure uniform microbial count in soil

Microbial count in the soil is the most important element of the microbial remediation because of the biome sensitivity and uniformity to attain the target. If the microbial application will not be uniform and sufficient for the target it will not work efficiently. So, to make sure the uniformity and microbial count we consider different techniques such as; plate loop count, Standard plate count, pectin gel plate count, Petrifilm aerobic count, and spiral plate count (Hayes et al., 2001). While Oehler et al. (2019) did a lot of work on that and revealed that array of flatbed scanners that are coupled or used with automatic image analyzer are highly recommended for the thousands of bacterial counts meanwhile he also stated that growth properties of a specific member of the whole community can also be investigated. This system allows us to differentiate the mixed samples that are taken from the environment based on the properties of the microbial communities provided through this system.

11.7.2 Endophytic microbes

Endophytic microorganisms are the plants' internal inhabitants; the ultimate fate of these microbes is in the soil. Endophytic microbes are usually helpful for the plant and the soil microbiota as well as for the phytoremediation and for the control of soil pathogens (Thomas et al., 2016). Endophytic microorganisms migrate towards the host (plant tissues) from the root zone. The survival of endophytic microorganisms in the field for a longer period of time under different climatic conditions of the soil is very important to successfully reach the host (Hardoim et al., 2015; Puri et al., 2016). Endophytes enhance plant growth and play an important role in nitrogen fixation (Santi et al., 2013). *Pseudomonas stutzeri* A1501 is the most studied endophyte that was first isolated in China and known for its efficiency in nitrogen fixation (Tkacz and Poole, 2015). *Herbaspirillum seropedicae* enhanced the growth of its host especially in the soils that are nitrogen and oxygen-deficient (Pedrosa et al., 2011). Two endophytic strains isolated from *Capsicum annuum L.* (pepper plant), *Arthrobacter* EZB-4 and *Bacillus* EZB-8 increased the plant biomass and IAA in the drylands (Sziderics et al., 2007). *Achromobacter piechaudii* E6S enhanced the plant development in the saline soils by producing IAA, 1-aminocyclopropane-1-carboxylate (ACC) deaminase, and solubilizing the phosphate (Anand et al., 2016).

11.7.3 Biofertilizer development, scope and application

Growing population is challenging our food requirements and food security, to meet our requirements chemical fertilizers have been used rapidly to meet the food requirements. So very soon we may have to face food scarcity and endangered food security at the same time because of land degradation due to extensive farming. To overcome the problem of chemical applications that are degrading our soils, we need to explore the microorganisms as biofertilizers in the field (Mahanty et al., 2017).

Rhizobium, *Bradyrhizobium*, *Azospirillum*, and *Azotobacter* function as nitrogen-fixing biofertilizers (Bhattacharjee et al., 2008; Franche et al., 2009). *Bacillus*, *Pseudomonas*, and *Aspergillus* used as phosphate solubilizing biofertilizers (Khan et al., 2005). Rate of Single-leaf photosynthesis is enhanced by inoculation of *Rhizobium leguminosarum*, *Rhizobium* sp. IRBG 74, and *Bradyrhizobium* sp. IRBG 271 named biofertilizers. Among these three inoculants IRBG strain gave maximum (14%) photosynthetic rate as compared to the controls (uninoculated) (Mahanty et al., 2017). Shinwari et al. (2015) listed some PGPRs that remediate the heavy metals (*Achromobacter xylosoxidans*, *P. putida*, *Mesorhizobium*, *P. aeruginosa*, *Kluyvera ascorbata*, *Ralstonia metallidurans*, *Brevibacillus* sp., *Rhizobium*, *Pseudomonas* sp., *Sinorhizobium* sp., *Bradyrhizobium*, *Variovox paradoxus*, *Bacillus megaterium*, *Ochrobactrum* sp., *Bacillus subtilis*, *Psychrobacter* sp., *Azotobacter chroococcum*, and *Xanthomonas* sp.) from a wide range of PGPRs existing.

11.7.4 Bacterial isolates extraction, enrichment and application

Among the microorganisms' plant growth-promoting bacteria (PGPB) have a key role in plant growth. Phytohormones production, control of phytopathogens and enhancement of nutrient availability to the plants are the major functions of PGPB (Alonso-Mora et al., 2017). However, enzymes are secreted by the microorganisms that are helpful in the degradation of the herbicides. These enzymes of different microorganisms are also helpful in the restoration of the fertility status of the soil. Short-lived enzyme activities result in the early denaturation and or inhibition of the enzymes. Enzymes that can persist in soil (free enzymes) are more resistant to denaturation/degradation because they can be bounded with humic substances or clay (Tejada and Benítez, 2017). Masciandaro et al. (2013) give a school of thought that exploring the complexes of humus enzymes will lead to a better understanding

Table 11.7 Genetic studies.

Gene used	Description	References
Metallothionein MTA1	Gene isolated from *Pisum sativum* was used in the *Populus alba* for the increased tolerance of copper and zinc	Balestrazzi et al. (2009)
Phytochelatin synthase PCS1 + γ-glutamylcysteine synthetase GSH	These two genes were isolated from *Arabidopsis thaliana* and introduced into the *P. alba* for increased copper and zinc tolerance	Balestrazzi et al. (2009)
γ-Glutamylcysteine synthetase ECS	This gene is used in hydroponic cultures that isolated from *Escherichia coli* and was seen neutralizing the high amounts of cadmium	He et al. (2015)
γ-Glutamylcysteine synthetase glutathione synthetase GCS-GS operon	*Beta vulgaris* was genetically engineered by these two genes isolated from *Streptococcus thermophilus* that enhance the accumulation and tolerance of the cadmium copper and zinc	Liu et al. (2010)
G protein DEP1	Insertion of this gene into *A. thaliana* that was isolated from *Oryza sativa* that results in the increased cadmium accumulation	Young et al. (2013)

of the herbicide's toxicity on soil microbial community. Isolation of the new bacterial strains or microbial strains is the necessity for green agriculture that can only be possible by new strains that control or remediate the marginal lands.

Alonso-Mora et al. (2017) conducted a study for the isolation of new bacterial strains that are tolerant to aluminum. To overcome the toxicity of Al^{3+}, soil in acidic volcanic conditions was used from the bacterial consortium that also increased the plant growth under Chilean volcanic soil conditions. Efficient results were given by the bacterial strains of the following genera *Serratia, Enterobacter, Klebsiella*, and *Stenotrophomonas* that were isolated from the endo and rhizosphere of ryegrass and selected based on their tolerance and expression of plant growth-promoting traits.

11.7.5 Genetic engineering

A new strain can be developed after the incorporation of genes into a transgenic plant that is built up or cloned with the genes that are involved efficiently in the combined pollutant transformation. This is because transgenic plants show reduced levels of ethylene while activating or improving the appearance of ACC deaminase genes, beneficial for the extension of roots for the potential uptake of toxic metals and dissipate the xenobiotics within the rhizosphere (Ye et al., 2017). Some of the genetic studies are listed below in Table 11.7.

11.8 Conclusion

Marginal soils are a major concern nowadays due to their role in compromising food sustainability. Out of all available options, using soil microbes, microbial-plant interactions, and microbial products, are the most viable options and must be studied further to ensure food sustainability for the world. It is evident from the literature that microbial associations such as symbiotic and nonsymbiotic play a significant role in plant nutrition and soil fertility management. Apart from associations, there are types of microbes that help in the solubilization of nutrients to enhance their availability to plants. The role of microorganisms in degraded land management is evident from the fact that the bioremediation of soils polluted with inorganic, organic, and biological pollutants is considered one of the best techniques, due to its environment-friendly nature and cost-effectiveness.

To increase the efficient use of microbes in degraded land management, it is necessary to develop techniques to enhance or maintain the microbial population in soil uniformly. Strategies should be used to increase awareness regarding the use of biofertilizers. Apart from that steps should be taken in the efficient development of biofertilizers that can be stable for a longer period without compromising the performance. The use of genetic engineering should be considered to increase the efficiency of the existing microbial population.

References

Abid, N., Maqbool, A., Malik, K.A., 2014. Screening commercial wheat (*Triticum aestivum* L.) varieties for agrobacterium mediated transformation ability. Pak. J. Agric. Sci. 51 (1), 83–89.

Achard, P., Cheng, H., De Grauwe, L., Decat, J., Schoutteten, H., Moritz, T., et al., 2006. Integration of plant responses to environmentally activated phytohormonal signals. Science 311 (5757), 91–94.

Ahmad, Z., Anjum, S., Waraich, E.A., Ayub, M.A., Ahmad, T., Tariq, R.M.S., Ahmad, R., Iqbal, M.A., 2018. Growth, physiology, and biochemical activities of plant responses with foliar potassium application under drought stress–a review. Journal of Plant Nutrition 41 (13), 1734–1743.

Ahmad, K., Bhatti, I.A., Muneer, M., Iqbal, M., Iqbal, Z., 2012. Removal of heavy metals (Zn, Cr, Pb, Cd, Cu and Fe) in aqueous media by calcium carbonate as an adsorbent. Int. J. Chem. Biol. Sci. 2, 48–53.

Ahmad, H.R., Sabir, M., ur Rehman, M.Z., Aziz, T., Maqsood, M.A., Ayub, M.A., Shahzad, A., 2020. Wastewater irrigation-sourced plant nutrition: concerns and prospects. In: *Plant Micronutrients*, Springer, Cham, pp. 417–434.

Akhtar, M.S., Chali, B., Azam, T., 2013. Bioremediation of arsenic and lead by plants and microbes from contaminated soil. Res. Plant Sci. 1 (3), 68–73.

Akıncı, H., Özalp, A.Y., Turgut, B., 2013. Agricultural land use suitability analysis using GIS and AHP technique. Comput. Electron Agric. 97, 71–82.

Akinyele, O.A., Ewemoje, O.E., Bamgboye, A., 2019. Effects of anthropogenic activities on some engineering properties of two selected soils. Eur. J. Adv. Eng. Technol. 6 (6), 1–9.

Ali, Q., Zahir, Z.A., Asghar, H.N., Jamil, A., 2017. Inoculation with rhizobial consortium for improving the growth, yield and quality of maize under salt-stressed conditions. Pak. J. Agric. Sci. 54 (1).

Alonso-Mora, J., Samaranayake, S., Wallar, A., Frazzoli, E., Rus, D., 2017. On-demand high-capacity ride-sharing via dynamic trip-vehicle assignment. PNAS 114 (3), 462–467.

Alvarado-Gutiérrez, M.L., Ruiz-Ordaz, N., Galíndez-Mayer, J., Curiel-Quesada, E., Santoyo-Tepole, F., 2020. Degradation kinetics of carbendazim by *Klebsiella oxytoca*, *Flavobacterium johnsoniae*, and *Stenotrophomonas maltophilia* strains. Environ. Sci. Pollut. Res. 1–9.

Anand, K.U.M.A.R., Kumari, B.A.B.Y., Mallick, M.A., 2016. Phosphate solubilizing microbes: an effective and alternative approach as biofertilizers. J. Pharm. Sci. 8, 37–40.

Anjum, S., Jaffari, G.H., Rumaiz, A.K., Rafique, M.S., Shah, S.I., 2010. Role of vacancies in transport and magnetic properties of nickel ferrite thin films. J. Phys. D. Appl. Phys. 43 (26), 265001.

Arora, S., Liang, Y., Ma, T., 2016. A simple but tough-to-beat baseline for sentence embeddings. Published as a conference paper at ICLR 2017.

Ashaiekh, M.A., Eltayeb, M.A., Ali, A.H., Ebrahim, A.M., Salih, I., Idris, A.M., 2019. Spatial distribution of total and bioavailable heavy metal contents in soil from agricultural, residential, and industrial areas in Sudan. Toxin Rev. 38 (2), 93–105.

Aslam, M.U., Maqbool, M.A., Zaman, Q.U., Latif, M.Z., Ahmad, R.M., 2013. Responses of mungbean genotypes to drought stress at early growth stages. Int. J. Sci.: Basic. Appl. 13 (5), 22–27.

Aslam, S., Qazi, J.I., Hussain, A., Ali, S., 2014. Vertical zonation and seed germination indices of chromium resistant cellulolytic and nitrogen fixing bacteria from a chronically metal exposed land area. Pak. J. Bot. 46 (6), 2257–2268.

Ayub, M.A., Ahmad, H.R., Ali, M., Rizwan, M., Ali, S., ur Rehman, Waris, A.A., 2020. Salinity and its tolerance strategies in plants. In: *Plant Life Under Changing Environment*, Academic Press, pp. 47–76.

Ayub, M.A., Farooqi, Z.U.R., Umar, W., Nadeem, M., Ahmad, Z., Fatima, H., Iftikhar, I., Anjum, M.Z., 2021. Role of Urban Vegetation: Urban Forestry in Micro-Climate Pollution Management. In: *Examining International Land Use Policies, Changes, and Conflicts*, IGI Global, pp. 231–251.

Ayub, M.A., ur Rehman, M.Z., Umar, W., Adnan, M., Farooqi, Z.U.R., Naveed, M., Aslam, M.Z., Ahmad, H.R., 2021. Physiological mechanisms and adaptation strategies of plants under heavy metal micronutrient deficiency/toxicity conditions. In: *Frontiers in Plant-Soil Interaction*, Academic Press, pp. 413–458.

Ayub, M.A., Usman, M., Faiz, T., Umair, M., ul Haq, M.A., Rizwan, M., et al., 2020. Restoration of degraded soil for sustainable agriculture. Soil Health Restoration and Management. Springer, Singapore, pp. 31–81.

Azhar, M., ur Rehman, M.Z., Ali, S., Qayyum, M.F., Naeem, A., Ayub, M.A., ul Haq, M.A., Iqbal, A., Rizwan, M., 2019. Comparative effectiveness of different biochars and conventional organic materials on growth, photosynthesis and cadmium accumulation in cereals. Chemosphere 227, 72–81.

Baas, B.C., Mancini, L.M., Lee, M.N., Conant, R.T., Wallenstein, M.D., 2016. Phosphorus mobilizing consortium Mammoth P™ enhances plant growth. Peer J. 4, 2121.

Balestrazzi, A., Botti, S., Zelasco, S., Biondi, S., Franchin, C., Calligari, P., et al., 2009. Expression of the PsMT A1 gene in white poplar engineered with the MAT system is associated with heavy metal tolerance and protection against 8-hydroxy-2′-deoxyguanosine mediated-DNA damage. Plant Cell Rep. 28 (8), 1179–1192.

Bechtold, U., Field, B., 2018. Molecular mechanisms controlling plant growth during abiotic stress. J. Exp. Bot. 69 (11), 2753–2758.

Bhattacharjee, R.B., Singh, A., Mukhopadhyay, S.N., 2008. Use of nitrogen-fixing bacteria as biofertiliser for non-legumes: prospects and challenges. Appl. Microbiol. Biotechnol. 80 (2), 199–209.

Bierne, H., Nielsen-LeRoux, C., 2017. Is there a transgenerational inheritance of host resistance against pathogens? lessons from the *Galleria mellonella, Bacillus thuringiensis* interaction model. Virulence 8 (8), 1471–1474.

Blanco, D., Hardtke, L.A., Rostagno, C.M., Del Valle, H.F., Metternicht, G.I., 2017. Soil degradation in Peninsula Valdes: causes, factors, processes, and assessment methods. Late Cenozoic of Península Valdés, Patagonia, Argentina. Springer, Cham, pp. 191–213.

Chaves, M.M., Maroco, J.P., Pereira, J.S., 2003. Understanding plant responses to drought—from genes to the whole plant. Funct. Plant Biol. 30 (3), 239–264.

Chen, X., Kumari, D., Cao, C.J., Plaza, G., Achal, V., 2019. A review on remediation technologies for nickel-contaminated soil. Hum. Ecol. Risk Assess.: An. Int. J. 1–15.

Clocchiatti, A., Hannula, S.E., van den Berg, M., Korthals, G., De Boer, W., 2020. The hidden potential of saprotrophic fungi in arable soil: patterns of short-term stimulation by organic amendments. Appl. Soil Ecol. 147, 103434.

Collins, B.D., Montgomery, D.R., 2002. Forest development, wood jams, and restoration of Floodplain rivers in the Puget Lowland, Washington. Restor. Ecol. 10 (2), 237–247.

Delang, C.O., 2017. Causes and distribution of soil pollution in China. Environ. Socioecon. Stud. 5 (4), 1–17.

Deredjian, A., Alliot, N., Blanchard, L., Brothier, E., Anane, M., Cambier, J.C., et al., 2016. Occurrence of *Stenotrophomonas maltophilia* in agricultural soils and antibiotic resistance properties. Res. Microbiol. 167 (4), 313–324.

Domingo-Calap, P., Delgado-Martínez, J., 2018. Bacteriophages: protagonists of a post-antibiotic era. Antibiotics 7 (3), 66.

Du, H., Zuo, X., Li, S., Wang, T., Xue, X., 2019. Wind erosion changes induced by different grazing intensities in the desert steppe, Northern China. Agric. Ecosyst. Environ. 274, 1–13.

Ebadi, A.G., Hisoriev, H., 2017. Metal pollution status of Tajan River–Northern Iran. Toxicol. Environ. Chem. 99 (9–10), 1358–1367.

Ebadi, A.G., Hisoriev, H., 2018. *Cladophora glomerata* L. as indicator of chemical pollution in coastal areas of Farahabad Region. Iran. Revolut. Publ. 5 (14 (2)), 183–191.

Eleftherianos, I., Yadav, S., Kenney, E., Cooper, D., Ozakman, Y., Patrnogic, J., 2018. Role of endosymbionts in insect–parasitic nematode interactions. Trends Parasitol. 34 (5), 430–444.

FAO Policy Brief, 2006. Food security. FAO Agriculture and Development Economics Division. Issue 2. Available from: http://www.fao.org/fileadmin/templates/faoitaly/documents/pdf/pdf_Food_Security_Cocept_Note.pdf. (accessed 15 August 2021).

Food and Agriculture Organization of United Nations (FAO), 2017. The Future of Food and Agriculture. Available from: http://www.fao.org/3/i6583e/i6583e.pdf. (accessed 15 August 2021).

Farooqi, Z.U.R., Ayub, M.A., Nadeem, M., Shabaan, M., Ahmad, Z., Umar, W., Iftikhar, I., 2021. Precision agriculture to ensure sustainable land use for the future: precision agriculture and arable land use. In *Examining International Land Use Policies, Changes*. In: *and Conflicts*, IGI Global, pp. 210–230.

Farooqi, Z.U.R., Mubarak, U., Zeeshan, N., Hussain, M.M., Ayub, M.A., 2021. Agricultural Wastes and Its Applications in Plant-Soil Systems. In: *Agricultural Waste*, Apple Academic Press, pp. 15–34.

Finlayson, M., Cruz, R.D., Davidson, N., Alder, J., Cork, S., de Groot, R.S., et al., 2005. Millennium Ecosystem Assessment: Ecosystems and Human Well-being: Wetlands and Water Synthesis. Island Press.

Fontaine, L., Thiffault, N., Paré, D., Fortin, J.A., Piché, Y., 2016. Phosphate-solubilizing bacteria isolated from ectomycorrhizal mycelium of *Picea glauca* are highly efficient at fluorapatite weathering. Botany 94 (12), 1183–1193.

Franche, C., Lindström, K., Elmerich, C., 2009. Nitrogen-fixing bacteria associated with leguminous and non-leguminous plants. Plant Soil 321 (1), 35–59.

Gore, N.S., Navale, A.M., 2017. Effect of consortia of potassium solubilizing bacteria and fungi on growth, nutrient uptake and yield of banana. Indian. J. Hortic. 74 (2), 189–197.

Gouda, S., Kerry, R.G., Das, G., Paramithiotis, S., Shin, H.S., Patra, J.K., 2018. Revitalization of plant growth promoting rhizobacteria for sustainable development in agriculture. Microbiol. Res. 206, 131–140.

Gregory, R., Henn, C., McDonald, B., Saito, M., 2010. Trade and the crisis: protect or recover. J. Int. Commer. Econ. Policy 1 (02), 165–181.

Grimm, N.B., Faeth, S.H., Golubiewski, N.E., Redman, C.L., Wu, J., Bai, X., et al., 2008. Global change and the ecology of cities. Science 319 (5864), 756–760.

Grover, V.I., 2016. Climate change and land degradation in China: challenges for soil conservation, Global Warming and Climate Change, 2. CRC Press, pp. 749–768.

Gupta, R., Singh, A., Srivastava, M., Shanker, K., Pandey, R., 2019. Plant-microbe interactions endorse growth by uplifting microbial community structure of *Bacopa monnieri* rhizosphere under nematode stress. Microbiol. Res. 218, 87–96.

Hameed, M.A., Farooqi, Z.U.R., Hussain, M.M. and Ayub, M.A., 2021. PGPR-Assisted Bioremediation and Plant Growth: A Sustainable Approach for Crop Production Using Polluted Soils. *Plant Growth Regulators: Signalling Under Stress Conditions*, p.403.

Hardoim, R., Overbeek, Van, Berg, L.S., Pirttilä, G., Compant, A.M., Campisano, S., et al., 2015. The hidden world within plants: ecological and evolutionary considerations for defining functioning of microbial endophytes. Microbiol. Mol. Biol. Rev. 79 (3), 293–320.

Hayes, M.C., Ralyea, R.D., Murphy, S.C., Carey, N.R., Scarlett, J.M., Boor, K.J., 2001. Identification and characterization of elevated microbial counts in bulk tank raw milk. Int. J. Dairy Sci. 84 (1), 292–298.

Hazell, P., Poulton, C., Wiggins, S., Dorward, A., 2010. The future of small farms: trajectories and policy priorities. World Dev. 38 (10), 1349–1361.

He, J., Li, H., Ma, C., Zhang, Y., Polle, A., Rennenberg, H., et al., 2015. Overexpression of bacterial γ-glutamylcysteine synthetase mediates changes in cadmium influx, allocation and detoxification in poplar. New Phytol. 205 (1), 240–254.

Huang, L., 2016. Enhancement of antifungal activity of *Burkholderia pyrrocinia* JKâ SH007 genetically modified with *Bacillus subtilis* chi113 gene. For. Pathol. 46 (6), 632–642.

Hussain, M.I., 2018. Restoration of salt-degraded marginal lands through cultivation of nutrient enriched quinoa in UAE: impact on growth yield traits and carbon isotope fractionation. In: Qatar Foundation Annual Research Conference Proceedings, vol. 2018, No. 1, EEPP382. Hamad Bin Khalifa University Press (HBKU Press).

Hu, S., Jiang, T., Zhou, Y., Ming, D., Gao, H., Wang, M., 2017a. Genomic analysis of the multi-drug-resistant clinical isolate *Myroides odoratimimus* PR63039. Mol. Genet. Genomics 292 (1), 133–144.

Hu, R., Yuan, J., Meng, Y., Wang, Z., Gu, Z., 2017b. Pathogenic *Elizabethkingia miricola* infection in cultured black-spotted frogs, China, 2016. Emerg. Infect. Dis. 23 (12), 2055.

Hussain, M.M., 2021. Application of Bioremediation as sustainable approach to remediate heavy metal and pesticide polluted environments. Plant and Environment 1 (02), 62–92.

Igiehon, N.O., Babalola, O.O., 2018. Rhizosphere microbiome modulators: contributions of nitrogen fixing bacteria towards sustainable agriculture. Int. J. Environ. Res. Public. Health 15 (4), 574.

Ijaz, M., Ahmad, H.R., Bibi, S., Ayub, M.A., Khalid, S., 2020. Soil salinity detection and monitoring using Landsat data: a case study from Kot Addu, Pakistan. Arabian Journal of Geosciences 13 (13), 1–9.

Irshad, M., Inoue, M., Ashraf, M., Al-Busaidi, A., 2007. The management options of water for the development of agriculture in dry areas. Res. J. Appl. Sci. 7 (11), 1551–1557.

Jin, S., Feng, G., 2013. Large-scale variations of global groundwater from satellite gravimetry and hydrological models, 2002–2012. Glob. Planet. Change 106, 20–30.

Joo, J., Park, Y.G., Baek, J., Moon, Y.E., 2015. Haloperidol dose combined with dexamethasone for PONV prophylaxis in high-risk patients undergoing gynecological laparoscopic surgery: a prospective, randomized, double-blind, dose-response and placebo-controlled study. BMC Anesthesiol. 15 (1), 99.

Kang, J.C., 2015. Our Demand Is Simple: Stop Killing Us. The New York Times, p. 4.

Kang, C.H., Kwon, Y.J., So, J.S., 2016. Bioremediation of heavy metals by using bacterial mixtures. Ecol. Eng. 89, 64–69.

Karamage, F., Shao, H., Chen, X., Ndayisaba, F., Nahayo, L., Kayiranga, A., et al., 2016. Deforestation effects on soil erosion in the Lake Kivu Basin. Dr. Congo-Rwanda. Forests 7 (11), 281.

Kaur, C., Selvakumar, G., Ganeshamurthy, A.N., 2017. *Burkholderia* to *Paraburkholderia*: the journey of a plant-beneficial-environmental bacterium. Recent Advances in Applied Microbiology. Springer, Singapore, pp. 213–228.

Khan, S., Joshi, N., 2019. Development of salt tolerant mutants of *pseudomonas*. Emerging technologies towards agriculture. Food Environ. 223.

Khan, S., Khan, R.A., 2017. Chronic stress leads to anxiety and depression. Ann. Psychiatry Ment. Health 5 (1), 1091.

Khan, N.M., Rastoskuev, V.V., Sato, Y., Shiozawa, S., 2005. Assessment of hydrosaline land degradation by using a simple approach of remote sensing indicators. Agric. Water Manag. 77 (1–3), 96–109.

King, P., Grealish, G., Shahid, S.A., Abdelfattah, M.A., 2013. Land evaluation interpretations and decision support systems: soil survey of Abu Dhabi Emirate. Dev. Soil Classif, Land Use Plan. Policy Implic. 147–164.

Kochian, L.V., Hoekenga, O.A., Pineros, M.A., 2004. How do crop plants tolerate acid soils? mechanisms of aluminum tolerance and phosphorous efficiency. Annu. Rev. Plant Biol. 55, 459–493.

Kong, K., Nandintsetseg, B., Shinoda, M., 2021. How plant production in the Mongolian grasslands is affected by wind-eroded coarse-textured topsoil. Journal of Arid Environments 189 (104443), 2.

Lai, H.W., Shen, Y.H., Chien, L.J., Tseng, S.H., Mu, J.J., Chan, Y.J., et al., 2016. Outbreak of *Ralstonia pickettii* bacteremia caused by contaminated saline solution in Taiwan. Am. J. Infect. Control 44 (10), 1191–1192.

Lakshmi, K.S., Sailaja, V.H., Reddy, M.A., 2017. Phytoremediation-a promising technique in waste water treatment. Int. J. Sci. Res. Manage. 5, 5480–5489.

Lal, A.M., Anouncia, S.M., 2015. Semi-supervised change detection approach combining sparse fusion and constrained k means for multi-temporal remote sensing images. Egypt J. Remote. Sens. Space Sci. 18 (2), 279–288.

Li, X., Peng, W., Jia, Y., Lu, L., Fan, W., 2016. Bioremediation of lead contaminated soil with Rhodobacter sphaeroides. Chemosphere 156, 228–235.

Lin, H., Fang, C., Li, Y., Lin, W., He, J., Lin, R., et al., 2016. Effect of silicon on grain yield of rice under cadmium-stress. Acta Physiol. Plant. 38 (7), 186.

Liu, W., Luo, Y., Teng, Y., Li, Z., Ma, L.Q., 2010. Bioremediation of oily sludge-contaminated soil by stimulating indigenous microbes. Environ. Geochem. Health 32 (1), 23–29.

Liu, D., An, Z., Mao, Z., Ma, L., Lu, Z., 2015a. Enhanced heavy metal tolerance and accumulation by transgenic sugar beets expressing *Streptococcus thermophilus* StGCS-GS in the presence of Cd, Zn and Cu alone or in combination. PLoS One 10 (6), e0128824.

Liu, Y., Wang, Y., Peng, J., Du, Y., Liu, X., Li, S., et al., 2015b. Correlations between urbanization and vegetation degradation across the world's metropolises using DMSP/OLS nighttime light data. Remote Sens. 7 (2), 2067–2088.

Lozano-Durán, R., Bourdais, G., He, S.Y., Robatzek, S., 2014. The bacterial effector HopM1 suppresses PAMP-triggered oxidative burst and stomatal immunity. New Phytol. 202 (1), 259–269.

Mahanty, T., Bhattacharjee, S., Goswami, M., Bhattacharyya, P., Das, B., Ghosh, A., et al., 2017. Biofertilizers: a potential approach for sustainable agriculture development. Environ. Sci. Pollut. Res. 24 (4), 3315–3335.

Malik, Z., Ahmad, M., Abassi, G.H., Dawood, M., Hussain, A., Jamil, M., 2017. Agrochemicals and soil microbes: interaction for soil health. Xenobiotics in the Soil Environment. Springer, Cham, pp. 139–152.

Masciandaro, G., Macci, C., Peruzzi, E., Ceccanti, B., Doni, S., 2013. Organic matter–microorganism–plant in soil bioremediation: a synergic approach. Rev. Environ. Sci. Biotechnol. 12 (4), 399–419.

Mehmood, U., Inam-ul-Haq, M., Saeed, M., Altaf, A., Azam, F., Hayat, S., 2018. A brief review on plant growth promoting rhizobacteria (PGPR): a key role in plant growth promotion. Plant Prot. 2 (2), 77–82.

Meng, Z., Dang, X., Gao, Y., Ren, X., Ding, Y., Wang, M., 2018. Interactive effects of wind speed, vegetation coverage and soil moisture in controlling wind erosion in a temperate desert steppe, Inner Mongolia of China. J. Arid Land 10 (4), 534–547.

Mondal, M., Halder, G., Oinam, G., Indrama, T., Tiwari, O.N., 2019. Bioremediation of organic and inorganic pollutants using microalgae. New and Future Developments in Microbial Biotechnology and Bioengineering. Elsevier, pp. 223–235.

Mora, M.L., Alfaro, M.A., Jarvis, S.C., Demanet, R., Cartes, 2006. Soil aluminium availability in Andisols of southern Chile and its effect on forage production and animal metabolism. Soil Use Manage 22 (1), 95–101.

Naeem, A., Zafar, M., Khalid, H., Zia-ur-Rehman, M., Ahmad, Z., Ayub, M.A., Qayyum, M. F., 2019. Cadmium-induced imbalance in nutrient and water uptake by plants. In: *Cadmium Toxicity and Tolerance in Plants*, Academic Press, pp. 299–326.

Nath, A.J., Rattan, L.A.L., 2017. Effects of tillage practices and land use management on soil aggregates and soil organic carbon in the north Appalachian region. USA. Pedosphere 27 (1), 172–176.

Nations, U., 2007. World Population Prospects: The 2006 Revision and World Urbanization Prospects. United Nations, New York.

Oehler, C., Janett, F., Schmitt, S., Malama, E., Bollwein, H., 2019. Development of a flow cytometric assay to assess the bacterial count in boar semen. Theriogenology 133, 125–134.

Oliveira-Longatti, S.M., Marra, L.M., Soares, B.L., Bomfeti, C.A., Da Silva, K., Ferreira, A. A., et al., 2014. Bacteria isolated from soils of the western Amazon and from rehabilitated bauxite-mining areas have potential as plant growth promoters. World J. Microbiol. Biotechnol. 30 (4), 1239–1250.

Orhan, F., 2016. Alleviation of salt stress by halotolerant and halophilic plant growth-promoting bacteria in wheat (*Triticum aestivum*). Braz. J. Microbiol. 47 (3), 621–627.

Ostrem Loss, E.M., Yu, J.H., 2018. Bioremediation and microbial metabolism of benzo(a) pyrene. Mol. Microbiol. 109 (4), 433–444.

Pan, X., Achal, V., Zhao, C., Yang, J., Kumari, D., 2018. Microbial remediation of heavy metals, arsenic-contaminated environments in the arid zone of northwest China. Twenty Years of Research and Development on Soil Pollution and Remediation in China. Springer, Singapore, pp. 477–486.

Pedrosa, F.O., Monteiro, R.A., Wassem, R., Cruz, L.M., Ayub, R.A., Colauto, N.B., et al., 2011. Genome of *Herbaspirillum seropedicae* strain SmR1, a specialized diazotrophic endophyte of tropical grasses. PLoS Genet. 7 (5).

Peng, A., Liu, J., Gao, Y., Chen, Z., 2013. Distribution of endophytic bacteria in *Alopecurus aequalis* Sobol. and *Oxalis corniculata* L. from soils contaminated by polycyclic aromatic hydrocarbons. PLoS One 8 (12).

Plaut, Z., Edelstein, M., Ben-Hur, M., 2013. Overcoming salinity barriers to crop production using traditional methods. Crit. Rev. Plant Sci. 32 (4), 250–291.

Prashar, P., Shah, S., 2016. Impact of fertilizers and pesticides on soil microflora in agriculture. Sustain. Agric. Rev. 331–361.

Prosdocimi, M., Cerdà, A., Tarolli, P., 2016. Soil water erosion on Mediterranean Vineyards: a review. Catena 141, 1–21.

Puri, A., Padda, K., Chanway, C., 2016. Evidence of nitrogen fixation and growth promotion in canola (*Brassica napus* L.) by an endophytic diazotroph *Paenibacillus polymyxa* P2b-2R. Biol. Fertil. Soils 52 (1), 119–125.

Puri, A., Padda, K.P., Chanway, C.P., 2020. In vitro and in vivo analyses of plant-growth-promoting potential of bacteria naturally associated with spruce trees growing on nutrient-poor soils. Appl. Soil Ecol. 149, 103538.

Qadir, M.I., Abbas, K., Hamayun, R., Ali, M., 2014. Analgesic, anti-inflammatory and anti-pyretic activities of aqueous ethanolic extract of *Tamarix aphylla* L. (Saltcedar) in mice. Pak. J. Pharm. Sci. 27 (6), 1985–1988.

Qasim, M., Abideen, Z., Adnan, M.Y., Gulzar, S., Gul, B., Rasheed, M., et al., 2017a. Antioxidant properties, phenolic composition, bioactive compounds and nutritive value of medicinal halophytes commonly used as herbal teas. S. Afr. J. Bot. 110, 240–250.

Qasim, S., Shrestha, R., Qasim, M., 2017b. A national review of land degradation in Pakistan. Curr. Polit. Econ. S., Southeast., Centr. Asia. 26 (1), 91–108.

Qureshi, R., Bhatti, G.R., 2008. Diversity of micro-habitats and their plant resources in Nara desert, Pakistan. Pak. J. Bot. 40 (3), 979–992.

Qu, S., Wang, L., Lin, A., Zhu, H., Yuan, M., 2018. What drives the vegetation restoration in Yangtze river basin, China: climate change or anthropogenic factors? Ecol. Indic. 90, 438–450.

Raposo, R.S., de Almeida, M.C.M., da Fonseca, M.M.R., Cesário, M.T., 2017. Feeding strategies for tuning poly (3-hydroxybutyrate-co-4-hydroxybutyrate) monomeric composition and productivity using *Burkholderia sacchari*. Int. J. Biol. Macromol. 105, 825–833.

Rashid, M.I., Mujawar, L.H., Shahzad, T., Almeelbi, T., Ismail, I.M., Oves, M., 2016. Bacteria and fungi can contribute to nutrients bioavailability and aggregate formation in degraded soils. Microbiol. Res. 183, 26–41.

Rath, K.M., Fierer, N., Murphy, D.V., Rousk, J., 2019. Linking bacterial community composition to soil salinity along environmental gradients. ISME J. 13 (3), 836–846.

Ravi, S., Breshears, D.D., Huxman, T.E., D'Odorico, P., 2010. Land degradation in drylands: interactions among hydrologic–aeolian erosion and vegetation dynamics. Geomorphology 116 (3–4), 236–245.

Rehman, M.Z., Rizwan, M., Ali, S., Ok, Y.S., Ishaque, W., Nawaz, M.F., et al., 2017. Remediation of heavy metal contaminated soils by using *Solanum nigrum*: a review. Ecotoxicol. Environ. Saf. 143, 236–248.

Rehman, M.Z., Rizwan, M., Rauf, A., Ayub, M.A., Ali, S., Qayyum, M.F., Waris, A.A., Naeem, A., Sanaullah, M., 2019. Split application of silicon in cadmium (Cd) spiked alkaline soil plays a vital role in decreasing Cd accumulation in rice (Oryza sativa L.) grains. Chemosphere 226, 454–462.

Rehman, M.Z., Zafar, M., Waris, A.A., Rizwan, M., Ali, S., Sabir, M., et al., 2020. Residual effects of frequently available organic amendments on cadmium bioavailability and accumulation in wheat. Chemosphere 244, 125548.

Rodrigo-Comino, J., Ponsoda Carreres, M., Salesa, D., Terol, E., Gyasi Agyei, Y., Cerdà, A., 2019. Soil erosion processes in subtropical plantations (*Diospyros kaki*) managed under flood irrigation in Eastern Spain. Singap. J. Trop. Geogr.

Saleem, M., Arshad, M., Hussain, S., Bhatti, A.S., 2007. Perspective of plant growth promoting rhizobacteria (PGPR) containing ACC deaminase in stress agriculture. J. Ind. Microbiol. Biotechnol. 34 (10), 635–648.

Santi, C., Bogusz, D., Franche, C., 2013. Biological nitrogen fixation in non-legume plants. Ann. Bot. 111 (5), 743–767.

Sarkar, D.J., Mukherjee, I., Shakil, N.A., Rana, V.S., Debnath, S., 2018. Antibiotics in Agriculture: Use and Impact. Indian journal of ethno-phytopharmaceuticals 4 (1), 4–19.

Sarkar, A., Reinhold-Hurek, B., 2014. Transcriptional profiling of nitrogen fixation and the role of NifA in the diazotrophic endophyte *Azoarcus* sp. strain BH72. PLoS One 9 (2).

Savov, V., Chakalov, K., Popova, T., 2011. Comparative performance of the microbial supplements voodoo juice, piranha liquid, and tarantula liquid on vegetative growth and flowering. Sofia: Sofia Univ. St. Kliment Ohridski 1–13.

Savov, V., Chakalov, K., Popova, T., 2012. Comparative performance of the microbial supplements voodoo juice, orca, and great white on flowering. Sofia: Sofia Univ. St. Kliment Ohridski 1–9.

Sethunathan, N., Megharaj, M., Chen, Z.L., Williams, B.D., Lewis, G., Naidu, R., 2004. Algal degradation of a known endocrine disrupting insecticide, α-endosulfan, and its metabolite, endosulfan sulfate, in liquid medium and soil. J. Agric. Food Chem. 52 (10), 3030–3035.

Shahane, A.A., Shivay, Y.S., Prasanna, R., Kumar, D., 2019. Nitrogen nutrition and use efficiency in rice as influenced by crop establishment methods, cyanobacterial and phosphate solubilizing bacterial consortia and zinc fertilization. Commun. Soil Sci. Plant Anal. 50 (12), 1487–1499.

Shahid, M., Dumat, C., Pourrut, B., Sabir, M., Pinelli, E., 2014. Assessing the effect of metal speciation on lead toxicity to *Vicia faba* pigment contents. J. Geochem. Explor. 144, 290–297.

Shahid, S.A., Zaman, M., Heng, L., 2018. Soil salinity: historical perspectives and a world overview of the problem. Guideline for Salinity Assessment, Mitigation and Adaptation Using Nuclear and Related Techniques. Springer, Cham, pp. 43–53.

Shaygan, M., Hilton, M., McIntyre, N., Edraki, M., 2019. The effect of particle size on weathering and salt release from coal mine spoils. Geophys. Res. Abstr. 21.

Shinwari, K.I., Shah, A.U., Afridi, M.I., Zeeshan, M., Hussain, H., Hussain, J., et al., 2015. Application of plant growth promoting rhizobacteria in bioremediation of heavy metal polluted soil. Asian J. Multidiscip. Stud. 3 (4).

Simarmata, T., Setiawati, M.R., Herdiyantoro, D., Edriana, I., Kamaludin, N.N., Fitriatin, B. N., 2019. Application of ameliorant and microbials fertilizer as bioagent for enhancing the health of rhizomicrobiome and yield of soybean on marginal soils ecosystem. In: IOP Conference Series: Earth and Environmental Science, vol. 393, No. 1, 012044. IOP Publishing.

Simpson, R.J., Oberson, A., Culvenor, R.A., Ryan, M.H., Veneklaas, E.J., Lambers, H., et al., 2011. Strategies and agronomic interventions to improve the phosphorus-use efficiency of farming systems. Plant Soil 349 (1–2), 89–120.

Singh, B., Singh, J., Kaur, A., Singh, N., 2016. Bioactive compounds in banana and their associated health benefits—a review. Food Chem. 206, 1–11.

Singh, R., Srivastava, P., Singh, P., Upadhyay, S., Raghubanshi, A.S., 2019. Human overpopulation and food security: challenges for the agriculture sustainability. Urban Agriculture and Food Systems: Breakthroughs in Research and Practice. IGI Global, pp. 439–467.

Slemp, C., Davenport, M.A., Seekamp, E., Brehm, J.M., Schoonover, J.E., Williard, K.W., 2012. "Growing too fast": local stakeholders speak out about growth and its consequences for community well-being in the urban–rural interface. Landsc. Urban Plan. 106 (2), 139–148.

Solgi, E., Sheikhzadeh, H., Solgi, M., 2018. Role of irrigation water, inorganic and organic fertilizers in soil and crop contamination by potentially hazardous elements in intensive farming systems: case study from Moghan agro-industry, Iran. J. Geochem. Explor. 185, 74–80.

Song, O.R., Lee, S.J., Lee, Y.S., Lee, S.C., Kim, K.K., Choi, Y.L., 2008. Solubilization of insoluble inorganic phosphate by *Burkholderia cepacia* DA23 isolated from cultivated soil. Braz. J. Microbiol. 39 (1), 151–156.

Sun, Z., Cao, H., Zhang, X., Lin, X., Zheng, W., Cao, G., et al., 2017. Spent lead-acid battery recycling in China—a review and sustainable analyses on mass flow of lead. Waste Manage 64, 190—201.

Sziderics, A.H., Rasche, F., Trognitz, F., Sessitsch, A., Wilhelm, E., 2007. Bacterial endophytes contribute to abiotic stress adaptation in pepper plants (*Capsicum annuum* L.). Can. J. Microbiol. 53 (11), 1195—1202.

Tejada, M., Benítez, C., 2017. Flazasulfuron behavior in a soil amended with different organic wastes. Appl. Soil Ecol. 117, 81—87.

Thakur, A.K., Ojha, C.S., 2010. Variation of turbidity during subsurface abstraction of river water: a case study. Int. J. Sediment. Res. 25 (4), 355—365.

Thakur, N., Nath, A.K., Chauhan, A., Parmar, S.C., Pandey, H., Thakur, K., 2019. Chitinases from microbial sources, their role as biocontrol agents and other potential applications. J. Entomol. Zool. Stud. 7 (1), 837—843.

Thomas, P., Sekhar, A.C., 2016. Effects Due to Rhizospheric Soil Application of an Antagonistic Bacterial Endophyte on Native Bacterial Community and Its Survival in Soil: A Case Study with Pseudomonas aeruginosa from Banana. Front Microbiol. 1664-302X7, 493. 10.3389/fmicb.2016.00493.

Thomashow, L.S., Kwak, Y.S., Weller, D.M., 2019. Root associated microbes in sustainable agriculture: models, metabolites and mechanisms. Pest. Manage. Sci. 75 (9), 2360—2367.

Tilman, D., Fargione, J., Wolff, B., D'Antonio, C., Dobson, A., Howarth, R., et al., 2001. Forecasting agriculturally driven global environmental change. Science 292 (5515), 281—284.

Tkacz, A., Poole, P., 2015. Role of root microbiota in plant productivity. J. Exp. Bot. 66 (8), 2167—2175.

Tripathi, V., Edrisi, S.A., Abhilash, P.C., 2016. Towards the coupling of phytoremediation with bioenergy production. Renew. Sust. Energ. Rev. 57, 1386—1389.

Umar, W., Ayub, M.A., ur Rehman, M.Z., Ahmad, H.R., Farooqi, Z.U.R., Shahzad, A., Rehman, U., Mustafa, A., Nadeem, M., 2020. Nitrogen and phosphorus use efficiency in agroecosystems. In: *Resources Use Efficiency in Agriculture*, Springer, Singapore, pp. 213—257.

USDA, 2008. The United States Department of Agriculture (USDA) National Soil Survey Handbook, Part 622. Available from: http://soils.usda.gov/technical/handbook/contents/part622.html. (accessed 15 August 2021).

Vaid, S.K., Srivastava, C., Pachauri, S., Sharma, A., Rawat, D., Mathpal, B., et al., 2018. Residual effect of zinc applied to rice on zinc nutrition of succeeding wheat crop inoculated with zinc solubilizing microbial consortium. Isr. J. Plant. Sci. 1(aop) 1—11.

Verma, J.P., Jaiswal, D.K., Singh, S., Kumar, A., Prakash, S., Curá, J.A., 2017. Consequence of phosphate solubilising microbes in sustainable agriculture as efficient microbial consortium: a review. Clim. Change Environ. Sustain. 5 (1), 1—19.

Weiner, M., Tröndle, J., Albermann, C., Sprenger, G.A., Weuster-Botz, D., 2014. Improvement of constraint-based flux estimation during L-phenylalanine production with *Escherichia coli* using targeted knock-out mutants. Biotechnol. Bioeng. 111 (7), 1406—1416.

Wu, W.T., Zhou, Y.X., Tian, B., 2017. Coastal wetlands facing climate change and anthropogenic activities: a remote sensing analysis and modelling application. Ocean Coast. Manage 1 (38), 1—10.

Wu, B., Wang, Z., Zhang, Q., Shen, N., 2018. Distinguishing transport-limited and detachment-limited processes of interrill erosion on steep slopes in the Chinese loessial region. Soil Till. Res. 177, 88—96.

Xie, Q., Brady, V., Pooryousefy, E., Zhou, D., Liu, Y., Saeedi, A., 2017. The low salinity effect at high temperatures. Fuel 200, 419–426.

Xue, Y., Duan, H., Liu, L., Wang, Z., Yang, J., Zhang, J., 2013. An improved crop management increases grain yield and nitrogen and water use efficiency in rice. Crop Sci. 53 (1), 271–284.

Yang, Y., Wang, Z., Li, J., Gang, C., Zhang, Y., Zhang, Y., et al., 2016. Comparative assessment of grassland degradation dynamics in response to climate variation and human activities in China, Mongolia, Pakistan and Uzbekistan from 2000 to 2013. J. Arid Environ. 135, 164–172.

Yang, Z., Wu, Z., Liao, Y., Liao, Q., Yang, W., Chai, L., 2017. Combination of microbial oxidation and biogenic schwertmannite immobilization: a potential remediation for highly arsenic-contaminated soil. Chemosphere 181, 1–8.

Yang, Z., Liang, L., Yang, W., Shi, W., Tong, Y., Chai, L., et al., 2018. Simultaneous immobilization of cadmium and lead in contaminated soils by hybrid bio-nanocomposites of fungal hyphae and nano-hydroxyapatites. Environ. Sci. Pollut. Res. 25 (12), 11970–11980.

Yan, L., Helguera, M., Kato, K., Fukuyama, S., Sherman, J., Dubcovsky, J., 2004. Allelic variation at the VRN-1 promoter region in polyploid wheat. Theor. Appl. Genet. 109 (8), 1677–1686.

Yan, D.Z., Gan, Y.T., Zhou, H., Liu, J., Li, X., 2018. Draft genome sequence of cyclohexylamine-degrading strain *Acinetobacter* sp. YT-02 isolated. Curr. Microbiol. 75 (3), 284–287.

Yeboah, F.R.A.N.C.I.S., Awotwi, A.L.F.R.E.D., Forkuo, E.K., Kumi, M.I.C.H.A.E.L., 2017. Assessing the land use and land cover changes due to urban growth in Accra, Ghana. J. Basic Appl. Res. Int. 22 (2), 43–50.

Ye, S., Zeng, G., Wu, H., Zhang, C., Dai, J., Liang, J., et al., 2017. Biological technologies for the remediation of co-contaminated soil. Crit. Rev. Biotechnol. 37 (8), 1062–1076.

Young, L.S., Hameed, A., Peng, S.Y., Shan, Y.H., Wu, S., 2013. Endophytic establishment of the soil isolate *Burkholderia* sp. CC-Al74 enhances growth and P-utilization rate in maize (*Zea mays* L.). Appl. Soil Ecol. 66, 40–47.

Yuliar, Nion, Y.A., Toyota, K., 2015. Recent trends in control methods for bacterial wilt diseases caused by *Ralstonia solanacearum*. Microbes Environ. 30 (1), 1–11.

Yusoff, A.H.M., Nading, M.E., Salimi, M.N., 2017. Extraction of gold (Au) particles from sea water by *Delftia Acidovorans* microbes. In: Journal of Physics: Conference Series, vol. 908, No. 1, 012045. IOP Publishing.

Yu, M., Li, Y., Banakar, S., Liu, L., Shao, C., Li, Z., et al., 2019. New metabolites from the co-culture of marine-derived actinomycete *Streptomyces rochei* MB037 and fungus *Rhinocladiella similis* 35. Front. Microbiol. 10, 915.

Zhang, H., Hu, C., Jia, X., Xu, Y., Wu, C., Chen, L., et al., 2012. Characteristics of γ-hexachlorocyclohexane biodegradation by a nitrogen-fixing cyanobacterium, *Anabaena azotica*. J. Appl. Phycol. 24 (2), 221–225.

Zhang, T., Zhan, X., He, J., Feng, H., Kang, Y., 2018. Salt characteristics and soluble cations redistribution in an impermeable calcareous saline-sodic soil reclaimed with an improved drip irrigation. Agric. Water Manage. 197, 91–99.

Assessment of microbial biomass for production of ecofriendly single-cell protein, bioenergy, and other useful products

12

Raman Kumar Ravi[1,2], *Ajay Neeraj*[2] *and R. Hiranmai Yadav*[2]
[1]Department of Agriculture, Uttaranchal College of Science and Technology, Dehradun, India, [2]School of Environment and Sustainable Development, Central University of Gujarat, Gandhinagar, India

12.1 Introduction

Increased population growth has led to a decrease in the quality of life, poverty and hunger. Therefore human beings have been trying to overcome the crisis through technological advancement, which can help them to get more food. In this era of globalized society, it is finally possible to make a reliable estimate of the available resources for human survival through the technological revolution. The global shortage of food and feed has prompted researchers to find ways to increase protein production. Increasing the accessibility of protein in agricultural waste in single-cell protein (SCP) through fermentation is one of the most advantageous methods. SCP is the dried cells of microorganisms and is used as a protein supplement in human foods or animal feeds (Sakine et al., 2009). Microorganisms have been used to produce high-protein foods, such as cheese and fermented soy products for many years. Protein is the main nutrient in both products (Haider et al., 1989).

Microorganisms have the ability to upgrade low-protein organic substances into high-protein foods, which have been used by the food industry. This phenomenon was used in Germany during World War I, when people used the growth of *Saccharomyces cerevisiae* for human consumption. Further, *Candida arborea* and *Candida utilis* were used during World War II and about 60% of the country's pre-war food was replaced (Litchfield, 1983; Arora et al., 1991). The development of microbial systems for use in the food industry has firstly, growth of microorganism is faster than of animals. Secondly, a broader range of materials may be considered as suitable substrates depending on the microorganism chosen. The inexpensive feedstocks and waste products are utilized as sources of carbon and energy. The microbes such as bacteria, fungi, yeast, and algae can be used to grow biomass, protein concentrates, or amino acids (Ivarson and Morita 1982; Aggelopoulos et al., 2014). Since protein contents are the quantitatively important part of microbial cells, these microorganisms, also called SCPs, are a natural protein concentrate

Microbes in Land Use Change Management. DOI: https://doi.org/10.1016/B978-0-12-824448-7.00015-2
© 2021 Elsevier Inc. All rights reserved.

(Esabi, 2001). In recent years, the production of SCP has been widely available globally. *Candida* sp. is one of the most widely used microbial population in SCP production (Adoki, 2002). Besides SCP, many microorganisms are also applied for the production of different products such as biofuel, enzymes, etc. for human welfare. This chapter mainly focused on different types of microbial biomass and their application in production of SCP, bioenergy, and other useful products to fulfill human needs.

12.2 Microbial biomass

Microbes are a versatile group of organisms, found in diversified habitats, and have been serving humanity in various aspects for sustaining life. The microbial biomass is a suitable supplemental source of protein and bioenergy obtained through processing in which different microorganisms such as bacteria, yeasts, fungi, or algae are cultivated in large quantities.

12.2.1 Characteristics of microbial biomass

The microbial biomass is the main natural source of proteins and bioenergy. Mainly four types of microbial population are utilized for the production of biomass, that is, bacteria, yeasts, fungi, and algae. The characteristics preferred for selection of microorganisms are high specific growth rate, high substrate to biomass yield, high cell density, ability to use complex substrates, affinity towards substrate, and low nutrient requirements. Besides these, nature of the raw material available, nutrient availability such as energy values, protein contents, amino acid balance, and technological properties, that is, type of culture, nutritional toxicity may also affect the biomass production.

12.2.2 Substrates for microbial biomass production

It is estimated that the agricultural sector doesn't seem to be sufficient to fulfill and satisfy the food demand for the growing human population due to several reasons. Microorganisms are one of the most important sources of food and can be exploited to fulfill the future food requirements of the increasing human population. As all microorganisms need a suitable substrate for their growth and development, some of the microorganisms with suitable substrate are summarized in Table. 12.1 for biomass production. The substrate may be of different ranges, that is, may be waste materials. The microorganisms utilized these substrate as raw material for their growth and convert it into useful byproducts. Thus microorganisms may not only be cultivated properly on different cost effective substrates to fulfill the requirements of our daily diet but also may be mediators of environmental renovation.

Table 12.1 Microorganisms utilizing different substrate for biomass production.

Substrate	Microorganisms	References
Potato starch processing waste	*Bacillus licheniformis*	Liu et al. (2014)
	Bacillus pumilus, Candida utilis	Liu et al. (2013a,b)
	Aspergillus niger	Liu et al. (2013a,b, 2014)
Ram horn	*Bacillus cereus* *Escherichia coli* *Bacillus subtilis*	Kurbanoglu and Algur (2002)
	Penicillium citrinum, Aspergillus flavus, Fusarium semitectum, Monascus ruber, Aspergillus ochraceus, Cladosporium cladosporioides, A. niger	Valentino et al. (2016)
	Aspergillus oryzae	Ravinder et al. (2003)
Soybean hull	*B. subtilis*	Wongputtisin et al. (2012, 2014)
Glucose + fructose	*Corynebacterium ammoniagenes*	Wang et al. (2013)
Latex rubber sheet wastewater	*Rhodopseudomonas palustris*	Kornochalert et al. (2014)
Brewery wastewater	*Rhizospheric diazotrophs*	Lee et al. (2015)
Methane	*Methylococcus capsulatus, Ralstonia* sp., *Brevibacillus agri, Aneurunibacillus* sp.	Overland et al. (2010)
Methane salt broth	*Methylomonas* sp.	Yazdian et al. (2005)
Petrochemical wastewater	*Haloarcula* sp. IRU1	Taran and Asadi (2014)
Wheat straw	*Pleurotus florida*	Ahmadi et al. (2010)
Apple pomace	*A. niger*	Bhalla and Joshi (1994)
Banana wastes		Baldensperger et al. (1985)
Stickwater		Kam et al. (2012)
Waste liquor		Chiou et al. (2001)
Cheese whey	*Candida crusei*	Yadav et al. (2014)
Glucose	*Fusarium venenatum*	Wiebe (2002)
Molasses	*Candida tropicalis*	Gao et al. (2012)
Bagasse		Pessoa et al. (1996)
Poultry litter	*C. utilis*	Zhao et al. (2010)
Waste capsicum powder		Jalasutram et al. (2013)

(Continued)

Table 12.1 (Continued)

Substrate	Microorganisms	References
Spoiled date palm fruits	*Hanseniaspora uvarum*	Hashem et al. (2014)
Cheese whey	*Kluyveromyces marxianus*	Yadav et al. (2014, 2016)
Whey, molasses, potato pulp, orange pulp, brewer's spent grain		Aggelopoulos et al. (2014)
Methanol	*Methylotrophus*	Rudravaram et al. (2009), Johnson (2013)
Molasses, orange pulp, brewer's spent grain	*Saccharomyces cerevisiae*	Aggelopoulos et al. (2014)
Citrus pulp	*Trichoderma virideae*	De Gregorio et al. (2002)
Cheese whey filtrate	*Trichoderma harzianum*	Sisman et al. (2013)
Inulin, crude oil, glycerol waste hydrocarbons	*Yarrowia lipolytica*	Cui et al. (2011), Zinjarde (2014)
Cheese whey	*Kefir* sp.	Paraskevopoulou et al. (2003)
Molasses, orange pulp, brewer's spent grain, whey, potato pulp, malt spent rootlets		Aggelopoulos et al. (2014)
CO_2 and light	*Chlorella vulgaris*	Li et al. (2013a,b), Safafar et al. (2016)
	Spirulina	Ritala et al. (2017)
	Euglena gracilis	Rodríguez-Zavala et al. (2010)
	Scenesdesmus obliquus	Duong et al. (2015), Apandi et al. (2017)

12.2.3 Development of microbial biomass

Microbial biomass is developed by using different microbial population. The microorganisms are able to grow on waste materials and be used as substrate for biomass development. These microorganisms use various metabolic pathways as enzymes for the substrate catabolism. The different physiological aspects, growth parameters, energy, and nutritional requirements are the influencing factors during biomass development.

12.2.4 Factors affecting microbial biomass development

Microbial growth and development is affected by several factors that affect the production of biomass on a given substrate significantly. These factors may include:

- Selection of suitable strain of microbial culture
- Genetic stability of microbial strain
- Suitable pH of growth medium
- Temperature
- Oxygen availability to provide aerobic condition
- Organic substrate and nitrogen concentration
- Maintenance of sterile conditions throughout growth period

12.3 Different types of microbial biomass (bacterial, fungal, yeast, algae)

Bacteria, fungi, yeast, and algae are the four different types of microbial population that are commonly used for microbial biomass production.

12.3.1 Bacterial biomass

The features that make bacteria appropriate for biomass generation include rapid growth and a shorter production time. They can also grow on several raw materials ranging from carbohydrates (starch, sugar) to gaseous and liquid hydrocarbons (methane, petroleum) (Bamberg, 2000) to petrochemicals such as methanol and ethanol, nitrogen sources necessary for bacterial growth, that is, urea, ammonia, nitrates, ammonium salts, and organic nitrogen in wastes. The mineral nutrients are also added to culture to fulfill the deficiency of nutrients in natural waters in sufficient amounts to support bacterial growth. The potential phototrophic bacterial species are suggested for biomass production such as SCP production. The other researchers also recommended methanotrophic bacteria as well as some other bacterial species for SCP production (Arora et al., 1991).

12.3.2 Fungal biomass

A number of fungi are known to be used as a source of protein rich food (Bhalla et al., 2007). The filamentous fungi are also utilized as microbial biomass source. Earlier, in 1973, in second international conference organized at Massachusetts Institute of Technology, some *Actinomycetes* and filamentous fungi were confirmed as protein producer from different raw materials. Later, during World War II, *Fusarium* and *Rhizopus* (Yousuf, 2012) grown in fermentation were utilized as a source of protein. The *Aspergillus oryzae* (Anupama and Ravindra, 2000) or *Rhizopus arrhizus* were applied due to nontoxic nature. Different saprophytic fungal species grow on complex organic compounds and convert them into simpler forms. The higher growth of fungi may result larger production of fungal biomass. The mycelial productivity varies significantly, which mainly depends on organisms and substrates.

12.3.3 Yeast biomass

Yeast biomass is a high protein content feed substitute and commonly used for single cell production (Burgents et al., 2004). Although, there are so many yeast species used for development of biomass, some of them are very popular such as *Candida* (Bozakouk, 2002), *Hansenula, Pitchia, Torulopsis*, and *Saccharomyces*. These species are commonly used for single cell production. The oily yeast genera include *Yarrowia, Candida, Rhodotorula, Rhodosporidium, Cryptococcus, Trichosporon, and Lipomyces*. The SCP are produced using *S. cerevisiae* grown on various fruit waste (Tanveer, 2010). Orange peels and cucumber were assessed for the SCP production using *S. cerevisiae* by submerged fermentation (Sengupta et al., 2006).

12.3.4 Algal biomass

Algae are important microbes that can exist in different ecosystems, that is, marine and fresh water, deserts, hot springs, and even in snow and ice. Since ancient times, algae are used for human welfare. The algal biomass is the most vital source of protein and energy. *Spirulina* has long been cultivated by people of Africa from Chad Lake and by Aztecs from Tescoco Lake in Mexico. They dried it and used it as food. *Spirulina* is the most widely used algae. Astronauts also bring it into space during space travel. Similarly, tribal communities around the world have harvested biomass obtained from *Chlorella* and *Scenedesmus* and used it as a food source. Algal biomass is used as food in many different ways. Algae biomass has many advantages, including simple cultivation, efficient use of solar energy, faster growth rate, and higher protein content. Algae *Spirulina* has been considered as a protein supplement (Raja et al., 2008). It is a blue green algae with strong antioxidant activity and can stimulate free radical scavenging enzyme system. *Spirulina maxima* can prevent the development of fatty liver induced by carbon tetrachloride (CCl_4). *Spirulina* is rich in nutrients and can be used for patients with malnutrition, immunosuppression, liver, and nerve function impairment. Although there is a strong need for further research on the antiviral effect of this algae and its clinical significance. Under the influence of different environmental factors, the biomass of five *Chlorella* strains isolated from different habitats is used to produce SCP (Mahasneh, 2005).

12.4 Application of microbial biomass

The microbial biomass is an important and natural source for production of different useful byproducts such as SCP, bioenergy, enzymes, etc. for human welfare (Fig. 12.1).

12.4.1 For single-cell protein production

SCP is the protein produced from biomass originating from different microbial population, that is, bacteria, yeast, fungi, and algae, which can be used as feed additives. A large amount of SCP can be produced using bacteria like *Lactobacillus* sp., *Methylomonas methylotrophus*, *Cellulomonas* sp. (Piper, 2004), *Brevibacterium* (Adedayo et al., 2011), *Bacillus subtilis, Methylophilus methylotrophus*, *Aeromonas hydrophila, Achromobacter delvaevate, Acinetobacter calcoaceticus, Bacillus megaterium*, (Gomashe et al., 2014), *Thermomonospora fusca, Rhodopseudomonas capsulate, Pseudomonas fluorescens, Flavobacterium* sp. (Dhanasekaran et al., 2011). Similarly, various species of fungi, yeast, and algae are also known to produce SCP. Some of the examples of microorganisms that are applied in SCP production are summarized in Table 12.2.

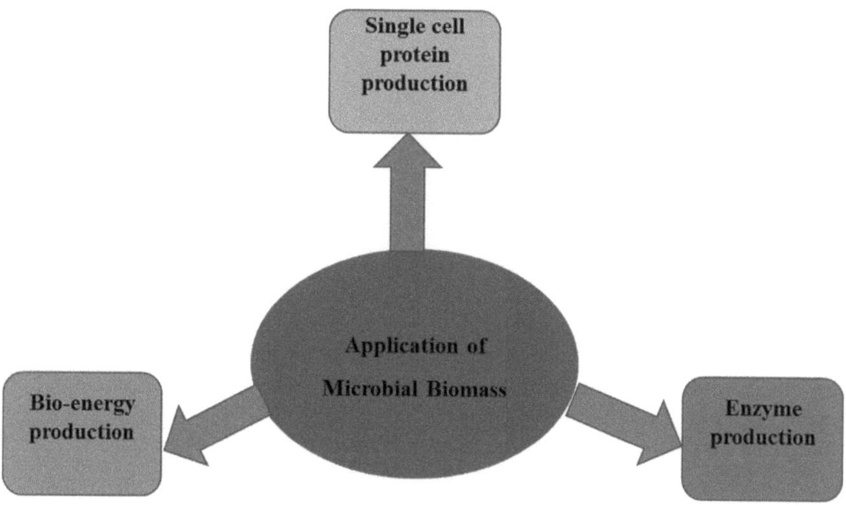

Figure 12.1 Application of microbial biomass.

Table 12.2 Microorganisms involved in single cell production.

Bacteria	Fungi	Yeast	Algae
Aeromonas, *Achromobacter,* *Acinetobacter,* *Bacillus,* *Flavobacterium,* *Lactobacillus,* *Methylomonas,* *Pseudomonas,* *Rhodopseudomonas*	*Aspergillus,* *Cephalosporium,* *Chaetomium,* *Penicillium,* *Rhizopus,* *Scytalidium,* *Trichoderma,* *Fusarium*	*Candida,* *Saccharomyces,* *Kluyveromyces,* *Thermomyces,* *Methylomonas,* *Rhodotorula,* *Trichosporon,* *Mucor*	*Anabaena,* *Nostoc,* *Spirulina, Tolypothrix,* *Chlorella, Oedogonium* *Scenedesmus,* *Spirogyra, Caulerpa,* *Coelastrum,* *Enteromorpha,* *Porphyra,* *Ulva*

Table 12.3 Microorganisms involved in biofuel production.

Biofuel	Microorganisms	References
Butanol	*Clostridium acetobutylicum*	Lütke-Eversloh and Bahl (2011)
	Escherichia coli	Shen et al. (2011)
	Pseudomonas putida	Nielsen et al. (2009)
Isobutanol	*Clostridium thermocellum*	Lin et al. (2015)
Ethanol	*Zymomonas mobilis*	Kremer et al. (2015)
	Caldicellulosiruptor bescii	Chung et al. (2014)
	Trichoderma reesei	Huang et al. (2014)
	E. coli	Romero-García et al. (2016)
Fatty acids	*Saccharomyces cerevisiae*	Yu et al. (2016)
	Yarrowia lipolytica	Beopoulos et al. (2009)
Isoprenoid based-biofuel	*S. cerevisiae*	Westfall et al. (2012)
1, 3-propanediol	*Synechococcus elongatus*	Hirokawa et al. (2016)
Limonene	*Synechococcus* sp.	Davies et al. (2014)
Lipids	*Cryptococcus vishniaccii*	Deeba et al. (2016)
2, 3-butanediol	*Zymomonas mobilis*	Yang et al. (2016)

12.4.2 For bioenergy production

Bioenergy is the world's largest contributor to the renewable and sustainable energy sector, and it plays a significant role in many energy sectors. These days a number of technologies are developed for cultivation of different microbial biomass for energy generation. Microorganisms produce biofuel using specific catabolic enzyme and metabolic pathway. For example, the direct decarboxylation of pyruvate leads to the production of ethanol in *S. cerevisiae*, while in the case of *Escherichia coli*, CoA activates the acyl group during the decarboxylation of pyruvate and is then reduced to ethanol. Some of the microorganisms used for bioenergy production are listed in Table 12.3.

In the field of bioenergy production, microalgae are an important biomass resource. The microalgae can provide different types of biofuels viz. methane produced by anaerobic digestion of algal biomass (Hernández et al., 2014), biodiesel derived from the microalgae oil (Martins et al., 2010), and biohydrogen generated photobiologically (Ghirardi et al., 2000). Several microalgae can store significant amount of lipids and contribute to high oil production. It is reported that, average lipid content of microalgae varies from 1% to 70%, and some can store up to 90% of dry weight under certain conditions (Martins et al., 2010). The lipids obtained from microalgal biomass contains higher amount of triglycerides, which is converted to biodiesel by transesterification reactions. In this reaction, three molecules of fatty acid are esterified with a glycerol molecule. The artificial consortia developed using microalgae cocultured with heterotrophs have advantageous from their mutualistic interactions.

In the case of cocultivation with microalgae, several oleaginous yeasts can also accumulate large amounts of lipids. In mixed culture, yeast uses different kinds of organic matter and provides CO_2 for microalgae for photosynthesis. At the same time, microalgae also act as an oxygen generator, which is conducive to the growth of yeast. Mixed cultures generally provide higher lipid production as well as higher microalgae growth rate and biomass concentration. Some examples of cocultivation of oleaginous yeast and microalgae are yeast *Rhodotorula glutinis* with microalgae *Chlorella vulgaris* (Cheirsilp et al., 2011) and yeast *R. glutinis* with cyanobacterium *Spirulina platensis* (Xue et al., 2010). In these cases, lipid production and algae growth are reported to be greater than in pure cultures. Similarly, cocultivation of filamentous fungus *Aspergillus fumigatus* and microalgae has also been reported with increased biomass production and lipid yield (Wrede et al., 2014). The filamentous fungi can support the flocculation and harvesting of microalgal biomass as they act as bioflocculating agents. On the other hand, in the cocultivation of oleaginous yeast and microalgae, organic carbon raw materials are required as a source of energy for yeast to produce biofuel products. *Synechococcus elongatus*, a phototrophic sucrose secreting cyanobacteria was engineered and cocultured with oleaginous yeast strain *Cryptococcus curvatus* or *R. glutinis* which is capable of producing biofuel products (Li et al., 2017). In this case, *S. elongatus* uses CO_2 and sunlight to produce sucrose, a carbon source for yeasts, and in return yeasts assist cyanobacterial growth and survival by eliminating oxidative stress. It was found that the lipid production of oleaginous yeast strains increased in the coculture and could be engineered to produce biofuels. This artificial cocultivation system provides a potential sustainable platform for the direct use of sunlight and carbon dioxide to produce biofuels. In addition, an algae-bacteria-archaea complex has been developed for the production of oily mixtures under anaerobic, thermophilic and atmospheric conditions. Under anaerobic conditions, the thermostable bacteria *Thermosipho globiformans* and archaeon *Methanocaldococcus jannaschii* were cocultured with different types of microalgal species at 68°C, and then pyrolyzed at 300°C. The consortium produces n-alkane rich biofuels and isoprenoids (Yamane et al., 2013). It was found that the composition and quantity of n-alkanes produced by pyrolysis were closely related to the lipid content and composition of microalgae.

12.4.3 For enzyme production

Enzymes are biological catalysts and play an important role at different stages of metabolic and biochemical activities. Microbial enzymes are considered to be one of the best enzymes obtained from different microbial populations, especially for industrial applications. Enzymes obtained from microbial sources have been used in different commercial processes. Different microorganisms such as bacteria, fungi, and yeast have been extensively studied for the biosynthesis of various enzymes for commercial purposes (Pandey et al., 1999). For example, a commercially available protease is alkaline and is mainly produced by bacteria such as *Pseudomonas*, *Bacillus*, and *Clostridium*, and certain fungi are also reported to produce this

Table 12.4 Microorganisms involved in enzyme production.

Enzymes	Microorganisms	References
Protease	*Bacilli, Pseudomonas, Clostridium, Rhizopus, Penicillium, Aspergillus*	Mukherjee et al. (2008), Nigam and Pandey (2009), Chudasama et al. (2010)
Keratinase	Bacteria, Actinomycetes, Fungi	Pandey et al. (1999), Nigam and Pandey (2009), Indhuja et al. (2012), Gupta et al. (2013), Baihong et al. (2013), Liu et al. (2013a,b)
Amylase	*Bacillus* sp., *Geobacillus*	Nigam and Pandey (2009), Gurumurthy and Neelagund (2012), Li et al. (2013a,b)
Xylanase	*Thermoactinomyces thalophilus, Bacillus* sp., *Humicola insolens, Bispora* sp.	Pandey et al. (1999), Nigam and Pandey (2009), Mamo et al. (2009), Luo et al. (2012), Jhamb and Sahoo (2012), Du et al. (2013)
Ligninase	*Steccherinum ochraceum, Polyporus versicolor, Panus tigrinus*	Pandey et al. (1999), Nigam and Pandey (2009), Mishra and Kumar (2009), Wongwilaiwalin et al. (2010), Kumar and Srikumar (2012)
Cellulase	*Polyporus* sp., *Pleurotus* sp., *Trichoderma* sp., *Aspergillus* sp.	Nigam and Prabhu (1988), Pandey et al. (1999), Nigam and Pandey (2009), Hardiman et al. (2010)
Lipase	*Candida* sp., *Aspergillus* sp., *Penicillium* sp., *Rhizopus, Mucor*	Pandey et al. (1999), Muralidhar et al. (2001), Reddivari et al. (2002), Nigam and Pandey (2009)

enzyme (Kumar and Takagi, 1999). Similarly, fungal species of genera *Trichoderma, Penicillium,* and *Aspergillus* synthesize xylanases that have important applications in the bioindustry. It was found that the xylanases produced by these microorganisms have high activity in a wide temperature range (40°C–60°C) (Ahmed et al., 2009). Some the enzymes with their producing microorganisms are summarized in Table 12.4.

12.5 Conclusion

The rapidly increasing population has created challenges for human livelihood. Shortages in food and other necessities have led us to other alternatives and microbial biomass as a possible solutions to fulfill these needs. The microbial biomass of bacteria, fungi, yeast, and algae are rich sources of proteins, minerals, vitamins, and essential fatty acids. These microorganisms can be used to replace a valuable amount of conventional vegetables and animal protein feed. The microbes are able to grow on waste materials and be used as substrate for the production of proteins and other useful products. SCP is one of the natural sources of protein produced by various microbial biomass. This microbial protein may be a sustainable future protein source to fulfill the requirements of human and animal food supplements. Other useful byproducts such as biofuels and enzymes are also significantly produced by different microbes to fulfill demands. Looking over the various important properties of microbes and their capability to grow on waste materials successfully and producing high amount of quality products, green clean technology is needed for the large-scale production in order to meet the needs of the future within a sustainable environment.

References

Adedayo, M.R., Ajiboye, E.A., Akintunde, J.K., Odaibo, A., 2011. SCP: as nutritional enhancer. J. Microbiol. 2 (5), 396–409.

Adoki, A., 2002. Culture characteristics of *Candida* sp. in waste conversion. J. Appl. Sci. 6, 49–58.

Aggelopoulos, T., Katsieris, K., Bekatorou, A., Pandey, A., Banat, I.M., Koutinas, A.A., 2014. Solid state fermentation of food waste mixtures for single cell protein, aroma volatiles and fat production. Food Chem. 15, 710–716.

Ahmadi, A.R., Ghoorchian, H., Hajihosaini, R., Khanifar, J., 2010. Determination of the amount of protein and amino acids extracted from the microbial protein (SCP) of lignocellulosic wastes. Pak. J. Biol. Sci. 13, 355–361.

Ahmed, S., Riaz, S., Jamil, A., 2009. Molecular cloning of fungal xylanases: an overview. Appl. Microbiol. Biotechnol. 84, 19–35.

Anupama, R., Ravindra, P., 2000. Value added food: single cell protein. Biotechnol. Adv. J. Microbiol. 18, 459–479.

Apandi, N.M., Mohamed, R.M.S.R., Latiffi, N.A.A., Rozlan, N.F.M., AlGheethi, A.A.S., 2017. Protein and lipid content of microalgae *Scenedesmus* sp. biomass grown in wet market wastewater. MATEC Web Conf. 103, 06011.

Arora, D., Mukerji, K., Marth, E., 1991. Single cell protein in hand book of applied mycology. J. Am. Sci. 18, 499–539.

Baihong, L., Juan, Z., Zhen, F., Lei, G., Xiangru, L., Guocheng, D., et al., 2013. Enhanced thermostability of keratinase by computational design and empirical mutation. J. Ind. Microbiol. Biotechnol. 40, 697–704.

Baldensperger, J., Le Mer, J., Hannibal, L., Quinto, P.J., 1985. Solid state fermentation of banana wastes. Biotechnol. Lett. 7, 743–748.

Bamberg, J., 2000. British Petroleum and Global Oil 1950–1975: The Challenge of Nationalism. Cambridge University Press, Cambridge.

Beopoulos, A., Cescut, J., Haddouche, R., Uribelarrea, J.L., Molina-Jouve, C., Nicaud, J.M., 2009. *Yarrowia lipolytica* as a model for bio-oil production. Prog. Lipid Res. 48, 375–387.

Bhalla, T.C., Joshi, M., 1994. Protein enrichment of apple pomace by coculture of cellulolytic moulds and yeasts. World J. Microbiol. Biotechnol. 10, 116–117.

Bhalla, T.C., Sharma, N.N., Sharma, M., 2007. Production of metabolites, industrial enzymes, amino acids, organic acids, antibiotics, vitamins and single cell proteins. J. Environ. (6), 34–78.

Bozakouk, A.H., 2002. Acid hydrolysis of *Phragmites austral*: is powder for production of single cell protein by *Candida utilis*. J. Res. 98, 876–897.

Burgents, J.E., Burnett, K.G., Burnett, L.E., 2004. Disease resistance of Pacific white shrimp, *Litopenaeus vannamei*, following the dietary administration of a yeast culture food supplement. Aquacult. J. Microbiol. 231, 1–8.

Cheirsilp, B., Suwannarat, W., Niyomdecha, R., 2011. Mixed culture of oleaginous yeast *Rhodotorula glutinis* and microalga *Chlorella vulgaris* for lipid production from industrial wastes and its use as biodiesel feedstock. N. Biotechnol. 28, 362–368.

Chiou, P.W.S., Chiu, S.W., Chen, C.R., 2001. Value of *Aspergillus niger* fermentation product as a dietary ingredient for broiler chickens. Anim. Feed Sci. Technol. 91, 171–182.

Chudasama, C.J., Jani, S.A., Jajda, H.M., Pate, H.N., 2010. Optimization and production of alkaline protease from *Bacillus thuringiensis* CC7. J. Cell Tissue Res. 10, 2257–2262.

Chung, D., Cha, M., Guss, A.M., Westpheling, J., 2014. Direct conversion of plant biomass to ethanol by engineered *Caldicellulosiruptor bescii*. Proc. Natl. Acad. Sci. U. S. A. 111, 8931–8936.

Cui, W., Wang, Q., Zhang, F., Zhang, S.C., Chi, Z.M., Madzak, C., 2011. Direct conversion of inulin into single cell protein by the engineered *Yarrowia lipolytica* carrying inulinase gene. Process. Biochem. 46, 1442–1448.

Davies, F.K., Work, V.H., Beliaev, A.S., Posewitz, M.C., 2014. Engineering limonene and bisabolene production in wild type and a glycogen-deficient mutant of *Synechococcus* sp. PCC 7002. Front. Bioeng. Biotechnol. 2, 21.

De Gregorio, A., Mandalari, G., Arena, N., Nucita, F., Tripodo, M.M., Lo Curto, R.B., 2002. SCP and crude pectinase production by slurry-state fermentation of lemon pulps. Bioresour. Technol. 83, 89–94.

Deeba, F., Pruthi, V., Negi, Y.S., 2016. Converting paper mill sludge into neutral lipids by oleaginous yeast *Cryptococcus vishniaccii* for biodiesel production. Bioresour. Technol. 213, 96–102.

Dhanasekaran, D., Lawanya, S.S., Saha, N.T., Panneerselvam, A., 2011. Production of single cell protein from pineapple waste using yeast. Innov. Rom. Food Biotechnol. 8, 26–32.

Du, Y., Shi, P., Huang, H., Zhang, X., Luo, H., Wang, Y., et al., 2013. Characterization of three novel thermophilic xylanases from *Humicola insolens* Y1 with application potentials in the brewing industry. Bioresour. Technol. 130, 161–167.

Duong, V.T., Ahmed, F., Thomas-Hall, S.R., Quigley, S., Nowak, E., Schenk, P.M., 2015. High protein- and high lipid-producing microalgae from northern Australia as potential feedstock for animal feed and biodiesel. Front. Bioeng. Biotechnol. 3, 53.

Esabi, B.K., 2001. Production of single cell protein from ram horn hydrolysate. Turk. J. Biol. 25, 371–377.

Gao, Y., Ki, D., Liu, Y., 2012. Production of single cell protein from soy molasses using *Candida tropicalis*. Ann. Microbiol. 62, 1165–1172.

Ghirardi, M.L., Zhang, L., Lee, J.W., Flynn, T., Seibert, M., Greenbaum, E., et al., 2000. Microalgae: a green source of renewable H2. Trends Biotechnol. 18, 506–551.

Gomashe, A.V., Pounikar, M.A., Gulhane, P.A., 2014. Liquid whey: a potential substrate for single cell protein production from *Bacillus subtilis* NCIM 2010. Int. J. Life Sci. 2 (2), 119–123.

Gupta, R., Sharma, R., Beg, Q.K., 2013. Revisiting microbial keratinases: next generation proteases for sustainable biotechnology. Crit. Rev. Biotechnol. 33, 216–228.

Gurumurthy, D.M., Neelagund, S.E., 2012. Molecular characterization of industrially viable extreme thermostable novel alpha-amylase of *Geobacillus* sp. Iso5 isolated from geothermal spring. J. Pure Appl. Microbiol. 6, 1759–1773.

Haider, M., AL-Barhawi, R., Hassan, H., 1989. Educ. Sci. 7, 123–135.

Hardiman, E., Gibbs, M., Reeves, R., Bergquist, P., 2010. Directed evolution of a thermophilic beta-glucosidase for cellulosic bioethanol production. Appl. Biochem. Biotechnol. 161, 301–312.

Hashem, M., Hesham, A.E.-L., Alamri, S.A., Alrumman, S.A., 2014. Production of single-cell protein from wasted date fruits by *Hanseniaspora uvarum* KKUY-0084 and *Zygosaccharomyces rouxii* KKUY-0157. Ann. Microbiol. 64, 1505–1511.

Hernández, D., Solana, M., Riaño, B., García-González, M.C., Bertucco, A., 2014. Biofuels from microalgae: lipid extraction and methane production from the residual biomass in a biorefinery approach. Bioresour. Technol. 170, 370–378.

Hirokawa, Y., Maki, Y., Tatsuke, T., Hanai, T., 2016. Cyanobacterial production of 1,3-propanediol directly from carbon dioxide using a synthetic metabolic pathway. Metab. Eng. 34, 97–103.

Huang, J., Chen, D., Wei, Y., Wang, Q., Li, Z., Chen, Y., et al., 2014. Direct ethanol production from lignocellulosic sugars and sugarcane bagasse by a recombinant *Trichoderma reesei* strain. Sci. World J. 2014, 798683.

Indhuja, S., Shiburaj, S., Pradeep, N.S., Thankamani, V., Abraham, T.K., 2012. Extracellular keratinolytic proteases from an alkalophilic *Streptomyces albidoflavus* TBG-S13A5: enhanced production and characterization. J. Pure Appl. Microbiol. 6, 1599–1607.

Ivarson, K.C., Morita, H., 1982. Single-cell protein production by the acid-tolerant fungus *Scytalidium acidophilum* from acid hydrolysates of waste paper. Appl. J. Environ. Microbiol. 43, 643–647.

Jalasutram, V., Kataram, S., Gandu, B., Anupoju, G.R., 2013. Single cell protein production from digested and undigested poultry litter by *Candida utilis*: optimization of process parameters using response surface methodology. Clean Technol. Environ. Policy 15, 265–273.

Jhamb, K., Sahoo, D.K., 2012. Production of soluble recombinant proteins in *Escherichia coli*: effects of process conditions and chaperone co-expression on cell growth and production of xylanase. Bioresour. Technol. 123, 135–143.

Johnson, E.A., 2013. Biotechnology of non-*Saccharomyces* yeasts – the ascomycetes. Appl. Microbiol. Biotechnol. 97, 503–517.

Kam, S., Abedian Kenari, A., Younesi, H., 2012. Production of single cell protein in stick-water by *Lactobacillus acidophilus* and *Aspergillus niger*. J. Aquat. Food Prod. Technol. 21, 403–417.

Kornochalert, N., Kantachote, D., Chaiprapat, S., Techkarnjanaruk, S., 2014. Use of *Rhodopseudomonas palustris* P1 stimulated growth by fermented pineapple extract to treat latex rubber sheet wastewater to obtain single cell protein. Ann. Microbiol. 64, 1021–1032.

Kremer, T.A., LaSarre, B., Posto, A.L., McKinlay, J.B., 2015. N_2 gas is an effective fertilizer for bioethanol production by *Zymomonas mobilis*. Proc. Natl. Acad. Sci. U. S. A. 112, 2222–2226.

Kumar, C.G., Takagi, H., 1999. Microbial alkaline proteases: from a bioindustrial viewpoint. Biotechnol. Adv. 17, 561–594.

Kumar, G.N., Srikumar, K., 2012. Characterization of xerophytic thermophilic laccase exhibiting metal ion-dependent dye decolorization potential. Appl. Biochem. Biotechnol. 167, 662–676.

Kurbanoglu, E.B., Algur, O.F., 2002. Single-cell protein production from ram horn hydrolysate by bacteria. Bioresour. Technol. 85, 125–129.

Lee, J.Z., Logan, A., Terry, S., Spear, J.R., 2015. Microbial response to single-cell protein production and brewery wastewater treatment. Microb. Biotechnol. 8, 65–76.

Li, C., Yang, H., Li, Y., Cheng, L., Zhang, M., Zhang, L., et al., 2013a. Novel bioconversions of municipal effluent and CO_2 into protein riched *Chlorella vulgaris* biomass. Bioresour. Technol. 132, 171–177.

Li, Y., Niu, D., Zhang, L., Wang, Z., Shi, G., 2013b. Purification, characterization and cloning of a thermotolerant isoamylase produced from *Bacillus* sp. CICIM 304. J. Ind. Microbiol. Biotechnol. 40, 437–446.

Li, T., Li, C.T., Butler, K., Hays, S.G., Guarnieri, M.T., Oyler, G.A., et al., 2017. Mimicking lichens: incorporation of yeast strains together with sucrose-secreting cyanobacteria improves survival, growth, ROS removal, and lipid production in a stable mutualistic co-culture production platform. Biotechnol. Biofuels 10, 1–11.

Lin, P.P., Mi, L., Morioka, A.H., Yoshino, K.M., Konishi, S., Xu, S.C., et al., 2015. Consolidated bioprocessing of cellulose to isobutanol using *Clostridium thermocellum*. Metab. Eng. 31, 44–52.

Litchfield, H., 1983. Science 219, 740–746.

Liu, B., Song, J., Li, Y., Niu, J., Wang, Z., Yang, Q., 2013a. Towards industrially feasible treatment of potato starch processing waste by mixed cultures. Appl. Biochem. Biotechnol. 171, 1001–1010.

Liu, B., Zhang, J., Li, B., Liao, X., Du, G., Chen, J., 2013b. Expression and characterization of extreme alkaline, oxidation-resistant keratinase from *Bacillus licheniformis* in recombinant *Bacillus subtilis* WB600 expression system and its application in wool fiber processing. World J. Microbiol. Biotechnol. 29, 825–832.

Liu, B., Li, Y., Song, J., Zhang, L., Dong, J., Yang, Q., 2014. Production of single-cell protein with two-step fermentation for treatment of potato starch processing waste. Cellulose 21, 3637–3645.

Luo, H., Wang, K., Huang, H., Shi, P., Yang, P., Yao, B., 2012. Gene cloning, expression, and biochemical characterization of an alkali-tolerant β-mannanase from *Humicola insolens* Y1. J. Ind. Microbiol. Biotechnol. 39, 547–555.

Lütke-Eversloh, T., Bahl, H., 2011. Metabolic engineering of *Clostridium acetobutylicum*: recent advances to improve butanol production. Curr. Opin. Biotechnol. 22, 634–647.

Mahasneh, I.A., 2005. Production of SCP from five strains of microalgae species. Biotechnol. Bioeng. J. 90, 153–161.

Mamo, G., Thunnissen, M., Hatti-Kaul, R., Mattiasson, B., 2009. An alkaline active xylanase: insights into mechanisms of high pH catalytic adaptation. Biochimie 91, 1187–1196.

Martins, A., Caetano, N.S., Mata, T.M., 2010. Microalgae for biodiesel production and other applications: a review. Renew. Sustain. Energy Rev. 14, 217–232.

Mishra, A., Kumar, S., 2009. Kinetic studies of laccase enzyme of *Coriolus versicolor* MTCC 138 in an inexpensive culture medium. Biochem. Eng. J. 46, 252–256.

Mukherjee, A.K., Adhikari, H., Rai, S.K., 2008. Production of alkaline protease by a thermophilic *Bacillus subtilis* under solid-state fermentation (SSF) condition using *Imperata* cylindrical grass and potato peel as low-cost medium: characterization and application of enzyme in detergent formulation. J. Biochem. Eng. 39, 353–361.

Muralidhar, R., Chirumamilla, R.R., Marchant, R., Nigam, P.A., 2001. Response surface approach for the comparison of lipase production by *Candida cylindraceae* using two different carbon sources. Biochem. Eng. J. 9, 17–23.

Nielsen, D.R., Leonard, E., Yoon, S.H., Tseng, H.C., Yuan, C., Prather, K.L.J., 2009. Engineering alternative butanol production platforms in heterologous bacteria. Metab. Eng. 11, 262–273.

Nigam, P., Pandey, A. (Eds.), 2009. Biotechnology for Agro-Industrial Residues Utilisation. Publisher Springer Science Business Media B.V, pp. 1–466.

Nigam, P., Prabhu, K.A., 1988. Isolation and recovery of cellulase and ligninase from crude enzymes produced by two basidiomycetes cultures in submerged bagasse fermentation. Sharkara 27, 40–46.

Overland, M., Tauson, A.H., Shearer, K., Skrede, A., 2010. Evaluation of methane-utilising bacteria products as feed ingredients for monogastric animals. Arch. Anim. Nutr. 64, 171–189.

Pandey, A., Selvakumar, P., Soccol, C.R., Nigam, P., 1999. Solid-state fermentation for the production of industrial enzymes. Curr. Sci. 77, 149–162.

Paraskevopoulou, A., Athanasiadis, I., Kanellaki, M., Bekatourou, A., Blekas, G., Kiosseoglou, V., 2003. Functional properties of sigle cell protein produced by kefir microflora. Food Res. Int. 36, 431–438.

Pessoa Jr., A., Mancilha, I.M., Sato, S., 1996. Cultivation of *Candida tropicalis* in sugar cane hemicellulosic hydrolyzate for microbial protein production. J. Biotechnol. 51, 83–88.

Piper, S., 2004. Continuous Cultures of Methylococcus capsulatus. Center of Microbial Biotechnology (Biocentrum) - Technical University of Denmark, 2004. Master's thesis.

Raja, R., Kumar, N.A., Sridhar, S., 2008. A perspective on biotechnological potential of microalgae. Cr. Revis. Microbiol. 34, 77–88.

Ravinder, R., Venkateshwar, R.L., Ravindra, P., 2003. Studies on *Aspergillus oryzae* mutants for the production of single cell proteins from deoiled rice bran. Food Technol. Biotechnol. 41, 243–246.

Reddivari, M., Chirumamilla, R., Nigam, P., 2002. Understanding lipase stereoselectivity. World J. Microbiol. Biotechnol. 18, 81–97.

Ritala, A., Häkkinen, S.T., Toivari, M., Wiebe, M.G., 2017. Single cell protein-state-of-the-art, industrial landscape and patents 2001–2016. Front. Microbiol. 8, 2009.

Rodríguez-Zavala, J.S., Ortiz-Cruz, M.A., Mendoza-Hernández, G., Moreno-Sánchez, R., 2010. Increased synthesis of α-tocopherol, paramylon and tyrosine by *Euglena gracilis* under conditions of high biomass production. J. Appl. Microbiol. 109, 2160–2172.

Romero-García, J.M., Martínez-Pati-o, C., Ruiz, E., Romero, I., Castro, E., 2016. Ethanol production from olive stone hydrolysates by xylose fermenting microorganisms. Bioethanol 2, 51–65.

Rudravaram, R., Chandel, A.K., Rao, L.V., Hui, Y.Z., Ravindra, P., 2009. Bio (single cell) protein: issues of production, toxins and commercialisation status. In: Ashworth, G.S., Azevedo, P. (Eds.), Agricultural Wastes, 1. Hauppage, New York, pp. 29–153.

Safafar, H., Norregaard, P.U., Ljubic, A., Møller, P., Holdt, S.L., Jacobsen, C., 2016. Enhancement of protein and pigment content in two *Chlorella* species cultivated on industrial process water. J. Mar. Sci. Eng. 4, 84.

Sakine, Y., Fatma, O., Berrin, G., Suzan, Y., 2009. The effect of dietary dried baker's yeast on the performance, egg traits and blood parameters in laying quails. Trop. Anim. Health Prod. 41, 5–10.

Sengupta, S., Bhowal, J., Bhattacharya, U., 2006. The Association of Official Analytical Chemists. The Official Methods of Analysis of AOAC International, 18th ed. J. Environ. Issues, Arlington, TX, 6, pp. 99–126.

Shen, C.R., Lan, E.I., Dekishima, Y., Baez, A., Cho, K.M., Liao, J.C., 2011. Driving forces enable high-titer anaerobic 1-butanol synthesis in *Escherichia coli*. Appl. Environ. Microbiol. 77, 2905–2915.

Sisman, T., Özlem, G., Dogan, N., Özdal, M., Algur, Ö.F., Ergon, T., 2013. Single-cell protein as an alternative food for zebrafish, *Danio rerio*: a toxicological assessment. Toxicol. Ind. Health 29, 792–799.

Tanveer, A., 2010. Production of single cell protein from *Saccharomyces cerevisiae* by utilizing fruit wastes. J. Environ. Issues 1 (2), 127–132.

Taran, M., Asadi, N., 2014. A novel approach for environmentally friendly production of single cell protein from petrochemical wastewater using a halophilic microorganism in different conditions. Pet. Sci. Technol. 32, 625–630.

Valentino, M.J.G., Ganado, L.S., Undan, J.R., 2016. Single cell protein potential of endophytic fungi associated with bamboo using rice bran as substrate. Adv. Appl. Sci. Res. 7, 68–72.

Wang, J.P., Kim, J.D., Kim, J.E., Kim, I.H., 2013. Amino acid digestibility of single cell protein from *Corynebacterium ammoniagenes* in growing pigs. Anim. Feed Sci. Technol. 180, 111–114.

Westfall, P.J., Pitera, D.J., Lenihan, J.R., Eng, D., Woolard, F.X., Regentin, R., et al., 2012. Production of amorphadiene in yeast, and its conversion to dihydroartemisinic acid, precursor to the antimalarial agent artemisinin. Proc. Natl. Acad. Sci. U. S. A. 109, 111–118.

Wiebe, M.G., 2002. Myco-protein from *Fusarium venenatum*: a well-established product for human consumption. Appl. Microbiol. Biotechnol. 58, 421–427.

Wongputtisin, P., Khanongnuch, C., Khongbantad, W., Niamsup, P., Lumyong, S., 2012. Screening and selection of *Bacillus* spp. for fermented corticated soybean meal production. J. Appl. Micobiol. 113, 798–806.

Wongputtisin, P., Khanongnuch, C., Kongbuntad, W., Niamsup, P., Lumyong, S., Sarkar, P.K., 2014. Use of *Bacillus subtilis* isolates from Tua-nao towards nutritional improvement of soya bean hull for monogastric feed application. Lett. Appl. Microbiol. 59, 328–333.

Wongwilaiwalin, S., Rattanachomsri, U., Laothanachareon, T., Eurwilaichitr, L., Igarashi, Y., Champreda, V., 2010. Analysis of a thermophilic lignocellulose degrading microbial consortium and multi-species lignocellulolytic enzyme system. Enzyme. Microb. Technol. 47, 283–290.

Wrede, D., Taha, M., Miranda, A.F., Kadali, K., Stevenson, T., Ball, A.S., et al., 2014. Co-cultivation of fungal and microalgal cells as an effcient system for harvesting microalgal cells, lipid production and wastewater treatment. PLoS One 9, 1–22.

Xue, F., Miao, J., Zhang, X., Tan, T., 2010. A new strategy for lipid production by mix cultivation of *Spirulina platensis* and *Rhodotorula glutinis*. Appl. Biochem. Biotechnol. 160, 498–503.

Yadav, J.S.S., Bezawada, J., Ajila, C.M., Yan, S., Tyagi, R.D., Surampalli, R.Y., 2014. Mixed culture of *Kluyveromyces marxianus* and *Candida krusei* for single-cell protein production and organic load removal from whey. Bioresour. Technol. 164, 119–127.

Yadav, J.S.S., Yan, S., Ajila, C.M., Bezawada, J., Tyagi, R., Surampalli, R.Y., 2016. Food-grade single-cell protein production, characterization and ultrafiltration recovery of residual fermented whey proteins from whey. Food Bioprod. Process. 99, 156–165.

Yamane, K., Matsuyama, S., Igarashi, K., Utsumi, M., Shiraiwa, Y., Kuwabara, T., 2013. Anaerobic coculture of microalgae with *Thermosipho globiformans* and *Methanocaldococcus jannaschii* at 68°C enhances generation of n-alkane-rich biofuels after pyrolysis. Appl. Environ. Microbiol. 79, 924–930.

Yang, S., Mohagheghi, A., Franden, M.A., Chou, Y.C., Chen, X., Dowe, N., et al., 2016. Metabolic engineering of *Zymomonas mobilis* for 2,3-butanediol production from lignocellulosic biomass sugars. Biotechnol. Biofuels 9, 189.

Yazdian, F., Hajizadeh, S., Shojaosadati, S.A., Khalilzadeh, R., Jahanshahi, M., Nosrati, M., 2005. Production of single cell protein from natural gas: parameter optimization and RNA evaluation. Iran. J. Biotech. 3, 235–242.

Yousuf, M.K., 2012. To determine protein content of single cell protein produced by using various combinations of fruit wastes in the production of SCP by using two standard food fungi *Aspergillus oryzae* and *Rhizopus oligospora*. Int. J. Adv. Biotechnol. Res. 3 (1), 533–536.

Yu, A.Q., Juwono, N.K.P., Foo, J.L., Leong, S.S.J., Chang, M.W., 2016. Metabolic engineering of *Saccharomyces cerevisiae* for the overproduction of short branched-chain fatty acids. Metab. Eng. 34, 36–43.

Zhao, G., Zhang, W., Zhang, G., 2010. Production of single cell protein using waste capsicum powder produced during capsanthin extraction. Lett. Appl. Microbiol. 50, 187–191.

Zinjarde, S., 2014. Food-related applications of *Yarrowia lipolytica*. Food Chem. 152, 1–10.

Participation and understanding of plant microbes interaction in plant health and growth by combating mercury stress: a sustainable approach towards agricultural practices

13

Shalini Singh and Vipin Kumar
Laboratory of Applied Microbiology, Department of Environmental Science and Engineering, Indian Institute of Technology (Indian School of Mines), Dhanbad, India

13.1 Introduction

In the current scenario, heavy metal pollution has become one of the most concernsable issue worldwide. Among all heavy metals, mercury is one of the most toxic heavy metals with no favorable functions for living beings. It is one of the most concerned environmental pollutants due to its bioaccumulative, persistent nature, high mobility, easily biomagnified (organic form) by biota and its transboundary movement (Kumari, 2011). It causes significant problems to human beings as well as the environment. Soil acts as a sink and source for global biogeochemical cycle of mercury as it acts as a barrier for transfer of mercury to biota, atmosphere, and hydrosphere (Yin et al., 2013). Also, biogeochemical cycle of mercury in the environment heavily relies upon regional environment factors Fig. 13.1. Biogeochemical cycle of mercury leads to the set out of highly toxic mercury compounds like methylated mercury through biotic (sulfur and iron reducing bacteria) and abiotic sources (sunlight). Once formed, methyl mercury (Me-Hg) has the ability to bioaccumalate in the food chain mainly via aquatic environment and reach to the top of predator. It can bind with sulfhydryl groups in protein of the living beings. Sulfur (SRB) and iron reducing bacteria (FeRB), methanogens and firmicutes comes under the category of methylating microbes (anaerobes) (Singh and Kumar, 2019). Application of mercury in various industries has been progressively reduced. But still many industries like artisanal mining, pesticide application, gold mining, wastewater irrigation, fossil fuel combustion, atmospheric deposition, battery, restoration process, erosion, chemical production, and medical equipment are using mercury knowingly after being aware of its toxic effects on human health and environment.

Natural sources of mercury are volcanic eruption, forest fire, biomass burning and low temperature volatilization Fig. 13.2. Both anthropogenic and natural activities

Microbes in Land Use Change Management. DOI: https://doi.org/10.1016/B978-0-12-824448-7.00016-4
© 2021 Elsevier Inc. All rights reserved.

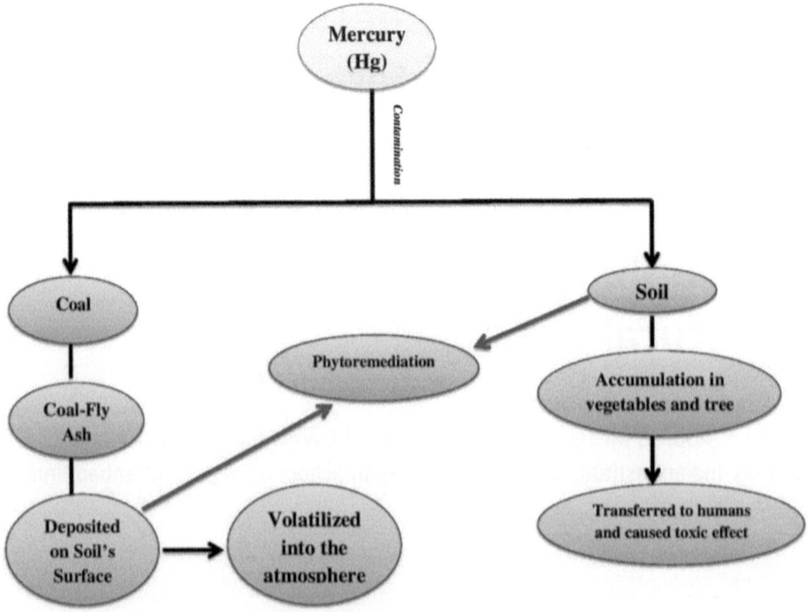

Figure 13.1 Flow chart of mercury contamination in coal and soil.

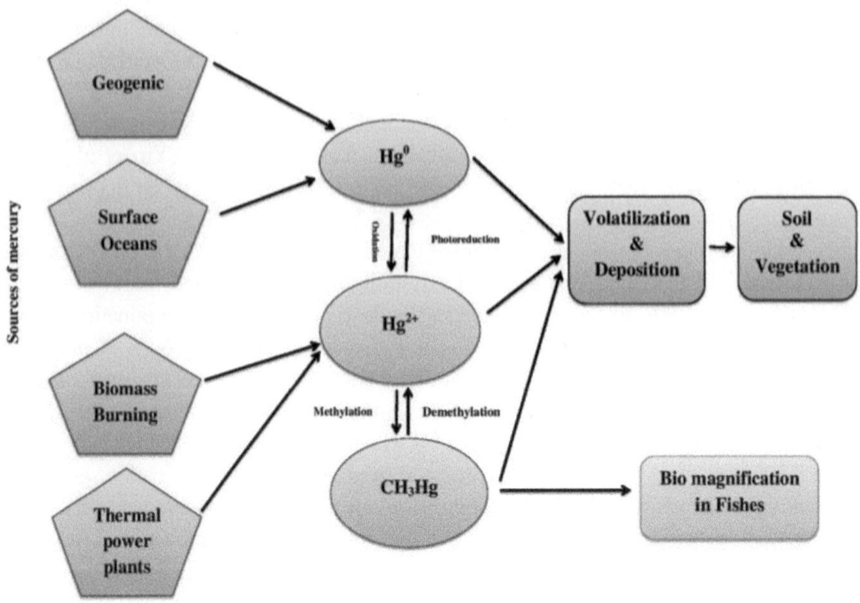

Figure 13.2 Mercury cycle in the environment.

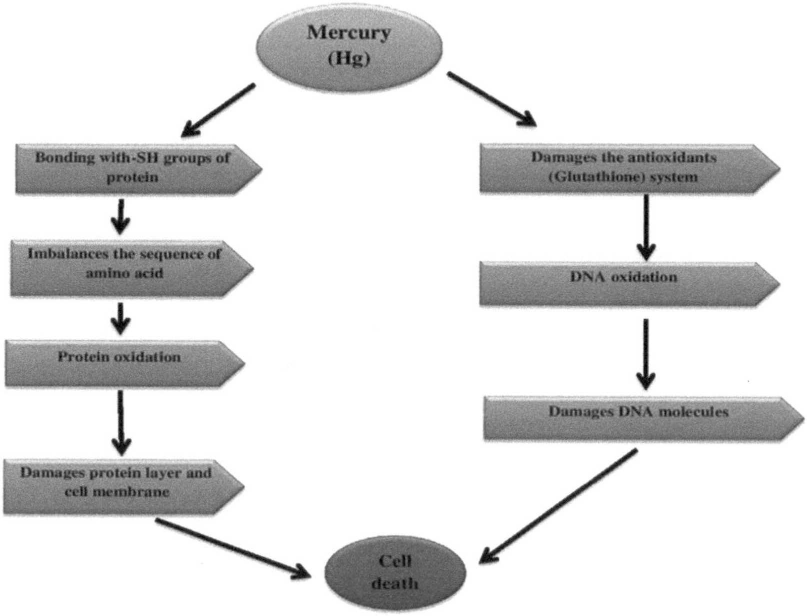

Figure 13.3 Mechanism of Hg toxicity in living beings.

release mercury directly into the atmosphere, promote the accumulation in the aquatic animals. Mercury compound mainly exists three in major forms, that is, inorganic (mercury chloride), organic (methyl mercury) and element mercury (liquid/gaseous or only liquid at room temperature) capable of traveling across regional and global distances. The conversion of mercury from one state to other induces its scattering and subsequent toxic effects on aquatic system and biosphere. In literature, it is clearly mentioned that in a wetland environment, microbial methylation elevates the production of methyl mercury form inorganic mercury (Windham-Myers et al., 2014). Methyl mercury is known as the most potent toxic form because of its ability to penetrate or bioconcentrate into the food chain easily via dietary consumption. However, other forms of mercury also cause harmful health effect depending upon dose and route of exposure (Hsu-Kim et al., 2013). Minamata disaster (Japan, 1956) is one of the lighting examples of mercury accumulation in aquatic system. Major diseases caused by mercury are Parkinson's disease, Alzheimer's disease, atherosclerosis, and cancer. Concentration of all the heavy metals in soil and plants is given in Fig. 13.3.

13.2 Biogeochemical cycle of mercury

Mercury cycle in the environment is facilitated by abiotic and biotic factors between aquatic system, land, and atmosphere. It mainly exits in three forms, that is, organic, inorganic, and elemental or gaseous form. Gaseous form of mercury can stay in the atmosphere for a prolonged period (lifetime 0.5–1 years), it represents about 95% of atmospheric mercury. Gaseous form (Hg^0) of mercury is released from oceans, lands, and from geological sources naturally. Anthropogenic sources may lead to the release of the gaseous form of mercury with two other different forms, that is, inorganic mercury and mercury associated particulate matters. The gaseous form of mercury further is divided into three different compounds, that is, gaseous oxidized mercury, gaseous elemental mercury, and particle bound mercury. The sum of all three compounds is known as total atmospheric mercury. In atmosphere, particle bound mercury exits in solid state. Gaseous oxidized mercury represents 95% of total atmospheric mercury, and is considered the most abundant compound (Esbrí et al., 2018).

Inorganic and particulate forms of mercury are more soluble than the gaseous form so these are predominant forms accumulated in the ecosystem by dry and wet deposition. The life time of gaseous mercury is much higher than that of inorganic and particulate mercury (a few weeks only).

13.3 Harmful impact of mercury

13.3.1 Living beings

There are no adverse effects of mercury when it is in Hg^0 form, but once this monovalent form is being digested by humans, the monovalent form of mercury get oxidized and converted into Hg^{2+} and this divalent form is highly toxic for the human body and can cause imbalance coordination with body functions and health effects such as improper functioning of gastrointestinal tract, kidney, lungs, and brain (De et al., 2014). The redox state is a primary important factor in terms of mercury toxicity, because the level of exposure differs in compound and thus their toxicity level. Mercury mainly affects kidney, brain development, lung, and nervous system in the human body through bloodstream while the gaseous form of mercury after inhalation huge dose of mercury can lead to interstitial pneumonitis and corrosive bronchitis (Plaza et al., 2011; Niane et al., 2019). Acute exposure of mercury may lead to insomnia, anxiety, and mental depression to humans (Raj and Maiti, 2019).

13.3.2 Plants

The high concentration of mercury favors enhanced formation of reactive oxygen species (ROS) which further induce oxidative stress in plant cells. It is mentioned in many reports that mercury exposure leads to oxidative stress in plants. In terms of genotoxicity, Hg ion forms a covalent bond with DNA molecules, and also induces sister chromatid exchanges in plant nuclei. Other harmful effects are membrane structural integrity disruption, mineral nutrient uptake reduction, photosynthesis, and transpiration reduction (Singh and Kumar, 2019). In literature, it has been mentioned that transgenic species are more efficient in mercury removal than non-transgenic species. In plants and small ferns, the common mode of mercury removal is to get accumulated by the plant tissues.

13.3.3 Aquatic species

High concentrations of mercury in birds affect most often their immune, detoxification, and nervous systems but the reproductive system is most sensitive endpoint. Most of the water bodies mainly affected fish by the discharge of mercury direct discharge of industrial wastes, mining wastes, or naturally occurring mercury minerals. Methyl mercury is the most toxic form of mercury formed during the methylation process, and transferred via food chain to human and animals by fish consumption. Semiaquatic animals and birds accumulate mercury from their diet through the consumption of aquatic species having mercury in their body. Cormorant, also known as Piscivorous are a top predator in the aquatic system, have a much higher concentration of mercury due to the capacity to eat 20–25 cm long fishes while crested grebe usually accumulates less mercury due to consumption of small fishes, insects, amphibians, and shellfish.

13.4 Biological approaches for remediation of mercury

The biological method of remediation is very safe and cost effective as per environmental point of view. The use of microorganisms and plants for remediation offers an effective solution as a green technology against degradation of contaminants. Biological remediation is further divided into two categories:

1. Intrinsic remediation
2. Engineered bioremediation

Intrinsic remediation technologies includes the degradation of contaminants with the help of microorganisms with human intervention, suitable for low contaminated zones while engineered bioremediation technologies include the human modification and manipulation in genetic material of microorganism and their environment to promote enhanced remediation potential and their activities, usually suitable for highly contaminated regions.

13.4.1 Phytoremediation of mercury

Working with plants to detox mercury in soil and water has been reported in many times. In this technique plant and rhizospheric microorganism is used for organic pollutant and heavy metal removal. Phytoremediation is cost effective, eco-friendly, and effective means of remediation and restoration especially in the case of heavy metals. Factors affecting phytoremediation include plant type, nutrient ion, valency of heavy metal, soil characteristics, climate condition, and soil-microbial community.

Toxicity of heavy metal elevates the synthesis of ROS by reducing the antioxidants (superoxide dismutase, glutathione, etc.) which defend cells' system. Heavy metals such as mercury, silver, lead, cadmium, chromium, selenium, zinc, gold, uranium, arsenic, and nickel, all these nonessentials do not play any role in plants' growth and promotion, lead to abrupt to metabolic activity in plants cause reduced plant mineral nutrition, photosynthetic activities and essential enzymes, also effect soil quality, crop yield and human health. If such situation continues for long, the

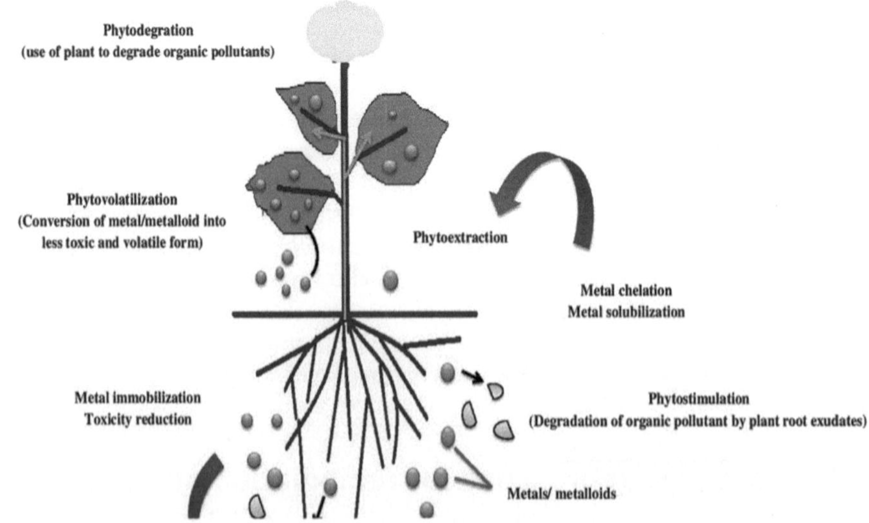

Figure 13.4 Steps involved in heavy metal phytoremediation.

functions of living system is adversely affected and cause cell death. There are many methods which come under phytoremediation mechanism such as phytovolatilization, phytostabilization, and phytoextraction Fig. 13.4.

Phytostabilization helps in mobilization of mercury from the contaminated sited through bioaccumulation and biosorption through the root of the plants or through Hg precipitation, this method averts mercury migration by soil erosion and deflation. In Phytoextraction, translocation and absorption of mercury occurs through plant root and shoot that further could be reaped and burned. However, in phytovolatilization, mercury is absorbed through plant root which further transported through the xylem and finally move into the atmosphere via cellular tissues. Light intensity and air temperature play strong role during the mercury emission from the plant tissues. The major pros of phytovolatilization technique are that without harvest and burnt, removal of mercury is easily achievable. Although there could be some skepticism regarding the safety of this strategy, safety assessment studies on Hg phytovolatilization have indicated that the wide dispersion and dilution in the atmosphere outweighs the potential risks (Moreno et al., 2005).

13.4.2 Hyperaccumulators for remediation of mercury

Hyperaccumulators are the plants with much potential to grow and tolerate high concentration/level of heavy metals, absorb through roots, transfer rapidly to their shoots and accumulate in their leaves and tissues. These hyperaccumulators can grow in contaminated soil and water and shows extraction of metals is much higher about 100–1000 times than non-hyperaccumulator plants. Because of deferential gene regulation and expression in two related plants species, the capacity of

Table 13.1 List of hyperaccumulator plants used for detoxification of mercury.

Heavy metal	Hyperaccumulator plant	Metal conc. (mg/kg)	References
Mercury	*Triticum aestivum L.*	0.14	Rodríguez and Fraga (1999)
	Helianthus tuberosus L.	1.89	Sas-Nowosielska et al. (2008)
	Armoracia lapathifolia	0.97	Sas-Nowosielska et al. (2008)
	Cicer arietinum L.	0.2	Wang et al. (2012)
	Arthraxon hispidus	70 μg/g	Qian et al. (2018)
	Eremochloa ciliaris	12 μg/g	
	Clerodendrum bungei	3.0 μg/g	
	Ixeris sonchifolia	20 μg/g	

hyperaccumulation is higher in one species compared to other species. Now, these days, hyperaccumulators are of interest for the potential to extract heavy metals from the contaminated regions and return back less toxic ecosystem. Higueras et al. (2004) reported in that the *Polypogon monspeliensis* hyperaccumulates 1500 ppm of mercury in their leaves. However, the other tax such as *Asparagus acutifolius, Dittrichia graveolens, Dodonaea viscosa, Helichrysum stoechas, Marrubium vulgare, P. monspeliensis, Rubus ulmifolius, Typha domingensis, and Verbascum sinuatum* were also reported as Hg hyperaccumulator plant species. Israr et al. (2006) observed the accumulation of mercury in root and shoot of *Sesbania drummondii* with higher accumulation of 41,403 mg/kg of mercury in roots while in shoot the accumulation was 998 mg/kg of mercury (Table 13.1).

13.4.3 Plant growth promoting rhizobacteria

Plant growth promoting rhizobacteria (PGPR) are group of bacteria, known for help in the promotion of plant growth rate and also in detoxification of contaminated sites under heavy metal stress condition. It further divided into two types based on coordination with the plant; free-living rhizobacteria (bacteria live outside the plant cells, do not produce nodules) and symbiotic bacteria (live inside the plant cell, give rise to nodules) (Zhuang et al., 2007) Table 13.2. PGPR promote growth of plant under heavy metal stress by producing phytohormones such as indole 3-acetic acid (IAA), also provide defense against pathogens. These bacteria provide nutrient for plant growth and development. Rhizopheric bacterial community can change the properties of metals present in their nearby proximity by synthesizing chelating substances, redox modification and soil acidification. Some major genera of PGPR are *Flavobacterium, Bacillus, Burkholderia, Pseudomonas, Klebsiella, Enterobacter, Azospirillum, Rhizobium, Serratia, Aeromonas, Alcaligenes, Acinetobacter, Variovorax, Arthrobacter*, and *Erwinia*. Strains of various bacteria mentioned in literature have increased elongation in root and shoot and increased the crop yield. Genus of *Pseudomonas* have enhance

Table 13.2 Types of PGPR.

PGPR form	Type of mechanism	References
Biofertilizer	**Direct mechanism**	Kumar et al. (2018)
	Biological nitrogen fixation	Oberson et al. (2013)
	Phosphate solubilization	Rodríguez and Fraga (1999)
Phytostimulator	Production of phytohormones	Glick (2014)
Biopesticide	**Indirect mechanism**	Glick (2014)
	Acquired and induced systemic resistance	Bhattacharyya and Jha (2012)
	Production of antibiotics, siderophores, production of lytic enzymes	

Table 13.3 Concentration of heavy metals in soil and plants (Alloway, 1990; Asad et al., 2019).

Soil			Plants	
Elements	**Normal range (μg/g)**	**Concentration in soil (μg/g)**	**Normal range (μg/g)**	**Concentration in soil (μg/g)**
Mercury	0.002–0.2	0.3–5	0.005–0.5	1–3
Zinc	10–300	70–400	8–400	100–400
Chromium	5–1000	75–1000	0.03–15	5–30
Cadmium	0.01–7	3–8	0.2–0.8	5–30
Lead	2–200	100–400	0.1–10	30–300
Nickel	10–1000	100	0.02–5	10–100
Manganese	200–2000	1500–3000	15–1000	300–500
Selenium	<5	50	0.2–6.8	60
Copper	2–100	60–125	4–15	20–100
Cobalt	1–70	25–50	0.05–0.5	5–30

the root and shoot in tomato, canola and lettuce and increase the crop yield in wheat, rice, sugar, beet, potato, lettuce, citrus, beans, apple and ornamental plants (Hall et al., 1996; Glick et al., 1997; Suslow, 1982; Lemanceau, 1992; Kloepper, 1988, 1994). These groups of bacteria function in three different manners; by release of specific compound for the plants; availing the uptake of specific nutrient and mineral form the nearby environment; decrease or deaden the risk of diseases to plants (Table 13.3).

13.5 Function of PGPR against biotic stress

Usually, plants are prone to various biotic and abiotic stresses, thereby all the plants have developed contrasting defense approaches to plants pathogens. Plants have the ability to discover the attack of pathogens and actuate complex signaling network, leading to induced defenses that confer a more tolerant state. Afterward the induced innate immune mechanism include hormone signaling, phosphorylation events, callose deposition, accumulation of reactive oxygen species, cell wall rigidification, defense hormone signaling and expression of genes encoding pathogen-related proteins. Conventionally, competition for nutrients, niche exclusion, ISR, and allelochemical production are the chief mode of biocontrol activity in PGPR. The allelochemicals synthesized by PGPR includes antibiotics, siderophores, volatile organic compound (VOC), enzymes and phytohormones such as 1-aminocyclopropane-1-carboxylic deaminase enzyme, cytokinnins, indole-3-acetic acid, and Gibberellic acid behaves as elicitors for defense system of plants. Many studies has been carried out in past to know the capacity of rhizobacteria to syntheize siderophores and metabolites contributing to antibiosis. Many bacteria have been reported for siderophores synthesis, among all, *pseudomonads* are reported to have high affinity for ferric ions. Also, many PGPR have the ability to synthesize peptide antibiotics. These oligopeptides discourage the cell wall synthesis of pathogen walls, influence membrane structures of cells, and inhibit the formation of initiation complex on small subunit of ribosomes. Various antibiotics have been identified, including compounds such as amphisin, 2,4-diacetylphloroglucinol, hydrogen cyanide, oomycin A, phenazine, pyoluteorin, pyrrolnitrin, tensin, tropolone, and cyclic lipopeptides produced by *pseudomonads* and oligomycin A, kanosamine, zwittermicin A, and xanthobaccin produced by *Bacillus*, *Streptomyces*, and *Stenotrophomonas* sp. (Ma et al., 2016). More than 12 antibiotics are synthesized by *Bacillus subtilis* strains: bacillomycin, mycobacillin, fungistatin, iturin, phengicin, plipastatin, surfactin, bacilizin, etc. The majority of *Bacillus* sp. antibiotics are active with both gram-positive and gram-negative bacteria (e.g., polymyxin, circulin, and colistin) and pathogenic fungi *Alternaria solani*, *Aspergillus flavus*, *Botryosphaeria ribis*, *Colletotrichum gloeosporioides*, *Fusarium oxysporum*, *Helminthosporium maydis*, and *Phomopsis gossypii* (Maksimov et al., 2011).

13.6 Role of PGPR in phytoremediation of contaminated sites

In the current situation, manmade activities have increased significant quantity of heavy metal into the aquatic and soil environment. Sources of heavy metals are mining activities, coal burning, fossil fuel burning, fertilizer, pesticides, herbicides and sewage sludge. Addition of such large quantity of heavy metal is accountable for reduced and contaminated crop yields, contaminated food chain, alteration in physical and chemical properties of soil and soil microbial interactions. For removal of metals and organic pollutant, physical and methods like thermal treatment, stabilization, encapsulation, electro reclamation, vapor extraction and soil washing is highly costly and not effective as that of biological remediation does. Biological methods have usually low cost, eco-friendly and nondestructive nature, these qualities attract towards biological remediation over physical and chemical remediation.

PGPR facilitate the phytoremediation of polluted sites due to having properties of growth-enhancing mechanism for plants such as phosphate solubilization, nitrogen fixation, siderophore production, hormone production such as IAA and Gibberellins (GA). These metabolites help in the precipitation of heavy metals, as they have ability to bind effectively with the heavy metals, thereby reduce the heavy metal simply by forming complex substances or by forming an operative barrier surrounding the cell (Etesami, 2018). Siderophores is one of the metabolites that can form complexes with heavy metals. These metabolites are synthesized by certain bacteria and fungi under Fe-limiting situations; they have strong affinity for Fe-chelating compounds (Nies, 1999).

13.7 Case studies of PGPR bacteria-assisted detoxification

The PGPR seemed to have successful work profile in plant growth and promotion of nutrition level by enhancing anti oxidative and photosynthetic activities as many studies have successfully completed their experiment in the presence of PGPR and attained desirable result in pilot scale as well as agriculture too. Swain et al. (2007) mentioned the positive role of *B. subtilis* on *Dioscorea rotundata* L. During experiment, the suspension of *B. subtilis* was applied on the surface of *D. rotundata* L, which leads to increase in root, shoot and sprouts due to the production of IAA. However, no such growth was observed in non-inoculated plants. In another case, As-resistant micro bacterium species strain isolated form gold mine worked as phytobial agent for the removal of As, as the strain help in siderophore production (Bahari et al., 2017). Ghosh et al. (2018) conducted an experiment using *Bacillus aryabhattai* which attenuate As induced toxicity which further leads to reduction of As in rice plant. Wu et al. (2006) conducted an experiment with *Brassica juncea* for the removal of lead-zinc. Inoculation of PGPR helps in the plant and alleviates the phytoextraction efficiency. Zaidi et al. (2006) used *B. subtilis* in their experiment for Nickel removal by using *B. juncea* in presence of metal tolerant Rhizobacteria and result showed that the rhizobacteria have the ability to produce IAA and solubilizing inorganic phosphate in culture medium and help in plant growth and Ni biosorption and bioaccumulation. *Kluyvera ascorbata*, a siderophore producing bacterium was used inoculate in the plantation of tomato, canola, and mustard seeds in presence of nickel, lead and zinc and the results indicated that the strains showed protective against inhibitory effects of high metal concentration and promote plant growth and yield. In another case, Vibhuti et al. (2017) conducted an experiment with endophtytic *Bacillus megaterium*, which help in the significant growth promotion of *Arabidopsis thaliana* root and shoot length. This bacterium also had shown great antagonistic activity contrary to varied plant pathogens. In another experiment, Gontia-Mishra et al. (2016) reported the PGPR activity on *Triticum aestivum* by using *Enterobacter ludwigii* and *Klebsiella pneumonia*. Both these strains showed greater growth pattern in *T. aestivum* under mercury stress condition. Significant higher growth was observed under stress condition in shoot and root length, shoot and root dry weight, water content in inoculated plants in comparison to that of uninoculated plant.

13.8 Agriculture and sustainability

Now these day, intense agriculture practices in one of the major issue for an accelerated "greenhouse gases production," as a consequence of that the temperature of earth is increasing day by day, thus disturbing the biosphere stability of Earth. Because of that, numerous stress conditions are triggered by intense agricultural practices, overall damaging on the functionality/ productivity of the ecosystem, obstruct their work, food, and safety. Major accountable stress factors are soil erosion, diseases, plant, invasion, pests, nutritional deficiency, salinity and drought. Besides, the use of agrochemicals also accelerates contamination and effect human health. The entire above mentioned factor surely provokes agricultural production losses, landscape fragmentation, water scarcity, soil erosion, biodiversity damage. In current scenario, sustainable way of agricultural practices is utterly important to meet economic, social and environmental development with the ultimate aim of balance the yield and conserve the biosphere. As stated by Altieri (2004), sustainable development is the outcome of crossing among three elementary components: environment, society, and economy, which in turn interact between each two of them. Hence, the junction in between the environment-society (environmental awareness), society-economy (life standard) and economy-environment (agroecology) certainly establishes the final action of sustainable development. In this high tech world, there are multiple keys that can yield high agriculture yield like applications of transgenic crops, agricultural land management, and massive use of PGPR. In all these, the application of transgenic crops is contentious and arise health risk issue to the public (Singh and Dubey, 2016; Sriprang et al., 2003). Living and microbial uses are the secure and sustainable practices to minimize constraints originated by man-made sources, thereby augmenting in the agricultural yield.

Classification of rhizosphere-associated bacteria depends on their habitat, function and structure. The area that comes close to plants is known as rhizosphere, which delivers beneficial symbiotic plant-microbes interactions. Different forms and mechanism associated with PGPR are universally harnessed to increase agricultural yield (Table 13.1). Any process that directly elevate the growth of plant by supplying nutrients or by accommodating growth regulator is characterized as direct mechanism while the process that defend plants from biotic stress (infection, diseases) or assist plant to grow healthily under abiotic stress (environmental stress) is portrayed as indirect mechanism. Myriad of report have been reported illustrating the function of PGPR in escalating the agricultural yield via recovery of soil nutrients or plant growth regulators.

13.9 Future prospects

Highly toxic heavy metal such as mercury is the major threat for the environment when present in excess quantity. Various anthropogenic and natural sources are the major cause of heavy metal pollution and their increased quantity in the environment. To the focus on remediation, bioremediation for highly toxic contaminant seems to have workable solution to clean up the environment.

The use of PGPR for the heavy metal toxicity reduction with different mechanisms is greener techniques now these days. These bacteria help in the reduction of heavy metals by accumulation of these heavy metals into the plant body, thereby decreasing the metal concentration in nearby contaminated areas. Now these days, scientists have developed different way of PGPR enhanced phytoremediation to conquer over provocation of plant stress in the contaminated regions. By altering metal bioavailability in the affected regions, this method showed an effective application in metal translocation. PGPR assisted remediation of Hg-contaminated soil is greener and highly favorable approach but with some limitation.

On the basis of collected knowledge up to now, here are some future avenues of research approaches:

- More deep research will be required in order to explore the mechanism of heavy metal toxicity on plant biochemistry.
- Comparative studies between plant, fungi, and heavy metal interaction would be an important aspect in conferring tolerance against stress condition.
- Microbiological, genetic, and chemical-assisted PGPR remediation should be performed in systematically combined way to get deeper understanding and mechanism.
- The tolerance and accumulation of mercury through PGPR should be studied thoroughly with major focus would be the role of different genes.
- Remediation time, cost, effect, and harmful impact over environment should be assessed.
- All the earlier performed experiment were mainly carried out in the lab or in specific greenhouse chambers, so the role of remediation over environmental change were not clearly understood, hence field study or in situ experiment is also important to know the impact of PGPR over complex ecosystem.
- Further research will be based on celluloid biological practices which make favorable interaction between root-microbes and leads to improve the output of the productivity of crops and bioenergy crop under both different environment conditions.

Acknowledgment

The authors would like to acknowledge the Department of Environmental Science and Engineering, Indian Institute of Technology (ISM), Dhanbad for availing assistance, encouragement, and needed facilities during construction of this chapter.

References

Alloway, B.J., 1990. Heavy Metals in Soils. Blackie and Son Ltd., Glasgow and London, and John Wiley and Sons, Inc., New York, pp. 125–146.

Altieri, M.A., 2004. Linking ecologists and traditional farmers in the search for sustainable agriculture. Front. Ecol. Environ. 2 (1), 35–42.

Asad, S.A., Farooq, M., Afzal, A., West, H., 2019. Integrated phytobial heavy metal remediation strategies for a sustainable clean environment—a review. Chemosphere 217, 925–941.

Bahari, Z.M., Ibrahim, Z., Jaafar, J., Shahir, S., 2017. Draft genome sequence of arsenic-resistant *Microbacterium* sp. strain SZ1 isolated from arsenic-bearing gold ores. Genome Announc 5 (43).

Bhattacharyya, P.N., Jha, D.K., 2012. Plant growth-promoting rhizobacteria (PGPR): emergence in agriculture. World J. Microbiol. Biotechnol. 28 (4), 1327–1350.

De, J., Dash, H.R., Das, S., 2014. Mercury pollution and bioremediation—a case study on biosorption by a mercury-resistant marine bacterium. Microbial Biodegradation and Bioremediation. Elsevier, pp. 137–166.

Esbrí, J.M., Izquierdo, C., Martínez-Coronado, A., Miteva, I., Higueras, P.L., 2018. Particulate matter and particulate-bound mercury in a heavily polluted site related to ancient mining and metallurgy: a proposal for dry deposition modeling based on micrometeorological conditions. Environ. Sci. Pollut. Res. 25 (35), 35312–35321.

Etesami, H., 2018. Bacterial mediated alleviation of heavy metal stress and decreased accumulation of metals in plant tissues: mechanisms and future prospects. Ecotoxicol. Environ. Saf. 147, 175–191.

Ghosh, P.K., Maiti, T.K., Pramanik, K., Ghosh, S.K., Mitra, S., De, T.K., 2018. The role of arsenic resistant *Bacillus aryabhattai* MCC3374 in promotion of rice seedlings growth and alleviation of arsenic phytotoxicity. Chemosphere 211, 407–419.

Glick, B.R., 2014. Bacteria with ACC deaminase can promote plant growth and help to feed the world. Microbiol. Res. 169 (1), 30–39.

Glick, B.R., Liu, C., Ghosh, S., Dumbroff, E.B., 1997. Early development of canola seedlings in the presence of the plant growth-promoting rhizobacterium *Pseudomonas putida* GR12-2. Soil Biol. Biochem. 29 (8), 1233–1239.

Gontia-Mishra, I., Sapre, S., Sharma, A., Tiwari, S., 2016. Alleviation of mercury toxicity in wheat by the interaction of mercury-tolerant plant growth-promoting rhizobacteria. J. Plant Growth Regul. 35 (4), 1000–1012.

Hall, J.A., Peirson, D., Ghosh, S., Glick, B., 1996. Root elongation in various agronomic crops by the plant growth promoting rhizobacterium *Pseudomonas putida* GR12–2. Isr. J. Plant Sci. 44 (1), 37–42.

Higueras, P.A.B.L.O., Molina, J.A., Oyarzun, R.O.B.E.R.T.O., Lillo, J., Esbrí, J.M., 2004. Identification of the plant-communities and hyperaccumulators in mercury contaminated sectors of the Almadén district, Spain. Mat. Geoenviron. 51, 103–107.

Hsu-Kim, H., Kucharzyk, K.H., Zhang, T., Deshusses, M.A., 2013. Mechanisms regulating mercury bioavailability for methylating microorganisms in the aquatic environment: a critical review. Environ. Sci. Technol. 47 (6), 2441–2456.

Israr, M., Sahi, S., Datta, R., Sarkar, D., 2006. Bioaccumulation and physiological effects of mercury in *Sesbania drummondii*. Chemosphere 65 (4), 591–598.

Kloepper, J.W., 1994. Plant growth-promoting rhizobacteria (other systems) *Azospirillum*. Plant Assoc. 187, 137–166.

Kloepper, J.W., Lifshitz, R., Schroth, M.N., 1988. *Pseudomonas* inoculants to benefit plant production. ISI Atlas Sci. Anim. Plant Sci. 1 (1), 60–64.

Kumar, A., Singh, V.K., Tripathi, V., Singh, P.P., Singh, A.K., 2018. Plant growth-promoting rhizobacteria (PGPR): perspective in agriculture under biotic and abiotic stress. Crop Improvement Through Microbial Biotechnology. Elsevier, pp. 333–342.

Kumari, R., 2011. Preliminary mercury emission estimates from non-ferrous metal smelting in India. Atmos. Pollut. Res. 2 (4), 513–519.

Lemanceau, P., 1992. Beneficial effects of rhizobacteria on plants: example of fluorescent *Pseudomonas* spp. Agronomie 12 (6), 413–437.
Ma, Y., Rajkumar, M., Zhang, C., Freitas, H., 2016. Beneficial role of bacterial endophytes in heavy metal phytoremediation. J. Environ. Manage 174, 14–25.
Maksimov, I.V., Abizgil'Dina, R.R., Pusenkova, L.I., 2011. Plant growth promoting rhizobacteria as alternative to chemical crop protectors from pathogens. Appl. Biochem. Microbiol. 47 (4), 333–345.
Moreno, F.N., Anderson, C.W., Stewart, R.B., Robinson, B.H., 2005. Mercury volatilisation and phytoextraction from base-metal mine tailings. Environ. Pollut. 136 (2), 341–352.
Niane, B., Devarajan, N., Poté, J., Moritz, R., 2019. Quantification and characterization of mercury resistant bacteria in sediments contaminated by artisanal small-scale gold mining activities, Kedougou region, Senegal. J. Geochem. Explor. 205, 106353.
Nies, D.H., 1999. Microbial heavy-metal resistance. Appl. Microbiol. Biotechnol. 51 (6), 730–750.
Oberson, A., Frossard, E., Bühlmann, C., Mayer, J., Mäder, P., Lüscher, A., 2013. Nitrogen fixation and transfer in grass-clover leys under organic and conventional cropping systems. Plant Soil 371 (1–2), 237–255.
Plaza, J., Viera, M., Donati, E., Guibal, E., 2011. Biosorption of mercury by *Macrocystis pyrifera* and *Undaria pinnatifida*: Influence of zinc, cadmium and nickel. J. Environ. Sci. 23 (11), 1778–1786.
Qian, X., Wu, Y., Zhou, H., Xu, X., Xu, Z., Shang, L., et al., 2018. Total mercury and methylmercury accumulation in wild plants grown at wastelands composed of mine tailings: insights into potential candidates for phytoremediation. Environ. Pollut. 239, 757–767.
Raj, D., Maiti, S.K., 2019. Sources, toxicity, and remediation of mercury: an essence review. Environ. Monit. Assess. 191 (9), 566.
Rodríguez, H., Fraga, R., 1999. Phosphate solubilizing bacteria and their role in plant growth promotion. Biotechnol. Adv. 17 (4–5), 319–339.
Sas-Nowosielska, A., Galimska-Stypa, R., Kucharski, R., Zielonka, U., Małkowski, E., Gray, L., 2008. Remediation aspect of microbial changes of plant rhizosphere in mercury contaminated soil. Environ. Monit. Assess. 137 (1-3), 101–109.
Singh, A.K., Dubey, S.K., 2016. Current trends in Bt crops and their fate on associated microbial community dynamics: a review. Protoplasma 253 (3), 663–681.
Singh, S., Kumar, V., 2019. Mercury detoxification by absorption, mercuric ion reductase, and exopolysaccharides: a comprehensive study. Environ. Sci. Pollut. Res. 1–21.
Sriprang, R., Hayashi, M., Ono, H., Takagi, M., Hirata, K., Murooka, Y., 2003. Enhanced accumulation of Cd^{2+} by a *Mesorhizobium* sp. transformed with a gene from *Arabidopsis thaliana* coding for phytochelatin synthase. Appl. Environ. Microbiol. 69 (3), 1791–1796.
Suslow, T.V., 1982. Role of root-colonizing bacteria in plant growth. Phytopathogenic Prokaryotes 1, 187–223.
Swain, M.R., Naskar, S.K., Ray, R.C., 2007. Indole-3-acetic acid production and effect on sprouting of yam (*Dioscorea rotundata* L.) minisetts by *Bacillus subtilis* isolated from culturable cow dung microflora. Pol. J. Microbiol. 56, 103–110.
Vibhuti, M., Kumar, A., Sheoran, N., Nadakkakath, A.V., Eapen, S.J., 2017. Molecular basis of endophytic *Bacillus megaterium*-induced growth promotion in *Arabidopsis thaliana*: revelation by microarray-based gene expression analysis. J. Plant Growth Regul. 36 (1), 118–130.
Wang, J., Feng, X., Anderson, C.W., Xing, Y., Shang, L., 2012. Remediation of mercury contaminated sites—a review. J. Hazard. Mater. 221, 1–18.

Windham-Myers, L., Ackerman, J.T., Fleck, J.A., Marvin-DiPasquale, M., Stricker, C.A., Bachand, P., et al., 2014. Mercury cycling in agricultural and managed wetlands: a synthesis of observations from an integrated field study of methylmercury production, hydrologic export, and bioaccumulation. Sci. Total Environ. Available from: http://ca.water.usgs.gov/pubs/2014/WindhamMyersEtAl2014.html.

Wu, S.C., Cheung, K.C., Luo, Y.M., Wong, M.H., 2006. Effects of inoculation of plant growth-promoting rhizobacteria on metal uptake by *Brassica juncea*. Environ. Pollut. 140 (1), 124–135.

Yin, R., Feng, X., Wang, J., Bao, Z., Yu, B., Chen, J., 2013. Mercury isotope variations between bioavailable mercury fractions and total mercury in mercury contaminated soil in Wanshan Mercury Mine, SW China. Chem. Geol. 336, 80–86.

Zaidi, S., Usmani, S., Singh, B.R., Musarrat, J., 2006. Significance of *Bacillus subtilis* strain SJ-101 as a bioinoculant for concurrent plant growth promotion and nickel accumulation in *Brassica juncea*. Chemosphere 64 (6), 991–997.

Zhuang, X., Chen, J., Shim, H., Bai, Z., 2007. New advances in plant growth-promoting rhizobacteria for bioremediation. Environ. Int. 33 (3), 406–413.

The role of plant growth promoting bacteria in mineralization of endosulfan and its metabolites

14

Rupa Rani, Vipin Kumar, Pratishtha Gupta and Avantika Chandra
Laboratory of Applied Microbiology, Department of Environmental Science and Engineering, Indian Institute of Technology (Indian School of Mines), Dhanbad, India

14.1 Introduction

Endosulfan is a chlorinated hydrocarbon insecticide that has a broad spectrum of activity. It is being used for protection against pest infestations on rice, cotton, tobacco, cashew, vegetables, cashew, and tea. India is the largest manufacturer of endosulfan followed by Germany and China (Li et al., 2007; Li and Macdonald, 2005). Hindustan Insecticide Limited, Coromandel Fertilizers, and Excel Crop Care are the three major endosulfan manufacturing companies (Venkatraman, 2004).

14.2 Physicochemical properties of endosulfan

In pure form, endosulfan exists as white crystalline solid, however the technical product is a yellow crystalline solid. The physical and chemical properties of endosulfan are shown in Table 14.1.

14.3 Persistence of endosulfan

Due to the presence of relatively reactive cyclic sulfite diester group, endosulfan is more moderately persistent than other organochlorine compounds (Giri and Rai, 2012). The physicochemical properties of endosulfan such as high octanol/water partition coefficient and low water solubility also make endosulfan persistent (Mukherjee and Kumar, 2012). Endosulfan has the tendency to get adsorbed in soil particles because of its hydrophobic nature (Harikumar et al., 2014).

Microbes in Land Use Change Management. DOI: https://doi.org/10.1016/B978-0-12-824448-7.00017-6
© 2021 Elsevier Inc. All rights reserved.

Table 14.1 Physicochemical properties of endosulfan.

Chemical name	6,7,8,9,10,10-Hexachloro-1,5,5a,6,9,9a-hexahydro-6,9-methano-2,4,3-benzodioxathiepin-3-oxide		
Chemical formula	$C_9H_6C_{16}O_3S$		
Trade names	Cyclodan, Thiodan, Thiofar, Thimol, Malix, Endofan, Agrisulfan, Ensure, Axis, Misulfan, Red Sun, Golden Leaf Tobacco Spray		
Molecular weight	406.9		
Relative density	1.745 g cm^{-3}		
Formulation	Emulsifiable concentrate, wettable powder, ultra-low volume liquid, smoke tablets		
Solubility in organic solvents g 100 mL^{-1} at 20°C			
Toluene	226		
Dichloromethane	200.7		
Acetone	116.4		
p-Xylene	104.4		
Ethyl acetate	100.9		
Acetonitrile	51.3		
Carbon tetrachloride	42.8		
n-Heptane	20.6		
Methanol	16.8		
n-Octanol	13.1		
	α	*B*	Technical endosulfan[a]
Melting point	109.2°C	213.3°C	70°C–100°C
Vapor pressure	1.9×10^{-3} Pa	9.2×10^{-5} Pa	1.7×10^{-3} Pa
Volatility	1.48 Pa × m^3 × mol^{-1} at 24°C	0.07 Pa × m^3 × mol^{-1} at 24°C	
Solubility in water (mg L^{-1})	0.41	0.23	0.32

[a]Technical endosulfan (mixture of isomers).
Source: Singh, S.N. (Ed.), 2016. Microbe-Induced Degradation of Pesticides. Springer.

14.4 Toxicity of endosulfan

Endosulfan is highly lipophilic in nature and exposure to it can lead to biomagnifications and bioaccumulation, which cause toxic effects on target and nontarget organisms. The persistence of endosulfan in aquatic and soil environments leads to accumulation in phytoplankton and zooplankton (Chowdhury et al., 2007; Chen et al., 2002; Arrebola et al., 2001; Ramaneswari and Rao, 2000). Exposure to endosulfan causes acute and chronic (mutagenic, genotoxic, carcinogenic, organ, reproductive effects) toxicity to humans. It is extremely toxic to aquatic and terrestrial ecosystems (Table 14.2). Endosulfan has a strong affinity with gamma-amino butyric acid receptors, which are inhibitory neuro-receptors in humans (Silva and Beauvais, 2010; UNEP/FAO, 2007).

Table 14.2 Toxic effects of endosulfan.

Toxicity	Effects	References
Acute toxicity	Convulsions, diarrhea, hyperactivity, hunching, tremors, piloerection, respiratory distress, salivation, nausea, vomiting, dizziness, headache, altered sensorium	Singh (2016)
Chronic toxicity		
Mutagenic effects	Damage human DNA, which leads to mutation, affect the cell-cycle kinetics, causes chromosomal aberrations in hamster and mouse and sex-linked mutations in *Drosophila*	Quijano (2000), Chowdhuri et al. (2001)
Genotoxic effects	Genotoxic to HepG2 cells, sister chromatid exchanges, micronuclei	Lu et al. (2000), Yuquan (2000)
Carcinogenic effects	Causes reproductive and central nervous system disorders	Joshi (2001)
Organ toxicity	Affects kidney, liver, blood, parathyroid gland	Amizadeh and Saryazdi (2011)
Reproductive effects	Interferes with sex-hormone synthesis as exposure in male child leads to delay sexual maturity	Saiyed et al. (2003)
Terrestrial ecosystem	Act as an immunosuppressive agent in birds, affects the olfactory learning in bees, exposure of chicken egg results in affecting the liver and brain enzymes, suppresses the immune system and decreases DNA and RNA in the brain	Garg et al. (2004), NIOH (2003), Pushpanjali et al. (2005)
Toxicity to aquatic ecosystem		
Fish	Causes changes in sodium and potassium levels, reduces blood magnesium and calcium concentration, and inhibits brain ATPase and acetyl cholinesterase, genotoxic effects	Dutta and Arends (2003), Sharma et al. (2007)
Aquatic invertebrates	Affects growth rates, impairment of feeding, reproduction rate and embryonic development	GFEA-U (2007)
Amphibians	Increases the vulnerability of tadpoles to the predators	Broomhall (2002)

14.5 Biodegradation of endosulfan

Bioremediation is an economical and ecofriendly technique that remediates organic pollutants, including pesticides from soil using various microbial interactions. Degradation of endosulfan through biological treatment is receiving serious attention as compared to existing treatments, for example, incineration and landfill (Siddique et al., 2003;

Sutherland et al., 2000) because the latter involves the possibility of causing secondary pollution problems and also are cost-effective. Microorganisms such as fungi and bacteria use pesticides as a source of carbon or sulfur. Due to the presence of complex, dynamic, and complicated enzymatic systems, microorganisms have the ability to degrade xenobiotic compounds (Singh, 2016). Microorganisms may possess genetic material which encodes the biochemical mechanisms essential to deal with the xenobiotic compounds. The degradation rate depends on temperature, pH, and organic matter content of the soil as well as the pesticide application rate (Gavrilescu, 2005). Many researchers have reported endosulfan degradation by several microbial strains (Table 14.3).

14.6 Metabolites formed during endosulfan degradation

The biodegradation of endosulfan can occur by attacking the sulfite group and form endosulfan sulfate (a toxic metabolite) via oxidation and endosulfan diol (nontoxic) by hydrolysis. Endosulfan diol can be further transformed into endosulfan lactone, endosulfan ether, endosulfan hydroxyether, endosulfan dialdehyde (Sutherland et al., 2000, 2002). Several researchers have reported the formation of metabolites during endosulfan degradation by different microbial strains (Table 14.4).

14.7 Factors regulating the degradation of endosulfan

The degradation rate of endosulfan is strongly influenced by several factors such as pH, temperature, the presence of the additional carbon source during biodegradation of pesticides.

14.7.1 Effect of pH

pH is one of the most important factors affecting the degradation of endosulfan. Awasthi et al. (2000) found that the optimal pH for endosulfan degradation was 8.5 because of the bioavailability of endosulfan and optimal biotic activity of bacterial cells at this pH. Guerin (1999) reported that endosulfan was transformed to endosulfan diol by hydrolysis at alkaline pH. Several researchers demonstrated a decrease in pH of the medium during endosulfan degradation by bacterial strains. This might be because of the metabolites formed or due to the organic acids and HCl produced by the microorganisms (Castillo et al., 2011; Li et al., 2009). Narkhede et al. (2015) reported that during endosulfan degradation by *Pseudomonas aeruginosa*, a reduction in pH from 7.2 to 4.98 in 288 h was observed. Siddique et al. (2003) observed the reduction in the pH of the medium to 3.3.

14.7.2 Effect of supplementary carbon source

Several researchers have reported that the addition of external carbon source increased the degradation potential of bacterial cultures. Kumar and Philip (2006b)

Table 14.3 Bacterial degradation of endosulfan (α- and β-endosulfan) by several microbial strains.

Initial endosulfan concentration	Bacteria strain	Degradation	References
100 mg L^{-1}	*Pseudomonas* sp. TAH	>80%	Zaffar et al. (2018)
50 mg L^{-1}	*Pseudomonas mendocina* ZAM1	>64.5%	Mir et al. (2017)
100 mg L^{-1}	*Stenotrophomonas maltophilia* OG2	81.53% (α-endosulfan)	Ozdal et al. (2017)
0.5 mM	*Pseudomonas* sp. RPT 52	α-96.6% β-92.7%	Gupta et al. (2016)
50 ppm	*Klebsiella* sp. M3	α-74.5% β-67.5%	Singh and Singh (2014)
	Alcaligenes faecalis strain JBW4	α-75.8% and 87.0% ß-58.5% and 69.5% (in sterilized and natural soils)	Kong et al. (2014)
100 ppm	*Bacillus subtilis*	94.2%	Kumar et al. (2014)
100 mg L^{-1}	*Agrobacterium tumefaciens*	>99%	Thangadurai and Suresh (2014)
100 mg L^{-1}	*A. faecalis* JBW4	α-87.5% ß-83.9%	Kong et al. (2013)
77 μg L^{-1} 119 μg L^{-1}	*Pseudomonas* *Bacillus*	78.73% (77 μg L^{-1}) 98.02% (119 μg L^{-1}) 56.62% (77 μg L^{-1}) 50.00% (119 μg L^{-1})	Harikumar et al. (2013)
	Pseudomonas fluorescens	α-92.80% β-79.35%	Giri and Rai (2012)
50 ppm	*Achromobacter xylosoxidans* C8B	α-94.12% β-84.52%	Singh and Singh (2011)
2 mg L^{-1} 10 mg L^{-1}	*Azotobacter* sp.	57.6% 72.3% (remaining concentrations)	Castillo et al. (2011)
50 μg mL^{-1}	*Pseudomonas* sp. IITR01	α-86.0% β-70.0%	Bajaj et al. (2010)
	Pseudomonas aeruginosa	α-50.25% β-69.77%	Kalyani et al. (2009)
100 μg mL^{-1}	*Bordetella* sp. B9	α-80% β-86%	Goswami and Singh (2009)

(*Continued*)

Table 14.3 (Continued)

Initial endosulfan concentration	Bacteria strain	Degradation	References
100 mg L^{-1}	*A. xylosoxidans* CS5	MM medium- α-24.8 mg L^{-1} β-10.5 mg L^{-1} NSM medium- α-30.8 mg L^{-1} β-13.65 mg L^{-1} MSM medium α-31.6 mg L^{-1} β-13.8 mg L^{-1} were removed.	Li et al. (2009)
50 mg kg^{-1}	*Ochrobacterum* sp. *Arthrobacter* sp. *Burkholderia* sp.	α-57%, β-04% α-88%, β-60% α-91%, β-67%	Kumar et al. (2008a)
100 mg L^{-1}	*P. aeruginosa*	85%	Arshad et al. (2008)
100 mg L^{-1}	*Pseudomonas spinosa* *P. aeruginosa* *Burkholderia cepacia*	α-92.9%, β-88.2% α-93%, β-87% α-93%, β-87%	Hussain et al. (2007)
100 mg L^{-1}	*S. maltophilia* *Rhodococcus erythropolis*	α-46%, β-51% α-24%, β-26%	Kumar et al. (2007)
5 mg g^{-1}	*Bacillus* sp.	50%	Shivaramaiah and Kennedy (2006)
50 mg L^{-1}	*Staphylococcus* sp. *Bacillus circulans I* *B. circulans II*	87.8% 88.1% 88.4%	Kumar and Philip (2006a)
23.5 μg mL^{-1}	*Pseudomonas* sp.	Concentration decreases to 6.0 μg mL^{-1}	Lee et al. (2006)
150 mg L^{-1}	*Klebsiella oxytoca*		Kwon et al. (2005)
100 mg L^{-1}	BF2 BF4 Lin-3	α-37.2%, β-50.3% α-79.6%, β-83.9% α-81.8%, β-86.8%	Siddique et al. (2003)
20 μg mL^{-1}	Bacterial coculture	α-94% β-82%	Awasthi et al. (2003)
93.9 μg mL^{-1}	*Klebsiella pneumoniae*	93.6%	Kwon et al. (2002)

MM, minimal medium; *MSM*, minimal salt medium; *NSM*, non sulfur medium.

Table 14.4 Endosulfan degrading microbial strains and the metabolites formed.

Bacteria strain	Metabolites	References
Pseudomonas sp.	Endosulfan lactone	Zaffar et al. (2018)
Klebsiella sp. M3	Endosulfan diol, endosulfan lactone endosulfan hydroxyether, endosulfan sulfate	Singh and Singh (2014)
Alcaligenes faecalis JBW4	Endosulfan ether and endosulfan lactone	Kong et al. (2014)
Bacillus subtilis	Endodiol and endosulfan lactone	Kumar et al. (2014)
A. faecalis JBW4	Endosulfan diol and endosulfan lactone	Kong et al. (2013)
Pseudomonas fluorescens	Endosulfan diol and endosulfan ether	Giri and Rai (2012)
Azotobacter sp.	Endosulfan diol	Castillo et al. (2011)
Pseudomonas sp. IITR01	Endosulfan diol, Endosulfan sulfate, endosulfan lactone, endosulfan ether	Bajaj et al. (2010)
Bordetella sp. B.9	Endosulfan ether, endosulfan lactone, endosulfan sulfate	Goswami and Singh (2009)
Achromobacter xylosoxidans CS5	Endosulfan diol and endosulfan ether	Li et al. (2009)
Ochrobacterum sp. *and Burkholderia* sp. *Arthrobacter* sp.	Endosulfan diol Endosulfan sulfate	Kumar et al. (2008a)
Pseudomonas spinosa, Pseudomonas aeruginosa, Burkholderia cepacia	Endosulfan diol, endosulfan ether	Hussain et al. (2007)
Stenotrophomonas sp; *Rhodococcus* sp.	Endosulfan diol	Kumar et al. (2007)
Bacterium S3	Endosulfan sulfate	Shivaramaiah and Kennedy (2006)
Klebsiella oxytoca KE-8	Endosulfan diol, endosulfan lactone, Endosulfan sulfate	Kwon et al. (2005)
Bacterial coculture	Endosulfan diol, endosulfan lactone	Awasthi et al. (2003)
Klebsiella pneumoniae KE-1	Endosulfan diol, endosulfan diol	Kwon et al. (2002)
Mycobacterium sp.	Endosulfan hydroxyether, endosulfan monoaldehyde	Sutherland et al. (2002)
Mixed culture	Endosulfan sulfate, endosulfan monoaldehyde, endosulfan diol, endosulfan ether	Sutherland et al. (2000)

observed that at 1 g L^{-1} of dextrose, bacterial growth and degradation of endosulfan increased, whereas further increase in the concentration of dextrose results in no improvement in the degradation of endosulfan. Similarly, Goswami and Singh (2009) found that endosulfan degradation increases in the presence of sucrose. However, Awasthi et al. (2000) found that no effect on the degradation of endosulfan was observed with the addition of glucose.

14.7.3 Effect of moisture content

Moisture content is an important parameter in endosulfan degradation (Ghadiri, 2001). Several researchers have reported that minimum moisture content of 30% is necessary for the degradation of endosulfan (Ali et al., 2014; Kataoka et al., 2011; Kumar et al., 2008b; Kumar and Philip 2006c) to make it bioavailable through film diffusion. Awasthi et al. (2000) found that the degradation of endosulfan was at maximum in flooded soils as compared to a non-flooded soils. However, Sethunathan et al. (2002) reported that degradation of endosulfan was faster in non-flooded condition (45%) as compared to the flooded condition (15%–18%). Tiwari and Guha (2013) observed that endosulfan degradation was at maximum in aqueous phase as compared to solid phase.

14.8 Plant growth promoting rhizobacteria

Plant growth promoting rhizobacteria (PGPR) are naturally occurring soil bacteria capable of promoting plant growth by colonizing plant roots (Kloepper, 1994). For a decade, PGPR were mainly used for the promotion of plant growth, such as biological nitrogen fixation, of iron sequestration by siderophores, solubilizing minerals such as phosphorous, production of plant hormones, decreasing the inhibitory effects of various pathogens in the forms of biocontrol agent (Glick et al., 1999). Further, much attention has recently been paid to remediate pesticides contaminated soils by using PGPR (Myresiotis et al., 2012; Abraham and Silambarasan, 2015).

14.9 Mechanism of plant growth promotion

PGPR promote plant growth through solubilization of nutrients in soils, production of phytohormones and siderophore, protecting plants from phytopathogens by controlling or inhibiting them, lowering of ethylene levels, improving soil structure, and bioremediating the polluted soils by degrading xenobiotic compounds (like pesticides) (Glick et al., 1999). Examples of PGPR and their beneficial traits cial traits are shown in Table 14.5.

Table 14.5 Examples of plant growth promoting rhizobacteria and their benefits.

PGPR	Plant	Results of addition of bacteria to plants	Reference
Bacillus sp.	*Peucedanum japonicum*	Increase in shoot growth by 10.9% Increase in root dry weight by 30.8%	Hong and Lee (2014)
Bacillus sp., *Azotobacter* sp.	*Vigna unguiculata*	Increase in yield by 30% Phosphate solubilization index 38.55 Produced IAA 24.2 $\mu g\ mL^{-1}$	Sivasankari and Anandharaj (2014)
Bacillus circulans CB7	*Malus domestica*	Increase in seed germination by 22.32% Increase in shoot length by 15.91% Increase in root length by 25.10% Increase in shoot dry weight by 52.92% Increase in root dry weight by 31.4% Increases nitrogen by 18.75% Increases potassium by 57.69% Increases phosphorous by 22.22%	Mehta et al. (2015)
Bacillus cereus	*Arabidopsis thaliana*	Increase in shoot weight by 68.8% Increase in root weight by 57.2% Increase in biomass by 67.4% Increase in seed production by 34.8%	Niu et al. (2011)
Bacillus sp. PSB10	*Cicer arietinum*	Significantly improved growth, nodulation, chlorophyll, leghemoglobin, seed yield and grain protein; reduced the uptake of chromium in roots, shoots and grains	Wani and Khan (2010)
Bacillus sp., *Arthrobacter* sp.	*Solanum lycopersicum*	Produced IAA 20.3 $\mu g\ mL^{-1}$ Produced IAA 3 $\mu g\ mL^{-1}$ Solubilization of phosphate (165.37 $\mu g\ mL^{-1}$)	Banerjee et al. (2010)

(*Continued*)

Table 14.5 (Continued)

PGPR	Plant	Results of addition of bacteria to plants	Reference
Bacillus weihenstephanensis strain SM3	*Helianthus annuus*	Increased plant biomass and the accumulation of Cu and Zn in the root and shoot systems, also augmented the concentrations of water soluble Ni, Cu and Zn in soil with their metal mobilizing potential	Rajkumar and Freitas (2008)
Bacillus subtilis, *Pseudomonas aeruginosa*	*S. lycopersicum* L., *Abelmoschus esculentus*, *Amaranthus* sp.	Dry biomass increased 31% for tomato, 36% for okra, and 83% for African spinach	Adesemoye et al. (2008)
Bacillus spp.	*Hordeum vulgare*	Increased root weight up to 16.7% and shoot weight up to 347%	Canbolat et al. (2006)
Bacillus sp. RJ31, *Xanthomonas* sp. RJ3, *Azomonas* sp. RJ4, *Pseudomonas* sp. RJ10	*Brassica napus*	Stimulated plant growth and increased cadmium accumulation	Sheng and Xia (2006)
B. subtilis SJ-101	*Brassica juncea*	Facilitated Ni accumulation	Zaidi et al. (2006)
Bacillus sp. SN9, *Pseudomonas* sp. SRI2, *Psychrobacter* sp. SRS8	*B. juncea, Brassica oxyrrhina*	Increased the biomass of the test plants and enhanced Ni accumulation in plant tissues	Ma et al. (2009a)
B. cereus SRA10, *Psychrobacter* sp. SRA1	*B. juncea*, *B. oxyrrhina*	Enhanced the metal accumulation in plant tissues by facilitating the release of Ni from the non-soluble phases in the soil	Ma et al. (2009b)
Bacillus sp., *Paenibacillus* sp.	Rice	Promoted the root and shoot growth	Beneduzi et al. (2008)
Bacillus edaphicus	*B. juncea*	Stimulated plant growth, facilitated soil Pb mobilization, enhanced Pb accumulation	Sheng et al. (2008)

B. subtilis CC-pg104, *Pseudomonas putida* CC-R2–4	*Lactuca sativa* L.	Significant increase in shoot length and root length achieved through encapsulated inoculant	Rekha et al. (2007)
Brevibacillus	*Trifolium repens*	Enhanced plant growth and nutrition of plants and decreased zinc concentration in plant tissues	Vivas et al. (2006)
Pseudomonas sp., *Bacillus* sp.	Mustard	Stimulated plant growth and decreased Cr (VI) content	Rajkumar et al. (2006)
B. cereus, *Ochrobactrum* sp.	Mung bean	Lowers the toxicity of chromium to seedlings by reducing Cr (VI) to Cr (III)	Faisal and Hasnain (2006)
Bacillus megaterium HKP-1, *Bacillus mucillaginosus* HKK-1, *Azotobacter chroococcum* HKN-5	*B. juncea*	Protected plant from metal toxicity, stimulated plant growth	Wu et al. (2006)
Bacillus pantothenticus, Azospirillum brasilense, Pseudomonas pickettii	*Oryza sativa*	Increased rice grain yield maximum up to 76.9%	Thakuria et al. (2004)
Paenibacillus polymyxa	Pepper	Significantly increased the biomass of plants and elicited induced systemic resistance against bacterial spot pathogen *Xanthomonas axonopodis* pv. *vesicatoria* untreated plants	Phi et al. (2010)
Klebsiella pneumoniae	*Triticum aestivum*	Significantly increased the root length and shoot length	Sachdev et al. (2009)
Klebsiella oxytoca 10MKR7, *Enterobacter sakazakii* 8Mr5, *Pseudomonas* sp. 4MKS8	*Zea mays* L	Inoculation increased growth parameters	Babalola et al. (2003)
P. putida, *Azospirilium* sp., *Azotobacter* sp.	*Cynara scolymus*	Phosphate solubilizing bacteria along with nitrogen fixing bacteria led to significant increase in radicle and shoot length, shoot weight, coefficient of velocity of germination, seedling vigority index, and significant decrease in mean time of germination	Jahanian et al. (2012)

(*Continued*)

Table 14.5 (Continued)

PGPR	Plant	Results of addition of bacteria to plants	Reference
Pseudomonas sp. PS1	*Vigna radiata* (L.)	Significantly increased plant dry weight, nodule numbers, total chlorophyll content, leghemoglobin, root N, shoot N, root P, shoot P, seed yield and seed protein	Ahemad and Khan (2010, 2011, 2012a)
Pseudomonas sp. A3R3	*Alyssum serpyllifolium, B. juncea*	Increased significantly the biomass (*B. juncea*) and Ni content (*A. serpyllifolium*) in plants grown in Ni-stressed soil	Ma et al. (2011)
Pseudomonas sp.	Soybean, wheat	Significantly increased soil enzyme activities, total productivity, and nutrient uptake	Sharma et al. (2011)
Pseudomonas sp., *Ochrobactrum cystitis* *Bradyrhizobium* sp. 750	*Lupinus luteus*	Increased both biomass, nitrogen content, accumulation of metals (improved phytostabilization potential)	Dary et al. (2010)
P. putida strain R-168, *Pseudomonas fluorescens* strain R-93, *P. fluorescens* DSM 50090, *P. putida* DSM291, *Azospirillum lipoferum* DSM 1691, *A. brasilense* DSM 1690	*Z. mays* L.	Plant height, seed weight, number of seeds per ear and leaf area, shoot dry weight significantly increased	Gholami et al. (2009)
P. aeruginosa, *P. fluorescens*, *Ralstonia metallidurans*	Maize	Promoted plant growth, facilitated soil metal mobilization, enhanced Cr and Pb uptake	Braud et al. (2009)
Pseudomonas sp.	Chickpea	Enhanced fresh and dry weight of plants, even at 2 mM nickel concentration	Tank and Saraf (2009)
Pseudomonas species	*O. sativa, Z. mays*	*Pseudomonas* isolated from rice showed a higher ability to control bacterial and fungal root pathogens than that obtained from maize	Lawongsa et al. (2008)

P. aeruginosa strain MKRh3	Black gram	Plants showed lessened cadmium accumulation, extensive rooting, and enhanced plant growth	Ganesan (2008)
P. aeruginosa	Indian mustard and pumpkin	Stimulated plant growth, reduced Cd uptake	Sinha and Mukherjee (2008)
Pseudomonas tolaasii ACC23, *P. fluorescens* ACC9, *Alcaligenes* sp. ZN4, *Mycobacterium* sp. ACC14	*B. napus*	Protected canola plant against the inhibitory effects of cadmium	Dell'Amico et al. (2008)
Pseudomonas jessenii PS06, *Mesorhizobium ciceri* C-2/2	*C. arietinum*	The co-inoculation treatment increased the seed yield (52% greater than the uninoculated control treatment) and nodule fresh weight	Valverde et al. (2007)
P. putida KNP9	Mung bean	Stimulated the plant growth, reduced Pb and Cd uptake	Tripathi et al. (2005)
P. fluorescens	Soybean	Increased plant growth	Gupta et al. (2005)

14.9.1 Phosphate (P) solubilization

Phosphorus is an important nutrient essential for plant growth and development. Phosphorus is abundantly present in soil in both organic and inorganic forms, but 95%–99% phosphate is present in the immobilized, insoluble, and precipitated form in the soil, so it becomes unavailable to plants. The amounts of P available to soil are lesser than 1 μMol L^{-1} (Rajasankar et al., 2013). Plants can absorb P only as monobasic ($H_2PO_4^-$) and the diabasic (HPO_4^{2-}) ions (Bhattacharyya and Jha, 2012). Phosphorous solubilization in soil is shown in Fig. 14.1.

Microorganisms that have phosphate solubilizing potential are termed as phosphate solubilizing microorganisms. It increases the availability of P in the soil environment as they solubilize the insoluble form of P to soluble form through secretion of organic acids or protons, acidification (Richardson et al., 2009) and chelation, and exchange reactions (Hameeda et al., 2008). Solubilization of insoluble P had been previously studied in rhizosphere soil bacteria like, *Rhizobium*, *Pantoea*, *Serratia*, *Bacillus*, *Pseudomonas*, *Azospirillum*, *Enterobacter* those bacteria also showed PGP activities (Ahemad and Khan, 2012b; Vyas and Gulati 2009; Park et al., 2009; Hameeda et al., 2008; Son et al., 2006; Rodriguez et al., 2004; Hwangbo et al., 2003).

14.9.2 Indole acetic acid

Auxins are phytohormones; the most common is indole acetic acid (IAA), which is known to increase cell division and differentiation, cell elongation, root growth, root elongation, stimulates seed and tuber germination; initiates lateral and adventitious root formation; control processes of vegetative growth; mediates responses to

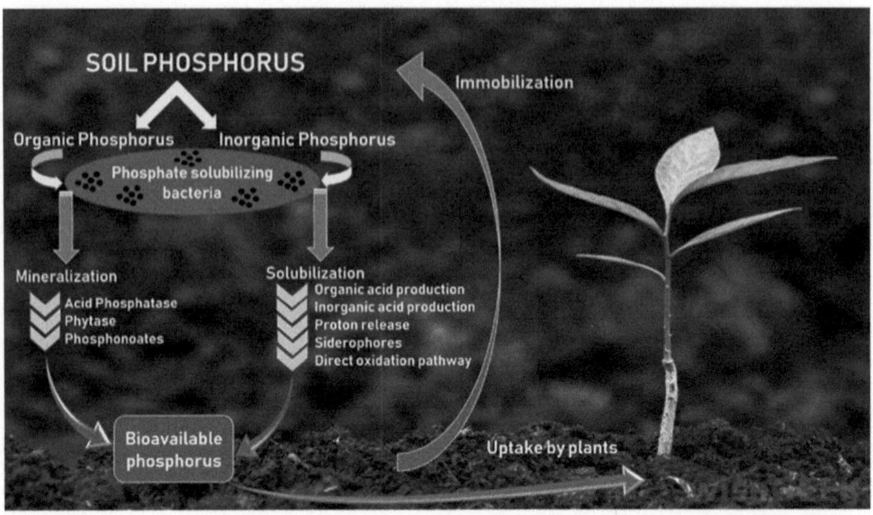

Figure 14.1 Soil phosphorous solubilization and immobilization by bacteria.

light, gravity, affects photosynthesis, pigment formation, resistance to stressful conditions and biosynthesis of various metabolites (Miransari and Smith, 2014). Plant as well as PGPR can synthesize IAA. IAA secreted by rhizobacteria interferes with the plant development by changing the endogenous pool plant auxin (Glick, 2012; Spaepen et al., 2007). Therefore IAA acts a significant role in plant-rhizobacteria interactions (Spaepen and Vanderleyden, 2011). Starting with tryptophan, rhizobacteria produce IAA through five different pathways; IAA biosynthesis via indole-3-pyruvic acid and indole-3-acetic aldehyde, tryptophan conversion into indole-3-acetic aldehyde through the formation of tryptamine, IAA formation via indole-3-acetamide production, IAA biosynthesis that involves conversion of tryptophan into indole-3-acetonitrile, the tryptophan-independent pathway (Spaepen et al., 2007).

14.9.3 Siderophore production

Siderophores are low molecular weight (400–1000 Da), iron-chelating agents with a high association constants ($K_d = 10^{-20}–10^{-50}$) for binding with Fe^{3+} (Gray and Smith, 2005). These are synthesized by bacteria in soil rhizosphere under neutral to alkaline pH conditions in iron limiting conditions due to low iron solubility at elevated pH. Iron is one of the essential components, occur as Fe^{3+} and is susceptible to form insoluble hydroxides and oxyhydroxides, which makes it unavailable to microorganisms and plants (Rajkumar et al., 2010). Bacteria utilize iron by producing siderophores, which plays a vital role in estimating bacterial ability to colonize plant roots and also to compete with other microbes for iron (Naik et al., 2013; Johri et al., 2003). Plants assimilate iron from bacterial siderophores through various mechanisms, such as, chelate and release of iron, the direct uptake of siderophore-Fe complexes, or by a ligand exchange reaction (Schmidt, 1999).

14.9.4 Aminocyclopropane-1-carboxylate deaminase

Ethylene also known as stress hormone, is produced endogenously by plants (Saleem et al., 2007). Under stress conditions, such as heavy metals, drought, pathogenicity and salinity, the endogenous level of endosulfan is significantly increased. PGPR which secretes the enzyme, 1-aminocyclopropane-1-carboxylate (ACC) deaminase, a precursor to plant ethylene levels, is one of the major mechanisms by which bacteria exhibited beneficial effects on plants under stress, and hence enhances plant growth and development (Zahir et al., 2008; Nadeem et al., 2007).

14.10 Conclusion

Endosulfan, an organochlorine pesticide has been widely used as an insecticide for different crops throughout the world. Due to its indiscriminate and unregulated use its contamination in food, air, soil, and water has also been detected. These environmental and health concerns have led to an interest in degradation of endosulfan using ecofriendly and efficient remediation techniques. Treatment of endosulfan contaminated soil through bioremediation is receiving serious attention as compared to physicochemical methods. PGPR are playing a vital role in promoting plant growth by colonizing the plant root through several mechanisms for instance phosphate solubilization, production of IAA and ACC deaminase, siderophore secretion. Common PGPR are *Bacillus* sp., *Pseudomonas* sp., *Klebsiella* sp., *Paenibacillus* sp., *Azospirillum* sp., which are playing a very important role in plant growth and promotion and also identified as endosulfan degrading bacteria.

References

Abraham, J., Silambarasan, S., 2015. Plant growth promoting bacteria *Enterobacter asburiae* JAS5 and *Enterobacter cloacae* JAS7 in mineralization of endosulfan. Appl. Biochem. Biotechnol. 175, 3336–3348.

Adesemoye, A.O., Obini, M., Ugoji, E.O., 2008. Comparison of plant growth-promotion with *Pseudomonas aeruginosa* and *Bacillus subtilis* in three vegetables. Braz. J. Microbiol. 39 (3), 423–426.

Ahemad, M., Khan, M.S., 2010. Phosphate-solubilizing and plant-growth-promoting *Pseudomonas aeruginosa* PS1 improves green gram performance in quizalafop-p-ethyl and clodinafop amended soil. Arch. Environ. Contam. Toxicol. 58 (2), 361–372.

Ahemad, M., Khan, M.S., 2011. *Pseudomonas aeruginosa* strain PS1 enhances growth parameters of greengram [*Vigna radiata* (L.) Wilczek] in insecticide-stressed soils. J. Pest. Sci. 84 (1), 123–131.

Ahemad, M., Khan, M.S., 2012a. Alleviation of fungicide-induced phytotoxicity in greengram [*Vigna radiata* (L.) Wilczek] using fungicide-tolerant and plant growth promoting *Pseudomonas* strain. Saudi J. Biol. Sci. 19 (4), 451–459.

Ahemad, M., Khan, M.S., 2012b. Effect of fungicides on plant growth promoting activities of phosphate solubilizing *Pseudomonas putida* isolated from mustard (*Brassica campestris*) rhizosphere. Chemosphere 86 (9), 945–950.

Ali, M., Kazmi, A.A., Ahmed, N., 2014. Study on effects of temperature, moisture and pH in degradation and degradation kinetics of aldrin, endosulfan, lindane pesticides during full-scale continuous rotary drum composting. Chemosphere 102, 68–75.

Amizadeh, M., Saryazdi, G.A., 2011. Effects of endosulfan on human health. Web. Cent. Toxicol. 2 (12).

Arrebola, F.J., Egea-González, F.J., Moreno, M., Fernández-Gutiérrez, A., Hernández-Torres, M.E., Martínez-Vidal, J.L., 2001. Evaluation of endosulfan residues in vegetables grown in greenhouses. Pest. Manage. Sci. 57 (7), 645–652.

Arshad, M., Hussain, S., Saleem, M., 2008. Optimization of environmental parameters for biodegradation of alpha and beta endosulfan in soil slurry by *Pseudomonas aeruginosa*. J. Appl. Microbiol. 104 (2), 364–370.

Awasthi, N., Ahuja, R., Kumar, A., 2000. Factors influencing the degradation of soil-applied endosulfan isomers. Soil Biol. Biochem. 32 (11–12), 1697–1705.

Awasthi, N., Singh, A.K., Jain, R.K., Khangarot, B.S., Kumar, A., 2003. Degradation and detoxification of endosulfan isomers by a defined co-culture of two *Bacillus* strains. Appl. Microbiol. Biotechnol. 62 (2–3), 279–283.

Babalola, O.O., Osir, E.O., Sanni, A.I., Odhiambo, G.D., Bulimo, W.D., 2003. Amplification of 1-amino-cyclopropane-1-carboxylic (ACC) deaminase from plant growth promoting rhizobacteria in *Striga*-infested soil. Afr. J. Biotechnol. 2 (6), 157–160.

Bajaj, A., Pathak, A., Mudiam, M.R., Mayilraj, S., Manickam, N., 2010. Isolation and characterization of a *Pseudomonas* sp. strain IITR01 capable of degrading α-endosulfan and endosulfan sulfate. J. Appl. Microbiol. 109 (6), 2135–2143.

Banerjee, S., Palit, R., Sengupta, C., Standing, D., 2010. Stress induced phosphate solubilization by '*Arthrobacter*' sp. and '*Bacillus*' sp. isolated from tomato rhizosphere. Aust. J. Crop. Sci. 4 (6), 378.

Beneduzi, A., Peres, D., Vargas, L.K., Bodanese-Zanettini, M.H., Passaglia, L.M.P., 2008. Evaluation of genetic diversity and plant growth promoting activities of nitrogen-fixing bacilli isolated from rice fields in South Brazil. Appl. Soil Ecol. 39 (3), 311–320.

Bhattacharyya, P.N., Jha, D.K., 2012. Plant growth-promoting rhizobacteria (PGPR): emergence in agriculture. World J. Microbiol. Biotechnol. 28 (4), 1327–1350.

Braud, A., Jézéquel, K., Bazot, S., Lebeau, T., 2009. Enhanced phytoextraction of an agricultural Cr- and Pb-contaminated soil by bioaugmentation with siderophore-producing bacteria. Chemosphere 74 (2), 280–286.

Broomhall, S., 2002. The effects of endosulfan and variable water temperature on survivorship and subsequent vulnerability to predation in *Litoriacitropa* tadpoles. Aquat. Toxicol. 61 (3–4), 243–250.

Canbolat, M.Y., Bilen, S., Çakmakçı, R., Şahin, F., Aydın, A., 2006. Effect of plant growth-promoting bacteria and soil compaction on barley seedling growth, nutrient uptake, soil properties and rhizosphere microflora. Biol. Fertil. Soils 42 (4), 350–357.

Castillo, J.M., Casas, J., Romero, E., 2011. Isolation of an endosulfan-degrading bacterium from a coffee farm soil: persistence and inhibitory effect on its biological functions. Sci. Total Environ. 412, 20–27.

Chen, Y., Zhang, X., Liu, H., Wang, Y., Xia, X., 2002. Study on *Pseudomonas* sp. WBC-3 capable of complete degradation of methyl parathion. Wei Sheng Wu Xue Bao 42 (4), 490–497.

Chowdhuri, D.K., Nazir, A., Saxena, D.K., 2001. Effect of three chlorinated pesticides on hsr ω stress gene in transgenic *Drosophila melanogaster*. J. Biochem. Mol. Toxicol. 15 (4), 173–186.

Chowdhury, A.G., Das, C., Kole, R.K., Banerjee, H., Bhattacharyya, A., 2007. Residual fate and persistence of endosulfan (50 WDG) in Bengal gram (*Cicer arietinum*). Environ. Monit. Assess. 132 (1–3), 467–473.

Dary, M., Chamber-Pérez, M.A., Palomares, A.J., Pajuelo, E., 2010. "In situ" phytostabilisation of heavy metal polluted soils using *Lupinus luteus* inoculated with metal resistant plant-growth promoting rhizobacteria. J. Hazard. Mater. 177 (1–3), 323–330.

Dell'Amico, E., Cavalca, L., Andreoni, V., 2008. Improvement of *Brassica napus* growth under cadmium stress by cadmium-resistant rhizobacteria. Soil Biol. Biochem. 40 (1), 74–84.

Dutta, H.M., Arends, D.A., 2003. Effects of endosulfan on brain acetylcholinesterase activity in juvenile bluegill sunfish. Environ. Res. 91 (3), 157–162.

Faisal, M., Hasnain, S., 2006. Growth stimulatory effect of Ochrobactrum intermedium and *Bacillus cereus* on *Vigna radiata* plants. Lett. Appl. Microbiol. 43 (4), 461–466.

Ganesan, V., 2008. Rhizoremediation of cadmium soil using a cadmium-resistant plant growth-promoting rhizopseudomonad. Curr. Microbiol. 56 (4), 403–407.

Garg, U.K., Pal, A.K., Jha, G.J., Jadhao, S.B., 2004. Haemato-biochemical and immuno-pathophysiological effects of chronic toxicity with synthetic pyrethroid, organophosphate and chlorinated pesticides in broiler chicks. Int. Immunopharmacol. 4 (13), 1709–1722.

Gavrilescu, M., 2005. Fate of pesticides in the environment and its bioremediation. Eng. Life Sci. 5 (6), 497–526.

GFEA-U, E., 2007. Draft Dossier prepared in support of a proposal of endosulfan to be considered as a candidate for inclusion in the CLRTAP protocol on persistent organic pollutants. German Federal Environment Agency–Umweltbundesamt, Berlin.

Ghadiri, H., 2001. Degradation of endosulfan in a clay soil from cotton farms of western Queensland. J. Environ. Manage. 62 (2), 155–169.

Gholami, A., Shahsavani, S., Nezarat, S., 2009. The effect of plant growth promoting rhizobacteria (PGPR) on germination, seedling growth and yield of maize. Int. J. Biol. Life Sci. 1 (1), 35–40.

Giri, K., Rai, J.P.N., 2012. Biodegradation of endosulfan isomers in broth culture and soil microcosm by *Pseudomonas fluorescens* isolated from soil. Int. J. Environ. Stud. 69 (5), 729–742.

Glick, B.R., 2012. Plant Growth-Promoting Bacteria: Mechanisms and Applications. Hindawi Publishing Corporation, Scientifica.

Glick, B.R., Patten, C.L., Holguin, G., Penrose, D.M., 1999. Biochemical and Genetic Mechanisms Used by Plant Growth-Promoting Bacteria. Imperial College Press, London.

Goswami, S., Singh, D.K., 2009. Biodegradation of α and β endosulfan in broth medium and soil microcosm by bacterial strain *Bordetella* sp. B9. Biodegradation 20 (2), 199–207.

Gray, E.J., Smith, D.L., 2005. Intracellular and extracellular PGPR: commonalities and distinctions in the plant–bacterium signaling processes. Soil Biol. Biochem. 37 (3), 395–412.

Guerin, T.F., 1999. The anaerobic degradation of endosulfan by indigenous microorganisms from low-oxygen soils and sediments. Environ. Pollut. 106 (1), 13–21.

Gupta, A., Rai, V., Bagdwal, N., Goel, R., 2005. In situ characterization of mercury-resistant growth-promoting fluorescent pseudomonads. Microbiol. Res. 160 (4), 385–388.

Gupta, M., Mathur, S., Sharma, T.K., Rana, M., Gairola, A., Navani, N.K., et al., 2016. A study on metabolic prowess of *Pseudomonas* sp. RPT 52 to degrade imidacloprid, endosulfan and coragen. J. Hazard. Mater. 301, 250–258.

Hameeda, B., Harini, G., Rupela, O.P., Wani, S.P., Reddy, G., 2008. Growth promotion of maize by phosphate-solubilizing bacteria isolated from composts and macrofauna. Microbiol. Res. 163 (2), 234–242.

Harikumar, P.S., Jesitha, K., Sreechithra, M., 2013. Remediation of endosulfan by biotic and abiotic methods. J. Environ. Prot. 4 (05), 418.

Harikumar, P.S., Jesitha, K., Megha, T., Kokkal, K., 2014. Persistence of endosulfan in selected areas of Kasaragod district, Kerala. Curr. Sci. 1421–1429.

Hong, S.H., Lee, E.Y., 2014. Vegetation restoration and prevention of coastal sand dunes erosion using ion exchange resins and the plant growth-promoting rhizobacteria *Bacillus* sp. SH1RP8 isolated from indigenous plants. Int. Biodeterior. Biodegradation 95, 262–269.

Hussain, S., Arshad, M., Saleem, M., Khalid, A., 2007. Biodegradation of α- and β-endosulfan by soil bacteria. Biodegradation 18 (6), 731–740.

Hwangbo, H., Park, R.D., Kim, Y.W., Rim, Y.S., Park, K.H., Kim, T.H., et al., 2003. 2-ketogluconic acid production and phosphate solubilization by *Enterobacter intermedium*. Curr. Microbiol. 47 (2), 0087–0092.

Jahanian, A., Chaichi, M.R., Rezaei, K., Rezayazdi, K., Khavazi, K., 2012. The effect of plant growth promoting rhizobacteria (PGPR) on germination and primary growth of artichoke (*Cynara scolymus*). Int. J. Agric. Crop Sci. 4 (14), 923–929.

Johri, B.N., Sharma, A., Virdi, J.S., 2003. Rhizobacterial diversity in India and its influence on soil and plant health. Biotechnology in India I. Springer, Berlin, Heidelberg, pp. 49–89.

Joshi, S., 2001. Children of endosulfan. Earth 19, 28.

Kalyani, S.S., Sharma, J., Singh, S., Dureja, P., Lata, 2009. Enrichment and isolation of endosulfan-degrading microorganism from tropical acid soil. J. Environ. Sci. Health B 44 (7), 663–672.

Kataoka, R., Takagi, K., Sakakibara, F., 2011. Biodegradation of endosulfan by *Mortieralla* sp. strain W8 in soil: influence of different substrates on biodegradation. Chemosphere 85 (3), 548–552.

Kloepper, J.W., 1994. Plant growth-promoting rhizobacteria (other systems). Azospirillum/ Plant Assoc. 187, 137–166.

Kong, L., Zhu, S., Zhu, L., Xie, H., Su, K., Yan, T., et al., 2013. Biodegradation of organochlorine pesticide endosulfan by bacterial strain *Alcaligenes faecalis* JBW4. J. Environ. Sci. 25 (11), 2257–2264.

Kong, L., Zhu, S., Zhu, L., Xie, H., Wei, K., Yan, T., et al., 2014. Colonization of *Alcaligenes faecalis* strain JBW4 in natural soils and its detoxification of endosulfan. Appl. Microbiol. Biotechnol. 98 (3), 1407–1416.

Kumar, M., Philip, L., 2006a. Endosulfan mineralization by bacterial isolates and possible degradation pathway identification. Bioremediat. J. 10 (4), 179–190.

Kumar, M., Philip, L., 2006b. Bioremediation of endosulfan contaminated soil and water—optimization of operating conditions in laboratory scale reactors. J. Hazard. Mater. 136 (2), 354–364.

Kumar, M., Philip, L., 2006c. Enrichment and isolation of endosulfan degrading mixed bacterial culture. J. Environ. Sci. Health B 41 (1), 81–96.

Kumar, K., Devi, S.S., Krishnamurthi, K., Kanade, G.S., Chakrabarti, T., 2007. Enrichment and isolation of endosulfan degrading and detoxifying bacteria. Chemosphere 68 (2), 317–322.

Kumar, M., Lakshmi, C.V., Khanna, S., 2008a. Biodegradation and bioremediation of endosulfan contaminated soil. Bioresour. Technol. 99 (8), 3116–3122.

Kumar, S., Habib, K., Fatma, T., 2008b. Endosulfan induced biochemical changes in nitrogen-fixing cyanobacteria. Sci. Total Environ. 403 (1–3), 130–138.

Kumar, A., Bhoot, N., Soni, I., John, P.J., 2014. Isolation and characterization of a *Bacillus subtilis* strain that degrades endosulfan and endosulfan sulfate. 3 Biotech 4 (5), 467–475.

Kwon, G.S., Kim, J.E., Kim, T.K., Sohn, H.Y., Koh, S.C., Shin, K.S., et al., 2002. *Klebsiella pneumoniae* KE-1 degrades endosulfan without formation of the toxic metabolite, endosulfan sulfate. FEMS Microbiol. Lett. 215 (2), 255–259.

Kwon, G.S., Sohn, H.Y., Shin, K.S., Kim, E., Seo, B.I., 2005. Biodegradation of the organochlorine insecticide, endosulfan, and the toxic metabolite, endosulfan sulfate, by *Klebsiella oxytoca* KE-8. Appl. Microbiol. Biotechnol. 67 (6), 845–850.

Lawongsa, P., Boonkerd, N., Wongkaew, S., O'Gara, F., Teaumroong, N., 2008. Molecular and phenotypic characterization of potential plant growth-promoting *Pseudomonas* from rice and maize rhizospheres. World J. Microbiol. Biotechnol. 24 (9), 1877–1884.

Lee, J.B., Sohn, H.Y., Shin, K.S., Jo, M.S., Kim, J.E., Lee, S.W., et al., 2006. Isolation of a soil bacterium capable of biodegradation and detoxification of endosulfan and endosulfan sulfate. J. Agric. Food Chem. 54 (23), 8824–8828.

Li, Y.F., Macdonald, R.W., 2005. Sources and pathways of selected organochlorine pesticides to the Arctic and the effect of pathway divergence on HCH trends in biota: a review. Sci. Total Environ. 342 (1−3), 87−106.

Li, J., Zhang, G., Guo, L., Xu, W., Li, X., Lee, C.S., et al., 2007. Organochlorine pesticides in the atmosphere of Guangzhou and Hong Kong: regional sources and long-range atmospheric transport. Atmos. Environ. 41 (18), 3889−3903.

Li, W., Dai, Y., Xue, B., Li, Y., Peng, X., Zhang, J., et al., 2009. Biodegradation and detoxification of endosulfan in aqueous medium and soil by *Achromobacter xylosoxidans* strain CS5. J. Hazard. Mater. 167 (1−3), 209−216.

Lu, Y., Morimoto, K., Takeshita, T., Takeuchi, T., Saito, T., 2000. Genotoxic effects of alpha-endosulfan and beta-endosulfan on human HepG2 cells. Environ. Health Perspect. 108 (6), 559−561.

Ma, Y., Rajkumar, M., Freitas, H., 2009a. Isolation and characterization of Ni mobilizing PGPB from serpentine soils and their potential in promoting plant growth and Ni accumulation by *Brassica* spp. Chemosphere 75 (6), 719−725.

Ma, Y., Rajkumar, M., Freitas, H., 2009b. Improvement of plant growth and nickel uptake by nickel resistant-plant-growth promoting bacteria. J. Hazard. Mater. 166 (2−3), 1154−1161.

Ma, Y., Rajkumar, M., Luo, Y., Freitas, H., 2011. Inoculation of endophytic bacteria on host and non-host plants—effects on plant growth and Ni uptake. J. Hazard. Mater. 195, 230−237.

Mehta, P., Walia, A., Kulshrestha, S., Chauhan, A., Shirkot, C.K., 2015. Efficiency of plant growth-promoting P-solubilizing *Bacillus circulans* CB7 for enhancement of tomato growth under net house conditions. J. Basic Microbiol. 55 (1), 33−44.

Mir, Z.A., Ali, S., Tyagi, A., Ali, A., Bhat, J.A., Jaiswal, P., et al., 2017. Degradation and conversion of endosulfan by newly isolated *Pseudomonas mendocina* ZAM1 strain. 3 Biotech 7 (3), 211.

Miransari, M., Smith, D.L., 2014. Plant hormones and seed germination. Environ. Exp. Bot. 99, 110−121.

Mukherjee, I., Kumar, A., 2012. Phytoextraction of endosulfan a remediation technique. Bull. Environ. Contam. Toxicol. 88 (2), 250−254.

Myresiotis, C.K., Vryzas, Z., Papadopoulou-Mourkidou, E., 2012. Biodegradation of soil-applied pesticides by selected strains of plant growth-promoting rhizobacteria (PGPR) and their effects on bacterial growth. Biodegradation 23 (2), 297−310.

Nadeem, S.M., Zahir, Z.A., Naveed, M., Arshad, M., 2007. Preliminary investigations on inducing salt tolerance in maize through inoculation with rhizobacteria containing ACC deaminase activity. Can. J. Microbiol. 53 (10), 1141−1149.

Naik, M.K., Rajalaxmi, K., Rangeshwaran, R., Amaresh, Y.S., Reddy, M.S., 2013. Search for 2, 4 DAPG positive gene in fluorescent *Pseudomonas* and their exploitation for sustainable disease management. In Recent Advances in Biofertilizers and Biofungicides (PGPR) for Sustainable Agriculture. Proceedings of third Asian Conference on Plant Growth-Promoting Rhizobacteria (PGPR) and other Microbials, Manila, Philippines, 21−24 April 2013. Asian PGPR Soc. Sustain. Agric. pp. 152−163.

Narkhede, C.P., Patil, A.R., Koli, S., Suryawanshi, R., Wagh, N.D., Patil, S.V., 2015. Studies on endosulfan degradation by local isolate *Pseudomonas aeruginosa*. Biocatal. Agric. Biotechnol. 4 (2), 259−265.

National Institute of Occupational Health (NIOH), 2003. Final report of the investigation of unusual illnesses allegedly produced by endosulfan exposure in Padre village of Kasargod District (N. Kerala). National Institute of Occupational Health, Ahmedabad. Indian Council for Medical Research.

Niu, D.D., Liu, H.X., Jiang, C.H., Wang, Y.P., Wang, Q.Y., Jin, H.L., et al., 2011. The plant growth–promoting rhizobacterium *Bacillus cereus* AR156 induces systemic resistance in *Arabidopsis thaliana* by simultaneously activating salicylate- and jasmonate/ethylene-dependent signaling pathways. Mol. Plant Microbe Interact. 24 (5), 533–542.

Ozdal, M., Ozdal, O.G., Algur, O.F., Kurbanoglu, E.B., 2017. Biodegradation of α-endosulfan via hydrolysis pathway by *Stenotrophomonas maltophilia* OG2. 3 Biotech 7 (2), 113.

Park, K.H., Lee, C.Y., Son, H.J., 2009. Mechanism of insoluble phosphate solubilization by *Pseudomonas fluorescens* RAF15 isolated from ginseng rhizosphere and its plant growth-promoting activities. Lett. Appl. Microbiol. 49 (2), 222–228.

Phi, Q.T., Park, Y.M., Seul, K.J., Ryu, C.M., Park, S.H., Kim, J.G., et al., 2010. Assessment of root-associated *Paenibacillus polymyxa* groups on growth promotion and induced systemic resistance in pepper. J. Microbiol. Biotechnol. 20 (12), 1605–1613.

Pushpanjali, Pal, A.K., Prasad, R.L., Prasad, A., Singh, S.K., Kumar, A., et al., 2005. In ovo embryotoxicity of α-endosulfan adversely influences liver and brain metabolism and the immune system in chickens. Pestic. Biochem. Physiol. 82 (2), 103–114.

Quijano, R.F., 2000. Risk assessment in a third-world reality: an endosulfan case history. Int. J. Occup. Environ. Health 6 (4), 312–317.

Rajasankar, R., Gayathry, G.M., Sathiavelu, A., Ramalingam, C., Saravanan, V.S., 2013. Pesticide tolerant and phosphorus solubilizing *Pseudomonas* sp. strain SGRAJ09 isolated from pesticides treated *Achillea clavennae* rhizosphere soil. Ecotoxicology 22 (4), 707–717.

Rajkumar, M., Freitas, H., 2008. Effects of inoculation of plant-growth promoting bacteria on Ni uptake by Indian mustard. Bioresour. Technol. 99 (9), 3491–3498.

Rajkumar, M., Nagendran, R., Lee, K.J., Lee, W.H., Kim, S.Z., 2006. Influence of plant growth promoting bacteria and Cr^{6+} on the growth of Indian mustard. Chemosphere 62 (5), 741–748.

Rajkumar, M., Ae, N., Prasad, M.N.V., Freitas, H., 2010. Potential of siderophore-producing bacteria for improving heavy metal phytoextraction. Trends Biotechnol. 28 (3), 142–149.

Ramaneswari, K., Rao, L.M., 2000. Bioconcentration of endosulfan and monocrotophos by *Labeo rohita* and *Channa punctata*. Bull. Environ. Contam. Toxicol. 65 (5), 618–622.

Rekha, P.D., Lai, W.A., Arun, A.B., Young, C.C., 2007. Effect of free and encapsulated *Pseudomonas putida* CC-FR2–4 and *Bacillus subtilis* CC-pg104 on plant growth under gnotobiotic conditions. Bioresour. Technol. 98 (2), 447–451.

Richardson, A.E., Barea, J.M., McNeill, A.M., Prigent-Combaret, C., 2009. Acquisition of phosphorus and nitrogen in the rhizosphere and plant growth promotion by microorganisms. Plant Soil 321 (1–2), 305–339.

Rodriguez, H., Gonzalez, T., Goire, I., Bashan, Y., 2004. Gluconic acid production and phosphate solubilization by the plant growth-promoting bacterium *Azospirillum* spp. Naturwissenschaften 91 (11), 552–555.

Sachdev, D.P., Chaudhari, H.G., Kasture, V.M., Dhavale, D.D., Chopade, B.A., 2009. Isolation and characterization of indole acetic acid (IAA) producing *Klebsiella pneumoniae* strains from rhizosphere of wheat (*Triticum aestivum*) and their effect on plant growth.

Saiyed, H., Dewan, A., Bhatnagar, V., Shenoy, U., Shenoy, R., Rajmohan, H., et al., 2003. Effect of endosulfan on male reproductive development. Environ. Health Perspect. 111 (16), 1958–1962.

Saleem, M., Arshad, M., Hussain, S., Bhatti, A.S., 2007. Perspective of plant growth promoting rhizobacteria (PGPR) containing ACC deaminase in stress agriculture. J. Ind. Microbiol. Biotechnol. 34 (10), 635–648.

Schmidt, W., 1999. Mechanisms and regulation of reduction-based iron uptake in plants. New Phytol. 141 (1), 1–26.

Sethunathan, N., Megharaj, M., Chen, Z., Singh, N., Kookana, R.S., Naidu, R., 2002. Persistence of endosulfan and endosulfan sulfate in soil as affected by moisture regime and organic matter addition. Bull. Environ. Contam. Toxicol. 68 (5), 725–731.

Sharma, S., Nagpure, N.S., Kumar, R., Pandey, S., Srivastava, S.K., Singh, P.J., et al., 2007. Studies on the genotoxicity of endosulfan in different tissues of fresh water fish *Mystus vittatus* using the comet assay. Arch. Environ. Contam. Toxicol. 53 (4), 617–623.

Sharma, S.K., Johri, B.N., Ramesh, A., Joshi, O.P., Prasad, S.S., 2011. Selection of plant growth-promoting *Pseudomonas* spp. that enhanced productivity of soybean-wheat cropping system in central India. J. Microbiol. Biotechnol. 21, 1127–1142.

Sheng, X.F., Xia, J.J., 2006. Improvement of rape (*Brassica napus*) plant growth and cadmium uptake by cadmium-resistant bacteria. Chemosphere 64 (6), 1036–1042.

Sheng, X.F., Jiang, C.Y., He, L.Y., 2008. Characterization of plant growth-promoting *Bacillus edaphicus* NBT and its effect on lead uptake by Indian mustard in a lead-amended soil. Can. J. Microbiol. 54 (5), 417–422.

Shivaramaiah, H.M., Kennedy, I.R., 2006. Biodegradation of endosulfan by a soil bacterium. J. Environ. Sci. Health B 41 (6), 895–905.

Siddique, T., Okeke, B.C., Arshad, M., Frankenberger, W.T., 2003. Enrichment and isolation of endosulfan-degrading microorganisms. J. Environ. Qual. 32 (1), 47–54.

Silva, M.H., Beauvais, S.L., 2010. Human health risk assessment of endosulfan. I: Toxicology and hazard identification. Regul. Toxicol. Pharmacol. 56 (1), 4–17.

Singh, S.N. (Ed.), 2016. Microbe-Induced Degradation of Pesticides. Springer.

Singh, N.S., Singh, D.K., 2011. Biodegradation of endosulfan and endosulfan sulfate by *Achromobacter xylosoxidans* strain C8B in broth medium. Biodegradation 22 (5), 845–857.

Singh, M., Singh, D.K., 2014. Biodegradation of endosulfan in broth medium and in soil microcosm by *Klebsiella* sp. M3. Bull. Environ. Contam. Toxicol. 92 (2), 237–242.

Sinha, S., Mukherjee, S.K., 2008. Cadmium–induced siderophore production by a high Cd-resistant bacterial strain relieved Cd toxicity in plants through root colonization. Curr. Microbiol. 56 (1), 55–60.

Sivasankari, B., Anandharaj, M., 2014. Isolation and molecular characterization of potential plant growth promoting *Bacillus cereus* GGBSTD1 and *Pseudomonas* spp. GGBSTD3 from vermisources. Adv. Agric. Article ID 248591.

Son, H.J., Park, G.T., Cha, M.S., Heo, M.S., 2006. Solubilization of insoluble inorganic phosphates by a novel salt-and pH-tolerant *Pantoea agglomerans* R-42 isolated from soybean rhizosphere. Bioresour. Technol. 97 (2), 204–210.

Spaepen, S., Vanderleyden, J., 2011. Auxin and plant-microbe interactions. Cold Spring Harb. Perspect. Biol. 3 (4), p.a001438.

Spaepen, S., Vanderleyden, J., Remans, R., 2007. Indole-3-acetic acid in microbial and microorganism-plant signaling. FEMS Microbiol. Rev. 31 (4), 425–448.

Sutherland, T.D., Horne, I., Lacey, M.J., Harcourt, R.L., Russell, R.J., Oakeshott, J.G., 2000. Enrichment of an endosulfan-degrading mixed bacterial culture. Appl. Environ. Microbiol. 66 (7), 2822–2828.

Sutherland, T.D., Horne, I., Harcourt, R.L., Russell, R.J., Oakeshott, J.G., 2002. Isolation and characterization of a *Mycobacterium* strain that metabolizes the insecticide endosulfan. J. Appl. Microbiol. 93 (3), 380–389.

Tank, N., Saraf, M., 2009. Enhancement of plant growth and decontamination of nickel-spiked soil using PGPR. J. Basic Microbiol. 49 (2), 195–204.

Thakuria, D., Talukdar, N.C., Goswami, C., Hazarika, S., Boro, R.C., Khan, M.R., 2004. Characterization and screening of bacteria from rhizosphere of rice grown in acidic soils of Assam. Curr. Sci. 978–985.

Thangadurai, P., Suresh, S., 2014. Biodegradation of endosulfan by soil bacterial cultures. Int. Biodeterior. Biodegradation 94, 38–47.

Tiwari, M.K., Guha, S., 2013. Kinetics of the biodegradation pathway of endosulfan in the aerobic and anaerobic environments. Chemosphere 93 (3), 567–573.

Tripathi, M., Munot, H.P., Shouche, Y., Meyer, J.M., Goel, R., 2005. Isolation and functional characterization of siderophore-producing lead- and cadmium-resistant *Pseudomonas putida* KNP9. Curr. Microbiol. 50 (5), 233–237.

UNEP/FAO, 2007. Inclusion of the chemical endosulfan in Annex III of the Rotterdam convention. United Nations Environment Programme and Food and Agriculture Organization, UNEP/FAO/RC/COP.4/9. http://www.pic.int/COPS/COP4/i9)/English/K0763331%20COP-4–9.pdf.

Valverde, A., Burgos, A., Fiscella, T., Rivas, R., Velazquez, E., Rodríguez-Barrueco, C., et al., 2007. Differential effects of coinoculations with *Pseudomonas jessenii* PS06 (a phosphate-solubilizing bacterium) and *Mesorhizobium ciceri* C-2/2 strains on the growth and seed yield of chickpea under greenhouse and field conditions. First International Meeting on Microbial Phosphate Solubilization. Springer, Dordrecht, pp. 43–50.

Venkatraman, L., 2004. Will banning endosulfan solve the problem? The Hindu business line. July 6. Online at: http://www.thehindubusinessline.com/2004/07/06/stories/2004070600790300.htm.

Vivas, A., Biro, B., Ruiz-Lozano, J.M., Barea, J.M., Azcon, R., 2006. Two bacterial strains isolated from a Zn-polluted soil enhance plant growth and mycorrhizal efficiency under Zn-toxicity. Chemosphere 62 (9), 1523–1533.

Vyas, P., Gulati, A., 2009. Organic acid production in vitro and plant growth promotion in maize under controlled environment by phosphate-solubilizing fluorescent *Pseudomonas*. BMC Microbiol. 9 (1), 174–188.

Wani, P.A., Khan, M.S., 2010. *Bacillus* species enhance growth parameters of chickpea (*Cicer arietinum* L.) in chromium stressed soils. Food Chem. Toxicol. 48 (11), 3262–3267.

Wu, C.H., Wood, T.K., Mulchandani, A., Chen, W., 2006. Engineering plant-microbe symbiosis for rhizoremediation of heavy metals. Appl. Environ. Microbiol. 72 (2), 1129–1134.

Yuquan, L., 2000. Genotoxic effects of alpha-endosulfan and beta-endosulfan on human HepG2 cells. Environ. Health Perspec. 108 (6), 559–561.

Zaffar, H., Ahmad, R., Pervez, A., Naqvi, T.A., 2018. A newly isolated *Pseudomonas* sp. can degrade endosulfan via hydrolytic pathway. Pestic. Biochem. Physiol. 152, 69–75.

Zahir, Z.A., Munir, A., Asghar, H.N., Shaharoona, B., Arshad, M., 2008. Effectiveness of rhizobacteria containing ACC deaminase for growth promotion of peas (*Pisum sativum*) under drought conditions. J. Microbiol. Biotechnol. 18 (5), 958–963.

Zaidi, S., Usmani, S., Singh, B.R., Musarrat, J., 2006. Significance of *Bacillus subtilis* strain SJ-101 as a bioinoculant for concurrent plant growth promotion and nickel accumulation in *Brassica juncea*. Chemosphere 64 (6), 991–997.

Influence of land use change on native microbial community and their response to the variations in micro environment

15

Snigdha Singh and R. Hiranmai Yadav
School of Environment and Sustainable Development, Central University of Gujarat, Gandhinagar, India

15.1 Introduction

A landscape can be considered as a geographical and/or an ecological space where organisms thrive. It can be heterogeneous in terms of resource availability. In order to explore the dynamics of land forms, human activities (i.e., land uses), and organisms, a landscape can be employed as a powerful model. "Landscape dynamics" refers to every single alteration that happens in the physical, biological, and cognitive assets of a landscape. Variation is one distinctive characteristic feature of landscapes. These changes lead to many ecological processes such as species movement and physical flows. Furthermore, anthropogenic activities are significant drivers of change. The prerequisite for the management of natural resources is the observation and understanding of the changes occurring in the ecosystem. Photos, census, and maps are the different analytical tools that bring in various information in order to detect changes in the landscape dynamics (Baudry, 1989).

Landscape dynamics are compelled by external agitations, such as fires, volcanic eruptions, earthquakes, extreme climate events, erosion, and human intrusion—and by internal perturbations, such as physical flows of energy and matter, succession, population fluctuations, and community turnover. Landscape dynamics comprise properties of the landscape, such as stability, persistence, resistance, resilience, and recovery, that operate along a broad range of temporal and spatial scales, such as shifting mosaic steady-state, and equilibrium spatial properties. Landscape dynamics are very significant for land management and conservation. The analysis of landscape dynamics must be conducted with robust conceptualization and explicit modeling taking into consideration the complexity of the processes that occur in the landscape and the different scales at which individuals, populations, and communities respond to environmental constraints. In particular, landscape dynamics can be studied at discontinuities or ecotones, or by evaluating the level of connectivity.

Landscapes changes differ in space and time. No landscape remains static, however the time scales of changes may show a discrepancy from month to millennia.

Microbes in Land Use Change Management. DOI: https://doi.org/10.1016/B978-0-12-824448-7.00018-8
© 2021 Elsevier Inc. All rights reserved.

Those changes are interesting and are important as drivers of different aspects of ecological processes, including biodiversity. Therefore attaining information on landscape changes and factors affecting those changes is an important aspect for the understanding and better management of our environment. In the past few years, the environmental effect of land-use change on different scales has gradually become a research hotspot (Zhang et al., 2019).

15.2 Urban transportation development

The transport sector of India is very diverse and huge catering to the transport needs of 1.1 billion people nationwide. The sector contributed about 5.2% of the nation's GDP, in 2012–13 (Agarwal, 2006). For the sake of economic growth, good physical connectivity among the rural as well as urban areas is essential. The growing economy of India has witnessed an upsurge in demand for transport infrastructure and services since the early 1990s. In order to sustain high economic growth, competent and reliable urban transportation services are crucial for the country (Estache, 2007).

The microbial communities existing in the urban soil have great significance in biogeo-transformation of urban waste emitted as they are the principal drivers of soil organic matter dynamics and perform a positive role under the enormous anthropogenic pressure (Zhang et al., 2020). The microbial community of the urban soil is very sensitive to the soil micro-environment. Yet, the response of soil microbial community to urbanization and the anthropogenic load remains unclear. The soil microorganisms in urban conditions can be influenced by plenty of factors that can cause stress. These factors include altered water regime, elevated temperatures, and high contamination levels of both inorganic as well as organic compounds. The qualitative and quantitative parameters of soil microbial communities can be influenced by the anthropogenic impacts.

For instance, a study carried out by Matinian et al. (2020), in Saint Petersburg, Russia a noteworthy variation in the structure of the microbial community was observed between the natural soil of the recreational and the industrial zone. Together with their heterogeneity, unique organic matter inputs, physical structure, chemical composition, temperature, and moisture regimes, and exposure to past and present anthropogenic activities urban soils are dissimilar to agricultural or pasture soils in numerous ways (Bray and Wickings, 2019; Lehmann and Stahr, 2007). For that reason, even slight disturbances in urban soil can significantly influence the microbial community.

15.3 Land use changes

15.3.1 Influence of land use change on environment

For the degradation and changes in the ecosystem, perhaps the change in the land use pattern is the most prevalent socioeconomic factor. The landscape of the Earth has been significantly transformed by several processes such as deforestation, agricultural practices, urban development as well as other anthropogenic activities. These instabilities of the land have affected the important ecosystem processes and services which can have wide-ranged long-term consequences. Farmland utilizes open space and valuable habitat for various wildlife species. On the other hand, intensive agriculture has potentially severe consequences on the ecosystem. For instance, it has been observed that agricultural practices and the land use pattern can cause water pollution which is further influenced by government policies.

Agricultural runoff has become a leading source of water pollution both in coastal as well as inland waters. Many of the wildlife species have gone on the verge of extinction due to the conversions of wetlands to agricultural lands. Many of the ecosystem services have been provided by forests as they support biodiversity by providing critical habitat for wildlife, slowing down the surface runoff, reducing flooding and soil erosion, intercepting precipitation and removing carbon dioxide from the atmosphere. Consequently the conversion of the forest land to urban development and agricultural practices results in the destruction and reduction of the important ecosystem services. For instance several anthropogenic activities along with agriculture, deforestation, and urban sprawl, etc. have resulted in substantial alteration and disintegration of the Earth's vegetative cover. The outcome of these disturbances affect the local, regional, and global climate along with resulting in the variation in the principal heat-trapping gas and global atmospheric carbon dioxide concentration.

Land use and land management practices have a major impact on natural resources including water, soil, air, nutrients, plants, and animals as listed below:

- Intensive farming and deforestation may cause soil erosion, salinization, desertification, and other soil degradations.
- Deforestation adds to the greenhouse effect, destroys habitats that support biodiversity, affects the hydrological cycle and increases soil erosion, runoff, flooding and landslides.
- Habitat destruction, fragmentation, and alteration associated with urban development are a leading cause of biodiversity decline and species extinctions.

Natural and anthropogenic changes both are included in the assessment of the landscape dynamics of the ecosystem. The relationship between the land use, species diversity, ecosystem health, and interactions is described with the inherent disturbances processes and biophysical capabilities.

15.3.2 Influence of land use change on biotic environment

In recent decades major changes in biodiversity are occurring due to the expansion and intensification of human land use. The biodiversity is altered at all the levels as a result of land use change. As an example, a reduction in population size and decrease in the genetic diversity within a species is observed as a result of reduction in the habitat from land use changes. On the other end, land use change commonly results in more homogenous landscapes leading to reduced ecosystem diversity (Flather et al., 1998). Rates of land-use change around the world and consequences on biodiversity are quantified by the emergence of the remote sensing data collected from satellites.

These studies have revealed that in recent decades, land use in the formerly occupied areas has been intensified and has expanded into the formerly natural habitats. The most evident repercussions of land use change are perhaps the loss of habitat along with its fragmentation and degradation. Transformation of the natural habitats to agriculture or other intensive human land use may cause these areas to become inhospitable for many native species. This conversion also results in the reduction of the area of natural habitats.

A few of the consequences of land use change are much less visible, since they encompass not habitats but the organisms within the habitats. Anthropogenic activities often result in the introduction of alien species and pathogens along with the variation in the distribution as well as the population of the native species. As a result there is a change in the ecosystem traits as well as biotic interactions among the species. The biotic interactions that are affected includes predation, competition, parasitism, and disease (Marzluff, 2001).

There is an emergent acknowledgment of the prominence of understanding the response of bacterial communities to land use change. Soil bacteria encompass the majority of global biodiversity and facilitate most biogeochemical cycles (Hug et al., 2016).

The biochemical cycles are often intensely altered when the ecosystem goes through conversion like from forest to agriculture (DeFries et al., 1999; Neill et al., 1997, 2005; Verchot et al., 2000). Land use can alter through water, heat, and other environmental conditions, the soil nutrient cycle and soil microbial activity and hence results in the variation of the structure of the soil microbial community (Bi et al., 2010). According to a study carried out by Deng et al. (2018) the results showed that under different land-use patterns, the metabolic activity of the soil microbial community was significantly different. The study also showed that the Shannon index of the soil microbial community in a maize farmland was lesser than that in other land-use patterns (Deng et al., 2018).

15.4 Soil as a microhabitat

Soil is vital and irreplaceable as it administers the productivity of plants in the terrestrial ecosystems. It also plays a key role in maintaining the biogeochemical cycles since the inhibiting microorganisms in the soil helps in the degradation of virtually all organic compounds such as persistent xenobiotics and naturally occurring polyphenolic compounds. Mircoflora, micro fauna, macrofauna, and mesofauna are the living population inhibiting the soil in general (Nannipieri et al., 2003).

Soil acts as a primary medium on which various living and nonliving components flourish and is an important part of the environment. The degradation of the fertile soil is the most sudden effect of development of roads on the soil

(Tsunokawa and Hoban, 1997). The delicate stabilizing factors like slopes, vegetation, micro fauna and flora is destabilized due to road activities. The soil may be contaminated due to many reasons as from heavy traffic activities on roads. Several metals such as Pb, Zn, and Cr persist for very long time in the soil atmosphere. The contaminants accumulated in the roadside soil have adverse effects on the growth and life span of soil organisms as well as vegetation and thus affecting the degradation of the soil. These effects customarily take place at confined levels affecting only a restricted band (Bogemans et al., 1989).

The de-icing salts usually calcium chloride, magnesium chloride, sodium chloride, and potassium chloride increases ion concentration in the soil resulting in the ultimate alteration in the chemical composition and pH of the soil (Tsunokawa and Hoban, 1997; Siddqiui and Iqbal, 1994; Garcia-Miragaya et al., 1981; Clift et al., 1983; Gjessing et al., 1984; Araratyan and Zakharyan, 1988).

15.5 Soil micro-floral population

There is a complex and enormous microbial diversity present in the soil (Nannipieri et al., 2003; Torsvik and Øvreås, 2002). There are a number of studies carried out in order to determine the effects of biotic and abiotic factors on soil microbes. To understand the diversity of the soil bacterial community identification of the dominant factors affecting the soil bacterial community is crucial. Soil pH is generally considered to be an important factor in determining the distribution of soil bacterial communities in various studies. (Fierer and Jackson, 2006; Shen et al., 2013; Lauber et al., 2009; Chu et al., 2010; Rousk et al., 2010). Soil pH predominantly determines the bacterial diversity as shown in the recent studies (Sun et al., 2015; Zhalnina et al., 2015), along with other soil parameters that also influences the composition and diversity of the bacterial community. Soil organic carbon and C/N ratios are considered as a significant influencers of soil microbial communities (Fierer et al., 2007; Tripathi et al., 2012; Sul et al., 2013).

Availability of the soil nutrient has been found to have a noteworthy influence on microbial communities present in the soil (Lauber et al., 2008). A significant correlation has been found between the structure of the soil microbial community with the salinity and soil moisture (Brockett et al., 2012; Rajaniemi and Allison, 2009). In some specific ecosystems several studies have inferred that the climate-change factors including temperature, pressure and elevated level of CO_2 can have an effect on the composition of the soil microbial communities (Rinnan et al., 2009; Kuffner et al., 2012; Yergeau et al., 2012; Chen et al., 2015). As stated by Yuan et al. (2014), an important factor for the change of soil bacterial communities would be precipitation. In this way, soil microbial communities can be influenced by a large number of factors, but the driving factor may be different in different ecosystems.

The quality and functionality of the urban environment can be determined by the biological characteristics of the urban soil that acts as bioindicators (Matinian et al.,

2020). The living organisms like plants and microbes are considered as bioindicators which are used for the screening of the urban health (Parmar et al., 2016). Bacteria, fungi, algae, and actinomycetes are the principal groups of the microbial communities inhibiting the urban soil. Microbes are easily used for assessment for instance, stress-response proteins are synthesized when the microbes are exposed to Cd. This stress-response protein can be used as primary signs of Cd pollution (Khatri and Tyagi, 2015). In the urban ecosystem, as compared to the other pollutants, bacteria are readily available due to its abundance (Tan et al., 2019; Torres et al., 2019).

The ecosystem processes at all scales are regulated by the soil microorganisms along with contribution in providing power for the carbon and nitrogen cycles. Many factors can affect the microbial diversity of the soil (Don and Kalbitz, 2005). Since the soil microbial traits are more sensitive as compared to the chemical and physical properties of soil, these are potentially the early indicators of the variations in soil quality as shown by the increasing evidences (Brookes, 2006; Geisseler and Horwath, 2009). The biomass of the microbes has been regarded as a sensitive index for variations in the soil organic carbon that results from variation in management and land use.

The microbial biomass undergoes variability in the primary phases, until it reaches a new equilibrium (Powlson et al., 1987). The biogeography is driven by special soil chemical properties that fluctuate among land-use types and is signified by the difference in the microbial community composition (Wal et al., 2006). Studies carried out by Johnson et al. (2003) and Lauber et al. (2008) have shown that pronounced effects on microbial communities has been observed due to the alterations in land-use type.

Natural ecosystems have been strongly altered by the destruction of habitat, conversion in land use, climate change and other intensive anthropogenic activities (Pabst et al., 2013). In local ecosystems, thorough soil agitations can bring change in the inherent soil properties by stimulating soil nutrient cycles and amending soil chemical, physical, and microbiological properties along with the structures of dwelling microbial communities (Lauber et al., 2008; Lin et al., 2011; Kuramae et al., 2012). For that reason, land-use or land-cover change is a vital factor affecting soil quality. The structure of the soil community and the microbial diversity functions as a sensitive indicator of soil quality and health. Hence, it acts quickly in response to the conversion of land use (He et al., 2008).

The conversion of the land-use can encourage long-term and significant effects on the microbial biomass, moisture content, temperature, pH, aeration, nutrient status and enzymatic activities of the soil. The conversion are stimulated due to the fluctuations in the composition of plant community and different management practices across land use type (Murty et al., 2002; Nishimura et al., 2008; Rahman et al., 2008; Wallenius et al., 2011).

The change in the land-use pattern has been confirmed to have a significant effect on the microbial community structure of the soil (Steenwerth et al., 2002; Johnson et al., 2003; Lauber et al., 2008). Alterations in plant cover and soil properties remain poorly understood whereas variations in the richness of soil-specific taxonomic groups is reformed during land-use conversion. The microbial communities of soil are designed by geology, land use, climate and soil chemical and physical factors like pH, porosity, and soil organic matter (Thomson et al., 2010, 2015; Griffiths et al., 2011; Hartmann et al., 2012). The conversion of the land use may lead to adverse alteration of quite a few environmental variables, which can indirectly or directly affect soil microbial diversity.

For that reason, during land-use conversion, the consequences of deviations in soil microbial diversity and community structure should be explored to provide a significant evidence on microbial responses to changes in the soil environment and interactions between soil microbial communities and soil chemical and physical properties. The main drivers of decline and deterioration of biodiversity and ecosystem services are the changes in land use pattern, that is, use from natural forest to commercial plantation and conventional agricultural landscape (Barnes et al., 2014, 2017; Kerfahi et al., 2014; McGuire et al., 2015; Ocampo-Peñuela et al., 2018; Rodrı́guez-Echeverry et al., 2018; Schneider et al., 2015). The soil community composition and soil microbial diversity including protozoa, fungi, bacteria and archaea are considerably transformed due to the effects of land use transition from the natural forest to other land uses (Falcucci et al., 2007; Newbold et al., 2015; Ding et al., 2013; Zhao et al., 2005; Krashevska et al., 2018). Numerous studies have been carried out in order to understand the association between land use changes and the shift in soil chemical attributes.

In the tropical rainforest of Southeast Asia and Latin America, numerous studies have been conducted to determine the effect of expansion of commercial plantation at the expense of natural forests (Ocampo-Peñuela et al., 2018; Sul et al., 2013; Cao et al., 2017). Studies have shown that changes in land use pattern results in substantial and often continuing influences on soil nutrient cycles and soil texture (Lauber et al., 2008; Murty et al., 2002). There is a shift observed in the composition of the soil bacterial and fungal communities due to these transformation in the physicochemical parameters of the soil (Elliott, 1986; DeGryze et al., 2004; Drenovsky et al., 2010; Wu et al., 2003). At large, alteration of natural habitats rigorously affects the natural vegetation diversity. Through addition a variety of organic matter it plays a momentous role in regulating the soil quality which further helps in the composition of the soil microbial community (Batten et al., 2006; Wardle, 2006).

15.6 Population changes and associated factors

15.6.1 Stress caused by transportation infrastructure on soil microbes

The main activity for the development of industrial units is the road construction. It has led to the green belt loss and subsequent biodiversity loss in the region. Microbial population and their associated activities in the soil is the resultant of complex interactions existing among the biotic as well as abiotic factors. For holding the complex soil system in a dynamic equilibrium, the interactions of physical, chemical, and biochemical factors are responsible. Vegetation and soil characteristics have largely regulated the population dynamics of the microbes inhibiting the soil (Joshi et al., 2010).

In the urban environment, a number of factors influence the soils and its microbial communities. A large number of anthropogenic activities including roads and communication constructions, industrial production, building construction, transport use, waste management, introduction of new plant and animal species, recreation completely the environment and the soil as its foremost element. For that reason, the anthropogenic impact turn out to be the leading factor in determining the urban soil microbial communities, associated to the soils of rural and natural areas. Disturbance of soil profiles is often resulted as an impact of construction of roads, buildings and infrastructure objects. As a result of the income of various types of waste, cement, sand, and dust into the upper horizons these disturbances can be intended (new soil profiles construction for reclamation, or soil sealing with asphalt or concrete) or unintended.

Soil profile changes in urban areas can be classified into several main types, sealing with an impermeable layer (concrete, asphalt). There are many soil contamination sources in the urban environment including transport, industrial enterprises, use of pesticides in parks and garden areas, and construction works (Müller et al., 2020; Sumampouw and Risjani, 2014; Torres et al., 2019). The soil can accumulate pollutants and acts as a natural filter that reaches high concentrations eventually and affects the soil microbial communities adversely. Synthetic nonnatural compounds are of particular threat inflowing in the nature as a result of various anthropogenic activities. Following groups of organic pollutants are most often considered in the urban soil; polycyclic aromatic hydrocarbons, polychlorinated biphenyls, dioxins, dibenzofurans, and phthalates, along with alicyclic and aliphatic hydrocarbons contained in petroleum products.

Inorganic pollutants are also common in urban soils, and heavy metals are the most studied among those. In the road construction and maintenance, many cities use a technical salt containing up to 99% sodium chloride to combat road icing. The increased concentration of various salts including chloride and sodium salts can affect the soil structure, impair water, and air permeability. It also causes osmotic stress in several microbial species (Černohlávková et al., 2008). The weakly sorbed metals from soil sorption sites are displaced by the readily soluble sodium ions leading to an increase in their toxicity and bioavailability (Linde et al., 2007). Excessive intake of calcium ions results in significant increase in the pH of the urban soil thus affecting the soil properties and its microbial community structure (Puskás and Farsang, 2009). The ratio and species composition of microbial communities in the soil is disturbed under salinization conditions, whereas the metabolic activity is suggestively reduced (Yuan et al., 2007). Ke et al. (2013) have reported similar results which shows that chloride de-icing salts reduces the diversity of the fungal community and has changed the dominant bacterial species in the urban soil.

15.7 Vehicular emission impact on soil and microbes

Urban development has been linked to many environmental problems, including air pollution, water pollution, and loss of wildlife habitat. Urban runoff often contains nutrients, sediment and toxic contaminants, and can cause not only water pollution but also large variation in stream flow and temperatures. Habitat destruction, fragmentation, and alteration associated with urban development have been identified as the leading causes of biodiversity decline and species extinctions (Czech et al., 2000; Soulé, 1991). Urban development and intensive agriculture in coastal areas and further inland are a major threat to the health, productivity, and biodiversity of the marine environment throughout the world.

15.7.1 NH development and its influence - case study of Gujarat

Transport network of India is one of the most extensive in the world. For unremitting and inclusive evolution of the economy and consolidation of social diversification of the nation road network is crucial. The road transport is considered to be more resilient and consumer friendly when compared to other transport means, as it provides door-to-door service to the people even in the remote parts of the country. With emerging progress, people are more and more reliant on transport system for accomplishing their needs and expansions. Thus these road activities greatly affect the environment qualities (Bhardwaj et al., 2019). Transportation has become an essential part of our daily lives and plays a fundamental role in the development and economy of the country. Transportation leads to growth by increasing access to basic requirements of living, that is, education and health and by providing opportunity of trade both at national and international fronts.

Transport activities cause adverse impacts on the human health and cause degradation to the environmental qualities irrespective of the wide range of benefits. Biodiversity along with air, soil and water qualities have been affected by the expansion of roads. The Indian national highways network is a network of highways that connects major ports, cities, big industrial and tourists center and state capitals. It spreads throughout the country and is managed by numerous governments of India agencies. The connectivity with the district headquarters, tourist centers, and national highways is provided by the state highways.

Profound disturbances are caused to the naturally existing vegetation and soil by the construction of roads in Australian Alps. Within a landscape, anthropogenic activities often result in land fragmentation with the partition of the landscape by the elements such as roads contributing to the loss of habitat and proliferation of exotic species (Reed et al., 1996; Saunders et al., 2002).

Roadside disturbances impart the alteration of the physical and chemical properties of the soil and even degrades its structure, stability, permeability, erodability, and porosity. However, it also indirectly influences biological and chemical properties of the soil including organic content, soil biota, chemical conditions, and the microclimatic conditions of the soil (Kuss et al., 1990). Most of these impacts lead to compromise primary ecosystem function, hydrological cycles, nutrient movement and availability, and surface stability (Johnston and Johnston, 2004).

15.8 Conclusion

In the urban environment, a number of factors influence the soils and its microbial communities. A large number of anthropogenic activities including roads and communication constructions, industrial production, building construction, transport use, waste management, introduction of new plant and animal species, recreation completes the environment and the soil as its foremost element. Disturbance is an important ecological process, and studies of disturbance have made key contributions to the development of landscape and ecosystem ecology. However, the soil microclimate along with the biological and chemical properties including organic content, chemical conditions, and soil biota have been indirectly influenced by the roadside disruption. A lot of these influences have the probability to compromise primary ecosystem function, hydrological cycles, nutrient movement and availability, and surface stability. It is imperative that we think confidently about how to better understand and adapt to these changes. Forthcoming trends in disturbance size, frequency, and severity are challenging to predict, and changes in disturbance will differ between regions. However, amongst the many persuasive challenges that command consideration, ecologists must escalate efforts to comprehend and anticipate the causes and consequences of fluctuating disturbance regimes and engage in the policy process.

References

Agarwal, O.P., 2006. Urban transport. In: Rastogi, A. (Ed.), India Infrastructure Report 2006: Urban Infrastructure. Oxford University Press, pp. 106–129.

Araratyan, L.A., Zakharyan, S.A., 1988. On the contamination of snow along main highways. 41, pp. 514–519.

Barnes, A.D., Jochum, M., Mumme, S., Haneda, N.F., Farajallah, A., Widarto, T.H., et al., 2014. Consequences of tropical land use for multitrophic biodiversity and ecosystem functioning. Nat. Commun. 5, 5351.

Barnes, A.D., Allen, K., Kreft, H., Corre, M.D., Jochum, M., Veldkamp, E., et al., 2017. Direct and cascading impacts of tropical land-use change on multi-trophic biodiversity. Nat. Ecol. Evol. 1 (10), 1511.

Batten, K.M., Scow, K.M., Davies, K.F., Harrison, S.P., 2006. Two invasive plants alter soil microbial community composition in serpentine grasslands. Biol. Invasions 8 (2), 217–230.

Baudry, J., 1989 Biodiversity: structure and function, vol. 2—Landscape Dynamics.

Bhardwaj, A., Aggarwal, R.K., Bhardwaj, S.K., 2019. A review on impacts of road activities and vehicular emissions on native ecosystems in mountainous region. Curr. World Environ. 14 (2), 194–204.

Bi, M., Yu, W., Jiang, Z., Ma, Q., Zhang, L., Xu, Y., 2010. Study on the effects of different land use patterns on microbial community structure in aquic brown soil by utilizing PLFA method. Sci. Agric. Sin. 43, 1834–1842.

Bogemans, J., Nierinck, L., Stassart, J.M., 1989. Effect of deicing chloride salts on ion accumulationin spruce (*Picea abies (L.)* sp.). Plant Soil 113 (1), 3–11.

Bray, N., Wickings, K., 2019. The roles of invertebrates in the urban soil microbiome. Front. Ecol. Evol. 7.

Brockett, B.F., Prescott, C.E., Grayston, S.J., 2012. Soil moisture is the major factor influencing microbial community structure andenzyme activities across seven biogeoclimatic zones in western Canada. Soil Biol. Biochem. 44, 9–20.

Brookes, P.C., 2006. The soil microbial biomass: concept, measurement and applications in soil ecosystem research. Microbes Environ. 16, 131–140.

Cao, C., Zhang, Y., Qian, W., Liang, C., Wang, C., Tao, S., 2017. Land-use changes influence soil bacterial communities in a meadow grassland in Northeast China. Solid Earth 8 (5), 1119–1129.

Černohlávková, J., Hofman, J., Bartoš, T., Sanka, M., Anděl, P., 2008. Effects of road deicing salts on soil microorganisms. Plant Soil Environ. 54 (11), 479–485.

Chen, D., Mi, J., Chu, P., Cheng, J., Zhang, L., Pan, Q., et al., 2015. Patterns and drivers of soil microbial communities along a precipitation gradient on the Mongolian Plateau. Landsc. Ecol. 30, 1669–1682.

Chu, H., Fierer, N., Lauber, C.L., Caporaso, J.G., Knight, R., Grogan, P., 2010. Soil bacterial diversity in the Arctic is not fundamentally different from that found in other biomes. Environ. Microbiol. 12, 2998–3006.

Clift, D., Dickson, I.E., Roos, T., Collins, P., Jolly, M., Klindworth, A., 1983. Accumulation of lead beside the Mulgrave Freeway, Victoria. Water Resour. Bull. 14, 155–157.

Czech, B., Krausman, P.R., Devers, P.K., 2000. Economic associations among causes of species endangerment in the United States. BioScience 50, 593–601.

DeFries, R.S., Field, C.R., Fung, I., Collatz, G.J., Bounoua, L., 1999. Combining satellite data and biogeochemical models to estimate global effects of human-induced land cover change on carbon emissions and primary productivity. Global Biogeochem. Cycles 13 (3), 803–815.

DeGryze, S., Six, J., Paustian, K., Morris, S.J., Paul, E.A., Merckx, R., 2004. Soil organic carbon pool changes following land-use conversions. Glob. Change Biol. 10 (7), 1120–1132.

Deng, J.J., Zhu, W.X., Zhou, Y.B., Yin, Y., Bai, X.J., Zhang, H.Z., et al., 2018. Effect of different land use patterns on the soil microbial community diversity in montane region of eastern Liaoning province, China. Chin. J. Appl. Ecol. 29, 2269–2276.

Ding, G.C., Piceno, Y.M., Heuer, H., Weinert, N., Dohrmann, A.B., Carrillo, A., et al., 2013. Changes of soil bacterial diversity as a consequence of agricultural land use in a semiarid ecosystem. PLoS One 8 (3), e59497.

Don, A., Kalbitz, K., 2005. Amounts and degradability of dissolved organic carbon from foliar litter at different decomposition stages. Soil Biol. Biochem. 37, 2171–2179.

Drenovsky, R.E., Steenwerth, K.L., Jackson, L.E., Scow, K.M., 2010. Land use and climatic factors structure regional patterns in soil microbial communities. Glob. Ecol. Biogeogr. 19 (1), 27–39.

Elliott, E., 1986. Aggregate structure and carbon, nitrogen, and phosphorus in native and cultivated soils. Soil Sci. Soc. Am. J. 50 (3), 627–633.

Estache, A., 2007. Infrastructure: a survey of recent and upcoming issues. In Annual World Bank Conference on Development Economics. World Bank, Washington, D.C., pp. 47–82.

Falcucci, A., Maiorano, L., Boitani, L., 2007. Changes in land-use/land-cover patterns in Italy and their implications for biodiversity conservation. Landsc. Ecol. 22 (4), 617–631.

Fierer, N., Jackson, R.B., 2006. The diversity and biogeography of soil bacterial communities. Proc. Natl. Acad. Sci. U. S. A. 103, 626–631.

Fierer, N., Bradford, M.A., Jackson, R.B., 2007. Toward an ecological classification of soil bacteria. Ecology 88, 1354–1364.

Flather, C.H., Knowles, M., Kendall, I.A., 1998. Threatened and endangered species geography: characteristics of hot spots in the conterminous United States. BioScience 48 (5), 365–376.

Garcia-Miragaya, J., Castro, S., Paolini, J., 1981. Lead and zinc levels and chemical fractionation inroad-side soils of Caracas, Venezuela. Water Air Soil Pollut. 15, 285–297.

Geisseler, D., Horwath, W.R., 2009. Short-term dynamics of soil carbon microbial biomass and soil enzyme activitivies compared to long-term effects of tillage in irrigated row-crops. Biol. Fertil. Soils 46, 65–72.

Gjessing, E., Lygren, E., Berglind, L., Gulbrandsen, T., Skanne, R., 1984. Effect of highway runoff on lake water quality. Sci. Total Environ. 33, 247–257.

Griffiths, R.I., Thomson, B.C., James, P., Bell, T., Bailey, M., Whiteley, A.S., 2011. The bacterial biogeography of British soils. Environ. Microbiol. 13, 1642–1654.

Hartmann, M., Howes, C.G., VanInsberghe, D., Yu, H., Bachar, D., Christen, R., et al., 2012. Significant and persistent impact of timber harvesting on soil microbial communities in northern coniferous forests. ISME J. 6, 2199–2218.

He, J.Z., Zheng, Y., Chen, C.R., He, Y.Q., Zhang, L.M., 2008. Microbial composition and diversity of an upland red soil under long-term fertilization treatments as revealed by culture dependent and culture-independent approaches. J. Soil Sediment 8, 349–358.

Hug, L.A., Baker, B.J., Anantharaman, K., Brown, C.T., Probst, A.J., Castelle, C.J., et al., 2016. A new view of the tree of life. Nat. Microbiol 1 (5), 16048.

Johnston, F.M., Johnston, S.W., 2004. Impacts of road disturbance on soil properties and on exotic plant occurrence in subalpine areas of the Australian Alps. Arct. Antarct. Alp. Res. 36 (2), 201–207.

Johnson, M.J., Lee, K.Y., Scow, K.M., 2003. DNA fingerprinting reveals links among agricultural crops, soil properties, and the composition of soil microbial communities. Geoderma 114, 279–303.

Joshi, S.R., Kumar, R., Saikia, P., Bhagobaty, R.K., Thokchom, S., 2010. Impact of roadside pollution on microbial activities in sub-tropical forest soil of North East India. Res. J. Environ. Sci. 4 (3), 280–287.

Ke, C., Li, Z., Liang, Y., Tao, W., Du, M., 2013. Impacts of chloride de-icing salt on bulk soils, fungi, and bacterial populations surrounding the plant rhizosphere. Appl. Soil Ecol. 72, 69–78.

Kerfahi, D., Tripathi, B.M., Lee, J., Edwards, D.P., Adams, J.M., 2014. The impact of selective-logging and forest clearance for oil palm on fungal communities in Borneo. PLoS One 9 (11), e111525.

Khatri, N., Tyagi, S., 2015. Influences of natural and anthropogenic factors on surface and groundwater quality in rural and urban areas. Front. Life Sci. 8, 23–39.

Krashevska, V., Malysheva, E., Klarner, B., Mazei, Y., Maraun, M., Widyastuti, R., et al., 2018. Micro-decomposer communities and decomposition processes in tropical lowlands as affected by land use and litter type. Oecologia 187 (1), 255–266.

Kuffner, M., Hai, B., Rattei, T., Melodelima, C., Schloter, M., Boltenstern, S.Z., et al., 2012. Effects of season and experimental warming on the bacterial community in a temperate mountain forest soil assessed by 16S rRNA gene pyrosequencing. FEMS Microbiol. Ecol. 82, 551–562.

Kuramae, E.E., Yergeau, E., Wong, L.C., Pijl, A.S., van Veen, J.A., Kowalchuk, G.A., 2012. Soil characteristics more strongly influence soil bacterial communities than land-use type. FEMS Microbiol. Ecol. 79, 12–24.

Kuss, F.R., Graefe, A.R., Vaske, J.J., 1990. Visitor Impact Management: A Review of Research. National Parks and Conservation Service, Washington, D.C., Xxx pp.

Lauber, C.L., Strickland, M.S., Bradford, M.A., Fierer, N., 2008. The influence of soil properties on the structure of bacterial and fungal communities across land-use types. Soil Biol. Biochem. 40, 2407–2415.

Lauber, C.L., Hamady, M., Knight, R., Fierer, N., 2009. Pyrosequencing-based assessment of soil pH as a predictor of soil bacterial community structure at the continental scale. Appl. Environ. Microbiol. 75, 5111–5120.

Lehmann, A., Stahr, K., 2007. Nature and significance of anthropogenic urban soils. J. Soils Sediment 7, 247–260.

Lin, Y.T., Jangid, K., Whitman, W.B., Coleman, D.C., Chiu, C.Y., 2011. Change in bacterial community structure in response to disturbance of natural hardwood and secondary coniferous forest soils in central Taiwan. Microb. Ecol. 61, 429–437.

Linde, M., Öborn, I., Gustafsson, J.P., 2007. Effects of changed soil conditions on the mobility of trace metals in moderately contaminated urban soils. Water Air Soil Pollut. 183 (1), 69–83.

Marzluff, J.M., 2001. Worldwide urbanization and its effects on birds. In: Marzluff, J.M., Bowman, R., Donnelly, R. (Eds.), Avian Ecology and Conservation in an Urbanizing World. Kluwer Academic Publishers, Boston, MA, pp. 19–48.

Matinian, N.N., Gusareva, A.L., Bakhmatova, K.A., Sheshukova, A.A., 2020. Microbiological indicators and heavy metal concentration in ecological assessment of urban soils of Saint Petersburg, Russia. Geogr. Environ. Sustain. 13 (1), 214–223.

McGuire, K.L., D'Angelo, H., Brearley, F., Gedallovich, S., Babar, N., Yang, N., et al., 2015. Responses of soil fungi to logging and oil palm agriculture in Southeast Asian tropical forests. Microb. Ecol. 69 (4), 733–747.

Müller, A., Österlund, H., Marsalek, J., Viklander, M., 2020. The pollution conveyed by urban runoff: a review of sources. Sci. Total Environ. 709, 136125.

Murty, D., Kirschbaum, M.U.F., McMurtrie, R.E., McGilvray, H., 2002. Does conversion of forest to agricultural land change soil carbon and nitrogen? A review of the literature. Glob. Change Biol. 8 (2), 105–123.

Nannipieri, P., Ascher, J., Ceccherini, M.T., Landi, L., Pietramellara, G., Renella, G., 2003. Microbial diversity and soil functions. Eur. J. Soil Sci. 54, 655–670.

Neill, C., Piccolo, M.C., Cerri, C.C., Steudler, P.A., Melillo, J.M., Brito, M., 1997. Net nitrogen mineralization and net nitrification rates in soils following deforestation for pasture across the southwestern Brazilian Amazon Basin landscape. Oecologia 110 (2), 243–252.

Neill, C., Steudler, P.A., Garcia-Montiel, D.C., Melillo, J.M., Feigl, B.J., Piccolo, M.C., et al., 2005. Rates and controls of nitrous oxide and nitric oxide emissions following conversion of forest to pasture in Rondônia. Nutr. Cycl. Agroecosyst. 71, 1–15.

Newbold, T., Hudson, L.N., Hill, S.L., Contu, S., Lysenko, I., Senior, R.A., et al., 2015. Global effects of land use on local terrestrial biodiversity. Nature 520 (7545), 45–50.

Nishimura, S., Yonemura, S., Sawamoto, T., Shirato, Y., Akiyama, H., Sudo, S., et al., 2008. Effect of land use change from paddy rice cultivation to upland crop cultivation on soil carbon budget of a cropland in Japan. Agric. Ecosyst. Environ. 125, 9–20.

Ocampo-Peñuela, N., Garcia-Ulloa, J., Ghazoul, J., Etter, A., 2018. Quantifying impacts of oil palm expansion on Colombia's threatened biodiversity. Biol. Conserv. 224, 117–121.

Pabst, H., Kuhnel, A., Kuzyakov, Y., 2013. Effect of land-use and elevation on microbial biomass and water extractable carbon in soils of Mt. Kilimanjaro ecosystems. Appl. Soil Ecol. 67, 10–19.

Parmar, T.K., Rawtani, D., Agrawal, Y.K., 2016. Bioindicators: the natural indicator of environmental pollution. Front. Life Sci. 9, 110–118.

Powlson, D.S., Brookes, P.C., Christensen, B.T., 1987. Measurement of soil microbial biomass provides an early indication of changes in total soil organic-matter due to straw incorporation. Soil Biol. Biochem. 19 (2), 159–164.

Puskás, I., Farsang, A., 2009. Diagnostic indicators for characterizing urban soils of Szeged, Hungary. Geoderma 148 (3), 267–281.

Rahman, M.H., Okubo, A., Sugiyama, S., Mayland, H.F., 2008. Physical, chemical and microbiological properties of an Andisol as related to land use and tillage practice. Soil Till. Res. 101, 10–19.

Rajaniemi, T., Allison, V., 2009. Abiotic conditions and plant cover differentially affect microbial biomass and community composition on dune gradients. Soil Biol. Biochem. 41, 102–109.

Reed, R.A., Johnson-Barnard, J., Baker, W.L., 1996. Contribution of roads to forest fragmentation in the Rocky Mountains. Conserv. Biol. 10, 1098–1106.

Rinnan, R., Rousk, J., Yergeau, E., Kowalchuk, G.A., Bååth, E., 2009. Temperature adaptation of soil bacterial communities along an Antarctic climate gradient: predicting responses to climate warming. Glob. Change Biol. 15, 2615–2625.

Rodríguez-Echeverry, J., Echeverría, C., Oyarzún, C., Morales, L., 2018. Impact of land-use change on biodiversity and ecosystem services in the Chilean temperate forests. Landsc. Ecol. 33 (3), 439–453.

Rousk, J., Baath, E., Brookes, P.C., Lauber, C.L., Lozupone, C., Caporaso, G.J., et al., 2010. Soil bacterial and fungal communities across a pH gradient in an arable soil. ISME J. 4, 1340–1351.

Saunders, S.C., Mislivets, M.R., Chen, J., Cleland, D.T., 2002. Effects of roads on landscape structure within nested ecological units of the northern Great Lakes Region, USA. Biol. Conserv. 103, 209–225.

Schneider, D., Engelhaupt, M., Allen, K., Kurniawan, S., Krashevska, V., Heinemann, M., et al., 2015. Impact of lowland rainforest transformation on diversity and composition of soil prokaryotic communities in Sumatra (Indonesia). Front. Microbiol. 6, 1339.

Shen, C., Xiong, J., Zhang, H., Feng, Y., Lin, X., Li, X., et al., 2013. Soil pH drives the spatial distribution of bacterial communities along elevation on Changbai Mountain. Soil Biol. Biochem. 57, 204–211.

Siddqiui, A.D., Iqbal, M.Z., 1994. Growth reduction in some roadside plants. Ekol. (Bratisl.) 13 (1), 155–159.

Soulé, M.E., 1991. Conservation: tactics for a constant crisis. Science 253, 744–750.

Steenwerth, K.L., Jackson, L.E., Calderon, F.J., Stromberg, M.R., Scow, K.M., 2002. Soil microbial community composition and land use history in cultivated and grassland ecosystems of coastal California. Soil Biol. Biochem. 34, 1599–1611.

Sul, W.J., Stella, A.B., Wang, Q., Tourlousse, D., Penton, C.R., Deng, Y., et al., 2013. Tropical agricultural land management influences on soil microbial communities through its effect on soil organic carbon. Soil Biol. Biochem. 65, 33–38.

Sumampouw, O.J., Risjani, Y., 2014. Bacteria as Indicators of Environmental Pollution: Review.

Sun, R., Zhang, X.X., Guo, X., Wang, D., Chu, H., 2015. Bacterial diversity in soils subjected to long-term chemical fertilization can be more stably maintained with the addition of livestock manure than wheat straw. Soil Biol. Biochem. 88, 9–18.

Tan, X., Kan, L., Su, Z., Liu, X., Zhang, L., 2019. The composition and diversity of soil bacterial and fungal communities along an urban-to-rural gradient in South China. Forests 10, 797.

Thomson, B.C., Ostle, N., McNamara, N., Bailey, M.J., Whiteley, A.S., Griffiths, R.I., 2010. Vegetation affects the relative abundances of dominant soil bacterial taxa and soil respiration rates in an upland grassland soil. Microb. Ecol. 59, 335–343.

Thomson, B.C., Tisserant, E., Plassart, P., Uroz, S., Griffiths, R.I., Hannula, S.E., et al., 2015. Soil conditions and land use intensification effects on soil microbial communities across a range of European field sites. Soil Biol. Biochem. 88, 403–413.

Torres, G.G., Figueroa-Galvis, I., Muñoz-García, A., Polanía, J., Vanegas, J., 2019. Potential bacterial bioindicators of urban pollution in mangroves. Environ. Pollut. 255 (2), 113293.

Torsvik, V., Øvreås, L., 2002. Microbial diversity and function in soil: from genes to ecosystems. Curr. Opin. Microbiol. 5, 240–245.

Tripathi, B.M., Kim, M., Singh, D., Larisa, L.C., Ang, L.H., Nuruddin, A.A., et al., 2012. Tropical soil bacterial communities in Malaysia: pH dominates in the equatorial tropics too. Microb. Ecol. 64, 474–484.

World Bank Technical Paper no. 376 Impact on soils. In: Tsunokawa, K., Hoban, C. (Eds.), Roads and the Environment: A Handbook. The World Bank, Washington, D.C., pp. 70–79.

Verchot, L.V., Davidson, E.A., Cattânio, J.H., Ackerman, I.L., 2000. Land-use change and biogeochemical controls of methane fluxes in soils of eastern Amazonia. Ecosystems 3 (1), 41–56.

Wal, A.V.D., Veen, J.A.V., Smant, W., Boschker, T.S., Bloem, J., Kardol, P., et al., 2006. Fungal biomass development in a chronosequence of land abandonment. Soil Biol. Biochem. 38, 51–60.

Wallenius, K., Rita, H., Mikkonen, A., Lappi, K., Lindstrom, K., Hartikainen, H., et al., 2011. Effects of land use on the level, variation and spatial structure of soil enzyme activities and bacterial communities. Soil Biol. Biochem. 43, 1464–1473.

Wardle, D.A., 2006. The influence of biotic interactions on soil biodiversity. Ecol. Lett. 9 (7), 870–886.

Wu, H., Guo, Z., Peng, C., 2003. Land use induced changes of organic carbon storage in soils of China. Glob. Change Biol. 9 (3), 305–315.

Yergeau, E., Bokhorst, S., Kang, S., Zhou, J., Greer, C.W., Aerts, R., et al., 2012. Shifts in soil microorganisms in response to warming are consistent across a range of Antarctic environments. ISME J. 6, 692–702.

Yuan, B.C., Li, Z.Z., Liu, H., Gao, M., Zhang, Y.Y., 2007. Microbial biomass and activity in salt affected soils under and conditions. Appl. Soil Ecol. 35 (2), 319–328.

Yuan, Y., Si, G., Wang, J., Luo, T., Zhang, G., 2014. Bacterial community in alpine grasslands along an altitudinal gradient on the Tibetan Plateau. FEMS Microbiol. Ecol. 87, 121–132.

Zhalnina, K., Dias, R., Quadros, P.D.D., Richardson, A.D., Camargo, F.A.O., Clark, I.M., et al., 2015. Soil pH determines microbial diversity and composition in the park grass experiment. Microb. Ecol. 69, 395–406.

Zhang, H., Gao, Z., Shi, M., Fang, S., Xu, H., Cui, Y., et al., 2019. Study of the effects of land use on hydrochemistry and soil microbial diversity. Water 11 (3), 466.

Zhang, F., Zhang, H., Yuan, Y., Liu, D., Zhu, C., Zheng, D., et al., 2020. Different response of bacterial community to the changes of nutrients and pollutants in sediments from an urban river network. Front. Environ. Sci. Eng. 14, 28.

Zhao, W.Z., Xiao, H.L., Liu, Z.M., Li, J., 2005. Soil degradation and restoration as affected by land use change in the semiarid Bashang area, northern China. Catena 59 (2), 173–186.

Soil-plant-microbial interactions for soil fertility management and sustainable agriculture

16

R. Hiranmai Yadav
School of Environment and Sustainable Development, Central University of Gujarat, Gandhinagar, India

16.1 Introduction

The soil is regarded as a living system in sustainable agriculture due to the presence and activity of microbes (Janvier et al., 2007). Soil and plant microbial interactions including secretion of plant exudates and complex biological and ecological processes benefit the plant growth and suppress soil borne pathogens (Liu et al., 2016). Microflora and fauna including nitrogen fixing, phosphate solubilizing and plant growth promoting organisms are part of the rhizosphere community. The soil borne pathogens that break the protective mechanism of plants immune system occur there. A complex and diverse population and activity is observed in this microenvironment (Raaijmakers et al., 2009; Philippot et al., 2013) (Fig. 16.1).

The concept of rhizosphere was introduced by Hiltner (1904). He described the narrow zone of soil around roots with microbe populations and activities that are stimulated by root. According to McCully (2005) this concept is now expanded to the physical, chemical, and biological properties influenced by the root exudates and microbial activities. The thin layer of soil immediately surrounding plant roots is an active area with root activity and metabolism and other physiological processes like mineral uptake and nutrient availability. This zone and its activities vary with plant genotype and soil conditions with abundance and diversity (Nielsen et al., 2015).

The soil rhizosphere system is governed by the roots, soil colloid and solution, particles and organic matter. Soil microbial ecology helps to unveil the decomposition of organic matter and how the small amount of organic matter triggers the reaction in nature. The biogeochemistry involves the plant- soil interaction that could be assessed through stochiometric reactions. These reactions describe nutrients reactions, their interactions, and demand by the microbes that decompose the organic litter (Dijkstra et al., 2013; Mooshammer et al., 2014; Rinkes et al., 2013). The stochiometric reactions in nutrient limited conditions help to identify different biogeochemical cycles for nitrogen fixation in ecosystems (Rousk et al., 2013). Isotopic tracer methods also help to systematically estimate the microbial contributions to nutrient transformations and turnover of low molecular weight components (Rinkes et al., 2013). The microbial role in ecosystem level processes is of significance.

Microbes in Land Use Change Management. DOI: https://doi.org/10.1016/B978-0-12-824448-7.00019-X
© 2021 Elsevier Inc. All rights reserved.

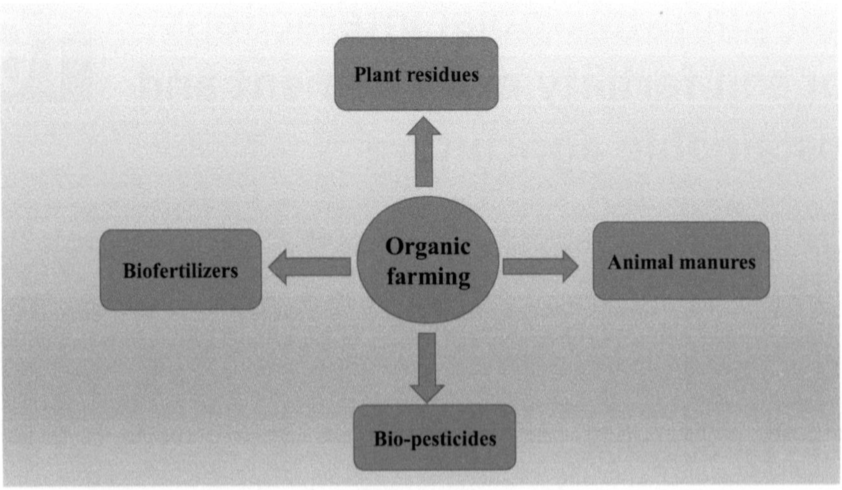

Figure 16.1 Components of organic farming.

Soil fertility management components influence the soil pH and soil organic carbon in addition to nutrient management. The organic matter decomposition is carried out by a large number of microbial species while xenobiotic degradation capability is restricted to more-specialized species. Using functional measures, stable-isotope probing, and molecular methods, it was observed that microbial (community diversity and function) response to land use change is stable over time. There exists a strong correlation between plants, animals and microbial community in maintaining biogeochemical cycle in any ecosystem (Nazaries et al., 2013). The metabolites act as molecular signals to microbes (in the form of nutrients) and support microbial growth that in turn provides nutrition and protection to plants. These beneficial microbes suppress plant pathogens and support direct plant growth (Nihorimbere et al., 2011).

Agrochemicals adversely affect the indigenous soil microbial population. Development of sustainable cultivation methods involves symbiotic microorganisms, diazotrophs and spore forming bacteria as they are significant for nutrient transformation and uptake. Their activity is influenced by physicochemical characteristics and mineral composition of soil (Paszt et al., 2017). Schmidt et al. (2019) reported that management practices and crop selection in agriculture highly influences microbial community.

Organic farming practices including crop rotation and addition of organic matter changes the microbial population, reduces pest incidence and weed control. Soil fertility is significant in crop production and ecosystem balance. Utilization of chemical fertilizers is unsustainable leading to degradation of soil and productivity. Integrated nutrient management system is eco-friendly and have the potential for sustainable agriculture. Biofertilizers are widely used for agricultural production that reduces the pathogen incidence and improves biological activity of soil (Cao et al., 2011). Application of biofertilizers are found to improve rhizosphere soil quality and

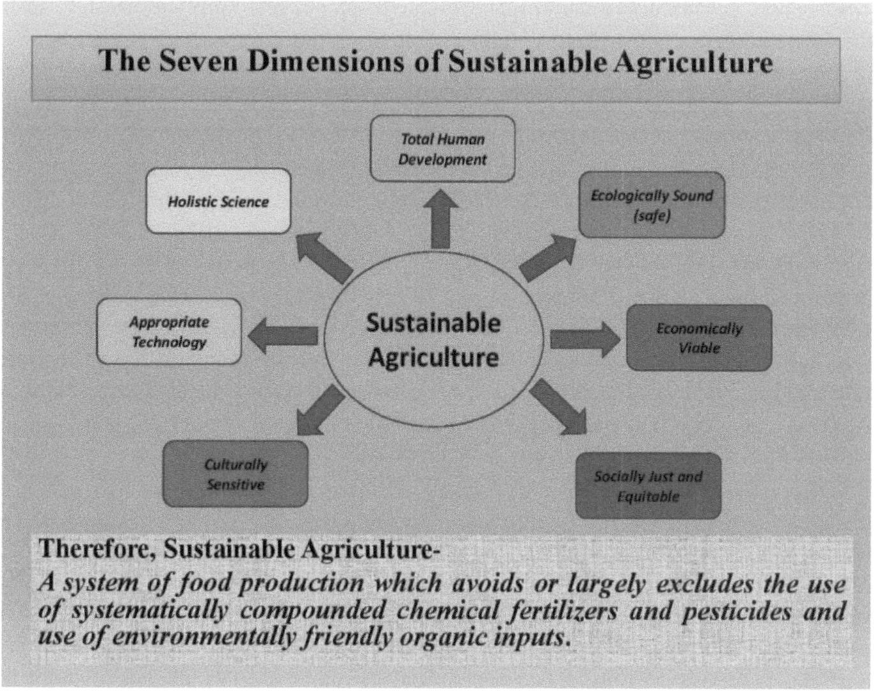

Figure 16.2 Dimensions of organic farming.

nutrient transformation and thereby plant nutrition and productivity. These are achieved by biological nitrogen fixation, nutrient solubilization, biocontrol activities and production of plant growth promoting substances (Raimi et al., 2017) (Fig. 16.2).

16.2 Organic matter and microbes in biogeochemical cycling

Microbes play an essential role in ecosystem functions, including biogeochemical cycles and considered to be black box in predictive models. This is due to the perceived temporal and spatial variations in microbial communities and lack of ecological theory explaining how microbes regulate ecosystem functions. Providing evidence of the microbial regulation of biogeochemical cycles is key for predicting ecosystem functions, including greenhouse gas fluxes, under current and future climate scenarios. Findings of Nazaries et al. (2013) provided support for the explicit consideration of microbial data in ecosystem/climate models to improve predictions of biogeochemical cycles. The organic matter in the soil forms the pool of carbon and other nutrients which is influenced by the environmental factors and microorganisms present in the soil (Schmidt et al., 2011).

Microbial communities and biogeochemical cycles are highly related to each other. The complex substrates including plant litter and chitin are degraded by soil microbes and utilized as a source of nutrients. These microbial activities are influenced by environmental factors and climate change. Production and activity of microbial enzymes are an important component of biogeochemical cycles. Biogenic volatile organic compounds, methanogens, and sulfate reducing bacteria form the community composition for reduction of substrate (Beier and Bertilsson, 2013; Moorhead et al., 2013; Rinkes et al., 2013; Aronson et al., 2013; Basiliko et al., 2013; Kappler and Nouwens, 2013; Aronson et al., 2013; Steinweg et al., 2013; de Vries and Shade, 2013; Thomson et al., 2013; Rinnan et al., 2013; Zeleke et al., 2013) (Fig. 16.3).

In agriculture, sustainability could be achieved by management of soil organic matter and nutrient cycling. Vermicompost production occurs through interaction of microbial and nonmicrobial interaction resulting in organic material. The product contains macro and micro elements and biological properties (Arancon et al., 2004) through mucus secretion from intestine and cuticle (Le Bayon and Binet, 2006) and enhances the transformation of minerals and microbial activity that supports nutrient cycling (Fracchia et al., 2006).

Nutrient transformation by microbes during vermicomposting is evidenced by improved biochemical changes and release of nutrients in available forms that could be utilized by plants (Kumari and Ushakumari, 2002). Bringing modification in the microbial composition during decomposition also changes the nutrient content of compost (Padmavathiamma et al., 2008) as reported by (Gutiérrez-Miceli et al., 2008a,b; Aria et al., 2010). These microbes mineralize the phosphorus by increased soil phosphatase activity (Saha et al., 2008; Scervino et al., 2010). Release of

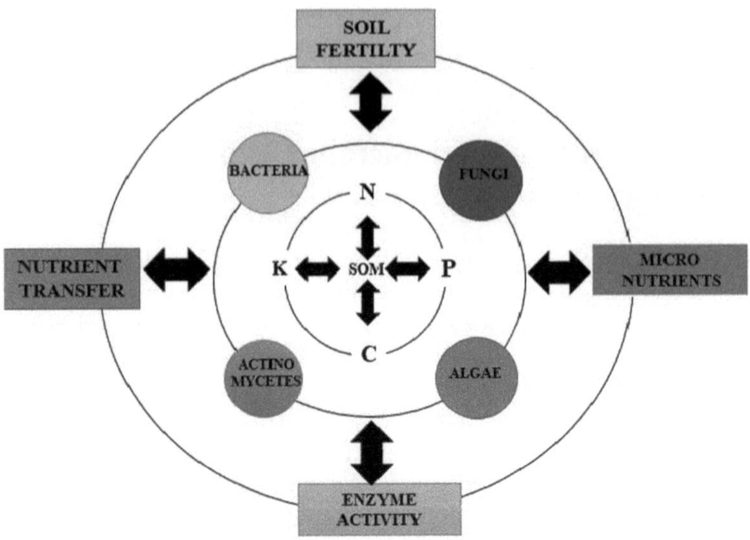

Figure 16.3 Soil organic matter microbial interactions.

phosphorus which is considered to be one of the major nutrients required for plant growth is increased during vermicomposting. It is achieved through P mineralization by phosphatase enzyme (both acid phosphatase and phytase enzymes) during vermicomposting. P mineralization in the first 50 days was by acid phosphatase enzyme and in the latter part of vermicomposting by phytase enzyme (Ghosh et al., 2018).

Increase in nutritional properties and electrical conductivity during composting was observed by different workers (Francou et al., 2005; Varma et al., 2014). Decomposition of organic materials increases the mineral cation exchange and improvement in quality of end product might be due to loss of dry mass (Amir et al., 2005; Singh and Kalamdhad, 2013). Decreased moisture content in end product indicates compost aeration and changes in temperature, bulk density due to lesser amount of available moisture in end compost implies that higher availability of air. This could also be attributed to microbial activity and less oxygen consumption by them (Jain et al., 2018). During vermicomposting earthworms consume organic substrates and the carbon content is either assimilated as energy or released as CO_2. The decrease in carbon content of end product is attributed to the consumption rate and nature of a substrate (Hait and Tare, 2011; Pramanik, 2012). Epigeic worms prefer cattle manure as a feed due to its palatable substrates (Atiyeh et al., 2000; Pramanik et al., 2010).

Formation of bicarbonates and humic acids during organic matter decomposition and stabilization of pH and changes in water soluble phosphorus was observed by Komilis and Ham (2006). Mineral phosphate solubilizing activity due to exudation of organic acids by *Burkholderia* and activation of phosphatases proves the ability of microorganisms to solubilize P complexes acidification, chelation, exchange reactions, and production of organic acids (Gaume et al., 2001; Gulati et al., 2010; Patel et al., 2011). The enzyme associated with phosphorus cycle phosphatase was increased in vermicomposting substrate 15 days after introduction of earthworms (Kızılkaya et al., 2010). Hydrolysis of aryl sulfate esters by oxygen-sulfur bond fission is catalyzed by arylsulphatase enzyme and helps in mineralization of ester sulfate in soil (Tabatabai, 1994). Changes in microbial population and requirements for phosphate is indicated by the increased phosphatase activity at the initial stage followed by phosphodiesterase activity and by the end of decomposition it was alkaline and acid phosphatase activity. Weak van der Waals interactions or hydrophobic forces aggregate the organic matter with phosphorus compounds (Piccolo, 2002) that may lead to organic matter fraction insoluble in water and poor nutrient extractability. Inoculation of selected microbes during vermicompost could be beneficial in modifying the nutrient and resulting in value added organic product (Busato et al., 2012).

Tea factory waste mixed with cattle manure in proper combination could produce quality end product with rich nutrient content and microbial population and enzyme activity (Ghosh et al., 2018). The palatability and rate of vermicomposting is influenced by the crude fiber, polyphenol content and other physicochemical characteristics of the substrate. Earthworms increase the total nitrogen content (Suthar, 2006). This could be attributed to the loss of carbon from substrate and can be correlated with increasing nitrogen, phosphorus, and potassium concentrations (Garg and Kaushik, 2005). Increase in the population of nitrogen fixing bacteria, mixing of earthworm mucus, and gut microbes enhances nitrogen content (Padmavathiamma et al., 2008; Hussain et al., 2016).

The biochemical changes in the vermicompost occurs as the substrate passes through the gut of worms where the gut microflora mineralizes the nutrient by enzyme and organic acid secretions (Atkins and de Paula, 2006; Le Bayon and Binet, 2006; Patel et al., 2011; Gaume et al., 2001). Enzyme activity by the microbes present in vermicompost significantly improves its quality. Enzyme urease and phosphatase are reported to improve the quality of compost and its application to crops beneficial. Humic enzyme complex in vermicompost is related to the microbial respiration and high enzyme activity (Banitez et al., 2000). Organic phosphorus contents in substrates are hydrolyzed by exocellular phosphatase enzymes (Pramanik et al., 2009). The organic wastes containing lignin are degraded by lignolytic strains like *Phanerochaete chrysosporium* and *Trichoderma viridae* by production of enzymes. The presence of nitrogen fixing microbes improved the nitrogen content of the end product (Manna et al., 2003; Anand et al., 2006) depending upon the waste used for decomposition (Beauchamp et al., 2006). The compost quality is determined by estimating the physical, chemical, and biological characteristics of the end product. Introducing earthworm *Eisenia foetida* to substrate increased urease activity that hydrolyzes urea to ammonia and carbon dioxide (Kızılkaya et al., 2010). During vermicomposting, earthworms feed upon microbes along with substrates and they consume the intestinal mucus. The mucus of worms contains protein, glyco protein, and glucoside which is taken up by microbes in the gut. The consumption of these nitrogenous compounds by microbes increases their activity and enzyme activity that contributes to substrate digestion (Zhang et al., 2000). The mineralization of complex substance to plant available form through a series of biological activity. Inoculation of *P. chrysosporium* and *T. viridae* to vermicompost enhanced the nutrient content and enzymatic activities of vermicompost (Pramanik et al., 2007).

Soil microbes enhance the enzymatic activity (cellulase, amylase, invertase, protease, peroxidase, urease, phosphatase, and dehydrogenase) that depicts the decomposition of organic matter and nitrogen transformations (Tiquia and Tam, 1998). Dehydrogenase enzyme related to the oxidative phosphorylation is an indicator of microbial activity in soil and other biological ecosystems (Garcia et al., 1997). The vermicompost produced from fruit pulp, vegetable waste, groundnut husk, and cow dung had abundant microbial numbers and their extracellular enzyme (Haritha Devi et al., 2009).

Temperature and moisture content during decomposition plays a significant role defining the quality of the end product. The varying temperature helps to improve reaction rates, kill pathogens, and sanitize the process (Haug, 1993; Agnew and Leonard, 2003). Temperature variation also helps in maintaining the degrading microbes that maintain the mineralization process (Mohee and Mudhoo, 2005; Varma and Kalamdhad, 2014).

Microbial activity is improved in organic farming that could be measured with enzyme dehydrogenase in terms of microbial redox processes, basal respiration, and metabolic quotient. Extracellular enzymes secreted by microbes help to break down organically bound nitrogen, proteins into peptides, dipeptides and single amino acids, urea into carbon dioxide and ammonia, and these enzymes act as indicators for mineralization and transformation.

16.3 Nutrient availability

The soil functions and processes in an agricultural system involve an active microbial community. They are the active part of nutrient release and mobilization involving ecosystem services like reducing soil erosion by aggregate formation and structure modification (Rillig and Mummey, 2006). Nutrient availability and transformation depend on various factors involving microbial metabolism and biogeochemical cycles. Acquisition of the essential nutrient nitrogen in ecosystem is controlled by various microbial processes in association with environmental factors. Mineralization, fixation, and transformation of nitrogen involves active rhizosphere mechanism with exudates and soil microbes. Plant roots are key regulators in nitrogen metabolism. Nitrogen cycling is influenced by agricultural and environmental factors (Stark et al., 2008; Hartmann et al., 2014; Fu et al., 2013).

Microorganisms in the root zone and present in adhering soil particles interact with the plant and improve their growth, provide nutrients, and protect from pathogens. Changes in pH and redox potential due to microbial activity are reported in the rhizosphere zone (Römheld and Neumann, 2005, 2006) (Fig. 16.4).

The length of the root and root hair are dependent upon the nitrogen and phosphorus available in the rhizosphere. Root growth and phosphorus solubilization are achieved by mycorrhizal association depending on the plant species. The phytohormones and exudates released by the plant initiate microbial activity and nutrient transformations. Rhizosphere acidification by plant biochemicals also induces an exchange of different nitrogen forms. Ammoniacal nitrogen supply is induced by rhizosphere acidification that in turn improves acid soluble calcium phosphate and micronutrients. Rhizosphere alkalinization in the acidic soil is beneficial for phosphorus solubilization in addition to potassium and micronutrient availability in legume soil as reported by Neumann and Römheld (2002). Plant health and defense is influenced by manganese, copper, and boron in the rhizosphere, and this microzone is an active site of various processes. Management of fertilizer requirements and agrochemicals application based on soil rhizosphere system help to improve productivity.

Nitrogen fertilizer is used by farmers in different forms and it helps to modify the pH and redox potential with exudates that sometimes also lead to manganese toxicity or deficiency. Accumulation or depletion of nutrients in the rhizosphere minerals based on solubility and uptake is reported by (Römheld and Neumann, 2006). According to Ma and Kalb (2006) slow release of nutrients by organic fertilizer leads to reduced plant growth and yields. This could be managed by utilization of starter

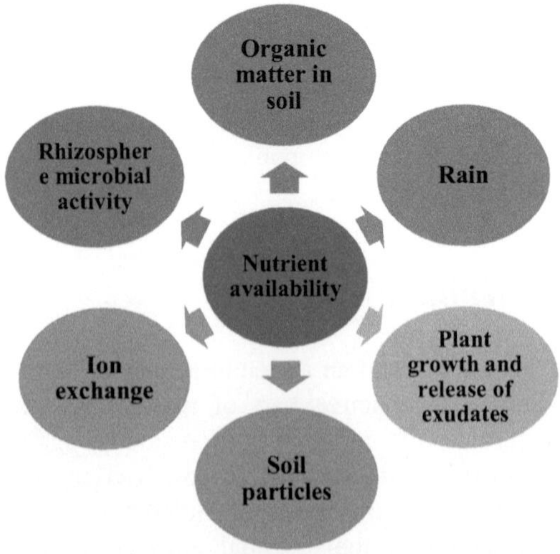

Figure 16.4 Interaction between soil particles and rhizosphere activity.

solution to sustain yields. Rhizosphere soil was depleted in NO_3-N, P, and K in both in the organic and conventional management systems (Schmidt et al., 2019).

16.4 Organic manure for soil buffering

Organic manure improves the water holding capacity of soil and thereby nutrient release and organic matter. The broadcasted organic fertilizers decompose quickly and release nutrients, which if not up taken by plants will get leached. In the rhizosphere zone they concentrate in the soil solution and retain for longer duration. This helps to improve soil fertility and carbon sequestration in terrestrial ecosystem. Lehmann and Rondon (2006) also observed that the addition of biocharcoal enhances the growth of the plant and helps to retain nutrients and make it available to plants. This could be attributed to increased surface area and higher exchange capacity (Glaser et al., 2002). Rice husk charcoal in addition to containing carbon entrapped in silica, includes micronutrients like calcium, magnesium, iron, manganese and potassium. Its addition to plants improves soil fertility by amelioration of soil nutrients in soil colloid particles (Fig. 16.5).

16.5 Rhizosphere management technologies

Researches in addition to focusing on microbes in rhizosphere, rhizosphere management technologies are also considered and promoted. This system identifies methods to decelerate soil quality deterioration. Rhizosphere soils can be amended first

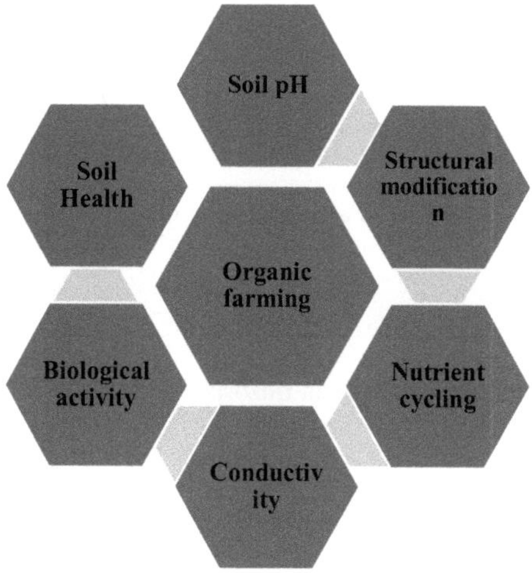

Figure 16.5 Organic manure for soil buffering.

Table 16.1 Details of different types of soil type.

S.no.	Type of soil	Application
1.	Acid soils	Liming
2.	Alkaline soils	Adding sulfur powder
3.	Sandy soils	Apply organic fertilizers
4.	Highly weathered soils	Apply lime and phosphorus fertilizers
5.	Saline soils	Leach away salts
6.	Sodic soils	Apply gypsum exchange Na with Ca
7.	Heavy textured soils	Incorporate sand or slowly decomposed residues

in problematic soil considering the nature of soil and treatment required. Details of different soil types are given in Table 16.1.

These treatments help to improve soil characters, drainage, and aeration. Flooding should be avoided in rhizospheric soils (Fig. 16.6).

16.6 Soil microbial communities

Ecosystem processes of significance in ecologic and economic factors are highly influenced by soil microbial communities. The influence of microbes on biogeochemical cycles has a positive effect on diversity of organisms, productivity, green house gases (GHG) emissions and climate change. The spatial and temporal interactions influence the microbial biodiversity-ecosystem functioning. There exists a systematic link between the metabolic processes of the microorganisms and

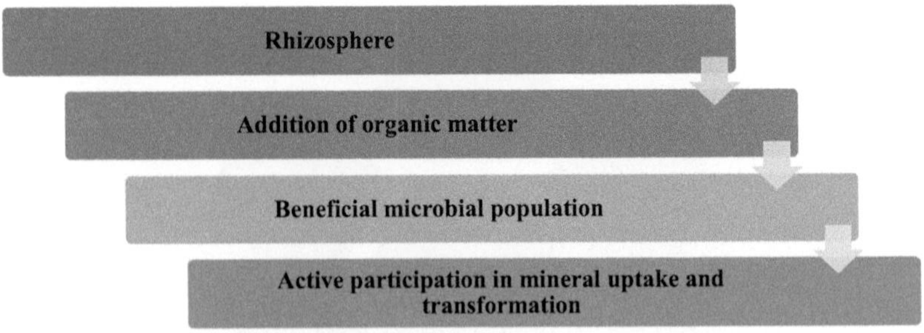

Figure 16.6 Organic matter for microbes in rhizosphere.

biogeochemistry of soil. There is also a need to recognize the relationship between mechanism that alter microbial activity and their interaction with primary producers (Rousk and Baath, 2011; Stein and Nicol, 2011; Treseder et al., 2012).

Microorganisms like *Bacillus, Pseudomonas, and Streptomyces* present in soil produces secondary metabolites that protects against pathogens (Pathma et al., 2011). Bioavailability of nutrients is enhanced by gut microflora of earthworms and the activity of earthworms influences soil biochemical processes as they ingest plant growth-promoting rhizospheric bacteria such as *Pseudomonas, Rhizobium, Bacillus, Azosprillium, Azotobacter*, etc. along with rhizospheric soil, and they might get activated or increased due to the ideal microenvironment of the gut. The earthworm activity increases the population of plant growth-promoting rhizobacteria (Sinha et al., 2010).

Earthworms live in association with free living bacteria in soil. Gut of various species of earthworm is found to be inhabited by aerobic bacteria. They mineralize organic matter and facilitate chelation of metal ions. Solubilization of nutrients by soil microbes increases availability and uptake by the plants and production of growth hormone and nitrogen fixation (Han et al., 2005; Ayyadurai et al., 2007; Ravindra et al., 2008). Anaerobic nitrogen fixing *Clostridium butyricum, Clostridium beijerinckii*, and *Clostridium paraputrificum* are also reported from earthworm gut. The number of proteolytic and oxalate-degrading bacteria and actinomycetes are high in earthworm gut and thereby the substrate that passes through gut carries them along with (Singleton et al., 2003; Karsten and Drake, 1997).

Conventional compost and vermicompost contains various microbial population that can change the soil quality (Vivas et al., 2009; Suhane, 2007). α-Proteobacteria, β-Proteobacteria, γ-Proteobacteria, Actinobacteria, Planctomycetes, Firmicutes, and Bacteroidetes (Yasir et al., 2009) are isolated from vermicompost. Parthasarathi and Ranganathan (1998) and Haritha Devi et al. (2009) reported considerable increase in total viable counts of actinomycetes and bacteria in the worm treated compost. The increase in microbial population could be due to the nutrient availability including cellulose, sugars, chitin, lignin, starch and polylactic acids from organic materials in earthworm gut (Tiwari et al., 1989; Zhang et al., 2000; Aira et al., 2007; Vivas et al., 2009). Details of beneficial soil microbial communities are given in Table 16.2.

Table 16.2 Details of beneficial soil microbial communities.

Microorganisms	References
Pseudomonas, Paenibacillus, Azoarcus *Burkholderia, Spiroplasm, Acaligenes, Acidobacterium* *Bacillus benzoevorans, Bacillus cereus, Bacillus licheniformis, Bacillus megaterium, Bacillus pumilus* *Bacillus subtilis, Bacillus macrolides, Actinobacteria*, namely, *Cellulosimicrobium cellulans* *Microbacterium* spp., *Microbacterium oxydans* **Proteobacteria such as** *Pseudomonas* spp., *Pseudomonas libaniensis* **Ungrouped genotypes** *Sphingomonas* sp., *Kocuria palustris and* **Yeasts, namely** *Geotrichum* spp. *and Williopsis californica*	Singleton et al. (2003) Vaz-Moreira et al. (2008)
Verminephrobacter eiseniae from *Eisenia foetida*	Pinel et al. (2008)
Ochrobactrum sp., *Massilia* sp., *Leifsonia* sp.,	Byzov et al. (2009)
Bacteroidetes, Alphaproteobacteria, *Betaproteobacteria* and **Representatives of classes** *Flavobacteria, Sphingobacteria* (Bacteroidetes) *Pseudomonas* spp. and *Alcaligenes* spp.	Nechitaylo et al. (2010)

16.7 Rhizosphere soils and microbial community in organic farming

The 10 most dominant phyla were identified and reported by Zhao et al. (2018) across all rhizosphere soil samples containing organic treatment.

1. Proteobacteria
2. Firmicutes
3. Planctomycetes
4. Actinobacteria
5. Bacteroidetes
6. Gemmatimonadetes
7. Acidobacteria
8. Chloroflexi
9. Verrucomicrobia
10. Nitrospirae

The organic management practices improved soil microbial populations with abundant microbial population in tea plantation soil. In agro ecosystems microbial population and their occurrence are influenced by agricultural management and plant species (Schmidt et al., 2019) and shaped by diverse, interacting forces. Agricultural practices such as fertilizer application and tillage changes the physic chemical characters and microbial community. Rhizosphere creates its environment with varying chemical supply through exudates. This zone acts as the major nutrient cycling area compared to other soil. Organic fertilizers and farming practices improves bacterial and fungal community composition (Fan et al., 2017; Li et al., 2017).

Organic farming including crop rotation practices have positive effect on soil microbiota that in turn improves soil based ecosystem services. Organic farming system influences microbial biomass nitrogen and carbon, phospholipid fatty acid analysis (PLFA) and enzymes. Eco-friendly agricultural practices could be achieved by organic farming that help to combat climate change and management of limited natural resources and reduce environmental degradation. Soil based ecosystem services are sustainable with organic farming practices and improving microbial diversity. A functionally diverse microbial population is required to manage the changes in climate and provide resilience to climate change. Adapting to climate changes is a critical factor for agriculturists and pastoralist community. As organic farming relies on nutrient cycling with plant residues and animal manures along with crop rotation it requires microorganisms actively involved in nutrient cycling. They have the capability to convert complex molecules to simpler ones that can be utilized by plants (Wagg et al., 2014).

Both liquid and solid farm yard manure and green manure crops provided as organic sources influenced the microbial biomass carbon with a positive correlation between soil organic carbon and microbial biomass carbon (Gattinger et al., 2012). Quality and quantity of inputs also influences the microbial abundance and activity (Tu et al., 2005). Inclusion of legumes in crop rotation improves microbial biomass nitrogen in organic management. Bio-organic fertilizers containing fermented cow plus chicken manure compost suppresses soil-borne diseases and promotes plant growth by altering the composition of the rhizosphere microbial community (Zhao et al., 2018).

A meta-analysis by McDaniel et al. (2014), who synthesized crop rotation effects on soil carbon and nitrogen pools. They found that the addition of one or more crops in the rotation to be beneficial in regard to microbial biomass carbon, microbial biomass nitrogen, total carbon and total nitrogen. According to Venter et al. (2016) inclusion of cover crops enhances total nitrogen and carbon. Diversity in crop rotation could give a positive effect on microbial richness and diversity. Similarly Bünemann et al. (2006) synthesized the varying effects of several pesticides on microbial communities and other soil organisms and observed that organic pesticides are less harmful to the environment.

Farming practices, land use, plant life cycle, and climatic conditions influences microbial activity and abundance. Microbial biomass is positively correlated with pH of soils (Baath et al., 2003). Microbial properties are also influenced by pH and organic systems with high biomass at pH 5.7 (Pietri Aciego and Brookes, 2008). Tillage affects the microbial population in addition to soil aggregate structure and soil chemical properties and they are microbial indicators. The combination of organic farming with reduced or no-tillage is challenging improve biological but also physical and chemical soil parameters considering climate change induced rainfall variability and soils' enhanced need for water infiltration and storage (Larsen et al., 2014).

16.8 Efficient management practices for soil quality improvement

Bidabadi et al. (2017) have reported the beneficial effects of vermicompost leachate on saline soil. Foliar spray with vermicompost leachate was found to develop salt tolerance in pomegranate seedlings with a reduction in oxidative stress and electrolyte leakage that could be due to improved activity of antioxidant enzymes. Efficiency of plant is improved by antioxidants and limited the agglomeration of sodium. Vermicomposting is an efficient nutrient recycling process involves harnessing earthworms as versatile natural bioreactors for organic matter decomposition (Fig. 16.7).

The vermicompost is obtained by the bio oxidative process that involves earthworms and associated microbes and biological organic waste decomposition process is termed as vermicomposting. It is a peat like material that enhances soil biodiversity and microbial activity in addition to improving physiochemical characters of soil. Its buffering capacity with high porosity, good aeration, drainage, water holding capacity and excellent nutrient status promoting the beneficial microbes which in turn enhances plant growth. By producing plant growth-regulating hormones and enzymes and indirectly by controlling plant pathogens, nematodes, and other pests enhances plant health and used to promote sustainable agriculture. The safe management of agricultural, industrial, domestic and hospital wastes which may otherwise pose serious threat to life and environment (Pathma and Sakthivel, 2012). Earthworms influences microbial community, physical and chemical properties of soil by breaking down large soil particles and leaf litter and thereby increase the availability of organic matter for microbial degradation and transforms organic wastes into valuable vermicomposts, a rich source of micro and macro-nutrients and microbial enzymes (Lavelle and Martin, 1992; Maboeta and van Rensburg, 2003).

Association of earthworms with microbes (few yeasts, protozoa and certain groups of fungi such as *Fusarium oxysporum*, *Alternaria solani*, and micro fungi) improves quality of Organic fertilizers. It improves the soil health by sustaining the balance of organic matter, thereby helping to amend physical, chemical, and biological properties of the soil structure and enhance nutrient availability (Mir et al., 2013; Ayyobi et al., 2014). Vermicompost is formed by crushing of substrate through worm gut and the liquid phase that comes from vermicompost called as vermicompost leachate has a very positive impact on soil improvement and plant growth (Chinsamy et al., 2013). This could be attributed to prolonged availability of nutrients from vermicomposts (Nandi et al., 2013). Vermicompost leachate also

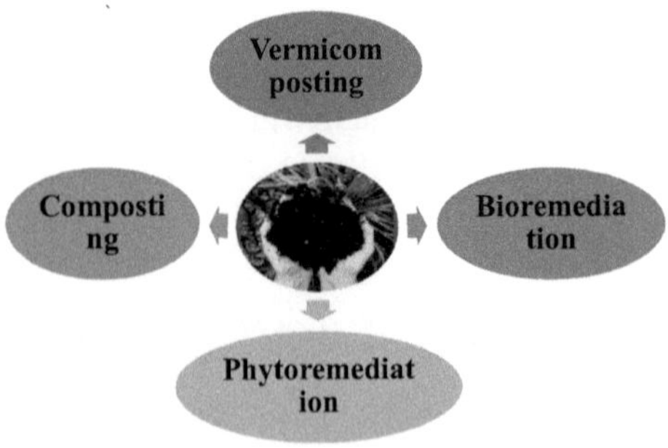

Figure 16.7 Soil management practices.

help to improve the negative effects caused by potassium and phosphorus deficiency (Fathi et al., 2012). Arthur et al. (2012) and also enhanced photosynthesis efficiency.

The large particular surface area provides many microsites for microbial activity thereby improving biological fertility of soil which adds beneficial microbes to soil. The mucus of earthworms accelerates and enhances decomposition of organic matter composing stabilized humic substances which embody water-soluble phytohormonal elements (Edwards and Arancon, 2004). The strong retention of nitrates, exchangeable phosphorus, soluble potassium, calcium, and magnesium in plant available forms vitamins, enzymes, and hormones are also recorded in vermicompost (Sinha et al., 2009). During vermicomposting the heavy metals forms complex, aggregates with humic acids, and other polymerized organic fractions resulting in lower availability of heavy metals to the plant, which are otherwise phytotoxic (Dominguez and Edwards, 2004). Solid pulp and paper mill sludge is a lignocellulosic waste is very difficult to manage as it is not easily degraded. Thyagarajan et al. (2010) and Hazarika et al. (2017) on composting of pulp and paper mill industry sludge added with saw dust and cow dung revealed improvement in chemical and nutritional properties and reduction in bioavailability and leachability of toxic metals such as copper, iron, nickel, lead, chromium, zinc, mercury, and manganese after a 20 days composting period.

16.9 Conclusion

The promotion of microbial abundance and activity enhances ecosystem services by crop rotations and organic farming. Soil microorganism play a pivotal role in maintaining the soil fertility. These organisms secrete enzymes that help in degrading the organic wastes and converting them into organic matter. They have an integral role in mineralization and biogeochemical cycle that maintain the ecosystem balance.

References

Agnew, J.M., Leonard, J.J., 2003. The physical properties of compost. Compost Sci. Util. 11 (3), 238–264.

Aira, M., Monroy, F., Dominguez, J., 2007. Earthworms strongly modify microbial biomass and activity triggering enzymatic activities during vermicomposting independently of the application rates of pig slurry. Sci. Total Environ. 385, 252–261.

Amir, S., Hafidi, M., Merlina, G., Revel, J.C., 2005. Sequential extraction of heavy metals during composting of sewage sludge. Chemosphere 59 (6), 801–810.

Anand, P., Isar, J., Saran, S., Saxena, R.K., 2006. Bioaccumulation of copper by *Trichoderma viride*. Bioresour. Technol. 97, 1018–1025.

Arancon, N.Q., Edwards, C.A., Bierman, P., Welch, C., Metzger, J.D., 2004. The influence of vermicompost applications to strawberries: Part1. Effects on growth and yield. Bioresour. Technol. 93, 145–153.

Aria, M.M., Lakzian, A., Haghnia, G.H., Berenji, A.R., Besharati, H., Fotovat, A., 2010. Effect of *Thiobacillus*, sulfur, and vermicompost on the water-soluble phosphorus of hard rock phosphate. Bioresour. Technol. 101, 551–554.

Aronson, E.L., Allison, S.D., Helliker, B.R., 2013. Environmental impacts on the diversity of methane-cycling microbes and their resultant function. Front. Microbiol. 4, 225. Available from: https://doi.org/10.3389/fmicb.2013.00225.

Arthur, G.D., Aremu, A.O., Kulkarni, M.G., Staden, J.V., 2012. Vermicompost leachate alleviates deficiency of phosphorus and potassium in tomato seedlings. HortScience 47 (9), 1304–1307.

Atiyeh, R.M., Subler, S., Edwards, C.A., Bachman, G., Metzger, J.D., Shuster, W., 2000. Effect of vermicomposts and composts on plantgrowth in horticultural container and soil. Pedobiologia 44 (5), 579–590. Available from: https://doi.org/10.1078/S0031-4056(04)70073-6.

Atkins, P., de Paula, J., 2006. Physical Chemistry for the Life Sciences. W.H. Freeman and Company, New York, pp. 363–371.

Ayyadurai, N., Ravindra Naik, P., Sakthivel, N., 2007. Functional characterization of antagonistic fluorescent pseudomonads associated with rhizospheric soil of rice (*Oryza sativa* L.). J. Microbiol. Biotechnol. 17, 919–927.

Ayyobi, H., Olfati, J.A., Peyvast, G.A., 2014. The effects of cow manure vermicompost and municipal solid waste compost on peppermint (*Mentha piperita* L.) in Torbat-e-Jam and Rasht regions of Iran. Int. J. Recycl. Org. Waste Agric. 3, 147–153.

Baath, E., Anderson, T.H., 2003. Comparison of soil fungal/bacterial ratios in a pH gradient using physiological and PLFA-based techniques. Soil Biol. Biochem. 35, 955–963.

Banitez, E., Nogales, R., Masciandraro, G., Ceccanti, B., 2000. Isolation of isoelectric focusing of humic-urease complexes from earthworm (*Eisenia foetida*)-processed sewage sludges. Biol. Fertil. Soils 31, 489–493.

Basiliko, N., Henry, K., Gupta, V., Moore, T.R., Driscoll, B.T., Dunfield, P.F., 2013. Controls on bacterial and archaeal community structure and green-house gas production in natural, mined, and restored Canadian peatlands. Front. Microbiol. 4, 215. Available from: https://doi.org/10.3389/fmicb.2013.00215.

Beauchamp, C.J., Levesque, G., Prevost, D., Chalifour, F.P., 2006. Isolation of free-living dinitrogen-fixing bacteria and their activity in compost containing de-inking paper sludge. Bioresour. Technol. 97, 1002–1011.

Beier, S., Bertilsson, S., 2013. Bacterial chitin degradation—mechanisms and ecophysiological strategies. Front. Microbiol. 4, 149. Available from: https://doi.org/10.3389/fmicb.2013.00149.

Bidabadi, S.S., Dehghanipoodeh, S., Wright, G.C., 2017. Vermicompost leachate reduces some negative effects of salt stress in pomegranate. Int. J. Recycl. Org. Waste Agric. 6, 255–263. Available from: https://doi.org/10.1007/s40093-017-0173-7.

Bünemann, E.K., Schwenke, G.D., Van Zwieten, L., 2006. Impact of agricultural inputs on soil organisms—a review. Aust. J. Soil. Res. 44, 379–406.

Busato, J.G., Lima, L.S., Aguiar, N.O., Canellas, L.P., Olivares, F.L., 2012. Changes in labile phosphorus forms during maturation of vermicompost enriched with phosphorus-solubilizing and diazotrophic bacteria. Bioresour. Technol. 110, 390–395.

Byzov, B.A., Nechitaylo, T.Y., Bumazhkin, B.K., Kurakov, A.V., Golyshin, P.N., Zvyagintsev, D.G., 2009. Culturable microorganisms from the earthworm digestive tract. Microbiology 78, 360–368.

Cao, Y., Zhang, Z., Ling, N., Yuan, Y., Zheng, X., Shen, B., 2011. *Bacillus subtilis* SQR 9 can control Fusarium wilt in cucumber by colonizing plant roots. Biol. Fertil. Soils 47, 495–506.

Chinsamy, M., Kulkarni, M.G., Staden, J.V., 2013. Garden-waste vermicompost leachate alleviates salinity stress in tomato seedlings by mobilizing salt tolerance mechanisms. Plant Growth Regul. 71, 41–47. Available from: https://doi.org/10.1007/s10725-013-9807-6.

de Vries, F.T., Shade, A., 2013. Controls on soil microbial community stability under climate change. Front. Microbiol. 4, 265. Available from: https://doi.org/10.3389/fmicb.2013.00265.

Dijkstra, F.A., Carrillo, Y., Pendall, E., Morgan, J.A., 2013. Rhizosphere priming:a nutrient perspective. Front. Microbiol. 4, 216. Available from: https://doi.org/10.3389/fmicb.2013.00216.

Dominguez, J., Edwards, C.A., 2004. Vermicomposting organic wastes: a review. In: Shakir Hanna S.H., Mikhail W.Z.A. (Eds.), Soil Zoology for Sustainable Development in the 21st Century. Cairo, pp. 369–395.

Edwards, C.A., Arancon, N.Q., 2004. Vermicomposts suppress plant pest and disease attacks. BioCycle 45, 51–53.

Fan, P., Ouyang, Z., Basnou, C., Pino, J., Park, H., Chen, J., 2017. Nature-based solutions for urban landscapes under post-industrialization and globalization: Barcelona versus Shanghai. Environ. Res. 156, 272–283.

Fathi, T., Golchin, A., Safikhani, F., 2012. Effect of drought stress and vermicompost on clary sage. Ann. Biol. Res. 3 (7), 3346–3349.

Fracchia, L., Dohrmann, A.B., Martinotti, M.G., Tebbe, C.C., 2006. Bacterial diversity in a finished compost and vermicompost: differences revealed by cultivation independent analyses of PCR-amplified 16S rRNA genes. Appl. Microbiol. Biotechnol. 71, 942–952.

Francou, U., Poitrenaud, M., Houot, S., 2005. Stabilization of organic matter during composting: influence of process and feedstocks. Compost Sci. Util. 13 (1), 72–83.

Fu, R., Vandermeer, B.W., Shamliyan, T.A., O'Neil, M.E., Yazdi, F., Fox, S.H., 2013. Handling continuous outcomes in quantitative synthesis. In: Methods Guide for Effectiveness and Comparative Effectiveness Reviews. Agency for Healthcare Research and Quality; 2013.

Garcia, C., Hernandez, T., Costa, F., 1997. Potential use of dehydrogenase activity as an index of microbial activity in degraded soils. Commun. Soil Sci. Plant Anal. 28, 123–134.

Garg, V., Kaushik, P., 2005. Vermistabilization of textile mill sludge spiked with poultry droppings by an epigeic earthworm *Eisenia foetida*. Bioresour. Technol. 1063–1071.

Gattinger, A., Muller, A., Haeni, M., Skinner, C., Fliessbach, A., Buchmann, N., 2012. Enhanced top soil carbon stocks under organic farming. Proc. Natl. Acad. Sci. 109 (44), 18226–18231. Available from: https://doi.org/10.1073/pnas.1209429109. PMID: 23071312.

Gaume, A., Mächler, F., Frossard, E., 2001. Aluminum resistance in two cultivars of *Zea mays* L.: root exudation of organic acids and influence of phosphorus nutrition. Plant Soil 73–81.

Ghosh, S., Goswami, A.J., Ghosh, G.K., Pramanika, P., 2018. Quantifying the relative role of phytase and phosphatase enzymes in phosphorus mineralization during vermicomposting of fibrous tea factory waste. Ecol. Eng. 116, 97–103.

Glaser, B., Lehmann, J., Zech, W., 2002. Ameliorating physical and chemical properties of highly weathered soils in the tropics with charcoal — a review. Biol. Fertil. Soils 35, 219–230.

Gulati, A., Sharma, N., Vyas, P., Sood, S., Rahi, P., Pathania, V., et al., 2010. Organic acid production and plant growth promotion as a function of phosphate solubilization by *Acinetobacter rhizosphaerae* strain BIHB 723 isolated from the cold deserts of the trans-Himalayas. Arch. Microbiol. 192, 975–983.

Gutiérrez-Miceli, F.A., Moguel-Zamudio, B., Abud-Archila, M., Gutiérrez-Oliva, V.F., Dendooven, L., 2008a. Sheep manure vermicompost supplemented with a native diazotrophic bacteria and mycorrhizas for maize cultivation. Bioresour. Technol. 99, 7020–7026.

Gutiérrez-Miceli, F.A., Gracia-Gomez, R.C., Rincon, R.R., Abud-Archila, M., MariaAngela, O.L., Gullin-Cruz, M.J., et al., 2008b. Formulation of liquid fertilizer for sorghum (*Sorghum bicolour* (L.) Moench) using vermicompost leachate. Bioresour. Technol. 99, 6174–6180.

Hait, S., Tare, V., 2011. Vermistabilization of primary sewage sludge. Bioresour. Technol. 2812–2820.

Han, J., Sun, L., Dong, X., Cai, Z., Yang, H., Wang, Y., et al., 2005. Characterization of a novel plant growth-promoting bacteria strain *Delftia tsuruhatensis* HR4 both as a diazotroph and a potential biocontrol agent against various pathogens. Syst. Appl. Microbiol. 28, 66–76.

Haritha Devi, S., Vijayalakshmi, K., Pavana Jyotsna, K., Shaheen, S.K., Jyothi, K., Surekha Rani, M., 2009. Comparative assessment in enzyme activities and microbial populations during normal and vermicomposting. J. Environ. Biol. 30, 1013–1017.

Hartmann, M., Frey, B., Mayer, J., MaÈder, P., Widmer, F., 2014. Distinct soil microbial diversity under long-term organic and conventional farming. ISME J. 2014, 1–18.

Haug, R.T., 1993. The Practical Handbook of Compost Engineering. Lewis Publishers, Boca Raton, FL.

Hazarika, J., Ghosh, U., Kalamdhad, A., Khwairakpam, M., Singh, J., 2017. Transformation of elemental toxic metals into immobile fractions in paper mill sludge through rotary drum composting. Ecol. Eng. 101, 185–192.

Hussain, N., Singh, A., Saha, S., Kumar, M.V.S., Bhattacharyya, P., Bhattacharya, S.S., 2016. Excellent N-fixing and P-solubilizing traits in earthworm gut-isolated bacteria: a vermicompost based assessment with vegetable market waste and rice straw feed mixtures. Bioresour. Technol. 165–174.

Jain, M.S., Daga, M., Kalamdhad, A.S., 2018. Composting physics: a degradation process-determining tool for industrial sludge. Ecol. Eng. 116, 14–20.

Janvier, C., Villeneuve, F., Alabouvette, C., Edel-Hermann, V., Mateille, T., Steinberg, C., 2007. Soil health through oil disease suppression: Which strategy from descriptors to indicators? Soil Biol. Biochem. 39, 1–23. Available from: https://doi.org/10.1016/j.soilbio.2006.07.001.

Kappler, U., Nouwens, A.S., 2013. Metabolic adaptation and trophic strategies of soil bacteria—C1- metabolism and sulfur chemolithotrophy in *Starkeya novella*. Front. Microbiol. 4, 304. Available from: https://doi.org/10.3389/fmicb.2013.00304.

Karsten, G.R., Drake, H.L., 1997. Denitrifying bacteria in the earthworm gastrointestinal tract and in vivo emission of nitrous oxide (N_2O) by earthworms. Appl. Environ. Microbiol. 63, 1878–1882.

Kızılkaya, R., Hepşen Türkay, F.Ş., Bayraklı, B., Aşkın, T., Türkmen, C., Akça, İ., et al., 2010. Enzyme and earthworm activities during vermicomposting in sewage sludge. In: Proceedings of the International Soil Science Congress on Management of Natural Resources to Sustain Soil Health and Quality. Ondokuz Mayis University, Samsun, Turkey, pp. 1047–1054.

Komilis, D.P., Ham, R.K., 2006. Carbon dioxide and ammonia emissions during composting of mixed paper, yard waste and food waste. Waste Manag 26, 62–70.

Kumari, S.S.M., Ushakumari, K., 2002. Effect of vermicompost enriched with rock phosphate on the yield and uptake of nutrients in cowpea (*Vigna unguiculata* L. Walp). J. Trop. Agric. 40, 27–30.

Larsen, E., Grossman, J., Edgell, J., Hoyt, G., Osmond, D., Hu, S., 2014. Soil biological properties, soil losses and corn yield in long-term organic and conventional farming systems. Soil Tillage Res 139, 37–45.

Lavelle, P., Martin, A., 1992. Small-scale and large-scale effects of endogeic earthworms on soil organic matter dynamics in soils of the humid tropics. Soil Biol. Biochem. 12, 1491–1498.

Le Bayon, R.C., Binet, F., 2006. Earthworms change the distribution and availability of phosphorous in organic substrates. Soil Biol. Biochem. 38, 235–246.

Lehmann, J., Rondon, M., 2006. Bio-char soil management on highly weathered soils in the humid tropics. In: Uphoff, N., et al., ed. Biological Approaches to Sustainable Soil Systems. CRC Press, Boca Raton, FL, pp. 517–530.

Li, F., Chen, L., Zhang, J., Yin, J., Huang, S., 2017. Bacterial community structure after long-term organic and inorganic fertilization reveals important associations between soil nutrients and specific taxa involved in nutrient transformations. Front Microbiol. 8, 187.

Liu, X., Zhang, S., Jiang, Q., Bai, Y., Shen, G., Li, S., 2016. Using community analysis to explore bacterial indicators for disease suppression of tobacco bacterial wilt. Sci. Rep. 6, 36773. Available from: https://doi.org/10.1038/srep36773. PMID: 27857159.

Ma, C.H., Kalb, T., 2006. Development of starter solution technology as a balanced fertilization practice in vegetable production. Acta Horticult. 700, 167172.

Maboeta, M.S., van Rensburg, L., 2003. Vermicomposting of industrially produced wood chips and sewage sludge utilizing *Eisenia foetida*. Ecotoxicol. Environ. Saf. 56, 265–270.

Manna, M.C., Jha, S., Ghosh, P.K., Acharya, C.L., 2003. Comparative efficacy of three epigeic earthworms under different deciduous forest litters decomposition. Bioresour. Technol. 88, 197–206.

McCully, M., 2005. The rhizosphere: the key functional unit in plant/soil/microbial interactions in the field. Implications for the understanding of allelopathic effects. In: Proceedings of the 4th World Congress on Allelopathy, eds. J. Harper, M. An, H. Wu and J. Kent, Charles Sturt University, Wagga Wagga, NSW, Australia. 21–26.

McDaniel, M.D., Tiemann, L.K., Grandy, A.S., 2014. Does agricultural crop diversity enhance soil microbial biomass and organic matter dyanmics? A meta-analysis. Ecol. Appl. 24 (3), 560–570. PMID: 24834741.

Mir, M., Hassan, G.I., Mir, A., Hassan, A., Sulaimani, M., 2013. Effects of bio-organics and chemical fertilizers on nutrient availability and biological properties of pomegranate orchard soil. Afr. J. Agric. Res. 8 (37), 4623–4627.

Mohee, R., Mudhoo, A., 2005. Analysis of the physical properties of an in-vessel composting matrix. J. Powder Technol. 155, 92–99.

Moorhead, D.L., Rinkes, Z.L., Sinsabaugh, R.L., Weintraub, M.N., 2013. Dynamic relationships between microbial biomass, respiration, inorganic nutrients and enzyme activities: informing enzyme-based decomposition models. Front. Microbiol. 4, 223.

Mooshammer, M., Wanek, W., Zechmeister-Boltenstern, S., Richter, A.A., 2014. Stoichiometric imbalances between terrestrial decomposer communities and their resources: mechanisms and implications of microbial adaptations to their resources. Front. Microbiol. 5, 22.

Nandi, B., Bhandari, S.C., Meena, R.H., Meena, R.R., 2013. Effect of vermicompost on plant growth, fruit yield and quality of pomegranate cv. Ganesh. Environ. Ecol. 31 (1A), 322–324.

Nazaries, L., Pan, Y., Bodrossy, L., Baggs, E.M., Millard, P., Murrell, J.C., et al., 2013. Evidence of microbial regulation of biogeochemical cycles from a study on methane flux and land use change. Appl. Environ. Microbiol. 4031–4040. Available from: aem.asm.org4031.

Nechitaylo, T.Y., Yakimov, M.M., Godinho, M., Timmis, K.N., Belogolova, E., Byzov, B. A., et al., 2010. Effect of the earthworms *Lumbricus terrestris* and *Aporrectodea caliginosa* on bacterial diversity in soil. Microb. Ecol. 59, 574–587.

Neumann, G., Römheld, V., 2002. Root induced changes in the availability of nutrients in the rhizosphere. In: Waisel, Y., Eshel, A., Kafkafi, U. (Eds.), Plant Roots The Hidden Half, 3rd ed. Marcel Dekker, New York, pp. 617–649.

Nielsen, U.N., Wall, D.H., Six, J., 2015. Soil biodiversity and the Environment. Annu. Rev. Environ. Resour. 40, 63–90.

Nihorimbere, V., Ongena, M., Smargiassi, M., Thonart, P., 2011. Focus on: beneficial effect of the rhizosphere microbial community for plant growth and health. Biotechnol. Agron. Soc.Environ. 15, 327–337.

Padmavathiamma, P.K., Li, L.Y., Kumari, U.R., 2008. An experimental study of vermi-biowaste composting for agricultural soil improvement. Bioresour. Technol. 99, 1672–1681.

Parthasarathi, K., Ranganathan, L.S., 1998. Pressmud vermicast are hot spots of fungi and bacteria. Ecol. Environ. Conserv. 4, 81–86.

Paszt, L.S., 2017. Beneficial microorganisms improving organic cultivation of horticultural crops and soil fertility. J. Bacteriol. Parasitol. 8:6 (Suppl). Available from: https://doi.org/10.4172/2155-9597-C1-041.

Patel, D.K., Murawala, P., Archana, G., Kumar, G.N., 2011. Repression of mineral phosphate solubilizing phenotype in the presence of weak organic acids in plant growth promoting fluorescent pseudomonads. Bioresour. Technol. 102, 3055–3061.

Pathma, J., Sakthivel, N., 2012. Microbial diversity of vermicompost bacteria that exhibit useful agricultural traits and waste management potential. http://www.springerplus.com/content/1/1/26.

Pathma, J., Rahul, G.R., Kamaraj Kennedy, R., Subashri, R., Sakthivel, N., 2011. Secondary metabolite production by bacterial antagonists. J. Biol. Control 25, 165–181.

Philippot, L., Raaijmakers, J.M., Lemanceau, P., Putten, W.H.V.D., 2013. Going back to the roots: the microbial ecology of the rhizosphere. Nat. Rev. Microbiol. 11 (11), 789±99. Available from: https://doi.org/10.1038/nrmicro3109. PMID: 24056930.

Piccolo, A., 2002. The supramolecular structure of humic substances: a novel understanding of humus chemistry and implications in soil science. Adv. Agron. 75, 57–133.

Pietri Aciego, J.C., Brookes, P.C., 2008. Relationships between soil pH and microbial properties in a UK arable soil. Soil Biol. Biochem. 40, 1856–1861.

Pinel, N., Davidson, S.K., Stahl, D.A., 2008. *Verminephrobacter eiseniae* gen. nov., sp. nov., a nephridial symbiont of the earthworm *Eisenia foetida* (Savigny). Int. J. Syst. Evol. Microbiol. 58, 2147–2157.

Pramanik, P., 2012. Effect of composting and vermicomposting on recycling of three aquatic weeds treated with rock phosphate on P-dynamics, phosphatase activity and biomass P. Dynamic Soil Dynamic Plant. Global Science Books, pp. 51–56.

Pramanik, P., Ghosh, G.K., Ghosal, P.K., Banik, P., 2007. Changes in organic - C, N, P and K and enzyme activities in vermicompost of biodegradable organic wastes under liming and microbial inoculants. Bioresour. Technol. 98, 2485–2494.

Pramanik, P., Ghosh, G., Banik, P., 2009. Effect of microbial inoculation during vermicomposting of different organic substrates on microbial status and quantification and documentation of acid phosphatase. Waste Manag 574–578.

Pramanik, P., Ghosh, G.K., Chung, Y.R., 2010. Changes in nutrient content, enzymatic activities and microbial properties of lateritic soil due to application of different vermicomposts: comparative study of ergosterol and chitin to determine fungal biomass in soil. Soil Use Manag. 508–515.

Raaijmakers, J.M., Paulitz, T.C., Steinberg, C., Alabouvette, C., MoeÈnne-Loccoz, Y., 2009. The rhizosphere: a playground and battlefield for soilborne pathogens and beneficial microorganisms. Plant Soil 321 (1 ± 2), 341–361.

Raimi, A., Adekele, R., Ashira, R., 2017. Soil fertility challenges and biofertiliser as a viable alternative for increasing smallholder farmer crop productivity in sub-Saharan Africa. Cogent Food Agric. 3, 1400933. Available from: https://doi.org/10.1080/23311932.2017.1400933.

Ravindra, N.P., Raman, G., Badri Narayanan, K., Sakthivel, N., 2008. Assessment of genetic and functional diversity of phosphate solubilizing fluorescent pseudomonads isolated from rhizospheric soil. BMC Microbiol. 8, 230.

Rillig, M.C., Mummey, D.L., 2006. Mycorrhizas and soil structure. New Phytol. 171, 41–53. Available from: https://doi.org/10.1111/j.1469-8137.2006.01750.x. PMID: 16771981.

Rinkes, Z.L., Sinsabaugh, R.L., Moorhead, D.L., Grandy, A.S., Weintraub, M.N., 2013. Field and lab conditions alter microbial enzyme and biomass dynamics driving decomposition of the same leaf litter. Front. Microbiol. 4, 260.

Rinnan, R., Gierth, D., Bilde, M., Rosenørn, T., Michelsen, A., 2013. Off-season biogenic volatile organic compound emissions from heath mesocosms: responses to vegetation cutting. Front. Microbiol. 4, 224.

Römheld, V., Neumann, G., 2005. The rhizosphere: definition and perspectives. In: Hartmann, A., et al., (Eds.), Rhizosphere 2004: A Tribute to Lorenz Hiltner. GSFReport, Munich, Neuherberg, Germany, pp. 47–49.

Römheld, V., Neumann, G., 2006. The rhizosphere: contributions of the soil-root interface to sustainable soil systems. In: Uphoff, N., et al., (Eds.), Biological Approaches to Sustainable Soil Systems. CRC Press, Boca Raton, FL, pp. 91–107.

Rousk, J., Baath, E., 2011. Growth of saprotrophic fungi and bacteria in soil. FEMS Microbiol. Ecol. 78, 17–30.

Rousk, K., Jones, D.L., DeLuca, T.H., 2013. Moss-cyanobacteria associations as biogenic sources of nitrogen in boreal forest ecosystems. Front. Microbiol. 4, 150.

Saha, S., Mina, B.L., Gopinath, K.A., Kundu, S., Gupta, H.S., 2008. Relative changes in phosphatase activities as influenced by source and application rate of organic composts in field crops. Bioresour. Technol. 99, 1750–1757.

Scervino, J.M., Mesa, M.P., Mónica, I.D., Recchi, M., Moreno, N.S., Godeas, A., 2010. Soil fungal isolates produce different organic acid patterns involved in phosphate salts solubilization. Biol. Fertil. Soils 46, 755–763.

Schmidt, M.G., Attaway, H.H., Fairey, S.E., Howard, J., Mohr, D., Craig, S., 2019. Self-disinfecting copper beds sustain terminal cleaning and disinfection effects throughout patient care. Appl. Environ. Microbiol. 86 (1). Available from: https://doi.org/10.1128/AEM.01886-19.

Schmidt, J.E., Angela, D.K., Vanessa, L.B., Amélie, C.M.G., 2019. Agricultural management and plant selection interactively affect rhizosphere microbial community structure and nitrogen cycling. Microbiome 7, 146. Available from: https://doi.org/10.1186/s40168-019-0756-9.

Singh, J., Kalamdhad, A.S., 2013. Assessment of bioavailability and leachability of heavy metals during rotary drum composting of green waste (water hyacinth). Ecol. Eng. 52, 59–69.

Singleton, D.R., Hendrixb, P.F., Colemanb, D.C., Whitmana, W.B., 2003. Identification of uncultured bacteria tightly associated with the intestine of the earthworm *Lumbricus rubellus* (Lumbricidae; Oligochaeta). Soil Biol. Biochem. 35, 1547–1555.

Sinha, R.K., Herat, S., Valani, D., Chauhan, K., 2009. Vermiculture and sustainable agriculture. Am-Euras. J. Agric. Environ. Sci. 5, 1–55. IDOSI Publication.

Sinha, R.K., Agarwal, S., Chauhan, K., Valani, D., 2010. The wonders of earthworms and its vermicompost in farm production: Charles Darwin's 'friends of farmers', with potential to replace destructive chemical fertilizers from agriculture. Agric. Sci. 1, 76–94.

Stark, C.H., Condron, L.M., O'Callaghan, M., Stewart, A., Di, H.J., 2008. Differences in soil enzyme activities, microbial community structure and short-term nitrogen mineralisation resulting from farm management history and organic matter amendments. Soil Biol. Biochem. 40, 1352–1363.

Stein, L.Y., Nicol, W.N., 2011. Grand challenges in terrestrial microbiology. Front. Microbiol. 2, 6. Available from: https://doi.org/10.3389/fmicb.2011.00006.

Steinweg, J.M., Dukes, J.S., Paul, E.A., Wallenstein, M.D., 2013. Microbial responses to multi-factor climate change: effects on soil enzymes. Front. Microbiol. 4, 146. Available from: https://doi.org/10.3389/fmicb.2013.00146.

Suhane, R.K., 2007. Vermicompost. Publication of Rajendra Agriculture University, Pusa, p. 88.

Suthar, S., 2006. Potential utilization of guar gum industrial waste in vermicompost production. Bioresour. Technol. 2474–2477.

Tabatabai, M.A., 1994. Soil enzymes. In: Mickelson, S.H., Bighan, J.M. (Eds.), Methods of Soil Analysis, Part 2: Microbiological and Biochemical Properties. Soil Science Society of America, Madison, pp. 775–826.

Thomson, B.C., Ostle, N.J., McNamara, N.P., Oakley, S., Whiteley, A.S., Bailey, M.J., et al., 2013. Plant soil interactions alter carbon cycling in an upland grassland soil. Front. Microbiol. 4, 253. Available from: https://doi.org/10.3389/fmicb.2013.00253.

Thyagarajan, L.P., Meenambal, T., Mangaleshwaran, L., Lashminarasimaiah, N., 2010. Recycling of pulp and paper industry sludge with sawdust by aerobic composting method. Nat. Environ. Pollut. Technol. 1, 149–154.

Tiquia, S.M., Tam, N.F.Y., 1998. Composting of spent pig litter in turned and forced aerated piles. Environ. Pollut. 99, 329–337.

Tiwari, S.C., Tiwari, B.K., Mishra, R.R., 1989. Microbial populations, enzyme activities and nitrogen, phosphorous, potassium enrichment in earthworm casts and in the surrounding soil of pineapple plantation. Biol. Fertil. Soils 8, 178–182.

Treseder, K.K., Balser, T.C., Bradford, M.A., Brodie, E.L., Dubinsky, E.A., Eviner, V.T., 2012. Integrating microbial ecology into ecosystem models: challenges and priorities. Biogeochemistry 109, 7–18. Available from: https://doi.org/10.1007/s10533-0119636-5.

Tu, C., Ristaino, J.B., Hu, S., 2006. Soil microbial biomass and activity in organic tomato farming systems: effects of organic inputs and straw mulching. Soil Biol. Biochem. 38, 247.

Varma, V.S., Kalamdhad, A.S., 2014. Effects of leachate during vegetable waste composting using rotary drum composter. Environ. Eng. Res. 19 (1), 1–7.

Varma, V.S., Ramu, K., Kalamdhad, A.S., 2014. Effects of waste lime sludge on nitrogen dynamics and stability of mixed organic waste using rotary drum composter. Int. J. Environ. Res. 9 (1), 395–404.

Vaz-Moreira, I., Maria, E., Silva, C.M., Manaia Olga, C., Nunes, 2008. Diversity of bacterial isolates from commercial and homemade composts. Microb. Ecol. 55, 714–722.

Venter, Z.S., Jacobs, K., Hawkins, H.-J., 2016. The impact of crop rotation on soil microbial diversity: a meta-analysis. Pedobiologia (Jena) 59 (4), 1–9.

Vivas, A., Moreno, B., Garcia-Rodriguez, S., Benitez, E., 2009. Assessing the impact of composting and vermicomposting on bacterial community size and structure, and microbial functional diversity of an olive-mill waste. Bioresour. Technol. 100, 1319–1326.

Wagg, C., Bender, S.F., Widmer, F., van der Heijden, M.G.A., 2014. Soil biodiversity and soil community composition determine ecosystem multifunctionality. Proc Natl Acad Sci. 111 (14), 5266–5270. Available from: https://doi.org/10.1073/pnas.1320054111. PMID: 24639507.

Yasir, M., Aslam, Z., Kim, S.W., Lee, S.W., Jeon, C.O., Chung, Y.R., 2009. Bacterial community composition and chitinase gene diversity of vermicompost with antifungal activity. Bioresour. Technol. 100, 4396–4403.

Zeleke, J., Sheng, Q., Wang, J.-G., Huang, M.-Y., Xia, F., Wu, J.-H., 2013. Effects of *Spartina alterniflora* invasion on the communities of methanogens and sulfate-reducing bacteria in estuarine marsh sediments. Front. Microbiol. 4, 243. Available from: https://doi.org/10.3389/fmicb.2013.00243.

Zhang, B.G., Li, G.T., Shen, T.S., Wang, J.K., Sun, Z., 2000. Changes in microbial biomass C, N, and P and enzyme activities in soil incubated with the earthworms *Metaphire guillelmi* or *Eisenia foetida*. Soil Biol. Biochem. 32, 2055–2062.

Zhao, J., Liu, J., Liang, H., Huang, J., Chen, Z., Nie, Y., 2018. Manipulation of the rhizosphere microbial community through application of a new bio-organic fertilizer improves watermelon quality and health. PLoS ONE 13, e0192967. Available from: https://doi.org/10.1371/journal.pone.0192967.

Endophytic microbial interaction with legume crop for developing resistance against nutrient stress

17

Priyanka Chauhan[1,2], *Pratibha Verma*[1,2], *Shipra Pandey*[1,2,3], *Arpita Bhattacharya*[1,2], *Ashutosh Tripathi*[1,2], *Ved Prakash Giri*[1,4], *Satyendra Pratap Singh*[1,5] *and Aradhana Mishra*[1,2]
[1]Division of Microbial Technology, CSIR-National Botanical Research Institute, Lucknow, India, [2]Academy of Scientific and Innovative Research (AcSIR), Ghaziabad, India, [3]Department of Chemical Engineering, Indian Institute of Technology Bombay, Mumbai, India, [4]Department of Botany, Lucknow University, Lucknow, India, [5]Division of Pharmacognosy, CSIR-National Botanical Research Institute, Lucknow, India

17.1 Introduction

In India the agriculture sector accounts for 18% of India's gross domestic product and accounts for about 50% of the country's labor force (Madhusudhan, 2015). The agriculture sector is utmost importance for humans and society as well as it makes food, fodder, fuel, and other ecosystem services available (Maitra and Ray, 2019). Various stresses including biotic stress such as bacteria, virus, fungi, nematodes, etc. causing diseases and abiotic stress such as salinity, drought, flooding, nutrient deficiency, heat, and cold significantly affect crop yield and production (Saijo, 2019).

Nutrients are essential for plant growth and development and its limitation causes a significant loss in yield as well as compromises plant's immunity against several biotic and abiotic stresses (Singh et al., 2019). Nutrient deficiency is one of the major concerns in plant productivity as an imbalance in nutrient concentration affects the entire metabolism of plants. Nutrient deficiency decreases photosynthesis, enhances accumulation of reactive oxygen species (ROS), reduces nodulation, nitrogen fixation, chlorophyll content, and causes hormonal imbalance (Sharma et al., 2012). Moreover, maintaining the soil fertility is an important aspect of plant nutrition, to provide an adequate amount of nutrients for healthy growth and plant development (Johnston and Bruulsema, 2014; Withers and Bowes, 2018). Adverse soil conditions affect the availability of mineral nutrients and increase the concentrations of toxic elements in soil solution which in turn also affects the plant growth (White and Greenwood, 2013).

Microbes in Land Use Change Management. DOI: https://doi.org/10.1016/B978-0-12-824448-7.00020-6
© 2021 Elsevier Inc. All rights reserved.

Although, chemical fertilizers increase the crop productivity, the excessive and prolong use of fertilizer affects the soil pH which becomes acidic, strengthen pesticides, and simultaneously the pathogens also become resistant (Amini and Sidovich, 2010). They also affect human health, harmful for soil fertility, and reduce the crop productivity by 15%–18% (Singh et al., 2019).

Legume crops, a sink of protein require macro and micronutrients (N, P, K, and S in higher concentrations and Fe, Cu, Mn, Zn, and other minerals in trace amounts) for normal physiological function and development (Nimac et al., 2020). In soybean plants, depending on plant variety, soil fertility and climatic factors, the deficiency of N, P, Fe, and S nutrients may cause yield loss up to 10%, 45%, 90%, and 30% respectively (Hellal and Abdelhamid, 2013). Among all macronutrients N, P, S and Fe are most important nutrient for the growth and development of leguminous crops. In agricultural production, lack of these nutrients causes adversely affects the plant growth and yield (Aleksandrov, 2019).

Nitrogen (N) is one of the major nutrients for leguminous crops, and a major component involved in the synthesis of amino acids, nucleic acid synthesis, and other nitrogenous compounds (Aleksandrov, 2019) considered as vital components of the living system. Phosphorus (P) is also an important nutrient for growth and development of leguminous crops which plays a central role in the diverse cellular processes (Kleinert et al., 2014), related to photosynthesis, energy production, carbohydrate metabolism, symbiotic nitrogen fixation (SNF), and redox reaction (Sulieman and Tran, 2015). Iron (Fe) is one of the most abundant element in soil and is involved in photosynthesis, respiration, chlorophyll synthesis, and N_2 fixation (Meharg, 2012). Lastly, sulfur (S) is the fourth major macro nutrient for the growth and development of the leguminous crops. Large amount of S is required in protein synthesis, mainly in the oil formation within the seeds of leguminous crops. Hence, S have also significant role in nutrient in influencing the nutritive quality of legume crops.

To cope up with this adverse environmental condition, beneficial microbes such as endophytes are new eco-friendly alternatives in place of chemical fertilizers to maintain sustainable crop production and reduce nutrient stress for plants (Miliute et al., 2015). The use of microorganisms is healthier alternative for plants during integrated plant disease management system along with nutrient acquisition (Jha et al., 2013).

Legumes being sensitive to the environmental stresses, the association with microbes enable them to persist under unfavorable conditions and help to cope up with the devastating effects of environmental stresses (Zahran, 2017). As the environmental condition changed, microbes and plants have co-evolved by an array of mechanisms to adapt with the changing environment. Various types of microbes are present in leguminous crops, solubilize the nutrients from soil and provides to the plants.

Among all the crops grown across the world, the leguminous crop occupies an important position being vital source of protein, calcium, phosphorus, iron, and other minerals. The pods of legumes (Family Fabaceae) are used in the human diet as well as the fodder for animal feed. In commercial agriculture, the production of legume crops ranks third after cereals and oil seeds, as they are rich source of proteins. Legume crops also have the ability to maintain soil fertility through biological nitrogen fixation (Mantri et al., 2013). Globally, the production of soybean, faba bean, and groundnut have the highest yields, which ranges between 1.5 and 2.4 tons/ha and while others legume crops like dry bean, pigeon pea, chickpea, and lentil have yield production levels is approximate 850 kg/ha. The main sources of dietary protein for vegetarians in legume crops are chickpea (*Cicer arietinum*), common bean (*Phaseolus vulgaris*), grass pea (*Lathyrus sativus*), lentil (*Lens culinaris*), mung bean (*Vigna radiata*), urad bean (*Vigna mungo*), pea (*Pisum sativum*), pigeon pea (*Cajanus cajan*), and soybean (*Glycine max*). Some legume crops such as soybean and groundnut are also a good source of vegetable oils (Bellaloui et al., 2013; Meena et al., 2015). As an important candidate for crop sequencing, legume crops can fix nitrogen biologically to maintain nitrogen fertility in the soil and thus improve soil's physical conditions and maintain environmental balance (Courty et al., 2015).

This chapter focused on effect of nutrient limitation on leguminous crop and what are the strategies adapted by endophytes to amelioration the nutrient stress.

17.2 Factors responsible for nutrient limitation and their effects

There are a variety of factors are associated with nutrient limitation in soil, which affects the crop yield and production. These include the parental rock material, particle size, humus and water content, pH, aeration, temperature, root surface area, the rhizoflora, and mycorrhizal development. Besides that nutrient stress may result from either low levels of availability of the element or by the presence of excess concentrations. In some cases the presence of one element in excess concentrations may induce the deficiency of another element (Reddy, 2006). Major factors that affect the soil nutrient availability includes soil pH, soil erosion, leaching, volatilization, denitrification, nitrogen immobilization and plant's nutrient uptake.

Nutrients are essential for plants, macro and micronutrients such as Ca, S, Mg, K, N, P and Fe have been recognized as crucial for plants. Plants cannot complete their life cycles and achieve their physiological functions in the absence of these micro and macronutrients. The deficiencies of these nutrients can reduce growth and yield of crops (Osman, 2013) (Fig. 17.1). This chapter discusses nutrient stresses of four major macronutrients such as P, N, S, and Fe which are primarily responsible for the growth, development and productivity of legume crops.

17.2.1 Phosphorus deficiency

The availability of soil phosphorus (P) may be one of the most salvaging environmental factors, and its deficiency limiting the physiology of legumes which involves association with rhizobia, engaged in symbiotic biological N_2 fixation (Gan et al., 2002). It is necessary for the development and functioning of the

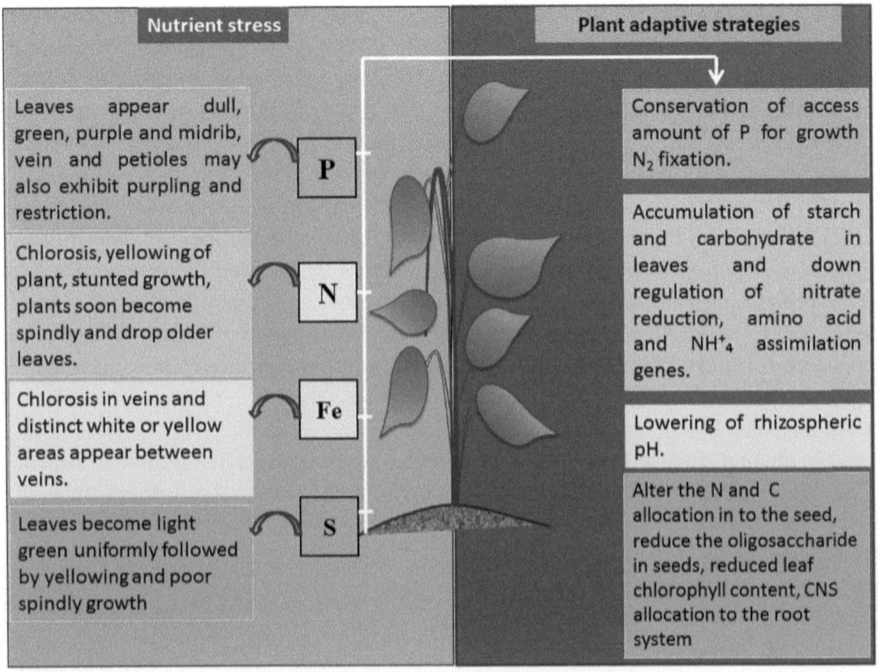

Figure 17.1 Effect of major nutrient stress on plants and their stress amelioration adaptive strategies.

nodule. The role of P is important in nodules mainly during the metabolic energy processes (ATP) that fixes N_2 to NH_3 (Le Roux et al., 2006) and then converts it into organic N compounds such as amino acids and ureides. Therefore the lack of P is critical for agriculture and natural ecosystems, posing a significant problem for leguminous plants.

However, the development of P deficiency within nodules occurs at a later stage, after long time roots have succumbed to P deficiency (Le Roux et al., 2006). Long-term P deficiency leads to a decrease in nodular inorganic P (Pi) levels (Kleinert et al., 2014), which can negatively affect the energy status of nodules and ATP availability (Plaxton and Podestá, 2006). This condition would certainly have an effect on the ability of nodules to assimilate the nitrogen. Phosphorus deficiency indirectly affects the plants such as it alters the utilization of inorganic nitrogen, affects the synthesis of the amino acid due to unavailability of organic C and bacteroid respiration in nodules. P deficiency can cause alterations in organic acids synthesis, especially of malate (C source) and its metabolism. Disruption in the supply of malate during P deficiency affects organic carbon available in nodules of bacteroids for respiration and also affects the amino acid synthesis and limits the biological N_2 fixation leading to feedback inhibition of organic acids such as malate which in turn reduces the amino acid synthesis (Le Roux et al., 2008). Carbon supply to respiration is limited in this situation, plants starts the recycling of organic N form amino acids. The switch to ureides export may reduce the carbon cost under Pi stress, which has a positive effect on legume's growth. This increases the participation of glutamate dehydrogenase (GDH) in deamination direction, another enzyme is aminotransferase may be key features of organic N recycling. Early literatures describes that the concentration of asparagine may increase due to mineral deficiency (Stewart and Larher, 1980). In legume white clover (*Trifolium repens*), an increase in asparagine concentration in legume roots and nodules induced a mechanism of organic N feedback, which affects the nodulation and the capacity to fix N_2 (Almeida et al., 2000). This enzyme has a vital role in the nitrogen and carbon metabolism of legume nodules. Under such circumstances GDH acts as an alternative enzyme for the recycling of organic N (Magadlela et al., 2016).

17.2.2 Iron deficiency

For all the higher plants iron is a transition metal that plants need for its survival. Iron plays a critical role in metabolic processes such as respiration, electron transport, and photosynthesis. Fe is a component of Fe-S cluster or heam containing enzymes involved in oxido-reduction reactions (Qin et al., 2015). Due to iron deficiency, crop's leaves starts yellowing and finally tunrs brown (Brittenham, 1994).

Chlorosis is a common nutritional disorder in plants under Fe deficiency, which affects crop yield and quality (Chen et al., 2018). In legumes SNF is adversely affected by various environmental limitations. Fe deficiency is devastating problem for legumes which limits SNF capacity, because Fe plays an important role in biosynthesis of several proteins which are involved in SNF, such as leghemoglobin. Leghemoglobin is highly abundant protein that buffers oxygen, nitrogenase involved in nitrogen fixation, and cytochromes is responsible for electron transport chain (Lüthje et al., 2018). Fe deficiency has been shown to limit the growth and SNF of legume crops (Slatni et al., 2011, 2014).

17.2.3 Sulfur deficiency

Sulfur is an essential nutrient for plants as a constituting element for the synthesis of amino acids, metal cofactors, coenzymes, and secondary metabolites. Sulfur is important for SNF, during symbiosis, bacteria associates with root organelles and differentiates into bacteroid. Bacteroid expresses nitrogenase enzymes complex (sulfur-rich complex) that reduces atmospheric nitrogen to ammonia. This reaction is highly sensitive for oxygen so leghemoglobin which is present in the cytoplasm of infected cell act as an O_2 diffusion barrier in the mid-inner cortex and this reaction requires high amount of ATP molecules as energy source (Minchin et al., 2008). Sulfur limitation can reduce N_2 fixation of legume crops by affecting nodule development and function (Divito and Sadras, 2014). There is no surprise that sulfur deficiency decreases the plant growth, photosynthesis and yield in both leguminous and nonleguminous crops. SNF is positively linked with sulfur supply and its deficiency affects SNF resulting in decrease in nodulation and nodule metabolism. Bacteroids provide ammonia to plant cells whereas plants provide sugar (sucrose) to nodules; here it is metabolized into phosphoenolpyruvate and then malate by malate dehydrogenase. Malate is taken up by bacteroids, oxidized into CO_2, and ATP obtained via the TCA cycle and used to reduce atmospheric nitrogen to ammonia (Becana et al., 2018). During sulfur deficiency, the impairment of nitrogenase biosynthesis leads to a decline in the accumulation of amino acids, leghemoglobin, ATP, and ferredoxin in the plant nodules.

17.2.4 Nitrogen deficiency

Nitrogen is also a major limiting nutrient, because it is the key constituent of amino acids, which are the building blocks of protein, nucleotides, chlorophyll and numerous other metabolites or cellular components (Gong et al., 2019). In plant leaf nearly 75% of nitrogen content is allocated in chloroplasts, and most of this proportion is used for the production of photosynthetic components. The capacity of photosynthesis is directly correlated with nitrogen content (Li et al., 2013). The radiation use efficiency and detection capacity reduces in plant's reproductive organs under nitrogen deficiency (Kakani and Reddy, 2010). Nitrogen deficiency constraints plant senescence, leaf area expansion by delaying reproductive and vegetative phenological development, which causes severe quality and yield decline in legume crops (Zhang et al., 2017).

17.3 Adaptive strategies of legume crops under nutrient deficient condition

Legume crops depend on symbiotic stable nitrogen, but the depletion of nitrogen is energetically more expensive than soil nitrogen. Due to abiotic stress viz. drought, heat stress, flooding, salinity, and nutrient deficiency (N, P, K, and S) are the major components causing about 50% of crop losses (Tiwari et al., 2020). Plants can be acclimatized against adverse conditions and adopt many strategies under stress conditions (Table 17.1). Under P deficiency legume nodules have evolved several strategies to counteract the initial effects of limitation, maintain the N metabolism (Valentine et al., 2017). In legume *Virgilia divaricata*, nodule reduces their N_2 acquisition from the atmosphere and preferred to uptake soil N from root during P stress (Magadlela et al., 2014).

There is a hypothesis that under Pi deficiency in soybean and common bean both activates the autoregulation of nodule (AON) pathway which leads to reduction in the number of nodules (Isidra-Arellano et al., 2020). According to Ferguson et al. (2019), SNF is a highly carbon costly process. Legumes tightly regulates the number of nodules per root system due to the high carbon cost of SNF. This regulation is mediated by the AON pathway, which allows legumes to maintain a balance between the gains and the costs of root nodule symbiosis (Ferguson et al., 2019). Plants getting N through chemical methods from NH_4 and NO_3 rather than

Table 17.1 Effect of different nutrient stress on plants and their adaptive strategies.

S. no.	Nutrient stress	Effect on plants	Plants adaptive strategies	References
1	Phosphorus	Leaves appear dull, dark green, blue green, or red-purple, especially on the underside, along the midrib and veins, petioles may also exhibit purpling and restriction in growth may be noticed	Nodules conserve the elevated P concentration to maintain growth and high rates of N_2 fixation	Qin et al. (2012)
2	Nitrogen	Chlorosis, progresses from light green to yellow, entire plant becomes yellow under prolonged stress, growth is immediately restricted and plants soon become spindly and drop older leaves	Genes involved in nitrate reduction and amino acid assimilation and ammonium assimilation are downregulated, starch and carbohydrate accumulation in leaves	Schlüter et al. (2012)
3	Iron	Symptoms are rare on mature leaves, veins become chlorotic and distinct white or yellow areas appear between veins	Increases the ability to decrease the rhizospheric pH through the activation of a plasms membrane–localized P-type H^+-ATPase (adenosine triphosphatase)	Santi and Schmidt (2009)
4	Sulfur	Chlorosis uniformly does not occur, leaves become light green uniformly followed by yellowing and poor spindly growth	Alter the N and C allocation into the seeds, reduces the oligosaccharide in seeds, reduces leaf chlorophyll content, C, N and S allocation to the root system	Zuber et al. (2013)

(Continued)

Table 17.1 (Continued)

S. no.	Nutrient stress	Effect on plants	Plants adaptive strategies	References
5	Zinc	Shortened of internode and leaves may be abnormally small and necrotic	Enhances the bioaccumulation of soil Zn for root uptake, translocation of Zn from the root to the shoot; alter the subcellular compartmentation of Zn in shoot cells, such that more Zn maintain in the cytoplasm; and enhance or more efficient biochemical utilization of Zn in shoot cells	Hacisalihoglu and Kochian (2003)
6	Boron	Internode may be short at shoot terminal and stems become rough, cracked, split along the vascular bundle and young leaves may be necrotic or distorted followed by death at growing point	Increases cytosolic Ca^{++} concentration, antioxidant enzymes and signal transduction, enhancement of DNA methylation and alternative splicing of mRNA	Yang et al. (2016)
7	Potassium	First appears the mild symptoms on recently matured leaves, leaf margins tanned, scorched, or have necrotic spots, become brown and cup downward and growth is restricted and die back may occur	Increases root length to enhance potassium uptake, specific isoforms of plasma membrane H^{+} ATPase are up regulated for unloading of K^{+} from the xylem stream, increases concentration of Na^{+}, Mg^{++}, and Ca^{++}	Hafsi et al. (2014)

(*Continued*)

Table 17.1 (Continued)

S. no.	Nutrient stress	Effect on plants	Plants adaptive strategies	References
8	Manganese	Symptoms appear later on older leaves, chlorosis is less marked near veins, some mottling occurs in interveinal areas, become brown, transparent, or necrotic	Enhances the accumulation of reactive oxygen species in chloroplast, length of chloroplast decreases and diameter of the grana thylakoids tends to increase	Schmidt et al. (2016)
9	Magnesium	Symptoms occur late in the growing seasons, chlorosis occur at margin of leaves, show yellow chlorotic interveinal tissue on some species, reddish purple progressing to necrosis on others, affected with continued stress, chlorotic areas may become necrotic, brittle, and curl upward	Increases the concentration of Ca, Fe, and Cu in leaves, root and K concentration decreases	Hermans et al. (2010)

biological N_2 fixation under P deficiency, under such conditions the plants accumulate more biomass (Magadlela et al., 2014).

In legumes, nitrate is the inhibitor of nodulation and biological N_2 fixation. The acquisition of inorganic nitrogen to nitrogen assimilation (amino acids) can vary to the legume host and also altered by the P status of the nodules. In most legumes from temperate regions such as lupins, pea, clovers transfer their fixed N_2 in xylem as amides (asparagine and glutamine) (Valentine et al., 2017). Whereas, legumes present in tropical regions (soybean, common bean, cowpea) the fixed nitrogen are transferred as arieds- exporters because they export purine derivatives, such as allantoin and allantoic acids in xylem during P deficiency owing to low carbon cost. Leguminous plants improve the activities of Fe(III)-chelate reductase (FC-R), H^+ ATPase, and the phosphoenolpyruvate carboxylase in nodules to promote Fe uptake from soil by the nodule epidermis during iron deficiency (Slatni et al., 2014). This strategy used by the legume crops is called as common Fe acquisition strategy.

17.4 Endophytes mediated mitigation of nutrient deficiencies

A wide range of microorganisms are found in different regions of plants including endophytic, epiphytic, rhizospheric, and phyllospheric which establish mutualistic or neutral associations with their host plants. In the form of microbial symbionts, endophytes are important parts of plants that live on the plant organs and in some occasions live within the tissue of host plants (Card et al., 2016). A beneficial microorganism that is associated with plants has immense potential to develop into microbial pesticide or bio-fungicides (Vryzas, 2016). Various microbes associated with plants have potential to provide biotic and abiotic stress tolerance in plants without causing any harmful effect (Fig. 17.2). Therefore the use of microbes provides an eco-friendly approach to improve nutrient use efficiency and stress tolerance in plants.

Leguminous crops and their associated microbes are found in different hostile habitats. Adapting to the changing environments, microbes have evolved various mechanisms and deal with such devastating environmental stresses, plants have traits that enable them to survive in unfavorable conditions. The plant has developed an

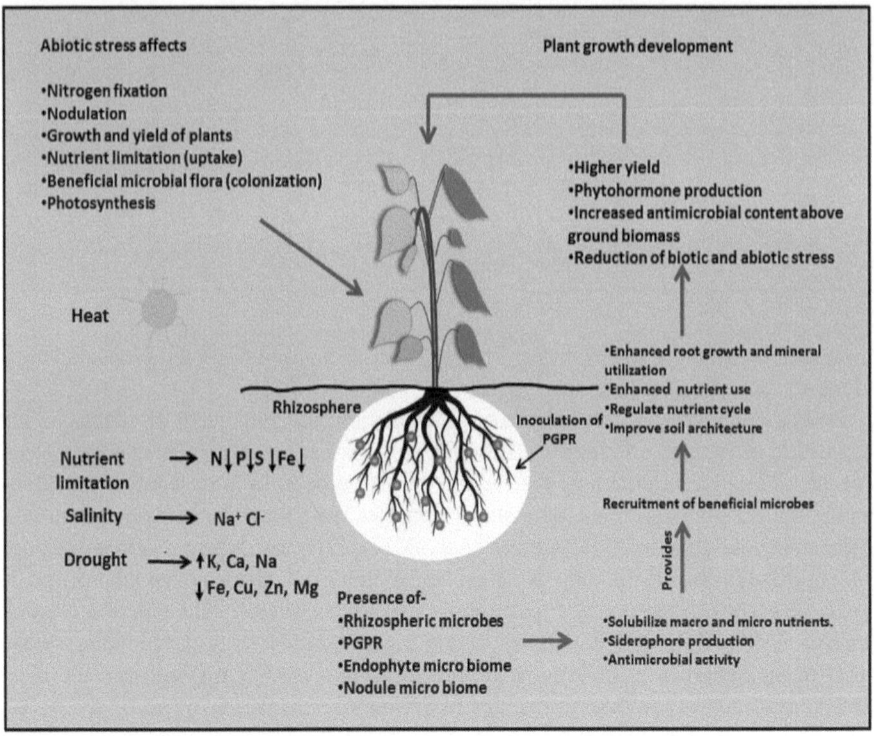

Figure 17.2 Abiotic and nutrient stress tolerance in plants through endophytic microbes.

array of symbiotic relationships with microbes to provide nutrients and revealed as best strategy to overcome from nutrient deficient stresses (Mabrouk et al., 2018). Legume crops secrete signaling molecules to establish a symbiotic interaction with soil bacteria to obtain nutrients. Only compatible microbes sense the specific chemical signals and induce the expression of nod genes, which stimulate the production of specific factors toward different host plant species for the identification of specific signal molecules such as Nod factors (Spaink, 2000; Poole et al., 2018). For instance, over 100 leguminous plants viz. cultivated soybean (*G. max*), wild soybean (*Glycine soja*), pigeon pea (*C. cajan*), and cowpea (*Vigna unguiculata*) (Brenner et al., 2009) establishes a symbiotic relationship with *Sinorhizobium fredii* strain NGR234. Various microorganisms such as Plant growth-promoting rhizobacteria (PGPR), arbuscular mycorrhizal fungi (AMF), and endophytic fungi or bacteria have been reported as beneficial microbes for plants, which mitigate the different abiotic stresses including nutrient deficiency stress in leguminous crops. Several mechanisms have been used by microorganism to mitigate the nutrient deficiency stress in plants (Table 17.2).

There are several mechanism adapted by endophytic bacteria that helps in abiotic stresses amelioration. Entophytic bacteria have two primary mechanisms for plants to overcome from nutrient deficiency stress, either by activating host stress response systems after exposure to stress, or through the biosynthesis of antistress biochemical compounds produced by endophytes (Schulz et al., 2002). According to Lata et al. (2018), there are various types of molecular mechanisms that are adopted by endophytes for increasing stress tolerance in plants such as induction and expression of stress-responsive genes, ROS scavenger molecules and synthesis of antistress metabolites. Endophytes even enhances the plant growth by providing oxygen limiting environment required for the activation of N_2 fixation and more efficient in transfer of the fixed nitrogen to the host plants (Dobbelaere et al., 2003).

17.4.1 Plant growth promoting rhizobacteria mediated mitigation of nutrient deficiencies

PGPR including both free-living and symbiotic, colonizes within the plant's root and promote growth and development either directly or indirectly (Khan et al., 2009a). Direct interaction involves N_2 metabolism, P-solubilization, phytohormones production (Indole-3-acetic acid, Gibberellic acid, and Abscisic acid) whereas indirect mechanisms are siderophore production for chelating iron, allelopathy, and competition with deleterious agents (Egamberdiyeva and Islam, 2008). PGPRs improves plant access to uptake more nutrients by promoting plant root growth through various mechanisms (Etesami and Adl, 2020). PGPRs are involved in plant's root elongation through the production of indole acetic acid (IAA) and 1-aminocyclopropane-1-carboxylate (ACC) deaminase (Saleem et al., 2018). Many PGPRs are able to fix atmospheric N_2 and make it available to plants. *Pseudomonas* and *Rhizobium* inoculated with Mung bean roots either alone or in consortia reduced the adverse effects of salinity and protein content in grain

Table 17.2 Effect of endophytic microbes in nutrient stress amelioration.

S. no.	Microbes	Benefits	Stress amelioration	Crops	References
1.	*Bradyrhizobium* sp. and *Pseudomonas aeruginosa*	Enhance the total N, P, and K contents of shoot	Nutrient deficiency	*Glycine max*	Kumawat et al. (2019)
2.	*Phomopsis liquidambaris*	Improve growth parameters, chlorophyll content, nitrate reductase (NR) activity, and Fe nutrition in plant	Iron deficiency	*Arachis hypogaea*	Su et al. (2019)
3.	*Bacillus megaterium*	Increase nodule number, root nitrogenase activity, leghemoglobin, and total nitrogen content in plant	Nitrogen deficiency	*G. max*	Subramanian et al. (2015)
4.	*Pantoea agglomerans*	Siderophore production, mineral phosphate solubilization	Phosphorus and nitrogen deficiency	*A. hypogaea*	Taurian et al. (2013)
5.	*Trichoderma* strain T-22	Enhance crops yield and productivity	Nitrogen deficiency	*G. max*	Harman (2000)
6.	*Trichoderma asperellum*	Increase seed germination, induce root and shoot development, increase carotenoid, photosynthetic pigment (chlorophyll), total sugar and protein content	NA	*Pisum sativum*	Singh et al. (2015)
7.	*Bradyrhizobium megaterium*, *Bradyrhizobium japonicum*, and *Methylobacterium oryzae*	Improve nodulation	Nitrogen deficiency	*G. max*	Subramanian et al. (2015)
8.	*Bacillus aryabhattai*	Improve uptake of Zn	Zinc deficiency	*G. max*	Ramesh et al. (2014)

9.	*Rhizobium leguminosarum*-PR1 and *Pseudomonas* sp. PGERs17	Improve N, P uptake, Fe and chlorophyll contents, nodulation, and leghemoglobin	Nitrogen and phosphorus deficiency	*P. sativum*	Mishra et al. (2012)
10.	*Pseudomonas* spp. and *R. leguminosarum*-PR1	Increase nodulation, leghemoglobin content, total iron chlorophyll content, P and N uptake	Nitrogen, iron and phosphorus deficiency	*Lens culinaris* L.	Mishra et al. (2011)
11.	Symbion-K (*Frauteria aurantia*), *Pseudomonas* sp. RA6, *Pelargonium citronella* (PC), *Serratia* sp. S2, and *Serratia marcescens* CDP-13	Enhance the macronutrients (calcium, nitrogen, and phosphorus) and micronutrients (boron, copper, iron, and zinc)	Nutrient deficiency	*Cicer arietinum*	Dogra et al. (2019)
12.	*B. aryabhattai* S10 and *Bacillus subtilis* ZM63	Increase nodules number, and nodules fresh and dry weight, enhance N, P, and K concentration in shoot	Nutrient deficiency	*Vigna radiata*	Ahmad et al. (2019)
13.	*Pantoea dispersa* MPJ9 and *Pseudomonas putida* MPJ6	Produce siderophore and increase iron uptake	Iron deficiency	*V. radiata*	Patel et al. (2018)

(Ahmad et al., 2013; Egamberdieva et al., 2013). When legume's root are co-inoculated with *Azospirillium* and rhizobia with other PGPR like Azotobacter, are more effective due to increasing nodulation and N_2 fixation of *Vicia faba* (Rodelas et al., 1996, 1999). Symbiotic relation between rhizobia and legumes plays a key role not only for enhancing N_2 uptake to plants but enhance productivity of other crops through crop rotations (Peoples et al., 2009; Herridge et al., 2008).

17.4.2 Arbuscular mycorrhizal fungi mediated mitigation of nutrient deficiencies

Arbuscular mycorrhizal fungi (AMF) colonizes with the root system of leguminous crops and transfers the nutrients such as phosphorus, potassium, nitrogen, sulfur, and other micronutrients to their host plants resulting improving the tolerance against biotic and abiotic stresses (Kafle et al., 2018). Besides that AM fungi also helps to improve tolerance activity of leguminous crops against arsenic and aluminium by reducing the uptake of these toxic metals (Zhang et al., 2015; Spagnoletti et al., 2016). AM fungi penetrate the root cortex and forms a highly branched specialized structure called arbuscular which is separated from plant symplast. AM fungi forms a huge extensive hyphal network in soil and radicle of mycelium extend to the roots which increases the absorption of nutrients and improves the soil microbial community. This condition provides plant-derived carbon to the microbial community (Kaiser et al., 2015). Arbuscules is the nutrient exchange site between plant and fungus (Brundrett, 2004). Strigolactones are the key chemical for the symbiotic relationship between plant and AM fungi. There are two pathways for mycorrhizal plants which involves in taking up nutrients from the soil. The first is plant pathway via low and high-affinity transporters in the root epidermis and root hairs. The second pathway is dominating for nutrient uptake known as the mycorrhizal pathway, in which the plants uptake nutrients with the help of fungus's ERM and then transport it to interfacial periarbuscular membrane through specialized transporters (Kafle et al., 2018)

17.4.3 Fungal mediated mitigation of nutrient deficiencies

Fungal species also helps in mitigation of nutrient deficiency. Many fungal species have been reported against nutrient deficiency stress and helps in enhancing the plant growth during unfavorable conditions (Khomari and Davari, 2017). Fungus secretes secondary metabolites which solubilizes soil nutrients from complex to simpler forms and make them available to plants (Altomare et al., 1999). A symbiotic association of fungus with roots could balance nutrient supply by increasing roots length and branching under nutrient deficiency conditions. When soybean plants inoculated with *Trichoderma* strain T-22, it colonizes within the roots of soybean and increases the yield (Harman, 2000). In the case of pea treated with *Trichoderma asperellum* increased 10%–20% seed germination (Singh et al., 2015). *Trichoderma* interaction with plant roots leads to enhanced nitrogen use efficiency (Singh et al., 2018). A piece of first evidence has been highlighted that *Trichoderma harzianum* T22 inoculated in maize crop, where it affects plant morphology under low soil nitrogen conditions (Harman, 2000). According to Marra et al. (2019), *Trichoderma* sp. releases siderophore and phosphate solubilizing metabolites to enhance nitrogen use efficiency, iron assimilation and phosphate accumulation.

17.5 Adaptive mechanism of endophytic microbes for nutrient stress amelioration

Some microorganisms benefited plants to uptake nutrient from soil such as actinomycetes, bacteria, and fungi of the genera viz., *Aspergillus*, *Scolecobasidium*, *Myrothecium*, and *Trichoderma* (Prakash et al., 2015) resulting improves their growth and development. Endophytes promote the plant growth through providing nutrient acquisition: nitrogen fixation, iron chelation, P-solubilization, and siderophores (Fig. 17.3). Likewise, endophytes also help the plants by preventing pathogen infection through the antimicrobial agents, competing with pathogens for nutrient uptake and activating the plant immune systemic resistance.

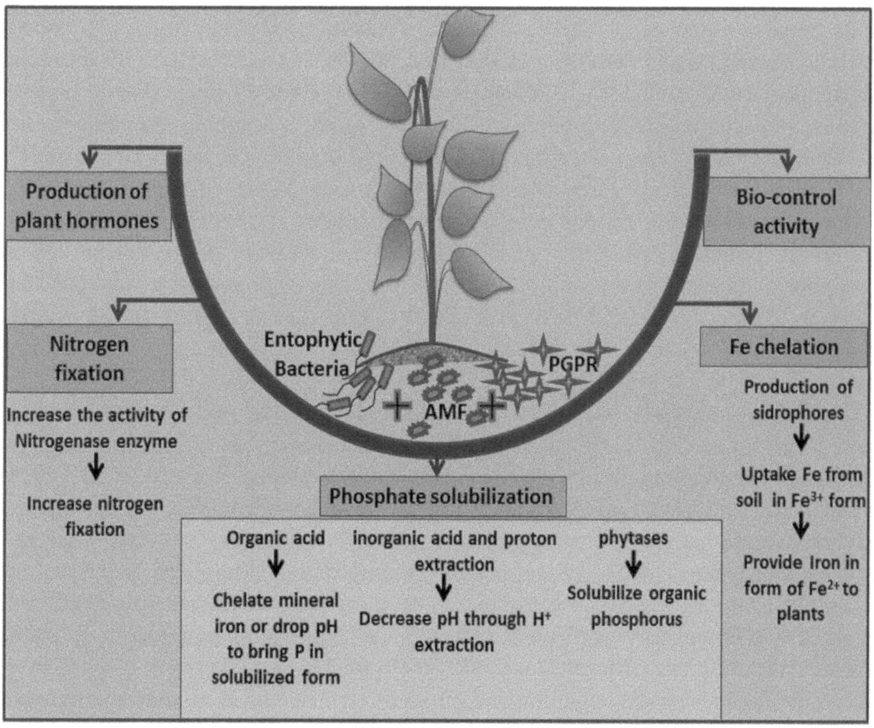

Figure 17.3 Endophytic microbes meditated nutrient stress amelioration.

17.5.1 Amelioration of nitrogen deficiency stress

In non-legume plants, the atmospheric nitrogen convert into another form that can be utilize by plants via chemical fixation and biological N fixation. While in case of legume plants, they are capable for biological N_2 fixation using nodules formed by symbiosis with rhizobia (Berman-Frank et al., 2003). This kind of symbiotic relationship between legume and rhizobia is positive to both the partners, the plants provide energy (carbohydrates) to symbiotic bacteria that convert gaseous N_2 to ammonia via the action of nitrogenase, and produces ammonia that is subsequently utilized for plant growth (Chakrabarti and Mukherji, 2003). Nitrate absorbed by the plant itself is also converted to ammonium by the action of nitrate reductase (NR) and nitrite reductase. Nitrogen fixation in legume plants is a partnership between plants and bacteria. Nitrogenase enzyme that is responsible for nitrogen fixation is sensitive for oxygen. Due to this several bacteria terminate production of the enzyme in presence of oxygen. At this situation many nitrogen fixing bacteria exist only in anaerobic condition, breathing to draw down oxygen levels, or binding the oxygen with leghemoglobin (Raza et al., 2020).

17.5.2 Amelioration of phosphorus deficiency stress

Endophytes solubilize the insoluble phosphorus and make it available to leguminous crops (Otieno et al., 2015). There are various endophytic microbial species such as *Bacillus* and *Pseudomonas* that have the ability to solubilize the insoluble phosphorus in soluble form (Grover et al., 2011). Endophytic microbes produce organic phosphates and solubilize insoluble inorganic phosphate compounds such as dicalcium phosphate, tricalcium phosphate, hydroxyl apatite and rock phosphate.

Several mechanisms have been reported for phosphate solubilization through endophytic microbes *viz.*: (1) Secretion of organic acids, protons, hydroxyl ions and anions for mineral dissolving; (2) Releasing of extracellular enzymes that help in phosphorus mineralization; and (3) Liberation of phosphorus during substrate degradation (Khan et al., 2009b). According to Pierzynski et al. (2005), the key step of soil phosphorus cycle is conversion of phosphorus from inorganic to organic forms through biological means.

Endophytic microbes, produces organic acids such as 2-ketogluconic acid, gluconic acid, (Otieno et al., 2015; Wagh et al., 2016) which lowers the pH of microbial cell and its surrounding to solubilizes the insoluble phosphorus (Naveed et al., 2014). Besides that, endophytic microbes helps to increase in acidification of the rhizospheric environment through exudation of proton or carboxylate ions such as citrate, malate, or oxalate which can immediately enhance the mobilization of P in plants by chelation or ligand exchange of P bound or complexed to Ca, Fe, or Al present in soil (Carrillo et al., 2002). Phosphatase or phytases can mobilize organic P through hydrolysis shown P availability in plants (Richardson et al., 2009). Phytate is major form of phosphorus which is stored in cereals, legumes, and oil seeds crop. It acts as an antinutrient factor, since it chelate to various type of metal ions such as Mg^{2+}, Ca^{2+}, Zn^{+}, $Fe^{2+/3+}$, and form complex with proteins (Cao et al., 2007; Liu et al., 2007). Endophytic microbes produce an enzyme that may help in to enhance phosphate utilization from phytate is called phytase. There are variety of endophytic microbes have been reported, which produce phytase enzyme. Endophytic fungus that produced phytase were isolated from leaf, stem and root of soybean and identified as *Rhizoctonia* sp. and *Fusarium verticillioides* (Gontia-Mishra et al., 2014).

17.5.3 Amelioration of iron deficiency stress

Endophytes helps leguminous crops by the production of siderophore, which chelates iron from soil under iron deficiency condition. Iron is most essential micronutrient and plays role in activation of several enzymes, in photosynthesis, nitrogen assimilation, chlorophyll biosynthesis, sulphate, and nitrate reduction (Rashid, 1996). Total concentration of iron in soil is generally greater than those consumed by crops, but iron is not present in available form in soil. Generally 10−18 M iron present in soil solution but it is not sufficient for crop sustainability and microbial growth. Endophytic microbes produce a low molecular weight compound (1−2 kDa) called siderophores, improves Fe uptake process. Siderophores can binds with Fe^{3+} with high specificity and connect to bacterial receptors. This process increases the Fe solubility and converts it into Fe^{2+} from Fe^{3+} (Boukhalfa and Crumbliss, 2002). Siderophores are classified into three groups, that is, hydroxamates, catecholates (phenolates), and carboxylates, based on the ligand molecule that bind to Fe^{2+} (Hider and Kong, 2010). Previous reports suggested that siderophores are involved in uptake of Fe strategy I by three different methods: (1) Affecting the behavior of plant in response to Fe deficiency conditions through production of IAA and ethylene; (2) Promote Fe transportation and reduction in plants through exchange of Fe^{2+} ions between siderophores and plants FC-R; and (3) Prevent the growth of phytopathogens through sequestering Fe (Meena et al., 2017). Siderophores are produced by several endophytic bacterial genera including *Agrobacterium, Bacillus, Burkholderia, Citrobacter, Enterobacter, Escherichia, Klebsiella, Pseudomonas, Rhizobium*, and *Stenotrophomonas* (Grobelak and Hiller, 2017).

17.6 A case study in soybean plant

Among all legume crops, soybean is a major valuable crop for domestic consumption. Soybean is rich source of protein and oil also an efficient in nitrogen fixation (Htwe et al., 2018). Many endophytic and plant growth promoting bacteria have been reported to improve soybean plant's growth after inoculation. Majorly it enhances the plants growth by increasing nodule number, nutrient accumulation and defense responses (Soe et al., 2012). In consequence, the endophytic microbes enhances the plant growth, improves the nutrient uptake, increases stress tolerance and induce systematic resistance in soybean plants by one of the following mechanisms: N_2 fixation, ACC deaminase enzyme, solubilization of insoluble minerals (e.g., phosphorus, potassium, etc.), phytohormones (e.g., IAA, cytokinin, gibberellin, etc.), and siderophores production etc. (Mishra et al., 2017).

Root nodules of soybean plants are colonized by numerous endophytic microbes (Dudeja et al., 2012). Several studies have reported the effect of endophytic microbes in soybean plants under nutrient stress condition. Subramanian et al. (2015), reported that the co-inoculation of *Bradyrhizobium megaterium*, *Bradyrhizobium japonicum*, and *Methylobacterium oryzae* improves the nodulation and nodule activity in soybean plants under nitrogen deficiency condition. Similarly, inoculation with two *Bacillus aryabhattai* strains (MDSR7 and MDSR14) improves the Zn uptake in soybeans under Zn-deficient soils (Ramesh et al., 2014). Co-inoculation of *B. japonicum* SAY3−7 with *Streptomyces griseoflavus* P4 improved N, P, K, Ca, and magnesium (Mg) uptake in soybean plants (Htwe et al., 2018). Similarly, treatment of *Glomus mosseae* increase in soybean root length and P uptake under phosphorus deficient condition (Thioub et al., 2019). Endophytes such as *Enterobacter sakazakii, Pseudomonas straminae, Acinetobacter calcoaceticus, Pseudomonas* sp. were isolated from root, shoot, and leaves of soybean plants which helps in Pi solubilization and IAA production (Kuklinsky-Sobral et al., 2004).

Future studies are necessary to evaluate the probability of using endophytic microbe at large scale and in field conditions for improvement of soybean production. Efforts for application of endophytic microbes should also give a better strategy to reduce the nutrient stress on soybean plants.

17.7 Conclusion

Based on the foregoing review, it can be concluded that nutrient deficient stresses are a major barrier to enhance crop growth and development. To minimize this adverse effect, endophytic microbes can be used as an eco-friendly approach to ameliorate nutrient deficiency and enhance tolerance against abiotic stresses in crops. Moreover, endophytes help in solubilizing soil nutrients to make them accessible to crops and increase the synthesis of antistress enzymes, nutrient supply, and phytohormones, which plays a vital role other adverse conditions. These endophytic microbes improves the soil fertility as well as nutrient use efficiency. Therefore there is a need to extend further study on the interaction of endophytes with legume crops, in order to gain a better insight and understanding to design effective farmer's practices to qualitatively and quantitatively enhance crop productivity while ameliorating nutrient deficiency stresses.

References

Ahmad, M., Zahir, Z.A., Khalid, M., Nazli, F., Arshad, M., 2013. Efficacy of *Rhizobium* and *Pseudomonas* strains to improve physiology, ionic balance and quality of mung bean under salt-affected conditions on farmer's fields. Plant Physiol. Biochem. 63, 170−176.

Ahmad, M., Adil, Z., Hussain, A., Mumtaz, M.Z., Nafees, M., Ahmad, I., et al., 2019. Potential of phosphate solubilizing *Bacillus* strains for improving growth and nutrient uptake in mung bean and maize crops. Pak. J. Agric. Sci. 56 (2), 283−289.

Aleksandrov, V., 2019. Identification of nutrient deficiency in bean plants by prompt chlorophyll fluorescence measurements and Artificial Neural Networks. arXiv preprint arXiv:1906.03312.

Almeida, J.F., Hartwig, U.A., Frehner, M., Nösberger, J., Luüscher, A., 2000. Evidence that P deficiency induces N feedback regulation of symbiotic N_2 fixation in white clover (*Trifolium repens* L.). J. Exp. Bot. 51 (348), 1289–1297.

Altomare, C., Norvell, W.A., Björkman, T., Harman, G.E., 1999. Solubilization of phosphates and micronutrients by the plant-growth-promoting and biocontrol fungus *Trichoderma harzianum* Rifai 1295–22. Appl. Environ. Microbiol. 65 (7), 2926–2933.

Amini, J., Sidovich, D., 2010. The effects of fungicides on *Fusarium oxysporum* f. sp. *lycopersici* associated with *Fusarium* wilt of tomato. J. Plant Prot. Res. 50 (2), 172–178.

Becana, M., Wienkoop, S., Matamoros, M.A., 2018. Sulfur transport and metabolism in legume root nodules. Front. Plant Sci. 9, 1434.

Bellaloui, N., Gillen, A.M., Mengistu, A., Kebede, H., Fisher, D.K., Smith, J.R., et al., 2013. Responses of nitrogen metabolism and seed nutrition to drought stress in soybean genotypes differing in slow-wilting phenotype. Front. Plant Sci. 4, 498.

Berman-Frank, I., Lundgren, P., Falkowski, P., 2003. Nitrogen fixation and photosynthetic oxygen evolution in cyanobacteria. Res. Microbiol. 154 (3), 157–164.

Boukhalfa, H., Crumbliss, A.L., 2002. Chemical aspects of siderophore mediated iron transport. Biometals 15 (4), 325–339.

second ed. Brenner, D.J., Krieg, N.R., Staley, J.T. (Eds.), 2009. Bergey's Manual of Systematic Bacteriology: The Proteobacteria, vol. 2. Springer, Berlin, 10.1007/978–0–387–68489-5.

Brittenham, G.Á., 1994. New advances in iron metabolism, iron deficiency, and iron overload. Curr. Opin. Hematol. 1 (2), 101–106.

Brundrett, M., 2004. Diversity and classification of mycorrhizal associations. Biol. Rev. 79 (3), 473–495.

Cao, L., Wang, W.M., Yang, C.T., Yang, Y., 2007. Effects of microbial phytase on phytate P contents in three plant ingredients [J]. Freshw. Fish. 4, 013.

Card, S., Johnson, L., Teasdale, S., Caradus, J., 2016. Deciphering endophyte behaviour: the link between endophyte biology and efficacious biological control agents. FEMS Microbiol. Ecol. 92 (8), fiw114.

Carrillo, A.E., Li, C.Y., Bashan, Y., 2002. Increased acidification in the rhizosphere of cactus seedlings induced by *Azospirillum brasilense*. Naturwissenschaften 89 (9), 428–432.

Chakrabarti, N., Mukherji, S., 2003. Effect of phytohormone pretreatment on nitrogen metabolism in *Vigna radiata* under salt stress. Biol. Plant 46 (1), 63–66.

Chen, H., Zhang, Q., Cai, H., Zhou, W., Xu, F., 2018. H_2O_2 mediates nitrate-induced iron chlorosis by regulating iron homeostasis in rice. Plant Cell Environ. 41 (4), 767–781.

Courty, P.E., Smith, P., Koegel, S., Redecker, D., Wipf, D., 2015. Inorganic nitrogen uptake and transport in beneficial plant root-microbe interactions. Crit. Rev. Plant Sci. 34 (1–3), 4–16.

Divito, G.A., Sadras, V.O., 2014. How do phosphorus, potassium and sulphur affect plant growth and biological nitrogen fixation in crop and pasture legumes? a meta-analysis. Field Crop Res. 156, 161–171.

Dobbelaere, S., Vanderleyden, J., Okon, Y., 2003. Plant growth-promoting effects of diazotrophs in the rhizosphere. Crit. Rev. Plant Sci. 22 (2), 107–149.

Dogra, N., Yadav, R., Kaur, M., Adhikary, A., Kumar, S., Ramakrishna, W., 2019. Nutrient enhancement of chickpea grown with plant growth promoting bacteria in local soil of Bathinda, North western India. Physiol. Mol. Biol. Plants 25 (5), 1251–1259.

Dudeja, S.S., Giri, R., Saini, R., Suneja-Madan, P., Kothe, E., 2012. Interaction of endophytic microbes with legumes. J. Basic Microbiol. 52 (3), 248–260.

Egamberdiyeva, D., Islam, K.R., 2008. Salt-tolerant rhizobacteria: plant growth promoting traits and physiological characterization within ecologically stressed environments. Plant-bacteria interactions: strategies and techniques to promote plant. Growth 257–281.

Egamberdieva, D., Jabborova, D., Wirth, S., 2013. Alleviation of salt stress in legumes by co-inoculation with *Pseudomonas* and *Rhizobium*. Plant Microbe Symbiosis: Fundamentals and Advances. Springer, New Delhi, pp. 291–303.

Etesami, H., Adl, S.M., 2020. Plant growth-promoting rhizobacteria (PGPR) and their action mechanisms in availability of nutrients to plants. Phyto-Microbiome in Stress Regulation. Springer, Singapore, pp. 147–203.

Ferguson, B.J., Minamisawa, K., Muñoz, N.B., Lam, H.M., 2019. Metabolic adjustments and gene expression reprogramming for symbiotic nitrogen fixation in legume nodules. Front. Plant Sci. 10, 898.

Gan, Y., Stulen, I., Van Keulen, H.E.R.M.A.N., Kuiper, P.J., 2002. Physiological changes in soybean (*Glycine max*) Wuyin9 in response to N and P nutrition. Ann. Appl. Biol. 140 (3), 319–329.

Gong, X., Li, J., Ma, H., Chen, G., Dang, K., Yang, P., et al., 2019. Nitrogen deficiency induced a decrease in grain yield related to photosynthetic characteristics, carbon–nitrogen balance and nitrogen use efficiency in proso millet (*Panicum miliaceum* L.). Arch. Agron. Soil Sci. 66 (3), 398–413.

Gontia-Mishra, I., Singh, V.K., Tripathi, N., Sasidharan, S., Tiwari, S., 2014. Computational identification, homology modelling and docking analysis of phytase protein from *Fusarium oxysporum*. Biologia 69 (10), 1283–1294.

Grobelak, A., Hiller, J., 2017. Bacterial siderophores promote plant growth: screening of catechol and hydroxamate siderophores. Int. J. Phytoremediation 19 (9), 825–833.

Grover, M., Ali, S.Z., Sandhya, V., Rasul, A., Venkateswarlu, B., 2011. Role of microorganisms in adaptation of agriculture crops to abiotic stresses. World J. Microbiol. Biotechnol. 27 (5), 1231–1240.

Hacisalihoglu, G., Kochian, L.V., 2003. How do some plants tolerate low levels of soil zinc? mechanisms of zinc efficiency in crop plants. New Phytol. 159 (2), 341–350.

Hafsi, C., Debez, A., Abdelly, C., 2014. Potassium deficiency in plants: effects and signaling cascades. Acta Physiol. Plant. 36 (5), 1055–1070.

Harman, G.E., 2000. Myths and dogmas of biocontrol changes in perceptions derived from research on *Trichoderma harzinum* T-22. Plant Dis. 84 (4), 377–393.

Hellal, F.A., Abdelhamid, M.T., 2013. Nutrient management practices for enhancing soybean (*Glycine max* L.) production. Acta Biol. Colomb. 18 (2), 239–250.

Hermans, C., Vuylsteke, M., Coppens, F., Cristescu, S.M., Harren, F.J., Inzé, D., et al., 2010. Systems analysis of the responses to long-term magnesium deficiency and restoration in *Arabidopsis thaliana*. New Phytol. 187 (1), 132–144.

Herridge, D.F., Peoples, M.B., Boddey, R.M., 2008. Global inputs of biological nitrogen fixation in agricultural systems. Plant Soil 311 (1–2), 1–18.

Hider, R.C., Kong, X., 2010. Chemistry and biology of siderophores. Nat. Prod. Rep. 27 (5), 637–657.

Htwe, A.Z., Moh, S.M., Moe, K., Yamakawa, T., 2018. Effects of co-inoculation of *Bradyrhizobium japonicum* SAY3–7 and *Streptomyces griseoflavus* P4 on plant growth, nodulation, nitrogen fixation, nutrient uptake, and yield of soybean in a field condition. Soil Sci. Plant Nutr. 64 (2), 222–229.

Isidra-Arellano, M.C., Pozas-Rodríguez, E.A., delRocíoReyero-Saavedra, M., Arroyo-Canales, J., Ferrer-Orgaz, S., del Socorro Sánchez-Correa, M., et al., 2020. Inhibition of legume nodulation by Pi deficiency is dependent on the autoregulation of nodulation (AON) pathway. Plant J. 103 (3), 1125–1139.

Jha, A., Saxena, J., Sharma, V., 2013. Investigation on Phosphate Solubilization Potential of Agricultural Soil Bacteria as Affected by Different Phosphorus Sources, Temperature, Salt, and pH. Communications in soil science and plant analysis 44 (16), 2443–2458.

Johnston, A.M., Bruulsema, T.W., 2014. 4R nutrient stewardship for improved nutrient use efficiency. Procedia Eng. 83, 365–370.

Kafle, A., Garcia, K., Peta, V., Yakha, J., Soupir, A., Bücking, H., 2018. Beneficial plant microbe interactions and their effect on nutrient uptake, yield, and stress resistance of soybeans. Soybean-Biomass, Yield and Productivity. Intech Open.

Kaiser, C., Kilburn, M.R., Clode, P.L., Fuchslueger, L., Koranda, M., Cliff, J.B., et al., 2015. Exploring the transfer of recent plant photosynthates to soil microbes: mycorrhizal pathway versus direct root exudation. New Phytol. 205 (4), 1537–1551.

Kakani, V.G., Reddy, K.R., 2010. Mineral deficiency stress: reflectance properties, leaf photosynthesis and growth of nitrogen deficient big bluestem (*Andropogon gerardii*). Agron. Crop Sci. 196 (5), 379–390.

Khan, A.A., Jilani, G., Akhtar, M.S., Naqvi, S.M.S., Rasheed, M., 2009a. Phosphorus solubilizing bacteria: occurrence, mechanisms and their role in crop production. J. Agric. Biol. Sci. 1 (1), 48–58.

Khan, M.S., Zaidi, A., Wani, P.A., Ahemad, M., Oves, M., 2009b. Functional diversity among plant growth-promoting rhizobacteria: current status. Microbial Strategies for Crop Improvement. Springer, Berlin, Heidelberg, pp. 105–132.

Khomari, S., Davari, M., 2017. *Trichoderma*-induced enhancement of soybean seedling performance in response to salt stress. J. Plant. Physiol. Breed. 7 (1), 27–39.

Kleinert, A., Venter, M., Kossmann, J., Valentine, A., 2014. The reallocation of carbon in P deficient lupins affects biological nitrogen fixation. J. Plant Physiol. 171 (17), 1619–1624.

Kuklinsky-Sobral, J., Araújo, W.L., Mendes, R., Geraldi, I.O., Pizzirani-Kleiner, A.A., Azevedo, J.L., 2004. Isolation and characterization of soybean-associated bacteria and their potential for plant growth promotion. Environ. Microbiol. 6 (12), 1244–1251.

Kumawat, K.C., Sharma, P., Sirari, A., Singh, I., Gill, B.S., Singh, U., et al., 2019. Synergism of *Pseudomonas aeruginosa* (LSE-2) nodule endophyte with *Bradyrhizobium* sp. (LSBR-3) for improving plant growth, nutrient acquisition and soil health in soybean. World J. Microbiol. Biotechnol. 35 (3), 47.

Lata, R., Chowdhury, S., Gond, S.K., White Jr, J.F., 2018. Induction of abiotic stress tolerance in plants by endophytic microbes. Lett. Appl. Microbiol. 66 (4), 268–276.

Le Roux, M.R., Ward, C.L., Botha, F.C., Valentine, A.J., 2006. Routes of pyruvate synthesis in phosphorus-deficient lupin roots and nodules. New Phytol. 169 (2), 399–408.

Le Roux, M.R., Khan, S., Valentine, A.J., 2008. Organic acid accumulation may inhibit N_2 fixation in phosphorus-stressed lupin nodules. New Phytol. 177 (4), 956–964.

Li, D., Tian, M., Cai, J., Jiang, D., Cao, W., Dai, T., 2013. Effects of low nitrogen supply on relationships between photosynthesis and nitrogen status at different leaf position in wheat seedlings. Plant Growth Regul. 70 (3), 257–263.

Liu, Z.H., Wang, H.Y., Wang, X.E., Zhang, G.P., Chen, P.D., Liu, D.J., 2007. Phytase activity, phytate, iron, and zinc contents in wheat pearling fractions and their variation across production locations. J. Cereal Sci. 45 (3), 319–326.

Lüthje, S., Meisrimler, C.N., Hopff, D., 2018. Proteomics of micronutrient deficiency and toxicity. Plant Micronutrient Use Efficiency. Academic Press, pp. 161–180.
Mabrouk, Y., Hemissi, I., Salem, I.B., Mejri, S., Saidi, M., Belhadj, O., 2018. Potential of rhizobia in improving nitrogen fixation and yields of legumes. Symbiosis 107.
Madhusudhan, L., 2015. Agriculture role on Indian economy. Bus. Econ. J. 6 (4), 1.
Magadlela, A., Kleinert, A., Dreyer, L.L., Valentine, A.J., 2014. Low-phosphorus conditions affect the nitrogen nutrition and associated carbon costs of two legume tree species from a Mediterranean-type ecosystem. Aust. J. Bot. 62 (1), 1–9.
Magadlela, A., Vardien, W., Kleinert, A., Steenkamp, E.T., Valentine, A.J., 2016. Variable P supply affects N metabolism in a legume tree, *Virgilia divaricata*, from nutrient-poor Mediterranean-type ecosystems. Funct. Plant Biol. 43 (3), 287–297.
Maitra, S., Ray, D.P., 2019. Enrichment of biodiversity, influence in microbial population dynamics of soil and nutrient utilization in cereal-legume intercropping systems: a review. Int. J. Bioresour. Sci. 6 (1), 11–19.
Mantri, N., Basker, N., Ford, R., Pang, E., Pardeshi, V., 2013. The role of micro-ribonucleic acids in legumes with a focus on abiotic stress response. Plant Genome 6 (3), 1–14.
Marra, R., Lombardi, N., d'Errico, G., Troisi, J., Scala, G., Vinale, F., et al., 2019. Application of *Trichoderma* strains and metabolites enhances soybean productivity and nutrient content. J. Agric. Food Chem. 67 (7), 1814–1822.
Meena, R.S., Yadav, R.S., Reager, M.L., De, N., Meena, V.S., Verma, J.P., et al., 2015. Temperature use efficiency and yield of groundnut varieties in response to sowing dates and fertility levels in western dry zone of India. J. Exp. Agric. Int. 170–177.
Meena, V.S., Meena, S.K., Verma, J.P., Kumar, A., Aeron, A., Mishra, P.K., et al., 2017. Plant beneficial rhizospheric microorganism (PBRM) strategies to improve nutrients use efficiency: a review. Ecol. Eng. 107, 8–32.
Meharg, A., 2012. Marschner's mineral nutrition of higher plantsEdited by P. Marschner. Amsterdam, Netherlands: Elsevier/Academic Press (2011), p. 684, United States $124.95. ISBN 978–0–12–384905-2 Exp. Agric. 48 (2), 305.
Miliute, I., Buzaite, O., Baniulis, D., Stanys, V., 2015. Bacterial endophytes in agricultural crops and their role in stress tolerance: a review. Zemdirbyste 102 (4), 465–478.
Minchin, F.R., James, E.K., Becana, M., 2008. Oxygen Diffusion, Production Of Reactive Oxygen And Nitrogen Species, And Antioxidants In Legume Nodules, Nitrogen-Fixing Leguminous Symbioses. Nitrogen-Fixation: Origins, Applications, and Research Progress, 7. Springer, pp. 321–362.
Mishra, P.K., Bisht, S.C., Ruwari, P., Joshi, G.K., Singh, G., Bisht, J.K., et al., 2011. Bioassociative effect of cold tolerant *Pseudomonas* spp. and *Rhizobium leguminosarum*-PR1 on iron acquisition, nutrient uptake and growth of lentil (*Lens culinaris* L.). Eur. J. Soil Biol. 47 (1), 35–43.
Mishra, P.K., Bisht, S.C., Mishra, S., Selvakumar, G., Bisht, J.K., Gupta, H.S., 2012. Coinoculation of *Rhizobium leguminosarum*-PR1 with a cold tolerant *Pseudomonas* sp. improves iron acquisition, nutrient uptake and growth of field pea (*Pisum sativum* L.). J. Plant Nutr. 35 (2), 243–256.
Mishra, J., Singh, R., Arora, N.K., 2017. Plant growth-promoting microbes: diverse roles in agriculture and environmental sustainability. Probiotics and Plant Health. Springer, Singapore, pp. 71–111.
Naveed, M., Mitter, B., Yousaf, S., Pastar, M., Afzal, M., Sessitsch, A., 2014. The endophyte *Enterobacter* sp. FD17: a maize growth enhancer selected based on rigorous testing of plant beneficial traits and colonization characteristics. Biol. Fertil. Soils 50 (2), 249–262.

Nimac, A., Štajcer, I., Šatović, Z., Carović-Stanko, K., Lazarević, B., 2020. Effects of nitrogen deficiency on some physiological parameters and root traits of three Croatian common bean landraces. *sa55*, p.70.

Osman, K.T., 2013. Plant nutrients and soil fertility management. Soils. Springer, Dordrecht, pp. 129–159.

Otieno, N., Lally, R.D., Kiwanuka, S., Lloyd, A., Ryan, D., Germaine, K.J., et al., 2015. Plant growth promotion induced by phosphate solubilizing endophytic *Pseudomonas* isolates. Front. Microbiol. 6, 745.

Patel, P., Trivedi, G., Saraf, M., 2018. Iron biofortification in mung bean using siderophore producing plant growth promoting bacteria. Environ. Sustain. 1 (4), 357–365.

Peoples, M.B., Brockwell, J., Herridge, D.F., Rochester, I.J., Alves, B.J.R., Urquiaga, S., et al., 2009. The contributions of nitrogen-fixing crop legumes to the productivity of agricultural systems. Symbiosis 48 (1–3), 1–17.

Pierzynski, G.M., McDowell, R.W., Thomas Sims, J., 2005. Chemistry, cycling, and potential movement of inorganic phosphorus in soils. Phosphorus: Agric. Environ. 46, 51–86.

Plaxton, W.C., Podestá, F.E., 2006. The functional organization and control of plant respiration. Crit. Rev. Plant Sci. 25 (2), 159–198.

Poole, P., Ramachandran, V., Terpolilli, J., 2018. Rhizobia: from saprophytes to endosymbionts. Nat. Rev. Microbiol. 16 (5), 291.

Prakash, O., Sharma, R., Rahi, P., Karthikeyan, N., 2015. Role of microorganisms in plant nutrition and health. Nutrient Use Efficiency: From Basics to Advances. Springer, New Delhi, pp. 125–161.

Qin, L., Zhao, J., Tian, J., Chen, L., Sun, Z., Guo, Y., et al., 2012. The high-affinity phosphate transporter GmPT5 regulates phosphate transport to nodules and nodulation in soybean. Plant Physiol. 159 (4), 1634–1643.

Qin, L., Wang, M., Chen, L., Liang, X., Wu, Z., Lin, Z., et al., 2015. Soybean Fe–S cluster biosynthesis regulated by external iron or phosphate fluctuation. Plant Cell Rep. 34 (3), 411–424.

Ramesh, A., Sharma, S.K., Sharma, M.P., Yadav, N., Joshi, O.P., 2014. Inoculation of zinc solubilizing *Bacillus aryabhattai* strains for improved growth, mobilization and biofortification of zinc in soybean and wheat cultivated in Vertisols of central India. Appl. Soil Ecol. 73, 87–96.

Rashid, A., 1996. Secondary and micronutrients. In: Rashid, A., Memon, K.S. (Eds.), Soil Science. National Book Foundation, Islamabad, Pakistan, pp. 341–386.

Raza, A., Zahra, N., Hafeez, M.B., Ahmad, M., Iqbal, S., Shaukat, K., et al., 2020. Nitrogen fixation of legumes: biology and physiology. The Plant Family Fabaceae. Springer, Singapore, pp. 43–74.

Reddy, K.J., 2006. Nutrient stress. Physiology and Molecular Biology of Stress Tolerance in Plants. Springer, Dordrecht, pp. 187–217.

Richardson, A.E., Barea, J.M., McNeill, A.M., Prigent-Combaret, C., 2009. Acquisition of phosphorus and nitrogen in the rhizosphere and plant growth promotion by microorganisms. Plant Soil 321 (1–2), 305–339.

Rodelas, B., Gonzalez-Lopez, J.E.S., Salmeron, V., Pozo, C., Martinez-Toledo, M.V., 1996. Enhancement of nodulation, Ni-fixation and growth of faba bean (*Vicia faba* L.) by combined inoculation with *Rhizobium leguminosarum* bv. *viceae* and *Azospirillum brasilense*. Symbiosis 21, 175–186.

Rodelas, B., González-López, J., Martínez-Toledo, M.V., Pozo, C., Salmerón, V., 1999. Influence of *Rhizobium*/*Azotobacter* and *Rhizobium*/*Azospirillum* combined inoculation on mineral composition of faba bean (*Vicia faba* L.). Biol. Fertil. Soils 29 (2), 165–169.

Saijo, T., 2019. Future design. The Future of Economic Design. Springer, Cham, pp. 253–260.

Saleem, A.R., Brunetti, C., Khalid, A., Della Rocca, G., Raio, A., Emiliani, G., et al., 2018. Drought response of *Mucuna pruriens* (L.) DC inoculated with ACC deaminase and IAA producing rhizobacteria. PLoS ONE 13 (2), e0191218.

Santi, S., Schmidt, W., 2009. Dissecting iron deficiency-induced proton extrusion in *Arabidopsis* roots. New Phytol. 183, 1072–1084.

Schlüter, U., Mascher, M., Colmsee, C., Scholz, U., Bräutigam, A., Fahnenstich, H., et al., 2012. Maize source leaf adaptation to nitrogen deficiency affects not only nitrogen and carbon metabolism but also control of phosphate homeostasis. Plant Physiol. 160 (3), 1384–1406.

Schmidt, S.B., Jensen, P.E., Husted, S., 2016. Manganese deficiency in plants: the impact on photosystem II. Trends Plant Sci. 21 (7), 622–632.

Schulz, B., Boyle, C., Draeger, S., Römmert, A.K., Krohn, K., 2002. Endophytic fungi: a source of novel biologically active secondary metabolites. Mycol. Res. 106 (9), 996–1004.

Sharma, P., Jha, A.B., Dubey, R.S., Pessarakli, M., 2012. Reactive oxygen species, oxidative damage, and antioxidative defense mechanism in plants under stressful conditions. J. Bot. 2012, Article ID 21703.

Singh, B.N., Singh, A., Singh, G.S., Dwivedi, P., 2015. Potential role of *Trichoderma asperellum* T42 strain in growth of pea plant for sustainable agriculture. J. Pure Appl. Microbiol. 9 (2), 1069–1074.

Singh, B.N., Dwivedi, P., Sarma, B.K., Singh, G.S., Singh, H.B., 2018. *Trichoderma asperellum* T42 reprograms tobacco for enhanced nitrogen utilization efficiency and plant growth when fed with N nutrients. Front. Plant Sci. 9, 163.

Singh, S.P., Pandey, S., Mishra, N., Giri, V.P., Mahfooz, S., Bhattacharya, A., et al., 2019. Supplementation of *Trichoderma* improves the alteration of nutrient allocation and transporter genes expression in rice under nutrient deficiencies. Plant Physiol. Biochem. 143, 351–363.

Slatni, T., Vigani, G., Salah, I.B., Kouas, S., Dell'Orto, M., Gouia, H., et al., 2011. Metabolic changes of iron uptake in N2-fixing common bean nodules during iron deficiency. Plant Sci. 181 (2), 151–158.

Slatni, T., Salah, I.B., Kouas, S., Abdelly, C., 2014. The role of nodules in the tolerance of common bean to iron deficiency. J. Plant Res. 127 (3), 455–465.

Soe, K.M., Bhromsiri, A., Karladee, D., Yamakawa, T., 2012. Effects of endophytic actinomycetes and *Bradyrhizobium japonicum* strains on growth, nodulation, nitrogen fixation and seed weight of different soybean varieties. Soil Sci. Plant Nutr. 58 (3), 319–325.

Spagnoletti, F.N., Balestrasse, K., Lavado, R.S., Giacometti, R., 2016. Arbuscular mycorrhiza detoxifying response against arsenic and pathogenic fungus in soybean. Ecotoxicol. Environ. Saf. 133, 47–56.

Spaink, H.P., 2000. Root nodulation and infection factors produced by rhizobial bacteria. Annu. Rev. Microbiol. 54 (1), 257–288.

Stewart, G.R., Larher, F., 1980. Accumulation of amino acids and related compounds in relation to environmental stress. Amino Acids and Derivatives. Academic Press, pp. 609–635.

Su, C.L., Zhang, F.M., Sun, K., Zhang, W., Dai, C.C., 2019. Fungal endophyte *Phomopsisliquidambari* improves iron and molybdenum nutrition uptake of peanut in consecutive monoculture soil. J. Soil Sci. Plant Nutr. 19 (1), 71–80.

Subramanian, P., Kim, K., Krishnamoorthy, R., Sundaram, S., Sa, T., 2015. Endophytic bacteria improve nodule function and plant nitrogen in soybean on co-inoculation with *Bradyrhizobium japonicum* MN110. Plant Growth Regul. 76 (3), 327–332.

Sulieman, S., Tran, L.S.P., 2015. Phosphorus homeostasis in legume nodules as an adaptive strategy to phosphorus deficiency. Plant Sci. 239, 36–43.

Taurian, T., Anzuay, M.S., Ludueña, L.M., Angelini, J.G., Muñoz, V., Valetti, L., et al., 2013. Effects of single and co-inoculation with native phosphate solubilising strain *Pantoea* sp. J49 and the symbiotic nitrogen fixing bacterium *Bradyrhizobium* sp. SEMIA 6144 on peanut (*Arachis hypogaea* L.) growth. Symbiosis 59 (2), 77–78.

Thioub, M., Ewusi-Mensah, N., Sarkodie-Addo, J., Adjei-Gyapong, T., 2019. Arbuscular mycorrhizal fungi inoculation enhances phosphorus use efficiency and soybean productivity on a Haplic Acrisol. Soil Till. Res. 192, 174–186.

Tiwari, R.K., Lal, M.K., Naga, K.C., Kumar, R., Chourasia, K.N., Subhash, S., et al., 2020. Emerging roles of melatonin in mitigating abiotic and biotic stresses of horticultural crops. Sci. Hortic. 272, 109592.

Valentine, A.J., Kleinert, A., Benedito, V.A., 2017. Adaptive strategies for nitrogen metabolism in phosphate deficient legume nodules. Plant Sci. 256, 46–52.

Vryzas, Z., 2016. The plant as metaorganism and research on next-generation systemic pesticides—prospects and challenges. Front. Microbiol. 7, 1968.

Wagh, J., Chanchal, K., Sonal, S., Praveena, B., Archana, G., Kumar, G.N., 2016. Inoculation of genetically modified endophytic *Herbaspirillum seropedicae* Z67 endowed with gluconic and 2-ketogluconic acid secretion, confers beneficial effects on rice (*Oriza sativa*) plants. Plant Soil 409 (1–2), 51–64.

White, P.J., Greenwood, D.J., 2013. Properties and management of cationic elements for crop growth. Soil Cond. Plant Growth 12, 160–194.

Withers, P.J., Bowes, M.J., 2018. Phosphorus the pollutant. Phosphorus: Polluter Resour. 1.

Yang, L.T., Lu, Y.B., Zhang, Y., Guo, P., Chen, L.S., 2016. Proteomic profile of *Citrus grandis* roots under long-term boron-deficiency revealed by iTRAQ. Trees 30 (4), 1057–1071.

Zahran, H.H., 2017. Legume-microbe interactions under stressed environments. Microbes for Legume Improvement. Springer, Cham, pp. 301–339.

Zhang, S., Zhou, J., Wang, G., Wang, X., Liao, H., 2015. The role of mycorrhizal symbiosis in aluminum and phosphorus interactions in relation to aluminum tolerance in soybean. Appl. Microbiol. Biotechnol. 99 (23), 10225–10235.

Zhang, Y., Wang, J., Gong, S., Xu, D., Sui, J., 2017. Nitrogen fertigation effect on photosynthesis, grain yield and water use efficiency of winter wheat. Agric. Water Manage. 179, 277–287.

Zuber, H., Poignavent, G., Le Signor, C., Aimé, D., Vieren, E., Tadla, C., et al., 2013. Legume adaptation to sulphur deficiency revealed by comparing nutrient allocation and seed traits in *Medicago truncatula*. Plant J. 76 (6), 982–996.

Function-driven microbial genomics for ecofriendly agriculture

18

G. Chethan Kumar, Jairam Chaudhary, Lalit Krishan Meena, Amrit Lal Meena and Amit Kumar
Indian Institute of Farming Systems Research, Meerut, India

18.1 Introduction

Present agricultural research is focused on addressing dual issues of feeding humongous population and maintaining sustainability in the wake of the changing climate. While the high yielding varieties, improved resource management practices and chemical fertilization have managed to achieve food security, they are also associated with negative effects of environmental damage, biodiversity loss, and land degradation. All these issues forced research attention to loss in functional diversity (FD) of agricultural lands and efforts to change land management practices by inclusion of integrated/organic nutrient management and ecological intensification practices. An inherent positive feature that promises reliability of these practices is the presence of huge soil microbiome which influences almost all plant metabolic processes apart from biogeochemical cycles.

Plant metabolism is affected by several biotic and abiotic environmental factors that lead to low yield in crops. A current research priority is to optimize plant–microbe interactions for more sustainable agricultural systems. Conventionally rhizospheric microbes are studied using traditional techniques and utilized in the form of biofertilizers, bioremediation/biocontrol agents, and recently as biostimulants. However, with the understanding of microbial community the richness in soil and complex ecosystems in which they function, it is now known that only a handful of microbes' functionality are known (as less than 1% are cultivable in lab) and the majority of microbes are largely uncharacterized along with the mechanisms of their plant growth promoting effects are yet to be deciphered.

Microbial genomics is a set of cultivation-independent methods accompanied by reductions in cost, technical advances in sequencing technologies, and new computational developments which have markedly increased the number of available microbial genomes. Genomic analyses are yielding unprecedented insights into microbial evolution, diversity, and their ecological roles. Elucidation of the functions of entire genome sequences to understand the inner workings of an organism to give biochemical, physiological, and ecological meaning to the information present in sequences is the objective of microbial functional genomics.

Microbes in Land Use Change Management. DOI: https://doi.org/10.1016/B978-0-12-824448-7.00021-8
© 2021 Elsevier Inc. All rights reserved.

Microbial genomic tools are also being used to study horizontal gene transfers (HGTs) to understand different evolutionary lineages and for functional annotation by reconstruction of metabolic pathways to shed light on the role of microbes in plant nutrition and biogeochemical cycles. Functional microbial genomics is powered by microarray-based genomic technology, "omics" (transcriptomics, proteomics and phenomics), cellular modeling, metagenomics, and single-cell genomics (SCG) as these are the most commonly used methodologies to gain genomic insights into cell functions in complex environments.

Functional genomics research is also trying to understand cell growth, differentiation, and evolution; pathogenesis, antibiotic resistance; agricultural issues of seed breeding and pesticide resistance; the biotechnological challenges of drug discovery; and the remediation of environmental contamination through a set of genomic and bioinformatics approaches. Recent genomic studies are beginning to reveal the specific microbial strains that contain metabolic pathways favorable for plant nutrition. Rhizosphere microbiome research describes the communities and their dynamics due to changes in environmental conditions. The understanding of microbiome functionality by using the highly promising genomic approaches will help in using microbes for more sustainable crop production. Genomic tools useful in characterization of beneficial microbes in agroecosystems so to improve their diversity and harness their potential for sustainable soil health management are briefly discussed in the following sections.

18.2 Genomics

The genome is the entire DNA content that is present within one cell of an organism and genomics is the study of this genome with the context of how the information in DNA is applied. Genomics takes into account all the DNA in an organism's genome – both the genes that code for proteins and the noncoding regions to understand how genes are expressed and the interplay between different genes. The term "genomics" was coined by T. H. Roderick (of Jackson Laboratory, Maine) and used for first time by Geisen et al. (2019a, 2019b) and McKusick and Ruddle (1987) for the discipline of mapping/sequencing/analysis of the information of an organism's genome. Weinstein (1998) extended "omics" terminology to transcriptomics and proteomics for transcriptomes and proteomes of organism, respectively. Phenomics is a recent addition that deals with plant phenology as governed by the environment.

Genomics is the study of whole genomes of organisms or more precisely the study of genes and their function. Genomics is an area within genetics that uses a combination of advanced molecular and cell biology tools (like recombinant DNA technology, DNA sequencing methods, gene mapping, gene expression/transcriptome profiling, DNA-protein, RNA-protein, and protein–protein interactions) aided by computational science/bioinformatics to sequence, assemble, and analyze the structure and function of genomes. Genomics differs from "classical genetics" as it considers an organism's full complement of hereditary material, rather than one gene or one gene product at a time. Genomics also focuses on interactions between loci and alleles within the genome and other interactions such as epistasis, pleiotropy, and heterosis.

Genomics is the most transformative technology of the 21st century that is intricate in almost all biological studies like functional genomics (the characterization of genes, their mRNA, and protein products), structural genomics (understanding of the architectural features of genes and chromosomes), comparative genomics (involving the evolutionary relationships between the genes and proteins of different species), epigenomics (or epigenetics dealing with DNA methylation patterns, imprinting and DNA packaging) and pharmacogenomics (new biological targets and new ways to design drugs/vaccines).

Genome sequencing project of any organism begins with the initial objective of getting complete contiguous DNA sequence of its chromosome(s). After whole genome has been sequenced, the next thing is to annotate it. Annotation is the process by which structural, functional, and other biological information is inferred from genes or proteins, and this is based on similarities of previously characterized sequence in public databases. This requires bioinformatic analyses, and bioinformatics tools such as basic local alignment search tool.

The first step in the annotation process is the identification of predicted protein coding sequences, generally referred to as open reading frames. Unlike eukaryotic genomes, identification of open reading frames in bacterial (and other prokaryotic) genomes is remarkably accurate and easier due to the absence of introns and the high gene density possessed by these organisms. Only a subset of all the open reading frames in the genomic sequence encodes proteins, and prediction of their functions by database comparison with similar genes of known functions is the next stage in the annotation process. However, this procedure can be problematic as the functions of a large percentage of genes in some organisms are unknown. Once a genome has been sequenced and annotated, basic information for understanding the biology of the organism has been obtained, and the next thing is to utilize the genome data (Donkor, 2013) for further molecular biological studies like updating taxonomic class, comparing relatedness, using potential organisms for genetic modifications, etc.

Almost all the sequence analysis including bacterial DNA sequences revealed that the biological functions of substantial portions of complete genomes are unknown. Dividing genomic DNA into coding and noncoding regions and subsequently defining the role of each gene in the complex cellular machine and network is a formidable task. In addition, genomes contain hundreds to thousands of genes, many of which encode multiple proteins that interact and function together as multicomponent systems or apparatuses for accomplishing specific cellular processes. The products of many genes are often coregulated in complex signal transduction networks, and understanding how the genome functions as a whole to give life to organism presents an even greater challenge.

Since the beginning of next-generation sequencing (NGS) methods, genomic sequencing and analysis grew exponentially. At present there are more than

260,000 sequenced prokaryote genomes (which includes both bacteria and archaea) and more than 12,000 eukaryote genomes publicly available in website of National Center for Biotechnology Information. There are more than 2000 metagenome projects. Determination of entire genome sequences, however, is only the first step in understanding the inner workings of an organism. The next critical step is to elucidate the functions of these sequences and give biochemical, physiological, and ecological meaning to the information.

18.3 Functional genomics

Functional genomics aims to look at gene functions, gene regulations, gene-gene/gene-proteins interactions in the genome and how they influence the phenotype of the cell or organism and interactions that regulate biological processes through genome-wide gene expression are studied. Knowledge of entire genetic sequences opens a whole new range of possibilities for more efficient research. Thus many laboratories are addressing important questions in functional genomics research by integrating genomic, proteomic, genetic, biochemical, and bioinformatic approaches. Consequently, areas in functional genomics and associated genomic technology are developing very rapidly.

FD is defined as a set of functional traits in a given community. Function-driven diversity describes the biological role of species or groups of species in an ecosystem. It is a description of the different ecological processes, performed by single organisms, populations, and communities. For example, "nitrogen fixation from the air" is a function performed by group of eubacteria called diazotrophs. In terms of diversity, this group is represented by symbiotic N-fixers, free living N-fixers and photosynthetic N-fixers (cyanobacteria). Also, FD of nitrogen fixing organisms not only includes species diversity and abundance, but also in the actual and maximal capacity of the ecosystem performing that function. FD is the diversity of species trait in an ecosystem as multifunctionality is another aspect of FD. For example, photosynthetic N-fixers (cyanobacteria) simultaneously fix atmospheric nitrogen and carbon along with release of oxygen. Hence, a degree of functionality of different genera is also studied under the term of FD.

FD influences ecosystem dynamics, stability, productivity, nutrient balance, and other aspects of ecosystem functioning. FD measures the performance of a given function by microbial communities and their distribution in given ecosystem. It tries to address aspects of impacts of activities performed by microbes on that particular ecosystem. For example, the impact of photosynthetic organisms in freshwater ecosystem (small lake) in terms of biomass production, life supporting activity through oxygen production and food generation for other trophic levels like fish, etc. FD also considers the complementarity and redundancy of cooccurring species. FD is usually predicted for its ecosystem productivity and vulnerability through its member's diversity.

Prokaryotes are small and undergo a rapid cell cycle coupled with a metabolic versatility that enables them to be key players in the functioning of all ecosystems, even the most extreme ones. The advent of metagenomic and SCG approaches has highlighted the metabolic capabilities of numerous members of uncultivable majority of microbiota through genome reconstruction. Linking the functions back to the species is revolutionizing our understanding of how ecosystem function is sustained by the microbial world. To improve our understanding of ecosystem functioning, it is necessary to obtain deeper insight into microbial processes and to establish a link between identified microbial populations and metabolic capabilities in the environment, even for its members present at very low abundance. Genome reconstruction of uncultured prokaryotes permits highlighting ecosystem functioning and revealing new metabolic potentialities. However, functional information obtained from genomic analyses should be considered as clues rather than strong evidence.

18.4 Biodiversity and microbial diversity of soil

The soil biodiversity is composed of size classes expressed as microflora (includes bacteria, archaea, fungi), microfauna (includes protozoa, nematodes), mesofauna (includes collembola, enchytraeids), and macrofauna (includes earthworms, isopods, diplopods). These size classes collectively involve the most important soil processes like decomposition of organic matter, nutrient transformations, biogeochemical cycling, carbon capturing, and soil structure formation. By influencing soil physical, chemical, and biological characteristics, the soil biota deliver ecosystem services and play A prominent role in agroecosystems. Among these size classes, the most versatile in terms of number, composition, and function are microbes (bacteria, archaea, cyanobacteria, green algae, and fungi).

Microorganisms inhabit almost every imaginable environment on Earth. In contrast to plant and animal diversity, however, the extent of microbial diversity is largely unknown. For example, a teaspoon of native grassland soil contains 600–800 million bacteria comprising ~10,000 species, approximately 5000 species of fungi with their mycelia stretched out for several miles, around 10,000 individual protozoa of over 1000 species, plus 20–30 different nematodes from as many as 100 different species. In any case, a gram of soil usually contains up to 10^9 bacterial cells and 10^6 distinct prokaryotic taxa. Current methods for isolation of soil microbiota can retrieve only 1% of such huge bacterial diversity present in soil. Similar is the case for fungal diversity and hence studies aiming to investigate the diversity and functions of microbial communities have been hampered by the inability of the vast majority of microorganisms to grow under laboratory conditions in defined culture media. This unculturable majority of microorganisms is referred to as "biological dark matter/microbial dark matter."

With understanding of contributions by terrestrial microbiota in terms of bio-geo-chemical cycling, biodiversity management, role in agroecosystems and aquatic (oceanic) microbiota in terms of atmospheric O_2-CO_2 balancing, bioremediation of oil spills or algal blooms, the ecosystem services of microbes has become clear. Hence interest to explore extent of role played by microbes in almost every aspect of human life has led to several genome projects and within a span of two decades an entirely new domain of microbial genomics has established. Studying relatedness microorganisms by comparing their genomes is microbial genomics. Here each genome is treated like a gene in functional genomics. Although microbial genome sequencing projects reveal enormous amounts of information about a particular microorganism, these projects only scratch the surface of microbial diversity in general. Insight into whole microbial diversity will require the sequencing of the genomes of many microbial species from various environments.

18.5 Microbial (functional) genomics

Microbial genomics is application of genomic tools like high-throughput (next-generation) sequencing technologies for the detection and functional characterization of different microbial species which are fundamentally important in understanding human infections, development of diagnostics and vaccines, bio-defense studies, antimicrobial target identification, and drug designing. Microbial genomic approaches have become regular in understanding the role of functional microbial diversity in agroecosystems, characterization of microbial communities, characterization of rhizospheric and plant microbiome communities, bioremediation of soil and water from organic pollutants, carbon management, and energy production.

Microbial diversity and the associated functional traits in soil are highly dynamic in time and space, constantly changing in response to natural factors or to human-induced perturbations. Changes in soil microbial biodiversity were assessed by traditional methods like classical taxonomic indices and morpho-physiological approaches. The taxonomic methods were successful in exploring and classifying organisms which helped us to understand vast role played by microbiota in soil processes. However as new approaches and techniques evolved, our understanding of immense microbial diversity grew further which lead to inclusion of methods that supplement taxonomy-based approaches with trait-based approaches (Moretti et al., 2017).

Application of cultivation-independent techniques based on biomolecules like DNA was introduced in the 1980s for the assessment of the soil microbiome which in the later years established as discipline of metagenomics. Today we are able to describe at least parts of the enormous diversity harbored in soils. Genetic material-based characterization of environmental samples helped us in studying complete diversity of a habitat, including archaea, bacteria, eukaryotes, viruses, and plasmids. Research into microbial communities has progressed from genotype–phenotype association studies to include experimental validation of bacterial species and functions (Forster, 2017). For this purpose, functional marker genes were identified and assigned to identify specific functional community in different environmental samples. Such analyses are based on the sequencing of a gene-specific region to reveal the diversity and composition of specific taxonomic groups present in an environmental sample.

Anthropogenic activities cause changes in soil affecting the soil microbiome. These changes consequently alter the functional profile of the microbiome, affecting the occurrence and abundance of essential metabolic pathways that maintain balance within the systems. Three interactions of microbes in the root zone soil (rhizosphere) are commensal, beneficial, and pathogenic. Microbiomes can activate complex signaling pathways in response to biotic or abiotic stress, leading to localized and systemic defenses in crop plants, but these functions may be impaired by the occurrence of pathogenic blooms. Therefore understanding the taxonomic and functional microbiome profiles caused by the occurrence of microbial pathogens in food production systems may provide valuable information to understand the negative effects of these phenomena and devise strategies to alleviate the effects or eradicate the pathogens. Also, crops can be engineered through their microbiota favoring desirable responses. Genomic disciplines, particularly functional metagenomics are being employed to provide adequate insights of microbiomes.

Functional trait-based approaches advanced our mechanistic understanding and predictive capabilities of the links between species traits and community responses, and thereby ecosystem processes. Thus genomic studies, especially metagenomics over the years paved way for microbial functional genomics. Functional genomics aims to look at the structure, function and regulation of genes and how these genes in the genome interact with each other to influence the phenotype of the cell or organism. Similarly comparing genomes of microorganisms to see how related they are can be used for classifying microorganisms, where each genome is treated like a gene in functional genomics.

18.6 Microbial genomic (molecular biological) tools useful in agricultural land management approaches

Present research methods try holistic approaches in including genetic data with functional traits data which not only help in detecting presence/absence of particular communities of organisms but also their role in shaping of the community along with its interactions with surroundings. The influence of abiotic factors and human interference (like agriculture or deforestation) in regulation of microbial community diversity is also being explored using function-driven microbial genomics approaches. Different agricultural land management practices affect soil biodiversity positively or negatively that can be studied using specific biotechnological tools with aim to develop strategy for sustainable land management practices. In this background, the basic genomics and microbial genomics approaches or system biology tools that help in studying agroecosystem microbial communities (soil microbial diversity) are discussed in below sections.

Unlike traditional cultural methods which used morphological and biochemical differences to distinguish microbes, genomic approach compares genetic material at various levels like genome level, transcriptome level and proteome level. In genomic studies DNA or RNA are sequenced to know their nucleotide arrangement which contains message for all the cellular processes of an organism. The first-generation DNA sequencing technology was chain termination sequencing or Sanger sequencing. In this sequencing, DNA is randomly fragmented, cloned into plasmid and transformed to *Escherichia coli*. The cloned fragment is amplified using flanking polymerase chain reaction (PCR) primer. Each PCR round is terminated using incorporation of fluorescently labeled dideoxyribonucleotide. The resultant terminated fragments are then separated in electrophoretic capillary containing polymer gel, following exposing capillary to excite the fluorescently labeled dye by argon laser, and then emitted spectrum is recorded in a form of chromatogram using charge-coupled device camera. This gives read length of 800 to 1000 bp with base call accuracy of 99.99%. However, its technology with very low output and high production cost limits the application.

In year 2005, massive parallel high-throughput sequencing technologies called NGS arrived that delivered tremendous output with high coverage and eventually became the essential tools for microbial genomics. The advantages of NGS over Sanger sequencing are construction of multiplexed sequencing library, clonal amplification of libraries, immobilization of amplified libraries on solid substrate and chip-based sequencing. Depending on the variation in methodology used to immobilize DNA on a solid substrate and detection NGS further comprised of pyrosequencing, sequencing by reversible termination (Illumina sequencing) and semiconductor sequencing.

The third-generation sequencing added improvements over previous sequencing techniques through direct DNA sequencing without utilizing the PCR amplification step, as amplification introduces a bias in read content and presence of high gas chromatography (GC) content affects depth and coverage. The major advantage of this technique is the longer read length with an average of 5–10 Kb. Commercially available third-generation sequencing are single-molecule real-time sequencing by Pacific Biosciences and Oxford Nanopore sequencing by Oxford Nanopore Technology. This technology delivers long reads at low cost with real-time nature of sequencing and thus an important tool in genomics and microbial community study.

Microbial community studies identify the presence or absence of species of microbes using marker genes or selected functional genes. Microbes are studied and compared at genome level, transcriptome level and proteome level to understand metabolic diversity of microbes in studied environment and their functions. Some molecular tools used in microbial studies are briefly discussed in following paragraphs.

18.6.1 Ribosomal RNA gene-based community analysis

To study microbes that are uncultivable in laboratory conditions and in comprehensive microbial community composition, diversity and microbe-habitat interaction studies, 16S ribosomal RNA (rRNA) gene sequencing tools are extensively employed. Early environmental genomic studies for bacteria were all from the analysis of the 16S rRNA sequence and 16S rRNA gene was targeted to obtain better picture of the species composing the community. The 16S sequence in the genome codes for the rRNA which forms part of the 30S subunit of the bacterial ribosome and is therefore highly likely to be present in all bacteria. The 16S rRNA sequence can be present multiple times in a bacterial genome and its function has been conserved. Hence changes in the 16S rRNA sequence are a good indication of evolutionary time.

The 16S rRNA gene consists of hypervariable regions of V1 to V9, with some conserveness between species to species, thus utilized as a molecular tool for bacterial characterization. The high-throughput 16S rRNA amplicon sequencing analysis of habitats such as the human gut, oral cavity and buffalo rumen microbiota have been characterized. High-throughput 16S rRNA amplicon sequencing has revealed the rare biosphere (composed of a large number of rare bacterial taxa that may be representatives of unidentified microbial lineages) and the members of these rare biosphere may become abundant depending on environmental conditions. Some studies focused on these populations have shown that a significant portion become active in the environment at some time intervals, sometimes even more active than abundant taxa and play crucial roles in biogeochemical cycles and overall metabolic fluxes. Such observations allowed the reclassification of the rare biosphere as not being simply a dormant seed bank and suggested that its ecological significance and roles should be further evaluated. The limitation of 16S rRNA sequence analysis is that it only shows what bacteria are present and does not give information on how their metabolism works. Thus whole genome shotgun sequencing using next-gen sequencing platforms have become alternatives to 16S rRNA based studies.

18.6.2 Metagenomics

Metagenomics is the study of microorganisms in an environment by extracting and analyzing the DNA (genomes) from that environmental sample. In general, the term metagenomics is used for genome sequencing projects in which many organisms are sequenced at once by shotgun cloning of all DNA present in a sample enabling to investigate microbes that are not culturable in pure form from any microbial ecosystem. Methods for sequencing the entire DNA from microbial communities (metagenomic sequencing) have the capacity to provide strain level resolution. The first and crucial step in a metagenomics study is the collection and processing of the environmental sample. Natural DNA-containing samples such as water, soil, cells, tissue can be collected and filtered. Filtering size should be chosen precisely to get correct target sequences from the environmental sample after removing large cells or debris.

In the early metagenomic studies, Sanger shotgun sequencing method was used which has now shifted to NGS technology with its ability to sequence thousands of organisms in parallel. The substantial improvements in Illumina throughput and read length have helped it dominate metagenomics studies and have promoted substantial increases in the number of metagenomic studies. Third-generation sequencing can create nearly complete genome assemblies of individual microbes directly from environmental samples without the need for cultivation methods. Faster and cheaper sequencing technologies combined with construction of sequencing libraries removed the barriers of the genomic study of uncultured microorganisms directly from environmental samples.

However, having genomic information does not necessarily mean that a lot of new discoveries are made to see which organisms are present in the environmental sample, genomic information needs to exist so that comparisons can be made. A problem with this approach is lack of availability of sequence data of unculturable majority as complete bacterial genomes available mostly represent species which could be grown as pure cultures in the laboratory. This means that the genomic database is biased to some extent towards bacteria which can be grown, and thus some species present in the environmental sample may be completely missed. Hence metagenomic approach is hindered by the numerous unsequenced microbial taxa that are contained within the environmental samples.

It is thought that similarities in the genomes of two organisms would lead to a similarity in their phenotypes, and this is often applied to microorganisms. However, several considerations are needed. Depending on genes that are being used as markers, the result may be different. Also, certain genes can be acquired by HGT – where genes are passed between microorganisms which are not directly related. This means that if a marker gene was horizontally transferred between two microorganisms it may appear that these two microorganisms are closely related, and this may in fact not be true.

18.6.3 Whole community shotgun metagenomics

Although 16S rRNA sequencing provided important insights into community diversity, this approach is frequently unable to achieve biologically meaningful resolutions. It has now become possible to sequence more than just the 16S rRNA sequence, and the development of large scale DNA sequencing means that it is now possible to sequence the whole environmental DNA rather than a few key genes and sequences. The first well-analyzed whole genome sequencing (WGS) metagenomic study of microbial genomes from the environment focused on acid mine drainage microbial biofilm and Sargasso Sea surface waters. During the last 10 years, many big projects and consortia have been launched for sequencing of metagenomes, such as the TerraGenome project for soil (Vogel et al., 2009).

The functional contribution of microorganism in various habitats is identifiable by performing the whole metagenome sequencing and their annotation determines the functional genes. The whole metagenome study revealed the correlation of genetic makeup of prevailing organism with its environment in terms of genome size, GC content, HGT, optimum growth temperature, antibiotic and metal ion resistance genes, and the presence or absence of DnaE2. The GC content of bacterial genomes has been shown to be positively correlated to optimal growth temperature and is a major barrier to HGT Researchers looking for the mechanisms involved in land colonization by bacteria have found DnaE2 is 55%–68% in soil bacteria as compared to 11%–21% in bacteria of water. The correlations between these factors and phylogenetic analysis have led them to further speculate that the duplication of dnaE1 to dnaE2 followed by an increase in GC were major steps leading to terrestrial bacteria. Metagenomic investigations also identified that microbes which thrive in soil generally have higher GC content with larger genome size compared to their aquatic counterparts (Wu et al., 2014).

18.6.4 Single-cell genomics

Isolate-based genome studies in microbial genomics are limited by the small number of microbes that can be cultured. SCG uncovers hereditary information at the most basic level of biological organization. It is emerging as a powerful complement to cultivation-based and microbial community-focused research approaches. SCG has been instrumental in identifying metabolic features, evolutionary histories and interorganismal interactions of the uncultured microbial groups that dominate many environments and biogeochemical cycles. The SCG approach also holds great promise in microbial microevolution studies and industrial bioprospecting.

Methods for SCG consist of a series of integrated processes, beginning with the collection and preservation of environmental samples, followed by physical separation, lysis and whole genome amplification of individual cells, and culminating in genomic sequencing and the inference of encoded biological features. When analyzing complex microbial communities, metagenomics is not well suited to deliver unequivocal information about the organization of discovered genes within genomes, obscuring the knowledge of the metabolic potential and evolutionary histories of specific microorganisms. To bridge this gap, SCG emerged as a powerful complement to cultivation and metagenomics, by providing genomic information from individual, uncultured cells.

18.6.5 Transcriptomics

The information content of an organism is present in its genomic DNA and is expressed through transcription into RNA. Transcription is the first phase of gene expression in which a particular segment of DNA is duplicated/transcribed into RNA language (as mRNA). The complete set of RNA transcripts produced by an organism at any one time is called the transcriptome and is the link between the genome, the protein complement and the phenotype. Because virtually all changes in cellular states are correlated with fluctuations in the mRNA levels, transcription patterns can be very useful in elucidating the function of unknown genes, regulatory pathways, and networks. Transcriptomics is the study of gene expression patterns and traditionally has used high throughput DNA microarray methodology. Transcriptomes allow us to study the differential expression pattern at a specific condition. Prior to 1990 serial analysis of gene expression (SAGE) based on Sanger sequencing was being used for gene expression studies. In 1990, the term transcriptome was first used as a synonym to SAGE.

Gene expression is affected by modification in DNA by DNA methylation without changing the DNA sequence. This can be determined by using methylation-dependent restriction enzyme or bisulfite conversion and its PCR. Other modifications like acetylation, phosphorylation, and ubiquitination also regulate the gene expression and these can be investigated using chromatin immunoprecipitation (CHIP) technique. NGS massive parallel sequencing of the genome using RNA sequencing (RNA-seq) with RNA intakes in nanogram is also used as high-throughput sequencing technology to study the genomes under specific conditions. It does not require the prior information of genomic sequencing whose sequencing/analysis has to be done and has high accuracy up to 90% in sequencing. NGS methods are also able to detect the single-nucleotide polymorphisms with high technical reproducibility of up to 99%.

18.6.6 Microarray

Microarrays (or microchips) are a recently developed genomic technology and are listed as one of 10 breakthrough technologies in 1998, along with genomics. Just as microprocessors have dramatically sped up computation, microarray-based genomic technologies have revolutionized the genetic analysis of biological systems. Microarray technology represents a powerful new tool that allows researchers to view the living cell under various physiological states from a comprehensive and dynamic molecular perspective. The widespread, routine use of such genomic technologies will shed light on a wide range of important research areas, such as how cells grow, differentiate, and evolve; the medical challenges of pathogenesis, antibiotic resistance, and cancer; agricultural issues of seed breeding and pesticide resistance; the biotechnological challenges of drug discovery; and the remediation of environmental contamination.

Although microarray-based genomic technology has great potential for overcoming the limitations of traditional molecular methods for studying microbial communities, adapting this technology to environmental studies from pure cultures is a challenge. Microarray has the ability to measure simultaneous expression of thousands of genes at particular stage of cells. This technique is used to investigate the diagnostic or prognostic biomarkers, disease associated genes, and the response of gene against a particular drug/stress in plants and animals to understand the mechanism involved in a particular condition. First step is mRNA isolation from a normal and experimental condition which are transcribed into cDNA in next step and labeled with dye and finally allowed to hybridize with a probe attached on chip to measure the level of mRNA in control and experimental condition.

Studies on various plants during biotic/abiotic stress using transcriptomics approach have revealed that many genes undergo differential expression during exposure to environmental stress. The genes which are upregulated and downregulated are a key player in several biological processes and molecular function. Stress-responsive gene which is over expressed is involved in different defense mechanism against the environmental stress while under expressed genes are generally involved in the storage process. For example, in metal toxicity conditions certain metabolic pathways are over expressed and under expressed for providing tolerance against heavy metal stress.

18.6.7 Proteomics

Cells can be considered biological factories that carry out and integrate thousands of discrete, highly specialized processes using molecular machines consisting of different proteins and other molecules. Many proteins assemble into larger complexes to execute fundamental cellular functions and metabolic processes (e.g., DNA replication, transcription, translation, and protein degradation), mediate information flow within and among cells and within their environment (e.g., signal transduction, energy conversion, and cell movement), and build cellular structures. Systematic identification of protein interactions in such complex molecular machines is very important in understanding how a given proteome works.

Proteins are the main catalysts, structural elements, signaling messengers, and molecular machines of biological systems. Recent progress in WGS has provided information on the composition of many proteins from a variety of organisms, but how proteins confer on cells their capabilities, structure, and higher-order properties remain to be clarified. In understanding the biological process, it is necessary to understand the function of the protein in a biological process.

Many times, transcriptomics studies do not correlate with proteomics study due to posttranslational modification which changes the function of the protein. Proteomics research, the large-scale analysis of the protein products of genes to provide a comprehensive identification of proteins to characterize the biochemical and cellular function of each protein and to analyze protein regulation. Two-dimensional gel electrophoresis can be used to analyze the protein content of a cell where proteins are first separated by size followed by mass spectrometry (MS). Translational regulation and posttranslational modification are also important in determining protein abundance in a cell.

To understand how a proteome (a complete set) is produced and the proteins interact with each other, high throughput proteomic approaches such as MS, phage display, and protein arrays are employed. Liquid chromatography-MS is also one of the approaches where proteins are first separated by one-dimensional sodium dodecyl sulphate–polyacrylamide gel electrophoresis (SDS PAGE), then protein is digested and separated by LC and analyzed by MS. Multidimensional protein identification technology (MudPIT) is high-throughput technology used for separating protein from complex mixture by digesting proteins into peptides followed by separation on the basis of charge and hydrophobicity and finally analyzed by MS. The signal obtained from MS data is compared with a database for identification of the protein.

Unlike high-throughput nucleic acid-based approaches, the development of high-throughput techniques for the analysis of proteins is a great challenge. MS is currently one of the most important tools in proteomics. With the availability of complete genomic sequences, proteins isolated from a given cell type no longer need to be identified by de novo sequencing but can instead be identified by correlating the molecular masses of short peptides with those predicted from sequence database. MS is the choice for peptide identification due to its exquisite sensitivity (femtomole level) and mass accuracy. The two-hybrid system is one of the most widely used approaches for generating comprehensive protein interaction maps. The main advantage of this approach is that no protein manipulation is required because proteins to be tested are expressed in vivo by the yeast cells. The two-hybrid system has been applied on a relatively small scale for generating yeast protein interaction maps.

18.6.8 Interactomics

The omics technology helps us to understand the biological system. Interaction of expressed proteins in a cell (the proteome) with genome encoded product along with its interaction in complex biological network is the interactome. The study of interactions in signal transduction, transcriptional regulation, metabolic pathways, and other biological processes is said as interactomics. It is a fast-growing area of systems biology for understanding the biological process and regulatory networks of plant metabolism.

Approaches developed for studying the interactome are in silico, in vivo, and in vitro. The first approach includes the computational analysis and text mining; second approach includes Y2H hybrid system, while the third approach involves experiments performed on the living organism to understand the biological functions interactome. Protein–protein interaction network study, protein interactome based on the interolog method, signaling and cellular function by enabling hypothesis generation. The union of informatics, biochemistry, and engineering technology enables the researcher to understand the interactions of proteins used to study under interactomics.

18.6.9 Metabolomics

The study of metabolites at a particular instant in an organism/system is known as metabolomics. Metabolomics is the large-scale study of metabolites produced by a species, again usually explored in terms of a particular cell, tissue or developmental stage. Broadly, metabolic analysis can be considered as a range of technical approaches: from the specific analysis of one or more target molecules, through metabolic profiling (determination of a group of components within a pathway), metabolite fingerprinting (direct comparison of plant samples) to metabolomics (qualitative/quantitative analysis of all metabolites present).

Metabolomic techniques are predicted to have major potential in providing molecular definition of complex quality attributes, in understanding metabolic control and metabolic diversity, the processes of plant development, differentiation, interaction with pathogens and resistance to stress. It is important to understand the plant stress response in terms of metabolites and deciphering the role of metabolite during different abiotic stress condition. Hence metabolomics approaches are used to understand the production of metabolite during different stress conditions in plants like water salinity, oxidative stress and heavy metal stress. For example, metabolic profiling of plants during drought, heat, and combined stress reported accumulation of sucrose and other sugar like starch.

Nuclear magnetic resonance (NMR) based metabolic fingerprinting of plants during heavy metal stress are performed and responsible metabolic pathways are explored. The metabolomics data is like transcriptomics and proteomics data. It requires more computational work including file handling, data mining and finally comparative analysis. Online server/databases/tools are available for analysis and visualization of metabolic pathway for understanding the agroecosystem. Once the function of metabolite will be known, one can directly correlate with the function of the gene during particular stress condition. It is a rapidly growing technology to understand the metabolic pathways in plants by using different strategies like targeted analysis, metabolite profiling, and metabolic fingerprinting.

18.6.10 Lipidomics

A diversity of lipids are found across organisms and environments, each playing critical roles in membrane organization, structure, and signaling, responding to environmental conditions such as desiccation or light, sporulation, as well as being important precursors for production of biofuels or involved in secondary metabolite biosynthesis. Often, species of lipids (e.g., phosphatidyl-choline, or PC lipid) are distinguished by a common head group attached to fatty acid chains of various carbon length and degree of saturation. Although several hundred forms of each species may exist, fragmentation during MS analysis often leads to a characteristic fragmentation ion or neutral loss allowing for identification. This enables us to identify common lipid species that have characteristic fragmentation spectra which are later annotated by lipidomics analyses.

18.6.11 Phenomics

Generally, plants interact with the biotic and abiotic component of the ecosystem. As a result, there is a change in genotype and phenotype of plants. The alteration in phenotype is due to combined interaction of genome with the environment. The study of phenotype during the interaction of biotic and abiotic factor of the environment at a particular instant is known as phenomics. Invasive and noninvasive techniques are available to understand the change in phenotype in model plants.

Phenomics technology is used in basic plant research during different stress conditions and crop breeding. Digital image of plants is taken by image acquisition to study the effect of biotic and abiotic effect on plants. It is done by several approaches like tomography imaging, thermography imaging, time-of-flight (TOF) camera and Light Detection and Ranging (LIDAR). After image acquisition, processing of the image is done for removing the noise and improving the contrast. Image cropping, contrast improvement, and dimensional reduction are major operation done during image processing. After the image preprocessing, segmentation of image is done, in which objects are identified and isolated by removing irrelevant background present in the objects. In the feature extraction process, the image of interest is used to extract the numerical value where different algorithms can be applied to understand the phenotype. Finally, the generated data is analyzed by a machine-learning approach.

All the above listed genomic tools are routinely used in plant biology are recently being used for soil microbial community analyses and functional characterization of soil microbial diversity.

18.7 Genomic tools to address issues of agricultural lands

18.7.1 Study of stress biology in agroecosystems using genomic tools

The natural ecosystems which are modified to produce food and fiber are known agroecosystems. Metabolic activities of crop plants are influenced by several biotic and abiotic environmental factors. Under suboptimal conditions these factors affect normal plant metabolism leading to low yield in crops. Generally, plants show symptoms to common abiotic factors like cold, drought, heat, salinity and heavy metals and biotic stress like pathogens. Appearance of stress symptoms lead to stress alleviating land management practices.

In general, some productivity changes are gradual and owing to climate variability crop expression of stress is not visible through observable symptoms ("hidden hunger type"). The time of realization of crop stress may in many cases become too late that land management have little or no effect as crops might have surpassed flowering and reproductive phase leading to below normal yields. Systems biology tools are useful for discovering the marker genes associated with stress and to develop the tolerance varieties of crops. Omics approaches and systems biology tools help in identifying the stress-responsive genes, pathways, secondary metabolites and crop plants response strategy to particular stress which help us to build plant stress resistance through suitable interventions.

Graphical view of biomolecules and their interactions in the form of a biological network represents the systems biology of organism. In such biological network, gene/protein/DNA/RNA are denoted as nodes of the graph and edges as way of interaction. The interaction may be direct (flow of information from one biomolecule to another) and undirected (having interaction but the direction is not sure). The way of interaction may be physical interaction, metabolic reaction, and regulatory connection. The most connected node in networks is termed as hub node, which is a key player in biological network. During stress condition in plants, the interaction between biomolecule changes and it can be well studied and understood by using systems biology by constructing or reconstructing gene to metabolite network, protein–protein interaction networks, gene regulatory networks, and transcriptional regulatory network. The functional genomics and systems biology can be used to develop a strategy to escape the plants from stress condition and during the development of new varieties.

18.7.2 Microbial biodiversity characterization and management

Soil is the main site of food production and peculiar to support life functionality. Soil plays an essential role for plant growth, cycling of carbon, and other nutrients which are mainly mediated by soil microbiota. Microbes play essential role in various metabolic activities like complex carbohydrate, nitrogen, sulfur and xenobiotic metabolism etc. The first report on soil microbial community using DNA-based study revealed that soil microbiota composition is enormously diversified (Torsvik et al., 1996). It is estimated that more than 10,000 bacterial species are present in one gram of soil, with strongly correlated complex network. The advancements in microbial genomics facilitated the soil microbiome study at various levels such as genus and species with abundance estimation (Nannipieri, 2014, 2019), including the functional gene content and actively involved genes.

The first-generation sequencing extensively facilitated the molecular characterization of microorganism and functional gene with expense of high cost with low throughput. Presently vast numbers of genomes (genomic sequential data) are available in public database generated through various NGS and third-generation sequencing contributing to the growth of genome and metagenome sequencing programs and associated projects helping in deducing microbial taxonomy and functional characterization through bioinformatic analysis.

Many important ecological niches like rumen of cattle, soil and human gut have been functionally characterized by such metagenomic sequencing programs. The extensive data generation and efficient computational resource development facilitated the finishing of various complete genomes and draft genomes with remarkable growth of complete genomes available on public databases. Bacterial genomes sequencing projects of agriculture were 7% of total genome projects along with 7% projects on environment samples and highest genome characterization was in medical sector (59%) for obvious reason of their importance in human health. There are some examples of big soil microbiome projects such as Earth Microbiome Project, Brazilian Microbiome Project, TerraGenome, China Soil Microbiome Initiative, MicroBlitz, and EcoFINDERS which characterized the soil microbiota community structure and FD.

In the beginning, providing a picture of the taxonomic profile and structure through 16S rRNA sequencing was considered as the limit to where this technique could lead. However, the feasibility of connecting databases containing information from simple biomarkers to those containing information on complete genomes made possible to infer the functional capabilities of the microbial community based in 16S rRNA sequencing. The correlation between taxonomy and metabolic function allows carrying out work focused on the identification of community structure and composition through analysis of taxonomic biomarker genes. There are bioinformatics tools that construct an approach of functional metagenomics of a microbiome.

Bioinformatics use the information from databases about the microbes constituting any niche and infer the metagenome of such microbial community and constructing a gene catalog with the abundance of different gene families. For example, tools like Phylogenetic Investigation of Communities by Reconstruction of Unobserved States (PICRUSt) are used for this purpose. In this approach, the efficiency of functional predictions depends on the taxonomic classification accuracy, the genomes used as reference, and the information collected in databases from the closest ancestors. The databases of genes used for the metagenome construction are the Kyoto Encyclopedia of Genes and Genomes and Clusters of Orthologous Groups of proteins. This can provide valuable information to evaluate factors influencing the functionality of microbial communities. One of the disadvantages of this approach is that this software uses the Green genes database, which has not received actualizations since several years ago. However, there are other options such as Tax4fun, an R package using the SILVA database as reference and reported to be more accurate for some microbial communities; however, neither PICRUSt nor Tax4fun has extensions to consider 18S rRNA data, and therefore the study of the microbiome may be incomplete.

Soil metagenomics studies use user-friendly computing software allowing the analysis of metagenomic data by most biologists. For example, MG-RAST and CLC Genomics Workbench (Microbial Metagenomics Module) and other programs involve user-friendly end-user systems to perform complete mandatory pipeline processes. HUMAnN is another computational pipeline developed to determine the relative abundance of gene families and metabolic pathways from short-read sequence datasets, while STAMP is a friendly software to perform statistics of data obtained from targeted or shotgun metagenomics. There are plenty of free tools available, however, the sets of tools to use to establish a valid pipeline depend on the objectives of the study and the computing skills of the biologist. Even though targeted metagenomics (16S rRNA) is an approach for the metagenome, shotgun sequencing offers not only more complete and precise results but also allows digging more into the functional potential of any microbial community, obtaining the genomic information from different taxa, and inferring metaproteomes from raw metagenomics sequences.

18.7.3 Functional metagenomics analysis of agricultural rhizospheric soil

Soil rhizosphere consists of complex interactions of microbes and plant root and these microbes play a vital role in plant growth and health promotion. Knowing the taxonomic profile and functional capabilities of soil and rhizospheric microbiota will provide greater certainty about the parameters or conditions to be manipulated. Microbes display increased activity in soil hot spots such as rhizosphere, mycosphere, drilosphere, and detritusphere. Rhizospheric soils are reservoirs of opportunistic pathogens also. This opportunistic biota can disrupt the normal host plant microbiota by competition for resources, release of antimicrobial compounds, or antagonism against beneficial microorganisms. Pathogen outbreaks usually occur when the conditions allow their proliferation which means that pathogens can either introduced into the systems or is part of the concerned environmental microbiota in a dormant or fully active state but not being virulent (Mendes et al., 2013).

All genes in a metagenome can be described by whole-metagenome shotgun sequencing. Here functional activities can be inferred by listing the abundance of each gene. But it is time-consuming and high level of experience is required. Alternatively, the amplification and high-throughput sequencing of the 16S rRNA gene allow describing the microbial composition. Few well known plant beneficial microbes are symbiotic nitrogen-fixing rhizobia, the phosphate-solubilizing bacteria, and pathogen suppressing such as *Pseudomonads* and *Bacilli*. One of the highly studied genes of soil microbiota is nitrogen fixing (*nif*) genes of various types. Among those, *nifH* was extensively targeted with different PCR primers for identification of N-fixing bacteria. The deep metagenome study explored the microbial community functional capacity for carbon cycling and correlation among community's functional genes.

Targeted metagenomics and shotgun metagenomics are commonly used for rhizospheric diversity studies. Through performing shotgun metagenomics investigations on the taxonomic and functional profiles of microbial communities in the bulk soil and in the rhizosphere of soybean plants were studied and the validity of neutral and niche theories were tested to explain the rhizosphere community assembly processes. This approach demonstrated that the assembly of the microbial community in the rhizosphere is based on niche-based processes because of selective pressure exerted by the plant itself and other environmental factors. Unlike targeted metagenomics, this approach can collect information about all the microbial taxa thriving in the rhizosphere but requires additional processes for cleaning, ordering, and annotating sequences.

Shotgun metagenomics has issue to consider that despite the easy generation of biomarker libraries, the manual exploration of the metabolisms associated with all observed operational taxonomic unit (OTUs) is impractical. Therefore the researcher has no other option but to trust the gene family ordering process performed by the software. Nevertheless, evidence suggests that despite the bias introduced by this imperfect correlation, shotgun metagenomics has validated the results obtained through targeted metagenomics using taxonomic biomarkers. Studies evaluating the rhizosphere have used these tools. Evaluation of the effect of urea-based fertilizer rate on maize root exudation, the associated rhizosphere microbial community, and nitrogen-use-efficiency were studied and explored via PICRUSt, the metagenomics contribution of bacterial OTUs obtained from 16S rRNA sequencing. These results revealed that some nitrifying and denitrifying genes were significantly influenced by the N rate, whereas some nitrogen-fixing and urease genes were not. However, on a total abundance basis, all N-cycle genes increased significantly as the N rate increased.

Metatranscriptomics is another approach being tried in soil biodiversity studies of agricultural soils. All the living microbes thriving in any niche that is metabolically active are contributing to the activities of the microbiome. These activities/functions can be estimated and tracked through metatranscriptomics. The identification and quantification of microbiome transcripts are now possible with metatranscriptomics which determines the profile and changes in gene expression occurring in a microbial community in response to environmental or experimental variations. Metatranscriptomics estimates the function and activity of complete sets of transcripts (RNA-seq) obtained from any sample. Therefore the "metatranscriptome" (messenger and noncoding RNAs) offers a vision framework about the regulatory networks and gene expression occurring at the time of sampling.

Biological processes depending on small RNAs such as virulence, quorum sensing, and stress/immune responses are accessed through metatranscriptome approach. It allows understanding some of the biomolecular networks dictating emergent phenotypes in the microbiome and their roles in the agricultural systems. As type of microbes and their mechanisms involved in disease suppression in soil and microbiome-plant interaction are still not completely elucidated, techniques like metatranscriptome analysis may unveil the role of genes responsible for these mechanisms. Comparative metatranscriptomic approach are performed for assessing the taxonomic and functional characteristics of the rhizosphere microbiome of wheat plants grown in adjacent fields previously known as suppressive and nonsuppressive to the pathogen *Rhizoctonia solani* AG8 (major pathogen of grain crops). The study indicated suppressive samples showing greater expression of a polyketide cyclase, a terpenoid biosynthesis backbone gene, and several cold shock proteins, whereas nonsuppressive samples showing higher expression of antibiotic genes (Hayden et al., 2018).

The combination of data obtained from metagenomics and metatranscriptomics can provide information about of the changes in gene expression that are accompanied by the changes in the microbiome structure, enabling to know what species are present in a given niche, their functional capabilities, and their real-time activities under particular scenarios. Both techniques complement and strengthens the results obtained from each other.

18.7.4 Genomics analysis of plant microbiome

The study of plant microbiota is recent research interest and is being done using modern genomic tools due to influencing role of these plant associated microbes in shaping various aspects of plant health. These plant associated microbial communities have vast genetic capacity to modulate the phenotypic expression of plants, its health, physiology, and productivity. Plant microbiome includes microbiota associated with plants both in above ground and below ground. While below ground microbes are well studied as rhizospheric microbes, above ground microbes are phyllosphere communities which have come to limelight recently due to their silent beneficial effects on plant health and development. The study and characterization of the total genetic stock of microorganisms (rhizo and phyllosphere) has revealed fundamental role they play in plant phenotype. The plant microbiome acts as an extension of the host genome through endophytes and epiphytes (in and on plants, respectively).

Microbes associated to plants influence seed germination, seedling vigor, plant growth and development, nutrition, diseases, and development. These microbiota function as an annexed organ or a host's genome extension, and therefore the entire comprehension of the biology of a plant also depends on the knowledge gathered about its microbiota. Microbes including bacteria, archaea, fungi, protozoa, and algae are usually part of these microbial consortia. This is a symbiotic relationship where plants depend on specific functions of the microbes while producing photosynthetically fixed carbon and other exudate components for the microbes thriving in the phyllosphere, rhizosphere, and mycorrhizosphere. Despite a considerable number of studies having demonstrated these associations, the diversity and complexity of these communities and interaction networks suggest that our current knowledge about this subject is still limited.

Plants are colonized by several microbes that together can reach higher cell loads than the plant itself. Most of these microbes thriving in the rhizosphere and nearby are directly or indirectly associated to the plant which can be considered as a harboring host and neighbor mutualist body for those microbes that are not in contact with the plant but interacting with it. Understanding the plant microbiome will comprehend the biology of plants and help in identifying microbes that can be exploited for biotechnological purposes aiming to improve plant growth and to avoid undesirable scenarios like stress or disease. Productivity of important agriculture crops can be improved by elucidating the taxonomic diversity and functional potential of their respective microbiota.

Beneficial rhizospheric microbes can alter the morphology of plants while enhancing their growth and increasing mineral availability. Rhizosphere microbiome is also fundamental because it can secrete growth hormones while inducing the immune system. Diseases in crops are sometimes strongly linked with the changes in their microbiome. When crops are affected by pathogens, alleviating strategies include the use of antibiotics. Functional metagenomics can provide information about potential resistance by pathogens to particular kinds of antibiotics and suggest a more targeted and efficient strategy (Santos et al., 2017), allowing the use of less antibiotic while alleviating the collateral damage to the environment. Achieving this level of understanding about particular features of microbiome requires highly specialized sets of tools.

18.7.5 Microbial genomics in carbon management

Role of microbes in carbon management is being studied by interlinking studies on microbes in their original ecological niche, their interaction with ecosystem and their influence on carbon cycle which are all intertwined with climatic change. There are countless numbers of ways in which different microbial communities, especially the soil microbes, influence the carbon cycle through their metabolic activities. These ways are often classified into two groups: one is known to affect the concentration of carbon dioxide and the uptake of methane in the ecosystem and the other controls the loss of carbon from soil due to respiration and methane production.

Climatic change is known to affect the soil microbiome activities and their response toward the emission of greenhouse gases in the atmosphere which further leads to global warming. Climate change affects the soil microbiome directly through change in drastic climatic events like precipitation, temperature, and greenhouses gases. Indirectly affect through changes in soil's physiochemical nature which alters the productivity of plants. It also alters the carbon content of soil, thus affecting those microbiomes, which are responsible in the decomposition activities.

Soil microbiome plays significant role in carbon cycling as remarkable enzyme system of soil microbes operate the Earth's biogeochemical processes or cycles. Microbes extract and break down the dead organic matter and transfer it into soil in usable form for the other living organisms. One of the major biogeochemical cycles is the global carbon cycle, is known to depend on microbial communities. Microbes fix atmospheric carbon (autotrophs), stimulate plant growth, and degrade/transform dead organic matter.

Huge amount of carbon is stored in grassland soil, permafrost, tropical forest, and many other ecosystems. Terrestrial carbon is cycling through two biological metabolic processes, the photosynthesis, and the respiration. The autotrophic and chemoautotrophic microbes fix atmospheric carbon via carbon fixation process into the soil. Microbes are responsible for the decomposition of specific carbon sources like cellulose, lignin, etc. in the soil. Soil microbiome through its decomposition process is also responsible for heterotrophic respiration.

Thus shift in microbial communities in the soil from autotrophic to heterotrophic communities will directly influence the carbon cycle. According to the prediction models of different ecosystem, global temperature or global warming process will cause increase in microbial decomposition of carbon sources present in the soil. Thus increased efficiency of using carbon compounds by soil microbes contributes toward the climatic change or global warming.

CO_2 is one of the major green house gas (GHGs) influencing the global environmental change. The carbon (C) emitted into the atmosphere is primarily due to anthropogenic activities in the past 200 years, and as a result, the global atmospheric concentration of CO_2 has risen from 280 to 410 ppm in 2020, with a current annual increase of 0.88 ppm. The annual rate of increase in atmospheric carbon dioxide over the past 60 years is about 100 times faster than previous natural increases. For comparison, between 2009–18, the growth rate (annual increase) has been 2.3 ppm per year. Thus there is an immediate need to find strategies for minimizing anthropogenic CO_2 emissions.

Terrestrial carbon sequestration is an important step towards mitigating anthropogenic CO_2 emissions. Increased CO_2 concentrations in the atmosphere are thought to be partly contributed by the soil under agriculture. According to an approximation, soils have contributed 55–878 billion tons (GT) of carbon to the total atmospheric CO_2. Soil microorganisms are of great importance in soil carbon cycling. Carbon sources are added externally to soil as crop residues undergo microbial decomposition which results in loss of 2/3 of the carbon. A small portion of it is absorbed as the microbial biomass, and a major part of it is released into the environment as CO_2. As a result of this process accumulation of comparatively resistant Soil Organic Carbon (SOC) fraction is less.

Carbon management requires the channelizing of CO_2 in the atmosphere into long lived pools to mitigate or reduce their immediate remittance. Soil is a reservoir of huge stock of potentially volatile C which act both as a buffer against atmospheric CO_2 increase and as a possible sink for additional C depending on the balance between photosynthesis, the respiration of decomposer organisms, and stabilization of C in soils (Lal, 2004). Soil C sequestration can significantly contribute to the idea of mitigation and potentially offset a remarkable amount of diffuse CO_2 sources for which direct capture is not yet reasonable. In terrestrial habitats, plants dominate the uptake of CO_2 from the atmosphere by net primary production, but microorganisms contribute largely to ecosystem C budgets with their roles as decomposers, plant symbionts, or pathogens, hence modifying nutrient availability and affecting C turnover and absorption in soil (Lal, 2004).

Methane (CH_4), the second most important greenhouse gas after CO_2, contributes to about 20% of the warming effects. Methane oxidation in aerobic soils is mediated by methane-oxidizing bacteria (methanotrophs), a subset of a physiological group of bacteria (methylotrophs). These bacteria utilize CH_4 as the sole carbon and energy source. The role of methanotrophs can never be underestimated as these are the important contributors to attenuate CH_4 flux at the oxic-anoxic interfaces. The WGS of methanotrophic bacteria such as *Methylocystis* sp. Strain Rockwell (ATCC 49242) and *Methylomonas methanica* MC09 provide insights into the genomic and physiological information that can utilize further to optimize the use of these bacteria in industry or biotechnological purposes.

Land management and land use practices can manipulate distinct microbial communities that support C sequestration in terrestrial ecosystems. The fungal to bacterial ratio in soils has been related with C sequestration capability with greater fungal population being related to greater C storage. Recently, a cross-biome metagenomic research has revealed that the ratio of fungal/bacterial rRNA reads showed variation across different soils, among temperate and boreal forests having the highest fungal/bacterial ratios. Higher C storage in fungal-dominated soils can be attributed to higher C use efficiency; longer retention of C in living biomass; and recalcitrant necromass resulting in longer resident time of C.

Challenges in manipulating microbial community for enhanced C sequestration arise from the enormous diversity and unculturability of soil microbial communities, which have precluded their comprehensive characterization and limited our understanding on their ecological functions. The new generation of omics methods (e.g., genomics, transcriptomics, proteomics, metabolomics, metagenomics) is proving instrumental in providing valuable information about the taxonomic, genetic, and functional properties of soil microbial communities. These techniques have begun to allow investigation of functional processes of terrestrial microbial communities involved in C cycling that can be incorporated into mechanistic and predictive ecological models.

Based on the information gathered from full genome sequences, we infer that bacteria belonging to *Acidobacteria* and *Actinobacteria* possess an impressive array of genes allowing breakdown, utilization, and biosynthesis of diverse structural and storage polysaccharides and resilience to stressful soil conditions making them truly ubiquitous in terrestrial ecosystems. These findings support the metagenomic evidence of higher SOC in *Acidobacteria*- and *Actinobacteria*-dominated communities and suggest that these groups promote soil C storage not only due to lifestyle (slow growth and lower metabolic activities) but also by producing polysaccharides for soil structural stability.

Metagenomic analysis has revealed that structure of microbial communities was markedly different between ambient CO_2 (aCO_2) and elevated CO_2 (eCO_2) as indicated by detrended correspondence analysis of gene-based pyrosequencing data and functional gene array (FGA) data. While the abundance of genes involved in decomposing recalcitrant C remained unchanged, those involved in labile C degradation and C and N fixation were significantly increased under eCO_2. Metagenomic analyses showed that 10 years of field exposure of a grassland ecosystem to eCO_2 altered the structural and functional potential of soil microbial communities.

18.7.6 Genomic tools to study bioremediation

Unscientific disposal of chemical substances after use by industrial, domestic and agricultural activities is the main cause of environmental pollution. Contaminants which pollute environment are petroleum oil, hydrocarbons [aliphatic, aromatic, and polycyclic aromatic hydrocarbons (PAHs); benzene, toluene, ethylbenzene, and xylenes; chlorinated hydrocarbons like polychlorinated biphenyls, trichloroethylene (TCE), and perchloroethylene; nitroaromatic compounds; and organophosphate compounds], solvents, pesticides and heavy metals. Due to their mobile and water-soluble nature, many organic and inorganic chemicals such as pesticides, fertilizers, and heavy metals are present in drainage and runoffs from the contaminated sites.

Regular cycling via volatilization and condensation of many organic chemicals such as volatile organic carbons are found in rain and fog. When the pollutants migrate from contaminated site to non-contaminated place as vapors or leachate or dust, they further contribute to the pollution of natural ecosystems. Pollutants from soil are reaching into underground and surface water by leaching and runoff, respectively, thus polluting water bodies. Pesticides are inherent in intensive agriculture promising high production and profits. But the environmental effects of these chemical substances necessitate the remediation of these chemicals after their deliberated use.

Uses of superphosphate fertilizers enrich soil with cadmium, while calcium nitrate can contain significant amounts of nickel. Certain fungicides containing copper and zinc increase the presence of these elements in the upper soil horizons. Most fertilizers, pesticides, and eroded soil reach the water bodies through runoff after rain and flood, leading to eutrophication in freshwater bodies. Phosphate plays a major role in eutrophication and promotes cyanobacterial growth when present in higher concentration, which ultimately decreases the dissolved oxygen concentration in water. These cyanobacterial blooms release some harmful toxins that get accumulated in the food chain. Nitrogen-rich fertilizer causes dissolved oxygen deficiency in rivers, lakes, and coastal zones which has hazardous effects on oceanic fauna. These fertilizers have high water solubility and increased runoff and leaching rate which results in groundwater pollution. Sandy soil favors leaching, thus promoting pollutant flow from the soil. Selenium (Se) is a heavy metal that is naturally found in soil, but irrigation practices favor its accumulation in the soil which ultimately reaches to water reservoirs and is very toxic for animals and humans.

Bioremediation process is an effective and economical way to curb out such environmental problems by enhancing the natural degradation processes. Bioremediation uses plants and/or microorganisms, such as bacteria, protozoa, and fungi to degrade contaminants into a less toxic or nontoxic compounds. Techniques of genomics, transcriptomics, proteomics, metabolomics, phenomics, and lipidomics are used to examine the system biology in different environments including contaminated ecosystems. It is used to unravel, optimize, propose, and scrutinize the function and survival tactics for bioremediation in the desired ecosystem.

The functional genomic approach is useful in identification of novel microbes and their strategies adopted for bioremediation and to uncover the biological function of a gene and their products in a cell. Metagenomics and WGS create possibility for identification and function of novel genes and their regulatory elements involved in biodegradation of various types of contaminants from various types of cultivable and noncultivable microbes from their respective environments.

Biological activities are complex and need a better understanding of interactions and networks at molecular, cellular, community, and ecosystem level, which can be achieved by system biology approach. The knowledge of these interactions clarifies the strategies adopted by the microbes at a polluted site. Bioinformatic tools with genomics, transcriptomics, proteomics, metabolomics, etc. are used to gather the knowledge of cellular interaction at different levels on a polluted site. Similarly, functional genomics also use the knowledge of genomics and transcriptomics to describe the gene functions and interactions in bioremediation process and helpful in phylogenetic identification of microbes involved in bioremediation.

The system biology approach for bioremediation requires the identification and characterization of microbial communities and the molecular processes involved, but it becomes difficult since the toxic contaminants present are influencing the normal activity of microbial communities. The main objective of bioremediation is to degrade or detox the toxic chemicals, which is not possible without knowing all plausible environmental factors that influence the cell-to-cell interactions.

Various bioremediation processes need different system biology approaches. If the purpose is to identify the microbial community structure, the DNA-based genomics tool such as 16S rRNA clone library, PhyloChip, or sequencing can be useful. If the aim is to identify the functional genes involved and to understand the associated cellular pathways with microbial bioremediation, various tools such as RNA-seq, GeoChip (for RNA), and several MS methods (for proteins) are used. Similarly, if we want to identify and characterize the metabolites produced by the microbes during bioremediation, the metabolomics tools such as high-performance liquid chromatography (HPLC) and GC coupled with MS, matrix-assisted laser desorption/ionization (MALDI), NMR, and desorption electrospray ionization are helpful. For the development of a useful conceptual model based on system biology, continuous monitoring of limiting nutrients, electron acceptors, electron donors, and hydrology is also needed.

It is important to understand the complex on-site bioremediation activities to apply the system biology approach. This can be achieved by various monitoring techniques that record and monitor terminal electron donors and electron acceptors, enzyme probes to access the functional activity in an ecosystem, functional genomic microarray, phylogenetic analysis, and metabolomics, proteomics, and quantitative PCR. These tools provide a greater understanding of microbial processes which are involved in bioremediation. An ecosystem generally consists of communities, populations, cells, proteins, lipids, and nucleic acids.

The nucleic acids (DNA and RNA), proteins, and lipids can be analyzed at the cellular level to know their effect on the cells and are also used to analyze populations and community to figure out the consequences of bioremediation on structural and functional relationships in an ecosystem. The detoxification of some hazardous metals and organic compounds can be achieved by simply changing their redox state. This method is particularly useful for heavy metals and radioactive nuclei. Degradation of these chemicals is not possible, so biotransformation is an effective strategy against these contaminants as it reduces their bioavailability, their mobility, and thus their harmful effect. In the presence of suitable electron donors, microbes can immobilize such contaminants by altering their redox state and using them as electron acceptor.

The measurement of microbial respiration and metabolism provides a better insight during the bioremediation process. The enzyme activity measurement during bioremediation is a cost-effective, reliable way to monitor the microbial respiration and metabolism. triphenyl tetrazolium chloride assays and iodonitrotetrazolium are the commonly used dehydrogenase enzyme-based assays which are successfully implied in measuring microbial respiration during bioremediation of metals, PAHs, and oils. Several other molecular probes (based on enzyme or DNA/RNA) from key bioremediation pathways can also be used to track the metabolic pathways. Such probes are widely used to sense the degradation processes, for example, in TCE, and of petroleum hydrocarbon degradation.

Oligonucleotide microarrays are used to detect genes involved in biodegradation and biotransformation of naphthalene, PAHs, and nitrotoluene in microbial communities (Rhee et al., 2004). The qPCR is a molecular technique used for identification of phylogenetic and metabolic genes linked to bioremediation process by microbes and can be used as a monitoring tool for metal detoxification and hydrocarbon degradation. To identify and characterize the microbial communities associated with bioremediation process, metagenomic approach such as 16S rRNA clone libraries is proved to be very useful. This approach is used to identify the microbial communities involved in bioremediation of metals, hydrocarbons, and chlorinated solvents.

In recent years, it has been found that high-throughput microarrays (PhyloChip and GeoChip) are implied to study the bioremediation process of metal and organic.

The PhyloChip (16S rRNA phylogenetic microarray) is used to identify, characterize, and monitor the fluctuation in microbial community, while the GeoChip (functional gene microarray) is used to monitor catabolic gene activity during bioremediation. Presently, proteomic and metabolomic approaches are helpful in detailed understanding of microbial cellular processes and gene products involved in catabolism of contaminants in the environment. Advances in genomics, transcriptomics, proteomics, and metabolomics enable the precise prediction of microbial responses toward the pollutants and their controlling factors in totality (from cell to community level). The vast amount of useful data gathered by system biology approach is helpful in designing a cost effective, sustainable, and low labor-intensive model for an effective bioremediation of organic and inorganic pollutants from soil and water.

Earlier most of the microbiological investigations of bioremediation were done by using treatability study in which the samples from contaminated sites were brought in the laboratory and the rates of degradation or immobilization are documented. Such studies provide an estimate of the degradation potential of the microbial community. The attempts are then made to isolate and purify the main degrader at the laboratory scale. The recovery of the isolates responsible for bioremediation process provides the opportunity to know not only their biodegradation reactions but also their physiology that may control their growth and activity rate in the contaminated site.

The main genomic approaches applied in the field of bioremediation include protein engineering, transcriptomics, WGS, and metagenomics. Identification of whole protein and their expression level in a cell through various methods such as two-dimensional gel electrophoresis, HPLC, and MS is known as proteomics. The membrane proteins play a crucial role in PAH biodegradation, where any mutation can affect membrane protein receptor. Two-dimensional polyacrylamide gel electrophoresis (2-DE) like MudPIT is used to detect such changes. MS techniques now enable the rapid identification of small peptides and proteins. MALDI-TOF-MS is the commonly used method for peptide/protein identification after separation by 2-DE through peptide mass fingerprinting. For identification of contaminants in water, liquid chromatography coupled with MS is used.

The recent advances in proteomics open a whole new possibility where identification as well as changes in composition of proteins during the process of bioremediation can be monitored. The regulation of genes related to bioremediation can be traced by using proteomic approach.

In protein engineering, the amino acid sequence of the target protein with desired activity is selected and altered. The active sites that are close to the substrate are selected for mutational analysis. Hybrid protein is then derived by site-directed mutagenesis, site saturation mutagenesis, gene shuffling, and generation of insertions and deletion. The bioremediation can be enhanced by successful alteration of specific amino acid. *Polaromonas* sp. strain JS666 is well studied by using transcriptomic approach for bioremediation of chlorinated solvents such as tetrachloroethylene and TCE under aerobic conditions. Transcriptome profiling of this strain has also been done and led to the identification of new genes and enzymes involved in cis-dichloroethene (cDCE) degradation. This type of studies paves the way to identify the main candidate for degradation of specific compounds such as cDCE.

Microarray systems named as PhyloChip used for community analysis in any environment allows detection of bacteria and archaeal taxa simultaneously. PhyloChip analysis offers a range of advantages over conventional techniques like Denaturing gradient gel electrophoresis (DGGE), Single-strand conformation polymorphism (SSCP), Restriction fragment length polymorphism (RFLP), Random amplification of polymorphic DNA (RADP), etc. Another set of arrays named as GeoChip used for functional gene analysis that target the genes indulge in the geochemical cycling of N, C, and P; sulfate reduction, metal resistance and reduction, and contaminant degradation have also been found. Nowadays, a new gene array known as FGA having probes for genes containing enzymes for specific degradative pathways are in the limelight. The GeoChip 2.0 is a broad range FGA now used for the study of various environmental metabolic processes. GeoChip 2.0 is successfully applied in the bioremediation of UV(I); diesel contamination; gene identification degradation of cellulose, phthalate, biphenyl, cyclohexanol, benzoate, and naphthalene; microbial N and C cycling in Antarctic sediment; metal resistance genes; stable isotope-probing experiments for microbial profiling; etc.

Bioremediation is transforming the bioavailable contaminants or pollutants in an optimized environment to make it less toxic or other forms that is not hazardous for the environment and human beings. Various bacteria and fungi have been searched out that are playing a vital role in bioremediation of harmful chemical present in the soil and the water, but it still needs to be explored to achieve effective and reliable cleaning up of contaminants. Genomics and metagenomics have potential applications in environmental bioremediation. For effective eradication of contaminants, diversification of microorganism at the contaminated site, identification of new enzymes and pathways, this technology seems to be very appropriate. Future concern on genomics update required in the field of bioremediation includes the development of methodologies for getting co-metabolic capability of the enzymes involved in degradation, identification of conditions that are conducive and facilitate the upregulation of degradative enzymes, and improvement of high bioavailability of mixed microbial populations.

18.8 Soil community composition and function-driven soil biodiversity

Although much is known about specific functions conducted by specific functional groups, it is unclear how widely functions are distributed among different taxa. Present soil microbial ecology studies are addressing research issue on the extent to which functional redundancy occurs in soil communities. It has been a long-held view that soil microbial communities comprise such high diversity and such a high level of functionally redundant organisms that changes in microbial community composition would not translate into changes in functioning. However, recent studies showed that community composition is particularly important.

Studies found that geographically distinct microbial communities have distinct rates of carbon mineralization. Other studies investigated the process of denitrification (where nitrogen oxides are reduced to the strong greenhouse gas N_2O and subsequently to unreactive N_2 gas) showed that it was linked to the composition and abundance of denitrifying soil communities. The proportion of denitrifying bacteria lacking the genes encoding proteins involved in N_2O reduction was shown to relate to the proportion of N_2O from total soil denitrification. Moreover, a reduction in the diversity of denitrifying communities could be shown to reduce total denitrification activity in soil. These examples provide proof of principle that microbial community composition can affect ecosystem functions.

Studies investigating soil biodiversity–ecosystem functioning relations found that at low levels of soil biodiversity, additional species often improves ecosystem functioning, while at higher-diversity levels, effects of species richness on functioning are less frequent. Community composition often has stronger effects compared with species richness. For proper functioning of ecosystem processes, a basic toolbox of organisms with certain functional characteristics is necessary, while further increases in soil biodiversity give no direct benefits (indicating presence of functional redundancy among species). However, agricultural land-use systems already harbor lower (sometimes much lower) level of soil biodiversity compared with less intensively used or natural ecosystems. Therefore the loss of a small number of species or functional groups in such systems could more easily hamper ecosystem functions compared with natural ecosystems. Functional capabilities of less-diverse soil communities are less resistant to stress compared with more diverse communities. Some ecosystem functions are provided by microbial consortia and different functional groups of soil biota complement each other in supporting plant productivity. Hence, simplification of soil food webs and the loss of particular soil biota can directly and indirectly affect the functioning of remaining soil biota.

In an experiment to see contributions of different components of soil biota, soil biodiversity and community composition were manipulated by filtering soil through different meshes of declining size. A successive reduction in soil biodiversity led to the successive decline in some of the measured ecosystem functions, such as plant diversity. Other functions, such as litter decomposition, were maintained at a constant degree at higher levels of biodiversity but declined sharply after a certain mesh size, indicating that the performance of this function depends on particular groups of organisms (i.e., "keystone species") (Wagg et al., 2017).

Categorization of ecosystem functions into physiologically and phylogenetically "narrow" (like nitrification) and "broad" processes (like organic matter decomposition) is also helpful (Schimel, 2018). For some functions, soil biodiversity per se seems to be important, while for others, the presence of certain organism groups (i.e., soil community composition) is crucial. This highlights the difficulties of applying a unifying concept (of soil biodiversity) to different ecosystem functions, because the underlying mechanisms are likely to vary from function to function.

However, the stochastic effects of soil biodiversity can directly provide benefits for ecosystem functioning.

It has been shown that increasing levels of soil biodiversity can reduce the ability of a pathogen to colonize the soil. With changing environmental conditions, such as through climate change, the ability of a particular organism to perform its function might be hampered. With high biodiversity, the probability is higher that a partly redundant organism can take over the function under the new environmental conditions (the insurance effect of biodiversity) (sensu Yachi et al., 2020). Therefore like above ground biodiversity, soil biodiversity might have direct implications for ecosystem stability under environmental change, such as global warming.

To maximize the beneficial effects of soil organisms on ecosystem functioning, both soil biodiversity per se and the presence of specific key organism groups have important roles, depending on the function considered. The variety of interactions and precisely evaluating which organisms provide which functions under which conditions and how different organisms interact, is a daunting task. While technology to detect and quantify soil organisms in high phylogenetic resolution is developing rapidly, knowledge of the ecological significance of this information is incomplete and ecological theory is lagging behind.

However, advanced analytical methods like interaction networks, might help us to understand the underlying processes leading to specific net ecosystem effects. By adopting conceptual advances from above ground ecology, important conclusions about the importance of soil biota for overall ecosystem functioning (multifunctionality) can be drawn. Experiments investigating effects of single groups of soil organisms on specific functions are important to understand mechanistic backgrounds.

Manipulation of soil biological complexity in microcosms, larger mesocosms, and in open fields helps in studying role of soil biodiversity on ecosystem multifunctionality. Combining ecological network analysis and food-web theory with biodiversity–ecosystem functioning research to detect correlative patterns in soil communities and ecosystem functions helps to elucidate these complex processes. Research showed that more diverse species communities often lead to a higher level of functioning for specific processes. It has been recognized that diverse species communities affect several ecosystem functions simultaneously.

Intensive agricultural practices are considered to lead to simpler soil food webs comprising smaller-bodied organisms and fewer functional groups. Agricultural management practices, such as intensive soil tillage, repeated and intensive fertilization, application of pesticides, and low plant diversity, have been shown to have adverse effects on several groups of soil organisms, including Arbuscular Mycorrhizal Fungus (AMF), earthworms and microarthropods, and to reduce overall soil microbial biomass. Blindly enhancing soil biodiversity infers random inclusions of more species. Maintaining more of everything in an unspecified manner might also include a greater diversity of undesired organisms, such as pathogens or weeds. Thus ecological intensification and targeted soil biological engineering approach helps to improve ecosystem functioning and services (Bender et al., 2016).

18.9 Agricultural intensification and biodiversity loss

Land degradation is negatively affecting the population and threatening sustained human well-being apart from contributing to climate change and biodiversity loss (Bach et al., 2020). Soil degradation leads to loss of land's production capacity due to loss of soil fertility and soil biodiversity. Accelerating factors of soil degradation are land use changes and climate change. Under land use change the main land degrading forces are pollution by agricultural, industrial, and commercial activities, arable land loss due to urban expansion and unsustainable agricultural practices. Under climate change factors are climate variability like intense rainfalls, extreme temperatures, and long-term climatic changes.

For food production alone, about one billion hectares of land could be turned into agriculturally managed land by 2050. United Nations reports disappearance of almost one-third of the world's farmable land in the last four decades and predicts that the entire world's topsoil could become unproductive within 60 years if current rates of loss continue. Furthermore, agricultural practices are being intensified to support the growing human population. Agricultural intensification results in a reduction in soil biodiversity leading to a loss of function which is detrimental to resilience and sustained productivity. The diversity and abundance of structures created by soil organisms is essential to the conservation and dynamics of soil organic matter turnover, nutrient release and soil physical properties essential for sustained productivity.

Soil biodiversity is the foundation of sustainable agriculture. Biodiversity is an essential part of proper ecosystem structure and functioning, and to the services ecosystems provide. Speciation (formation of new species) and extinction (permanent loss of species) are the major components of changes in the Earth's biological diversity. The background rate of extinction (or normal extinction rate) is a measure of the rate of species disappearance due to natural processes operating under normal conditions. The rate has been determined to approximately five extinctions per year averaged over geologic time. The current extinction rate is estimated to be anywhere from 10 to 1000 times the background rate of extinction, with the rate varying across taxonomic groups. This is the result of a myriad of human activities and their impacts. The rate at which the climate is changing is accelerating the rate of species extinction by reducing the time plants and animals have to adapt to changes in ecosystem conditions (Maximillian et al., 2019).

Communities of soil biota are substantially altered by direct human activities such as soil sealing, agricultural land-use intensification, and biological invasions resulting from the introduction of non-native species. In addition, altered abiotic conditions resulting from climate change also impact soil biodiversity. These changes in soil biodiversity can alter ecosystem functions performed by the soil biota. Overall, we need to better understand the determinants of soil biodiversity and how they function, plan to avoid further losses, and restore soil biodiversity where possible.

Intensified agriculture heavily depends on irrigation, the use of heavy machines and increased application of chemical fertilizers and pesticides. These practices cause habitat modifications that immediately change soil structure and physicochemical properties, which, together with changes in plant biodiversity, affect and commonly reduce soil biodiversity. Intensified agriculture triggers a spatial and temporal decoupling of organisms which alters the regulation of soil structure development and nutrient cycling. Increased inputs through fertilizer application is a form of pollution and can shift soil bacterial communities and decrease overall microbial activity. Planned above-ground diversity of crop plants and animals (specialization through monoculture or limiting animal types) and soil disturbance (tillage) result in reductions in the biodiversity of soil organisms.

One of the main fluctuating conditions affecting the microbiota is the soil nutrient profile. For example, N-fertilization is commonly used for modern varieties which reduce microbial biomass and diversity, and consequently some functional capabilities may be lost. These crop beneficial strategies have short-time effect on production but have consequences in the long term because the soil microbiota lack adequate functional profile to stimulate the responses in the plants of future crops or inclusively may fail to suppress pathogens. Microbial communities are becoming bacteria-dominated, earthworms are getting killed, and mycorrhizal fungal associations with plants are being disrupted by soil tillage and other agricultural practices.

Herbicides reduce plant diversity with unpredictable effects on the species diversity of various groups of soil fauna. Pesticide use, such as neonicotinoids, and herbicides like glyphosate, which can remain in soils for years impact nontarget organisms and affect soil biodiversity. Common nontarget effects are evident for neonicotinoids, as they are assumed to be causative agents of above ground insect declines, but also can kill soil invertebrates including insects and earthworms. Pollution by heavy metals emerging from agricultural practices, as well as runoff pollution emerging from mining sites and smelters, can kill microbial taxa and change the community composition of soil.

Human induced invasions can lead to an increase in soil biodiversity in the short term, as not everything occurs everywhere, but can reduce biodiversity of native species on a longer term. Plant species host their own species-specific microbiomes and many plants associated species will disappear when their host plant species go extinct. Modern plant varieties do not support microbiomes that degrade organic nitrogen and solubilize mineral nutrients such as phosphorus. More research is needed in order to determine the magnitude of soil biodiversity declines. In this regard, taxonomic and functional metagenomics insights could offer clues for performing strategies minimizing this collateral damage to the rhizospheric microbiota. Engineering rhizospheres through inoculants that form connections with the native microbiome or soil amendments that stimulate microbial activity is helpful.

Human-induced introductions of exotic plants, animals, and microbes change the composition and functioning of soil biota. Several introduced exotic plant species become invasive, for example when released from specialized pathogens in their native range. In turn, invasions of soil borne oomycetes (in particular) or fungal species are considered the main sources of emerging infectious diseases of plants.

Other anthropogenic impacts on soil biodiversity are associated with unintended changes to the environment from ongoing global climate change. Climate change factors that affect soil biodiversity include increases of extreme events such as drought and heavy rainfall, but also more continuous changes such as increases in CO_2 levels and temperature. Extended periods of drought negatively impact most groups of soil life, for example by reducing the abundance and diversity of protists and larger soil animals. Heavy rain events are occurring increasingly. Although these events can promote the abundance and diversity of soil biota through increasing moisture levels, the associated waterlogging and increased soil erosion serve to reduce soil biodiversity.

Increased atmospheric CO_2 levels can enhance microbial biomass by increasing plant production, but might reduce food-web complexity, for example by decreasing the abundance of larger, omnivorous and predacious nematodes. Warming affects soil biodiversity, for example by promoting fungi over bacteria, thus affecting the composition of higher trophic-level consumers. Another unintended anthropogenic change that causes soil biodiversity declines is acid rain, resulting from increased sulfur dioxide and nitrogen oxide emissions, which change vegetation and soil chemistry.

Decrease in soil biodiversity is associated with a simultaneous decrease of several soil functions and loss of soil biodiversity will lead to reduction of soil multifunctionality. Although the exact mechanisms often remain unknown, soil biodiversity plays a pivotal role in providing key ecosystem functions and services. Knowledge about the composition and functions of soil biota are increasing, particularly due to new methods such as high-throughput sequencing. Safeguarding this rich biotic reservoir is essential for soil sustainability and, ultimately, the sustainability of human society.

18.10 Restoration of functional soil biodiversity and its ecosystem services

Soil contains an unknown repertoire of organisms that are directly involved in biochemical nutrient cycling, such as in the global carbon cycle and other ecosystem processes and services. Therefore soil biota is directly involved in climate warming-related processes also, such as the control of greenhouse gas emissions. Soil biodiversity offers major and nearly infinite opportunities as it serves as a reservoir of novel antibiotics apart from being reservoir of biocontrol agents and biofertilizers. To use this immense biotic reservoir and mitigate negative anthropogenic changes that threaten below ground biodiversity, we must understand more about the complex communities of the soil. Protecting and sustainably using soil biodiversity are key challenges to maintain a functional ecosystem and to increase ecosystem health for soil, plants, animals, and humans.

In terms of functioning, the soil community contains major decomposers, playing key roles in carbon and nutrient cycling. Pathogens, parasites, and mutualists in the soil can directly control the performance of plants, which in turn change the performance of above ground biota. Although our understanding of soil biodiversity is increasing, the taxonomic diversity of the soil biota remains undescribed, and ecological functioning at high taxonomic resolution, such as at the species level, remains unknown for most soil organisms, particularly microorganisms (viruses, bacteria, fungi, and protists).

Despite the functional importance of soil biodiversity for major soil functions and ecosystem services- such as provisioning of clean drinking water, prevention of greenhouse gas production, and control of diseases- soil organisms are not typically did not include in assessments of soil quality or biodiversity declines, nor are they incorporated in Earth-system models. Considering and integrating soil biodiversity into large-scale analyses and models will help us to better understand the importance of soil biodiversity at a global scale.

Land managers and farmers can reduce their negative footprint on soil biodiversity by increasing their efforts in ecologically intensifying agricultural practices. Planting arable land margins with tree plants/bushes/perennial grasses will influence the soil biodiversity underneath the strips and help in capturing carbon. Reducing soil tillage will enhance soil organic matter, which positively affects soil carbon storage, soil drainage, nutrient provisioning and, possibly, crop sensitivity to extreme weather events. Organic agriculture comes at some cost of a reduced yield in the short-term, but through increases in the taxonomic and functional soil biodiversity, improved soil functions it results in increased plant production over longer time spans.

Nutrient stresses alleviate as agricultural systems intensify and stress dominate as agricultural systems go fully conservative way. Thus relationships of productivity, soil disturbance and stress show that the greatest diversities among soil organisms may be found at intermediate stages of agricultural intensification. Sustainable agricultural actions focused on soil biology are needed across cultures and farming contexts.

Soil biodiversity only slowly recovers once industrial and agricultural sites are allowed for restoration. Without collaborative, scientifically informed land management it takes decades or more to restore. In many areas, restoration projects may become necessary to speed up recovery and to increase soil biodiversity as in the case of abandoned mining sites or to counteract global change-induced desertification. Microbial inoculation together with seeding of target plant species might help speed up these processes.

Even though soil biodiversity loss can be mitigated through many options of restoration of biodiversity to some extent, the best option is to preserve it in the first place. Soil biodiversity is not infinite and once species go extinct, they cannot be replaced. Even the previously assumed infinite diversity of bacteria seems limited-instead of trillions there are likely only millions of species, which illustrates the limitations of the reservoir of life in soils on Earth.

The potential of soil biodiversity to increase soil functions can be exploited in an ecologically intensified and thereby a more sustainable type of agriculture. Increased soil biodiversity makes soil productivity less dependent on external inputs including fertilizers and pesticides. Cost reductions to buy and apply fertilizers and pesticides may be achieved through a more sustainable manner of soil management, as a diverse soil biota can act as biofertilizers by increasing nutrient turnover and as biocontrol by inhibiting pathogens. This would also result in decreased environmental pollution and insect die-off, while minimizing yield losses.

With increased evidence pointing to negative side effects of many pesticides, including loss of biodiversity, reducing their application will require alternatives, such as biocontrol agents, to sustain agricultural production. Many currently applied biocontrol agents that control plant pathogens and plant pests are soil-borne and act in the plant's rhizosphere, such as plant growth-promoting rhizobacteria, biocontrol fungi, and entomopathogenic nematodes. Similarly, many so-far unknown soil organisms contain a reservoir of genes of direct human interest, such as novel antibiotics that are potentially more effective than currently used antibiotics.

Soil biodiversity as a whole, or individual groups of soil biota, have been proposed to be reliable bioindicators for soil quality, as they quickly respond to changes and perturbations, such as those induced by contaminants. The diversity of life in soils might contain taxa or communities that can potentially indicate the presence of nearly all imaginable compounds, but the potential of soil biodiversity for the indication of (toxic) compounds has yet to be exploited. Similarly, microorganisms and soil animals may actively contribute to the degradation of environmental pollutants and can be used for bioremediation of pesticides and other toxic or poorly degradable compounds.

Establishment of permanent cultures and conservation sites of unique ecosystems may help to protect local soil biodiversity wherever it is threatened. Although research efforts are increasing, an even more complete understanding of the taxon-level diversity in soils is needed. Functions performed by individual species or entire soil communities should be understood fully to make efforts to improve targeted species and enrich overall diversity as per the needs. Linking soil taxonomic group and FD to ecosystem functioning and services will benefit in ascertaining changes occurring in soil with respect to richness of particular community and its respective ecological function. Understanding soil biodiversity at species/subspecies level will help us to re-connect ecosystem services to soil biodiversity, thereby enhancing the sustainable provisioning of those services. Some strategies to enhance microbial diversity of agroecosystems are discussed below.

18.10.1 Soil biodiversity intensification

Interventions in natural processes are necessary to maximize benefits from ecosystems for human requirements. Blindly enhancing soil biodiversity infers random inclusions of more species and maintaining more of everything in an unspecified manner might also include greater diversity of undesired organisms, such as pathogens or weeds. Thus instead of general approach targeted enhancement of consortia of microbes capable of organic matter decomposition, plant growth promotion, plant hormone production, siderophore production, metal transformation (to plant available form), nitrogen fixation and pathogen suppression should be followed. To achieve maximum effects, management strategies must apply at multiple scales, from soil and plant community management to plant genetic and rhizosphere microbiome management. At entry point for management crop, plant choice should consider enhanced crop diversity, intercropping, living mulch, including N-fixing legumes, diverse rooting patterns, mixing perennial and annual and other factors. At another entry level of soil management mulching, reduced soil tillage and no-till should be given priority. At microbiome level, rhizosphere microbiome management like nitrification inhibition or denitrification inhibition, disease suppression, inoculation with beneficial soil organisms (such as plant growth-promoting bacteria), fostering indigenous AMF communities or inoculation with AMF shall be considered.

18.10.2 Soil tillage management

Practices that conserve the soil biological potential while allowing economic farm management are important in promoting biodiversity. Conventional mechanized/heavy tillage often has adverse effects on soil biota and promotes the decomposition and mineralization of organically bound nutrients even though it eases weed control. Hence, practices that minimize negative effects on soil biota while providing the desired agricultural benefits (like no or reduced soil tillage, and strip tillage) need to be followed. Such conservation tillage practices are often most successful in combination with other measures, such as cover crops and mulches. The application of organic residues and composts has also been shown to reduce pest incidence and weed pressure, and to favor soil biota. In addition, in the past decade applications of biochar to agricultural soils to improve soil properties and plant performance received much attention and are showing enhanced nitrogen fixation and AMF abundance. Practices conserving the soil biological potential can enhance or maintain soil organic matter content and, therefore, can contribute to long-term soil preservation.

18.10.3 Crop diversification

Species mixtures and crop rotations favor higher soil biodiversity and make use of complementary biodiversity effects. Spatial crop diversification (intercropping) or temporal crop diversification (crop rotation or cover crops or relay crops) has proven to have several beneficial effects on ecosystem processes. Specific cover crops enhance the abundance of soil biota that increase the yield of the subsequent crop. Planting diverse mixtures of cultivars reduces pathogen incidence. There is a lack of studies addressing beneficial effects on soil biota and below ground ecosystem processes through enhanced crop diversity in an applied agronomic context except few studies comparing effect of grass family crops (rice, wheat, maize, etc.) and legume family crops (cow pea, alfalfa, soybean, etc.). The targeted combinations of crop varieties with different traits in relation to ecosystem functioning also exploiting niche complementarity effects theoretically opens up a range of possibilities to manage ecosystem services in cropping systems and to reduce the dependence on external resource inputs.

18.10.4 Plant selection for rhizosphere microbiome enrichment

It is now well understood that plants can shape their root endophytic and rhizospheric microbial communities through root exudates and selective promotion and/or inhibition of microbial species. This feature theoretically provides the option to breed crops to acquire soil microbes that provide specific services. For example, some plants can inhibit the transformation of ammonium into nitrate (nitrification) by affecting nitrifying microbes. This potentially improves the nitrogen availability of the plant and can also reduce nitrogen losses from soil through leaching and denitrification. The integration of knowledge about how plants regulate the composition of the root microbiome into crop breeding strategies could greatly contribute to agricultural sustainability. For example, the recruitment of root symbionts, such as AMF or rhizobia, is mediated through carbon allocation and root exudation of specific compounds. Selecting crops for the high production of such compounds could maximize symbiotic benefits.

Selection for the plant microbiome can also contribute to disease suppression or to altered plant traits, such as flower time. Plant breeders have largely ignored such processes and it is now a key challenge to integrate root traits and associate microbiomes in future breeding programs, especially since several modern plant cultivars have partly lost their ability to associate with beneficial soil biota. Action is urgent because the development of new cultivars is often time consuming and can take decades, especially if multiple traits are involved. The adoption of transgenic methods could also be an option to engineer plant effects on rhizosphere communities.

18.10.5 Biofertilizer/effective microbial consortia application and biocontrol

Plant nutrition contributing or supplementing microorganisms are well recognized and are being used to also to considerable extent. Bio-stimulating stress enhancing microbes are being promoted as "effective microbes" very recently. Several groups of such organisms are identified, and their mechanisms of plant growth promotion (other than direct nutrient supplementation) and biotic/abiotic stress tolerance are being elucidated regularly. Maximum utilization of these groups of microbes along with regular well known biofertilizers (N-fixers, PGPR, P-solubilizers) as consortia will help plant establishment and enhanced production. Microbial redundancy protects key functional groups of organisms from reduction to crucial thresholds below which agroecosystem function is impaired.

Inoculation of seedlings or soils with AMF propagules, in addition to fostering indigenous AMF communities, is an option to profit from beneficial effects provided by these fungi. The use of biocontrol agents to control agricultural pests is much appreciated approach. However, due to lack of biocontrol agents for vast number of insect pests and easy availability of broad-spectrum insecticides, these chemicals are being used excessively to save crops from pest and disease attacks. A range of microbial phyla, such as *Pseudomonas*, *Bacillus*, and *Trichoderma*, can induce systemic resistance of plants against pathogen attacks. Recent developments in sequencing-based methods from soil or root samples to identify species and even specific isolates will make it possible to follow the fate of microorganisms introduced into soils and will allow risk assessment (Bender et al., 2016). Application of consortia of soil biota as biofertilizers and biopesticides and in storing carbon will be the important step in sustainable agriculture through the potential of reducing fertilizer and pesticide inputs (Geisen et al., 2019a,b).

18.11 Conclusion and future perspectives

The soil biodiversity function at global, regional and local levels and hence has become an environmental asset. At global level soil biota are important for future generations as a reservoir of carbon, novel products (like antibiotics) or for scientific discoveries associated with some aspect of soil biodiversity. At national scale, soil biodiversity contribute to ensure sustained food security through high crop yields, livestock production and by increasing aesthetic value of landscape. At local level (farming system scale), soil biodiversity contribute to productive capacity of the system by ensuring the mineralization of nutrients from organic resources and buffer the functions of the soil and their resilience to environmental risks (e.g., pest attacks, drought and fire). By performing these roles across different land management approaches and climates, soil biodiversity has become the heart of sustainable agriculture.

Microbial communities, among soil biota, play an important role for agriculture progress. In recent years, interest in understanding the activities of these microbial communities has increased. Hence, UN Convention on Biodiversity is actively creating awareness across the globe about the ecosystem functions of soil biota through participatory approaches like conferences and strategic plans. One such plan in the year 2020 is composed of 20 targets: to address the underlying causes of biodiversity loss by mainstreaming biodiversity across government and society; reduce the direct pressures on biodiversity and promote sustainable use; improve the status of biodiversity by safeguarding ecosystems, species, and genetic diversity; enhance the benefits to all from biodiversity and ecosystem services; enhance implementation through participatory planning, knowledge management and capacity building.

Increasing food demand together with the environmental changes represent a challenge that requires more intensive yet environmentally safe production, as well as the sustainability of the resources involved in agricultural activity. Considering the beneficial role of microbial communities in soil environments, it is important increasing the understanding of plant-microbe relationships to provide biotechnological information for control and management with sustainable practices.

Molecular methods show promise for assessment of the diversity of soil organisms in which morphological taxonomy is difficult or even impossible. Since the vast majority of organisms living in the soil have never been described, advances in the systematics of the soil biota using systems biology and molecular genomics tools are essential. Genomics harnesses the availability of complete DNA sequences for entire organisms and was made possible by both the pioneering work of Fred Sanger and the more recent NGS technology. Bioinformatics tools have become indispensable in metagenomics research, allowing to know the structure and metabolic potential of environmental microbes.

Main challenges in using genomic tools is discriminating and/or interpreting massive amount of information it generates and challenging task of linking functional groups of soil microbes to molecular data to species (rather than broad taxonomic groups) and to performance in the field rather than just potential physiological capabilities. Functional groups are an indispensible aid in studying the role of the soil biota in maintaining ecosystem services. The identification of functional groups of the soil biota that occupy key positions in ecosystem processes is a very high priority. A fundamental problem, however, is the limited knowledge available at the species level to assign species to such groups (Brussaard et al., 1997).

In systems biology, improving the accuracy of the results is a constant desire and hence the discipline is progressing regularly by providing cost-effective, simplified, thorough methods to study microbial metabolic networks. A few metagenome studies by sequence-based metagenomics approaches have been performed for plants and associated microorganisms. Study of the interaction of plant-microbes in a global context could lead to find a natural mechanism facilitated by microbes in controlling diseases and their role in bio-geo-chemical cycles by their ecological process.

Complex ecosystems need to be tackled using cultivation-independent methods and both metagenomics and SCG are today's commonly used methodologies to gain genomic insights into complex environments. Novel approaches that efficiently enrich or isolate, prior to sequencing, specific environmental cells that catalyze a function of interest or exhibit a wanted phenotype is needed in future to conserve and utilize ecofriendly natural resource of biodiversity.

References

Bach, E.M., Ramirez, K.S., Fraser, T.D., Wall, D.H., 2020. Soil biodiversity integrates solutions for a sustainable future. Sustainability 12 (2662), 1–20.

Bender, S.F., Wagg, C., van der Heijden, M.G.A., 2016. An underground revolution: biodiversity and soil ecological engineering for agricultural sustainability. Trends Ecol. Evol. 31 (6), 440–452.

Brussaard, L., Behan-Pelletier, V.M., Bignell, D.E., Brown, V.K., Didden, W., Folgarait, P., et al., 1997. Biodiversity and ecosystem functioning in soil. Ambio 26 (8), 563–570.

Donkor, E.S., 2013. Sequencing of Bacterial Genomes: Principles and Insights into Pathogenesis and Development of Antibiotics. Genes 4, 556–572.

Forster, S.C., 2017. Illuminating microbial diversity. Nat. Rev. Microbiol. 15, 1.

Geisen, S., Briones, M.J.I., Gan, H., Behan-Pelletier, V.M., Friman, V.P., de Groot, G.A., et al., 2019a. A methodological framework to embrace soil biodiversity. Soil Biol. Biochem. 136, 107536.

Geisen, S., Wall, D.H., van der Putten, W.H., 2019b. Challenges and opportunities for soil biodiversity in the anthropocene. Curr. Biol. 29, 1036–1044.

Hayden, H.L., Savin, K.W., Wadeson, J., Gupta, V.V.S.R., Mele, P.M., 2018. Comparative metatranscriptomics of wheat rhizosphere microbiomes in disease suppressive and non-suppressive soils for *Rhizoctonia solani* AG8. Front. Microbiol. 9, 859.

Lal, R., 2004. Soil carbon sequestration impacts on global climate change and food security. Science 304 (5677), 1623–1627.

Maximillian, J., Brusseau, M.L., Glenn, E.P., Matthias, A.D., 2019. Pollution and environmental perturbations in the global system. Glob. Syst. Hum. Dimens. Environ. Pollut. 457–476.

McKusick, V.A., Ruddle, F.H., 1987. A new discipline, a new name, a new journal. Genomics 1, 1–2.

Mendes, R., Garbeva, P., Raaijmakers, J.M., 2013. The rhizosphere microbiome: significance of plant beneficial, plant pathogenic, and human pathogenic microorganisms. FEMS Microbiol. Rev. 37, 634–663.

Moretti, M., Dias, A.T.C., de Bello, F., Altermatt, F., Chown, S.L., Azcárate, F.M., 2017. Handbook of protocols for standardized measurement of terrestrial invertebrate functional traits. Funct. Ecol. 31, 558–567.

Nannipieri, P., 2014. Soil as a biological system and omics approaches. Int. J. Environ. Qual. 13, 61–66.

Nannipieri, P., Penton, C.R., Purahong, W, 2019. Recommendations for soil microbiome analyses. Biol Fertil Soils 55, 765–766.

Rhee, S., Liu, X., Wu, L., Chong, S., Wan, X., Zhou, J., 2004. Detection of genes involved in biodegradation and biotransformation in microbial communities by using 50-mer oligonucleotide microarrays. Appl. Environ. Microbiol. 70, 4303–4317.

Santos, M.S., Hungria, M., Nogueira, M.A., 2017. Production of polyhydroxybutyrate (PHB) and biofilm by *Azospirillum brasilense* aiming at the development of liquid inoculants with high performance. Afr. J. Biotechnol. 16, 1855–1862.

Schimel, J.P., 2018. Life in dry soils: effects of drought on soil microbial communities and processes. Annu. Rev. Ecol. Evol. Syst. 49 (1), 409–432.

Torsvik, V., Sørheim, R., Goksøyr, J., 1996. Total bacterial diversity in soil and sediment communities—a review. J. Ind. Microbiol. Biotechnol. 17, 170–178.

Vogel, T., Hirsch, P., Simonet, P., et al., 2009. Advantages of the metagenomic approach for soil exploration: reply from Vogel et al. Nat. Rev. Microbiol. 7, 756–757.

Wagg, C., O'Brien, M.J., Vogel, A., Scherer-Lorenzen, M., Eisenhauer, N., Schmid, B., Weigelt, A., 2017. Plant diversity maintains long-term productivity under frequent drought by increasing short-term variation. Ecology 98, 2952–2961.

Weinstein, J.N., 1998. Fishing expeditions. Science 282, 627–628.

Wu, X.H., Ge, T.D., Yuan, H.Z., Li, B.Z., Zhu, H.H., Zhou, P., et al., 2014. Changes in bacteria CO_2 fixation with depth in agricultural soils. Appl. Microbiol. Biotechnol. 98, 2309–2319.

Yachi, S., Loreau, M., 1999. Biodiversity and ecosystem productivity in a fluctuating environment: The insurance hypothesis. Proceedings of the National Academy of Sciences of the United States of America 96 (4), 1463–1468. Available from: https://www.ncbi.nlm.nih.gov/genome/https://gold.jgi.doe.gov/.

Further reading

Alkorta, I., Amezaga, I., Albizu, I., Aizpurua, A., Onaindia, M., Büchner, V., et al., 2003. Molecular microbial biodiversity assessment: a biological indicator of soil health. Rev. Environ. Health 18 (2), 131–151.

Barea, J.M., Pozo, M.J., Azcón, R., Azcón-Aguilar, C., 2005. Microbial co-operation in the rhizosphere. J. Exp. Bot. 56 (417), 1761–1778.

Bhattacharjee, S., Prakash, T., 2019. Microbial genomics in carbon management and energy production. In: Tripathi, V., Kumar, P., Tripathi, P., Kishore, A. (Eds.), Microbial Genomics in Sustainable Agroecosystems, vol. 1. Springer Nature, Singapore.

Breure, A.M., 2004. Soil biodiversity: measurements, indicators, threats and soil functions. In: International Conference on Soil and Compost Eco-Biology, pp. 83–96.

Coleman, D.C., Callaham Jr, M.A., Crossley Jr, D.A., 2018. Future Developments in Soil Ecology. Fundamentals of Soil Ecology, third ed. Academic Press, pp. 255–282.

Fernando, T.M., Delgado-Baquerizo, M., Jeffries, T.C., Eldridge, D.J., Ochoa, V., Gozalo, B., et al., 2015. Increasing aridity reduces soil microbial diversity and abundance in global drylands. PNAS 1–6.

Fierer, N., Jackson, R.B., 2006. The diversity and biogeography of soil bacterial communities. PNAS 103 (3), 626–631.

Frąc, M., Hannula, S.E., Bełka, M., Jędryczka, M., 2018. Fungal biodiversity and their role in soil health. Front. Microbiol. 9 (707), 1–9.

Fraser, C.M., Dujon, B., 2000. The genomics of microbial diversity, Editorial overview. Curr. Opin. Microbiol. 3, 443–444.

Garibay-Valdez, E., Calderón, K., Vargas-Albores, F., Lago-Lestón, A., Martínez-Córdova, L.R., Martínez-Porchas, M., 2019. Functional metagenomics for rhizospheric soil. In: Tripathi, V., Kumar, P., Tripathi, P., Kishore, A. (Eds.), Microbial Genomics in Sustainable Agroecosystems, vol. 1. Springer Nature, Singapore.

Giller, K.E., Beare, M.H., Lavelle, P., Izac, A.M.N., Swift, M.J., 1997. Agricultural intensification, soil biodiversity and agroecosystem function. Appl. Soil Ecol. 6, 3–16.

Handelsman, J., 2004. Metagenomics: application of genomics to uncultured microorganisms. Microbiol. Mol. Biol. Rev. 68 (4), 669–685.

Hugenholtz, P., 2002. Exploring prokaryotic diversity in the genomic era. Genome Biol. 3 (03), 1.

Janda, J.M., Abbott, S.L., 2007. 16S rRNA gene sequencing for bacterial identification in the diagnostic laboratory: pluses, perils and pitfalls. J. Clin. Microbiol. 45 (9), 2761–2764.

Kanianska, R., 2016. Agriculture and its impact on land-use, environment, and ecosystem services. Landscape Ecology: The Influences of Land Use and Anthropogenic Impacts of Landscape Creation. Intech Open Publishing, pp. 4–26.

Lutz, S., Anesio, A.M., Field, K., Benning, L.G., 2015. Integrated "omics," targeted metabolite and single-cell analyses of arctic snow algae functionality and adaptability. Front. Microbiol. 6, 1323.

Maheshwari, D.K., Agarwal, M., Dheeman, S., 2014. Trends and prospects of microbial diversity in rhizosphere. In: Maheshwari, D.K. (Ed.), Bacterial Diversity in Sustainable Agriculture, Sustainable Development and Biodiversity. Springer International Publishing, Switzerland, pp. 1–22.

Markowitz, V.M., et al., 2008. The integrated microbial genomes (IMG) system in 2007: data content and analysis tool extensions. Nucleic Acids Res. 36, D528–D533 (database issue).

Morgado, R.G., Loureiro, S., González-Alcaraz, M.N., 2018. Changes in Soil Ecosystem Structure and Functions Due to Soil Contamination, Soil Pollution from Monitoring to Remediation. Academic Press, Elsevier Inc., pp. 59–87.

Ramsey, P., Rillig, M.C., Feris, K.P., Gordon, N.S., 2005. Relationship between communities and processes: new insights from a field study of a contaminated ecosystem. Ecol. Lett. 8 (11), 1201–1210.

Rinke, C., et al. Insights into the phylogeny and coding potential of microbial dark matter. Nature 499(7459), pp. 431–437.

Saha, N., Roy, S.S., Biswas, S., Datta, S., 2017. Adaptive Soil Management: A Tool for Plant Fitness in Stressful Environment Through Microbial Integrity. In: A., Rakshit et al. (Eds.), Adaptive Soil Management: From Theory to Practices, pp. 277–299.

Sanger, F., Coulson, A.R., 1975. A rapid method for determining sequences in DNA by primed synthesis with DNA polymerase. J. Mol. Biol. 94 (3), 441–448.

Santos, M.S., Nogueira, M.A., Hungria, M., 2019. Microbial inoculants: reviewing the past, discussing the present and previewing an outstanding future for the use of beneficial bacteria in agriculture. AMB Express 9, 205.

Schimel, J.P., Schaeffer, S.M., 2012. Microbial control over carbon cycling in soil. Front. Microbiol. 3.

Schimel, J., Balser, T.C., Wallenstein, M., 2007. Microbial stress-response physiology and its implications for ecosystem function. Ecology 88, 1386–1394.

Sehgal, N., Singh, D.K., Chiary, H.R., Rao, K.S., 2019. Community characteristics, types of biodiversity, diversity index, abundance, species richness. Vert. Horiz. Stratification: II Princ. Ecol. 1–12.

Shih, et al., 2013. Improving the coverage of the cyanobacterial phylum using diversity-driven genome sequencing. PNAS 110 (3), 1053–1058.

Thompson, L., Sanders, J., McDonald, D., et al., 2017. A communal catalogue reveals Earth's multiscale microbial diversity. Nature 551, 457–463.

Thoms, C., Gattinger, A., Jacob, M., Thomas, F.M., Gleixner, G., 2010. Direct and indirect effects of tree diversity drive soil microbial diversity in temperate deciduous forest. Soil Biol. Biochem. 42, 1558–1565.

Tsai, M.H., et al., 2019. A new genome-to-genome comparison approach for large-scale revisiting of current microbial taxonomy. Microorganisms 7 (6), 161.

van der Heijden, M.G.A., Bardgett, R.D., van Straalen, N.M., 2008. The unseen majority: soil microbes as drivers of plant diversity and productivity in terrestrial ecosystems. Ecol. Lett. 11, 269–310.

Wagg, C., Bendera, S.F., Widmer, F., van der Heijden, M.G.A., 2014. Soil biodiversity and soil community composition determine ecosystem multifunctionality. PNAS 111 (14), 5266–5270.

Wagg, C., Schlaeppi, K., Banerjee, S., Kuramae, E.E., van der Heijden, M.G.A., 2019. Fungal-bacterial diversity and microbiome complexity predict ecosystem functioning. Nat. Commun. 10, 4841.

Wallenstein, M.D., 2017. Managing and manipulating the rhizosphere microbiome for plant health: a systems approach. Rhizosphere 3, 230–232.

Washburn, M.P., Wolters, D., Yates, J.R., 2001. Large-scale analysis of the yeast proteome by multidimensional protein identification technology. Nat. Biotechnol. 19 (3), 242–247.

Wu, D., Hugenholtz, P., Mavromatis, K., et al., 2009. A phylogeny-driven genomic encyclopaedia of Bacteria and Archaea. Nature 462, 1056–1060.

Yachi, S., Loreau, M., 1999. Biodiversity and ecosystem productivity in a fluctuating environment: the insurance hypothesis. Proc. Natl. Acad. Sci. U. S. A. 96 (4), 1463–1468.

Yadav, B.S., Mani, A., 2019. Functional genomics and systems biology approach for understanding agroecosystems in microbial genomics. In: Tripathi, V., Kumar, P., Tripathi, P., Kishore, A. (Eds.), Microbial Genomics in Sustainable Agroecosystems, vol. 1. Springer Nature, Singapore.

Yadav, S., Bhardwaj, Y., Neha, Singh, A., 2019. Functional genomics and system biology approach in bioremediation of soil and water from organic and inorganic pollutants. In: Tripathi, V., Kumar, P., Tripathi, P., Kishore, A. (Eds.), Microbial Genomics in Sustainable Agroecosystems, vol. 1. Springer Nature, Singapore.

Yin, R., Kardol, P., Thakur, M.P., Gruss, I., Wu, G.L., Eisenhauer, N., et al., 2020. Soil functional biodiversity and biological quality under threat: intensive land use outweighs climate change. Soil Biol. Biochem. 147, 107847.

Zhou, Z., Wang, C., Luo, Y., 2020. Meta-analysis of the impacts of global change factors on soil microbial diversity and functionality. Nat. Commun. 11 (3072), 1–10.

Role of microbial communities in restoration disturbed lands

19

Monu Jariyal[1] *and Ruchi Soni, PhD (Microbiology), Jr. Scientific Officer*[2,*]
[1]Regional Centre of Organic Farming Panchkula, Ministry of Agriculture & Farmers Welfare, Government of India, India, [2]Regional Centre of Organic Farming, Nagpur, India, Ministry of Agriculture & Farmers Welfare, Government of India, India
*Corresponding author. Regional Centre of Organic farming (HQ) Ghaziabad, Ministry of Agriculture & Farmers Welfare, Government of India, India. e-mail address: ruchihpau1987@gmail.com

19.1 Introduction

The rapid increase in the population, urbanization, and sprawling is reducing the availability of the agricultural land day by day (Tiwari and Singh, 2017; Kumar et al., 2018) while the world population is growing intensively at such a high rate. To accomplish worldwide food requirements for the increasing population, food production will have to rise rapidly in the upcoming decades. Consequently, to meet the urgency of food demand and due to less availability of agricultural land, cultivation on the marginal lands is the need of the hour. Marginal lands are having huge potential for crop production. As per available literature, in Asian countries, a large portion of land belongs to marginal lands and can be exploited for food production. The marginal land is having very little agricultural importance because of the production of the less worthy crop. These lands comprise of soil having high salinity and less rainfall, an undrained soil system, and improper physical appearance. Plants are the basic unit of life on this earth. These are various environmental parameters such as biotic and abiotic factors which can impact plants life cycle and thus affecting the production of agricultural crop. Bacteria, viruses insects, and fungi can affect plant growth using biotic factors (Singh et al., 2018; Tiwari et al., 2018). Salinity, flooding, drought, extreme temperatures, acidity, heavy metals, and UV radiations are various abiotic factors which affect plant and soil health (Vimal et al., 2018; Singh and Gupta, 2018). This chapter discusses the role of different microbes in order restore marginal lands so that their great potential can be exploited for agriculture use and crop production.

19.2 Why marginal lands

At the global level, salinity is known as one of the main factors that adversely affects crop productivity and soil health (Mittler, 2006; Mahajan and Tuteja, 2005). Hence, there is an extensive need for the development of agriculture-based technologies to reclaim salt-affected lands. The reason for the high amount of salt in some

Microbes in Land Use Change Management. DOI: https://doi.org/10.1016/B978-0-12-824448-7.00022-X
© 2021 Elsevier Inc. All rights reserved.

soil areas is mainly due to various natural processes. There are many countries worldwide which are having soils affected due to high salt content. Globally, Near to 831 million hectares of land are affected due to high salt content (Martinez-Beltran and Manzur, 2005). Besides the primary reason for the salinity of natural processes, human activities and increasing levels of groundwater are also some of the factors responsible for the high salinity (Metternicht and Zink, 2003). It was found that at the global level, in arid and semiarid areas, nearly 4000 ha of irrigated land is affected by high salinity every day and hence can be used as agriculture suitable land (Shabala and Cuin, 2008). High salinity is the main concern for the environment and sustainable agriculture because it affects crop productions and also adversely affects the soil system (Giri et al., 2003; Al-Karaki, 2006). High salinity mainly affects soil physical and chemical properties and soil microflora, hence badly affecting crop productivity. Plant growth is not properly, whenever they are under influence of any abiotic and biotic adversely affecting factors (Swarbrick et al., 2006; Massad et al., 2012), and thus at last negatively affecting crop production (Shao et al., 2008). In response to stress, plant activates its basic defense system which involves a series of complex signaling cascades (Chinnusamy et al., 2004; Abou Qamar et al., 2009; Andreasson and Ellis, 2010).

19.3 Types of marginal lands and their managements

Globally, Drought is known as one of the major factors affecting agriculture. As the rate of increase of heat events and rainfall reduction because of global warming, in temperate areas crop production is in danger, plant growth is also reduced due to water scarcity and ultimately affecting seed maturation, leaf wilting, and unbalanced fruit composition. Another major concern is a salinity that is also called as some of the soil is having a high amount of salt and known as saline soil (Szabolcs, 1989). A saline soil contains a high amount of sodium together with negatively charged clay particles. It is a natural process that makes high salt accumulation in soil water and affects crop productivity negatively. It is due to the high salt content for a long period. Another reason is the release of the high salt soluble salts such as magnesium, sodium chloride, carbonates, sulfates, and calcium due to rocks weathering. Sometimes, due to high wind flow and heavy rains, oceanic salt is also deposited which ultimately increases salinity. Some of the other factors responsible for salinity are the production of the annual crop in replace of perennial vegetation, land clearing. Others are the use of saline water for irrigation and the drainage system is also insufficient (Munns and Tester, 2008).

In agriculture, soil salinity is known as one of the major concerns for affecting crop productivity. In a certain environment, a high amount of salts also badly affects living organisms and thus affecting plant productivity in a substantial way. The main reason behind less crop productivity is high amounts of ions in the soil such as calcium, carbonate, chloride, sodium, bicarbonate, sulfate, and magnesium. There is a reduction of essential nutrient activity in the soil due to the high amount of salt ions and which will ultimately reduce accessibility and nutrient uptake of plants. There has been a report of declination of phosphorous content in plants due to high salinity which will lead to P-deficiency in the plants. The major effects of high salinity are imbalance in the nutrient and that affects the uptake of nutrients to the shoots and ultimately causes a deficiency of ions in the plants (Munns, 2002). Some parameters such as exchangeable sodium percentage (ESP), electrical conductivity of the saturated extract (ECE), pH, and sodium adsorption ratio (SAR) are used to classified saline soils (Rengasamy, 2010). Saline soils are classified as saline-sodic soils, having a high amount of exchangeable sodium and soluble salts; sodic soils, containing high levels of exchangeable sodium (SAR.13.0 and ESP.15.0) and saline soils, having high levels of soluble salts (ECE.4.0 dS m21) (Ghasemi et al., 1995).

The main mechanism behind drought and salinity affecting plant growth is by blocking photosynthetic pathway thus increasing photorespiration, which ultimately leads to disturbing cell homeostasis and high amount of reactive oxygen species (ROS) production such as hydroxyl radical, superoxide radical, and hydrogen peroxide (Das and Roychoudhury, 2014) whereas in the case of normal environmental conditions, less amount of ROS is produced in mitochondria, chloroplasts and peroxisomes (Apel and Hirt, 2004). Due to the high amount of ROS production under such stress conditions, plant defense pathway genes, and stress-responsive genes are activated (Pitzschke et al., 2006). Extensive research has been done about plants under stress conditions, specifically on the accumulation of metabolites photosynthesis, and water relations. Due to increasing population rate at an explosive level with urgency to maintain land resources and to meet the future generation needs, there is an urgent need to find out the reasons for the degradation of land and drought and salinity. Also, it is important to find novel methods to deal with these adverse conditions. A biological way can be one of the emerging options to deal with such adverse environmental conditions.

Marginal lands provide essential and suitable land resources for food production because they account for 1.3 billion hectares, that is, about 36% of global agricultural land which support around one-third of the world's population. Although the generalized understanding about the concept and knowledge of marginal land has been broadly applied but still their assessment and management are restricted and diverse. A holistic approach for redeeming marginal land in order to optimize its maximum use, sustain crop productivity, lower environmental stress and enhance ecosystem services should be encouraged as depicted in Fig. 19.1.

19.4 Phytoremediation as a potential approach

Under drought conditions, there will be an imbalance in the uptake of nutrients by plants that can adversely affect plant health (Vassilev et al., 2012; Qi and Zhao, 2013). The agricultural lands are mainly used for food crop cultivation and therefore it is necessary to reclaim poor quality lands for agriculture to meet current food demand. There are several reports of using salinity and drought-tolerant plants to reclaim marginal lands for sustainable agriculture. Under drought conditions,

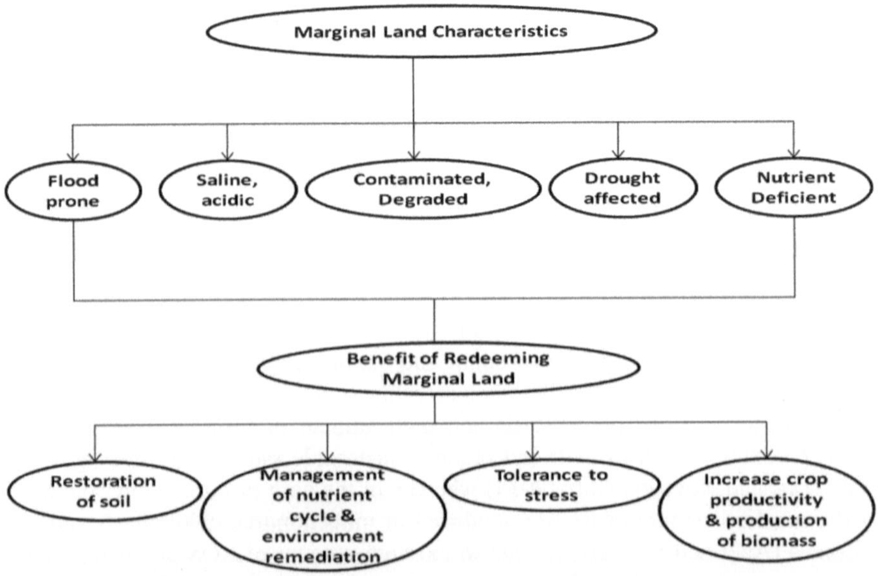

Figure 19.1 Picture depicting different characteristics of marginal land and benefit of redeeming it in order to increase crop productivity.

plant root and leaf system undergo certain modifications. These changes can occur between annual plants (cereals) and perennial trees.

Though in agriculture, drought is one of the major problems, still there is a scarcity of studies related to rhizospheric microorganism's contribution in response to water stress. Recently, there has been extensive research on the cultivation of medicinal and aromatic plants to reclaim marginal lands because of not only its flavor, clinical use, and aroma but it also has been proven good cash crops and ultimately contributes to increasing farmer's income. These plants are grown up for many derived beneficial products from the processing of their crude materials such as cosmetics, insecticides, pharmaceuticals, dietary supplements, sweets, polishes, the aroma industry, and herbal remedies (Lange, 2004). There are many reports on the use of grasses, particularly aromatic grasses to reclaim drought and salinity affected areas, for example, lemongrass (*Cymbopogon flexuous*), vetiver (*Chrysopogon zizanioides*), and palmarosa (*Cymbopogon martini*), (Patra and Singh, 1998; Tomar and Minhas, 2004).

19.5 Arbuscular mycorrhizal fungi to reclaim marginal land

In 1881, Franciszek Kamienski, the Polish botanist found that there is a symbiotic relationship between the fungus and the roots of higher plants. After that in 1885, he called mycorrhizal, Greek word to this mutual process between the fungi and roots, implicated "mycorrhiza (fungus-root)," earlier called as endomycorrhiza, or endotrophic mycorrhizas. The arbuscular mycorrhizal fungi (AMF) association is the most common one (Sjoberg, 2005; Tahat and Sijam, 2012) among all other mycorrhizal associations. The AMF reproduces asexually.

Around 80% of higher plant roots show symbiotic association with AMF. There are various reports in the literature regarding the integral role of AMF for sustainable agriculture and they can even survive under adverse conditions such as drought and saline environments (Giri et al., 2003; Evelin et al., 2009). The AMF can overcome the negative effects of drought and salinity on plant growth by various mechanisms. The AMF can lessen the harmful effects of salinity to plant by colonizing its roots. Under the saline environment, the AMF association can enhance plant growth by increasing nutrients and water (Al-Khaliel, 2010). In abiotic stress conditions, the AMF association promotes the germination of seeds and also improves the physiological characters of the plants under abiotic stress. It also enhances drought and sails uptake tolerance capacity of many crops such as lettuce, maize, and tomato (Jahromi et al., 2008; Ruiz-Lozano and Azcón, 1995). The AMF helps plants to grow under stress conditions by balancing osmotic pressure, removal of ROS, stabilizing protein, and by maintaining membrane enzyme and integrity (Abdel Latef and Chaoxing, 2011). They produce osmoprotectants that protect plant cells under any osmotic stress conditions. The examples of osmoprotectants are soluble sugars, trehalose, proline, quaternary ammonium compounds (alanine betaine, proline, glycine betaine, betaine,), and polyols (Hamdia et al., 2004).

The rhizosphere around mycorrhiza involves unique interaction among plant-microbe and microbe-microbe. Sometimes these interactions around mycorrhizosphere can affect either positively or negatively (Barea et al., 2002; Awasthi et al., 2011). In mycorrhizosphere, these interactions need to be studied in a more extensive way for better exploitation for sustainable agriculture in the environment (Krishnamoorthy et al., 2016). Table 19.1 depicts the role of mycorrhizal fungi in improving growth of various crops in the marginal soils. Moreover, this is a vital step for improving plant growth and enhancing crop productivity under adverse environmental conditions (Johansson et al., 2004). The PGPR's (plant growth promoting rhizobacteria) activities of microbes depend on many factors such as type of soil, host plant, cultivation practices, and strains. Also, to some extent, the specificity depends among plants and their associated microbes in such a way that plant plays an important role in the selection of microorganisms to reclaim poor quality soil for sustainable agriculture.

19.6 Restoration of marginal lands through microbes

Drought has adverse effects on the plants and as well as on their microorganisms. Generally, some microorganisms have a unique capacity to tolerate extreme environmental conditions such as temperature variation, salinity, and osmotic stress. Osmotolerant microbes can grow in low moisture conditions and hence beneficial to plant growth. Plant growth-promoting bacteria can help in the uptake of

Table 19.1 Role of mycorrhizal fungi in improving plants growth in marginal soils.

Type of mycorrhizal inoculum used	Name of mycorrhizal species used	Type of marginal land	Name of plant inoculated with mycorrhizae	Effect on plant growth	References
Vesicular arbuscular mycorrhizae	-	Barren lands	Forest	Food, fiber, forage, fuel wood, fertilizer (root nodule), and timber production	Chamola et al. (1999)
Arbuscular mycorrhizal fungi (AMF)	-	Phosphorous deficient sandy loam soil	*Zea mays*, *Medicago sativa*, *Trifolium alexandrinum*, *Avena sativa*, and *Sorghum vulgare*	Significantly increased yield in terms of shoot and root dry weights and total uptake of P and N of all the test plants	Gaur and Adholeya (2002)
AMF	*Glomus intraradices* and *Scutellospora calospora*	Red loam P-deficient and low levels of root pathogens	Maize and pea	Significantly enhanced uptake of Zn and no increase in P uptake and yield	Ryan and Angus (2003)
Mycorrhizae propagules and *Rhizobium*	-	-	*Arachis hypogeal*	Increased plant height, dry weight, and nodule weight	Puspito et al. (2013)
AMF	-	Organic matter and nutrient-poor, dry, thin of solum soil and hilly topography	Soyabean	Increased biodiversity of microbial population and increased yield	Samanhudi et al. (2017)

AMF	*Glomus manihotis, Glomus etunicatum, Gigaspora margarita*, and *Acaulospora tuberculata*	Acidic soil and low phosphorous level	*Pennisetum purpureum*	Shoot dry weight production, crude protein content, phosphorus uptake (40%), N content, N uptake, and protein production	Karti et al. (2018)
Arbuscular and ectomycorrhizal	*Rhizoglomus irregular* and *Hebeloma longicaudum*	Sandy, rocky, and dry	Willows (*Salix miyabeana* Seeman)	No significant growth in plants but reported viability of mycorrhizal inoculums used	Pray et al. (2018)
AMF	*Glomus* sp.	Low fertility soils	*Z. mays*	Significant increase in plant height, root length, dry weight, crude protein, fat, moisture and ash content	Sarah et al. (2019)

micronutrients and produce phytohormone which activates plant immune system and hence improve soil properties. The PGPR's are having the various number of beneficial effects on plant and as well as on soil. The PGPR's can produce the ACC-deaminase enzyme (Glick et al., 2007) to inhibit ACC (1-aminocyclopropane-1-carboxylase) which is an ethylene precursor, hence lower the ethylene level against stress in the plant (Glick, 2004). The PGPRs also produce siderophore and help in inorganic phosphate compounds solubilization. Besides this, PGPRs also secrete certain exogenous compounds that aid in bacteria in root colonization and adhesion and even contribute to soil stability via various ways such as by increasing root adhering soil and water holding capacity thus, minimizing water loss under desiccation conditions. It also stimulates root exudates and protects roots from any kind of mechanical damage.

Globally, saline soil is one of the major reasons affecting plant growth and crop productivity which ultimately leads to marginal land area production. The adverse effects of salinity comprise of osmotic imbalance, nutrient stress, and destabilization of the membrane which ultimately affects plant growth and crop productivity (Zhu, 2007). It is necessary to control the harmful effects of soil salinity for reclaiming marginal lands and sustainable agriculture. Some halophilic microbes can grow well in extremely high salt concentrations (Qin et al., 2016). One of the major alternatives to overcome this problem is enhancing the growth of rhizobacteria which improves plant growth and crop productivity. Under saline conditions, to have better chances of survival of PGPRs, both good halophiles and PGPR strain having auxin and siderophore producer are necessary (Bharti et al., 2013). The PGPRs can also contribute to plant productivity by protecting plant borne pathogens due to the release of certain compounds such as hydrogen cyanide. The plant growth-promoting rhizobacteria can enter within root cortex cells; hence these endorhizobacteria also contribute to plant growth. Good PGPR's should have certain properties such as the capability of root colonization, can grow around and within root cells, and should have a positive relationship with other beneficial rhizospheric microflora (Kloepper et al., 1991). The PGPR can enhance plant growth by producing essential nutrients such as minerals, phosphorus, and nitrogen to plants and controlling hormone levels of plants. Various reports are available regarding the role of PGPRs in salinity and drought improvement.

19.7 Mechanism of restoration of marginal lands through these microbial processes

The plant can grow under adverse environmental conditions mainly due to interactions among PGPRs and AMF. There are various studies conducted by combining different strains of PGPRS's and AMF to improve crop production. The AMF lessens the effects of stress on plants by enhancing the rate of photosynthesis, uptake of nutrient, water, and organic solutes (Abdel Latef and Chaoxing, 2011).

In a case study, when *Rhizophagus intraradices* was inoculated in a maize plant, it was found that tolerance level of plant against stress has been improved by increasing nutrient uptake and thus lowers the stress level in mycorrhizae treated plants by lowering the accumulation of proline (Estrada et al., 2013). Iron plays an important role in plant growth by participating in metabolic pathways such as photosynthesis, DNA synthesis, and respiration. Many pants can produce phytosiderophores (Fe specific chelating compounds). Furthermore, in some plant, there is a release of organic acids which form complexes with iron ions. In an adverse environmental condition, there have been reports of the significant role of PGRR's to alleviate the effects of stress by producing siderophores and iron siderophores complexes (Kloepper et al., 1980). Also, Siderophores have been known for desorption of certain metals, for example, Pb and Cd (Hepinstall et al., 2005).

When two or more microbes live together, called symbiosis and it can be endosymbiosis or endosymbiosis. In 1872, the consortium model was proposed by Johannes Reinke. Recently there has been a focus on the use of microbial consortium to deal with the drought or salinity issues for better plant growth and crop productivity (Krishnamoorthy et al., 2016). There have been many reports available in the literature showing interactions microbes and rhizosphere under stressed environmental conditions (Qin et al., 2016). There are three different sections (rhizosphere, rhizoplane, and endosphere) for microbe's colonization. To maintain plant growth hormones, both bacteria and AMF releases ACC-deaminase and other hormones under adverse stress environmental conditions. The main function of ACC-deaminase is to lower the ethylene level in plants in stressful environmental conditions. Under stress conditions, there is an imbalance in the ratio of the ions, so to sustain ion homeostasis, bacterial cells produce certain exopolysaccharides that bind with toxic Na ion and restricts its entry to the root system. The PGPR's can produce volatile organic compounds that induce production of high-affinity K transporter (HKT) in shoots and reduction of HKT in roots, and thus restricting Na influx in the roots and ultimately maintaining Na recirculation from shoot-to-root. The AMF fungi enhance the K/Na ratio by specifically increasing uptake of K and Ca and thus restricting the toxic Na in high salinity environmental conditions (Elhindi et al., 2016).

The AMF fungi and PGPR's can control genes expression for plasma membrane integral proteins to aquaporin activity for better uptake of water and under saline environmental conditions. Moreover, all the microorganisms associated with the plant rhizosphere can produce antioxidants to protect plants from ROS scavenging. In some cases, plant microbes can produce osmolytes such as polyamines and proline to protect plants under stressed conditions. The entophytes can produce some plant hormones, micro RNAs, and signaling molecules to activate many important metabolic pathways. They can enhance the efficiency of metabolic pathways and maintain the down regulation and up regulation genes. Under abiotic stress conditions, in some in vitro case studies, endophytes can produce certain bioactive compounds such as trehalose, melanin, and mannitol to reduce the plant stress level.

The intensive use of chemical fertilizers and modification in traditional cultivation patterns has affected the physical, chemical, and biological properties of land, converting fertile land into marginal land. As we have already discussed that the microbes facilitate the rejuvenation of degraded ecosystems by extenuating abiotic and biotic stresses on plants. In recent years, PGPR and AMF have drawn considerable attention towards alleviation of various environmental stresses such as salinity, drought, and poor nutrient soil. PGPR help in promoting plant growth and restoring soil fertility by various mechanisms that involve biological nitrogen fixation, phosphate solubilization, siderophore production, and synthesis of plant hormones such as indole acetic acid, cytokinins, or gibberellins. Besides that, AMF contributes in agricultural assistance by numerous mechanisms including nutrient acquirement and transport to the plant, preventing crop diseases, and increasing soil carbon storage. There are some examples of microbes possessing potential role in restoration of marginal lands and contribution in crop productivity in varying stressful conditions of drought, salinity, extreme temperature, pH and heavy metals as shown in Table 19.2 and Fig. 19.2.

19.8 Amendment of organic manure

In soil, organic matter can enhance plant growth and crop productivity by increasing the water holding capacity of the soil. Also, organic matter can enhance the soil aeration, release, and absorption of nutrients to prevent soil leaching and erosion (Reijntjes et al., 1992; Sekhon and Meelu, 1994). It was found that the amendment of different microbial strains with organic fertilizers can improve the quality of postharvest soil of many gown plants (Singh et al., 2013). In a case study of Hussain et al. (2001), amendments of farmyard manure (FYM) with saline soil enhances many physical and chemical properties of soil such as soil porosity, hydraulic conductivity, and bulk density.

The application of microbial strains *Dietzia natronolimnaea* and *Glomus intraradices* with organic manure (vermicompost) improves the *Ocimum* (*Ocimum basilicum*) plant growing in saline conditions (Bharti et al., 2016). In another similar case study, the application of *R. intraradices* and *Massilia* spp. improves the growth of maize plants (Krishnamoorthy et al., 2016). The AMF plays an important role in the carbon cycle, thus enhancing carbon flow in the soil and ultimately improves poor quality soil (Cardoso and Kuyper, 2006). The interaction between AMF fungi and plants can affect the carbon storage of soil (Zhu and Miller, 2003); around 70% of organic carbon can be lost from the host plant to the AMF mycorrhizosphere (Jakobsen and Rosendahl, 1990). Furthermore, AMF fungi can produce an iron-containing glycoproteinaceous, glomalin, which is helpful in the aggregation of soil and carbon storage of soil (Driver et al., 2005). The concentration of glomalin is directly linked with the soil aggregate water stability in the soil system.

The soil enzyme plays a vital role in determining the extent of soil stresses. Generally, enzymes are the indicator of soil quality and contribute to soil microbial activity, thus ultimately affect the soil nutrient cycling. There are various studies available in the literature regarding the interaction between soil properties and enzyme activities in the presence of different organic and inorganic amendments.

Table 19.2 Some examples of potential microbes showing functional ecological roles in marginal soils.

Type of microorganism	Name of microorganism	Type of marginal land	Functional role in marginal soil	Technique	References
Fungi	*Aspergillus* sp., *Penicillium* sp., *Acrophialophora* sp., *Alternaria* sp.	Arid soils of western Rajasthan (India)	Release of bioavailable phosphorous from organic phosphorous compounds	Culturing	Tarafdar et al. (1988)
	Arbuscular mycorrhizal fungi (AMF)	Sonoran Desert (United States)	As colonists of both roots and soils, they contribute to the creation of "resource-Island" soils by stabilizing windborne soil under plant canopies and enhancing plant colonization	Spore count/ culturing/ propagule density/ mycelia formation	Carrillo-Garcia et al. (2002)
	AMF	Semiarid Mediterranean steppes (Spain)	Soil structuring and aggregation	Soil hyphal length and glomalin content	Rillig et al. (2003)
	Acaulospora sp., *Archaeospora* sp., *Entrophospora* sp., *Glomus* sp., *Paraglomus* sp.	Junggar Basin (northwest China)	AMF associated with desert ephemerals that probably play an important role in their development and maintenance	Culturing	Shi et al. (2007)
	N.D.	Negev (Israel)	Biodegradation of recalcitrant organic matter (i.e., plant polymer)	Substrate utilization patterns	Oren and Steinberger (2008)
	Phylotypes from the Ascomycota (mainly *Alternaria* sp. and *Acremonium* sp.) and Basidiomycota divisions	BSCs of the Colorado Plateau (United States)	The dark-pigmented *Alternaria* sp. may be involved in UV resistance of the BSC	DGGE	Bates and Garcia-Pichel (2009)

(*Continued*)

Table 19.2 (Continued)

Type of microorganism	Name of microorganism	Type of marginal land	Functional role in marginal soil	Technique	References
Bacteria	*Sinorhizobium meliloti* and *Sinorhizobium Medicae*	Land affected by drought, extremes of temperature and soil pH, soil salinity and heavy metals (Morocco)	Improved root nodules of forage legume alfalfa (*Medicago sativa* L.) under salt and drought soils	16S rRNA gene sequencing	Elboutahiri et al. (2010)
	Bacillus megaterium	Saline land	Regulation of osmotic balance by balancing water potential and stomatal opening by affecting hydraulic conductivity and transpiration rate. Maize showed increased root hydraulic conductivity	16S rRNA gene sequencing	Marulanda et al. (2010)
	Pseudomonas simiae AU	Saline land	Decrease in *Glycine max* root Na^+ accumulation and increase in proline and chlorophyll content	16S rRNA gene sequencing	Vaishnav et al. (2016)
	Planomicrobium chinense strain P1 and *Bacillus cereus* strain P2	Hot semiarid climate, Mardan (Pakistan)	Increased leaf proline content, lipid peroxidation, and antioxidant enzyme activity in wheat under moisture stress	16S rRNA gene sequencing	Khan and Bano (2019)

BSCs of the Colorado Plateau, BASICs of the Colorado Plateau (United States); *DGGE*, denaturing gradient gel electrophoresis; *N.D.*, not defined.

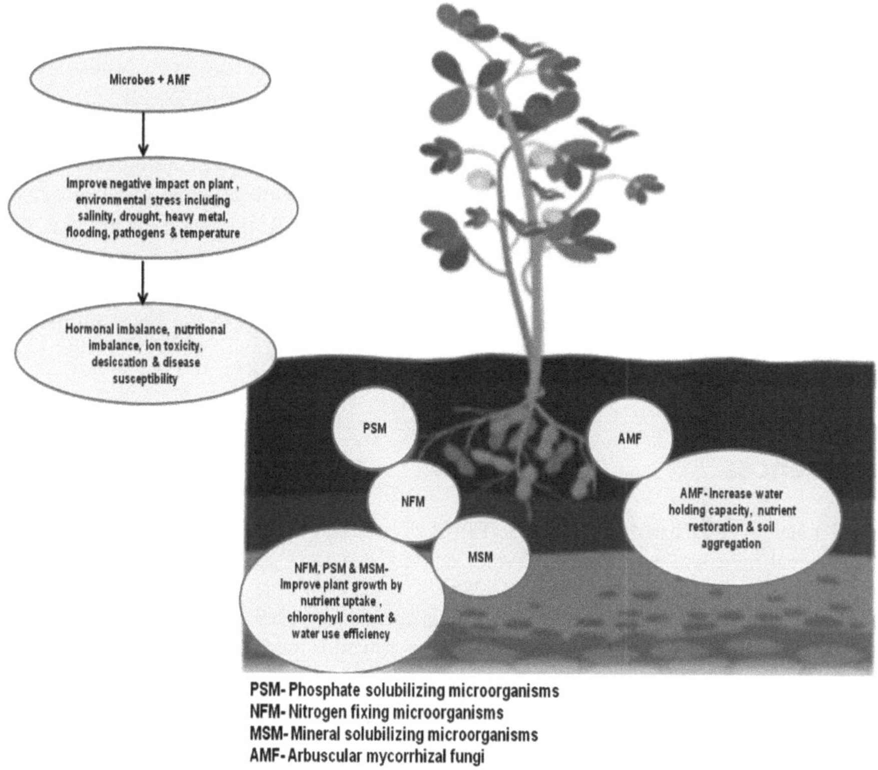

Figure 19.2 Contribution of microbes and arbuscular mycorrhizal fungi in improving plant growth to revegetate marginal lands.

The soil quality depends on different soil enzymes such as urease, catalase, dehydrogenase, protease, and phosphatase (Mariela et al., 2016). The main enzyme (alkaline phosphatase) responsible for the conversion of organic P into inorganic P in the P cycle and male it available for the plants (Dick and Burns, 2011). The enzyme alkaline phosphatase is very crucial for the biological activity of soil as it converts the organic P available in the soil (Zhang et al., 2014). In marginal lands, the application of different PGPR and organic fertilizers such as vermicompost and FYM can improve the soil enzyme activity (Singh, 2016). The PGPR's and AMF fungi are a good source of enzymes for metabolic reactions and thus increase soil enzyme activity even in adverse environmental stressed conditions.

19.9 Conclusion

Microorganisms are generally exploited for reclaiming the soils from various agro-environmental stress conditions as beneficial microbes can improve the availability and uptake of nutrients and ultimately enhances the plant growth under adverse conditions. The AMF fungi and PGPR's can lessen the harsh environmental conditions to pants and thus improves plant growth, germination, and crop productivity. In severe adverse abiotic conditions, the plant growth can be enhanced by a biological model which ultimately improves the enzyme activity and nutrient uptake. The application of organic fertilizers such as vermicompost and FYM along with AMF fungi and PGPR's can be the best combination to deal with adverse environmental stressed conditions such as high salinity and drought soil. Since to reduce the stressed environmental conditions, the biological mode can be used this involves the use of drought and salt-tolerant genotypes and better-integrated nutrient systems along with the addition of organic fertilizers, ultimately leads to the best results and increases crop yield and productivity.

References

Abdel Latef, A.A., Chaoxing, H., 2011. Effect of arbuscular mycorrhizal fungi on growth, mineral nutrition, antioxidant enzymes activity and fruit yield of tomato grown under salinity stress. Sci. Hortic. 127, 228–233.

Abou Qamar, S., Luo, H., Laluk, K., Mickelbart, V.M., Mengiste, T., 2009. Crosstalk between biotic and abiotic stress responses in tomato is mediated by AIM1 transcription factor. Plant J. 58, 1–13.

Al-Karaki, G.N., 2006. Nursery inoculation of tomato with arbuscular mycorrhizal fungi and subsequent performance under irrigation with saline water. Sci. Hortic. 109, 1–7.

Al-Khaliel, A.S., 2010. Effect of salinity stress on mycorrhizal association and growth response of peanut infected by *Glomus mosseae*. Plant Soil Environ. 56 (7), 318–324.

Andreasson, E., Ellis, B., 2010. Convergence and specificity in the *Arabidopsis* MAPK nexus. Trends Plant Sci. 15, 106–113.

Apel, K., Hirt, H., 2004. Reactive oxygen species: metabolism, oxidative stress, and signal transduction. Annu. Rev. Plant Biol. 55, 373–399.

Awasthi, A., Bharti, N., Nair, P., Singh, R., Shukla, A.K., Gupta, M.M., et al., 2011. Synergistic effect of *Glomus mosseae* and nitrogen fixing *Bacillus subtilis* strain Daz26 on artemisinin content in *Artemisia annua* L. Appl. Soil Ecol. 49, 125–130.

Barea, J.M., Pozo, M.J., Azcón, R., Azcón-Aguilar, C., 2002. Microbial cooperation in the rhizosphere. J. Exp. Bot. 56 (417), 1761–1778.

Bates, S.T., Garcia-Pichel, F., 2009. A culture-independent study of free-living fungi in biological soil crusts of the Colorado Plateau: their diversity and relative contribution to microbial biomass. Environ. Microbiol. 11, 56–67.

Bharti, N., Yadav, D., Barnawal, D., Maji, D., Kalra, A., 2013. *Exiguobacterium oxidotolerans*, a halotolerant plant growth promoting rhizobacteria, improves yield and content of secondary metabolites in *Bacopa monnieri* (L.) Pennell under primary and secondary salt stress. World J. Microbiol. Biotechnol. 29 (2), 379–387.

Bharti, N., Pandey, S.S., Barnawal, D., Patel, V.K., Kalra, A., 2016. Plant growth promoting rhizobacteria *Dietzia natronolimnaea* modulates the expression of stress responsive genes providing protection of wheat from salinity stress. Sci. Rep. 6, 347–368.

Cardoso, I.M., Kuyper, T.W., 2006. Mycorrhizas and tropical soil fertility. Agric. Ecosyst. Environ. 116, 72–84.

Carrillo-Garcia, Á., La Luz, D., León, J.L., Bashan, Y., Bethlenfalvay, G.J., 2002. Nurse plants, mycorrhizae, and plant establishment in a disturbed area of the Sonoran Desert. Restor. Ecol. 7, 321.

Chamola, B.P., Giri, B., Mukerji, K.G., 1999. Vesicular arbuscular mycorrhizal fungi: biofertilizer for the future. Ethnomycol Fungal Biotechnol. 225–234.

Chinnusamy, V., Schumaker, K., Zhu, J.K., 2004. Molecular genetics perspectives on cross-talk and specificity in abiotic stress signalling in plants. J. Exp. Bot. 55, 225–236.

Das, K., Roychoudhury, A., 2014. Reactive oxygen species (ROS) and response of antioxidants as ROS-scavengers during environmental stress in plants. Front. Environ. Sci. Available from. Available from: https://doi.org/10.3389/fenvs.2014.00053.

Dick, R.P., Burns, R.G., 2011. A Brief History of Soil Enzymology Research. Soil Science Society of America, Madison, WI, pp. 1–34.

Driver, J.D., Holben, W.E., Rillig, M.C., 2005. Characterization of glomalin as a hyphal wall component of arbuscular mycorrhizal fungi. Soil Biol. Biochem. 37, 101–106.

Elboutahiri, N., Thami-Alami, I., Udupa, S., 2010. Phenotypic and genetic diversity in *Sinorhizobium meliloti* and *Sinorhizobium medicae* from drought and salt affected regions of Morocco. BMC Microbiol. 10. Available from: https://doi.org/10.1186/1471-2180-10-15.

Elhindi, K.M., El-din, A.S., Elgorban, A.M., 2016. The impact of arbuscular mycorrhizal fungi in mitigating salt-induced adverse effects in sweet basil (*Ocimum basilicum* L.). Saudi J. Biol. Sci. 24 (1), 170–179.

Estrada, B., Aroca, R., Maathuis, F.J., Barea, J.M., Ruiz-Lozano, J.M., 2013. Arbuscular mycorrhizal fungi native from a Mediterranean saline area enhance maize tolerance to salinity through improved ion homeostasis. Plant Cell Environ. 36, 1771–1782.

Evelin, H., Kapoor, R., Giri, B., 2009. Arbuscular mycorrhizal fungi in alleviation of salt stress: a review. Ann. Bot. 104, 1263–1280.

Gaur, A., Adholeya, A., 2002. Arbuscular-mycorrhizal inoculation of five tropical fodder crops and inoculum production in marginal soil amended with organic matter. Biol. Fertil. Soils 35, 214–218.

Ghasemi, F., Jakeman, A.J., Nix, H.A., 1995. Salinisation of Land and Water Resources: Human Causes, Extent, Management, and Case Studies. NSW University Press, Sydney.

Giri, B., Kapoor, R., Mukerji, K.G., 2003. Influence of arbuscular mycorrhizal fungi and salinity on growth, biomass and mineral nutrition of *Acacia auriculiformis*. Biol. Fertil. Soils 38, 170–175.

Glick, B.R., 2004. Bacterial ACC-deaminase and the alleviation of plant stress. Adv. Appl. Microbiol. 56, 291–312.

Glick, B.R., Cheng, Z., Czarny, J., Duan, J., 2007. Promotion of plant growth by ACC deaminase-producing soil bacteria. Eur. J. Plant Pathol. 119, 329–339.

Hamdia, M.A.E.S., Shaddad, M.A.K., Doaa, M.M., 2004. Mechanisms of salt tolerance and interactive effects of *Azospirillum brasilense* inoculation on maize cultivars grown under salt stress conditions. Plant Growth Regul. 44 (2), 165–174.

Hepinstall, S.E., Turner, B.F., Maurice, P.A., 2005. Effects of siderophores on Pb and Cd adsorption to kaolinite. Clays Clay Miner. 53, 557–563.

Hussain, N., Hassan, G., Arshadullah, M., Mujeeb, F., 2001. Evaluation of amendments for the improvement of physical properties of sodic soil. Int. J. Agric. Biol. 3, 319–322.

Jahromi, F., Aroca, R., Porcel, R., Ruiz-Lozano, J.M., 2008. Influence of salinity on the in vitro development of *Glomus intraradices* and on the in vivo physiological and molecular responses of mycorrhizal lettuce plants. Microbiol. Ecol. 55, 45–53.

Jakobsen, I., Rosendahl, L., 1990. Carbon flow into soil and external hyphae from roots of mycorrhizal cucumber plants. New Phytol. 115, 77–83.

Johansson, J.F., Paul, L.R., Finlay, R.D., 2004. Microbial interactions in the mycorrhizosphere and their significance for sustainable agriculture. FEMS Microbiol. Ecol. 48, 1–13.

Karti, P.D.M.H., Prihantoro, I., Setiana, M.A., 2018. Evaluation of arbuscular mycorrhizal fungi inoculum on production and nutrient content of *Pennisetum Purpureum*. Trop. Anim. Sci. J. 41 (2), 114–120.

Khan, N., Bano, A., 2019. Exopolysaccharide producing rhizobacteria and their impact on growth and drought tolerance of wheat grown under rainfed conditions. PLoS One 14 (9), e0222302.

Kloepper, J.W., Leong, J., Teintze, M., Schroth, M.N., 1980. Enhancing plant growth by siderophores produced by plant growth-promoting rhizobacteria. Nature 286, 885–886.

Kloepper, J.W., Rodrguez-kbana, R., Mcinroy, J.A., Collins, D.J., 1991. Analysis of populations and physiological characterization of microorganisms in rhizospheres of plants with antagonistic properties to phytopathogenic nematodes. Plant Soil 136, 95–102.

Krishnamoorthy, R., Kim, K., Subramanian, P., Senthilkumar, M., Anandham, R., Sa, T., 2016. Arbuscular mycorrhizal fungi and associated bacteria isolated from salt-affected soil enhances the tolerance of maize to salinity in coastal reclamation soil. Agric. Ecosyst. Environ. 231, 233–239.

Kumar, A., Kaushal, S., Saraf, S.A., Singh, J.S., 2018. Microbial bio-fuels: a solution to carbon emissions and energy crisis. Front. Biosci. (Landmark ED.) 1 (23), 1789–1802.

Lange, D., 2004. Medicinal and aromatic plants: trade, production, and management of botanical resources. Acta Hortic. 629, 177–197.

Mahajan, S., Tuteja, N., 2005. Cold, salinity and drought stresses: an overview. Arch. Biochem. Biophys. 444 (2), 139–158.

Mariela, F.P., IvańPavel, M.E., Luis Manuel, S.R., MaríaJesús, F.G., Reyes, L.O., 2016. Dehydrogenase and mycorrhizal colonization: tools for monitoring agrosystem soil quality. Appl. Soil Ecol. 100, 144–153.

Martinez-Beltran, J., Manzur, C.L., 2005. Overview of salinity problems in the world and FAO strategies to address the problem. Paper Presented at the International Salinity Forum, Riverside.

Marulanda, A., Azcon, R., Chaumont, F., Ruiz-Lozano, J.M., Aroca, R., 2010. Regulation of plasma membrane aquaporins by inoculation with a *Bacillus megaterium* strain in maize (*Zea mays* L.) plants under unstressed and salt-stressed conditions. Planta 232, 533–543.

Massad, T.J., Dyer, L.A., Vega, C.G., 2012. Cost of defense and a test of the carbon-nutrient valance and growth-differentation balance hypotheses for two co-occurring classes of plant defense. PLoS One 7, e7554.

Metternicht, G., Zink, J., 2003. Remote sensing of soil salinity: potentials and constraints. Remote Sens. Environ. 85, 1–20.

Mittler, R., 2006. Abiotic stress, the field environment and stress combination. Trends Plant Sci. 11, 15–19.

Munns, R., 2002. Comparative physiology of salt and water stress. Plant Cell Environ. 25 (2), 239–250.

Munns, R., Tester, M., 2008. Mechanisms of salinity tolerance. Annu. Rev. Plant Biol. 59, 651–681.

Oren, A., Steinberger, Y., 2008. Catabolic profiles of soil fungal communities along a geographic climatic gradient in Israel. Soil Biol. Biochem. 40, 2578–2587.

Patra, P., Singh, D.V., 1998. Medicinal and aromatic crops. In: Tyagi, N.K., Minhas, P.S. (Eds.), Agricultural Salinity Management in India. Central Soil Salinity Research Institute, Karnal, pp. 499–506.

Pitzschke, A., Fornazi, C., Hirt, H., 2006. Reactive oxygen species signalling in plants. Antioxid. Redox Signal. 8, 1757–1764.

Pray, T.J., Nissim, W.G., St-Arnaud, M., Michel, L., 2018. Investigating the effect of a mixed mycorrhizal inoculum on the productivity of biomass plantation willows grown on marginal farm land. Forests 9, 185.

Puspito, N.A., Raditia, N., Nova, P., 2013. Introduction of mycorrhizae propagules and *Rhizobium* as growth promoter legumes on marginal lands. In: The 3rd International Conference on Biological Science 2013 (The 3rd ICBS-2013). Available from: https://doi.org/10.18502/kls.v2i1.246.

Qi, W.Z., Zhao, L., 2013. Study of the siderophore-producing *Trichoderma asperellum* Q1 on cucumber growth promotion under salt stress. J. Basic Microbiol. 53, 355–364.

Qin, Y., Druzhinina, I.S., Pan, X., Yuan, Z., 2016. Microbially mediated plant salt tolerance and microbiome-based solutions for saline agriculture. Biotechnol. Adv. 34, 1245–1259.

Reijntjes, C., Haverkort, B., Waters-Bayer, A., 1992. Farming for the Future: An Introduction to Low-External-Input and Sustainable Agriculture. Macmillan Press L, London.

Rengasamy, P., 2010. Soil processes affecting crop production in salt affected soils. Funct. Plant Biol. 37, 613–620.

Rillig, M.C., Maestre, F.T., Lamit, L.J., 2003. Microsite differences in fungal hyphal length, glomalin, and soil aggregate stability in semiarid Mediterranean steppes. Soil Biol. Biochem. 35, 1257–1260.

Ruiz-Lozano, J.M., Azcón, R., 1995. Hyphal contribution to water uptake in mycorrhizal plants as affected by the fungal species and water status. Physiol. Plant 95, 472–478.

Ryan, M.H., Angus, J.F., 2003. Arbuscular mycorrhizae in wheat and field pea crops on a low P soil: increased Zn-uptake but no increase in P-uptake or yield. Plant Soil 250, 225–239.

Samanhudi, Pujiasmanto, B., Yunus, A., Supyani, Suntoro, W.H., Prabowo, S.M., 2017. The effect of manure and mycorrhiza application to the soil microbes biodiversity in terms of increasing soybean yield in marginal land in Indonesia. Bulg. J. Agric. Sci. 23 (6), 994–1003.

Sarah, S., Muhammad, S., Ikram, M., Ikramullah, K., Farooq, J., Fida, H., et al., 2019. Symbiotic response of three tropical maize varieties to eco-friendly arbuscular mycorrhizal fungal inoculation in marginal soil. Biocell (43), 245–252.

Sekhon, G.S., Meelu, O.P., 1994. Organic matter management in relation to crop production in stressed rainfed systems. In: Virmani, S.M., Katyal, J.C., Eswaran, H., Abrol, I.P. (Eds.), Stressed Ecosystems and Sustainable Agriculture. Oxford University Press and IBH Publishing, New Delhi.

Shabala, S., Cuin, T.A., 2008. Potassium transport and plant salt tolerance. Physiol. Plant 133, 651–669.

Shao, H.B., Chu, L.Y., Shao, M.A., Abdul Jaleel, C., Hong-Mei, M., 2008. Higher plant antioxidants and redox signalling under environmental stresses. C. R. Biol. 331, 433–441.

Shi, Z., Zhang, L., Li, X., Feng, G., Tian, C.Y., Christie, P., 2007. Diversity of arbuscular mycorrhizal fungi associated with desert ephemerals in plant communities of Junggar Basin, northwest China. Appl. Soil Ecol. 35, 10–20.

Singh, K., 2016. Microbial and enzyme activities of saline and sodic soils. Land Degrad. Dev. 27 (3), 706–718.

Singh, J.S., Gupta, V.K., 2018. Soil microbial biomass: a key soil driver in management of ecosystem functioning. Sci. Total Environ. 634, 497–500.

Singh, R., Soni, S.K., Patel, R.P., Kalra, A., 2013. Technology for improving essential oil yield of *Ocimum basilicum* L. (sweet basil) by application of bioinoculant colonized seeds under organic field conditions. Ind. Crop Prod. 45, 335–342.

Singh, C., Tiwari, S., Gupta, V.K., Singh, J.S., 2018. The effect of rice husk biochar on soil nutrient status, microbial biomass and paddy productivity of nutrient poor agriculture soils. Catena 171, 485–493.

Sjoberg, J., 2005. Arbuscular mycorrhizal fungi-occurrence in Sweden and interaction with a pathogenic fungus in barley (Ph.D. Thesis). Swedish University of Agricultural Sciences, Uppsala, pp. 1–5.

Swarbrick, P.J., Schulze-Lefert, P., Scholes, J.D., 2006. Metabolic consequences of susceptibility and resistance in barley leaves challenged with powdery mildew. Plant Cell Environ. 29, 1061–1076.

Szabolcs, I., 1989. Salt-Affected Soils. CRC Press Inc, Boca Raton, FL.

Tahat, M.M., Sijam, K., 2012. Arbuscular mycorrhizal fungi and plant root exudates biocommunications in the rhizosphere. Afr. J. Microbiol. Res. 6, 7295–7301.

Tarafdar, J., Rao, A., Bala, K., 1988. Production of phosphatates by fungi isolated from desert soils. Folia Microbiol. (Praha) 33, 453–457.

Tiwari, P., Singh, J.S., 2017. A plant growth promoting rhizospheric *Pseudomonas aeruginosa* strain inhibits seed germination in *Triticum aestivum* (L) and *Zea mays* (L). Microbiol. Res. 8, 1–7.

Tiwari, S., Singh, C., Singh, J.S., 2018. Land use changes: a key ecological driver regulating methanotrophs abundance in upland soils. Energy Ecol. Environ. 3, 355–371.

Tomar, O.S., Minhas, P.S., 2004. Relative performance of some aromatic grasses under saline irrigation. Indian J. Agron. 49 (3), 207–208.

Vaishnav, A., Kumari, S., Jain, S., Tuteja, N., Choudhary, D., 2016. PGPR-mediated expression of salt tolerance gene in soybean through volatiles under sodium nitroprusside. J. Basic Microbiol. 56. Available from: https://doi.org/10.1002/jobm.201600188.

Vassilev, N., Eichler-Lobermann, B., Vassileva, M., 2012. Stress-tolerant P-solubilizing microorganisms. Appl. Microbiol. Biotechnol. 95, 851–859.

Vimal, S.R., Patel, V.K., Singh, J.S., 2018. Plant growth promoting *Curtobacterium albidum* strain SRV4: an agriculturally important microbe to alleviate salinity stress in paddy plants. Ecol. Indic. in press.

Zhang, T.B., Kang, Y.H., Liu, S.H., Liu, S.P., 2014. Alkaline phosphatase activity and its relationship to soil properties in a saline-sodic soil reclaimed by cropping wolfberry (*Lycium barbarum* L.) with drip irrigation. Paddy Water Environ. 12, 309–317.

Zhu, J.K., 2007. Plant Salt Stress. Encyclopedia of Life Sciences. John Wiley and Sons Ltd, pp. 1–3.

Zhu, Y.G., Miller, R.M., 2003. Carbon cycling by arbuscular mycorrhizal fungi in soil-plant systems. Trends Plant Sci. 8, 407–409.

Impact of microbial biofilm on crop productivity and agricultural sustainability

20

Muzamil Ahmad Rather, Kuldeep Gupta and Manabendra Mandal
Department of Molecular Biology and Biotechnology, Tezpur University, Tezpur, India

20.1 Introduction

The term "biofilm" is an assemblage of well-organized, structured, self-aggregating communities of microbes that adhere to surfaces and enclose within self-produced extracellular matrix (Costerton et al., 1978; Teschler et al., 2015; Di Martino, 2018; Decho and Gutierrez, 2017). The formation of biofilm is a complex and sequential multistep process involving reversible attachment of planktonic cells, irreversible adhesion to the surfaces, formation of extracellular matrix, maturation and dispersion as per the five-step model of biofilm formation (Sauer, 2003). Water, microbial cells, Exopolymeric Substances (EPSs) (polysaccharides, proteins, lipids, surfactants, eDNA) are the major biofilm constituents (De Natale et al., 2020; Gannesen et al., 2019; Karygianni et al., 2020; Dang and Lovell, 2016). The formation of biofilm is regarded as a result of the evolutionary success of microbes against environmental stresses like pH, temperature, nutrient deficiency, free radicals, detergents, and antibiotics (Muhammad et al., 2020; Costa et al., 2018). The complex structure of biofilm attributes some advantages to the embedded microbes within. It provides protection from the harsh environment, maintains availability of nutrients, acquisition of new genetic traits, reproductive fitness, and restriction to the penetration of antibiotics, detergents, and other chemicals (Rafique et al., 2015; de Carvalho, 2018). The differential gene expression and regulation in planktonic and biofilm microbes impart differences in terms of growth rate, metabolic activity, antibiotic resistance, rate of genetic material transfer, and other physiological processes (Bisht and Wakeman, 2019; Sharma et al., 2019). The antibiotic resistance and transformational frequencies are respectively about 1000 and 10–600 times greater in biofilms compared to their planktonic form (Donlan, 2002; Penesyan et al., 2019). Low metabolic activity and growth rate, horizontal gene transfer, persister cells, efflux pumps, and antibiotic modifying enzymes provide multifold resistance against antibiotics, pollutants, and other environmental conditions to biofilm forming microbes (Velmourougane et al., 2017). The attributes that regulate the formation and development of biofilm include environmental factors like temperature, pH, osmolarity, etc. (Toyofuku et al., 2016), nutrient availability, antimicrobials like antibiotics, biocides, and other chemicals (Penesyan et al., 2020), enzymes like lyase, chitinase,

Microbes in Land Use Change Management. DOI: https://doi.org/10.1016/B978-0-12-824448-7.00023-1
© 2021 Elsevier Inc. All rights reserved.

dispersin, nuclease (Tielen et al., 2010) and regulatory mechanisms like modulation of cyclic di-GMP (involved in expression of genes for motility, protein synthesis and adhesion) (Ahmad et al., 2020), cell-to-cell communication [quorum sensing (QS)], and mechanical and host-derived signals (cell-surface interactions) (Wang et al., 2017a). Thus, in general, the combination of bacterial cell surface components and extracellular components like EPS, motility factors, and lipopolysaccharides, along with environmental and cell-to-cell signaling are important for formation, development, and maintenance of biofilm (Liu et al., 2020; Rabin et al., 2015).

Although a vast literature is available on virulence, invasiveness, and pathogenicity of microbial biofilms in the health sector (Dewasthale et al., 2018) and food industry like dairy, aquaculture, poultry, meat, and ready-to-eat, yet limited knowledge is available regarding agriculturally important biofilms (Galie et al., 2018; Rana et al., 2020; Rather et al., 2021). It is well documented that pathogenic microbial biofilms are responsible for a rise in recurrent infections (Jamal et al., 2018), loss of human life (Landry et al., 2020), food spoilage (Speranza and Corbo, 2017), frequent outbreaks of food-borne diseases (Schirone et al., 2019), and burden on the economy (Muhammad et al., 2020). Now-a-days researchers are focusing on using microbial biofilms in agriculture for enhancing overall crop productivity, improving soil quality and fertility, providing plant protection, increasing availability of plant nutrients, stress tolerance, bioremediation, etc. (Velmourougane et al., 2017). As per the reports of Food and Agriculture Organization, the expected population of the world will be about 8.5×10^9 in 2025, requiring an enormous increase of agriculture productivity of about 2.4×10^9 tons per year (Timmusk et al., 2017). Keeping in view the existing adverse environmental conditions, increased stress levels, global climate change, decline in agricultural land (conversion of agricultural land to nonagricultural sectors) and frequent attacks of diseases and pests, the use of microbial biofilms for overall enhancement of crop productivity is gaining interest and popularity as well (Enebe and Babalola, 2018). Application of microbial biofilm in agriculture is a key strategy to develop positive interaction between agriculturally important biofilms and important crops for sustainable and eco-friendly agriculture in order to overcome the declining crop production to meet the increasing food requirements of extensively growing population (Lens, 2011). Fig. 20.1 is the representation of biofilm formation and its role in sustainable agriculture.

20.2 Plant microbe interactions and plant growth promoting rhizobacteria

It has been demonstrated that the transition of planktonic bacteria to biofilm formation is important in order to develop a beneficial relationship between microbes and plants which could be achieved through microbial interaction in or near the plant microenvironment (phylosphere or rhizosphere) (Dong et al., 2019) and

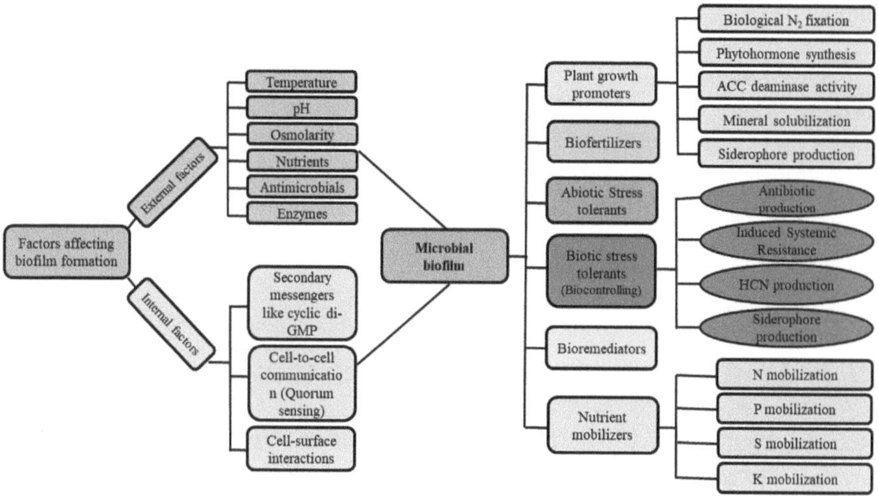

Figure 20.1 Overview of the factors affecting biofilm formation and its role in agriculture.

communication between plant and microbial surfaces (leaf, root, vessels) (Bogino et al., 2013). Rhizosphere, a soil compartment that is affected by root secretions and associated root microbial biome, is the region where numerous positive (beneficial) and negative (harmful) interactions have been reported between roots and microbes (Nadeem et al., 2013). It has been demonstrated that nonpathogenic plant growth promoting rhizobacteria (PGPR) are believed to enhance the crop production by contributing toward soil fertility, nutrient enrichment, boost in mineral uptake, synthesis of plant hormones, antibiotics mediated suppression of pathogens, and secondary-metabolite mediated resistance (Rudrappa et al., 2008; Bhattacharyya and Jha, 2012). Various physiochemical parameters within and throughout the surface of the root, nutrient availability, pH gradient, humidity, temperature, and most predominantly the nature and composition of root exudates are important for initiation of root colonization and development of both homogenous as well as heterogeneous biofilm associations (Ryu et al., 2004; Lorite et al., 2013). At the molecular level the microbial cell density dependent cell-to-cell-communication referred to as QS, employing autoinducers (AIs) along with diffusion and efficiency sensing, plays a significant role in micro-colony formation and subsequent colonization within plant roots (Hense and Schuster, 2015). It has been reported that specific root exudates within rhizosphere resemble AIs and serve as plant QS mimics, thus, amplify the process of microbial colonization (Olanrewaju et al., 2019; Walker et al., 2003). A study has been reported on the presence of reporter genes tagged to *LuxRI, AhyRI*, and *LasRI* QS promoters in root extract of pea that mimic the acyl homoserine lactons (AHLs) of gram-negative bacteria (Teplitski et al., 2000). The plants also influence microbial association by carrying out the structural changes of the plant-root-associated biofilm by varying the nutrient composition and concentration within the rhizosphere (Lareen et al., 2016). Thus, the participating microbes as well as plants regulate and influence the microbial colonization by secreting AHLs and AHL mimics, respectively within the rhizosphere (Tkacz et al., 2020). Furthermore, plant-root-associated microbial biofilms in PGPRs play a significant role in eradicating pathogenic microbes by producing antimicrobial agents against competing pathogenic microbes (Olanrewaju et al., 2017). A study described the role of root-surface biofilm and surfactants in biocontrolling potential of *Bacillus subtilis* against *Pseudomonas syringae* pv. tomato DC3000 (Bais et al., 2004).

20.3 Importance of microbial biofilms in agriculture

20.3.1 Biofilmed biofertilizers-novel alternative of chemical fertilizers

The excessive use of chemical fertilizers in agriculture to meet the increasing food demands led to the deterioration of quality and fertility of soil besides posing a high risk of health issues, groundwater contamination, eutrophication, and soil erosion (Rahman and Zhang, 2018; Patra et al., 2016). Keeping in view the growing demand for agricultural products and sustainability of biosphere as well, an alternative of chemical fertilizers in the form biofertilizers is gaining the attention of scientists (Gortari et al., 2019; Mazid and Khan, 2015; Reddy et al., 2020). A preparation consisting of living or latent microbes that enhances overall crop production by interacting with plants within the rhizosphere is known as biofertilizer (Ramasamy et al., 2020). Biofertilizers are monocultures or mixed cultures of agriculturally significant microbes *viz.* N_2-fixing bacteria, cellulolytic microbes, phosphorous solubilizers, mycorrhizal fungi, *Azolla*, and algae (Mahdi et al., 2010). It has been reported that the use of biofertilizer microbes as surface-attached biofilm, called biofilmed biofertilizers (BFBFs), show significant results in terms of crop productivity over using the microbes in planktonic mode (Marapatla, 2014). Furthermore, the introduction of BFBFs in modern agricultural practices reduced the use of chemical fertilizers by 50% in many crops like vegetables, cereals, pulses, rubber, tea, etc. which has not been obtained by nonbiofilm forming biofertilizers (Ghiasian, 2020). Some of the important BFBFs used in crop productivity and that maintain the agroecosystem have been described in Table 20.1.

The use of BFBFs restores the microbiota of the soil by breaking or shortening the dormancy period of the sexual and asexual spores which were depleted by using chemical fertilizers alone as the only conventional agricultural practice (Buddhika et al., 2013). The role of BFBFs is enormously large as they play a vital role in mobilization, solubilization, and availability of both organic and inorganic nutrients, enhance biological nitrogen fixation (Wang et al., 2017b), reduce nitrate availability, protect the plants from pathogens, promote seed germination, and increase the production of plant hormones to enhance plant growth (Ghiasian, 2020; Zakeel and

Table 20.1 Some of the biofilmed biofertilizers used in crop productivity.

S. no.	Crop	Biofilmed biofertilizers	Role	Reference
1	*Triticum* sp.	Cyanobacterium-based biofilms developed with *Azotobacter, Mesorhizobium, Serratia*, and *Pseudomonas*	Improved viability and N fixing potential	Swarnalakshmi et al. (2013)
2.	*Fragaria ananassa*	*Aspergillus* sp. and *Enterobactor* sp.	Enhanced crop production and economic benefits	Singhalage et al. (2019)
3	*Cicer arietinum* L.	*Anabaena laxa*	Improved plant growth, nutrient uptake, nitrogen fixation	Bidyarani et al. (2016)
4	*Triticum* sp.	*Providencia* sp.	Biofortification and improve quality of grains	Rana et al. (2012)
5	Tea	*Arachis pintoi, Gliricidia, Bacillus megaterium*	Improvements of soil exchangeable K, carbon, phosphorus, microbial biomass carbon, cation exchange capacity	De Silva et al. (2014)
6	Soya bean	*Bradyrhizobium japonicum* in combination with 33% of chemical fertilizers	Increased growth performance	Janagard et al. (2013)
7	Leguminous crops	*Rhizobium* with phosphobacteria	Enhances crop improvement by fixing N_2 and improves availability of P	Selvakumar et al. (2012)

Safeena, 2019; Galelli et al., 2015; Kanchan et al., 2019). Thus, BFBFs may be regarded as biofilmed microbial ameliorators since the role of BFBFs in agriculture is more than biofertilization as it restores the sustainability of the agroecosystem by enhancing revival of microbial diversity (Singh and Gupta, 2016), ecosystem functioning, energy flow, improving soil quality, plant growth, and reducing the use of agroenvironment deteriorating chemical fertilizers (Ramasamy et al., 2020; Okur, 2018; Sahoo et al., 2013).

20.3.2 Application of microbial biofilms as plant growth promoters

As mentioned earlier that plant growth and health, net productivity and agro-ecosystem fertility are determined mainly by interactions between plants and microbes within the rhizosphere (Yadav et al., 2015; Singh et al., 2017; Mhlongo et al., 2018; Mohanram and Kumar, 2019). PGPR are responsible for enhancing plant growth promotion (PGP) traits in order to amplify the agronomic production (Shaikh et al., 2016). The effect of microbes on plant growth depends upon the inoculant and shows variations among different species and strains as well (Souza et al., 2015). Direct and indirect mechanisms have been described through which microbes can be beneficial for plant growth (Lugtenberg and Kamilova, 2009). The direct mechanism includes: (1) rhizoremediation; (2) root growth stimulation; (3) plant stress control; and (4) biofertilization (Vejan et al., 2016) while as: (1) reduction of diseases; (2) declining food and space competition; (3) systemic resistance; and (4) antibiosis are included under indirect mechanisms of plant growth through biofilmed microbes (Suryadi et al., 2019; Ahemad and Kibret, 2014). The biofilm forming rhizobacteria regulate plant growth through enhanced N_2-fixation (Kuan et al., 2016), synthesis of indolic compounds (Olanrewaju et al., 2017), P and S solubilization (Sarikhani et al., 2020; Fox et al., 2014), production of siderophores (enterobactin, rhizobactin, ferrioxamine B, etc.) (Felestrino et al., 2017; Sharma and Johri, 2003), plant hormones, etc. (Kudoyarova et al., 2019; Egamberdieva et al., 2017; Gupta et al., 2017). Fungal-bacterial biofilms (FBBs) can correlate the production of indoleacetic acid like substances (IAAS) production and pH (Bandara et al., 2006). A prominent increase in IAAS production and a decline in pH in FBBs than nonbiofilm forming mono and mixed cultures was reported. The high concentration of IAAS plays a significant role in plant growth and development while high acidity protects the plants from pathogenic attack. Some important PGP biofilm forming microbes include *Azospirillum, B. subtilis, Azoarcus, Pseudomonas fluorescens, Enterobacter, Anabaena-Azotobacter chroococcum, Azotobacter, Trichoderma-Azotobacter chroococcum, Trichoderma-Bacillus* sp., *Azospirillum brasilense, Klebsiella, Bacillus*, and *Rhizobium* (Ramasamy et al., 2020; Ghiasian, 2020; Kumar and Meena, 2019; Ren et al., 2019; Kumar et al., 2020). *P. fluorescens* FAP2 and *Bacillus licheniformis* as PGPR in biofilm mode can enhance the PGP traits *viz.* P solubilization and synthesis of siderophores, Indole Acetic Acid (IAA), and ammonia (Ansari and Ahmad, 2019). Table 20.2 mentions some biofilm forming rhizobacteria used to synthesize plant hormones for enhanced plant growth. The extensive synthesis of volatile organic compounds (VOCs) by PGPRs in biofilm mode enhances the metabolic activity, physiological processes, and crop productivity of certain crops. Few VOCs *viz.*: (1) 2,3-butanediol and 3-hydroxy-2-butanone increases leaf surface area and systemic resistance in *Arabidopsis thaliana* by *B. subtilis* strain GB03, *Bacillus amyloliquefaciens* strain IN937 a inoculants; (2) indole production induced by *Escherichia coli* in *A. thaliana* enhanced lateral root biomass and number; and (3) β-caryophyllene production in lettuce promoted enhancement in length of root and shoot as well as the weight of seedlings (Ruzzi and Aroca, 2015).

Table 20.2 Biofilm forming rhizobacteria used to synthesize plant hormones for enhanced plant growth.

S. no.	Biofilm forming rhizobacteria	Phytohormone	Function	Reference
1	Pink-pigmented facultative methylotrophic bacteria	Enhance indoleacetic acid (IAA) synthesis	Increase root mass and length, decrease stomatal size and density	Omer et al. (2004)
2	*Rhizobium, Bradyrhizobium, Agrobacterium*	Enhance IAA synthesis	Enhance root length	Gupta et al. (2015)
3	*Bacillus pumilus* and *Bacillus licheniformis*	Increase Gibberellic Acid (GA) production	Enhance shoot development	Kumar et al. (2015)
4	*Bacillus subtilis*	Cytokinin synthesis	Cell division and tissue expansion	Liu et al. (2013)
5	*Pseudomonas* sp.	1-aminocyclopropane-1-carboxylate deaminase to lower plant ethylene synthesis	Root lengthening	Glick (2014)

20.3.3 Microbial biofilms as plant stress busters

The growing demand for agricultural products has depleted about 60% of soil ecosystem functioning due to soil erosion, reduction in nutrients availability, and biotic and abiotic stresses (Cohen and Leach, 2019; Panagos et al., 2018). The sessile nature of plants has made them more susceptible to stress conditions and, therefore, reduction in crop productivity as well (Gull et al., 2019; Zhu, 2016). The environmental stresses on the plants are divided into two groups: (1) abiotic: drought, salinity, extreme temperature, pH, and pressure; and (2) biotic: stress, pathogens, and diseases (Cohen and Leach, 2019; Rejeb et al., 2014). Since, the development of genetically engineered crop plants has ethical issues and inducing tolerance through conventional breeding is costly and time taking process, therefore, the role of beneficial biofilmed PGPRs are gaining interest in stress management (Backer et al., 2018; Mokrani et al., 2020).

20.3.4 Role of microbial biofilms in evading abiotic stress

Pseudomonas putida MTCC5279 biofilm improved the drought stress in *Cicer arietinum* by Reactive Oxygen Species (ROS) scavenging, regulating membrane integrity, and stockpiling of osmolytes (Tiwari et al., 2016). Biofilm forming

B. licheniformis and *Pseudomonas plecoglossicida* induced salt tolerance in *Helianthus annuus* by reducing the oxidative stress through upregulation of catalase, guaiacol peroxidase, and superoxide dismutase which enhanced biochemical and physiological conditions of the plant (Yasmeen et al., 2020). It has been demonstrated that *B. amyloliquefaciens*, a biofilm forming PGPR, induced salt tolerance in *Hordeum vulgare* L.by enhancing shoot and root weights (Kasim et al., 2016). *P. fluorescens* REN1 supported plants growing in flooded soils by root elongation through 1-aminocyclopropane-1-caboxylate deaminase production (Etesami et al., 2014). Upregulation of antioxidant activity and cold acclimation genes in low temperature exposed *Solanum lycopersicum* by inoculating *Burkholderia phytofirmans* PsJN in biofilm forming mode (Fernandez et al., 2012). Thus, biofilm PGPRs could enhance the ecosystem sustainability and crop productivity by ameliorating the stress conditions through enhancing the release of antistress hormones, chlorophyll content, root nodule biomass, root and shoot length, nutrient uptake, Abscisic Acid (ABA)-signaling cascades, ROS pathways, antioxidant enzymes, reducing cell destruction, etc. (Backer et al., 2018; Ilangumaran and Smith, 2017; Tsukanova et al., 2017; Wang et al., 2019).

20.3.5 Role of microbial biofilms in resisting biotic stress (biocontrolling)

To overcome the pest and disease attack, minimize the damage thereafter and to enhance the crop productivity (Nazir et al., 2019), biocontrolling has been regarded as an appropriate alternative tool in contrast to the conventional methods of using chemical pesticides and insecticides (Barratt et al., 2018). The studies have demonstrated that the microbes in the biofilm mode amplify the biocontrolling process by decreasing the sensitivity of plants to infections as compared to their planktonic form (Seneviratne et al., 2010; Köhl et al., 2019). To deter the chronic threat of phytopathogens, biofilm forming microbes play a significant role by producing antibiotics, stimulating induced systemic resistance (ISR) (Olowe et al., 2020), and production of siderophores (Olanrewaju et al., 2017). An increase in gossypol and jasmonic acid production and a reduction in pest attack of *Spodoptera exigua* larva have been reported when *Gossypium hirsutum* plants were grown in presence of biofilm forming *Bacillus* sp. (Zebelo et al., 2016). *Enterobacter asburiae* BQ9 inoculated tomato plants resisted the yellow leaf curl virus through activation of antioxidant enzymes *viz.*superoxide dismutase, peroxidase, and catalase (Li et al., 2016). The potential of root colonization of biofilm forming *Pleurotus ostreatus*−*P. fluorescens* was compared with the colonization when *P. fluorescens* was used alone. The study found that the endophytic root colonization was 10 times more in biofilm forming mode of *P. ostreatus*−*P. fluorescens* than when used alone, suggesting that biofilm forming plant microbes show amplified results toward crop productivity (Jayasinghearachchi and Seneviratne, 2010). *Paenibacillus polymyxa*, a biofilm forming PGPR provides protection to *A. thaliana* against pathogens by synthesizing several pathogen specific antibiotics (Timmusk et al., 2005). Activation of ISR by biofilmed-PGPRs is an advanced strategy to counter phytopathogens. The microbes secrete several components like lipid peroxidases, salicylic acid, and ROSs that induce ISR by activating pathogenesis-related genes and modulate the defense system of the plant (Gupta et al., 2017). Along with pronounced colonization, the strains of *B. subtilis* and *Bacillus cereus* form stable and extensive biofilms. Inoculants of *B. cereus* are present in commercially available biocontrol agent, Serenade, which effectively reduces the pathogenic bacteria such as *Erwina, Xanthomonas*, and *Pseudomonas*. Similarly, *B. subtilis* has been used as a biocontrol agent against pathogens including pathogenic fungi (Ghiasian, 2020). Root colonized biofilm forming *B. cereus* help in biocontrolling of alfalfa damping by synthesizing zwittermicin A and kanosamine antibiotics (Ongena and Jacques, 2008). Hydrogen cyanide is the most abundant volatile gas released in response to phytopathogen attack predominantly by *Bacillus* sp. (Gutiérrez-Luna et al., 2010). The lytic enzymes produced by biocontrolling agents such as β-glucanase, phosphatases, chitinases, dehydrogenase, lipases, and proteases are hyperpathogenic, thus, protect the crops from phytopathogens (Gupta et al., 2015; Jiang et al., 2020).

20.4 Microbial biofilm mediated bioremediation

The anthropogenic activities such as industrialization and extensive agricultural practices have added life threatening and soil depleting contaminants into the aquatic as well as terrestrial ecosystems (Khatri and Tyagi, 2015). Polychlorinated biphenyls, pesticides, radio elements, herbicides, metals, and polyaromatic hydrocarbons (PAHs) are some of the major ecosystem degrading contaminants released through industrial effluents and agricultural wastes (Das and Adholeya, 2012). Although a number of physical (electrokinesis, washing, encapsulation) (Chen et al., 2019), chemical (ion exchange, precipitation, solidification) (Du et al., 2011) and physiochemical (electrochemical) techniques (Niroumand et al., 2012) have been used in remediation, yet bioremediation through biofilm forming microbes are gaining interest due to cost-effective, safe, efficient, and environmental friendly nature (Streche et al., 2018). An enhanced Uranium (iv) remediation potential of *Staphylococcus aureus* V329 biofilm cells than the planktonic form has been reported by increasing the biosorption and activity of acid phosphatase (Shukla et al., 2020). Nutrient pollution is a major concern of aquaculture. In this context, a mycoalgae biofilm (*Mucor indicus* and *Chlorella vulgaris*) helped in bioremediation of aquaculture wastewaters and converted the nutrient pollutants into proteins (Barnharst et al., 2018). The bioremediation of PAHs demonstrated a significant degradation of phenanthrene and pyrene by *Pseudomonas aeruginosa* N6P6 and *Pseudomonas pseudoalcaligenes* NP103 in biofilm mode (Mangwani et al., 2017). The effluents form textile industries (mainly dyes) are also posing a global threat (Imtiazuddin et al., 2018). Bioremediation of dyes employing bacteria, fungi, and algae is advantageous over conventional chemical methods and the same is achieved by mineralization, degradation, and transformation. It has been reported that dyes crystal violet, reactive red 2, acid red 18, orange II, dorasyn red azo, reactive orange 13, cresol red, remazole brown, and cotton blue can be degraded by biofilm forming *Aeromonas hydrophila, Pseudomonas* sp. SUK1, *P. putida, B. cereus*, lactic acid bacteria, *Alcaligenes faecalis* PMS-1, *Streptomyces ipomoeae* CECT3341, *Aspergillus foetidus*, and *Penicillium ochrochloron*, respectively (Ihsanullah et al., 2020). It has been demonstrated that the bioremediation of cobalt-zinc-cadmium, organic and inorganic mercury, phenol, nickel-cobalt-cadmium, and chromate can be achieved by *Bacillus* sp., *P. aeruginosa, Pseudomonas* sp., *Bordetella* sp., and *S. aureus*, respectively (Das and Dash, 2014). The EPS encapsulation of cells within biofilm provides them better chances of adaptability (horizontal gene transfer) and survivability in contaminated soils as compared to the planktonic form, thus, acts as a persistent and ecofriendly approach of remediation (Ghiasian, 2020; Yin et al., 2019).

20.5 Role of microbial biofilms in nutrient mobilization

Root colonization of nonpathogenic microbial biofilms upregulate nutrient mobilization, integration and biogeochemical cycling, thus, concentrating the nutrients for plant growth, disease resistance and productivity (Bargaz et al., 2018; Ghiasian, 2020). Microbial biofilms play a significant role in decomposition of organic matter, nutrient cycling, and addition of polysaccharides, thus, are the determinants of soil structure, health, nutrient availability, water holding capacity, porosity, soil density, and erodibility (Wani et al., 2015). A study showed the prominent role of fungi *Glomus intraradices* in the mobilization of organic nitrogen (Thirkell et al., 2016). A study reviewed few phosphorous solubilizing microbes *viz. Pseudomonas* sp., *Azotobacter*, *Agrobacterium* sp., *Paenibacillus*, and *Bacillus* sp., and discussed their role in plant growth and crop yield by mineralizing and solubilizing insoluble P (Alori et al., 2017). The biofilm forming *Bacillus pumilus* and *P. pseudoalcaligenes*, two potassium mobilizing bacteria, upregulated the carbohydrate accumulation, and membrane permeability of paddy under salinity stress by increasing the solubility, mobility, and uptake of potassium (Jha, 2017). A comparison of the potential of biofilm forming hyphobacteria with their planktonic form in terms of potassium releasing and zinc solubilizing capability under in vitro conditions has been conducted (Nagaraju et al., 2017). The study demonstrated that there was a multifold increase in potassium and zinc mobilization when the microbes were allowed to form biofilm. Therefore, the use of microbial nutrient mobilizers provides an alternative approach to meet the deficiency of nutrients in soil, increasing food demands and restores the depleted agricultural lands since the excessive use of chemical fertilizers lead to environmental pollution, eutrophication and addition of toxic metals (Sharma et al., 2013).

20.6 Conclusion and future perspective

Agrochemical poisoning through the excessive use of chemicals in crop lands has devastated the agro-ecosystems by deteriorating the quality of soil, accumulation of hazardous chemicals in the crops, transfer of such toxic substances to each level of food chain and development of dreaded diseases including cancers. Thus, the sustainable and ecofriendly strategies need to be implemented to restore the contaminated, unfertilized, and depleted agricultural lands in order to meet the enormously growing demands for food and other agricultural products. The plant microbe interaction is a thoroughly researched field as it is regarded as an important and a beneficial solution to stressed agricultural lands. In this regard, the incorporation of beneficial biofilm forming microbes in the soil and commercially available fertilizers is gaining a global interest as it turns out be an efficient, reliable, persistent, and ecofriendly approach for sustainable agricultural practices. The contributions of microbes could be either direct by acting as biofertilizers, stress tolerant, PGPs and nutrient mobilizers or indirect by synthesizing antibiotics and activating ISR. This wide range of applications of biofilm PGPRs promises a crop and soil security as well as sustainability.

The future perspective of the novel approach of using microbes in agricultural practices lies in overcoming the already existing drawbacks *viz.* enhancing survivability and shelf life, development of multistrain inoculants and carriers, preparation of improved formulations, and commercialization as well.

References

Ahemad, M., Kibret, M., 2014. Mechanisms and applications of plant growth promoting rhizobacteria: current perspective. J. King Saud Univ. Sci. 26 (1), 1–20.

Ahmad, I., Nygren, E., Khalid, F., Myint, S.L., Uhlin, B.E., 2020. A cyclic-di-GMP signaling network regulates biofilm formation and surface associated motility of *Acinetobacter baumannii* 17978. Sci. Rep. 10 (1), 1–11.

Alori, E.T., Glick, B.R., Babalola, O.O., 2017. Microbial phosphorus solubilization and its potential for use in sustainable agriculture. Front. Microbiol. 8, 971.

Ansari, F.A., Ahmad, I., 2019. Fluorescent *Pseudomonas*-FAP2 and *Bacillus licheniformis* interact positively in biofilm mode enhancing plant growth and photosynthetic attributes. Sci. Rep. 9 (1), 1–12.

Backer, R., Rokem, J.S., Ilangumaran, G., Lamont, J., Praslickova, D., Ricci, E., et al., 2018. Plant growth-promoting rhizobacteria: context, mechanisms of action, and roadmap to commercialization of biostimulants for sustainable agriculture. Front. Plant Sci. 9, 1473.

Bais, H.P., Fall, R., Vivanco, J.M., 2004. Biocontrol of *Bacillus subtilis* against infection of *Arabidopsis* roots by *Pseudomonas syringae* is facilitated by biofilm formation and surfactin production. Plant Physiol. 134 (1), 307–319.

Bandara, W.M.M.S., Seneviratne, G., Kulasooriya, S.A., 2006. Interactions among endophytic bacteria and fungi: effects and potentials. J. Biosci. 31 (5), 645–650.

Bargaz, A., Lyamlouli, K., Chtouki, M., Zeroual, Y., Dhiba, D., 2018. Soil microbial resources for improving fertilizers efficiency in an integrated plant nutrient management system. Front. Microbiol. 9, 1606.

Barnharst, T., Rajendran, A., Hu, B., 2018. Bioremediation of synthetic intensive aquaculture wastewater by a novel feed-grade composite biofilm. Int. Biodeterior. Biodegradation 126, 131–142.

Barratt, B.I.P., Moran, V.C., Bigler, F., Van Lenteren, J.C., 2018. The status of biological control and recommendations for improving uptake for the future. Biol. Control 63 (1), 155–167.

Bhattacharyya, P.N., Jha, D.K., 2012. Plant growth-promoting rhizobacteria (PGPR): emergence in agriculture. World J. Microbiol. Biotechnol. 28 (4), 1327–1350.

Bidyarani, N., Prasanna, R., Babu, S., Hossain, F., Saxena, A.K., 2016. Enhancement of plant growth and yields in chickpea (*Cicer arietinum* L.) through novel cyanobacterial and biofilmed inoculants. Microbiol. Res. 188, 97–105.

Bisht, K., Wakeman, C.A., 2019. Discovery and therapeutic targeting of differentiated biofilm subpopulations. Front. Microbiol. 10, 1908.

Bogino, P.C., Oliva, M.D.L.M., Sorroche, F.G., Giordano, W., 2013. The role of bacterial biofilms and surface components in plant-bacterial associations. Int. J. Mol. Sci. 14 (8), 15838–15859.

Buddhika, U.V.A., Athauda, A.R.W.P.K., Seneviratne, G., Kulasooriya, S.A., Abayasekara, C.L., 2013. Emergence of diverse microbes on application of biofilmed biofertilizers to a maize growing soil. Ceylon J. Sci. 42 (2), 87–94.

Chen, X., Kumari, D., Cao, C.J., Plaza, G., Achal, V., 2019. A review on remediation technologies for nickel-contaminated soil. Hum. Ecol. Risk Assess. 26 (3), 571–585.

Cohen, S.P., Leach, J.E., 2019. Abiotic and biotic stresses induce a core transcriptome response in rice. Sci. Rep. 9 (1), 1–11.

Costa, O.Y., Raaijmakers, J.M., Kuramae, E.E., 2018. Microbial extracellular polymeric substances: ecological function and impact on soil aggregation. Front. Microbiol. 9, 1636.

Costerton, J.W., Geesey, G.G., Cheng, K.J., 1978. How bacteria stick. Sci. Am. 238, 86–95.

Dang, H., Lovell, C.R., 2016. Microbial surface colonization and biofilm development in marine environments. Microbiol. Mol. Biol. Rev. 80 (1), 91–138.

Das, M., Adholeya, A., 2012. Role of microorganisms in remediation of contaminated soil. Microorganisms in Environmental Management. Springer, Dordrecht, pp. 81–111.

Das, S., Dash, H.R., 2014. Microbial bioremediation: a potential tool for restoration of contaminated areas. Microbial Biodegradation and Bioremediation. Elsevier, pp. 1–21.

de Carvalho, C.C., 2018. Marine biofilms: a successful microbial strategy with economic implications. Front. Mar. Sci. 5, 126.

De Natale, A., Mele, B.H., Cennamo, P., Del Mondo, A., Petraretti, M., Pollio, A., 2020. Microbial biofilm community structure and composition on the lithic substrates of Herculaneum Suburban Baths. PLoS One 15 (5), e0232512.

De Silva, M.S.D.L., Jayasekera, A.P.D.A., Seneviratne, G., Abeysekera, U.P., Premathunge, E.W.T.P., Wijesekera, S.N., 2014. Soil fertility improvement through biofilmed biofertilizers: potential for field applications in tea cultivation. Sri Lanka J. Tea Sci. 79 (1/2), 46–61.

Decho, A.W., Gutierrez, T., 2017. Microbial extracellular polymeric substances (EPSs) in ocean systems. Front. Microbiol. 8, 922.

Dewasthale, S., Mani, I., Vasdev, K., 2018. Microbial biofilm: current challenges in health care industry. J. Appl. Biotechnol. Bioeng. 5 (3), 156–160.

Di Martino, P., 2018. Extracellular polymeric substances, a key element in understanding biofilm phenotype. AIMS Microbiol. 4 (2), 274.

Dong, C.J., Wang, L.L., Li, Q., Shang, Q.M., 2019. Bacterial communities in the rhizosphere, phyllosphere and endosphere of tomato plants. PLoS One 14 (11), e0223847.

Donlan, R.M., 2002. Biofilms: microbial life on surfaces. Emerg. Infect. Dis. 8 (9), 881.

Du, Y.J., Jin, F., Liu, S.Y., Chen, L., Zhang, F., 2011. Review of stabilization/solidification technique for remediation of heavy metals contaminated lands. Yantu Lixue/Rock Soil Mech. 32, 116–124.

Egamberdieva, D., Wirth, S.J., Alqarawi, A.A., Abd_Allah, E.F., Hashem, A., 2017. Phytohormones and beneficial microbes: essential components for plants to balance stress and fitness. Front. Microbiol. 8, 2104.

Enebe, M.C., Babalola, O.O., 2018. The influence of plant growth-promoting rhizobacteria in plant tolerance to abiotic stress: a survival strategy. Appl. Microbiol. Biotechnol. 102 (18), 7821–7835.

Etesami, H., Hosseini, H.M., Alikhani, H.A., 2014. Bacterial biosynthesis of 1-aminocyclopropane-1-caboxylate (ACC) deaminase, a useful trait to elongation and endophytic colonization of the roots of rice under constant flooded conditions. Physiol. Mol. Biol. Plants 20 (4), 425–434.

Felestrino, É.B., Santiago, I.F., Freitas, L.D.S., Rosa, L.H., Ribeiro, S.P., Moreira, L.M., 2017. Plant growth promoting bacteria associated with langsdorffia hypogaea-rhizosphere-host biological interface: a neglected model of bacterial prospection. Front. Microbiol. 8, 172.

Fernandez, O., Theocharis, A., Bordiec, S., Feil, R., Jacquens, L., Clement, C., 2012. *Burkholderia phytofirmans* PsJN acclimates grapevine to cold by modulating carbohydrate metabolism. Mol. Plant Microbe Interact. 25, 496–504.

Fox, A., Kwapinski, W., Griffiths, B.S., Schmalenberger, A., 2014. The role of sulfur- and phosphorus-mobilizing bacteria in biochar-induced growth promotion of *Lolium perenne*. FEMS Microbiol. Ecol. 90 (1), 78–91.

Galelli, M.E., Sarti, G.C., Miyazaki, S.S., 2015. *Lactuca sativa* biofertilization using biofilm from *Bacillus* with PGPR activity. J. Appl. Hortic. 17 (3).

Galie, S., García-Gutiérrez, C., Miguélez, E.M., Villar, C.J., Lombó, F., 2018. Biofilms in the food industry: health aspects and control methods. Front. Microbiol. 9, 898.

Gannesen, A.V., Zdorovenko, E.L.E., Botchkova, E.A., Hardouin, J., Massier, S., Kopitsyn, D.S., et al., 2019. Composition of biofilm matrix of *Cutibacterium acnes* acneic strain RT5. Front. Microbiol. 10 (1284).

Ghiasian, M., 2020. Microbial biofilms: beneficial applications for sustainable agriculture. New and Future Developments in Microbial Biotechnology and Bioengineering. Elsevier, pp. 145–155.

Glick, B.R., 2014. Bacteria with ACC deaminase can promote plant growth and help to feed the world. Microbiol. Res. 169 (1), 30–39.

Gortari, F., Nowosad, M.I.P., Laczeski, M.E., Onetto, A., Cortese, I.J., Castrillo, M.L., et al., 2019. Biofertilizers and biocontrollers as an alternative to the use of chemical fertilizers and fungicides in the propagation of yerba mate by mini-cuttings. Rev. Arvore 43 (4).

Gull, A., Lone, A.A., Wani, N.U.I., 2019. Biotic and abiotic stresses in plants. Abiotic and Biotic Stress in Plants. IntechOpen.

Gupta, G., Parihar, S.S., Ahirwar, N.K., Snehi, S.K., Singh, V., 2015. Plant growth promoting rhizobacteria (PGPR): current and future prospects for development of sustainable agriculture. J. Microb. Biochem. Technol. 7 (2), 096–102.

Gupta, G., Snehi, S.K., Singh, V., 2017. Role of PGPR in biofilm formations and its importance in plant health. Biofilms in Plant and Soil Health. p. 27.

Gutiérrez-Luna, F.M., López-Bucio, J., Altamirano-Hernández, J., Valencia-Cantero, E., de la Cruz, H.R., Macías-Rodríguez, L., 2010. Plant growth-promoting rhizobacteria modulate root-system architecture in *Arabidopsis thaliana* through volatile organic compound emission. Symbiosis 51 (1), 75–83.

Hense, B.A., Schuster, M., 2015. Core principles of bacterial autoinducer systems. Microbiol. Mol. Biol. Rev. 79 (1), 153–169.

Ihsanullah, I., Jamal, A., Ilyas, M., Zubair, M., Khan, G., Atieh, M.A., 2020. Bioremediation of dyes: current status and prospects. J. Water Process Eng. 38, 101680.

Ilangumaran, G., Smith, D.L., 2017. Plant growth promoting rhizobacteria in amelioration of salinity stress: a systems biology perspective. Front. Plant Sci. 8, 1768.

Imtiazuddin, S.M., Tiki, S., Chemicals, A.V.M., 2018. Impact of textile wastewater pollution on the environment. Pak. Text. J. 10, 38–39.

Jamal, M., Ahmad, W., Andleeb, S., Jalil, F., Imran, M., Nawaz, M.A., et al., 2018. Bacterial biofilm and associated infections. J. Chin. Med. Assoc. 81 (1), 7–11.

Janagard, M.S., Raei, Y., Gasemi-Golezani, K., Aliasgarzad, N., 2013. Soybean response to biological and chemical fertilizers. Int. J. Agric. Crop Sci. 5 (3), 261–266.

Jayasinghearachchi, H.S., Seneviratne, G., 2010. A mushroom-fungus helps improve endophytic colonization of tomato by *Pseudomonas fluorescens* through biofilm formation. Res. J. Microbiol. 5 (7), 689–695.

Jha, Y., 2017. Potassium mobilizing bacteria: enhance potassium intake in paddy to regulates membrane permeability and accumulate carbohydrates under salinity stress. Braz. J. Biol. Sci. 4 (8), 333–344.

Jiang, Y., Geng, M., Bai, L., 2020. Targeting biofilms therapy: current research strategies and development hurdles. Microorganisms 8 (8), 1222.

Kanchan, A., Simranjit, K., Ranjan, K., Prasanna, R., Ramakrishnan, B., Singh, M.C., et al., 2019. Microbial biofilm inoculants benefit growth and yield of *chrysanthemum* varieties under protected cultivation through enhanced nutrient availability. Plant Biosyst. 153 (2), 306–316.

Karygianni, L., Ren, Z., Koo, H., Thurnheer, T., 2020. Biofilm matrixome: extracellular components in structured microbial communities. Trends Microbiol. 28 (8), 668–681.

Kasim, W.A., Gaafar, R.M., Abou-Ali, R.M., Omar, M.N., Hewait, H.M., 2016. Effect of biofilm forming plant growth promoting rhizobacteria on salinity tolerance in barley. Ann. Agric. Sci. 61 (2), 217–227.

Khatri, N., Tyagi, S., 2015. Influences of natural and anthropogenic factors on surface and groundwater quality in rural and urban areas. Front. Life Sci. 8 (1), 23–39.

Köhl, J., Kolnaar, R., Ravensberg, W.J., 2019. Mode of action of microbial biological control agents against plant diseases: relevance beyond efficacy. Front. Plant Sci. 10, 845.

Kuan, K.B., Othman, R., Abdul Rahim, K., Shamsuddin, Z.H., 2016. Plant growth-promoting rhizobacteria inoculation to enhance vegetative growth, nitrogen fixation and nitrogen remobilisation of maize under greenhouse conditions. PLoS one 11 (3), e0152478.

Kudoyarova, G., Arkhipova, T.N., Korshunova, T., Bakaeva, M., Loginov, O., Dodd, I.C., 2019. Phytohormone mediation of interactions between plants and non-symbiotic growth promoting bacteria under edaphic stresses. Front. Plant Sci. 10, 1368.

Kumar, A., Meena, V.S., 2019. Plant Growth Promoting Rhizobacteria for Agricultural Sustainability, 10. Springer, Berlin, pp. 978–981.

Kumar, A., Bahadur, I., Maurya, B.R., Raghuwanshi, R., Meena, V.S., Singh, D.K., et al., 2015. Does a plant growth-promoting rhizobacteria enhance agricultural sustainability. J. Pure Appl. Microbiol. 9 (1), 715–724.

Kumar, A., Singh, S., Gaurav, A.K., Srivastava, S., Verma, J.P., 2020. Plant growth-promoting bacteria: biological tools for the mitigation of salinity stress in plants. Front. Microbiol. 11.

Landry, K.S., Morey, J.M., Bharat, B., Haney, N.M., Panesar, S.S., 2020. Biofilms—impacts on human health and its relevance to space travel. Microorganisms 8 (7), 998.

Lareen, A., Burton, F., Schäfer, P., 2016. Plant root-microbe communication in shaping root microbiomes. Plant Mol. Biol. 90 (6), 575–587.

Lens, P., 2011. Biofilms for Environmental Biotechnology in Support of Sustainable Development: A Report.

Li, H., Ding, X., Wang, C., Ke, H., Wu, Z., Wang, Y., 2016. Control of tomato yellow leaf curl virus disease by *Enterobacter asburiae* BQ9 as a result of priming plant resistance in tomatoes. Turk. J. Biol. 40, 150–159.

Liu, X., Liu, Y., Yang, L., Guo, T., Lv, L., Yin, C., et al., 2013. Classification of potassium-releasing bacteria isolated from four agricultural soil samples. J. Pure Appl. Microbiol. 7 (4), 3001–3008.

Liu, C., Sun, D., Zhu, J., Liu, J., Liu, W., 2020. The regulation of bacterial biofilm formation by cAMP-CRP: a mini-review. Front. Microbiol. 11, 802.

Lorite, G.S., Janissen, R., Clerici, J.H., Rodrigues, C.M., Tomaz, J.P., Mizaikoff, B., et al., 2013. Surface physicochemical properties at the micro and nano length scales: role on bacterial adhesion and *Xylella fastidiosa* biofilm development. PLoS One 8 (9), e75247.

Lugtenberg, B., Kamilova, F., 2009. Plant-growth-promoting rhizobacteria. Annu. Rev. Microbiol. 63, 541–556.

Mahdi, S.S., Hassan, G.I., Samoon, S.A., Rather, H.A., Dar, S.A., Zehra, B., 2010. Bio-fertilizers in organic agriculture. J. Phytol. 2, 42–54.

Mangwani, N., Kumari, S., Surajit, D.A.S., 2017. Marine bacterial biofilms in bioremediation of polycyclic aromatic hydrocarbons (PAHs) under terrestrial condition in a soil microcosm. Pedosphere 27 (3), 548–558.

Marapatla, N.P., 2014. Study of Biofilmed Biofertilizers to Improve Crop Production and Disease Control in Chickpea (Doctoral dissertation), Acharya N.G. Ranga Agricultural University, Rajendranagar, Hyderabad.

Mazid, M., Khan, T.A., 2015. Future of bio-fertilizers in Indian agriculture: an overview. Int. J. Agric. Food Res. 3 (3).

Mhlongo, M.I., Piater, L.A., Madala, N.E., Labuschagne, N., Dubery, I.A., 2018. The chemistry of plant–microbe interactions in the rhizosphere and the potential for metabolomics to reveal signaling related to defense priming and induced systemic resistance. Front. Plant Sci. 9, 112.

Mohanram, S., Kumar, P., 2019. Rhizosphere microbiome: revisiting the synergy of plant-microbe interactions. Ann. Microbiol. 69 (4), 307–320.

Mokrani, S., Nabti, E.H., Cruz, C., 2020. Current advances in plant growth promoting bacteria alleviating salt stress for sustainable agriculture. Appl. Sci. 10 (20), 7025.

Muhammad, M.H., Idris, A.L., Fan, X., Guo, Y., Yu, Y., Jin, X., et al., 2020. Beyond risk: bacterial biofilms and their regulating approaches. Front. Microbiol. 11, 928.

Nadeem, S.M., Naveed, M., Zahir, Z.A., Asghar, H.N., 2013. Plant–microbe interactions for sustainable agriculture: fundamentals and recent advances. Plant Microbe Symbiosis: Fundamentals and Advances. Springer, New Delhi, pp. 51–103.

Nagaraju, Y., Triveni, S., Subhashreddy, R., Jhansi, P., 2017. Biofilm formation of zinc solubilizing, potassium releasing bacteria on the surface of fungi. Int. J. Curr. Microbiol. Appl. Sci. 6 (4), 2037–2047.

Nazir, T., Khan, S., Qiu, D., 2019. Biological control of insect pest. Pests-Insects, Management, Control. IntechOpen.

Niroumand, H., Nazir, R., Kassim, K.A., 2012. The performance of electrochemical remediation technologies in soil mechanics. Int. J. Electrochem. Sci. 7 (6), 5708–5715.

Okur, N., 2018. A review: bio-fertilizers-power of beneficial microorganisms in soils. Biomed. J. Sci. Tech. Res. 4 (4), 4028–4029.

Olanrewaju, O.S., Glick, B.R., Babalola, O.O., 2017. Mechanisms of action of plant growth promoting bacteria. World J. Microbiol. Biotechnol. 33 (11), 197.

Olanrewaju, O.S., Ayangbenro, A.S., Glick, B.R., Babalola, O.O., 2019. Plant health: feedback effect of root exudates-rhizobiome interactions. Appl. Microbiol. Biotechnol. 103 (3), 1155–1166.

Olowe, O.M., Akanmu, A.O., Asemoloye, M.D., 2020. Exploration of microbial stimulants for induction of systemic resistance in plant disease management. Ann. Appl. Biol. 177 (3), 282–293.

Omer, Z.S., Tombolini, R., Broberg, A., Gerhardson, B., 2004. Indole-3-acetic acid production by pink-pigmented facultative methylotrophic bacteria. Plant Growth Regul. 43 (1), 93–96.

Ongena, M., Jacques, P., 2008. *Bacillus lipopeptides*: versatile weapons for plant disease biocontrol. Trends Microbiol. 16 (3), 115–125.

Panagos, P., Standardi, G., Borrelli, P., Lugato, E., Montanarella, L., Bosello, F., 2018. Cost of agricultural productivity loss due to soil erosion in the European Union: from direct cost evaluation approaches to the use of macroeconomic models. Land Degrad. Dev. 29 (3), 471–484.

Patra, S., Mishra, P., Mahapatra, S.C., Mithun, S.K., 2016. Modelling impacts of chemical fertilizer on agricultural production: a case study on Hooghly district, West Bengal, India. Model. Earth Syst. Environ. 2 (4), 1–11.

Penesyan, A., Nagy, S.S., Kjelleberg, S., Gillings, M.R., Paulsen, I.T., 2019. Rapid microevolution of biofilm cells in response to antibiotics. NPJ Biofilms Microb. 5 (1), 1–14.

Penesyan, A., Paulsen, I.T., Gillings, M.R., Kjelleberg, S., Manefield, M.J., 2020. Secondary effects of antibiotics on microbial biofilms. Front. Microbiol. 11, 2109.

Rabin, N., Zheng, Y., Opoku-Temeng, C., Du, Y., Bonsu, E., Sintim, H.O., 2015. Biofilm formation mechanisms and targets for developing antibiofilm agents. Future Med. Chem. 7 (4), 493–512.

Rafique, M., Hayat, K., Mukhtar, T., Khan, A.A., Afridi, M.S., Hussain, T., et al., 2015. Bacterial biofilm formation and its role against agricultural pathogens. The Battle Against Microbial Pathogens: Basic Science, Technological Advances and Educational Programs. Spain: Formatex Research Center, pp. 373–382.

Rahman, K.M., Zhang, D., 2018. Effects of fertilizer broadcasting on the excessive use of inorganic fertilizers and environmental sustainability. Sustainability 10 (3), 759.

Ramasamy, M., Geetha, T., Yuvaraj, M., 2020. Role of biofertilizers in plant growth and soil health. Nitrogen Fixation. IntechOpen.

Rana, A., Joshi, M., Prasanna, R., Shivay, Y.S., Nain, L., 2012. Biofortification of wheat through inoculation of plant growth promoting rhizobacteria and cyanobacteria. Eur. J. Soil Biol. 50, 118–126.

Rana, K.L., Kour, D., Yadav, A.N., Yadav, N., Saxena, A.K., 2020. Agriculturally important microbial biofilms: biodiversity, ecological significances, and biotechnological applications. New and Future Developments in Microbial Biotechnology and Bioengineering: Microbial Biofilms. Elsevier, pp. 221–265.

Rather, Muzamil, Gupta, Kuldeep, Bardhan, Pritam, Borah, Munmi, Sarkar, Anupama, Eldiehy, et al., 2021. Microbial biofilm: A matter of grave concern for human health and food industry. Journal of Basic Microbiology 61 (5), 380–395. https://doi.org/10.1002/jobm.202000678.

Reddy, G.C., Goyal, R.K., Puranik, S., Waghmar, V., Vikram, K.V., Sruthy, K.S., 2020. Biofertilizers toward sustainable agricultural development. Plant Microbe Symbiosis. Springer, Cham, pp. 115–128.

Rejeb, I.B., Pastor, V., Mauch-Mani, B., 2014. Plant responses to simultaneous biotic and abiotic stress: molecular mechanisms. Plants 3 (4), 458–475.

Ren, X., Guo, S., Tian, W., Chen, Y., Han, H., Chen, E., et al., 2019. Effects of plant growth-promoting bacteria (PGPB) inoculation on the growth, antioxidant activity, Cu uptake, and bacterial community structure of rape (*Brassica napus* L.) grown in Cu-contaminated agricultural soil. Front. Microbiol. 10, 1455.

Rudrappa, T., Biedrzycki, M.L., Bais, H.P., 2008. Causes and consequences of plant-associated biofilms. FEMS Microbiol. Ecol. 64 (2), 153–166.

Ruzzi, M., Aroca, R., 2015. Plant growth-promoting rhizobacteria act as biostimulants in horticulture. Sci. Hortic. 196, 124–134.

Ryu, C.M., Farag, M.A., Hu, C.H., Reddy, M.S., Kloepper, J.W., Paré, P.W., 2004. Bacterial volatiles induce systemic resistance in *Arabidopsis*. Plant Physiol. 134 (3), 1017–1026.

Sahoo, R.K., Bhardwaj, D., Tuteja, N., 2013. Biofertilizers: a sustainable eco-friendly agricultural approach to crop improvement. Plant Acclimation to Environmental Stress. Springer, New York, pp. 403–432.

Sarikhani, M.R., Aliasgharzad, N., Khoshru, B., 2020. P solubilizing potential of some plant growth promoting bacteria used as ingredient in phosphatic biofertilizers with emphasis on growth promotion of *Zea mays* L. Geomicrobiol. J. 37 (4), 327–335.

Sauer, K., 2003. The genomics and proteomics of biofilm formation. Genome Biol. 4 (6), 1–5.

Schirone, M., Visciano, P., Tofalo, R., Suzzi, G., 2019. Foodborne pathogens: hygiene and safety. Front. Microbiol. 10, 1974.

Selvakumar, G., Reetha, S., Thamizhiniyan, P., 2012. Response of biofertilizers on growth, yield attributes and associated protein profiling changes of blackgram (*Vigna mungo* L. Hepper). World Appl. Sci. J. 16 (10), 1368–1374.

Seneviratne, G., Weerasekara, M.L.M.A.W., Seneviratne, K.A.C.N., Zavahir, J.S., Kecskés, M.L., Kennedy, I.R., 2010. Importance of biofilm formation in plant growth promoting rhizobacterial action. Plant Growth and Health Promoting Bacteria. Springer, Berlin, Heidelberg, pp. 81–95.

Shaikh, S.S., Sayyed, R.Z., Reddy, M.S., 2016. Plant growth-promoting rhizobacteria: an eco-friendly approach for sustainable agroecosystem. Plant, Soil and Microbes. Springer, Cham, pp. 181–201.

Sharma, A., Johri, B.N., 2003. Growth promoting influence of siderophore-producing *Pseudomonas* strains GRP3A and PRS9 in maize (*Zea mays* L.) under iron limiting conditions. Microbiol. Res. 158 (3), 243–248.

Sharma, S.B., Sayyed, R.Z., Trivedi, M.H., Gobi, T.A., 2013. Phosphate solubilizing microbes: sustainable approach for managing phosphorus deficiency in agricultural soils. SpringerPlus. 2 (1), 587.

Sharma, D., Misba, L., Khan, A.U., 2019. Antibiotics versus biofilm: an emerging battleground in microbial communities. Antimicrob. Resist. Infect. Control 8 (1), 1–10.

Shukla, S.K., Hariharan, S., Rao, T.S., 2020. Uranium bioremediation by acid phosphatase activity of *Staphylococcus aureus* biofilms: can a foe turn a friend? J. Hazard. Mater. 384, 121316.

Singh, J.S., Gupta, V.K., 2016. Degraded land restoration in reinstating CH4 sink. Front. Microbiol. 7, 923.

Singh, D.P., Singh, H.B., Prabha, R. (Eds.), 2017. Plant-Microbe Interactions in Agro-Ecological Perspectives. Springer, New Delhi.

Singhalage, I.D., Seneviratne, G., Madawala, H.M.S.P., Wijepala, P.C., 2019. Profitability of strawberry (*Fragaria ananassa*) production with biofilmed biofertilizer application. Sci. Hortic. 243, 411–413.

Souza, R.D., Ambrosini, A., Passaglia, L.M., 2015. Plant growth-promoting bacteria as inoculants in agricultural soils. Genet. Mol. Biol. 38 (4), 401–419.

Speranza, B., Corbo, M.R., 2017. The impact of biofilms on food spoilage. The Microbiological Quality of Food. Woodhead Publishing, pp. 259–282.

Streche, C., Cocârţă, D.M., Istrate, I.A., Badea, A.A., 2018. Decontamination of petroleum-contaminated soils using the electrochemical technique: remediation degree and energy consumption. Sci. Rep. 8 (1), 1–14.

Suryadi, Y., Susilowati, D.N., Fauziah, F., 2019. Management of plant diseases by PGPR-mediated induced resistance with special reference to tea and rice crops. Plant Growth Promoting Rhizobacteria for Sustainable Stress Management. Springer, Singapore, pp. 65–110.

Swarnalakshmi, K., Prasanna, R., Kumar, A., Pattnaik, S., Chakravarty, K., Shivay, Y.S., et al., 2013. Evaluating the influence of novel cyanobacterial biofilmed biofertilizers on soil fertility and plant nutrition in wheat. Eur. J. Soil Biol. 55, 107–116.

Teplitski, M., Robinson, J.B., Bauer, W.D., 2000. Plants secrete substances that mimic bacterial N-acyl homoserine lactone signal activities and affect population density-dependent behaviors in associated bacteria. Mol. Plant Microbe Interact. 13 (6), 637–648.

Teschler, J.K., Zamorano-Sánchez, D., Utada, A.S., Warner, C.J.A., 2015. Living in the matrix: assembly and control of *Vibrio cholera* biofilms. Nat. Rev. Microbiol. 13, 255–268.

Thirkell, T.J., Cameron, D.D., Hodge, A., 2016. Resolving the 'nitrogen paradox'of arbuscular mycorrhizas: fertilization with organic matter brings considerable benefits for plant nutrition and growth. Plant Cell Environ. 39 (8), 1683–1690.

Tielen, P., Rosenau, F., Wilhelm, S., Jaeger, K.E., Flemming, H.C., Wingender, J., 2010. Extracellular enzymes affect biofilm formation of mucoid *Pseudomonas aeruginosa*. Microbiology 156 (7), 2239–2252.

Timmusk, S., Grantcharova, N., Wagner, E.G.H., 2005. *Paenibacillus polymyxa* invades plant roots and forms biofilms. Appl. Environ. Microbiol. 71 (11), 7292–7300.

Timmusk, S., Behers, L., Muthoni, J., Muraya, A., Aronsson, A.C., 2017. Perspectives and challenges of microbial application for crop improvement. Front. Plant Sci. 8, 49.

Tiwari, S., Lata, C., Chauhan, P.S., Nautiyal, C.S., 2016. *Pseudomonas putida* attunes morphophysiological, biochemical and molecular responses in *Cicer arietinum* L. during drought stress and recovery. Plant Physiol. Biochem. 99, 108–117.

Tkacz, A., Bestion, E., Bo, Z., Hortala, M., Poole, P.S., 2020. Influence of plant fraction, soil, and plant species on microbiota: a multikingdom comparison. mBio 11 (1).

Toyofuku, M., Inaba, T., Kiyokawa, T., Obana, N., Yawata, Y., Nomura, N., 2016. Environmental factors that shape biofilm formation. Biosci. Biotechnol. Biochem. 80 (1), 7–12.

Tsukanova, K.A., Meyer, J.J.M., Bibikova, T.N., 2017. Effect of plant growth-promoting rhizobacteria on plant hormone homeostasis. S. Afr. J. Bot. 113, 91–102.

Vejan, P., Abdullah, R., Khadiran, T., Ismail, S., Nasrulhaq Boyce, A., 2016. Role of plant growth promoting rhizobacteria in agricultural sustainability—a review. Molecules 21 (5), 573.

Velmourougane, K., Prasanna, R., Saxena, A.K., 2017. Agriculturally important microbial biofilms: present status and future prospects. J. Basic Microbiol. 57 (7), 548–573.

Walker, T.S., Bais, H.P., Grotewold, E., Vivanco, J.M., 2003. Root exudation and rhizosphere biology. Plant Physiol. 132 (1), 44–51.

Wang, S., Shi, W., Tang, T., Wang, Y., Zhi, L., Lv, J., et al., 2017a. Function of quorum sensing and cell signaling in the formation of aerobic granular sludge. Rev. Environ. Sci. Biotechnol. 16 (1), 1–13.

Wang, D., Xu, A., Elmerich, C., Ma, L.Z., 2017b. Biofilm formation enables free-living nitrogen-fixing rhizobacteria to fix nitrogen under aerobic conditions. ISME J. 11 (7), 1602–1613.

Wang, D.C., Jiang, C.H., Zhang, L.N., Chen, L., Zhang, X.Y., Guo, J.H., 2019. Biofilms positively contribute to *Bacillus amyloliquefaciens* 54-induced drought tolerance in tomato plants. Int. J. Mol. Sci. 20 (24), 6271.

Wani, F.S., Ahmad, L., Ali, T., Mushtaq, A., 2015. Role of microorganisms in nutrient mobilization and soil health—a review. J. Pure Appl. Microbiol. 9, 1401–1410.

Yadav, B.K., Akhtar, M.S., Panwar, J., 2015. Rhizospheric plant-microbe interactions: key factors to soil fertility and plant nutrition. Plant Microbes Symbiosis: Applied Facets. Springer, New Delhi, pp. 127–145.

Yasmeen, T., Ahmad, A., Arif, M.S., Mubin, M., Rehman, K., Shahzad, S.M., et al., 2020. Biofilm forming rhizobacteria enhance growth and salt tolerance in sunflower plants by stimulating antioxidant enzymes activity. Plant Physiol. Biochem. 156, 242–256.

Yin, W., Wang, Y., Liu, L., He, J., 2019. Biofilms: the microbial "protective clothing" inextreme environments. Int. J. Mol. Sci. 20 (14), 3423.

Zakeel, M.C.M., Safeena, M.I.S., 2019. Biofilmed biofertilizer for sustainable agriculture. Plant Health Under Biotic Stress. Springer, Singapore, pp. 65–82.

Zebelo, S., Song, Y., Kloepper, J.W., Fadamiro, H., 2016. Rhizobacteria activates (+)-delta-cadinene synthase genes and induces systemic resistance in cotton against beet armyworm (*Spodoptera exigua*). Plant Cell Environ. 39, 935–943.

Zhu, J.K., 2016. Abiotic stress signaling and responses in plants. Cell 167 (2), 313–324.

21 Valuing each patch of land: utilizing plant-microbe interactions for the betterment of agriculture

Sinha Sahab, Ibha Suhani and Rajeev Pratap Singh
Department of Environment and Sustainable Development, Institute of Environment and Sustainable Development, Banaras Hindu University, Varanasi, India

21.1 Introduction

For ages, land has been one of the essential parts for human livelihood (food and shelter). The fate of human development and expansion followed by urbanization, modernization, and industrialization, has created a major shift in the land use paradigm. The developments and ongoing environmental processes over time and space have made humans work in socioeconomic dimensions which have given rise to many types of land use changes. Land use changes have led to great transformations of major arable land cover into urban land covers. These land transformations gave rise to problems like biodiversity loss, deforestation, less availability of arable land, flooding, desertification, land degradation (directly or indirectly), etc. (Rattan, 2015). Overpopulation has made the food demands rise globally, creating immense pressure on agriculture. It is projected that by 2050 the global population will cross 9 billion counts and there will be a need to increase the food supply by 70% to withstand global food demands (Rattan, 2015). The unavailability of agricultural lands and the existence of marginal or degraded lands has become a great challenge. This has made science focus on utilizing the degraded and existing arable land resources in the best way possible, to meet the agricultural demand as well as to reduce further land degradation (Suhani et al., 2020). Also, the emphasis should be made on using marginal or degraded lands for urban expansions instead of using valuable arable land resources, to fix the major problem, that is, unavailability of agricultural lands. The linkage between landscape and their management practices reveals the impact on the available land area (Suhani et al., 2020). Due to the limitation of available arable land resources, we need to adopt strategies involving restoration, reclamation, and management of degraded lands to fulfill the present and future global food demands. For generating a good agricultural output, plants or crops need a suitable edaphic environment.

Microbes in Land Use Change Management. DOI: https://doi.org/10.1016/B978-0-12-824448-7.00024-3
© 2021 Elsevier Inc. All rights reserved.

The agricultural productivity depends upon the following (Suhani et al., 2020):

1. Environmental factors and topography—The region-specific environmental stresses (sunlight, temperature, humidity, wind, soil pH, soil salinity, flood, drought, etc.), geographical distribution and climatic conditions.
2. Land quality—Arable and barren land (natural/manmade).
3. Available land area—More the arable land area more is the agricultural output.
4. Anthropogenic activities—Farming modes (sustainable/extensive), use of chemicals (overuse of inorganic fertilizers and pesticides), improper agricultural techniques, irrigation types, soil pollution, etc.

The knowledge of root causes and the link between biophysical and socioeconomic dimension helps in understanding the challenge of utilizing each patch of land for the betterment of agriculture. Though we already know that land degradation is caused by natural as well as human induced processes. Not all causes of land degradation can be controlled especially natural causes, but we can minimize the harmful anthropogenic activities. According to Gibbs and Salmon (2015), the global estimate varies from 1 to 6 billion hectares, which is categorized as lightly, moderately, severely degraded lands. Among these categories of degradation, the severely degraded proportion of land is not in a condition of restoration or it can be said that the damage is irreversible. The lightly and moderately degraded lands can be used for agricultural practices by involving onsite management techniques, one being the microbial technique (Sahab et al., 2020; Suhani et al., 2020). Biological techniques which utilize microorganisms are given more significance now a days. This is mainly because of they are good stress regulators, root level workers, sustainable, magnifies resilience and well efficient to uplift soil nutrient profile. Not only for the present scenario but also for the future, adopting strategies which ensure management of land use patterns and agricultural practices on available (degraded/arable) land is the only way out to utilize each patch of land in its best way. In this chapter, we are looking forward for such microbial techniques which will help in restoring degraded lands by diminishing or regulating existing environmental stresses and increase agricultural outputs, to meet the upcoming global food demands.

21.2 Relationship between abiotic stresses and soil-plant health

Out of the total world's agricultural land area, 91% of the land area is under different types of stress (Sahab et al., 2020). About 50% of the agricultural losses are said to be because of abiotic stresses. Abiotic stresses being the most agricultural yield dipping factor for the major crop worldwide, around 50% of the average yield gets damaged by heat stress 20%, cold stress 7%, drought 9%, and other stress is 4% (Minhas et al., 2017). In nature the soil-plant health is mainly influenced by variety of abiotic factors. Some of the abiotic factors plays vital role in nurturing the soil ecosystem and plant growth. When these factors are unfavorable to maintain soil-plant health then they cause stress. As we know light is an essential factor which determines the growth of plants, but it can be transformed into stress if it exceeds than an optimal requirement of plants. According to Bechtold and Field (2018), excess of light reduces and overexcites the etc. (electron transport chain) of photosynthetic pathway. Thus it leads to production of reactive oxygen species (ROS) which disrupts membranes, DNA, and proteins of plants. Sometimes stresses are exaggerated by prevailing climatic conditions or anthropogenic activities. The environmental stress is said to be more destructive if its frequency and persistence increases within a time frame.

Some of the abiotic stresses which occurs due to natural or human interferences is mentioned below.

21.2.1 Temperature

Approximately all organisms present on earth including plants are sensitive to temperature fluctuations. In soil-plant system temperature plays significant role in maintaining rate of reactions, enzymatic activities, and viscosity of cytoplasm, xylem, phloem, and other solutions (Źróbek-Sokolnik, 2012). The soil temperature significantly affects the plant root system and its functionality (Abenavoli et al., 2012). Mainly the temperature stress can be classified into two types, that is, high temperature stress (HTS) and low temperature stress (LTS). Both HTS and LTS have different nature of effects on plants. In most of the plants HTS increases the cell membrane fluidity, whereas LTS toughens the biological membranes, hence increases membrane rigidity and activation energy requirement of membrane enzymes (Sung et al., 2003). HTS can create secondary water stress in plants, when the leaf transpiration rate is higher than the rate of root water uptake. On the other hand, in extreme LTS (mainly below 0°C) water gets converted into ice crystals inside and outside the plants. Intracellular ice crystal formed inside the cell (protoplast) can cause disruption or damage to the cell structures and extracellular crystals may cause dehydration of cell (Źróbek-Sokolnik, 2012). Most of the growing crop or plant species needs an optimal range of temperature (20°C–35°C) for photosynthesis. During HTS (temperature above 40°C) and LTS (temperature below 0°C), the photosynthetic rate gets ceased.

21.2.2 Drought

The global ambient temperature is rising due to prevailing climatic changes. Inter-Governmental Panel on Climate Change has predicted that by the end of 21st century there will be an increase of about 1°C–4°C in global mean air temperature (Change, 2007). Such kind of change in global mean air temperature can lead to many types of associated abiotic stresses like HTS, drought, salinity, etc., and this may cause to environmental stress exaggeration. In general, drought is a condition of soil drying in an area or region due to lack of precipitation, thus water availability become suboptimal, that is, below necessity (Abenavoli et al., 2012). It is considered a major threat that diminishes the crop productivity worldwide. Reduction in stem extension, leaf size, and root proliferation is observed in plants as the result of disturbed water use efficiency enhanced by drought (Farooq et al., 2009). In response to disturbed cell water status, exposed plants start producing 1-aminocyclopropane-1-carboxylic acid (ACC) (Kieber and Polko, 2019). As a result of elevated ACC concentration inside plant higher amount of ethylene synthesis takes place which leads to abscission, premature senescence, and growth inhibition (Gamalero and Glick, 2012). Drought disturbs plant root system by disrupting the mineral-nutrient relation of plants and severely affects the partition, transport and availability of nutrients (Hu and Schmidhalter, 2005). In subtropical arid and semi-arid regions, an increased span of rainless period and evapotranspiration events can intensify drought stress (Minhas et al., 2017). Different parts of South Asia have reported for heavy decline in productivity of paddy and wheat crops due to reduction in number of rainy days. Projections are made for arid and semiarid regions that for every 1°C rise in temperature the necessity of irrigational water will rise for about 10% (Grover et al., 2011).

21.2.3 Flood

Globally, agricultural area of 10%–12% has been estimated to be in condition of severe waterlogging (Shabala, 2011). Flooding is a condition of waterlogging in soil which leads to leaching, runoff or erosion of soil causing nutrient deficiency. Flooding causes soil nitrogen loss through denitrification, runoff and nitrate leaching. Flooding causes disturbance in root-respiration and causes anoxic condition. Flooding reduces soil oxygen content which causes reduction in nitrogenase activity and nodulation. Because of waterlogging there is decrease in water and nutrient uptakedue to lower root conductance eventually results in poor plant growth from nutrient deficiency (Suhani et al., 2020). Flooding or waterlogging situation causes submergence of plant parts under water, accountable for accelerating ethylene concentration, due to rise in the synthesis of ACC and ACC-oxidase in root as well as shoot. The waterlogged condition does not allow ethylene to escape the plant system in anyway thus causes number of stresses to the plant. Few studies were done to in order to reduce flooding stress on plant, plant growth promoting rhizobacteria (PGPR) with ACC deaminase activity were used in the situation as they help in rerouting ACC from ethylene biosynthesis pathway in the roots of host plants and thus causes low ethylene production (Barnawal et al., 2012; Li et al., 2013). There are several reports showing that water logging in soil resulted in rise of ACC-oxidase activity in plants and resulted in accelerated ethylene synthesis as well as abscisic acid (ABA) (Grover et al., 2011; Barnawal et al., 2012; Nadeem et al., 2014). The waterlogged condition reduces photosynthesis due to limitation of CO_2 concentration, as there is reduction in stomatal opening which results in accumulation of oxygen free radicals (Suhani et al., 2020). Cellular metabolism gets altered as the active oxygen species production causes oxidative damage to proteins, lipids, and nucleic acids and can further damage photosystem under severe stress circumstances.

21.2.4 Salinity

Around 15% of the world's total land has been degraded due to salinization and soil erosion. (Wild, 2003; Tóth et al., 2008). High concentrations of salts adversely affect soil's physiochemical properties and causes structural deterioration (slaking, swelling, hard setting, dispersion, and surface crust), high pH, organic matter loss, reduced nutrient cycling (C and N mineralization), inhibition of numerous soil enzymatic activities [alkaline phosphatase, benzoyl-argininamide, β-glucosidase, dehydrogenase, hydrolase, amylase, urease, invertase, and catalase (CAT)], and reduced availability of the micronutrients (Fe, Al, Cu, Zn, Mn) (Sahab et al., 2020). Salinity is one of the major abiotic stress, which is accountable for limiting plant growth and development by disturbing different plant physiology and biochemical processes such ashindering nutrient uptake, alters protein metabolism, enzyme activities, hormone balance, ion toxicity of salts, making ion inadequacy, and troubles symbiotic interaction between legumes and Rhizobium (Upadhyay et al., 2011; Egamberdieva and Lugtenberg, 2014; Nadeem et al., 2014). Salinity also disturbs photosynthesis by increasing photorespiration, fluctuating the normal homeostasis of cells and increases production of ROS as well as ethylene (Miller et al., 2010; Carmen and Defez, 2011). Seed germination and seedling growth are mostly affected by salinity (Egamberdieva et al., 2011). It has been reported that metal removal efficiency is decreased with increase in salinity (Fritioff et al., 2005). Soil salinity greatly impacts metabolic processes, population, abundance and life cycles of most of the soil organisms (earthworms, nematodes, etc.) (Owojori et al., 2009; Škarková et al., 2016; Sharif et al., 2016, Su et al., 2016).

21.2.5 Acidification

Acidic soils spread over 30% of the total land area worldwide, this approximately equals to 50% of the total world cultivable lands. Acidification leads to change in the chemical nature of the soil, which is caused by many factors like leaching, improper drainage practices, excessive land use practices, acid rain, microbial respiration, nitrification and use of inorganic fertilizers (Muthukumar et al., 2014). All of the mentioned processes produce CO_2 or H^+ ions, which later gets dissolved in soil water and forms carbonic acid. The acidification process leads to distribution of heavy metals and decline in pH. The soil nutrients decline as the soil particles and important minerals forms complexes. By acidification land gets damaged and not only the soil's nutrient profile but also the microbiome as well as the plants are adversely affected (Kumar, 2019). Plants are greatly damaged because of the hindered biochemical activities due to deficiency of nutrients. Low soil pH reduces organic matter decomposition and hinders the symbiotic relationship of native soil organisms and vegetation of the area. Low pH reduces the growth of beneficial microorganisms like rhizobia. It has been reported that Ca^{2+} and Mo becomes inaccessible and there is 40%–60% decrease in nodule formation per plant (like in soyabean) when the pH level is below 5. Whereas, at pH level below 5 the availability of Al and Mn gets enhanced which are toxic to symbiosis of legume-rhizobia (Bordeleau and Prévost, 1994).

21.2.6 Nutrient availability

Nutrient deficiency is one of the important effects which limits plants growth by reducing cellular expansion, plant development and photosynthesis. When there is deficiency in soil then plant translocate nutrients from mature leaves and stem for their functioning. Plant organs become chlorotic and necrotic during severe deficiency. Deleterious effects can be observed on the cellular and molecular effects. Nitrogen (N) and potassium (K) are the most important nutrients responsible for physiological processes, such as ion balance, enzyme activation, solute transport, photosynthetic rate, respiration rate, protein synthesis, growth and yields of many crops and also helps in alleviating abiotic and biotic stresses (Zhu et al., 2019). Among the macronutrients, N-deficiency decreases chlorophyll content, enzymatic activity, respiration rate, photosynthesis, and yield. N is very essential as it participates in Cd accumulation and uptake (Yang et al., 2020). K plays a vital role in osmotic adjustment, and when tissues accumulate more K then it helps in enhancing water acquisition under stress like drought (Wang et al., 2013). When K^+ limiting conditions arises in rhizospheric soil then there is decrease in K^+ activity, protein synthesis, pH in the cytosol of root cells and growth. At the metabolic level, the roots and shoots are subjected to have increased levels of sugars and decline in pyruvate content (Armengaud et al., 2009). Some other changes also take place such as maintaining carbon flux into amino acids and proteins, reduction in negative metabolic charge, and rise in CN ratio in amino acids. It has been reported by Walker et al. (1996) that barley shows decrease in protein synthesis when there is K^+ nutrient deficiency. Under adverse conditions plants are more susceptible to abiotic stresses like salinity, drought, fungal attacks and cold. Boron is involved in metabolism and transport of carbohydrates, nucleic acid synthesis, root elongation, photosynthetic activities and water absorption in plant parts. Its deficiency may cause shortening of the terminal growth, sterility, leaves become brittle with scorched appearance poor fruit set, small fruit size, growth seized, lower yield and cracking or distorted growth of fruits (Islam et al., 2018; Kumar et al., 2020). Zinc is essential for reproduction, carbohydrate formation, synthesis of plant hormones (auxin), structural and functional integrity of biomembranes and ribomsomes, enzyme activation, starch formation, protein metabolism, nitrogen and sulphur metabolism, photosynthetic metabolism, seed maturation, gene expression, growth and development of plant. Its deficiency causes Stunted growth, and clustering of leaves on short branches, distortion in shape of leaves, and interveinal chlorosis of older leaves. Zinc is essential for the synthesis of tryptophan (a precursor of auxin) thus aids in declining fruit drop. (Pankaj et al., 2018; Kumar et al., 2020). Iron is required for synthesis of chlorophyll molecule, protein synthesis, DNA synthesis, helps in the absorption of other elements, reduction of sulphates and nitrates, normal growth and plant function. Iron act as catalyst in synthesis of chlorophyll molecule and helps in the absorption of other elements. Furthermore, it helps in activation of metabolic pathways, essential structural component of porphyrin molecules (hemes, cytochrome, ferrichrome, hematin, and leg hemoglobin), and is a prosthetic group constituent of various enzymes (Kumar et al., 2020). Copper has pivotal role in several biochemical reactions, during chlorophyll synthesis it helps in utilization of iron, and indirectly effects nodule formation. Its deficiency causes yellowing of young leaves, distortion of the younger leaves, leaf edges ragged, curling of tips, plant top may wither, reduction in growth, and necrosis of apical meristem (Harris and Lavanya, 2016; Kumar et al., 2020). Nickel is important component required at reproductive phase as it is needed for nitrogen metabolism for urease activation, can substitute (for Fe and Zn) as a cofactor for some enzymes, iron uptake and control senescence. Its deficiency causes reduction in efficiency of nitrogen fixation in leguminous plants and leaf tip necrosis in some plants (Kumar et al., 2020).

21.2.7 Soil pollutants

Many sources are involved in soil pollution and some of these are pesticides, chemical fertilizers, smelting wastes, mining wastes, municipal wastes, industrial wastes, leakage of landfill leachate, accidentals spill, vehicular emissions, energy (or fuel) production and other sources like parent material (Nieder et al., 2018). The contaminated soil consequently disturbs plant health by creating hindrance in root growth. Toxic metals (Cd, Cr, Hg, Pb, Cu, Al, Zn, Ni, etc.) and metalloids (Sb, As, and Se) in soil is detrimental to plants and microflora (Bolan et al., 2014). The high concentration of these contaminants exerts stress which produces high level of ethylene that depletes nutrient uptake and restricts root growth (Sharma et al., 2017). Through different industrial as well as agricultural sources heavy metals are rising and getting concentrated into soil (Singh and Agrawal, 2007). Due to high concentration of heavy metals there is production of ethylene in plants which restricts root growth and reduces nutrient availability (Suhani et al., 2021). The buildup of metals (Cd, Cr, Hg, Pb, Cu, Al, Zn, Ni, etc.) and metalloids (Sb, As, and Se) is severely toxic and have detrimental effect on the environment, plants, animals and microbes (Suhani et al., 2021). Soil dwelling organisms get adversely affected as heavy metals causes functional disturbance, protein denaturation or the destruction of the integrity of cell membranes and this creates negative impact on morphology, growth, and metabolism of soil dwelling organisms.

Abiotic factors affected the human's livelihood since ages, but percussions were felt better, only when land-use patterns changed, nutritional demands increased, and food security became a peril. The existence of degraded lands is worldwide, which is increasing at a frightening rate. Causes of land degradation are occasionally natural, but increased anthropogenic activities have augmented the rate of degradation, which has become riskier. There are several researches available that shows the harmful effects of abiotic factors on soil properties, microflora, seed germination, plant growth, and soil-dwelling organisms.

21.3 Stress alleviating microbial communities

Soil microflora is an essential part of soil-plant environment, mainly because of their key roles in improving soil and plant health. Many of known beneficial microbes present in environment performs a number of tasks including environmental stresses alleviation, nutrient solubilization and mineralization, enhances plants systemic tolerance, bioremediation of contaminated environments, and improves nutrient uptake in plants (Sapre et al., 2019; Shah et al., 2020). These beneficial microorganisms include several species of bacteria, fungi, algae, actinomycetes, protozoa, nematodes, etc. (Grover et al., 2011; Selvakumar et al., 2012). Degraded soils can be restored by the means applying microbial inoculants, extracts, biofertilizers, and biostimulants, which is produced at a commercial scale worldwide for the betterment of agriculture.

21.3.1 Bacterial communities

A number of endophytic and exophytic plant growth promoting bacteria resides in soil-plant system. Rhizospheric soil microflora is dominated by beneficial bacterial communities which are either symbiotic (as rhizobia) or free living (like rhizobacteria). In comparison to rhizospheric and symbiotically living bacterial species, endophytes were seen to feature more stress tolerance and mitigation abilities to plant (Upadhyay et al., 2019). The application of these bacteria on seed, plant and soil influences plant growth and imparts tolerance against many abiotic stresses.

21.3.2 Rhizobia

Rhizobia are a group of symbiotic nitrogen fixing bacteria of family *Rhizobiaceae*. In agriculture, around 80% of total symbiotically fixed nitrogen accounted as a result of symbiosis between leguminous plants and bacteria of *Rhizobiaceae* (Sindhu et al., 2020). Symbiotically rhizobia form nodules (specialized nitrogen structure for nitrogen fixation) on the roots of host legume plants. The symbiotic establishment of rhizobia depends upon many variables like interaction to host plants, stress severity, soil type, plantation site conditions, local microflora and management tactics (Sindhu et al., 2020). The environmental abiotic stresses limit the nitrogen fixing attribute of rhizobia by affecting their survival, growth, life cycle and population (Sindhu et al., 2019). However, researchers identified many symbiotic rhizobia which were able to tolerate the harsh stress conditions of soil salinity, temperature, acidity, alkalinity, drought, and metal toxicities (Table 21.1) (Sindhu et al., 2019). Rhizobia promotes plant growth in abiotic stress conditions by fixing nitrogen in roots, producing phytohormones and vitamins, nutrient solubilization, mobilization, and mineralization (Yu et al., 2012; Prasad et al., 2015; Sindhu et al., 2019). They also contribute to biotic stress alleviation in plants by producing bioactive molecules which inhibits plant pathogens and also induces systemic tolerance (Datta and Chakrabartty, 2014). Some temperature tolerant rhizobial species are known for fixing nitrogen in plant roots at temperatures ranging from 25°C–65°C, like, *Mesorhizobium* sp., *Rhizobium* sp. (strain NBRI0102 and NBRI2505) (Kulkarni et al., 2000; Zahran et al., 2012). Zeenat et al. (2017) performed a laboratory experiment to assess the thermo-tolerance of rhizobial species isolated from root nodule of Chickpea plant. It was found that the growth performance of rhizobial isolates were optimum at 28°C–30°C, moderate at 35°C, and retarded at 40°C temperature. Damodara Chari (2018), collected rhizospheric soil samples from different salt and drought affected areas of Telangana state and isolated some rhizobial isolates from

Table 21.1 Rhizobia involved in abiotic stress alleviation under different conditions.

Rhizobia	Isolated from	Benefitted plant species/in vitro analysis	Stress/conditions	Remarks	Reference
Sinorhizobium meliloti	–	In vitro	Desiccation (a combined result of salt, osmotic, and temperature stress)	*S. meliloti* tolerated and survived in desiccation because of producing desiccation protectants, reactive oxygen species-scavenging compounds, heat shock proteins and accumulating osmoprotectants	Vriezen et al. (2007)
Bradyrhizobial sp. strains	Soybean	Soybean	Growth and survival of rhizobial isolate assessed at high temperatures (38°C and 41°C)	There were 19 rhizobial isolates found growing at 38°C and 10 isolates were capable of growing at 41°C. Isolates	Rahmani et al. (2009)
Rhizobium tropici CIAT899 strain	–	*Phaseolus vulgaris* L. (common bean)	Inoculated plants evaluated for nutrient use efficiency	Inoculation significantly increased nodulation, and elevated nitrogen content and phosphorus use efficiency of plants	Tajini et al. (2012)
Bradyrhizobium japonicum	Root nodules of soybean	Soybean	Performance of isolates evaluated under drought stress	Strain inoculated plants tolerated drought stress effectively. Strains and inoculated plants possessed good growth up to 15% polyethylene glycol (PEG) concentration. Inoculated plants also survived with reduced growth at 30% PEG concentration	Uma et al. (2013)
Sinorhizobium sp., and *Rhizobium* sp.	Root nodules of *Lens culinaris*, *Lathyrus sativus*, *Medicago truncatula*, *Medicago marina*, and *Medicago minima*	*L. culinaris*, *L. sativus*, *M. truncatula*, *M. marina*, and *M. minima*	Inoculated plants were evaluated for phytoremediation efficiency of heavy metal (Cd)	The rhizobial inoculation increased plant biomass and root nodulation in Cd stress	Guefrachi et al. (2013)

respective samples. All the isolated rhizobial isolates were screened for thermo-tolerance at 30°C, 35°C, 40°C, 45°C, and 50°C and drought tolerance at different concentrations of polyethylene glycol (PEG 6000) solution, −0.05, −0.15, −0.30, and −0.73 MPa, respectively. It was observed that the one rhizobial isolate (YR-21) showed thermo-tolerance from temperature 20°C to 50°C, three isolates (RR-1, GR-5, and GNR-8) showed thermo-tolerance from 20°C to 45°C, and one isolate (ASR-17) showed thermo-tolerance from 30°C to 40°C. Results revealed that, three isolates (GNR-8, ASR-19, and YR-21) showed drought tolerance up to −0.30 MPa, and no isolate was found tolerant to −0.73 MPa. Dhull and Gera (2017) assessed drought tolerance of 158 rhizobial strains isolated from root nodules of cluster bean. It was reported that most of isolated rhizobia were able to grow up to 20% PEG concentration, and a decrease in growth was observed with increasing PEG concentration of 30% and 40% only 54 and 8 rhizobial isolates were found growing, respectively. Salt stress is another major factor reducing crop yield and agricultural productivity worldwide. Some of the rhizobia were found to have ability of tolerating salt stress and imparting systemic tolerance to the plants. Chaudhary and Sindhu (2017) isolated 55 rhizobacterial strains from root nodule of chickpea plant and screened them for salt tolerance. It was found that 41.8% of isolates were growing at 3% salinity (NaCl concentration) and only 10.9% of isolates were able to tolerate 4% NaCl concentration. Coinoculation of rhizobacterial isolate (RHD18) and rhizobial isolate (*Mesorhizobium* isolate MBD26) tolerated salt stress more effectively than singly inoculated chickpea plants. In salt stress, coinoculation elevated the plant growth in terms of root dry weight, shoot dry weight, and root nodule number. Some of the tree root rhizobia is more effective in stress alleviation as compared to legume plant nodule rhizobial strains. Zhang et al. (1991) isolated rhizobial strains from root nodules of *Prosopis chilenis* and *Acacia senegal* and nodules of some woody trees, and screened the isolates for temperature, heavy metal as well as salt tolerance. Results revealed that the tree rhizobia exhibited high temperature (38°C–44°C), salt (normal growth at 3% NaCl) and heavy metal tolerance as compared with other legume isolated strains. According to Vriezen et al. (2007), rhizobia were able to tolerate abiotic stresses because of production of exopolysaccharides, ROS-scavenging compounds, heat shock proteins, chaperons, improved accumulation of desiccation protectants and osmoprotectants through expression of NaCl-responsive loci.

21.3.3 Endophytic, free living and rhizospheric plant growth promoting bacteria

Many of endophytic, free living, and rhizospheric bacterial strains are found significantly influencing the soil as well as plant health by direct and indirect effects against plethora of abiotic stresses (Haq et al., 1999; Singh et al., 2016; Komaresofla et al., 2019). Among the beneficial plant growth promoting bacteria there were many bacterial genera such as, *Burkholderia*, *Bacillus*, *Beijerinkia*, *Klebsiella*, *Enterobacter*, *Erwinia*, *Gluconacetobacterium*, *Staphylococcus*, *Flavobacterium*, *Pseudomonas*, and *Serratia*, which are known to influence plant growth (Tiwari et al., 2016; Ashraf et al., 2017; Komaresofla et al., 2019).

A diversity of endophytic bacteria resides inside entire plant system (that includes leaf, stem, seed, root, etc.), many of which enhances nutrient uptake, produces osmoprotectants, phytohormones and ROS-scavenging enzymes, etc. and imparts tolerance to growing plants (Singh et al., 2016; Komaresofla et al., 2019). In reclamation of degraded lands through phytoremediation, the application of endophytic bacteria is preferred more than use of rhizospheric bacteria. This is mainly because of the natural tendency of endophytes to inhabit inside plant structures, this reduces the competition between endophytes and other microbial strains (Upadhyay et al., 2019). A variety of plant growth promoting endophytic bacteria were found, which provided systemic tolerance against environmental stresses to many plants (Table 21.2) (Khan et al., 2020; Fan et al., 2020). Khan et al. (2020) found that an endophytic bacteria *Bacillus cereus* SA1 isolated from *Echinochloa crus-galli* L. is capable of tolerating high temperature up to 45°C and also producing PGP-traits (Indole-acetic acid, gibberellins, organic acids, etc.) in presence of stress. They inoculated soybean plants with the bacterium and found a better tolerance response to heat stress and significantly increased plant growth in terms of chlorophyll content, root length, shoot length, plant fresh and dry weight, as compared to noninoculated control plants. Fan et al. (2020) isolated 98 endophytic and ectophytic bacterial isolates from four different plants which is *Scorzoneroides autumnalis*, *Glycine max*, *Phalaris arundinacea*, and *Solanum dulcamara*. Out of 98 isolates, four endophytic bacterial isolates (which belongs to *Pseudomonas* sp., *Mucilaginibacter* sp., *Bacillus* sp., and *Rhizobium* sp.) were screened on basis of their plant growth promoting traits (PGP-traits) and salt tolerance capabilities. It was reported that all the four endophytes were able to tolerate salinity stress up to 200 mM NaCl concentration, but *Mucilaginibacter* sp., *Bacillus* sp., and *Rhizobium* sp. seen to be more effective against salt stress mitigation in inoculated plant (*Arabidopsis thaliana*). The inoculated *A. thaliana* plants resulted in better growth and vigor in terms of physiological (plant dry and fresh weight) and biochemical (antioxidative enzymes) responses in presence of salt stress. Komaresofla et al. (2019) reported an increase in PGP-traits production (IAA, ACC-deaminase, available phosphate), physiological [relative water content, total chlorophyll, superoxide dismutase (SOD), proline] and morphological (Plant height and biomass) parameters of plants (*Salicornia* sp.) against salt stress (different concentration of NaCl), when coinoculated with an endophytic and a rhizospheric bacterial strain. In some cases, endophytic bacterial strains isolated from medical plants are seen to reduce or inhibit biotic stress (pathogenesis) in certain plants. Abdelshafy Mohamad et al. (2020) isolated some endophytic strains of medicinal plant *Thymus vulgaris* and observed them for biotic (pathogen) and abiotic stress (NaCl salt stress) tolerance. Results revealed that some strains of *Bacillus* sp. promoted growth of inoculated tomato plants in salt stress up to 200 mM NaCl and strains of *Enterobacter* sp. showed antagonistic effects to plant pathogen (*Fusarium oxysporum*).

Nowadays, inoculation or use of PGPR is considered as an effective technique to reduce or alleviate wide range of abiotic and biotic stresses (Table 21.2). PGPR provides systemic tolerance and enables plant growth in many stress conditions by performing wide array of mechanisms such as, production of phytohormones

Table 21.2 Bacterial species involved in abiotic stress alleviation under different conditions.

Free living/endophytic/ rhizospheric bacterium	Isolated from	Benefitted plant species/in vitro analysis	Stress/conditions	Remarks	Reference
Drought stress					
Pseudomonas oryzihabitans (Ep4 strain), *Variovorax paradoxus* (5C-2 strain), and *Achromobacter xylosoxidans* (Cm4 strain) (rhizospheric)	Rhizosphere of *Brassica juncea* and *Pisum sativum*	*Solanum tuberosum* (potato)	Plants grown in water deficient conditions	Inoculation increased plant biomass and tuber yield against drought stress in both pot and field experiment. PGPR's also decreased rhizospheric ACC concentrations	Belimov et al. (2015)
Pseudomonas putida (strain MTCC5279)	Desert areas of Rajasthan	*Cicer arietinum L.* (chickpea)	Inoculated plants were evaluated for physiological and morphological parameters at different PEG stress levels (15% and 30%)	The inoculated seeds showed increased germination percentage against different PEG concentrations (15% and 30%). Root length, shoot length, nodes number, and total biomass is found increased in inoculated plants	Tiwari et al. (2016)
Flood stress					
Ochrobactrum rhizosphaerae, *A. xylosoxidans*, *Herbaspirillum seropedicae*, and *Serratia Ureilytica* (rhizospheric)	Rhizosphere of *Ocimum sanctum* (tulsi)	*O. sanctum*	Inoculated plants were tested for waterlogging stress tolerance	Rhizobacteria protected Tulsi plants from waterlogging stress and provided resistance to stress by producing ACC-deaminase, reducing proline concentration, lipid peroxidation and stress ethylene levels	Barnawal et al. (2012)

Temperature stress					
Pseudomonas sp. *strain* AKM-P6 (97% homology with *Pseudomonas aeruginosa*) (rhizospheric)	Rhizosphere of pigeon pea	*Sorghum*	Inoculated plants tested on higher temperatures, 47°C–50°C (day) and 30°C–33°C (night)	Enhanced biosynthesis of high molecular weight protein, improved metabolites level (proline, sugars, chlorophyll, proteins and amino acids), and higher plant biomass	Ali et al. (2009)
Pseudomonas frederiksbergensis OS211, and *Flavobacterium* sp. OR306 (free living)	Isolated from soil in winter season	*Solanum lycopersicum* (tomato plant)	Plants were subjected to grow in lower temperature or chilling stress (In day 12°C and night 10°C)	A significant reduction observed in stress ethylene concentrations, and ACC levels observed in inoculated tomato plants. Inoculation improved resistance against chilling stress in plants	Subramanian et al. (2015)
Salt stress					
Bacillus tequilensis, Bacillus endophyticus, V. paradoxus, Planococcus rifietoensis, and *Arthrobacter agilis* (endophytes)	Inside root of *Salicornia europaea*	In vitro	Endophytes were evaluated for PGP-trait production and salt tolerance (50–500 mM)	All the strains were found to solubilizing phosphate, producing IAA and ACC-deaminase at different NaCl concentration	Zhao et al. (2016)
Staphylococcus sp. E14 (endophyte)	Inside root of *Salicornia* sp.	*Salicornia* sp.	Inoculated plants were evaluated for morphological, physiological, and biochemical parameters against different NaCl concentrations (200, 400, 600 mM)	Plant height and root length observed increasing up to 600 mM. Plant weight found increased up to 400 mM and started decreasing at higher concentration when compared to uninoculated control	Komaresofla et al. (2019)

(*Continued*)

Table 21.2 (Continued)

Free living/endophytic/ rhizospheric bacterium	Isolated from	Benefitted plant species/in vitro analysis	Stress/conditions	Remarks	Reference
Bacillus thuringiensis, *Bacillus alcalophilus*, and *Gracilibacillus saliphilus* (endophytes)	Phyllosphere of halophyte *Arthrocnemum macrostachyum*	*A. macrostachyum*	Salinity tolerance assessed for the bacterial isolate and inoculated plants against different NaCl concentrations (0, 510, and 1030 mM)	Bacterial isolates exhibited siderophore, IAA, ACC-deaminase production, nitrogen fixation and phosphate solubilization. Inoculation resulted in improved germination rate, osmotic adjustments, and salinity tolerance in plants	Navarro-Torre et al. (2017)
Staphylococcus sp. R11 (rhizospheric)	Rhizosphere of *Salicornia* sp.	*Salicornia* sp.	Inoculated plants were evaluated for morphological, physiological, and biochemical parameters against different NaCl concentrations (200, 400, 600 mM)	Plants height found to be increased at all the tested NaCl concentrations. Increased root length and plant weight observed up to 400 mM as compared to uninoculated control	Komaresofla et al. (2019)
Soil pollutants stress					
P. aeruginosa strain BS2 (free living strain)	Isolated from oily sludge	In vitro	Bacterial strain evaluated for Cd and Pb removal efficiency	Strain possess biosurfactant production (di-rhamnolipid), and efficiently removed Cd (92%) and Pb (88%) contaminated soils after 36 h of leaching	Juwarkar et al. (2007)

Klebsiella sp., and *Enterobacter cloacae* (free living bacterial strains)	Industrial effluents (textile and chemical industry)	In vitro	Isolates were evaluated for growth in medium supplemented with variable concentrations of heavy metals like, cadmium (Cd), chromium (Cr), and lead (Pb)	*E. cloacae* isolate tolerated 220 μg/mL of Cd, 800 μg/mL of Cr, and 1400 μg/mL of Pb. While *Klebsiella* sp. isolates tolerated up to 110 μg/mL of Cd, 600 μg/mL of Cr, and 1200 μg/mL of Pb	Haq et al. (1999)
Enterobacter sp. C1D (rhizospheric)	Root surface of *Cajanus cajan*	*C. cajan*	Isolate evaluated for Cd tolerance and alleviating effects on plants	The isolate showed good tolerance to Cd stress (120 μg/mL). Inoculated plants have increased growth parameters, and antioxidant metabolites	Sharma et al. (2020)
Azotobacter sp., *Pseudomonas* sp., *and Bacillus* sp. (rhizospheric)	Rhizosphere of *Withania somnifera* (ashwagandha)	*W. somnifera*	Isolates tested for Ni and Cd tolerance	Isolates effectively tolerated both the heavy metal stress, also significantly increased the seed germination, PGP-traits (HCN, CAT, IAA, and ammonia), and root shoot growth	Rathaur et al. (2012)
Stenotrophomonas maltophilia, Stenotrophomonas acidaminiphila, Brevundimonas diminuta (rhizospheric)	Rhizosphere of *Cynodon dactylon* grown in effluent contaminated soil	In vitro	Bacterial strains were evaluated for tolerance at different concentrations (50–600 mg/L) of copper metal	All strains exhibited PGP-traits (phosphate solubilization, siderophore production, ACC-deaminase) production and Cu resistance	Rathi and Nandabalan (2017)

(*Continued*)

Table 21.2 (Continued)

Free living/endophytic/ rhizospheric bacterium	Isolated from	Benefitted plant species/in vitro analysis	Stress/conditions	Remarks	Reference
Bacillus sp. SBER3 (endophytic)	Roots of *Populus deltoides*	*P. deltoids*	Inoculated plants were grown in PAH (naphthalene and anthracene) contaminated soil	The bacterial isolate degraded PAH contamination by 45.6%. Inoculated plants showed significant increase in biomass and plant shoot root length	Bisht et al. (2014)
Sphingomonas sp., *Brevibacillus parabrevis* FCm9, and *Acinetobacter calcoaceticus* MCm5	Agricultural soil	In vitro	Isolates tested for pesticide (cypermethrin) degradation and PGP-traits production	The bacterial isolates were able to degrade cypermethrin up to 85% within 10 days. Isolates also exhibited PGP-traits production (IAA, ammonia, P-solubilization, etc.)	Akbar et al. (2015)

ACC, 1-aminocyclopropane-1-carboxylic acid; CAT, catalase; HCN, hydrogen cyanide; IAA, indole acetic acid; PAH, polycyclic aromatic hydrocarbons; PEG, polyethylene glycol; PGPR, plant growth promoting rhizobacteria.

(IAA, gibberellins, cytokinin, etc.), osmolyte, exopolysaccharide, siderophore, antioxidants, and mineral solubilization (K, P, Zn, etc.), as well as stress enzymes modulation (increase proline, reduce stress ethylene, etc.) (Tiwari et al., 2016; Sapre et al., 2019). Ethylene, a plant hormone synthesized by plants to regulate essential processes such as, root hairs development, root elongation, nodule development, fruit ripening, senescence, seed germination, etc. But as a result of biotic (plant pathogens, and nematodes) and abiotic (temperature extremes, drought, salt, flooding, metal toxicity, etc.) stress persistence, plant generates high ethylene concentrations by certain mechanisms which ultimately harm growth and health of plants (Sapre et al., 2019; Husain et al., 2020). Some of the ACC deaminase producing PGPR's were identified by researchers, which are capable of combating negative effects of high ethylene levels produced during stress in plants (Belimov et al., 2015; Sapre et al., 2019). Torbaghan et al. (2017), in their study on different saline soils of Khorasan Razavi province, Iran, isolated 48 rhizobacterial strains, out of which alkaliphilic, halophilic, and haloakaliphilic PGPRs were isolated. On further identification these isolates were found belonging to four genera of *Bacillaceae* family, that is, *Alkalibacillus*, *Virgibacillus, Haloalkaliphilic bacillus*, and *Bacillus*. The isolates helped in increasing salinity and alkalinity tolerance, better root growth, high yield, and development in wheat plants. Following mechanisms were observed for imparting tolerance and betterment of wheat plants-ACC-deaminase production, ammonia, IAA, gibberellic acid production, phosphate solubilization and nitrogen fixation. Excessive amount of metals and heavy metals such as cadmium (Cd), chromium (Cr), copper (Cu), lead (Pb), zinc (Zn), arsenic (As), and nickel (Ni) in agricultural soil system reduces the health and fertility of soil which leads to alter plant or crop health and agricultural productivity (Hassan et al., 2016). Rathaur et al. (2012) isolated some strains of PGPR from rhizosphere of *Withania somnifera* plant and evaluated them for ACC-deaminase production and certain heavy metal stress tolerance. Results revealed that strains were capable of tolerating Ni and Cd stress and also showed significant increase in germination and growth parameters of inoculated plant in presence of heavy metal stress. Hassan et al. (2016) isolated rhizobacterial strains from rhizosphere of sugarcane plant and tested their inoculation effects on Cd stress tolerance in wheat plants. It was reported that those strains which have both nitrogen fixing and ACC-deaminase producing activity are alleviated Cd stress effects more efficiently than the strain containing either nitrogen fixing or ACC-deaminase producing activity in inoculated wheat plants. The heavy metal stress tolerance mechanism of rhizospheric and other bacteria mainly involves biosurfactants and exopolysaccharides production, metal chelation or complexation, impermeability toward metals, enzymatic detoxification, and efflux (Ullah et al., 2015). The production of biosurfactants by microbes play a major role in solubilization and mobilization of soil heavy metals. Biosurfactants forms complexes with heavy metals, and this complexation initiates metals desorption from soil matrix, thus the process increases metal solubility and mobility (Ullah et al., 2015). Juwarkar et al. (2007) evaluated effectiveness rhamnolipid biosurfactant producing *Pseudomanas aeruginosa* strain on removal of Cd and Pb contaminated soils. Results revealed that the di-rhamnolipid biosurfactant producing strain effectively removed the Cd and Pb metals from contaminated soils.

21.4 Fungal communities

Many plant growth promoting fungal strains were identified belonging to genera-*Aspergillus, Alternaria, Achrothcium, Aurobassium, Arthrobotrys, Cephalosporium, Curvularia, Cunninghamella, Cladosporium, Chaetomium, Dothideomycetes, Glomus, Helminthosporium, Fusarium, Micromonospora, Myrothecium, Mortierella, Oidiodendron, Penicillium, Paecilomyces, Pichia fermentans, Phoma, Pythium, Populospora, Rhizopus, Rhizoctonia, Schizosaccharomyces, Saccharomyces, Sclerotium, Schwanniomyces, Trichoderma, Torula,* and *Yarrowiaare,* which are known to produce antioxidant enzymes, PGP-traits, phytohormones, mobilize organic and inorganic forms of soil phosphorus and other nutrients by means of mineralization and solubilization (Sharma et al., 2013; Alori et al., 2017; Sadeghi et al., 2020). There are many researches available about the benefits of using fungal species in improving plant growth and tolerance in stressed environments (Asrar and Elhindi, 2011; Yaseen et al., 2011; Shah et al., 2020; Khan et al., 2020).

21.4.1 Endophytic and free-living fungi

Many plant growth promoting endophytic and free living fungi are known to alleviate effects caused due to biotic and abiotic stress on growing plants. Endophytes synthesize growth promoting phytohormones (IAA, gibberellic acid, jasmonic acid, ABA, etc.), facilitates essential minerals (P, K, Ca, Mg, and S) absorption, and also play key role in uplifting the nutritional values (lipid, carbohydrate and total protein) against environmental stresses (drought, temperature, and salt stress) in host plants (Table 21.3) (Ismail et al., 2020; Shah et al., 2020; Sadeghi et al., 2020). There were many fungal strains identified by researchers which are capable of tolerating heavy metals and also for the removal of metals from polluted sites via leaching (Khan et al., 2020). In bioremediation of metals, fungal strains are highly suitable and more preferred than other microbial communities because of fungus mycelial attribute and well-recognized capabilities to accumulate all type of metals (Sanyaolu, 2018). In a study Khan et al. (2020), isolated some fungal strains from industrial effluent polluted soil samples and evaluated them for heavy metal tolerance and removal efficiency. The results revealed that the fungal strains *Aspergillus flavus* and *Aspergillus fumigatus* removed heavy metal Pb (lead) efficiently and strains *Aspergillus niger* and *Aspergillus terreus* efficiently removed Hg (mercury) metal. According to Khan et al. (2020), there are certain mechanisms done by fungal strains which governs the bioremediation of different heavy metals. The mechanisms include vacuole formation (metal ions get collected in it), chelation (complexes formation), active exclusion on metals outside the cell, and production of enzymes, pigments and metal binding compounds inside the cell. Endophytic

Table 21.3 Fungal species involved in abiotic stress alleviation under different conditions.

Endophytic/AM fungal strains	Isolated from	Benefitted plant/in vitro analysis	Stress conditions	Remarks	Reference
Drought Stress					
Dothideomycetes sp., *Aurobassium pullunts*, and *Penicillium citrinum* (endophytes)	Trunk, stems, leaves, and roots of *Citrus reticulata* L. (mandarin plant)	Mandarin plant	Drought stress	Inoculation induced ROS scavenging by producing antioxidant enzymes and improved drought tolerance in plants	Sadeghi et al. (2020)
Glomus versiforme	Strain procured from a facility	*Poncirus trifoliate* L. (trifoliate orange)	Drought stress	Inoculation imparted drought tolerance to plants by significantly increasing activities of antioxidants such as SOD, CAT, glaiacol peroxidase	Wu and Zou (2009)
Glomus constrictum Trappe (AMF)	Root of *Allium cepa* L. (White Lisbon)	*Tagetes erecta* (marigold)	Different levels of drought stress	Inoculation improved photosynthesis, root shoot length and total biomass of plants against drought stress by improving nutritional profile (mainly P) and plant water status	Asrar and Elhindi (2011)

(*Continued*)

Table 21.3 (Continued)

Endophytic/AM fungal strains	Isolated from	Benefitted plant/in vitro analysis	Stress conditions	Remarks	Reference
Salt stress					
Glomus intraradices, *Glomus etunicatum*, and *Glomus mosseae* (AMF)	*Acacia gerrardii* (talh tree)	*Cucumis sativus L.* (cucumber)	Salt stress (up to 200 mM NaCl)	Inoculation provided tolerance to cucumber plants by enhancing pigments synthesis, antioxidant activities, and improving accumulation of proline and phenols. Increased total biomass of plants was observed as a result of inoculation	Hashem et al. (2018)
Rhizophagus irregularis (AMF)	Isolate collected from experimental field of Zaidín, Granada	*Solanum lycopersicum* (tomato plants)	Salinity stress	AMF inoculation increased the levels of IAA, cytokinin, and ABA, thus increased the growth of plants in terms of biomass and yield	Khalloufi et al. (2017)
Nutrient deficiency					
G. etunicatum (AMF)	Acidic soil samples	*Panicum virgatum L.* (switchgrass)	Inoculated plants evaluated for nutrient uptake	Inoculated plants showed higher uptake of nutrients such as S, P, K, N, Mg, Cu, and Zn	Clark (2002)

Glomus aggregatum, *G. mosseae*, and *Glomus fasciculatum* (AMF)	Rhizosphere of wheat and maize plants	*Vigna unguiculata* (cowpea)	Inoculation effects on nutrient uptake is evaluated	AMF inoculation significantly improved the uptake of nutrients (such as Mg, K, Ca, Fe, P, and Si). Plants also showed better growth after inoculation	Yaseen et al. (2011)
G. intraradices		*Phaseolus vulgaris* L.	Strain inoculated plants evaluated for nutrient use efficiency	Inoculation significantly increased the phosphorus use efficiency, also fresh and dry weight of plants	Tajini et al. (2012)
Soil acidity					
Glomus sp. (*G. intraradices*, *Glomus diaphanum*, *Glomus clarum*, *G. etunicatum*), *Gigaspora* sp. (*Gigaspora albida*, *Gigaspora margarita*, *Gigaspora Rosea*) and *Acaulospora* sp. (AMF)	Collected by means of program and procurement	Switchgrass	Inoculated plants were observed for growth in acidic soils of different acidity (pH-4 and 5)	Nutrient profile of inoculated plants was more at pH 4 than pH 5. Inoculation improved resistance of plants by uplifting nutrient uptake status	Clark (2002)
Scutellospora pellucida SCT111, *Scutellospora heterogama* SCT113, *Acaulospora scrobiculata* SCT112, and *G. etunicatum* SCT110	Collected from orchard soils	*Malus prunifolia* (apple)	Inoculated plants grown in different pH conditions and evaluated for nutrient uptake	Fungal colonization observed more at lower pH. Shoot of AM plants had higher P concentration and normal Zn, Mn, Cu, and Fe at all pH levels	Cavallazzi et al. (2007)

(Continued)

Table 21.3 (Continued)

Endophytic/AM fungal strains	Isolated from	Benefitted plant/in vitro analysis	Stress conditions	Remarks	Reference
G. etunicatum and *Gi. margarita* (AMF)	-	Cowpea plant	Acidic soils (pH of 4.6, 4.9, 5.3)	Both AMF strains performed well on acidic pH. Inoculation of plants with strains increased root-shoot length, root-shoot weight, shoot phosphorus content and overall plant growth in acidic soil	Rohyadi (2008)
Temperature stress					
Rhizopus oryzae (Endophyte)	Roots of *Adiantum capillus-veneris* L.	*Helianthus Annuus* L., and *Glycine Max* L.	Normal and high temperature stress (25°C and 40°C)	Inoculation alleviated temperature stress in both plants. Plants were observed with high amount of CAT, sugars, proline, flavonoids, phenolic metabolites, lipids and proteins	Ismail et al. (2020)
Funneliformis mosseae, *Rhizophagus intraradices*, and *Funneliformis geosporum* (AMF)	-	*Zea mays* (maize)	High temperature stress (44°C)	AMF protected plants at high temperature by increasing photochemical reactions, as well as improved plant growth and development by	Mathur et al. (2018)

				enhancing Mg uptake, chlorophyll content, photosynthetic activity and plant biomass	
Soil pollutants stress					
Serratia sp. IU01 (endophyte)	*Solanum nigrum*	*Brassica napus* L.	Inoculated plants tested for Cd stress alleviation	Endophytic strain alleviated Cd stress and increased growth of inoculated plants	Shah et al. (2020)
R. intraradices and *G. intraradices* (AMF)	Procured from laboratory	*Robinia pseudoacacia* (black locust)	Inoculated plants evaluated for phytore-mediation efficiency of Pb polluted site	The AMF inoculation improved plant total biomass. Pb translocated in root and shoot more efficiently in inoculated plants	Yang et al. (2016)
Aspergillus flavus and *Aspergillus fumigatus* (free living)	Industrial effluent contaminated soil	In vitro	Isolates evaluated for heavy metal (Pb) removal efficiency	Both isolates tolerated and removed Pb efficiently. The removal efficiency of *A. flavus* ranged from 99%–99.3% and for *A. fumigatus* ranged from 98.5%–99.2%	Khan et al. (2020)
Aspergillus niger and *Aspergillus terreus* (free living)	Industrial effluent contaminated soil	In vitro	Isolates evaluated for heavy metal (Hg) removal efficiency	Both isolates tolerated and removed Hg efficiently. The removal efficiency of *A. niger* ranged from 92%–96% and for *A. terreus* ranged from 91.6%–95.5%	Khan et al. (2020)

(*Continued*)

Table 21.3 (Continued)

Endophytic/AM fungal strains	Isolated from	Benefitted plant/in vitro analysis	Stress conditions	Remarks	Reference
A. niger (free living)	Soil of refused dumping area	*Lactuca sativa* L.	Lambda-cyhalothrin pesticide and associated heavy metals (As, Pb, Cr, and Cd)	Inoculation effectively reduced pesticide contamination to a level below of the maximum residue level. A significant reduction is seen in all heavy metal levels (except As)	Sanyaolu (2018)
G. aggregatum, *R. intraradices*, *Rhizophagus fasciculatus*, and *F. mosseae* (AMF)	Procured from laboratory	*Z. mays L.*	Inoculated plants supplemented with heavy metal contaminated sludge	All fungal strains effectively accumulated metals such as Ni, Cd, Pb, Cr, Fe, Cu, and Zn	Singh et al. (2019)

ABA, Abscisic acid; ACC, 1-aminocyclopropane-1-carboxylic acid; AMF, Arbuscular Mycorrhizal Fungi; CAT, catalase; ROS, reactive oxygen species; SOD, superoxide dismutase.

fungal strains produce exogenous antioxidant enzymes that prevent the accrual of ROS and moderate the oxidative stress in plants, thus providing a defense mechanism to plants (Shah et al., 2020). Sadeghi et al. (2020) observed that the inoculation *Citrus reticulata* L. plants with some endophytic strains belonging to *Dothideomycetes, Aurobassium*, and *Penicillium* genera of fungus, effectively alleviated oxidative damage raised due to drought stress in plants.

21.4.2 Arbuscular mycorrhizal fungi

It is widely believed and accepted fact about AM fungal symbiosis that it prevents host plants from deleterious effects of environmental stresses and facilitates growth and development (Table 21.3) (Asrar and Elhindi, 2011; Rapparini and Peñuelas, 2014; Muthukumar et al., 2014; Hashem et al., 2018). The main mechanisms governed by arbuscular mycorrhizal fungi (AMF) which provide tolerance to host plants against stresses includes increased hydraulic conductivity of root, stomatal regulation, improved water uptake either at poor soil water content via extraradical hyphae, osmotic alterations, accumulation of carbohydrate and proline, improved photosynthesis, and increased nutritional profile (Clark, 2002; Rohyadi, 2008; Asrar and Elhindi, 2011). Soil nutrients uptake plays a major role in any plant growth and development, but its deficiency induces many changes such as growth patterns, lowers defense abilities and antioxidants of plants (Latef et al., 2016). Extraradical hyphae of AMF helps in nutrient acquisition of host plants by absorbing nutrients like N, P, Cu, Zn, etc., from soil and transferring them to plants (Smith and Read, 2010). Yaseen et al. (2011) detected maximum uptake of nutrients such as Mg, Ca, K, P, Si, and Fe in AMF inoculated plants. Soil pH is considered as a delicate factor which determines microfloral abundance and diversity in soil, microbe-plant interactions, vital nutrient availability (N, P, Ca, Mg, Mo, and Zn), and survival of a plant (Muthukumar et al., 2014). A slight decrease in pH can turn normal soil into acidic, which leads to reduced availability of vital nutrients (micro and macro nutrients) and increased availability of phytotoxic elements such as Mn (manganese), Al (aluminum), and Fe (Kochian et al., 2004). AMF inoculation is suggested as a best way to cope up deleterious effects of soil acidity and nutrient deficiency to plants or crops. Rohyadi (2008) reported increase in nutrient uptake and overall growth of cowpea plants in acidic soil (pH-4.6) as a response of AMF inoculation. Drought stress have deleterious effects on plant or crop life cycle, it disrupts plants many essential mechanisms such as root water relation, transpiration rates, osmotic adjustments, etc., Some known AMF strains have notable potential to alleviate deleterious effects of drought stress and promote growth of plants (Asrar and Elhindi, 2011). AMF have ability to modulate plant aquaporin genes under water deficient conditions (drought), this gene is responsible for symbiotic exchange of vital nutrients and water between AMF and host plant (Rapparini and Peñuelas, 2014). Drought exposure increases oxidative damage to plants, many AMF alleviates the damages by antioxidant enzymes activities such as CAT, SOD, peroxidase (POD), and guaiacol POD (Wu and Zou, 2009).

21.5 Algal communities

Biostimulants are group of organic compounds or materials and other diverse products that, improves crop or plant growth when applied in limited amounts (Sharma et al., 2014). Agricultural biostimulants includes plant growth promoters, protein hydrolysates, nitrogenous compounds, elements, fulvic and humic acids, microorganisms, and algal extracts (Sharma et al., 2014; Zou et al., 2019). Many species of locally found micro and macro algae such as *Arthrospira* (*Spirulina*) sp., *Sargassum* sp., *Ascophylum nodosum*, *Botryococcus braunii*, *Chlamydomonas reinhardtii*, *Durvillea antarctica*, *Dunaliella salina*, *Durvillea potatorum*, *Dunaliella tertiolecta*, *Fucus serratus*, *Isochrysis galbana*, *Ecklonia maxima*, *Laminaria digitata*, *Laminaria hyperborea*, *Macrocystis pyrifera*, *Porphyridium cruentum*, and *Himanthalia elongate* are known to produce biostimulants (Sharma et al., 2014; Arroussi et al., 2018). Algal extract biostimulants includes phytohormones, amino acids, phenolics, betains, brassinosteroids, carbohydrates, flavonoids, carotenoids, polyamines, polyols, and polysaccharides (Arroussi et al., 2018; Carillo et al., 2020). Application of many algal extracts are known for their usefulness in improving plant or crop productivity against abiotic (salt, drought, heavy metal stress, etc.) and biotic (pathogen) stresses (Table 21.4) (El-Eslamboly et al., 2019; El-Katony et al., 2020). El-Katony et al. (2020) applied extract of alga *Dictyota dichotoma* to growing plants (*Oryza sativa* L.) for the evaluation of efficiency to alleviate deleterious effects of salt and drought stress. It was observed that the aqueous extract constitutes a variety of beneficial components including PGP-hormones, flavonoids, phenolics, sterols, fucoidan vitamins and many essential micro and macronutrients. A significant increase in growth parameters of salt sensitive rice plants were found in salt and drought stressed soils amended with algal extracts. Similarly, Patel et al. (2018) evaluated efficiency of algae *Kappaphycus alvarezii* sap to improve plant growth and tolerance in drought and salt stress conditions. Findings suggest that algal sap significantly reduced negative effects of both the stresses and promoted plant growth. Zou et al. (2019) found that the application of polysaccharides produced from brown algae (*Lessonia nigrescens*) provides tolerance and enhances growth of wheat plants in salt stressed conditions. The polysaccharides application increased compartmentalization, exclusion of intracellular ions, chlorophyll content and antioxidant activities, also it decreased membrane peroxidation and thus led to improved tolerance and growth of plants. Hussein et al. (2020) reported higher nutrient uptake in sugarbeet plants in salt stress when supplemented with algal extract. The supplemented plants were seen to uptake more N, K, P, Mg, Mn, Zn, Cu, Fe, and Ca in comparison to those plants which were growing in salt stress without supplements. Recently, some species of micro and macro algae is considered as a big spectrum applicable material, these algal species and their extracts are produced commercially in the form of biofertilizers, food source, animal feeding material, stress alleviators, etc. (Hussein et al., 2020).

21.6 Conclusion

Decreasing arable land patches with increasing soil degradation and diminishing nutritional value of crops are key concerns in this shifting global land use paradigm.

Table 21.4 Application of algal extracts and their benefits to plants against abiotic stresses.

Algae	Isolated from	Biostimulants	Benefitted plant material	Remarks	Reference
Laurencia obtuse and *Sargassum muticum*	Red Sea at Hurghada coast in Egypt	Bioactive compounds of algal extract	*Triticum vulgare*	Extract application significantly enhanced the germination rate, germination percentage, growth parameters and antioxidant enzymes	Ibrahim (2016)
Dunaliella salina	Oualidia lagoon in the Atlantic Moroccan Coast	Exopolysaccharides	*Triticum aestivum* (wheat)	Extract application increased germination rate and growth of seedlings in salt stressed (3 and 6 g/L NaCl) soil	El Arroussi et al. (2015)
D. salina Ms002 (chlorophyte)	Oualidia lagoon at the Moroccan Atlantic	Exopolysaccharides	*Solanum lycopersicum* (tomato)	Algal extract application alleviated salt stress and increased growth parameters by altering proline, phenolic compounds, Na^+, and antioxidant enzymes (POD, SOD, CAT)	Arroussi et al. (2018)
Kappaphycus alvarezii	Procured form CSIR-CSMCRI Mandapam field station	Algal sap (K-sap)	*Triticum Durum* (pasta wheat)	Algal K-sap application improved plant growth parameters in drought as well as salt stress by maintaining Na^+/K^+ ratio, osmoprotectants (proteins, amino acids, proline, and soluble sugars) and phytohormones	Patel et al. (2018)

(*Continued*)

Table 21.4 (Continued)

Algae	Isolated from	Biostimulants	Benefitted plant material	Remarks	Reference
Grateloupia filicina	Beach in Qingdao, China	Polysaccharides	*Oryza sativa* L. (rice)	Polysaccharides treatment induced salt tolerance, elevated germination rates and growth parameters of plants. Treatments increased proline, SOD, POD and thus improved ROS scavenging and osmotolerance	Liu et al. (2019)
D. salina and *Phaeodactylum tricornutum*	Procured from laboratory	Extracts	*Capsicum annuum* L. (bell pepper)	Extracts application effectively mitigated salt stress in plants. Treatment increased germination rates by decreasing superoxide radicals and increasing antioxidant enzymes	Guzmán-Murillo et al. (2013)

CAT, catalase, POD, peroxidase; ROS, reactive oxygen species; SOD, superoxide dismutase.

Expanding human settlements made conversion of arable land resources into nonarable forms and the leftover agricultural lands are degrading due to improper techniques and management. On global scale, a vast proportion of land resources are degraded or facing some degree of deterioration. In agroecosystems, a growing plant needs optimal environmental conditions for their healthy growth and producing satisfactory or substantial amount of yields. But the presence of environmental stresses (biotic and abiotic) hinders growth and productivity of plants or crops. The natural causes of abiotic stress make situation more worsened when the negative anthropogenic activities are increased. The increasing food demands with decreasing arable lands have created pressure on agricultural systems to utilize each patch of land in the best way possible. Our curtailed perceptive of how some microbial communities through their different mechanisms such as PGP-traits production, pollutants detoxification, symbiosis, tolerance, nutrients solubilization and mobilization can help plants or crops to survive in environmental stress conditions. The strategy involving microorganisms could give multiple benefits like economic viability, sustainability, upgrading health and nutritional profile of soil-plant system, increase yield, improving ecosystem functioning and no damaging effect. The development done in understanding the field of microbes seems ready to use but further researches will enhance the knowledge and its worth will increase to the next level. Microbial utilizing strategy gives us the lost hope to reduce the stress and stop degradation of arable land masses, also the degraded lands can be reclaimed for the use of agriculture. Hence, the microbial community seems to serve the whole mankind and fulfill the future food demands.

Acknowledgment

The authors are grateful to the Dean & Head, Department of Environment and Sustainable Development and Director, Institute of Environment and Sustainable Development, Banaras Hindu University, for providing essential amenities. Rajeev Pratap Singh is grateful to the Science and Engineering Research Board (SERB), India for the project support (EMR/2017/002525).

Conflict of interest/declaration of interest

The authors declare that they have no known competing financial interests or personal relationships that could have appeared to influence the work reported in this paper.

Credit roles

Sinha Sahab: Visualization, Investigation, Writing—original draft, Writing—review & editing.

Ibha Suhani: Visualization, Investigation, Writing—original draft, Writing—review & editing.

Rajeev Pratap Singh: Conceptualization, Visualization, Writing—review & editing, Resources, Supervision, Funding acquisition.

References

Abdelshafy Mohamad, O.A., Ma, J.B., Liu, Y.H., Zhang, D., Hua, S., Bhute, S., et al., 2020. Beneficial endophytic bacterial populations associated with medicinal plant *Thymus vulgaris* alleviate salt stress and confer resistance to *Fusarium oxysporum*. Front. Plant Sci. 11, 47.

Abenavoli, M.R., Panuccio, M.R., Sorgonà, A., 2012. Root form and function in plant as an adaptation to changing climate. Environmental Adaptations and Stress Tolerance of Plants in the Era of Climate Change. Springer, New York, NY, pp. 175–198.

Akbar, S., Sultan, S., Kertesz, M., 2015. Determination of cypermethrin degradation potential of soil bacteria along with plant growth-promoting characteristics. Curr. Microbiol. 70 (1), 75–84.

Ali, S.Z., Sandhya, V., Grover, M., Kishore, N., Rao, L.V., Venkateswarlu, B., 2009. *Pseudomonas* sp. strain AKM-P6 enhances tolerance of sorghum seedlings to elevated temperatures. Biol. Fertil. Soils 46 (1), 45–55.

Alori, E.T., Glick, B.R., Babalola, O.O., 2017. Microbial phosphorus solubilization and its potential for use in sustainable agriculture. Front. Microbiol. 8, 971.

Armengaud, P., Sulpice, R., Miller, A.J., Stitt, M., Amtmann, A., Gibon, Y., 2009. Multilevel analysis of primary metabolism provides new insights into the role of potassium nutrition for glycolysis and nitrogen assimilation in *Arabidopsis* roots. Plant Physiol. 150 (2), 772–785.

Arroussi, H.E., Benhima, R., Elbaouchi, A., Sijilmassi, B., Mernissi, N.E., Aafsar, A., et al., 2018. *Dunaliella salina* exopolysaccharides: a promising biostimulant for salt stress tolerance in tomato (*Solanum lycopersicum*). J. Appl. Phycol. 30 (5), 2929–2941.

Ashraf, M.A., Hussain, I., Rasheed, R., Iqbal, M., Riaz, M., Arif, M.S., 2017. Advances in microbe-assisted reclamation of heavy metal contaminated soils over the last decade: a review. J. Environ. Manage. 198, 132–143.

Asrar, A.W.A., Elhindi, K.M., 2011. Alleviation of drought stress of marigold (*Tagetes erecta*) plants by using arbuscular mycorrhizal fungi. Saudi J. Biol. Sci. 18 (1), 93–98.

Barnawal, D., Bharti, N., Maji, D., Chanotiya, C.S., Kalra, A., 2012. 1-Aminocyclopropane-1-carboxylic acid (ACC) deaminase-containing rhizobacteria protect *Ocimum sanctum* plants during waterlogging stress via reduced ethylene generation. Plant Physiol. Biochem. 58, 227–235.

Bechtold, U., Field, B., 2018. Molecular mechanisms controlling plant growth during abiotic stress. J. Exp. Bot., 69, 2753–2758.

Belimov, A.A., Dodd, I.C., Safronova, V.I., Shaposhnikov, A.I., Azarova, T.S., Makarova, N. M., et al., 2015. Rhizobacteria that produce auxins and contain 1-amino-cyclopropane-1-carboxylic acid deaminase decrease amino acid concentrations in the rhizosphere and improve growth and yield of well-watered and water-limited potato (*Solanum tuberosum*). Ann. Appl. Biol. 167 (1), 11–25.

Bisht, S., Pandey, P., Kaur, G., Aggarwal, H., Sood, A., Sharma, S., et al., 2014. Utilization of endophytic strain *Bacillus* sp. SBER3 for biodegradation of polyaromatic hydrocarbons (PAH) in soil model system. Eur. J. Soil Biol. 60, 67–76.

Bolan, N., Kunhikrishnan, A., Thangarajan, R., Kumpiene, J., Park, J., Makino, T., et al., 2014. Remediation of heavy metal(loid)s contaminated soils–to mobilize or to immobilize? J. Hazard. Mater. 266, 141–166.

Bordeleau, L.M., Prévost, D., 1994. Nodulation and nitrogen fixation in extreme environments. Plant Soil 161 (1), 115–125.

Carillo, P., Ciarmiello, L.F., Woodrow, P., Corrado, G., Chiaiese, P., Rouphael, Y., 2020. Enhancing sustainability by improving plant salt tolerance through macro- and micro-algal biostimulants. Biology 9 (9), 253.

Carmen, B., Defez, R., 2011. Soil bacteria support and protect plants against abiotic stresses. In: Abiotic Stress in Plants: Mechanisms and Adaptations, pp. 143–171.

Cavallazzi, J.R.P., Klauberg Filho, O., Stürmer, S.L., Rygiewicz, P.T., de Mendonça, M.M., 2007. Screening and selecting arbuscular mycorrhizal fungi for inoculating micropropagated apple rootstocks in acid soils. Plant Cell Tissue Organ Cult. 90 (2), 117–129.

Change, I.P.O.C., 2007. Working Group III Report: Mitigation of Climate Change. In: Climate Change 2007, IPCC Fourth Assessment Report (AR4). <http://www.ipcc.ch/> (Consulta: Noviembre 2008).

Chaudhary, D., Sindhu, S.S., 2017. Amelioration of salt stress in chickpea (*Cicer arietinum* L.) by coinculation of ACC deaminase-containing rhizospheric bacteria with *Mesorhizobium* strains. Legume Research-An Int. J. 40 (1), 80–86.

Clark, R.B., 2002. Differences among mycorrhizal fungi for mineral uptake per root length of switchgrass grown in acidic soil. J. Plant Nutr. 25 (8), 1753–1772.

Damodara Chari, K., 2018. Screening and characterization of stress tolerant rhizobia from different soils of Telangana. African J. Microbiol. Res. 10, 1038–1042. *ISSN*, 0975–5276 & E-ISSN: 0975–9174.

Datta, B., Chakrabartty, P.K., 2014. Siderophore biosynthesis genes of *Rhizobium* sp. isolated from *Cicer arietinum* L. 3 Biotech 4 (4), 391–401.

Dhull, S., Gera, R., 2017. Assessing stress tolerant rhizobium isolates of clusterbean (*Cymopsis tetragonoloba* (L.) Taub.) retrieved from semi-arid regions of Haryana, India. Int. J. Curr. Microbiol. Appl. Sci. 6 (4), 744–753.

Egamberdieva, D., Kucharova, Z., Davranov, K., Berg, G., Makarova, N., Azarova, T., et al., 2011. Bacteria able to control foot and root rot and to promote growth of cucumber in salinated soils. Biol. Fertil. Soils 47 (2), 197–205.

Egamberdieva, D., Lugtenberg, B., 2014. Use of plant growth-promoting rhizobacteria to alleviate salinity stress in plants, Use of Microbes for the Alleviation of Soil Stresses, vol. 1. Springer, New York, NY, pp. 73–96.

El Arroussi, H., Elbaouchi, A., Benhima, R., Bendaou, N., Smouni, A., Wahby, I., 2015. Halophilic microalgae *Dunaliella salina* extracts improve seed germination and seedling growth of *Triticum aestivum* L. under salt stress. In: Proceedings of the Second World Congress on the Use of Biostimulants in Agriculture, 1148, pp. 13–26.

El-Eslamboly, A.A.S.A., Abd El-Wanis, M.M., Amin, A.W., 2019. Algal application as a biological control method of root-knot nematode *Meloidogyne incognita* on cucumber under protected culture conditions and its impact on yield and fruit quality. Egypt. J. Biol. Pest Control 29 (1), 18.

El-Katony, T.M., Deyab, M.A., El-Adl, M.F., Ward, F.M.E.N. 2020. The aqueous extract and powder of the brown alga *Dictyota dichotoma* (Hudson) differentially alleviate the impact of abiotic stress on rice (*Oryza sativa* L.). Physiol. Mol. Biol. Plants, 26, 1155–1171.

Fan, D., Subramanian, S., Smith, D.L., 2020. Plant endophytes promote growth and alleviate salt stress in *Arabidopsis thaliana*. Sci. Rep. 10 (1), 1–18.

Farooq, M., Wahid, A., Kobayashi, N., Fujita, D.B.S.M.A., Basra, S.M.A., 2009. Plant drought stress: effects, mechanisms and management. Sustainable Agriculture. Springer, Dordrecht, pp. 153–188.

Fritioff, Å., Kautsky, L., Greger, M., 2005. Influence of temperature and salinity on heavy metal uptake by submersed plants. Environ. Pollut. 133 (2), 265–274.

Gamalero, E., Glick, B.R., 2012. Ethylene and abiotic stress tolerance in plants. Environmental Adaptations and Stress Tolerance of Plants in the Era of Climate Change. Springer, New York, NY, pp. 395–412.

Gibbs, H.K., Salmon, J.M., 2015. Mapping the world's degraded lands. Appl. Geogr. 57, 12–21.

Grover, M., Ali, S.Z., Sandhya, V., Rasul, A., Venkateswarlu, B., 2011. Role of microorganisms in adaptation of agriculture crops to abiotic stresses. World J. Microbiol. Biotechnol. 27 (5), 1231–1240.

Guefrachi, I., Rejili, M., Mahdhi, M., Mars, M., 2013. Assessing genotypic diversity and symbiotic efficiency of five rhizobial legume interactions under cadium stress for soil phytoremediation. Int. J. Phytoremediation 15 (10), 938–951.

Guzmán-Murillo, M.A., Ascencio, F., Larrinaga-Mayoral, J.A., 2013. Germination and ROS detoxification in bell pepper (*Capsicum annuum* L.) under NaCl stress and treatment with microalgae extracts. Protoplasma 250 (1), 33–42.

Haq, R., Zaidi, S.K., Shakoori, A.R., 1999. Cadmium resistant *Enterobacter cloacae* and *Klebsiella* sp. isolated from industrial effluents and their possible role in cadmium detoxification. World J. Microbiol. Biotechnol. 15 (2), 283–290.

Harris, K.D., Lavanya, L., 2016. Influence of foliar application of boro the quality of tomato. Res. J. Agric. For. Sci. 4 (7), 1–5. E-ISSN 2320–6063.

Hassan, W., Bashir, S., Ali, F., Ijaz, M., Hussain, M., David, J., 2016. Role of ACC-deaminase and/or nitrogen fixing rhizobacteria in growth promotion of wheat (*Triticum aestivum* L.) under cadmium pollution. Environ. Earth Sci. 75 (3), 267.

Hashem, A., Alqarawi, A.A., Radhakrishnan, R., Al-Arjani, A.B.F., Aldehaish, H.A., Egamberdieva, D., et al., 2018. Arbuscular mycorrhizal fungi regulate the oxidative system, hormones and ionic equilibrium to trigger salt stress tolerance in *Cucumis sativus* L. Saudi J. Biol. Sci. 25 (6), 1102–1114.

Hu, Y., Schmidhalter, U., 2005. Drought and salinity: a comparison of their effects on mineral nutrition of plants. J. Plant Nutr. Soil Sci. 168 (4), 541–549.

Husain, T., Fatima, A., Suhel, M., Singh, S., Sharma, A., Prasad, S.M., et al., 2020. A brief appraisal of ethylene signaling under abiotic stress in plants. Plant Signal. Behav. 15 (9), 1782051.

Hussein, M.M., El-Saady, A., Gobarah, M., Abo El-Khier, A., 2020. Nutrient content and growth responses of sugar beet plants grown under salinity condition to citric acid and algal extract. Egypt. J. Agron. 42 (2), 209–224.

Ibrahim, W.M., 2016. Potential impact of marine algal extracts on the growth and metabolic activities of salinity stressed wheat seedlings. J. Appl. Sci. 16 (8), 388–394.

Islam, M.Z., Mele, M.A., Ki-Young, C.H.O.I., Ho-Min, K.A.N.G., 2018. The effect of silicon and boron foliar application on the quality and shelf life of cherry tomatoes. Zemdirbyste-Agric. 105 (2).

Ismail, A.H., Mehmood, A.S.I.F., Qadir, M.U.H.A.M.M.A.D., Husna, A.I., Hamayun, M.U. H.A.M.M.A.D., Khan, N.A.E.E.M., 2020. Thermal stress alleviating potential of endophytic fungus *Rhizopus oryzae* inoculated to sunflower (*Helianthus annuus* L.) and soybean (*Glycine max* L.). Pak. J. Bot. 52 (5), 1857–1865.

Juwarkar, A.A., Nair, A., Dubey, K.V., Singh, S.K., Devotta, S., 2007. Biosurfactant technology for remediation of cadmium and lead contaminated soils. Chemosphere 68 (10), 1996–2002.

Khalloufi, M., Martínez-Andújar, C., Lachaâl, M., Karray-Bouraoui, N., Pérez-Alfocea, F., Albacete, A., 2017. The interaction between foliar GA3 application and arbuscular mycorrhizal fungi inoculation improves growth in salinized tomato (*Solanum lycopersicum* L.) plants by modifying the hormonal balance. J. Plant Physiol. 214, 134–144.

Khan, M.A., Asaf, S., Khan, A.L., Jan, R., Kang, S.M., Kim, K.M., et al., 2020. Thermotolerance effect of plant growth-promoting *Bacillus cereus* SA1 on soybean during heat stress. BMC Microbiol. 20 (1), 1–14.

Kieber, J.J., Polko, J.K., 2019. 1-Aminocyclopropane 1-carboxylic acid and its emerging role as an ethylene-independent growth regulator. Front. Plant Sci. 10, 1602.

Kochian, L.V., Hoekenga, O.A., Pineros, M.A., 2004. How do crop plants tolerate acid soils? Mechanisms of aluminum tolerance and phosphorous efficiency. Annu. Rev. Plant Biol. 55, 459–493.

Komaresofla, B.R., Alikhani, H.A., Etesami, H., Khoshkholgh-Sima, N.A., 2019. Improved growth and salinity tolerance of the halophyte *Salicornia* sp. by co–inoculation with endophytic and rhizosphere bacteria. Appl. Soil Ecol. 138, 160–170.

Kulkarni, S., Surange, S., Nautiyal, C.S., 2000. Crossing the limits of *Rhizobium* existence in extreme conditions. Curr. Microbiol. 41 (6), 402–409.

Kumar, N., 2019. Natural and artificial soil amendments for the efficient phytoremediation of contaminated soil. Phyto and Rhizo Remediation. Springer, Singapore, pp. 1–32.

Kumar, S., Kumawat, P., Suman, M., 2020. A-review on role of micro-nutrients on banana, mango and pomegranate. Int. J. Pure Appl. Biosci. 8 (1), 190–198.

Latef, A.A.H.A., Hashem, A., Rasool, S., Abd_Allah, E.F., Alqarawi, A.A., Egamberdieva, D., et al., 2016. Arbuscular mycorrhizal symbiosis and abiotic stress in plants: a review. J. Plant Biol. 59 (5), 407–426.

Li, J., McConkey, B.J., Cheng, Z., Guo, S., Glick, B.R., 2013. Identification of plant growth-promoting bacteria-responsive proteins in cucumber roots under hypoxic stress using a proteomic approach. J. Proteomics 84, 119–131.

Liu, H., Chen, X., Song, L., Li, K., Zhang, X., Liu, S., et al., 2019. Polysaccharides from *Grateloupia filicina* enhance tolerance of rice seeds (*Oryza sativa* L.) under salt stress. Int. J. Biol. Macromol. 124, 1197–1204.

Mathur, S., Sharma, M.P., Jajoo, A., 2018. Improved photosynthetic efficacy of maize (*Zea mays*) plants with arbuscular mycorrhizal fungi (AMF) under high temperature stress. J. Photochem. Photobiol. B Biol. 180, 149–154.

Miller, G., Susuki, N., Ciftci-Yilmaz, S., Mittler, R., 2010. Reactive oxygen species homeostasis and signaling during drought and salinity stresses. Plant Cell Environ. 33, 453–467.

Minhas, P.S., Rane, J., Pasala, R.K., 2017. Abiotic stresses in agriculture: an overview. Abiotic Stress Management for Resilient Agriculture. Springer, Singapore, pp. 3–8.

Muthukumar, T., Priyadharsini, P., Uma, E., Jaison, S., Pandey, R.R., 2014. Role of arbuscular mycorrhizal fungi in alleviation of acidity stress on plant growth, Use of Microbes for the Alleviation of Soil Stresses, vol. 1. Springer, New York, NY, pp. 43–71.

Nadeem, S.M., Ahmad, M., Zahir, Z.A., Javaid, A., Ashraf, M., 2014. The role of mycorrhizae and plant growth promoting rhizobacteria (PGPR) in improving crop productivity under stressful environments. Biotechnol. Adv. 32 (2), 429–448.

Navarro-Torre, S., Barcia-Piedras, J.M., Mateos-Naranjo, E., Redondo-Gómez, S., Camacho, M., Caviedes, M.A., et al., 2017. Assessing the role of endophytic bacteria in the halophyte *Arthrocnemum macrostachyum* salt tolerance. Plant Biol. 19 (2), 249–256.

Nieder, R., Benbi, D.K., Reichl, F.X., 2018. Role of potentially toxic elements in soils. Soil Components and Human Health. Springer, Dordrecht, pp. 375–450.

Owojori, O.J., Reinecke, A.J., Voua-Otomo, P., Reinecke, S.A., 2009. Comparative study of the effects of salinity on life-cycle parameters of four soil-dwelling species (*Folsomia candida, Enchytraeus doerjesi, Eisenia fetida* and *Aporrectodea caliginosa*). Pedobiologia 52 (6), 351–360.

Pankaj, P., Kujur, P.K., Saravanan, S., 2018. Effect of different micronutrients on plant quality of broccoli (*Brassica oleracea* var. italica) cv. green magic. J. Pharmacogn. Phytochem. SP1 2825–2828. E-ISSN: 2278–4136; P-ISSN: 2349–8234.

Patel, K., Agarwal, P., Agarwal, P.K., 2018. *Kappaphycus alvarezii* sap mitigates abiotic-induced stress in *Triticum durum* by modulating metabolic coordination and improves growth and yield. J. Appl. Phycol. 30 (4), 2659–2673.

Prasad, R., Kumar, M., Varma, A., 2015. Role of PGPR in soil fertility and plant health. Plant-Growth-Promoting Rhizobacteria (PGPR) and Medicinal Plants. Springer, Cham, pp. 247–260.

Rahmani, H.A., Saleh-Rastin, N., Khavazi, K., Asgharzadeh, A., Fewer, D., Kiani, S., et al., 2009. Selection of thermotolerant bradyrhizobial strains for nodulation of soybean (*Glycine max* L.) in semi-arid regions of Iran. World J. Microbiol. Biotechnol. 25 (4), 591–600.

Rapparini, F., Peñuelas, J., 2014. Mycorrhizal fungi to alleviate drought stress on plant growth, Use of Microbes for the Alleviation of Soil Stresses, vol. 1. Springer, New York, NY, pp. 21–42.

Rathaur, P., Ramteke, P.W., Raja, W., John, S.A., 2012. Isolation and characterization of nickel and cadmium tolerant plant growth promoting rhizobacteria from rhizosphere of *Withania somnifera*. J. Environ. Sci. 6, 253–261. Corpus ID: 88826946.

Rathi, M., Nandabalan, Y.K., 2017. Copper-tolerant rhizosphere bacteria—characterization and assessment of plant growth promoting factors. Environ. Sci. Pollut. Res. 24 (10), 9723–9733.

Rattan, L., 2015. Restoring soil quality to mitigate soil degradation. Sustainability 7, 5875–5895.

Rohyadi, A., 2008. Growth responses of external hyphae of arbuscular mycorrhizal fungi to acidic soil conditions and their effects on cowpea growth. Microbiol. Indones. 2 (1), 5–5.

Sadeghi, F., Samsampour, D., Seyahooei, M.A., Bagheri, A., Soltani, J., 2020. Fungal endophytes alleviate drought-induced oxidative stress in mandarin (*Citrus reticulata* L.): toward regulating the ascorbate–glutathione cycle. Sci. Hortic. 261, 108991.

Sahab, S., Suhani, I., Srivastava, V., Chauhan, P.S., Singh, R.P., Prasad, V., 2020. Potential risk assessment of soil salinity to agroecosystem sustainability: Current status and management strategies. Science of The Total Environment 144164. Available from: https://doi.org/10.1016/j.scitotenv.2020.144164.

Sanyaolu, A.A.A., 2018. Verification of *Aspergillus niger* as a myco-remediation agent of lambda-cyhalothrin and associated heavy metals in *Lactuca sativa* (L.) leaf. J. Appl. Sci. Environ. Manage. 22 (5), 621–624.

Sapre, S., Gontia-Mishra, I., Tiwari, S., 2019. ACC deaminase-producing bacteria: a key player in alleviating abiotic stresses in plants. Plant Growth Promoting Rhizobacteria for Agricultural Sustainability. Springer, Singapore, pp. 267–291.

Selvakumar, G., Panneerselvam, P., Ganeshamurthy, A.N., 2012. Bacterial mediated alleviation of abiotic stress in crops. Bacteria in Agrobiology: Stress Management. Springer, Berlin, Heidelberg, pp. 205–224.

Shabala, S., 2011. Physiological and cellular aspects of phytotoxicity tolerance in plants: the role of membrane transporters and implications for crop breeding for waterlogging tolerance. New Phytol. 190 (2), 289–298.

Shah, A.U., Rajab, H., Jalal, A., Ajmal, M., Bangash, S.A.K., Ahmad, D., et al. (2020). Inoculation of *Brassica napus* L. genotypes with endophytic bacteria promote growth and alleviate cadmium toxicity. J. Animal Plant Sci., 30.

Sharif, F., Danish, M.U., Ali, A.S., Khan, A.U., Shahzad, L., Ali, H., et al., 2016. Salinity tolerance of earthworms and effects of salinity and vermi amendments on growth of *Sorghum bicolor*. Arch. Agron. Soil Sci. 62 (8), 1169–1181.

Sharma, S.B., Sayyed, R.Z., Trivedi, M.H., Gobi, T.A., 2013. Phosphate solubilizing microbes: sustainable approach for managing phosphorus deficiency in agricultural soils. Springer Plus 2 (1), 587.

Sharma, H.S., Fleming, C., Selby, C., Rao, J.R., Martin, T., 2014. Plant biostimulants: a review on the processing of macroalgae and use of extracts for crop management to reduce abiotic and biotic stresses. J. Appl. Phycol. 26 (1), 465–490.

Sharma, B., Sarkar, A., Singh, P., Singh, R.P., 2017. Agricultural utilization of biosolids: a review on potential effects on soil and plant grown. Waste Manage. 64, 117–132.

Sharma, R.K., Barot, K., Archana, G., 2020. Root colonization by heavy metal resistant *Enterobacter* and its influence on metal induced oxidative stress on *Cajanus cajan*. J. Sci. Food Agric. 100 (4), 1532–1540.

Sindhu, S.S., Sharma, R., Sindhu, S., Sehrawat, A., 2019. Soil fertility improvement by symbiotic rhizobia for sustainable agriculture. Soil Fertility Management for Sustainable Development. Springer, Singapore, pp. 101–166.

Sindhu, S., Dahiya, A., Gera, R., Sindhu, S.S., 2020. Mitigation of abiotic stress in legume-nodulating rhizobia for sustainable crop production. Agric. Res. 9, 444–459.

Singh, A., Shankar, A., Gupta, V.K., Prasad, V., 2016. Rhizobacteria: tools for the management of plant abiotic stresses. Microbial Biotechnology: An Interdisciplinary Approach. CRC Press, pp. 241–261.

Singh, G., Pankaj, U., Chand, S., Verma, R.K., 2019. Arbuscular mycorrhizal fungi-assisted phytoextraction of toxic metals by *Zea mays* L. from tannery sludge. Soil Sediment Contam. 28 (8), 729–746.

Singh, R.P., Agrawal, M., 2007. Effects of sewage sludge amendment on heavy metal accumulation and consequent responses of Beta vulgaris plants. Chemosphere 67 (11), 2229–2240. Available from: https://doi.org/10.1016/j.chemosphere.2006.12.019.

Škarková, P., Kos, M., Drobne, D., Vávrová, M., Jemec, A., 2016. Effects of food salinization on terrestrial crustaceans *Porcellio scaber*. Appl. Soil Ecol. 100, 1–7.

Smith, S.E., Read, D.J., 2010. Mycorrhizal Symbiosis. Academic Press, 978–0–12–370526-6.

Su, Y., Liu, T., Wang, X., Yang, R., 2016. Salinity effects on soil organic carbon and its labile fractions, and nematode communities in irrigated farmlands in an arid region, northwestern China. Sci. Cold Arid Reg. 8 (1), 46–53.

Subramanian, P., Krishnamoorthy, R., Chanratana, M., Kim, K., Sa, T., 2015. Expression of an exogenous 1-aminocyclopropane-1-carboxylate deaminase gene in psychrotolerant bacteria modulates ethylene metabolism and cold induced genes in tomato under chilling stress. Plant Physiol. Biochem. 89, 18–23.

Suhani, I., Sahab, S., Singh, R.P., 2020. Impacts of Land-use Paradigm Shifts on the Agricultural Economy: Sustainable Management. RASSA Journal of Science for Society 2 (3), 177–181.

Suhani, I., Sahab, S., Srivastava, V., Singh, R.P., 2021. Impact of Cadmium Pollution on Food Safety and Human Health. Current Opinion in Toxicology 27, 1–7. Available from: https://doi.org/10.1016/j.cotox.2021.04.004.

Suhani, I., Vaish, B., Singh, P., Singh, R.P., 2020. Restoration, construction, and conservation of degrading wetlands: a step toward sustainable management practices. In: *Restoration of Wetland Ecosystem: A Trajectory Towards a Sustainable Environment*, Springer, Singapore, pp. 1–16. Available from: https://doi.org/10.1007/978-981-13-7665-8_1.

Sung, D.Y., Kaplan, F., Lee, K.J., Guy, C.L., 2003. Acquired tolerance to temperature extremes. Trends Plant Sci. 8 (4), 179–187.

Tajini, F., Trabelsi, M., Drevon, J.J., 2012. Combined inoculation with *Glomus intraradices* and *Rhizobium tropici* CIAT899 increases phosphorus use efficiency for symbiotic nitrogen fixation in common bean (*Phaseolus vulgaris* L.). Saudi J. Biol. Sci. 19 (2), 157–163.

Tiwari, S., Lata, C., Chauhan, P.S., Nautiyal, C.S., 2016. *Pseudomonas putida* attunes morphophysiological, biochemical and molecular responses in *Cicer arietinum* L. during drought stress and recovery. Plant Physiol. Biochem. 99, 108–117.

Torbaghan, M.E., Lakzian, A., Astaraei, A.R., Fotovat, A., Besharati, H., 2017. Salt and alkali stresses reduction in wheat by plant growth promoting *haloalkaliphilic* bacteria. J. Soil Sci. Plant Nutr. 17 (4), 1058–1087.

Tóth, G., Adhikari, K., Várallyay, G., Tóth, T., Bódis, K., Stolbovoy, V., 2008. Updated map of salt affected soils in the European Union. Threats to Soil Quality in Europe EUR 23438 EN, 65–77.

Ullah, A., Heng, S., Munis, M.F.H., Fahad, S., Yang, X., 2015. Phytoremediation of heavy metals assisted by plant growth promoting (PGP) bacteria: a review. Environ. Exp. Bot. 117, 28–40.

Uma, C., Sivagurunathan, P., Sangeetha, D., 2013. Performance of bradyrhizobial isolates under drought conditions. Int. J. Curr. Microbiol. Appl. Sci. 2 (5), 228–232.

Upadhyay, S.K., Singh, J.S., Singh, D.P., 2011. Exopolysaccharide-producing plant growth-promoting rhizobacteria under salinity condition. Pedosphere 21 (2), 214–222.

Upadhyay, N., Vishwakarma, K., Singh, J., Verma, R.K., Prakash, V., Jain, S., et al., 2019. Plant-microbe-soil interactions for reclamation of degraded soils: potential and challenges. Phyto and Rhizo Remediation. Springer, Singapore, pp. 147–173.

Vriezen, J.A., Bruijn, De, Nüsslein, K., F.J., 2007. Responses of rhizobia to desiccation in relation to osmotic stress, oxygen, and temperature. Appl. Environ. Microbiol. 73 (11), 3451–3459.

Walker, D.J., Leigh, R.A., Miller, A.J., 1996. Potassium homeostasis in vacuolate plant cells. Proc. Natl. Acad. Sci. USA 93 (19), 10510–10514.

Wang, M., Zheng, Q., Shen, Q., Guo, S., 2013. The critical role of potassium in plant stress response. Int. J. Mol. Sci. 14 (4), 7370–7390.

Wild, A., 2003. Soils, Land and Food: Managing the Land During the Twenty-First Century. Cambridge University Press, ISBN: 0521527597.

Wu, Q.S., Zou, Y.N., 2009. Mycorrhiza has a direct effect on reactive oxygen metabolism of drought-stressed citrus. Plant Soil Environ. 55 (10), 436–442.

Yang, Y., Liang, Y., Han, X., Chiu, T.Y., Ghosh, A., Chen, H., et al., 2016. The roles of arbuscular mycorrhizal fungi (AMF) in phytoremediation and tree-herb interactions in Pb contaminated soil. Sci. Rep. 6, 20469.

Yang, Y., Xiong, J., Tao, L., Cao, Z., Tang, W., Zhang, J., et al., 2020. Regulatory mechanisms of nitrogen (N) on cadmium (Cd) uptake and accumulation in plants: a review. Sci. Total Environ. 708, 135186.

Yaseen, T., Burni, T., Hussain, F., 2011. Effect of arbuscular mycorrhizal inoculation on nutrient uptake, growth and productivity of cowpea (*Vigna unguiculata*) varieties. Afr. J. Biotechnol. 10 (43), 8593–8598.

Yu, X., Liu, X., Zhu, T.H., Liu, G.H., Mao, C., 2012. Co-inoculation with phosphate-solubilzing and nitrogen-fixing bacteria on solubilization of rock phosphate and their effect on growth promotion and nutrient uptake by walnut. Eur. J. Soil Biol. 50, 112–117.

Zahran, H.H., Abdel-Fattah, M., Yasser, M.M., Mahmoud, A.M., Bedmar, E., 2012. Diversity and environmental stress responses of rhizobial bacteria from Egyptian grain legumes. Aust. J. Basic Appl. Sci. 6 (10), 571–583. Corpus ID: 86256708.

Zeenat, W., Vivek, S., Raj, R., Tan, vi, Kanchan, M., Sumit, J., 2017. Isolation and characterization of *Rhizobium* from chickpea (*Cicer arietinum*). Int. J. Curr. Microbiol. Appl. Sci. 2880–2893.

Zhang, X., Harper, R., Karsisto, M., Lindström, K., 1991. Diversity of *Rhizobium* bacteria isolated from the root nodules of leguminous trees. Int. J. Syst. Evol. Microbiol. 41 (1), 104–113.

Zhao, S., Zhou, N., Zhao, Z.Y., Zhang, K., Wu, G.H., Tian, C.Y., 2016. Isolation of endophytic plant growth-promoting bacteria associated with the halophyte *Salicornia europaea* and evaluation of their promoting activity under salt stress. Curr. Microbiol. 73 (4), 574–581.

Zhu, B., Xu, Q., Zou, Y., Ma, S., Zhang, X., Xie, X., et al., 2019. Effect of potassium deficiency on growth, antioxidants, ionome and metabolism in rapeseed under drought stress. Plant Growth Regul. 1–12.

Zou, P., Lu, X., Zhao, H., Yuan, Y., Meng, L., Zhang, C., et al., 2019. Polysaccharides derived from the brown algae *Lessonia nigrescens* enhance salt stress tolerance to wheat seedlings by enhancing the antioxidant system and modulating intracellular ion concentration. Front. Plant Sci. 10, 48.

Źróbek-Sokolnik, A., 2012. Temperature stress and responses of plants. In: Environmental adaptations and stress tolerance of plants in the era of climate change, Springer, New York, NY, pp. 113–134. Available from: https://doi.org/10.1007/978-1-4614-0815-4_5.

Tripartite interactions: plant-*Pseudomonas putida*-microRNAs in agricultural productivity

22

Ram Jatan[1,2] *and Charu Lata*[2,3]
[1]CSIR-National Botanical Research Institute, Lucknow, India, [2]Academy of Scientific and Innovative Research (AcSIR), Ghaziabad, India, [3]CSIR-National Institute of Science Communication and Policy Research, New Delhi, India

22.1 Introduction

In the current scenario, food and energy are the two main emerging problems faced by humans due to the growing population, decreasing fossil energy, global warming, and environmental destruction events. It is very difficult to grow crops in such conditions and therefore, new approaches for increased productivity are needed. Indiscriminate use of chemical fertilizers, herbicides, and pesticides for improved production may prove detrimental for the environment. These hazardous materials may be absorbed by plants and introduced to the food chain (Seenivasan et al., 2016). Due to several plant growth promotion (PGP) attributes, beneficial microorganisms have the potential to enhance plant growth and productivity in a sustainable manner, and hence have the potential of becoming a promising agrotechnology and research area. Further, beneficial rhizospheric bacteria, also known as plant growth-promoting rhizobacteria (PGPR) attach to the root surface of plants (Beneduzi et al., 2012; Vacheron et al., 2013). Several studies have shown that PGPR has a number of PGP characteristics, such as P-solubilization, 1-aminocyclopropane-1-carboxylate (ACC) deaminase activity, indole acetic acid and siderophore production, that improve seed germination, growth and plant yield, as well as defense responses towards phytopathogens (Beneduzi et al., 2012; Glick, 2014; Jatan and Lata, 2019; Tiwari et al., 2017) Among different PGPR genus, soil-borne *Pseudomonas* spp. is one of the most predominant root microbiome genera and has the potential to facilitate plant growth through abscisic acid (ABA) signaling, auxin (AUX) production and phosphate solubilization (Srivastava et al., 2012; Zamioudis et al., 2013). *Pseudomonas putida* (RA) strain MTCC5279 is a PGPR that has been demonstrated to facilitate a number of beneficial attributes in plants, like hormone production, nutrient uptake, growth and development, and alleviating defense mechanisms against abiotic stresses (Jatan et al., 2018; Kumar et al., 2016; Srivastava et al., 2012; Tiwari et al., 2016).

Alterations in physiochemical, cellular, and molecular mechanisms and regulation of different gene expressions in plants have been associated with complex

Microbes in Land Use Change Management. DOI: https://doi.org/10.1016/B978-0-12-824448-7.00025-5
© 2021 Elsevier Inc. All rights reserved.

interactions among plants and PGPR (Chauhan et al., 2019; Jatan et al., 2018, 2019; Tiwari et al., 2017). MicroRNAs (miRNAs) are 20–24 nt noncoding small RNAs, which control their target gene expression either by the cleavage or translational inhibition (Bartel, 2004; Sunkar et al., 2012; Voinnet, 2009; Yu et al., 2017). miRNAs are known to regulate plant growth, development, and metabolism by regulating several cellular, biological, and molecular processes miRNAs have been reported to control the growth, development, and metabolism of plants by controlling multiple cellular, biological, and molecular processes (Xia et al., 2016; Yu et al., 2017). In addition, miRNAs are also involved in response to several environmental factors, like cold, drought, heat, salt, and CO_2, as well as against numerous phytopathogens in various plants (Liu et al., 2015, 2017; Yang et al., 2017; Yin et al., 2017). Previous studies have shown that miRNAs play a very pivotal role in the growth and development of plants, as well as in response to stress during interactions between plant and microbes (Jatan et al., 2018; Jatan and Lata, 2019; Khraiwesh et al., 2012). *P. putida*-mediated regulation of miRNAs and their target genes play important roles in the reduction of drought and salt stress in chickpea (Jatan et al., 2018). Moreover, significant changes in the expression pattern of conserved and putative novel miRNAs in *P. putida*-inoculated chickpea have been reported their involvement in the growth and development as well as drought stress amelioration in chickpea (Jatan et al., 2019). Recently, a genome-wide profiling of miRNAs suggested that the *P. putida*-mediated modulation of miRNAs and its targets are involved in the growth and development of Arabidopsis during *P. putida*-inoculation (Jatan et al., 2020). In this chapter, we address the current status of PGPR-mediated miRNA modulation as well as the growth and development of host plants during tripartite (plant-PGPR-miRNAs) interactions, and their role in the sustainable agricultural growth and productivity of plants.

22.2 Plant-microbe interaction

Plants and their associated microbes have a dramatic impact on lifestyle and health benefits for each other. On the basis of the zone of microbial adherence to plants, colonization may be either rhizospheric or endophytic. The rhizosphere is a narrow zone of soil and influenced by the root exudates of plants, which can establish favorable circumstances for the growth of microbes (Egamberdieva et al., 2008). Several microbes referred to as endophytes which reside in root endosphere (interior tissues of root) of the host plants in natural ecosystems (Card et al., 2016). The nutrients and metabolites of the plant root exudates attract several microbes allowing the rhizosphere a hotspot of microbial activity and interactions (Mendes et al., 2013).

Root exudates consist of variety of components, such as amino acids, carbohydrates, organic acids, phytohormones, flavonoids, phenols, etc., which play important roles in plant-microbe interactions (Badri and Vivanco, 2009). Further, the organic acids of plant root exudates act as a carbon source and chemoattractant for the microorganisms and also in association with secreted amino acids enhance the growth and diversity of microbes in the rhizosphere. For example, malic acid secreted from Arabidopsis roots that recognizes and recruits the beneficial microbes *Bacillus subtilis* FB17 (Rudrappa et al., 2008). Additionally, phenolics like benzoxanoids and salicylic acid (SA) have been reported to secrete from the roots of plants including maize and Arabidopsis, help to recruit and promote the growth and diversity of *P. putida* within the rhizosphere (Lebeis et al., 2015; Neal et al., 2012). Further, roots of plant have the ability to recognize and recruit suitable microbes to successfully grow in the rhizosphere from the reservoir (bulk soil) of microbial diversity. The soil type, along with the host plant genotype, is a major factor for the microbial population diversity in the rhizosphere (Berg and Smalla, 2009; Garbeva et al., 2008). Overall, this indicates that the diversity of rhizospheric microbiome is dramatically affected by different plant species and genotypes, as well as depending on the type and content of the root exudates.

The majority of microorganisms act as destructive plant pathogens and secrete phytotoxic metabolites that inhibit germination of seeds and have profound effect on the productivity of the plants (Hassani et al., 2018; Kamoun et al., 2015). Potentially detrimental effects of various microorganisms such as plant-pathogenic bacteria, fungi, nematodes and viruses have been previously studied in different plants (Dean et al., 2012; Jones et al., 2013; Mansfield et al., 2012; Scholthof et al., 2011). Moreover, the "disease triangle" is a well-known conceptual model in explaining plant-pathogen interactions between plants that are genetically susceptible to a plant diseases, genetically virulent pathogens, and favorable environmental conditions for pathogenicity and susceptibility to plants (Pangga et al., 2011). However, the neutral bacteria do not have any positive or negative impact on the growth and development of host plants. Besides, plants interact with a variety of beneficial microorganisms such as bacteria, fungi which enhance the growth and production of plants. The association of beneficial bacteria and their impact on the growth of plants will be addressed in the subsequent sections. The details of abiotic and biotic stress-responsive miRNAs in major crops have been presented in Table 22.1.

22.3 Plant growth promoting rhizobacteria

Beneficial bacteria are mainly composed of PGPR which attach to the plant roots and facilitate the growth and development either directly by producing and secreting diverse regulatory substances and controlling plant hormones or indirectly by suppressing phytopathogens (Navarro et al., 2006; Vacheron et al., 2013; Yang et al., 2009) (Fig. 22.1). The term PGPR was first used by Kloepper and Schroth for beneficial rhizospheric bacteria (Kloepper, 1978). There are various bacterial genera that belong to PGPR including, *Acinetobacter, Acetobacter, Arthrobacter, Azospirillum, Azotobacter, Bacillus, Burkholderia, Bradyrhizobium, Enterobacter, Klebsiella, Paenibacillus, Pseudomonas, Rhizobium, Stenotrophomonas*, and

Table 22.1 Abiotic and biotic stress-responsive miRNAs in major crops.

Types of stress	miRNA families	Response of miRNAs	Putative/ validated targets	References
Cold	miR393,	*Osa*↑	TIR1, AFB	Barrera-Figueroa et al. (2012)
	miR394,	*Osa*↑	F-box protein	(-do-)
	miR396,	*Osa*↑	GRF	(-do-)
	miR529,	*Osa*↑	SBP-box gene	(-do-)
	miR809	*Osa*↓	Glutaredoxin 2	(-do-)
Drought	miR156,	*Gma*↑, *Tae*↑	SPL TFs	Li et al. (2013), Akdogan et al. (2015)
	miR160,	*Osa*↑, *Tae*↑	ARF	Barrera-Figueroa et al. (2012), Akdogan et al. (2015)
	miR164,	*Tae*↑	NAC	Pandey et al. (2014)
	miR166,	*Osa*↓, *Ssp*↑, *Tae*↑	HD-ZIPIII	Barrera-Figueroa et al. (2012), Gentile et al. (2015), Akdogan et al. (2015)
	miR167,	*Gma*↑, *Osa*↑, *Tae*↑	ARF	Li et al. (2013), Barrera-Figueroa et al. (2012), Akdogan et al. (2015)
	miR168,	*Osa*↓	AGO1	Barrera-Figueroa et al. (2012)
	miR169,	*Ssp*↓	NFY	Gentile et al. (2015)
	miR170,	*Osa*↓	SCL	Zhou et al. (2010)
	miR171,	*Gma*↑, *Hvu*↓, *Ssp*↓, *Tae*↑	GRAS	Li et al. (2013), Kruszka et al. (2014), Gentile et al. (2015), Pandey et al. (2014)
	miR172,	*Osa*↓, *Ssp*↓, *Tae*↓	bZIP, AP2	Zhou et al. (2010), Gentile et al. (2015), Akdogan et al. (2015)
	miR393,	*Osa*↑	TIR1, AFB	Barrera-Figueroa et al. (2012)
	miR395,	*Gma*↑, *Tae*↑	APS/AST	Li et al. (2013), Akdogan et al. (2015)
	miR396,	*Gma*↑, *Osa*↑, *Ssp*↓, *Tae*↑	GRF	Li et al. (2013), Barrera-Figueroa et al. (2012), Gentile et al. (2015), Akdogan et al. (2015)
	miR408,	*Osa*↓	Plantacyanin	Zhou et al. (2010)
	miR444,	*Hvu*↓	MADS-box	Kruszka et al. (2014)
	miR474,	*Osa*↑	Protein kinase, PPR SBP-box	Zhou et al. (2010)

(*Continued*)

Table 22.1 (Continued)

Types of stress	miRNA families	Response of miRNAs	Putative/ validated targets	References
	miR529,	*Osa*↑	Glutaredoxin 2	Barrera-Figueroa et al. (2012)
	miR809,	*Osa*↓	Cytochrome c	(-do-)
	miR1428,	*Osa*↓	AAA ATPase	(-do-)
	miR1884,	*Osa*↓	Wpk4 protein kinase	(-do-)
	miR5049	*Tae*↑		Akdogan et al. (2015)
Heat	miR156,	*Tae*↓	SPL	Xin et al. (2010)
	miR159,	*Tae*↑	MYB domain	Pandey et al. (2014)
	miR160,	*Hvu*↑, *Tae*↑	ARF 13, 17	Hackenberg et al. (2015), Pandey et al. (2014)
	miR164,	*Tae*↓	NAC	Xin et al. (2010)
	miR166,	*Hvu*↑, *Tae*↑	HD-ZIPIII	Hackenberg et al. (2015), Xin et al. (2010)
	miR167,	*Tae*↓	ARF	Xin et al. (2010)
	miR168,	*Tae*↑	AGO	(-do-)
	miR169,	*Tae*↑	NF-YA	(-do-)
	miR172,	*Tae*↓	AP2-like	(-do-)
	miR396,	*Tae*↓	GRF3	(-do-)
	miR827,	*Tae*↑	NLA	(-do-)
	miR5175	*Hvu*↑	ACC-like oxidase	Hackenberg et al. (2015)
Heavy metals	miR156,	*Osa*↓	SPL TFs	Ding et al. (2016)
	miR162a,	*Osa*↓	DCL1	(-do-)
	miR166,	*Osa*↓	HD-ZIPIII	(-do-)
	miR168,	*Osa*↓	AGO	(-do-)
	miR171,	*Osa*↓	Scarecrow-like TF	(-do-)
	miR390,	*Osa*↓	LRR-RLK	(-do-)
	miR396,	*Osa*↓	GRF	(-do-)
	miR444,	*Osa*↓	MADS-box	(-do-)
	miR528,	*Osa*↑	DCL1	(-do-)
Salt	miR156,	*Ssp*↑, *Tae*↓	SPL TFs	Bottino et al. (2013), Pandey et al. (2014)
	miR159,	*Osa*↓, *Ssp*↑	MYB TFs	Barrera-Figueroa et al. (2012), Bottino et al. (2013)
	miR160,	*Osa*↓, *Tae*↓	ARF	Barrera-Figueroa et al. (2012), Pandey et al. (2014)

(*Continued*)

Table 22.1 (Continued)

Types of stress	miRNA families	Response of miRNAs	Putative/ validated targets	References
	miR164,	*Tae*↓	NAC	Pandey et al. (2014)
	miR166,	*Ssp*↑	Class III HD-Zip	Bottino et al. (2013)
	miR167,	*Ssp*↑, *Tae*↓	ARF	Bottino et al. (2013), Pandey et al. (2014)
	miR168,	*Hvu*↓, Ssp↑	AGO1	Deng et al. (2015), Bottino et al. (2013)
	miR169,	Ssp↑	HAP12-CCAAT-box	Bottino et al. (2013)
	miR171,	*Hvu*↑, *Tae*↓	Protein FAN, scarecrow-like TF	Deng et al. (2015), Pandey et al. (2014)
			TCP family	Barrera-Figueroa et al. (2012)
			TIR1, AFB	(-do-)
	miR319,	*Osa*↓		(-do-)
	miR393,	*Osa*↑	F-box protein GRF1,3	Bottino et al. (2013), Pandey et al. (2014)
	miR394,	*Osa*↓		Deng et al. (2015)
	miR396,	*Ssp*↑, *Tae*↓	MADS-box	
	miR444	*Hvu*↓		
Bacteria	miR169,	*Mtr*↓	HAP2/NF-YA	Reynoso et al. (2013)
	miR171,	*Mtr*↑	GRAS (NSP2)	(-do-)
	miR172,	*Mtr*↑	AP2 TF	Reynoso et al. (2013), Nova-Franco et al. (2015)
	miR2118	*Mtr*↑	NBS-LRRs	Reynoso et al. (2013)
Viruses	miR169,	*Grapevine*	Zinc finger	Singh et al. (2012)
	miR398,	*Grapevine*	Terpene synthase	(-do-)
	miR1885,	*Bra*	TIR-NBS-LRR	He et al. (2008)
	miR3623,	*Grapevine*	(-do-)	Singh et al. (2012)
	miR6019	*Nbe*	(-do-)	Li et al. (2012)

↑, upregulation; ↓, downregulation.
Bra, Brassica rapa; *Gma, Glycine max*; *Hvu, Hordeum vulgare*; *Mtr, Medicago truncatula*; *Nbe, Nicotiana benthamiana*; *Osa, Oryza sativa*; *Ssp, Saccharum* sp.; *Tae, Triticum aestivum*; *Vvi, Vitis vinifera.*
Source: Adapted from Jatan, R., Lata, C., 2019. Role of microRNAs in abiotic and biotic stress resistance in plants. In: Proc. Indian Natl. Sci. Acad. pp. 553–567.

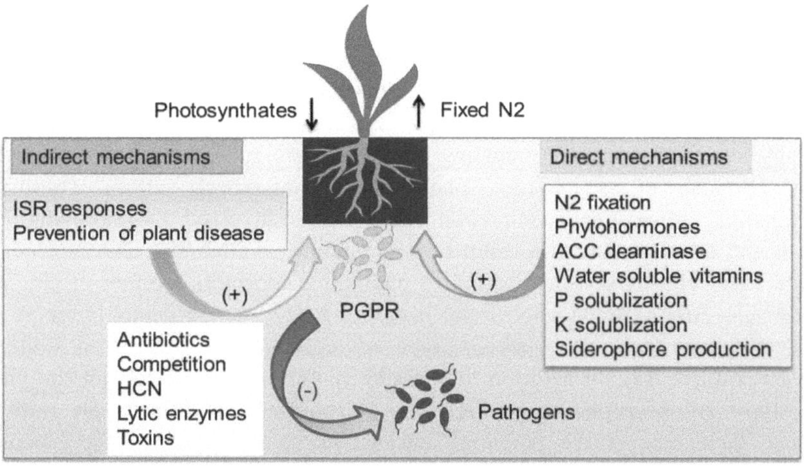

Figure 22.1 Mechanisms of plant growth promotion by PGPR.

Thiobacillus (Finkel et al., 2017; Katiyar et al., 2016). PGPR have direct and indirect PGP activities including: (1) nitrogen fixation; (2) siderophore production; (3) production of plant growth regulators; (4) nutrient availability; (5) production of protective enzymes; and (6) production of volatile organic compounds (VOCs); which improve seed germination, plant growth, and yield, as well as defense responses toward phytopathogens by synthesizing antibiotics and biofilm formation (Choudhary, 2011; García-Fraile et al., 2015). However, based on the type of soil and host plants, different PGPR have performed different mechanisms of action (Garbeva et al., 2008).

22.3.1 Direct mechanisms

22.3.1.1 Nitrogen fixation

Nitrogen (N_2) is a vital nutrient for the synthesis of biological macromolecules, enzymatic activities, energy transfer, and signal transduction as well as growth and productivity of plants. It is the most limiting essential element for plant growth and development. However, about 78% of the nitrogen is found in the atmosphere. The process of biological nitrogen fixation (BNF) is performed either by symbiotic or nonsymbiotic interactions among plants and nitrogen fixing microbes, by which atmospheric nitrogen transformed into available forms (NH_3) for plants (Bagali, 2012). Various PGPR including *Pseudomonas* spp. are able to colonize the root surface of plants and thus improve their growth-promoting impact by both the fixation of atmospheric nitrogen and effective transfer of the fixed nitrogen to the host plants (Lobo et al., 2019). Further, BNF is an economically advantageous and environmentally sustainable alternative to conventional (chemical) fertilizers.

22.3.1.2 Siderophore production

Siderophores are low-molecular-weight iron binding agents, which chelate ferric ion (Fe^{3+}) in a reversible way from the surrounding environment. In plant system, iron (Fe) acts as a cofactor in several enzymes that are essential for various important physiological processes including photosynthesis, nitrogen fixation, and respiration (Bogdan et al., 2016; Cohen, 2014; Hallenbeck, 2017). During Fe unavailability, rhizospheric microbial siderophores involved in improving plant growth and development as a result of iron binding, availability, and translocation of iron into cells by communicating to membrane receptors (Kumar et al., 2017; Saha et al., 2016). Among beneficial bacteria, *Pseudomonas* genus is reported to produce pyoverdine and pyochelin type of siderophores under iron limitation (Dumas et al., 2013). In addition to availability and uptake of iron for the plants, these types of siderophores produced by rhizobacteria also have the ability to enhance their competitive potential, because these compounds have biocontrol activity. For example, pseudobactin siderophore is secreted by *P. putida*, which inhibits the growth of *Fusarium oxysporum* in iron deficient soil (Kloepper et al., 1980a,b). The above studies have indicated that the microbial siderophores are participated in growth and development of plants by iron uptake as well as biocontrol activity.

22.3.1.3 Production of plant growth regulators

Plant growth regulators (also known as plant hormones) are small molecules that play a significant role in alleviation of abiotic stresses, plant growth, and regulate numerous developmental events during the plant life cycle at very low concentrations (Egamberdieva et al., 2017; Hu et al., 2013). Various phytohormones such as AUX, cytokinins (CKs), gibberellins (GAs), ABA, jasmonic acid (JA), ethylene (ET), brassinosteroids, oxide, SA, and strigolactone are involved in germination, growth, development, and flowering as well as in different signaling pathways in plants during responses to environmental stresses (Backer et al., 2018; Khan et al., 2015; Kudoyarova et al., 2019). Various beneficial PGPR have been found to enhance the plant growth by the biosynthesis of several phytohormones such as AUX, GAs, CKs, and ABA (Park et al., 2017; Pereira et al., 2016).

Auxins secreted by PGPR have been shown to enhance root length, root biomass, transcriptional changes in hormone, decrease size and densities of stomata, and activate defense-related, cell wall related, and AUX response genes that stimulate plant growth (Llorente et al., 2016; Ruzzi and Aroca, 2015; Spaepen et al., 2014). The root elongation in wheat was stimulated by AUX producing *Bacillus, Enterobacter, Moraxella*, and *Pseudomonas* genera, whereas inhibition of primary root elongation by bacterial AUX has been demonstrated in *Arabidopsis thaliana* (Raheem et al., 2018; Zamioudis et al., 2013). Cytokinin-releasing bacteria have been reported to decrease the root to shoot ratio and stimulate shoot growth of the host plants (Arkhipova et al., 2007). *Pseudomonads* isolated from rhizospheric soils of various crops such as *Pennisetum glaucum, Helianthus annuus, Zea mays* produced CKs during osmotic stress (Sandhya et al., 2010). Many beneficial bacteria including *Pseudomonas* have *acds* gene coding for an ACC deaminase, which lowers the ET levels by the conversion of ACC into ammonium and α-ketobutyrate (Singh et al., 2015). It have previously demonstrated that the application of ACC deaminase producing PGPR is involved in salt stress alleviation and growth promotion of host plants (Glick et al., 2007; Misra et al., 2017).

22.3.1.4 Nutrient availability

Nutritional imbalances have a major effect on plant growth, production, and productivity. PGPR have the ability to improve nutrient solubility and promote the accumulation of nutrients in soil, thus reducing the use of chemical fertilizers. In addition to nitrogen fixation, several PGPR including *Pseudomonas* can strongly influence plant growth by conversion of insoluble nutrient to soluble forms, increasing nutrient absorption and mineralizing organic phosphate, so plants can easily take in the nutrients (Vejan et al., 2016). *Pseudomonas* species have been reported to be involved in solubilization of several nutrient elements like phosphate, iron, potassium, and zinc and maintaining availability and absorption of these nutrients to the host plants (Srivastava and Srivastava, 2020; Vejan et al., 2016). Recently, a study has been reported that *P. putida* enhances the production of hormone and proline, phosphatase activity, and phosphate accumulation which results in improved growth of *A. thaliana* upon nutrient deficiency and salt stress condition (Srivastava and Srivastava, 2020).

22.3.1.5 Production of protective enzymes

PGPR promotes plant growth by the regulation of phytopathogens through the production of protective enzymes and metabolites, which have antifungal and antibiotic properties participating defense mechanisms. Various bacteria including *Pseudomonas* and *Bacillus* sp. are reported to secret enzymes such as chitinases and glucanase which inhibit the growth of filamentous fungi (Miljakovic et al., 2020; Vejan et al., 2016). For example, *Pseudomonas fluorescens* LPK2 and *Sinorhizobium fredii* KCC5 act as biocontrol agents against *Fusarium udum* by the production of beta-glucanase and chitinase (Kumar et al., 2010). The role of chitinases and β-glucanases producing *Pseudomonas* spp. has been reported to inhibit the growth of two most crop destructive pathogens, *Rhizoctonia solani* and *Phytophthora capsici* (Arora et al., 2008). The use of such PGPR has been widely exploited for biological protection of crops from pathogens containing glucans and chitin in their cell wall (Miljakovic et al., 2020; Vejan et al., 2016).

22.3.1.6 Production of volatile organic compounds

In addition to the production of protective enzymes, numerous PGPR like *Arthrobacter, Bacillus, Pseudomonas*, *Serratia*, and *Stenotrophomona* have the ability to produce VOCs that contribute to enhanced plant growth, induced systemic resistance (ISR) and inhibit the growth of plant-associated pathogenic agents (Park et al., 2015; Raza et al., 2016; Ryu et al., 2003). Naturally, VOC secretion is a common phenomenon of various rhizospheric microbes, while the amount and types of secreted VOC varies across species (Effmert et al., 2012; Kanchiswamy et al., 2015). *Pseudomonas chlororaphis* strain O6 reported to secrete 2,3-butanediol that can trigger ISR and exhibited induced drought tolerance in Arabidopsis (Liu and Zhang, 2015). Some other bacterial VOCs including 2-pentylfuran, 3-hydroxy-2-butanone, *N,N*-dimethyl-hexadecanamine, 2-butanone, 13-tetradecadien-1-ol, and 2-methyl-n-1-tridecene were found to be involved in PGP and root system architecture modulation (Delaplace et al., 2015; Park et al., 2015). In previous studies, VOCs from *P. fluorescens* WCS417r and *P. putida* enhanced the shoot fresh weight of *Mentha piperita* (Chu et al., 2020). Recently, application of *Pseudomonas* PS01 enhances the growth and modulates the root system architecture of Arabidopsis by the combination of VOCs and different diffusible compounds (Chu et al., 2020). Taken together, VOCs secreted by beneficial rhizospheric microbes play very important roles in defense against phytopathogens, as well as enhancing the PGP.

22.3.2 Indirect mechanisms

22.3.2.1 Biocontrol activity

Biocontrol activity is defined as the reduction of soil microbial density by one or more PGPR. The biocontrol activity of PGPR has been exhibited by different mechanism and production of antimicrobial agents including antibiotics, siderophores, VOCs, lytic enzymes and ISR (Kumudini et al., 2017; Olorunleke et al., 2015; Sajitha and Dev, 2016). Therefore PGPR have been used as biocontrol against various phytopathogens in agricultural practices in many regions of the world. Among PGPR, *Pseudomonas* sp. is the most frequently abundant bacteria in cultivated fields with potential for biocontrol activity. There are different types of antimicrobial compounds such as cyclic lipopeptides, hydrogen cyanide, phenazines, 2,4-diacetylphloroglucinol (DAPG), pyoluteorin, and pyrrolnitrin have been reported to produce by *Pseudomonas* sp. and these compounds have also been correlated with biocontrol activity against broad spectrum phytopathogens (Flury et al., 2017; Haas and Défago, 2005). Secretion of 2,4-DAPG by *P. fluorescens* CHA0 has been reported to protect against black root rot of tobacco triggered by *Thielaviopsis basicola* (Huang et al., 2004; Zhao et al., 2020). Phenazines secreted by plant-beneficial *Pseudomonas* sp. which are responsible for the growth inhibition of different phytopathogens including fungi and nematodes (Biessy and Filion, 2018; Yu et al., 2018). Beside antibiotics, pyoverdin siderophore has been reported to suppress the growth of several microbes like bacteria and fungi under in vitro conditions in iron-deficient medium (Beneduzi et al., 2012; Kloepper et al., 1980a,b).

22.3.2.2 Biofilm formation

PGPRs are also able to enhance plant growth indirectly by the formation of biofilms. In the process biofilm formation, a single microbial cell multiplies to produce multiple microcolonies. In microcolonies, all the cells are attached to each other through extracellular polymeric substances called exopolysaccharides (EPS). Therefore biofilms have a higher microbial density, and are capable to preventing the entry of toxic metals and pathogens into plant roots, and provide tolerance to biotic stresses, enhanced growth and yield of host plants (Ramey et al., 2004). There are several PGPR including *Pseudomonas* spp. have been reported to colonize the leaves or root surfaces of plant and have capacity to biofilm formation (Ueda and Saneoka, 2015). In a recent study, *Pseudomonas* sp. biofilm has a greater positive effect on height and root dry weight of tomato (Ricci et al., 2019).

22.4 PGPR: as biofertilizers

Biofertilizers are products with living microbial cells that have beneficial effects on growth and yield of host plant through several direct or indirect mechanisms as a result of their effective colonization on plant surfaces, seed sand/or rhizospheric soil (Mishra et al., 2013; Vessey, 2003). A study has reported that formulated biofertilizer products have one or more microbes that significantly improve nutrient level, subsequently growth and yield of host plants either by substituting soil nutrients or by allowing more available forms of nutrients to plants (Malusá and Vassilev, 2014).

Biofertilizer components are generally composed of PGP microbes such as nitrogen fixing rhizobia, arbuscular mycorrhizal fungi and PGPR (Franche et al., 2009; Jeffries et al., 2003). Additionally, PGPR have been recognized as effective biofertilizers globally leading to improved soil fertility and crop yields. Therefore with the potent beneficial capability of the PGPR, they contribute to better agricultural productivity (Khalid et al., 2009). Endospore forming ability of PGPR, particularly *Pseodomonas* spp. and *Bacillus* spp. facilitates to survive under variety of adverse environmental factors, thus enabling the efficient formulation of biofertilizers (Pérez-García et al., 2011). In addition, PGPR abundance in the biofertilizers has very important roles in establishing an appropriate rhizosphere supporting plant growth and increasing the accessibility of nutritionally essential elements, including N, P, K, and also inhibiting the growth of phytopathogens (Vejan et al., 2016; Yang et al., 2011).

22.5 Pseudomonas spp

The word *Pseudomonas* originated from the Greek terms *pseudes* = false and *monas* = a single unit (false unit) represents one of the most versatile and environmentally suited genus of bacteria (Palleroni, 2015). All the members of group *Pseudomonas* have been formerly described by the word "Pseudomonad." *Pseudomonas* spp. is a genus of Gram-negative, rod-shaped motile bacteria, considered to be omnipresent group in the rhizospheric soils, which has various PGP attributes and enables them to be highly suited as PGPR (Shaikh et al., 2018).

Previous studies have reported that most of the members of the *Pseudomonas* genus have participated in plant growth and productivity by different direct and indirect mechanisms including nitrogen fixation, nutrient solubilization, siderophore production, phytohormone production and production of enzymes and other metabolites (Panpatte et al., 2016; Srivastava et al., 2012; Tiwari et al., 2016). In addition to their beneficial attributes, some members have been exhibited biological control activity by various activities like production of antibiotics, siderophores, VOCs, lytic enzymes, EPS, biofilm formation and also ISR (Kumudini et al., 2017; Olorunleke et al., 2015; Sajitha and Dev, 2016). A previous study reported an increase in *Solanum tuberosum* yield by upon inoculation of potato seed pieces with suspension of *P. fluorescens* and *P. putida* (Burr et al., 1978). *P. putida* (RA) strain MTCC5279 has ability to enhance plant growth through phosphate solubilization, ABA signaling and AUX production (Srivastava et al., 2012). This strain also ameliorates drought stress in *desi* and *kabuli* chickpea genotypes by the regulation of morpho-physiological and biochemical parameters and also by controlling differential expression of stress-responsive genes (Jatan et al., 2018, 2019; Tiwari et al., 2016). Recently a study reported that *P. putida* involved in the modification of root architecture as well as regulation of several biochemical and physiological processes in *A. thaliana* upon nutrient deficiency and salt stress condition (Srivastava and Srivastava, 2020).

22.6 Small RNA: emerging master regulators of gene expression

Small RNAs are 19–30 nt, noncoding RNAs that regulate the expression of protein-coding genes at either transcriptional or posttranscriptional levels in plants and animals (Sunkar et al., 2012; Trindade et al., 2010). Small RNAs were first discovered from *Caenorhabditis elegance* (Lee et al., 1993) and are known to be involved in RNA interference, cosuppression, gene silencing, or quelling (de Carvalho et al., 1992; Hannon, 2002; Napoli et al., 1990; Romano and Macino, 1992). Numerous biological processes including development, maintenance of genome integrity, metabolism, abiotic stress responses, and immunity against pathogens are the crucial attributes for plant growth and development, regulated by small RNAs. On the basis of biogenesis, precursor structure and mode of action, the small RNAs are classified into two major groups: miRNAs and small interfering RNAs.

22.6.1 MicroRNAs

The miRNAs are 20–24 nt, noncoding, small regulatory RNAs reported from different plants and animals (Lee and Ambros, 2001; Llave et al., 2002; Reinhart et al., 2002). Functional analysis of miRNAs has shown their crucial roles in the growth, development, signal transduction, developmental timing, organ formation and development (formation and development of leaf, shoot and root) and cell fate determination of plants (Mallory et al., 2004; Sunkar, 2010; Sunkar et al., 2012). Beside cellular, biological and developmental processes (Li and Zhang, 2016), miRNAs also have been reported to play important roles in biotic (Khraiwesh et al., 2012) and abiotic stresses such as drought (Ferdous et al., 2017; Jatan et al., 2018, 2019), salinity (Jatan et al., 2018; Yin et al., 2017) heat (Mangrauthia et al., 2017; Pan et al., 2017), cold (Song et al., 2017), oxidation (Sunkar et al., 2006), nutrients (Liang et al., 2010; Pant et al., 2009) and heavy metals (HMs) (Zeng et al., 2012; Zhou et al., 2012).

22.6.2 MicroRNAs: biogenesis and mechanism of actions

miRNAs are produced by the transcription of endogenous MIR genes (Fig. 22.2). Metazoans miRNAs are mostly encoded by intergenic regions as independent transcription units, while some are represented by intronic region of the coding sequences (Olena and Patton, 2010; Tong et al., 2013). Moreover, MIR genes of plants are mainly located in the intergenic regions (Voinnet, 2009). Nuclear-encoded MIR genes are transcribed by Pol II to produce transcripts with a hairpin structure called primary miRNAs (pri-miRNAs) (Voinnet, 2009). The RNA-binding protein DAWDLE is thought to stabilize the hairpin structure of pri-miRNAs and

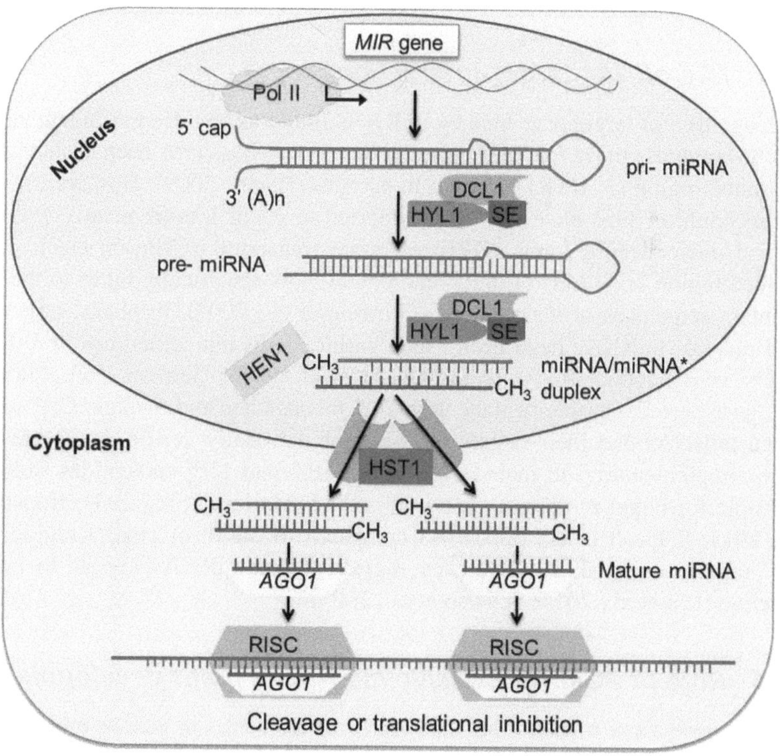

Figure 22.2 Biogenesis and function of microRNAs.

also involve in the formation of precursor miRNAs (pre-miRNAs). In plants, Dicer-like 1 (DCL1), HYPONASTIC LEAVES 1 (HYL1), SERRATE (SE) proteins and the cap-binding proteins complex are involved in the processing of pri-miRNAs into pre-miRNAs with a characteristic hairpin structure (Dolata et al., 2016; Iwata et al., 2013; Voinnet, 2009). Finally, the pre-miRNA hairpin is processed into 20–24 nt miRNA/miRNA* duplexes by HYL1, DCL1, and SE. Then the methylation of the miRNA/miRNA* duplexes at the 3′ terminus takes place by HUA ENHANCER 1 and the methylated miRNA/miRNA* duplexes are exported into the cytoplasm by an exportin protein, HASTY (Bartel, 2004; Boutet et al., 2003). In the cytoplasm, guide strand of the duplex is integrated into an ARGONAUTE (AGO), the catalytic component of RNA-induced silencing complex (RISC), and guides RISC to interact with the complementary sequence of the transcripts. The RISC protein complex inhibit the transcript expression either by the cleavage or translation inhibition based on sequence complementarity (Baumberger and Baulcombe, 2005; Brodersen and Voinnet, 2009). However, miR172 mediated regulation of the APETALA 2 (AP2) transcripts has been reported through both mRNA cleavage and translation inhibition (Schwab et al., 2005). Besides, posttranscriptional regulation of target genes, miRNAs also control the target expression by epigenetic modifications like DNA and histone methylation (Khraiwesh et al., 2010; Wu et al., 2010).

22.6.3 MicroRNAs target-interactions

The recognition of target sequence by miRNAs found to be different in animals and plants. In animals, most frequent target sites of miRNAs have been located in 3′-untranslated region (3′-UTR) of target transcripts (Bartel, 2009). However, miRNA target recognition sites have also been reported to occur less frequently within 5′-UTR and open reading frame (ORF) of target transcripts (Grimson et al., 2007). The "seed region" (2–7 nt) of the 5′ end of miRNAs specifically binds to the complementary sequences of their targets (Grimson et al., 2007). In plants, target sites for the most of miRNAs have been found within ORFs but sometimes in 3′-UTRs, 5′-UTRs, or noncoding RNAs (Addo-Quaye et al., 2008; German et al., 2008). In plants, near-perfect complementary with ≤ 5 mismatches and a single G:U wobble between miRNAs and their targets have been functionally verified. miRNAs have perfect complementarity to their targets at the 10th and 11th nucleotides which are responsible for target repression (Liu et al., 2014; Mallory et al., 2004; Rogers and Chen, 2013; Schwab et al., 2005). As compared to the 5′ or central region, mismatch alignments in the 3′ sequence region of the miRNAs appear to be less destructive (Liu et al., 2014; Schwab et al., 2005).

22.6.4 Role of PGPR-responsive miRNAs in plant development

Functional analysis of miRNAs has shown their crucial roles in various metabolic and biological processes including developmental pattering, phytohormone signaling, metabolism, genome integrity maintenance, flowering and sex determination, which are crucial to plant growth and development (Li and Zhang, 2016; Waheed and Zeng, 2020) (Fig. 22.3). miRNAs are not only involved in the regulation of plant growth processes, but also responsible for regulating plant response to nutrient deficiency, HM stress, biotic and abiotic stress (Ferdous et al., 2017; Islam et al., 2017; Khraiwesh et al., 2012; Song et al., 2019). In order to guide proper development, the expression of miRNAs has been found to vary within families and play different regulatory roles throughout several tissues and organs in their developmental perspective (Swarup and Denyer, 2018). For example, the miR165/166 family have seven members (vary in one nucleotide) and the target HD-Zip transcription factors including PHABULOSA (PHB), PHAVOLUTA (PHV), and REVOLUTA (REV), which are associated with plant developmental activities like embryo, leaf, shoot apical meristem, root and vascular development (Du and Wang, 2015; Sakaguchi and Watanabe, 2012; Swarup and Denyer, 2018; Wójcik et al., 2017).

In *Arabidopsis*, miR396-mediated control of GRF–GIF complex involved in the development of the pistil and the carpel number as well as in the determination of the meristematic cells of gynoecia and anthers (Lee et al., 2018). The upregulation of miR396a-5p has been shown to downregulate the GRF1 indicated their possible role in the growth and development of Arabidopsis on *P. putida*-inoculation (Jatan et al., 2020). The plant leaf actively participates in photosynthesis and thereby facilitates the growth and productivity of plants. miR319 targets *TEOSINTE BRANCHED1*, *CYCLOIDEA*, and *PROLIFERATING CELL NUCLEAR ANTIGEN BINDING FACTOR* (*TCP*) gene and miR319-mediated regulation of the TCP involved in the leaf development (Palatnik et al., 2003). Overexpression of miR319

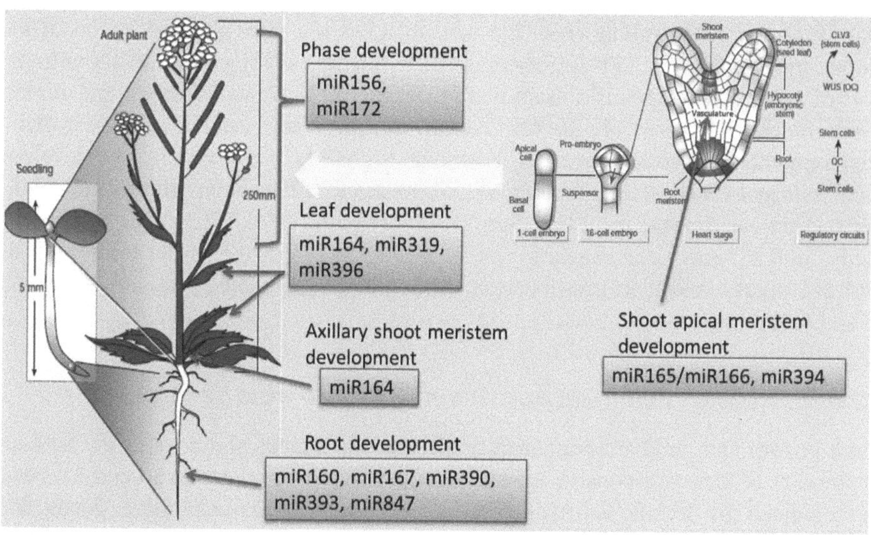

Figure 22.3 miRNA-mediated regulation of growth and development in plants.
Source: Adapted from Li, C., Zhang, B., 2016. MicroRNAs in control of plant development. J. Cell Physiol. 231, 303–313.

led to large leaflets, growth of leaf margins and a wider leaf blade in different transgenic plants including tomato, creeping bent grass and rice (Ori et al., 2007; Yang et al., 2013; Zhou et al., 2013). Further, analysis of different mutants of *miR319* and *TCP* reported that the miR319-TCP module is also responsible for determining cotyledon boundary, leaf senescence and leaf serration (Koyama et al., 2017; Yang et al., 2018).

A number of miRNAs influence the growth and expansion of root by controlling various genes and transcription factors (Swarup and Denyer, 2018). For example, the expression of miR166/165 has been reported to involve in root growth by the negative regulation of *HD-ZIP III* in the RAM (Singh et al., 2014). miR159 regulates the primary root growth in *A. thaliana* by the inhibition of MYB33, MYB65, and MYB101 (Xue et al., 2017). A recent study has shown that miR837−3p-LBD16 module may involve in lateral root formation by the upregulation of miR837−3p as well as significant downregulation of LBD16 in *P. putida*-inoculated *Arabidopsis* (Jatan et al., 2020).

Macronutrients including phosphorus (Pi), potassium (K), nitrogen (N2), and sulfur (S) are absolutely essential for normal growth and development of plants. miRNAs have also been reported to maintain the homeostasis of various nutrients. For example, induced expressions of miR395, miR398, and miR399 have been reported in response to sulfate, copper and phosphate deficiency, respectively (Khraiwesh et al., 2012). The target genes of miR399 belong to two different families: a phosphate transporter and ubiquitin conjugating enzyme and both genes involved in Pi homeostasis (Khraiwesh et al., 2012; Nath and Tuteja, 2016). miR399-PHO2-IPS1 module is reported to be an essential Pi-deprivation signaling pathway for the restoration of phosphate homeostasis. Plant roots absorb sulfur mainly as inorganic sulfate from the soil. miR395 regulates the expression of the sulfate transporter and ATP sulfurylases (APSs), which involved root to shoot transport of sulfate and the sulfur assimilation, respectively (Jones-Rhoades and Bartel, 2004; Jones-Rhoades et al., 2006). Overall, miRNAs play critical roles in growth, development, metabolic processes, hormonal signaling pathways, flower development, nutrient homeostasis and adaptive responses to different environmental conditions (Hong and Jackson, 2015; Sun, 2012; Tripathi et al., 2015). In a previous study, miR167-mediated regulation of NRAMP1 may involve in Fe homoeostasis and Mn uptake and also translocation, in *P. putida*-inoculated Arabidopsis (Jatan et al., 2020).

22.6.5 Role of PGPR-responsive miRNAs in abiotic stress

Plant growth and yield are dramatically affected by various abiotic stresses. Besides expression of stress-responsive genes, several miRNA families have been found to be regulated by abiotic stresses (Fig. 22.4). These stress-responsive genes and miRNAs are known to regulate the expression of several other genes related to growth, development and physiology of plants. In addition, various reports have demonstrated that miRNAs perform very crucial roles in adaptive response to different abiotic stresses (Khraiwesh et al., 2012; Sunkar et al., 2006). A large number

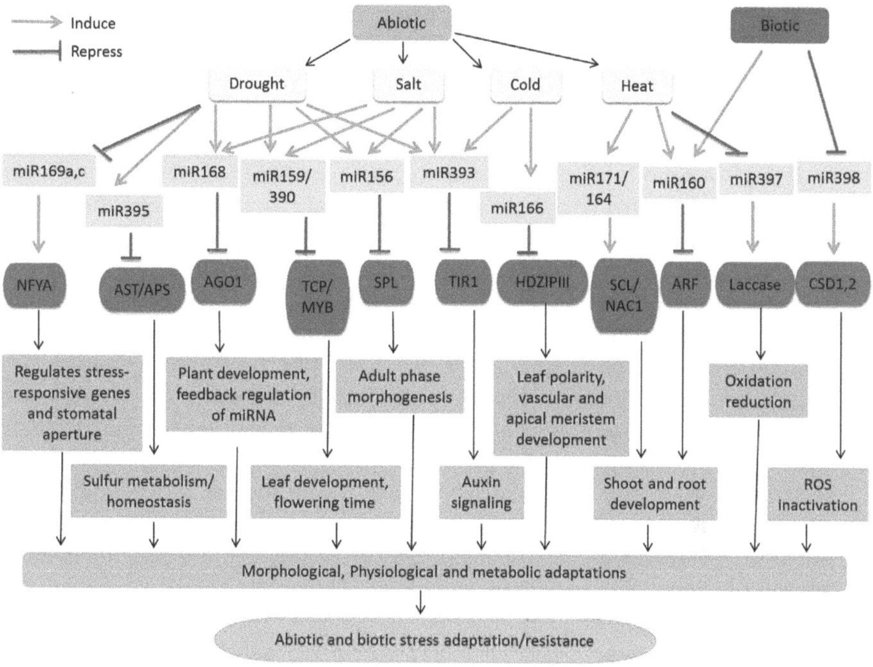

Figure 22.4 Distribution of novel miRNAs on chickpea.
Source: Adapted from Jatan, R., Lata, C., 2019. Role of microRNAs in abiotic and biotic stress resistance in plants. In: Proc. Indian Natl. Sci. Acad. pp. 553–567.

of miRNAs related to drought were identified in Arabidopsis (Liu et al., 2008), rice (Zhou et al., 2010), barley (Ferdous et al., 2017), common bean (Wu et al., 2017), maize (Aravind et al., 2017), tea (Guo et al., 2017), and tomato (Liu et al., 2017) through small RNAs sequencing. Overexpression of ath-miR408 in chickpea repressed the expression of plantacyanin leading to copper accumulation and enhanced drought tolerance in transgenic plants (Hajyzadeh et al., 2015). The overexpression of miR156 in *Arabidopsis* and rice down regulated the expression of *SPL9* genes which affected anthocyanin metabolism pathway by regulating the downstream genes such as *PRODUCTION OF ANTHOCYANIN PIGMENT 1* and *DIHYDROFLAVONOL-4-REDUCTASE* (*DFR*). The transgenic plants thus showed improved tolerance to drought and salt stresses through miR156-SPLs-DFR network (Cui et al., 2014). The regulation of NF-YA by miR169 in Arabidopsis and wheat under drought stress contributes significant role in regulating the expression of various drought-responsive genes, leading to an increased resistance to stress (Luan et al., 2014; Ma et al., 2015). High-throughput sequencing of chickpea inoculated with *P. putida* under drought stress led to the identification of 923 conserved and 216 putative novel miRNAs (Jatan et al., 2019).

Like drought stress, salinity also interferes with the normal growth and physiology of the plants. Numerous salt stress-associated miRNAs have been identified from different plants including rice (Macovei and Tuteja, 2012; Mondal et al., 2018), populus (Chen et al., 2017; Li et al., 2013), radish (Sun et al., 2015), *Salicornia europaea* (Feng et al., 2015), cotton (Yin et al., 2017), and maize (Fu et al., 2017). The overexpressing osa-miR319a exhibited enhanced drought and salt tolerance in transgenic creeping bent grass by targeting TCP (Zhou et al., 2013). Expression of numerous cold-associated miRNAs has been reported in different plants like *Prunus persica* (Barakat et al., 2012), Medicago (Shu et al., 2016), egg-plant (Yang et al., 2017), wheat (Song et al., 2017), and brassica (Zeng et al., 2018). In rice, overexpression of *OsmiR156k* exhibited less resistance to cold stress in transgenic plants by controlling SPL*3*, SPL*14*, and SPL*17* genes and inhibited the growth at the very early seedling stage by decreasing survival rates, chlorophyll and proline content (Cui et al., 2015). As like cold stress, the expression of miRNAs under heat stress have been identified in numerous plants including *Saccharina japonica* (Liu et al., 2015), cotton (Wang et al., 2016), *Betula luminifera* (Pan et al., 2017), and rice (Mangrauthia et al., 2017). Overexpression of Ta-miR159 in transgenic rice exhibited more sensitive to heat stress as compared to wild-type (Wang et al., 2012). Among abiotic stresses, HM toxicity also leads to deleterious effects on the growth, development and productivity of plant by modifying the physiological and metabolic processes (Gupta et al., 2014; Noman and Aqeel, 2017). It has been reported that the expression of miRNAs is also regulated by cadmium (Cd), mercury (Hg) and aluminum (Al) in Medicago (Zhou et al., 2008). miR390 overexpression in transgenic rice revealed reduced Cd tolerance with higher accumulation of Cd as compared to wild-type (Ding et al., 2016). *P. putida* inoculation led to the altered expression patterns of miR159, miR160, miR166, miR167, miR169, miR171, miR172, miR393, and miR396 at different time-points under drought and salt stress indicating their significant roles in abiotic stress alleviation in chickpea (Jatan and Lata, 2019).

22.6.6 Role of PGPR-responsive miRNAs in biotic stress

Besides abiotic stresses, miRNAs are also known involved in plant-microbe interactions and thus regulating the growth, development and physiology of host plants by the regulation of different genes and transcription factors (Fig. 22.4; Table 22.1). The first miRNA was reported as key players during plant-bacterial interactions when Arabidopsis treated with bacterial pathogen-associated molecular pattern peptide flg22 that leads to the induced expression of miR393 (Navarro et al., 2006). The high-throughput sequencing of small RNAs from Arabidopsis leaves infected with nonpathogenic *Pseudomonas syringae* pv. *tomato* strain DC3000 *hrc* (*Pst* DC3000 *hrc*, mutated in type III secretion system) at 1 and 3 h postinoculation (hpi) revealed that three miRNAs (miR160, miR167, and miR393) were highly upregulated and miR825 was downregulated at 3 hpi (Fahlgren et al., 2007). Overexpressing transgenic lines of miR393 led to the restricted growth of *Pst* DC3000 by the suppression of AUX signaling pathways by regulating TIR1 and other related proteins. Induced accumulation of miR863–3p by *Pst* DC3000 *avrRpt2* (*avrRpt2*, avirulence gene) infection, sequentially targets both negative and positive regulators of plant immunity depending on the timing and the magnitude of defense responses (Niu et al., 2006). miR863−3p enhances plant defense by targeting *a typical receptor-like pseudokinase1* (*ARLPK1*) and *ARLPK2* during early infection. However, miR863−3p positively regulates plant defense by targeting SE during later stage of infection. Most of bacterial-regulated miRNAs target genes were associated with hormone biosynthesis and signaling pathways such as AUX, JA, and ABA pathways, suggested that they play a crucial role in plant defense signaling.

The miRNAs are also play role in modulating nonpathogenic plant-microbe interaction rather than their role in plant-pathogen interactions. For example, symbiotic interaction between nitrogen-fixing bacteria (rhizobia) and leguminous plants is facilitated by miRNAs mediated regulatory systems (Simon et al., 2009). Symbiotic association of rhizobia and plant roots trigger the formation of specific nitrogen fixing organs called as root nodules in which reduction of nitrogen to ammonia is facilitated by rhizobia. miR169-mediated regulation of *MtHAP2−1* plays a significant role in the root nodule growth of *Medicago truncatula* (Combier et al., 2006). Interestingly, the expression level of *MtHAP2−1* confined to the nodule meristematic zone by miR169, suggesting that the spatial and temporal control of this TF by miR169 play critical role in differentiation of nodule cells. Reynoso et al. (2013) demonstrated that miRNAs were selectively recruited to the polyribosome in *M. truncatula* roots upon inoculation with *Sinorhizobium meliloti*. miR169 and miR172 targets the mRNAs of the *NF-YA/HAP2* and AP2 TFs, respectively. Expression of miR169 was downregulated in polysomal complexes led to the upregulation of the *NF-YA/HAP2* upon rhizobia inoculation, indicating that both mRNAs and miRNAs differentially recruited to polyribosomes. Expression of miR172c was found to be upregulated in common bean (*Phaseolus vulgaris*) by infection with *Rhizobium etli* and elevated during nodule development, with maximal expression in mature nodule while expression was declined in senescent nodules (Nova-Franco et al., 2015). miR172c targets the mRNA of APETALA2−1 (AP2−1), which is negatively correlated with the expression of miR172c. Phenotypic analysis of miR172c overexpressing lines or a mutated AP2−1 resistant to miR172c cleavage suggested the miR172 play important regulatory role in rhizobial symbiosis with the common bean.

22.7 Conclusion and future prospective

Overall, the growth promoting impact on host plants and also a suppressive effect on phytopatogens have been established by various PGP attributes of different PGPR under adverse conditions. Different *Pseudomonas* sp. are well known to increase the growth and productivity of host plants by the production of phytohormones, nutrient solubilization, and inhibition of phytopathogens during interactions with plants. Several PGPR, including *Pseudomonas* spp., have been actively recognized as an effective biofertilizers to enhance crop yields and biocontrol agents against different phytopathogens, and thereby reducing the use of chemical fertilizers, herbicides, and pesticides. Further, PGPR regulates the expression of stress-responsive genes, morpho-physiological and also biochemical parameters which lead to the overcoming of adverse effects against various abiotic stresses. In recent years, PGPR have also been reported to involve in amelioration of different abiotic stresses by regulating the expression of various stress-responsive miRNAs. Although tolerance to abiotic stresses is a complex phenomenon, PGPR-mediated modulation of expression of miRNAs and their target genes suggested that their crucial roles in the adaptive responses associated to sustainable growth and development of plants under various stresses. Further studies in PGPR-mediated regulation of miRNAs and their targets will uncover the molecular mechanism underlying stress tolerance and will provide a new strategy for crop improvement programs against global-warming in various agricultural crops.

Acknowledgment

CL acknowledges "Start-up Research Grant for INSA Young Scientist" (Grant No. INSA/SP/YSP/148/2017/1582) by the Indian National Science Academy (INSA), New Delhi. RJ acknowledges Junior Research Fellowship (Fellow No. DBT/2015/NBRI/322) from Department of Biotechnology, Government of India and Academy of Scientific and Innovative Research (AcSIR).

Conflict of interest

The authors declare that they have no conflict of interest.

References

Addo-Quaye, C., Eshoo, T.W., Bartel, D.P., Axtell, M.J.J.C.B., 2008. Endogenous siRNA and miRNA targets identified by sequencing of the *Arabidopsis* degradome. Curr. Biol. 18, 758–762.

Akdogan, G., Tufekci, E.D., Uranbey, S., Unver, T., 2015. miRNA-based drought regulation in wheat. Funct. Integr. Genomics 16, 221–233.

Aravind, J., Rinku, S., Pooja, B., Shikha, M., Kaliyugam, S., Mallikarjuna, M.G., et al., 2017. Identification, characterization, and functional validation of drought-responsive microRNAs in subtropical maize inbreds. Front. Plant Sci. 8, 941.

Arkhipova, T., Prinsen, E., Veselov, S., Martinenko, E., Melentiev, A., Kudoyarova, G.J.P., 2007. Cytokinin producing bacteria enhance plant growth in drying soil. Plant Soil 292, 305–315.

Arora, N.K., Khare, E., Oh, J.H., Kang, S.C., Maheshwari, D.K., 2008. Diverse mechanisms adopted by fluorescent Pseudomonas PGC2 during the inhibition of Rhizoctonia solani and Phytophthora capsici. World Journal of Microbiology and Biotechnology 24, 581–585.

Backer, R., Rokem, J.S., Ilangumaran, G., Lamont, J., Praslickova, D., Ricci, E., et al., 2018. Plant growth-promoting rhizobacteria: context, mechanisms of action, and roadmap to commercialization of biostimulants for sustainable agriculture. Front. Plant Sci. 9, 1473.

Badri, D.V., Vivanco, J.M., 2009. Regulation and function of root exudates. Plant Cell Environ. 32, 666–681.

Bagali, S.S., 2012. Review: nitrogen fixing microorganisms. Int. J. Microbiol. Res. 3, 46–52.

Barakat, A., Sriram, A., Park, J., Zhebentyayeva, T., Main, D., Abbott, A., 2012. Genome wide identification of chilling responsive microRNAs in *Prunus persica*. BMC Genomics 13, 481.

Bartel, D.P., 2004. MicroRNAs: genomics, biogenesis, mechanism, and function. Cell 116, 281–297.

Bartel, D.P., 2009. MicroRNAs: target recognition and regulatory functions. Cell 136, 215–233.

Barrera-Figueroa, B.E., Gao, L., Wu, Z., Zhou, X., Zhu, J., Jin, H., et al., 2012. High throughput sequencing reveals novel and abiotic stress-regulated microRNAs in the inflorescences of rice. BMC Plant Biol 12, 132.

Baumberger, N., Baulcombe, D.C., 2005. *Arabidopsis* ARGONAUTE1 is an RNA Slicer that selectively recruits microRNAs and short interfering RNAs. Proc. Natl. Acad. Sci. U. S. A. 102, 11928–11933.

Beneduzi, A., Ambrosini, A., Passaglia, L.M., 2012. Plant growth-promoting rhizobacteria (PGPR): their potential as antagonists and biocontrol agents. Genet. Mol. Biol. 35, 1044–1051.

Berg, G., Smalla, K., 2009. Plant species and soil type cooperatively shape the structure and function of microbial communities in the rhizosphere. FEMS Microbiol. Ecol. 68, 1–13.

Biessy, A., Filion, M., 2018. Phenazines in plant-beneficial *Pseudomonas* spp. biosynthesis, regulation, function and genomics. Environ. Microbiol. 20, 3905–3917.

Bogdan, A.R., Miyazawa, M., Hashimoto, K., Tsuji, Y., 2016. Regulators of iron homeostasis: new players in metabolism, cell death, and disease. Trends Biochem. Sci. 41, 274–286.

Boutet, S., Vazquez, F., Liu, J., Beclin, C., Fagard, M., Gratias, A., et al., 2003. *Arabidopsis* HEN1: a genetic link between endogenous miRNA controlling development and siRNA controlling transgene silencing and virus resistance. Curr. Biol. 13, 843–848.

Bottino, M.C., Rosario, S., Grativol, C., Thiebaut, F., Rojas, C.A., Farrineli, L., et al., 2013. High-throughput sequencing of small RNA transcriptome reveals salt stress regulated microRNAs in sugarcane. PLoS ONE 8, e59423.

Brodersen, P., Voinnet, O., 2009. Revisiting the principles of microRNA target recognition and mode of action. Nat. Rev. Mol. Cell Biol. 10, 141–148.

Burr, T., Schroth, M., Suslow, T., 1978. Increased potato yields by treatment of seed pieces with specific strains of *Pseudomonas fluorescens* and *Pseudomonas putida*. Phytopathology 68, 1377–1383.

Card, S., Johnson, L., Teasdale, S., Caradus, J., 2016. Deciphering endophyte behaviour: the link between endophyte biology and efficacious biological control agents. FEMS Microbiol. Ecol. 92.

Chauhan, P.S., Lata, C., Tiwari, S., Chauhan, A.S., Mishra, S.K., Agrawal, L., et al., 2019. Transcriptional alterations reveal *Bacillus amyloliquefaciens*-rice cooperation under salt stress. Sci. Rep. 9.

Chen, J., Zhang, J., Hu, J., Xiong, W., Du, C., Lu, M., 2017. Integrated regulatory network reveals the early salt tolerance mechanism of *Populus euphratica*. Sci. Rep. 7, 1–13.

Choudhary, D.K., 2011. Plant growth-promotion (PGP) activities and molecular characterization of rhizobacterial strains isolated from soybean (*Glycine max* L. Merril) plants against charcoal rot pathogen, *Macrophomina phaseolina*. Biotechnol. Lett. 33, 2287–2295.

Chu, T.N., Bui, L.V., Hoang, M.T.T., 2020. *Pseudomonas* PS01 isolated from maize rhizosphere alters root system architecture and promotes plant growth. Microorganisms 8 (4), 471.

Cohen, G., 2014. Microbial Biochemistry. Springer.

Combier, J.P., Frugier, F., de Billy, F., Boualem, A., El-Yahyaoui, F., Moreau, S., et al., 2006. MtHAP2–1 is a key transcriptional regulator of symbiotic nodule development regulated by microRNA169 in *Medicago truncatula*. Gene Dev. 20, 3084–3088.

Cui, L.G., Shan, J.X., Shi, M., Gao, J.P., Lin, H.X., 2014. The miR156-SPL 9-DFR pathway coordinates the relationship between development and abiotic stress tolerance in plants. Plant J. 80, 1108–1117.

Cui, N., Sun, X., Sun, M., Jia, B., Duanmu, H., Lv, D., et al., 2015. Overexpression of OsmiR156k leads to reduced tolerance to cold stress in rice (*Oryza Sativa*). Mol. Breed. 35, 214.

de Carvalho, F., Gheysen, G., Kushnir, S., Van Montagu, M., Inze, D., Castresana, C., 1992. Suppression of beta-1, 3-glucanase transgene expression in homozygous plants. EMBO J. 11, 2595–2602.

Dean, R., Van Kan, J.A., Pretorius, Z.A., Hammond-Kosack, K.E., Di Pietro, A., Spanu, P. D., et al., 2012. The top 10 fungal pathogens in molecular plant pathology. Mol. Plant Pathol. 13, 414–430.

Deng, P., Wang, L., Cui, L., Feng, K., Liu, F., Du, X., et al., 2015. Global identification of microRNAs and their targets in barley under salinity stress. PLoS ONE 10, 0137990.

Delaplace, P., Delory, B.M., Baudson, C., Mendaluk-Saunier de Cazenave, M., Spaepen, S., Varin, S., et al., 2015. Influence of rhizobacterial volatiles on the root system architecture and the production and allocation of biomass in the model grass *Brachypodium distachyon* (L.) P. Beauv. BMC Plant Biol. 15, 195.

Ding, Y., Ye, Y., Jiang, Z., Wang, Y., Zhu, C., 2016. MicroRNA390 is involved in cadmium tolerance and accumulation in rice. Front. Plant Sci. 7, 235.

Dolata, J., Bajczyk, M., Bielewicz, D., Niedojadlo, K., Niedojadlo, J., Pietrykowska, H., et al., 2016. Salt stress reveals a new role for ARGONAUTE1 in miRNA biogenesis at the transcriptional and posttranscriptional levels. Plant Physol. 172, 297–312.

Du, Q., Wang, H., 2015. The role of HD-ZIP III transcription factors and miR165/166 in vascular development and secondary cell wall formation. Plant Signal. Behav. 10, e1078955.

Dumas, Z., Ross-Gillespie, A., Kummerli, R., 2013. Switching between apparently redundant iron-uptake mechanisms benefits bacteria in changeable environments. Proc. Biol. Sci. 280, 20131055.

Effmert, U., Kalderás, J., Warnke, R., Piechulla, B., 2012. Volatile mediated interactions between bacteria and fungi in the soil. J. Chem. Ecol. 38, 665–703.

Egamberdieva, D., Kamilova, F., Validov, S., Gafurova, L., Kucharova, Z., Lugtenberg, B., 2008. High incidence of plant growth-stimulating bacteria associated with the rhizosphere of wheat grown on salinated soil in Uzbekistan. Environ. Microbiol. 10, 1–9.

Egamberdieva, D., Wirth, S., Jabborova, D., Räsänen, L.A., Liao, H., 2017. Coordination between *Bradyrhizobium* and *Pseudomonas* alleviates salt stress in soybean through altering root system architecture. J. Plant Int. 12, 100–107.

Fahlgren, N., Howell, M.D., Kasschau, K.D., Chapman, E.J., Sullivan, C.M., Cumbie, J.S., et al., 2007. High-throughput sequencing of *Arabidopsis* microRNAs: evidence for frequent birth and death of MIRNA genes. PLoS One 2, e219.

Feng, J., Wang, J., Fan, P., Jia, W., Nie, L., Jiang, P., et al., 2015. High-throughput deep sequencing reveals that microRNAs play important roles in salt tolerance of euhalophyte *Salicornia europaea*. Plant Biol. 15, 63.

Ferdous, J., Sanchez-Ferrero, J.C., Langridge, P., Milne, L., Chowdhury, J., Brien, C., et al., 2017. Differential expression of microRNAs and potential targets under drought stress in barley. Plant Cell Environ. 40, 11–24.

Finkel, O.M., Castrillo, G., Herrera, P.S., Salas Gonzalez, I., Dangl, J.L., 2017. Understanding and exploiting plant beneficial microbes. Curr. Opin. Plant Biol. 38, 155–163.

Flury, P., Vesga, P., Péchy-Tarr, M., Aellen, N., Dennert, F., Hofer, N., et al., 2017. Antimicrobial and insecticidal: cyclic lipopeptides and hydrogen cyanide produced by plant-beneficial *Pseudomonas* strains CHA0, CMR12a, and PCL1391 contribute to insect killing. Front. Microbiol. 8, 100.

Franche, C., Lindström, K., Elmerich, C., 2009. Nitrogen-fixing bacteria associated with leguminous and non-leguminous plants. J. Plant Soil 321, 35–59.

Fu, R., Zhang, M., Zhao, Y., He, X., Ding, C., Wang, S., et al., 2017. Identification of salt tolerance-related microRNAs and their targets in maize (*Zea mays* L.) using high-throughput sequencing and degradome analysis. Front. Plant Sci. 8, 864.

Garbeva, P., Van Elsas, J., Van Veen, J., 2008. Rhizosphere microbial community and its response to plant species and soil history. J. Plant Soil 302, 19–32.

García-Fraile, P., Menéndez, E., Rivas, R., 2015. Role of bacterial biofertilizers in agriculture and forestry. AIMS Bioeng. 2, 183.

Gentile, A., Dias, L.I., Mattos, R.S., Ferreira, T.H., Menossi, M., 2015. MicroRNAs and drought responses in sugarcane. Front. Plant Sci. 6, 58.

German, M.A., Pillay, M., Jeong, D.H., Hetawal, A., Luo, S., Janardhanan, P., et al., 2008. Global identification of microRNA–target RNA pairs by parallel analysis of RNA ends. Nat. Biotechnol. 26, 941–946.

Glick, B.R., 2014. Bacteria with ACC deaminase can promote plant growth and help to feed the world. Microbiol. Res. 169, 30–39.

Glick, B.R., Cheng, Z., Czarny, J., Duan, J., 2007. Promotion of plant growth by ACC deaminase-producing soil bacteria. New Perspectives and Approaches in Plant Growth-Promoting Rhizobacteria Research. Springer, pp. 329–339.

Grimson, A., Farh, K.K.H., Johnston, W.K., Garrett-Engele, P., Lim, L.P., Bartel, D.P., 2007. MicroRNA targeting specificity in mammals: determinants beyond seed pairing. Mol. Cell 27, 91–105.

Guo, Y., Zhao, S., Zhu, C., Chang, X., Yue, C., Wang, Z., et al., 2017. Identification of drought-responsive miRNAs and physiological characterization of tea plant (*Camellia sinensis* L.) under drought stress. BMC Plant Biol. 17, 211.

Gupta, O.P., Meena, N.L., Sharma, I., Sharma, P., 2014. Differential regulation of microRNAs in response to osmotic, salt and cold stresses in wheat. Mol. Biol. Rep. 41, 4623–4629.

Haas, D., Défago, G., 2005. Biological control of soil-borne pathogens by fluorescent pseudomonads. Nat. Rev. Microbiol. 3, 307–319.

Hackenberg, M., Gustafson, P., Langridge, P., Shi, B.J., 2015. Differential expression of microRNAs and other small RNAs in barley between water and drought conditions. Plant Biotechnol. J. 13, 2–13.

Hajyzadeh, M., Turktas, M., Khawar, K.M., Unver, T., 2015. miR408 overexpression causes increased drought tolerance in chickpea. Gene 555, 186–193.

Hallenbeck, P.C., 2017. Modern Topics in the Phototrophic Prokaryotes: Metabolism, Bioenergetics, and Omics. Springer.

Hannon, G.J., 2002. RNA interference. Nature 418, 244–251.

Hassani, M.A., Duran, P., Hacquard, S., 2018. Microbial interactions within the plant holobiont. Microbiome 6, 58.

He, X.F., Fang, Y.Y., Feng, L., Guo, H.S., 2008. Characterization of conserved and novel microRNAs and their targets, including a TuMV-induced TIR–NBS–LRR class R gene-derived novel miRNA in Brassica. FEBS letters 582, 2445–2452.

Hong, Y., Jackson, S., 2015. Floral induction and flower formation—the role and potential applications of miRNAs. Plant Biotechnol. J. 13, 282–292.

Hu, Y.F., Zhou, G., Na, X.F., Yang, L., Nan, W.B., Liu, X., et al., 2013. Cadmium interferes with maintenance of auxin homeostasis in *Arabidopsis* seedlings. J. Plant Physiol. 170, 965–975.

Huang, Z., Bonsall, R.F., Mavrodi, D.V., Weller, D.M., Thomashow, L.S., 2004. Transformation of *Pseudomonas fluorescens* with genes for biosynthesis of phenazine-1-carboxylic acid improves biocontrol of rhizoctonia root rot and in situ antibiotic production. FEMS Microbiol. Ecol. 49, 243–251.

Islam, W., Islam, S., Qasim, M., Wang, L., 2017. Host-pathogen interactions modulated by small RNAs. RNA Biol. 14, 891–904.

Iwata, Y., Takahashi, M., Fedoroff, N.V., Hamdan, S.M., 2013. Dissecting the interactions of SERRATE with RNA and DICER-LIKE 1 in *Arabidopsis* microRNA precursor processing. Nucleic Acids Res. 41, 9129–9140.

Jatan, R., Chauhan, P.S., Lata, C., 2018. *Pseudomonas putida* modulates the expression of miRNAs and their target genes in response to drought and salt stresses in chickpea (*Cicer arietinum* L.). Genomics 111 (4), 509–519.

Jatan, R., Lata, C., 2019. Role of microRNAs in abiotic and biotic stress resistance in plants. In: Proc. Indian Natl. Sci. Acad. pp. 553–567.

Jatan, R., Tiwari, S., Asif, M.H., Lata, C., 2019. Genome-wide profiling reveals extensive alterations in *Pseudomonas putida*-mediated miRNAs expression during drought stress in chickpea (*Cicer arietinum* L.). Environ. Exp. Bot. 157, 217–227.

Jatan, R., Chauhan, P.S., Lata, C., 2020. High-throughput sequencing and expression analysis suggest the involvement of *Pseudomonas putida* RA-responsive microRNAs in growth and development of *Arabidopsis*. Int. J. Mol. Sci. 21 (15), 5468.

Jeffries, P., Gianinazzi, S., Perotto, S., Turnau, K., Barea, J.M., 2003. The contribution of arbuscular mycorrhizal fungi in sustainable maintenance of plant health and soil fertility. Biol. Fertil. Soils 37, 1–16.

Jones-Rhoades, M.W., Bartel, D.P., 2004. Computational identification of plant microRNAs and their targets, including a stress-induced miRNA. Mol. Cell 14, 787–799.

Jones-Rhoades, M.W., Bartel, D.P., Bartel, B.J., 2006. MicroRNAs and their regulatory roles in plants. Annu. Rev. Plant Biol. 57, 19–53.

Jones, J.T., Haegeman, A., Danchin, E.G., Gaur, H.S., Helder, J., Jones, M.G., et al., 2013. Top 10 plant-parasitic nematodes in molecular plant pathology. Mol. Plant Pathol. 14, 946–961.

Kamoun, S., Furzer, O., Jones, J.D., Judelson, H.S., Ali, G.S., Dalio, R.J., et al., 2015. The top 10 oomycete pathogens in molecular plant pathology. Mol. Plant Pathol. 16, 413–434.

Kanchiswamy, C.N., Malnoy, M., Maffei, M.E., 2015. Chemical diversity of microbial volatiles and their potential for plant growth and productivity. Front. Plant Sci. 6, 151.

Katiyar, D., Hemantaranjan, A., Singh, B., 2016. Plant growth promoting rhizobacteria-an efficient tool for agriculture promotion. Adv. Plants Agric. Res. 4 (6), 426–434.

Khalid, A., Arshad, M., Shaharoona, B., Mahmood, T., 2009. Plant growth promoting rhizobacteria and sustainable agriculture. Microbial Strategies for Crop Improvement. Springer, pp. 133–160.

Khan, H.A., Siddique, K.H., Munir, R., Colmer, T.D., 2015. Salt sensitivity in chickpea: growth, photosynthesis, seed yield components and tissue ion regulation in contrasting genotypes. J. Plant Physiol. 182, 1–12.

Khraiwesh, B., Arif, M.A., Seumel, G.I., Ossowski, S., Weigel, D., Reski, R., et al., 2010. Transcriptional control of gene expression by microRNAs. Cell 140, 111–122.

Khraiwesh, B., Zhu, J.K., Zhu, J., 2012. Role of miRNAs and siRNAs in biotic and abiotic stress responses of plants. Biochim. Biophys. Acta 1819, 137–148.

Kloepper, J.W., 1978. Plant growth-promoting rhizobacteria on radishes. Proceedings of the Fourth International Conference on Plant Pathogenic Bacteria, Station de Pathologie Vegetale et Phytobacteriologie. INRA, Angers, France, pp. 879–882.

Kloepper, J.W., Leong, J., Teintze, M., Schroth, M.N., 1980a. Enhanced plant growth by siderophores produced by plant growth-promoting rhizobacteria. Nature 286, 885–886.

Kloepper, J.W., Leong, J., Teintze, M., Schroth, M.N., 1980b. *Pseudomonas* siderophores: a mechanism explaining disease-suppressive soils. Curr. Microbiol. 4, 317–320.

Koyama, T., Sato, F., Ohme-Takagi, M., 2017. Roles of miR319 and TCP transcription factors in leaf development. Plant Physiol. 175, 874–885.

Kruszka, K., Pacak, A., Swida-Barteczka, A., Nuc, P., Alaba, S., Wroblewska, Z., et al., 2014. Transcriptionally and post-transcriptionally regulated microRNAs in heat stress response in barley. J. Exp. Bot. 65, 6123–6135.

Kudoyarova, G., Arkhipova, T., Korshunova, T., Bakaeva, M., Loginov, O., Dodd, I.C., 2019. Phytohormone mediation of interactions between plants and non-symbiotic growth promoting bacteria under edaphic stresses. Front. Plant Sci. 10, 1368.

Kumar, H., Bajpai, V.K., Dubey, R., Maheshwari, D., Kang, S.C., 2010. Wilt disease management and enhancement of growth and yield of *Cajanus cajan* (L) var. Manak by bacterial combinations amended with chemical fertilizer. Crop. Prot 29, 591–598.

Kumar, M., Mishra, S., Dixit, V., Agarwal, L., Chauhan, P.S., Nautiyal, C.S., 2016. Synergistic effect of *Pseudomonas putida* and *Bacillus amyloliquefaciens* ameliorates drought stress in chickpea (*Cicer arietinum* L.). Plant Signal. Behav. 11, e1071004.

Kumar, R.K., Chu, H.H., Abundis, C., Vasques, K., Rodriguez, D.C., Chia, J.C., et al., 2017. Iron-nicotianamine transporters are required for proper long distance iron signaling. Plant Physiol. 175, 1254–1268.

Kumudini, B.S., Jayamohan, N.S., Patil, S.V., 2017. Integrated mechanisms of plant disease containment by rhizospheric bacteria: unraveling the signal cross talk between plant and fluorescent *Pseudomonas*. Agriculturally Important Microbes for Sustainable Agriculture. Springer, pp. 263–291.

Lebeis, S.L., Paredes, S.H., Lundberg, D.S., Breakfield, N., Gehring, J., McDonald, M., et al., 2015. PLANT MICROBIOME. Salicylic acid modulates colonization of the root microbiome by specific bacterial taxa. Science 349, 860–864.

Lee, R.C., Ambros, V.J., 2001. An extensive class of small RNAs in *Caenorhabditis elegans*. Science 294 (5543), 862–864.

Lee, R.C., Feinbaum, R.L., Ambros, V., 1993. The *Caenorhabditis elegans* heterochronic gene lin-4 encodes small RNAs with antisense complementarity to lin-14. Cell 75, 843–854.

Lee, S.J., Lee, B.H., Jung, J.H., Park, S.K., Song, J.T., Kim, J.H., 2018. GROWTH-REGULATING FACTOR and GRF-INTERACTING FACTOR specify meristematic cells of gynoecia and anthers. Plant Physiol. 176, 717–729.

Li, C., Zhang, B., 2016. MicroRNAs in control of plant development. J. Cell Physiol. 231, 303–313.

Li, B., Duan, H., Li, J., Deng, X.W., Yin, W., Xia, X., 2013. Global identification of miRNAs and targets in *Populus euphratica* under salt stress. Plant Mol. Biol. 81, 525–539.

Li, F., Pignatta, D., Bendix, C., Brunkard, J.O., Cohn, M.M., Tung, J., Sun, H., Kumar, P., Baker, B., 2012. MicroRNA regulation of plant innate immune receptors. Proceedings of the National Academy of Sciences 109 (5), 1790–1795.

Li, H., Dong, Y., Yin, H., Wang, N., Yang, J., Liu, X., Wang, Y., Wu, J., Li, X., 2011. Characterization of the stress associated microRNAs in Glycine max by deep sequencing. BMC Plant Biol 11 (1), 1–12.

Liang, G., Yang, F., Yu, D., 2010. MicroRNA395 mediates regulation of sulfate accumulation and allocation in *Arabidopsis thaliana*. Plant J. 62, 1046–1057.

Liu, X.M., Zhang, H., 2015. The effects of bacterial volatile emissions on plant abiotic stress tolerance. Front. Plant Sci. 6, 774.

Liu, H.H., Tian, X., Li, Y.J., Wu, C.A., Zheng, C.C., 2008. Microarray-based analysis of stress-regulated microRNAs in *Arabidopsis thaliana*. RNA 14, 836–843.

Liu, Q., Wang, F., Axtell, M.J., 2014. Analysis of complementarity requirements for plant microRNA targeting using a *Nicotiana benthamiana* quantitative transient assay. Plant Cell 26, 741–753.

Liu, F., Wang, W., Sun, X., Liang, Z., Wang, F., 2015. Conserved and novel heat stress-responsive microRNAs were identified by deep sequencing in *Saccharina japonica* (Laminariales, Phaeophyta). Plant Cell Environ. 38, 1357–1367.

Liu, M., Yu, H., Zhao, G., Huang, Q., Lu, Y., Ouyang, B., 2017. Profiling of drought-responsive microRNA and mRNA in tomato using high-throughput sequencing. BMC Genomics 18, 481.

Llave, C., Kasschau, K.D., Rector, M.A., Carrington, J.C., 2002. Endogenous and silencing-associated small RNAs in plants. Plant Cell 14, 1605–1619.

Llorente, B.E., Alasia, M.A., Larraburu, E.E., 2016. Biofertilization with *Azospirillum brasilense* improves in vitro culture of *Handroanthus ochraceus*, a forestry, ornamental and medicinal plant. N. Biotechnol. 33, 32–40.

Lobo, C.B., Juarez-Tomas, M.S., Viruel, E., Ferrero, M.A., Lucca, M.E., 2019. Development of low-cost formulations of plant growth-promoting bacteria to be used as inoculants in beneficial agricultural technologies. Microbiol. Res. 219, 12–25.

Luan, M., Xu, M., Lu, Y., Zhang, Q., Zhang, L., Zhang, C., et al., 2014. Family-wide survey of miR169s and NF-YAs and their expression profiles response to abiotic stress in maize roots. PLoS One 9, e91369.

Ma, X., Xin, Z., Wang, Z., Yang, Q., Guo, S., Guo, X., et al., 2015. Identification and comparative analysis of differentially expressed miRNAs in leaves of two wheat (*Triticum aestivum* L.) genotypes during dehydration stress. BMC Plant Biol. 15, 21.

Macovei, A., Tuteja, N., 2012. MicroRNAs targeting DEAD-box helicases are involved in salinity stress response in rice (*Oryza sativa* L.). BMC Plant Biol. 12, 183.

Mallory, A.C., Dugas, D.V., Bartel, D.P., Bartel, B., 2004. MicroRNA regulation of NAC-domain targets is required for proper formation and separation of adjacent embryonic, vegetative, and floral organs. Curr. Biol. 14, 1035–1046.

Malusá, E., Vassilev, N., 2014. A contribution to set a legal framework for biofertilisers. Appl. Microbiol. Biotechnol. 98 (15), 6599–6607.

Mangrauthia, S.K., Bhogireddy, S., Agarwal, S., Prasanth, V.V., Voleti, S.R., Neelamraju, S., et al., 2017. Genome-wide changes in microRNA expression during short and prolonged heat stress and recovery in contrasting rice cultivars. J. Exp. Bot. 68, 2399–2412.

Mansfield, J., Genin, S., Magori, S., Citovsky, V., Sriariyanum, M., Ronald, P., et al., 2012. Top 10 plant pathogenic bacteria in molecular plant pathology. Mol. Plant Pathol. 13, 614–629.

Mendes, R., Garbeva, P., Raaijmakers, J.M., 2013. The rhizosphere microbiome: significance of plant beneficial, plant pathogenic, and human pathogenic microorganisms. FEMS Microbiol. Rev. 37, 34–663.

Miljakovic, D., Marinkovic, J., Balesevic-Tubic, S., 2020. The significance of *Bacillus* spp. in disease suppression and growth promotion of field and vegetable crops. Microorganisms 8 (7), 1037.

Mishra, D., Rajvir, S., Mishra, U., Kumar, S.S., 2013. Role of bio-fertilizer in organic agriculture: a review. Res. J. Recent. Sci. 2, 39–41.

Misra, S., Dixit, V.K., Khan, M.H., Mishra, S.K., Dviwedi, G., Yadav, S., et al., 2017. Exploitation of agro-climatic environment for selection of 1-aminocyclopropane-1-carboxylic acid (ACC) deaminase producing salt tolerant indigenous plant growth promoting rhizobacteria. Microbiol. Res. 205, 25–34.

Mondal, T.K., Panda, A.K., Rawal, H.C., Sharma, T.R., 2018. Discovery of microRNA-target modules of African rice (*Oryza glaberrima*) under salinity stress. Sci. Rep. 8, 570.

Napoli, C., Lemieux, C., Jorgensen, R., 1990. Introduction of a chimeric chalcone synthase gene into petunia results in reversible co-suppression of homologous genes in trans. Plant Cell 2, 279–289.

Nath, M., Tuteja, N., 2016. NPKS uptake, sensing, and signaling and miRNAs in plant nutrient stress. Protoplasma 253, 767–786.

Navarro, L., Dunoyer, P., Jay, F., Arnold, B., Dharmasiri, N., Estelle, M., et al., 2006. A plant miRNA contributes to antibacterial resistance by repressing auxin signaling. Science 312, 436–439.

Neal, A.L., Ahmad, S., Gordon-Weeks, R., Ton, J., 2012. Benzoxazinoids in root exudates of maize attract *Pseudomonas putida* to the rhizosphere. PLoS One 7, e35498.

Niu, Q.W., Lin, S.S., Reyes, J.L., Chen, K.C., Wu, H.W., Yeh, S.D., et al., 2006. Expression of artificial microRNAs in transgenic *Arabidopsis thaliana* confers virus resistance. Nat. Biotechnol. 24, 1420–1428.

Noman, A., Aqeel, M., 2017. miRNA-based heavy metal homeostasis and plant growth. Environ. Sci. Pollut. Res. Int. 24, 10068–10082.

Nova-Franco, B., Íñiguez, L.P., Valdés-Lópe,z, O., Alvarado-Affantranger, X., Leija, A., Fuentes, S.I., et al., 2015. The micro-RNA172c-APETALA2–1 node as a key regulator of the common bean-*Rhizobium etli* nitrogen fixation symbiosis. Plant Physiol. 168, 273–291.

Olena, A.F., Patton, J.,G., 2010. Genomic organization of microRNAs. J. Cell Physiol. 222, 540–545.

Olorunleke, F.E., Hua, G.K.H., Kieu, N.P., Ma, Z., Höfte, M., 2015. Interplay between orfamides, sessilins and phenazines in the control of *Rhizoctonia* diseases by *Pseudomonas* sp. CMR12a. Environ. Microbiol. Rep. 7, 774–781.

Ori, N., Cohen, A.R., Etzioni, A., Brand, A., Yanai, O., Shleizer, S., et al., 2007. Regulation of LANCEOLATE by miR319 is required for compound-leaf development in tomato. Nat. Genet. 39, 787–791.

Palatnik, J.F., Allen, E., Wu, X., Schommer, C., Schwab, R., Carrington, J.C., et al., 2003. Control of leaf morphogenesis by microRNAs. Nature 425, 257–263.

Palleroni, N.J., 2015. Pseudomonas. Bergey's manual of systematics of archaea and bacteria 1.

Pan, Y., Niu, M., Liang, J., Lin, E., Tong, Z., Zhang, J., 2017. Identification of heat-responsive miRNAs to reveal the miRNA-mediated regulatory network of heat stress response in *Betula luminifera*. Trees 31, 1635–1652.

Pandey, R., Joshi, G., Bhardwaj, A.R., Agarwal, M., Katiyar-Agarwal, S., 2014. A comprehensive genome-wide study on tissue-specific and abiotic stress-specific miRNAs in *Triticum aestivum*. PLoS ONE 9, 95800.

Pangga, I.B., Hanan, J., Chakraborty, S., 2011. Pathogen dynamics in a crop canopy and their evolution under changing climate. Plant Pathol. 60, 70–81.

Panpatte, D.G., Jhala, Y.K., Shelat, H.N., Vyas, R.V., 2016. *Pseudomonas fluorescens*: a promising biocontrol agent and PGPR for sustainable agriculture. Microbial Inoculants in Sustainable Agricultural Productivity. Springer, pp. 257–270.

Pant, B.D., Musialak-Lange, M., Nuc, P., May, P., Buhtz, A., Kehr, J., et al., 2009. Identification of nutrient-responsive *Arabidopsis* and rapeseed microRNAs by comprehensive real-time polymerase chain reaction profiling and small RNA sequencing. Plant Physiol. 150, 1541–1555.

Park, Y.S., Dutta, S., Ann, M., Raaijmakers, J.M., Park, K., 2015. Promotion of plant growth by *Pseudomonas fluorescens* strain SS101 via novel volatile organic compounds. Biochem. Biophys. Res. Commun. 461, 361–365.

Park, Y.G., Mun, B.G., Kang, S.M., Hussain, A., Shahzad, R., Seo, C.W., et al., 2017. *Bacillus aryabhattai* SRB02 tolerates oxidative and nitrosative stress and promotes the growth of soybean by modulating the production of phytohormones. PLoS One 12, e0173203.

Pereira, S., Monteiro, C., Vega, A., Castro, P.M.L., 2016. Endophytic culturable bacteria colonizing *Lavandula dentata* L. plants: isolation, characterization and evaluation of their plant growth-promoting activities. Ecol. Eng. 87, 91–97.

Pérez-García, A., Romero, D., De Vicente, A.D.V., 2011. Plant protection and growth stimulation by microorganisms: biotechnological applications of *Bacilli* in agriculture. Curr. Opin. Biotechnol. 22, 187–193.

Raheem, A., Shaposhnikov, A., Belimov, A.A., Dodd, I.C., Ali, B., 2018. Auxin production by rhizobacteria was associated with improved yield of wheat (*Triticum aestivum* L.) under drought stress. Arch. Agron. Soil Sci. 64, 574–587.

Ramey, B.E., Koutsoudis, M., von Bodman, S.B., Fuqua, C., 2004. Biofilm formation in plant–microbe associations. Curr. Opin. Micrbiol. 7, 602–609.

Raza, W., Yousaf, S., Rajer, F.U., 2016. Plant growth promoting activity of volatile organic compounds produced by biocontrol strains. Sci. Lett. 4, 40–43.

Reinhart, B.J., Weinstein, E.G., Rhoades, M.W., Bartel, B., Bartel, D.P., 2002. MicroRNAs in plants. Genes Dev. 16, 1616–1626.

Reynoso, M.A., Blanco, F.A., Bailey-Serres, J., Crespi, M., Zanetti, M.E., 2013. Selective recruitment of mRNAs and miRNAs to polyribosomes in response to rhizobia infection in *Medicago truncatula*. Plant. J. 73, 289–301.

Ricci, E., Schwinghamer, T., Fan, D., Smith, D.L., Gravel, V.J., 2019. Growth promotion of greenhouse tomatoes with *Pseudomonas* sp. and *Bacillus* sp. biofilms and planktonic cells. Appl. Soil Ecol. 138, 61–68.

Rogers, K., Chen, X., 2013. Biogenesis, turnover, and mode of action of plant microRNAs. Plant Cell 25, 2383–2399.

Romano, N., Macino, G., 1992. Quelling: transient inactivation of gene expression in *Neurospora crassa* by transformation with homologous sequences. Mol. Microbiol. 6, 3343–3353.

Rudrappa, T., Czymmek, K.J., Pare, P.W., Bais, H.P., 2008. Root-secreted malic acid recruits beneficial soil bacteria. Plant Physiol. 148, 1547–1556.

Ruzzi, M., Aroca, R., 2015. Plant growth-promoting rhizobacteria act as biostimulants in horticulture. Sci. Hortic. 196, 124–134.

Ryu, C.M., Farag, M.A., Hu, C.H., Reddy, M.S., Wei, H.X., Pare, P.W., et al., 2003. Bacterial volatiles promote growth in *Arabidopsis*. Proc. Natl. Acad. Sci. U. S. A. 100, 4927–4932.

Saha, M., Sarkar, S., Sarkar, B., Sharma, B.K., Bhattacharjee, S., Tribedi, P., 2016. Microbial siderophores and their potential applications: a review. Environ. Sci. Pollut. Res. Int. 23, 3984–3999.

Sajitha, K., Dev, S.A., 2016. Quantification of antifungal lipopeptide gene expression levels in *Bacillus subtilis* B1 during antagonism against sapstain fungus on rubberwood. Biol. Control 96, 78–85.

Sakaguchi, J., Watanabe, Y., 2012. miR165/166 and the development of land plants. Dev. Growth Differ. 54, 93–99.

Sandhya, V., Ali, S.Z., Venkateswarlu, B., Reddy, G., Grover, M., 2010. Effect of osmotic stress on plant growth promoting *Pseudomonas* spp. Arch. Microbiol. 192, 867–876.

Scholthof, K.B., Adkins, S., Czosnek, H., Palukaitis, P., Jacquot, E., Hohn, T., et al., 2011. Top 10 plant viruses in molecular plant pathology. Mol. Plant Pathol. 12, 938–954.

Schwab, R., Palatnik, J.F., Riester, M., Schommer, C., Schmid, M., Weigel, D., 2005. Specific effects of microRNAs on the plant transcriptome. Dev. Cell 8, 517–527.

Seenivasan, S., Anderson, T.A., Muraleedharan, N., 2016. Heavy metal content in tea soils and their distribution in different parts of tea plants, *Camellia sinensis* (L). O. Kuntze. Environ. Monit. Assess. 188–428.

Shaikh, S., Wani, S., Sayyed, R., 2018. Impact of interactions between rhizosphere and rhizobacteria: a review. Plant Microbe Interact. 5, 1058.

Shu, Y., Liu, Y., Li, W., Song, L., Zhang, J., Guo, C.J.G.G., 2016. Genome-wide investigation of microRNAs and their targets in response to freezing stress in *Medicago sativa* L., based on high-throughput sequencing. G3 (Bethesda) 6, 755–765.

Simon, S.A., Meyers, B.C., Sherrier, D.J., 2009. MicroRNAs in the rhizobia legume symbiosis. Plant Physiol. 151, 1002–1008.

Singh, K., Talla, A., Qiu, W., 2012. Small RNA profiling of virus-infected grapevines: evidences for virus infection-associated and variety-specific miRNAs. Funct & Inegrative genomics 12, 59–669.

Singh, A., Singh, S., Panigrahi, K.C., Reski, R., Sarkar, A.K., 2014. Balanced activity of microRNA166/165 and its target transcripts from the class III homeodomain-leucine zipper family regulates root growth in *Arabidopsis thaliana*. Plant Cell Rep. 33, 945–953.

Singh, R.P., Shelke, G.M., Kumar, A., Jha, P.N., 2015. Biochemistry and genetics of ACC deaminase: a weapon to "stress ethylene" produced in plants. Front. Microbiol. 6, 937.

Song, G., Zhang, R., Zhang, S., Li, Y., Gao, J., Han, X., et al., 2017. Response of microRNAs to cold treatment in the young spikes of common wheat. BMC Genomics 18, 212.

Song, X., Li, Y., Cao, X., Qi, Y., 2019. MicroRNAs and their regulatory roles in plant–environment interactions. Annu. Rev. Plant Biol. 70, 489–525.

Spaepen, S., Bossuyt, S., Engelen, K., Marchal, K., Vanderleyden, J., 2014. Phenotypical and molecular responses of *Arabidopsis thaliana* roots as a result of inoculation with the auxin-producing bacterium *Azospirillum brasilense*. New Phytol. 201, 850–861.

Srivastava, S., Srivastava, S., 2020. Prescience of endogenous regulation in *Arabidopsis thaliana* by *Pseudomonas putida* MTCC 5279 under phosphate starved salinity stress condition. Sci. Rep. 10, 5855.

Srivastava, S., Chaudhry, V., Mishra, A., Chauhan, P.S., Rehman, A., Yadav, A., et al., 2012. Gene expression profiling through microarray analysis in *Arabidopsis thaliana* colonized by *Pseudomonas putida* MTCC5279, a plant growth promoting rhizobacterium. Plant Signal. Behav. 7, 235–245.

Sun, G., 2012. MicroRNAs and their diverse functions in plants. Plant Mol. Biol. 80, 17–36.

Sun, X., Xu, L., Wang, Y., Yu, R., Zhu, X., Luo, X., et al., 2015. Identification of novel and salt-responsive miRNAs to explore miRNA-mediated regulatory network of salt stress response in radish (*Raphanus sativus* L.). BMC Genomics 16, 197.

Sunkar, R., 2010. MicroRNAs with macro-effects on plant stress responses. Semin. Cell Dev. Biol. 21, 805–811.

Sunkar, R., Kapoor, A., Zhu, J.K., 2006. Posttranscriptional induction of two Cu/Zn superoxide dismutase genes in *Arabidopsis* is mediated by downregulation of miR398 and important for oxidative stress tolerance. Plant Cell 18, 2051–2065.

Sunkar, R., Li, Y.F., Jagadeeswaran, G., 2012. Functions of microRNAs in plant stress responses. Trends Plant Sci. 17, 196–203.

Swarup, R., Denyer, T., 2018. miRNAs in plant development. Annu. Plant Rev. 689–712.

Tiwari, S., Lata, C., Chauhan, P.S., Nautiyal, C.S., 2016. *Pseudomonas putida* attunes morphophysiological, biochemical and molecular responses in *Cicer arietinum* L. during drought stress and recovery. Plant Physiol. Biochem. 99, 108–117.

Tiwari, S., Prasad, V., Chauhan, P.S., Lata, C., 2017. *Bacillus amyloliquefaciens* confers tolerance to various abiotic stresses and modulates plant response to phytohormones through osmoprotection and gene expression regulation in rice. Front. Plant Sci. 8, 1510.

Tong, Y.-A., Peng, H., Zhan, C., Fan, L., Ai, T., Wang, S., 2013. Genome-wide analysis reveals diversity of rice intronic miRNAs in sequence structure, biogenesis and function. PLoS One 8, e63938.

Trindade, I., Capitao, C., Dalmay, T., Fevereiro, M.P., Santos, D.M., 2010. miR398 and miR408 are up-regulated in response to water deficit in *Medicago truncatula*. Planta 231, 705–716.

Tripathi, A., Goswami, K., Sanan-Mishra, N., 2015. Role of bioinformatics in establishing microRNAs as modulators of abiotic stress responses: the new revolution. Front. Physiol. 6, 286.

Ueda, A., Saneoka, H., 2015. Characterization of the ability to form biofilms by plant-associated *Pseudomonas* species. Curr. Microbiol. 70, 506–513.

Vacheron, J., Desbrosses, G., Bouffaud, M.L., Touraine, B., Moenne-Loccoz, Y., Muller, D., et al., 2013. Plant growth-promoting rhizobacteria and root system functioning. Front. Plant Sci. 4, 356.

Vejan, P., Abdullah, R., Khadiran, T., Ismail, S., Boyce, A.N., 2016. Role of plant growth promoting rhizobacteria in agricultural sustainability—a review. Molecules 21, 573.

Vessey, J.K., 2003. Plant growth promoting rhizobacteria as biofertilizers. Plant Soil 255, 571–586.

Voinnet, O., 2009. Origin, biogenesis, and activity of plant microRNAs. Cell 136, 669–687.

Waheed, S., Zeng, L., 2020. The critical role of miRNAs in regulation of flowering time and flower development. Genes (Basel) 11, 319.

Wang, Y., Sun, F., Cao, H., Peng, H., Ni, Z., Sun, Q., et al., 2012. TamiR159 directed wheat TaGAMYB cleavage and its involvement in anther development and heat response. PLoS One 7, e48445.

Wang, Q., Liu, N., Yang, X., Tu, L., Zhang, X., 2016. Small RNA-mediated responses to low-and high-temperature stresses in cotton. Sci. Rep. 6, 35558.

Wójcik, A.M., Nodine, M.D., Gaj, M.D., 2017. miR160 and miR166/165 contribute to the LEC2-mediated auxin response involved in the somatic embryogenesis induction in *Arabidopsis*. Front. Plant Sci. 8, 2024.

Wu, L., Zhou, H., Zhang, Q., Zhang, J., Ni, F., Liu, C., et al., 2010. DNA methylation mediated by a microRNA pathway. Mol. Cell 38, 465–475.

Wu, J., Wang, L., Wang, S., 2017. MicroRNAs associated with drought response in the pulse crop common bean (*Phaseolus vulgaris* L.). Gene 628, 78–86.

Xia, H., Zhang, L., Wu, G., Fu, C., Long, Y., Xiang, J., et al., 2016. Genome-wide identification and characterization of micrornas and target genes in *Lonicera japonica*. PLoS One 11, e0164140.

Xin, M., Wang, Y., Yao, Y., Xie, C., Peng, H., Ni, Z., et al., 2010. Diverse set of microRNAs are responsive to powdery mildew infection and heat stress in wheat (*Triticum aestivum* L.). BMC Plant Biol 10, 123.

Xue, T., Liu, Z., Dai, X., Xiang, F., 2017. Primary root growth in *Arabidopsis thaliana* is inhibited by the miR159 mediated repression of MYB33, MYB65, and MYB101. Plant Sci. 262, 182–189.

Yang, J., Kloepper, J.W., Ryu, C.M., 2009. Rhizosphere bacteria help plants tolerate abiotic stress. Trends Plant Sci. 14, 1–4.

Yang, X., Chen, L., Yong, X., Shen, Q., 2011. Formulations can affect rhizosphere colonization and biocontrol efficiency of *Trichoderma harzianum* SQR-T037 against *Fusarium* wilt of cucumbers. Biol. Fertil. Soil 47, 239–248.

Yang, C., Li, D., Mao, D., Liu, X., Ji, C., Li, X., et al., 2013. Overexpression of microRNA319 impacts leaf morphogenesis and leads to enhanced cold tolerance in rice (*Oryza sativa* L.). Plant Cell Environ. 36, 2207–2218.

Yang, X., Liu, F., Zhang, Y., Wang, L., Cheng, Y.F., 2017. Cold-responsive miRNAs and their target genes in the wild eggplant species *Solanum aculeatissimum*. BMC Genomics 18, 1000.

Yang, T., Wang, Y., Teotia, S., Zhang, Z., Tang, G., 2018. The making of leaves: how small RNA networks modulate leaf development. Front. Plant Sci. 9, 824.

Yin, Z., Han, X., Li, Y., Wang, J., Wang, D., Wang, S., et al., 2017. Comparative analysis of cotton small RNAs and their target genes in response to salt stress. Genes (Basel) 8.

Yu, Y., Jia, T.R., Chen, X.M., 2017. The "how" and "where" of plant microRNAs. New Phytol. 216, 1002–1017.

Yu, J.M., Wang, D., Pierson III, L.S., Pierson, E.A., 2018. Effect of producing different phenazines on bacterial fitness and biological control in *Pseudomonas chlororaphis*. Plant Pathol. J. 34 (1), 44–58.

Zamioudis, C., Mastranesti, P., Dhonukshe, P., Blilou, I., Pieterse, C.M., 2013. Unraveling root developmental programs initiated by beneficial *Pseudomonas* spp. bacteria. Plant Physiol. 162, 304–318.

Zeng, F., Qiu, B., Wu, X., Niu, S., Wu, F., Zhang, G., 2012. Glutathione-mediated alleviation of chromium toxicity in rice plants. Biol. Trace Elem. Res. 148, 255–263.

Zeng, X., Xu, Y., Jiang, J., Zhang, F., Ma, L., Wu, D., et al., 2018. Identification of cold stress responsive microRNAs in two winter turnip rape (*Brassica rapa* L.) by high throughput sequencing. BMC Plant Biol. 18, 52.

Zhao, M.M., Lyu, N., Wang, D., Wu, X.G., Zhao, Y.Z., Zhang, L.Q., et al., 2020. PhlG mediates the conversion of DAPG to MAPG in *Pseudomonas fluorescens* 2P24. Sci. Rep. 10, 4296.

Zhou, Z.S., Huang, S.Q., Yang, Z.M., 2008. Bioinformatic identification and expression analysis of new microRNAs from *Medicago truncatula*. Biochem. Biophys. Res. Commun. 374, 538–542.

Zhou, L., Liu, Y., Liu, Z., Kong, D., Duan, M., Luo, L., 2010. Genome-wide identification and analysis of drought-responsive microRNAs in *Oryza sativa*. J. Exp. Bot. 61, 4157–4168.

Zhou, Z.S., Zeng, H.Q., Liu, Z.P., Yang, Z.M., 2012. Genome-wide identification of *Medicago truncatula* microRNAs and their targets reveals their differential regulation by heavy metal. Plant Cell Environ. 35, 86–99.

Zhou, M., Li, D., Li, Z., Hu, Q., Yang, C., Zhu, L., et al., 2013. Constitutive expression of a miR319 gene alters plant development and enhances salt and drought tolerance in transgenic creeping bentgrass. Plant Physiol. 161, 1375–1391.

Next-generation omics technologies to explore microbial diversity

23

Ayushi Sharma, Jitendraa Vashistt and Rahul Shrivastava
Department of Biotechnology and Bioinformatics, Jaypee University of Information Technology, Waknaghat, Solan, India

23.1 Introduction

Studies into microbial complexity undergo several challenges. Low-abundance members increasingly require deep sampling procedures. Others, which have never been cultivated, demand still unknown growth conditions (Weinstock, 2012). Technological advances from culture-independent 5S and 16S rRNA sequencing to next-generation strategies allow extensive interventions into microbial communities (Weinstock, 2012). Metagenomic and several other omic approaches are indeed the most efficient technologies for characterizing microbes (Malla et al., 2019). These pave way for deciphering microbial physiology, improve annotations of the database, and help in understanding new biotechnological applications of microbes (Gutleben et al., 2018). Recent developments in omic technologies have drastically benefitted our understanding into biology of microbial communities and their interactions with the ecology. Omic advances and rapid progression in next-generation sequencing (NGS) associated computational and bioinformatic databases have provided significant breakthroughs in analyzing functional diversity of microbes (Jagadeesan et al., 2019). Yet, there are numerous mechanisms underlying the effect of land use changes on microbes, and their contribution toward management of these changes that need to be uncovered. This chapter reviews classical techniques of genome intervention and provides a comparative overview of modern NGS tools with their current and future applications relevant to microbial diversity. It aims at promoting use of NGS for exploring microbial diversity and reviews its usage for upcoming research.

23.1.1 Microbial diversity and metagenomics

Environmental metabolome and microbial metabolic potential are chief agents shaping the composition of microbial communities (Garza et al., 2018). Microbial activities are responsible for proper functioning of the ecosystem to a great extent. Microorganisms play crucial roles in maintaining habitats, facilitating biotic and abiotic interactions, enabling nutrient recycling, and being instrumental in waste degradation. Their contribution toward functional diversity is therefore a matter of

Microbes in Land Use Change Management. DOI: https://doi.org/10.1016/B978-0-12-824448-7.00026-7
© 2021 Elsevier Inc. All rights reserved.

assessment. Humans have impacted the physical, chemical, and biological microbial community structure. Anthropogenic interventions frequently lead to "land use changes," altering ecosystem functionality altogether. Most influential global disturbances which have been observed as an outcome of human activities include landscape transformation, increased croplands, degradation of forests, climatic changes, and loss in microbial diversity (Tiwari et al., 2019). Since microbes are the largest fraction of ecosystem inhabitants, the prevalent stresses force them to adapt to a new environment and gain new functions. Microbial diversity can very well be explored through techniques providing insights into microbial phylogeny (Rusch et al., 2007). Analysis into the nature and range of diversity assists in determining functionality and dynamics of microbes in environment (Zhou et al., 2008).

Revelation of the importance of soil microbes for generation of new natural compounds by Handelsman et al. (1998) enlightened the concept of "metagenome" (Alves et al., 2018). Metagenomics focuses on shotgun sequencing of complete genomes (Gilbert and Dupont, 2011). It enables identification of microbial genes via their functional aspects by cloning and subsequent expression in surrogate hosts (Weinstock, 2012), together with offering far greater genome coverage (Jagadeesan et al., 2019). Various genomes have been discovered from uncultured microbes through the use of metagenomic approach (Handelsman, 2004; Chivian et al., 2008). Metagenomic bioprospecting offers several beneficial biotechnological and industrial applications (Schloss and Handelsman, 2003; Lorenz and Eck, 2005; Hess et al., 2011). Its coupling with transcriptomics or proteomics facilitates identification of biochemical and functional aspects of microbial populations. Collectively, the main aim of metagenomics coupled NGS techniques is the resolution of data limitation and increasing understanding into microbial communities (Malla et al., 2019).

Another such approach for identifying complete microbial populations is metabarcoding or amplicon sequencing, which considers amplification and subsequent sequencing of marker gene families (Jagadeesan et al., 2019). DNA sequences resulting from high-throughput experiments are bar-coded for use in the strategy developed. The technique thereby efficiently helps in biodiversity monitoring, following taxonomic classification of species from mixed microbial samples (Zhang et al., 2018). Whole-genome sequencing (WGS) on the counter side allows investigations into individual strains, wherein multilocus sequence typing and analysis of single nucleotide polymorphism (SNP) help in discriminating between two strains by studying the generated phylogenetic clusters (Jagadeesan et al., 2019). The approaches utilizing NGS for exploring microbial diversity have been summarized and illustrated in Fig. 23.1.

23.2 Omics and ecology

"Omics" engulfs characterized and quantified associations of biological entities like genes, RNA, protein, and metabolites, in context with their dynamics, structure, and

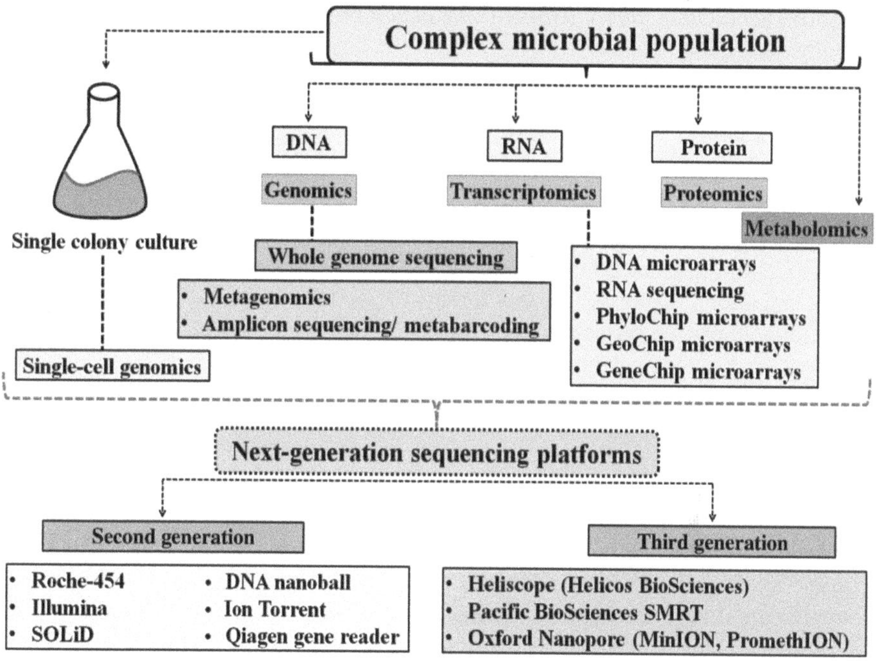

Figure 23.1 Illustration of next-generation sequencing technologies used for exploring microbial diversity.

functional attributes (Kulski, 2016; Zhang et al., 2018). Omic technologies, enlisted below, provide global coverage into cellular makeup of microbial communities.

23.2.1 Genomics

NGS has flourished genomic investigations for management and upbringing of human issues (Kulski, 2016). "Functional genomics" broadly indulges in high-throughput gene sequencing resultant changes in function of genes, either by knockdown (suppression) or knockout (deletion) (Zhang et al., 2018). The underlying technologies, RNA interference (RNAi) and clustered regularly interspaced short palindrome repeat (CRISPR)-Cas9 mediated editing of genomes, can either be incorporated individually or simultaneously. Considering complete knockout by CRISPR-Cas9 approach, functions of essential genes can be complementarily elucidated via RNAi partial knockdown.

23.2.1.1 RNA interference

The technique works by degradation of specific gene mRNA (Shalem et al., 2015). It finds numerous applications in functional screening of genomes for discovery of new drugs (Zhang et al., 2018). The approach however causes unexpected and irrelevant alterations in the phenotypes under study, on account of drawbacks like incomplete knockdown of genes and various off-target effects. The discovery of alternative techniques like CRISPR-Cas9 hence provided a leading edge for complete gene knockout studies.

23.2.1.2 Clustered regularly interspaced short palindrome repeats-Cas9

CRISPRs and Cas9 nuclease presents genome editing technique which relies on foreign nucleic acid cleavage capability of the prokaryotic immune system. A 20 bp sequence of single-guide RNA guides the nuclease appropriately for generating permanent mutations (cleavage) in the foreign DNA (Zhang et al., 2018). Impact of omic technologies on assessment of chemical consequences of genes, metabolic pathways, and ecological interventions can therefore be efficiently be studied by CRISPR-Cas9 screening of genomes (Zhang et al., 2018).

23.2.1.3 Single-cell genomics

Single-cell genomics can be conveniently applied to overcome constraints offered by complex microbial communities, the microbial dark matter (Woyke et al., 2010). The principle relies on isolation of microbial cells from the environment, their subjection to amplification, and subsequent sequencing. It incorporates techniques ranging from microfluidics (Leung et al., 2012) and flow cytometry to PCR (polymerase chain reaction) and genome sequencing (Kalisky and Quake, 2011). Separation of cells is done by fluorescence-activated cell sorting, following their rupture using alkaline lysis method (Raghunathan et al., 2005). Amplified fragments generated from multiple displacement amplification approach are used with NGS platforms (Lasken, 2007; Chitsaz et al., 2011). Complete genome from *Korachaeum cryptofilum* has been drafted using the approach (Elkins et al., 2008). The technique has been used widely for several environmental investigations. Novel microbial populations have been deducted in hypersaline aquatic environments (Ghai et al., 2011). Interactions between marine uncultivated protists are available (Yoon et al., 2011). Metagenomic analysis coupled with single-cell genomics has also targeted genomes of the most abundant Earth microbes, the ammonia-oxidizing archaea, in low-salinity ecosystems (Blainey et al., 2011). It has been deployed for investigating response of microbial communities to deep water oil spills (Mason et al., 2012). The platform efficiently recovered polyketide biosynthetic pathways from symbiotic sponges (Siegl et al., 2011) and helped in discovering degradative carbon-dioxide fixing macromolecules through chemoautotrophy (Martinez-Garcia et al., 2012). The technique on the whole focuses on exploitation of microbial diversity. It largely confronts investigations into the microbial world, since a large fraction of biotechnological applications hold possibility of sequencing only the microbes which can be cultured (Soon et al., 2013).

23.2.2 Transcriptomics

The dimension encompasses study of mRNA transcripts (Zhang et al., 2018). Initially employed for mammalian high-throughput sequencing ventures, the area has expanded its impact on a wide range of organisms. Transcriptomics increases understanding of genomic data and opens new dimensions for microbial functional biology (Kulski, 2016) and ecological diversity (Vieites et al., 2008).

23.2.2.1 DNA microarrays

DNA microarray is the fundamental transcriptome based approach which utilizes arrays of numerous probes (DNA oligonucleotides or cDNA fragments) on solid surface for interrogating abundance of thousands of target nucleic acid molecules analogously (Zhu, 2003). The targets undergo radioactive or fluorescent labeling before being detected. The fundamentals of complementary base pairing allow DNA probes to capture the targets, yield detectable signals, and provide quantitative estimation of relative abundances (Zhu, 2003). It has widely been used for investigating diversity and functions of microbial communities (Torsvik and Øvreås, 2002; Dennis et al., 2003; Majtan et al., 2004). A study utilized DNA microarray for monitoring bacteria having relevance to nitrogen cycling (Wu et al., 2001). Several other bacterial species, such as those belonging to *Pseudomonas* species have also been determined using the platform (Cho and Tiedje, 2001).

23.2.2.2 RNA sequencing

The analytical RNA-seq technique quantitatively samples an RNA population to generate significant reads using discrete templates for whole transcriptome shotgun sequencing (Li et al., 2014). It leads to identification of coding as well as noncoding transcriptional activities for providing sensitive and precise gene expression management (Wang et al., 2009; Ozsolak and Milos, 2011). RNA-seq generally utilizes cDNA libraries generated from either a fractionated or total RNA population. Molecules are sequenced without distinguishing between their amplification patterns (Wang et al., 2009). The approach provides insights into splicing events, small and noncoding RNAs, transcription factors, SNPs, and isoforms related to expression regulation (Kulski, 2016).

23.2.2.3 PhyloChip and GeoChip microarrays

Variations in microbial taxonomies can be best presented by PhyloChips. It incorporates thousands of DNA templates for diversity studies. 16S rRNA serves as the biomarker, owing to its conserved nature. A comprehensive phylogenetic oligonucleotide array PhyloChip is based on Affymetrix format (Santa Clara, CA). Second-generation of the arrays has been developed in 2002, and commercialized in 2006. These are advantageous on their sensitivity attributes, allowing detection of less dominant microbial populations. A major drawback however is the inability to discover and characterize new taxa (Nikolaki and Tsiamis, 2013). The approach has previously been used for describing microbial diversity in complex ecosystems like deep sea oil plumes (Hazen et al., 2010), olive-mill wastewater (Tsiamis et al., 2012), solar saltern (Tsiamis et al., 2008), and aircraft HEPA (high efficiency particulate air) filters (Korves et al., 2013). GeoChips on the counterside are advanced functional gene arrays used in exploration of microbial populations associated with agriculture, bioenergy, land use, climatic changes, biogeochemical cycles, ecosystem restoration, and human health (Nikolaki and Tsiamis, 2013). Diversity and metabolic potential of microbes in acid mine drainage have been studied using GeoChip microarrays (Xie et al., 2011). The drainage site is an ideal target for analyzing ecology, evolution and function of interacting microbial partners (Baker and Banfield, 2003). Composition, structure and function of microbial communities have been rapidly analyzed by these functional gene microarrays (He et al., 2010, 2012). Both the arrays can widely be used for monitoring and remediation of environmental contaminants on the whole.

23.2.2.4 GeneChip microarray

GeneChip microarrays are a standard tool for whole-genome expression analysis, opening new avenues into transcriptomic dynamics, and regulation. It utilizes *in silico* generated oligonucleotides for abundance estimation. Increased probe density owing to reduced area coverage on the chip provides broad coverage of the genome (Zhu, 2003). The technique has earlier been applied for characterizing complexity, abundance and stability of genomes for identification of associated biological pathways and treatment possibilities. Moreover, it has been utilized for prediction of biological performances and gene regulating determinants (Zhu, 2003). The technique has hence opened ways for several prokaryotic and eukaryotic expression studies (Kulski, 2016).

23.2.3 Epigenomics and methylomics

Epigenomics covers studies related to heritable gene regulation without involving DNA coding sequences (Hirst, 2013). It however considers DNA methylation, nucleosome positioning, posttranslational histone modifications, and target-transcription factor interactions (Landt et al., 2012). Methylation sequencing involves reactions of genomic DNA with bisulfate chemicals before proceeding towards library construction. Methylomics studies effect of DNA methylations on expression of genes and heredity (Kulski, 2016). This generates uracil from unmethylated cytosines. The treated sequences can then be efficiently sequenced with the alliance of NGS platforms. 5'-cytosine methylation increasingly regulates chromatin state and subsequent expression (Horner et al., 2010).

23.2.3.1 Chromatin immunoprecipitation-seq and RIP (RNA immunoprecipitation)-seq coupled next-generation sequencing

Chromatin immunoprecipitation (ChIP) profiles histones, DNA-binding proteins, and modifications in nucleosomes (Soon et al., 2013). It reveals associations between proteins and DNA (Mardis, 2008), yielding high resolution, minimum noise, and comparatively greater coverage than ChIP-chip array based method (Kulski, 2016).

modENCODE and ENCODE consortiums work together for maintaining ChIP-seq platform (Landt et al., 2012). Another technology for epigenomic and methylomic studies is RIP-seq. Both ChIP and RIP sequencing use antibody capture for enrichment of sequences prior to preparation of fragments for sequencing (Kulski, 2016). Although NGS has not been used widely for epigenomic investigations, the area is increasingly gaining scientific interest. A major paradigm related to the technique was reported in 2008 by Johnson and collaborators, who elucidated *Caenorhabditis elegans* nucleosome positioning using Roche platform (Mardis, 2008).

23.2.4 Proteomics

"Proteomics" refers to the study and characterization of gene products, the proteins (Zhang et al., 2018). This provides information of polymorphisms, abundance, cellular interactions, and related modifications. Sequences obtained are translated into open reading frames, and bioinformatic tools and algorithms are then applied for constructing proteomic profiles using NGS data. Other methods for gaining functional insights from protein studies use liquid chromatography along with mass spectrometry, yeast two-hybrid assays and affinity-tags (Kulski, 2016). Numerous public databases are available for proteomic studies. Two highly used portals providing protein related data are PRIDE (http://www.ebi.ac.uk/pride/archive/) and ExPASy (http://www.expasy.org/proteomics/protein-protein_interaction).

23.2.5 Metabolomics

This approach investigates an organism's metabolic response to genetic modifications or environmental stimuli. It is an important functional genomics aspect which helps in determining phenotypic consequences of mutations, insertions or deletions. The outcomes of metabolism are drawn from NGS data indirectly. Various physiological and biochemical pathways are studied on a higher scale through metabolomic interventions (Kulski, 2016).

23.3 Classical techniques of genome sequencing

Classical techniques for gaining insights into microbial ecology relied on culturing microbes to infer their associations with the host (Malla et al., 2019). Previous approaches relying on metagenomic analysis belonged to hybridization based signature sequences for distinguishing various taxa, restriction fragment or polymorphism centered PCR fingerprinting, or DNA sequencing categories (Weinstock, 2012). Ribosomal RNA sequences were widely investigated as targets for bacterial assays (Suzuki and Giovannoni, 1996; Lee et al., 2006). Cloning the entire 16S rRNA after amplification with primers used on numerous organisms, followed by Sanger sequencing of the cloned product was the most sorted approach before NGS (Rajendhran and Gunasekaran, 2011). PCR amplification of rRNA small subunit has undoubtedly increased the range of naming bacterial phyla from 12% to 92% over the previous 40 years (Gutleben et al., 2018). Uncultured bacteria residing within environmental niches have been largely classified by targeting 16S rDNA conserved regions (Wang and Qian, 2009). The approach however was culture-dependent, relied on narrow-range genomic sequencing, and provided limited sequence coverage (DeSantis et al., 2007). It was also inefficient in discriminating between organisms at the species level (Schmidt, 2017). Besides, sequencing errors often generated misclassified data. Metagenomic (sequence investigation of total DNA), metatranscriptomic (investigation of total RNA) and metaproteomic (functional associations between genotypes and phenotypes) interventions were the ones providing high-throughput details of taxon-specific gene expression (Gutleben et al., 2018). Alongside, these studies enabled identification of both disease-specific microbes and drug resistant genes between taxa shared by social contacting (Archie and Tung, 2015).

23.3.1 First-generation sequencing: Sanger sequencing

Sanger sequencing came into effect in 1977, and dominated the research arena for about three decades (Sanger et al., 1977). Automatic first-generation sequencers were initially developed by Applied Biosystems (ABI). These utilized fluorescent termination reagents based on Sanger method for sequencing in a single reaction (França et al., 2002). The sequencing utilizes capillary-based instrument and chain-terminating di-deoxynucleotides for sequencing single DNA fragments. It found major application in the 2001 Human Genome Project (HGP) (Malla et al., 2019). Global sponge microbiome profiling has also been done by 16S rRNA sequencing for studying structure, diversity and evolution of the phylum Porifera (Thomas et al., 2016). The platform is being used even today for low-throughput sequencing (Malla et al., 2019). PCR based traditional barcoding and Sanger sequencing needed advancements, since the earlier methodologies provided limited throughput data (Jagadeesan et al., 2019). Moreover, they suffered from volume limitations and increasingly high cost issues (Malla et al., 2019). Incurring cost of sequencing was 1000 times higher than NGS platform, allowing only 6 Mb output per day (Alves et al., 2018). A single genome sequencing by Sanger technique amounted to approximately $50,000 (Land et al., 2015).

23.4 Automated next-generation sequencing platforms for exploring microbial diversity

NGS has been routinely applied to various fields like diagnostics, forensics, antimicrobial resistance outbreaks, and exploring microbial diversity (Jagadeesan et al., 2019). It encompasses DNA, RNA, or methylated sequences for large-scale sequencing purposes (Kulski, 2016). Studies largely utilize NGS for WGS (Lam et al., 2012). Limitations of classical techniques are increasingly being addressed by NGS metagenomic approaches, which reveal detailed information about microbial populations (Jagadeesan et al., 2019). The advent of modern-day sequencing began with the end of HGP (Venter et al., 2001; Lander et al., 2001). The Venter Genome was completed with greater than $1 million by Sanger sequencing. It took around 15 years for the project completion. Watson genome however was sequenced by 454 pyrosequencer within 2 months, cutting down the cost 100 times (Kulski, 2016). A project on sequencing 1000 genomes has also come into effect owing to reduced NGS cost (McVean et al., 2012). NGS hence offers a cost efficient (Metzker, 2010) and highly parallel platform for independent sequencing of template molecules (Pareek et al., 2011).

NGS platforms have an enormous impact on biological sciences, owing to in-depth high-throughput screening of data, minimum technical expertise, and lower processing cost when compared to Sanger sequencing (Malla et al., 2019). NGS techniques aim at drastically reducing the cost and production of sequenced genomes (Weinstock, 2012). Table 23.1 describes the various classical and next-generation DNA sequencing platforms for easy briefing. The technologies rely either on amplicon sequencing or functional metagenomics (Malla et al., 2019). Preference of choosing the platforms is made as per aim, strategy, and purpose of the study.

23.4.1 Second-generation sequencing

Second-generation sequencing technologies do not rely on microbial cloning of fragments that need to be sequenced (Aird et al., 2011). Instead, the fragments are ligated to specific adapters and subjected to in vitro amplification (Niedringhaus et al., 2011). The library is amplified on a solid surface. Isolation is done within emulsified miniature droplets. Changes in luminescence or electrical charge are accordingly monitored for sequencing completion (Kulski, 2016). They however require generation of amplification libraries (Kulski, 2016). Numerous draft genomes have been produced by this sequencing generation (Land et al., 2015).

23.4.1.1 GS FLX pyrosequencing/Roche 454 platform

Roche-454 platform undoubtedly dominates analysis into microbial communities (Weinstock, 2012). It relies on pyrophosphate release principle; described in 1985 by Pal Nyrén (Ronaghi et al., 1996), and reported in 1988 by Hyman. The commercial Roche 454 sequencer released in 2005 utilized sequencing by synthesis principle (Malla et al., 2019). DNA synthesis proceeds by utilizing ATP sulfurylase, adenosine 5' phosphosulfate, luciferase and luciferin substrates (Mardis, 2008). DNA fragments undergo amplification by Emulsion PCR, where bead containing water droplets and PCR reagents are submerged in oil (Malla et al., 2019). Light received as output is proportional to the count of incorporated nucleotides (Mardis, 2008). The sequencer provides highly efficient reads of around 400–500 bp (Nossa et al., 2010), generating 20 Mb output per run (Malla et al., 2019). The platform is however costly and more error prone when considering homopolymer repeats (Kulski, 2016). Additional information is available at http://www.roche.com. The

Table 23.1 Classical and next-generation DNA sequencing platforms (Metzker, 2010; Kulski, 2016; Malla et al., 2019; Jagadeesan et al., 2019).

Sequencing platform	Principle/technique incorporated	Run time	Average read length (bp)	Output per run	Application	Error rate
First generation sequencing						
Sanger sequencing	Fluorescent dideoxy terminators	2 h	750–800	84 kb	Whole genome sequencing	Negligible
Second generation sequencing						
Roche-454	Pyrosequencing, sequencing by synthesis	18–48 h	500–700	~1 Gb	Identification of disease causing agents	Very low
Illumina	Stepwise fluorescent sequencing by synthesis, fluorescent terminators	2 h—3 days	~300	~2–900 Gb	Variant discovery	Low
SOLiD	Oligonucleotide ligation of adapters	~10 days	50–75	~120–320 Gb	Variant discovery	Low
DNA nanoball	Ligation	~14 days	50	3000 Gb	Decreased cost and increased throughput	
Ion Torrent	Detection of protons	2–20 h	200–400	100 Gb	Variant calling	Low
Third generation sequencing						
Heliscope	Real-time single molecule sequencing	~8 days	35	25 Gb	Unbiased sequencing	High
PacBio SMRT	Single molecule real-time fluorescent sequencing	0.5–6 h	~60,000	0.5–10 Gb	*De novo* genomic assembly	High
Oxford nanopore	Single molecule sequencing using electronic signals	48–72 h	~200,000	0.1–20 Gb	Complete genome sequencing, metagenomics	High

platform has been used for enumerating and contrasting soil microbial diversity (Roesch et al., 2007; Buee et al., 2009; Zhou et al., 2011). It also elucidated effects of land use and management on plethora of soil microbial inhabitants (Acosta-Martinez et al., 2008). Biogeography of microbial population facing long-term climatic and environmental changes has also been investigated using 454 pyrosequencing (Logares et al., 2013). The technique has been utilized for freshwater sampling (Monchy et al., 2011) demonstrating chemical, physical, and geographical biotopic factors influencing spatial variability. Another study identified aerosolized microbes by culture-independent universal (Sanger) and bacterial-specific (pyrosequencing) technologies (Robertson et al., 2013). Complex microbial communities can efficiently be characterized by 16S rRNA pyrosequencing (Jonasson et al., 2002; Kumar et al., 2011). Blocking the supply and servicing of 454 sequencing platforms by Roche after 2016 however directs researchers to unwillingly shift their preferences towards other NGS platforms (Kulski, 2016).

23.4.1.2 Illumina genome analyzer (HiSeq and MiSeq)

Solexa Limited originally developed the first commercial system utilizing sequencing technology (Balasubramanian, 2015). Illumina platform emerged in 2006 and expanded later in 2007, after acquiring Solexa (Malla et al., 2019). As for Roche 454 platform, Illumina also offered paired end sequencing. The scientific community shifted interest from Roche 454 to Illumina, owing to improved read length (300 bp by MiSeq 2) and much cost-effectiveness (Malla et al., 2019). MiSeq usually provides results in about 2 days (Liu et al., 2012). Sequencing through this platform relies on bridge-based amplification and the incorporation of reversible terminators (Pareek et al., 2011; Chun and Rainey, 2014). It utilizes laboratory-based library preparation, amplification of respective clusters, sequencing-by-synthesis followed by analysis of generated images, and data processing (Whiteford et al., 2009). Upgradation of the platform also came into light with the release of Genome Analyzer II, which magnified output to thrice as produced by the native form. Illumina HiSeq on the other hand generates up to 20 Gb sequences per read (Malla et al., 2019). Despite of producing enormous amount of data, the technique is underrepresented, owing to uneven coverage of extreme base constituents (Aird et al., 2011). It delivers short reads when compared to Roche-454 (Loman et al., 2012). Additional information about Illumina and its upgraded versions is available at http://solexaqa.sourceforge.net/.

23.4.1.3 Sequencing by oligonucleotide ligation and detection

ABI working in subsidiary with Life Technologies presented the sequencing by oligonucleotide ligation and detection (SOLiD) platform in 2007, and released it in 2008 (Shendure and Ji, 2008). It works on the principle of sequencing by oligonucleotide ligation of adapters (Tang et al., 2009; Cloonan et al., 2008). These products are immobilized on beads and suspended in emulsions of water and oil. Sequential hybridization using adapter complementary universal primers, followed by detection of emitted fluorescence completes the sequencing process (Malla et al., 2019).

23.4.1.4 *DNA nanoball sequencing*

The platform is a hybrid of hybridization and sequencing based sequencing (Niedringhaus et al., 2011). It has been developed by Complete Genomics and requires creation of DNA nanoballs by amplification of genomic DNA or cDNA by rolling-circle replication. Deposition of nanoballs on flow cells allows DNA reads from adapters. The chemistry happens to be related to that of exact cell in SOLiD sequencing. The technique however generates short read lengths and increases the time duration of sequencing. China based BGI-Shenzhen company introduced a highly upgraded Retrovelocity system by acquiring Complete Genomics in 2015. Several genomes have since been sequenced by the highly efficient platform (Kulski, 2016). Additional information is available at http://www.completegenomics.com and http://www.completegenomics.com/public-data/.

23.4.1.5 *Ion Torrent/Ion Proton personal genome machine*

Personal genome machine was introduced by Ion Torrent in 2010. An upgraded version Ion Proton, comprising of increasing output generation ability was accordingly launched in 2012. It utilizes enzyme based cascade for signal generation and has no relevance to optical signaling (Rothberg et al., 2011). The principle revolves around semiconductor technology and detects pH changes instead of light signals (Schadt et al., 2010; Chun and Rainey, 2014). Nucleotides are added in parallel to microwells containing different DNA templates. Ion sensors located beneath these wells detect protons released upon dNTP (deoxyribonucleotide triphosphate) addition. An ion-sensitive detector integrated to field-effect transistor measures the resultant pH change (Quail et al., 2012). A typical Ion Torrent has the capability of generating about 10 Gb output in a single run. An important highlight is the ability to distinguish between long fragments, which are generally omitted from the sequencing procedure owing to size-selection aspect (Malla et al., 2019). Additional information is available at https://www.thermofisher.com/in/en/home/brands/ion-torrent.html.

23.4.1.6 *Qiagen gene reader*

Gene reader was originally introduced as cyclic reversible termination platform by Qiagen Intelligent BioSystems. It was later available commercially in 2015. It is an all-in-one apparatus executing complete DNA sequencing steps. The platform is entirely same as Illumina, except that it considers only a few fluorophore labeled dNTPs. A run in gene reader considers four flow cells, each containing up to ten different samples (Malla et al., 2019). Additional information is available at http://www.qiagen.com.

23.4.2 Third-generation sequencing

Biasness in sequencing technologies is being overcome for balancing output yield, length of reads and efficient management of cost. Advancements have therefore been made towards establishment of third-generation sequencing platforms. These are single molecule based real-time technologies which deal with problems related to read length and amplification biasness (Malla et al., 2019). The area covers a complete genome within few hours, simultaneously adding methylation sites to chromosomes (Land et al., 2015). A significant reduction in cost is observed owing to nonrequirement of amplification libraries (Kulski, 2016).

23.4.2.1 Helicos BioSciences heliscope

Helicos BioSciences released heliscope, single-molecule sequencer in 2007 (Morozova et al., 2009). It was the first commercial platform using single-molecule templates (Metzker, 2010). It utilizes principal of sequencing by synthesis (Thompson and Steinmann, 2010). Several charge-coupled cameras help in obtaining reliable signals. The platform highly eliminates amplification biasness, average intensities, and problems related to synchronization. It is based on immobilization of around 200 oligonucleotides within microfluidic flow cell. Introduction of fluorescently labeled nucleotides into complementary strand by DNA polymerase increases the detection ability after ignoring DNA template amplification. A drawback however is the provision of just 35 nucleotides by the instrument (Malla et al., 2019). Even though Helicos BioSciences went bankrupt to present day, the Helioscope sequenced human genome for providing insights into disease signatures. It also sequenced RNA for quantitative transcriptomics of cells and tissues (Kulski, 2016). Additional information is available at http://www.helicosbio.com.

23.4.2.2 Pacific BioSciences single molecule real time

A DNA sequencer initially utilizing single molecule real-time (SMRT) approach was developed by Pacific BioSciences around 2010. It also utilizes principal of sequencing by synthesis for *de novo* genomic assembly (Eid et al., 2009). Florescent dyes used by several other NGS platforms are routinely used by this approach. Variation lies in the manner of signal emission and its detection in real-time. SMRT chip/cell containing zero mode wave (ZMW) carries out the sequencing. DNA polymerases are immobilized beneath ZMW guide along with single stranded templates. Incorporation of nucleotides proceeds by production of light signals, formulation of base calls and signal recordation (Malla et al., 2019). Additional information is available at http://www.pacificbiosciences.com.

23.4.2.3 Oxford Nanopore sequencing technologies (MinION and PromethION)

Oxford Nanopore was developed in 2005 for exploiting stochastic sequencing. The platform makes single-molecule detection possible (Bayley, 2015). Library preparation herein exempts amplification step, and allows direct sequencing of DNA template. Ion current is modulated as the template traverses through nanopore, providing insights into length, conformation, and diameter of the molecule under study. A conductive electrolytic solution within the pore generates potential gradient which is measured for completion of the sequencing protocol. Oxford Nanopore Technologies released MinION in 2014. A recent upgraded device developed on the foundation of nanopore is PromethION. The individual platform encompasses about 48 MinIONs and can efficiently be incorporated for analysis into very large genomes, as that of humans (Malla et al., 2019). Additional information is available at https://www.nanoporetech.com.

23.5 Computational biology interventions for analysis and storage of metagenomic data

Bioinformatics acts as rate-limiting arena for NGS applications (Horner et al., 2010). Advanced NGS technologies produce increasingly magnified data which needs continuous management. This can only be achieved through sophisticated computational and bioinformatic pipelines (Malla et al., 2019). Large amounts of generated data need to be analyzed, maintained, and interpreted from molecular genomic studies (Oliver et al., 2015). Shotgun reads can be compared using BLAST algorithm and metabolic pathways can therefore be constructed. Functional inferences from sequenced reads can be made from databases such as KEGG, which help in sorting gene products into relevant processes and pathways. This aims in providing insights into drug-resistant or virulent attributes of pathogens, and highlights antibiotic therapy consequences (Weinstock, 2012). Computational analysis further determines cooccurring microbial communities for yielding insights into their symbiotic associations or competitiveness. Collaborative use of supercomputers and efficient software, along with cloud-computing ventures provide an attractive means of managing vast datasets (Weinstock, 2012).

Various bioinformatic pipelines have been incorporated for metaomic analyses, metabarcoading, and investigations into ecological networks. These include QDME2 (http://qiime2.org/), MOTHUR (http://www.mothur.org/), Oligotyping (http://merenlab.org/software/oligotyping/) for metabarcoding; DADA2 (http://github.com/benjjneb/dada2), MG-RAST (http://metagenomics.anl.gov/) (Rath et al., 2011), MOCAT (http://www.bork.embl.de/mocat/), ANvro (http://merenlab.org/software/anvio/), IMP (http://r3lab.uni.lu/web/imp/) for metaomics; and Co Net (http://psbweb05.psb.ugent.be/conet/), sparCC (http://bitbucket.org/yonatanf/sparcc), SPIEC-EASI (http://github.com/zdk123/SpiecEasi), eLSA (http://bitbucket.org/charade/elsa) for network inference (Jagadeesan et al., 2019). Interactions between species can be studied and evaluated globally for reconstruction of ecological networks revealing involvement of new species in ecosystem functionality. Malla et al. (2019) describe various bioinformatic pipelines for metagenomic analysis, some of which are

FastQC (http://www.bioinformatics.babraham.ac.uk/projects/fastqc/), MEGAN (http://ab.inf.uni-tuebingen.de/software/megan/), PRINSEQ (http://prinseq.sourceforge.net/), QIIME (http://qiime.org/), CAMERA (https://omictools.com/camera-2-tool), and GlimmerMG (http://www.cbcb.umd.edu/software/glimmermg/). Tools for annotating noncoding sequence variants are GWAVA (http://www.sanger.ac.uk/sanger/StatGen_Gwava), CADD (http://cadd.gs.washington.edu), Phen-Gen (http://phen-gen.org), and RegulomeDB (http://regulome.stanford.edu) (Wang et al., 2015). Taxonomic investigations can be made using ANI (http://www.imedea.uib.es/jspecies/; http://www.ezbiocloud.net/ezgenome/ani), MUMi (http://genome.jouy.inra.fr/mumi/), GBDP (http://ggdc.dsmz.de/), and Nucleotide identity (http://www.bork.embl.de/software/specI/) (Chun and Rainey, 2014).

Metabolic associations can be revealed from assembled genomes using THE SEED. NetSeed modeling tool was developed by Carr and Borenstein in 2012 for prediction of compounds, the microbe obtains from the natural environment. Another tool, the Minimal ENvironment TOol helps in predicting nutritional demands of microbes at stake. These can be successfully incorporated for designing medium for laboratory cultivation of microbes that have not yet been cultured (Gutleben et al., 2018).

23.6 Advantages of next-generation sequencing in exploring microbial diversity

Multiomics revolution has astoundingly reshaped microbial ecology and provided understanding of the ecosystem. Studies have been conducted for characterizing microbial diversity within soil, wastewater and oceans (Gutleben et al., 2018). Omic advances offer increased efficiency, efficacy, and timeliness for assessment into the environment. New insights have been provided into genetic conservation and microbial diversity (Kulski, 2016). The technologies aim at generating novel knowledge and provide insights into new mechanisms and regulatory processes by collaborative support from superfast bioinformatic screening (Zhang et al., 2018). NGS involves cheap interventions, automation, less time consumption, and less labor intensive approaches. The platforms are increasingly creating paradigm-shifts in study of microbial diversity. Various pipelines are simultaneously being created for permitting high-throughput mechanistic interventions into molecular basis of microbial existence. While Illumina and Ion Torrent provide good coverage, SMRT platforms needs to be involved for detecting gene isoforms, genomic assemblies, and methylation patterns Decreased cost has fostered several collaborative ventures and lead to investigations into more specific scientific problems. It has impacted microbial diversity, medicine, genomics, biotechnology, microbiology, and environmental research beyond imagination. A good to the humanity example is seen at the Sanger Institute, where sequencing is routinely been carried out for cancer patients and those suffering from rare diseases (Malla et al., 2019). Humans are tremendously being blessed by the technologies as comprehensive sequencing provides details as to how DNA continuously struggles to keep people healthy. Human Microbiome Project and Metagenomics of the Human Intestinal Tract consortium have largely provided insights into the human microbiome (Morgan et al., 2013).

Illumina sequencing has successfully been applied for characterizing bacterial communities underlying Mississippi river (Staley et al., 2013). Soil microbial diversity resultant agricultural management also finds way from NGS investigations (Vasileiadis et al., 2012, 2013). The advanced platforms suggest importance of dinoflagellates and their relatives for the marine ecosystem (Stoeck et al., 2010). NGS conveniently offers possibilities of investigating the widespread marine microbial diversity (Pedrós-Alió, 2006). Besides, stable culturing has paved way for microbial use in ecosystem engineering and bioremediation (Greene and Voordouw, 2003). Suitable inferences have been made regarding toxic metabolite degradation pathways relative to gene abundances (Gutleben et al., 2018). Restoration of ecosystem and treatment of wastewater has also been made possible through multiomics ventures (Gutleben et al., 2018). NGS has also been incorporated for sequencing and characterizing archaeal, bacterial (Wu et al., 2009), algal, and cyanobacterial sequences (Wang et al., 2012; Fernández-Martínez et al., 2013).

23.6.1 Environmental genomic novelties

Metagenomic investigations have increasingly generated data from novel genomes since the availability of various tools for study (Medini et al., 2008). The most astounding example has been presented by Sargasso Sea short gun sequencing in 2004. This has lead to accumulation of around 70,000 new genes from hub of about 1.2 million. Three marine sites were sequenced, leading to discovery of 148 new bacterial rRNA phenotypes, ammonia oxidative pathway unrestrained by UV light, genes offering resistance towards trace metals copper and arsenate, and 782 proteorhodopsin-like receptors (Venter et al., 2004). Metagenomics has benefitted evolutionary biology in context of understanding the constraints presented by sexual reproduction on epigenetic and genomic variations. It revealed the nonclonal attributes of microbial species (Wang et al., 2015). Microbial dark matter was further investigated by Rinke et al. (2013) in a study involving single-cell genomics. The research considered around 20 new bacterial and archaeal lineages, providing outstanding details on eukaryotic to archaeal lateral transfer of genes. An evidence of rebuilding microbial communities from acid mine drainage utilizing fluoresce in situ hybridization based assay has also been reported (Tyson et al., 2004). The study provided abundant ferrous iron oxidizing genes amongst extreme environment inhabited low microbial diversity. In another report, Tringe et al. (2005) carried out investigation for comparing composition and functions of microbes from complex environments for illustrating and overcoming complications related to genomic analysis of the resident communities. The high-throughput technologies have allowed rapid access into microbial populations, than was possible using traditional techniques (Podar et al., 2007). Dinsdale et al. (2008) efficiently conducted comparative profiling of nine biomes for showing the specific role played by various biological traits in environment.

23.7 Next-generation sequencing future research directions

The increasing use of NGS for evaluation of microbial diversity generated through land use changes undoubtedly may lead to improvement in understanding the quality of ecosystems for developing powerful and constructive monitoring strategies. Although increasing metagenomic interventions provides exponentially increasing data, more sophisticated data analyzing platforms are required for overcoming shotgun sequencing drawbacks. PacBio and Oxford Nanopore are advanced alternatives yielding long reads, and distinguishing these between organism datasets. Their enormous power and speed, blessed with portability, definitely has wide industrial applications. Benchtop sequencers will help in maintaining small laboratories and allow a few scientists to ignite genomics revolution for providing new knowledge in the field of life sciences. Multidisciplinary collaborations and international teams will continue producing high-throughput comprehensive data in the future. Expansion of the reference genome catalog also needs consideration. Moreover, genomic ventures need accompanying transcriptomic or proteomic investigations, for having clear insights into data from living or dead microbial population. This would overcome variations in data from live microbiota, and that which has succumbed to host defense or drug action. Statistically validated data generated from ongoing projects will definitely manifest future experimental design and analysis.

References

Acosta-Martinez, V., Dowd, S., Sun, Y., Allen, V., 2008. Tag-encoded pyrosequencing analysis of bacterial diversity in a single soil type as affected by management and land use. Soil Biol. Biochem. 40 (11), 2762–2770.

Aird, D., Ross, M.G., Chen, W.S., Danielsson, M., Fennell, T., Russ, C., et al., 2011. Analyzing and minimizing PCR amplification bias in Illumina sequencing libraries. Genome Biol. 12 (2), 1–14.

Alves, L.D.F., Westmann, C.A., Lovate, G.L., de Siqueira, G.M.V., Borelli, T.C., Guazzaroni, M.E., 2018. Metagenomic approaches for understanding new concepts in microbial science. Int. J. Genomics 2018.

Archie, E.A., Tung, J., 2015. Social behavior and the microbiome. Curr. Opin. Behav. Sci. 6, 28–34.

Baker, B.J., Banfield, J.F., 2003. Microbial communities in acid mine drainage. FEMS Microbiol. Ecol. 44 (2), 139–152.

Balasubramanian, S., 2015. Solexa sequencing: decoding genomes on a population scale. Clin. Chem. 61 (1), 21–24.

Bayley, H., 2015. Nanopore sequencing: from imagination to reality. Clin. Chem. 61 (1), 25–31.

Blainey, P.C., Mosier, A.C., Potanina, A., Francis, C.A., Quake, S.R., 2011. Genome of a low-salinity ammonia-oxidizing archaeon determined by single-cell and metagenomic analysis. PLoS One 6 (2).

Buee, M., Reich, M., Murat, C., Morin, E., Nilsson, R.H., Uroz, S., et al., 2009. 454 Pyrosequencing analyses of forest soils reveal an unexpectedly high fungal diversity. New Phytol. 184 (2), 449–456.

Chitsaz, H., Yee-Greenbaum, J.L., Tesler, G., Lombardo, M.J., Dupont, C.L., Badger, J.H., et al., 2011. Efficient *de novo* assembly of single-cell bacterial genomes from short-read data sets. Nat. Biotechnol. 29 (10), 915–921.

Chivian, D., Brodie, E.L., Alm, E.J., Culley, D.E., Dehal, P.S., DeSantis, T.Z., et al., 2008. Environmental genomics reveals a single-species ecosystem deep within Earth. Science 322 (5899), 275–278.

Cho, J.C., Tiedje, J.M., 2001. Bacterial species determination from DNA-DNA hybridization by using genome fragments and DNA microarrays. Appl. Environ. Microbiol. 67 (8), 3677–3682.

Chun, J., Rainey, F.A., 2014. Integrating genomics into the taxonomy and systematics of the Bacteria and Archaea. Int. J. Syst. Evol. Microbiol. 64 (2), 316–324.

Cloonan, N., Forrest, A.R., Kolle, G., Gardiner, B.B., Faulkner, G.J., Brown, M.K., et al., 2008. Stem cell transcriptome profiling via massive-scale mRNA sequencing. Nat. Methods 5 (7), 613–619.

Dennis, P., Edwards, E.A., Liss, S.N., Fulthorpe, R., 2003. Monitoring gene expression in mixed microbial communities by using DNA microarrays. Appl. Environ. Microbiol. 69 (2), 769–778.

DeSantis, T.Z., Brodie, E.L., Moberg, J.P., Zubieta, I.X., Piceno, Y.M., Andersen, G.L., 2007. High-density universal 16S rRNA microarray analysis reveals broader diversity than typical clone library when sampling the environment. Microb. Ecol. 53 (3), 371–383.

Dinsdale, E.A., Edwards, R.A., Hall, D., Angly, F., Breitbart, M., Brulc, J.M., et al., 2008. Functional metagenomic profiling of nine biomes. Nature 452 (7187), 629–632.

Eid, J., Fehr, A., Gray, J., Luong, K., Lyle, J., Otto, G., et al., 2009. Real-time DNA sequencing from single polymerase molecules. Science 323 (5910), 133–138.

Elkins, J.G., Podar, M., Graham, D.E., Makarova, K.S., Wolf, Y., Randau, L., et al., 2008. A korarchaeal genome reveals insights into the evolution of the Archaea. Proc. Natl. Acad. Sci. U. S. A. 105 (23), 8102–8107.

Fernández-Martínez, M.A., de Los Ríos, A., Sancho, L.G., Pérez-Ortega, S., 2013. Diversity of endosymbiotic *Nostoc* in *Gunnera magellanica* (L) from Tierra del Fuego, Chile. Microb. Ecol. 66 (2), 335–350.

França, L.T.C., Carrilho, E., Kist, T.B.L., 2002. A review of DNA sequencing techniques. Q. Rev. Biophys. 35 (2), 169.

Garza, D.R., van Verk, M.C., Huynen, M.A., Dutilh, B.E., 2018. Towards predicting the environmental metabolome from metagenomics with a mechanistic model. Nat. Microbiol. 3 (4), 456–460.

Ghai, R., Pašić, L., Fernández, A.B., Martin-Cuadrado, A.B., Mizuno, C.M., McMahon, K. D., et al., 2011. New abundant microbial groups in aquatic hypersaline environments. Sci. Rep. 1, 135.

Gilbert, J.A., Dupont, C.L., 2011. Microbial metagenomics: beyond the genome. Annu. Rev. Mar. Sci. 3, 347–371.

Greene, E.A., Voordouw, G., 2003. Analysis of environmental microbial communities by reverse sample genome probing. J. Microbiol. Methods 53 (2), 211–219.

Gutleben, J., Chaib De Mares, M., van Elsas, J.D., Smidt, H., Overmann, J., Sipkema, D., 2018. The multi-omics promise in context: from sequence to microbial isolate. Crit. Nat. Rev. Microbiol. 44 (2), 212–229.

Handelsman, J., 2004. Metagenomics: application of genomics to uncultured microorganisms. Microbiol. Mol. Biol. Rev. 68 (4), 669–685.

Handelsman, J., Rondon, M.R., Brady, S.F., Clardy, J., Goodman, R.M., 1998. Molecular biological access to the chemistry of unknown soil microbes: a new frontier for natural products. Chem. Biol. 5 (10), R245–R249.

Hazen, T.C., Dubinsky, E.A., DeSantis, T.Z., Andersen, G.L., Piceno, Y.M., Singh, N., et al., 2010. Deep-sea oil plume enriches indigenous oil-degrading bacteria. Science 330 (6001), 204–208.

He, Z., Deng, Y., Van Nostrand, J.D., Tu, Q., Xu, M., Hemme, C.L., et al., 2010. GeoChip 3.0 as a high-throughput tool for analyzing microbial community composition, structure and functional activity. ISME J. 4 (9), 1167–1179.

He, Z., Deng, Y., Zhou, J., 2012. Development of functional gene microarrays for microbial community analysis. Curr. Opin. Biotechnol. 23 (1), 49–55.

Hess, M., Sczyrba, A., Egan, R., Kim, T.W., Chokhawala, H., Schroth, G., et al., 2011. Metagenomic discovery of biomass-degrading genes and genomes from cow rumen. Science 331 (6016), 463–467.

Hirst, M., 2013. Epigenomics: sequencing the methylome. Array Comparative Genomic Hybridization. Humana Press, Totowa, NJ, pp. 39–54.

Horner, D.S., Pavesi, G., Castrignano, T., De Meo, P.D.O., Liuni, S., Sammeth, M., et al., 2010. Bioinformatics approaches for genomics and post genomics applications of next-generation sequencing. Brief. Bioinfo. 11 (2), 181–197.

Jagadeesan, B., Gerner-Smidt, P., Allard, M.W., Leuillet, S., Winkler, A., Xiao, Y., et al., 2019. The use of next generation sequencing for improving food safety: translation into practice. Food Microbiol. 79, 96–115.

Jonasson, J., Olofsson, M., Monstein, H.J., 2002. Classification, identification and subtyping of bacteria based on pyrosequencing and signature matching of 16S rDNA fragments. APMIS 110 (3), 263–272.

Kalisky, T., Quake, S.R., 2011. Single-cell genomics. Nat. Methods 8 (4), 311–314.

Korves, T.M., Piceno, Y.M., Tom, L.M., DeSantis, T.Z., Jones, B.W., Andersen, G.L., et al., 2013. Bacterial communities in commercial aircraft high-efficiency particulate air (HEPA) filters assessed by PhyloChip analysis. Indoor Air 23 (1), 50–61.

Kulski, J.K., 2016. Next-generation sequencing—an overview of the history, tools, and "omic" applications. Next Generation Sequencing - Advances, Applications and Challenges. InTechOpen, pp. 3–60.

Kumar, P.S., Brooker, M.R., Dowd, S.E., Camerlengo, T., 2011. Target region selection is a critical determinant of community fingerprints generated by 16S pyrosequencing. PLoS One 6 (6), e20956.

Lam, H.Y., Clark, M.J., Chen, R., Chen, R., Natsoulis, G., O'huallachain, M., et al., 2012. Performance comparison of whole-genome sequencing platforms. Nat. Biotechnol. 30 (1), 78–82.

Land, M., Hauser, L., Jun, S.R., Nookaew, I., Leuze, M.R., Ahn, T.H., et al., 2015. Insights from 20 years of bacterial genome sequencing. Funct. Integr. Genomics 15 (2), 141–161.

Lander, E.S., et al., 2001. Initial sequencing and analysis of the human genomeInternational Human Genome Sequencing Consortium Nature 409, 860–921.

Landt, S.G., Marinov, G.K., Kundaje, A., Kheradpour, P., Pauli, F., Batzoglou, S., et al., 2012. ChIP-seq guidelines and practices of the ENCODE and modENCODE consortia. Genome Res. 22 (9), 1813–1831.

Lasken, R.S., 2007. Single-cell genomic sequencing using multiple displacement amplification. Curr. Opin. Microbiol. 10 (5), 510–516.

Lee, D.Y., Shannon, K., Beaudette, L.A., 2006. Detection of bacterial pathogens in municipal wastewater using an oligonucleotide microarray and real-time quantitative PCR. J. Microbiol. Methods 65 (3), 453–467.

Leung, K., Zahn, H., Leaver, T., Konwar, K.M., Hanson, N.W., Pagé, A.P., et al., 2012. A programmable droplet-based microfluidic device applied to multiparameter analysis of single microbes and microbial communities. Proc. Natl. Acad. Sci. U. S. A. 109 (20), 7665–7670.

Li, S., Tighe, S.W., Nicolet, C.M., Grove, D., Levy, S., Farmerie, W., et al., 2014. Multi-platform assessment of transcriptome profiling using RNA-seq in the ABRF next-generation sequencing study. Nat. Biotechnol. 32 (9), 915–925.

Liu, L., Li, Y., Li, S., Hu, N., He, Y., Pong, R., et al., 2012. Comparison of next-generation sequencing systems. Biomed. Res. Int. 2012.

Logares, R., Lindström, E.S., Langenheder, S., Logue, J.B., Paterson, H., Laybourn-Parry, J., et al., 2013. Biogeography of bacterial communities exposed to progressive long-term environmental change. ISME J. 7 (5), 937–948.

Loman, N.J., Misra, R.V., Dallman, T.J., Constantinidou, C., Gharbia, S.E., Wain, J., et al., 2012. Performance comparison of benchtop high-throughput sequencing platforms. Nat. Biotechnol. 30 (5), 434–439.

Lorenz, P., Eck, J., 2005. Metagenomics and industrial applications. Nat. Rev. Microbiol. 3 (6), 510–516.

Majtan, T., Bukovska, G., Timko, J., 2004. DNA microarrays—techniques and applications in microbial systems. Folia Microbiol. 49 (6), 635.

Malla, M.A., Dubey, A., Kumar, A., Yadav, S., Hashem, A., Abd_Allah, E.F., 2019. Exploring the human microbiome: the potential future role of next-generation sequencing in disease diagnosis and treatment. Front. Immunol. 9, 2868.

Mardis, E.R., 2008. Next-generation DNA sequencing methods. Annu. Rev. Genomics Hum. Genet. 9, 387–402.

Martinez-Garcia, M., Brazel, D.M., Swan, B.K., Arnosti, C., Chain, P.S., Reitenga, K.G., et al., 2012. Capturing single cell genomes of active polysaccharide degraders: an unexpected contribution of *Verrucomicrobia*. PLoS One 7 (4).

Mason, O.U., Hazen, T.C., Borglin, S., Chain, P.S., Dubinsky, E.A., Fortney, J.L., et al., 2012. Metagenome, metatranscriptome and single-cell sequencing reveal microbial response to Deepwater Horizon oil spill. ISME J. 6 (9), 1715–1727.

McVean, G., Altshuler (Co-Chair), D., Durbin (Co-Chair), R., et al., 2012. An integrated map of genetic variation from 1092 human genomes1000 Genomes Project Consortium Nature 491 (7422), 56–65.

Medini, D., Serruto, D., Parkhill, J., Relman, D.A., Donati, C., Moxon, R., et al., 2008. Microbiology in the post-genomic era. Nat. Rev. Microbiol. 6 (6), 419–430.

Metzker, M.L., 2010. Sequencing technologies—the next generation. Nat. Rev. Genet. 11 (1), 31–46.

Monchy, S., Sanciu, G., Jobard, M., Rasconi, S., Gerphagnon, M., Chabé, M., et al., 2011. Exploring and quantifying fungal diversity in freshwater lake ecosystems using rDNA cloning/sequencing and SSU tag pyrosequencing. Environ. Microbiol. 13 (6), 1433–1453.

Morgan, X.C., Segata, N., Huttenhower, C., 2013. Biodiversity and functional genomics in the human microbiome. Trends Genet. 29 (1), 51–58.

Morozova, O., Hirst, M., Marra, M.A., 2009. Applications of new sequencing technologies for transcriptome analysis. Annu. Rev. Genomics Hum. Genet. 10, 135–151.

Niedringhaus, T.P., Milanova, D., Kerby, M.B., Snyder, M.P., Barron, A.E., 2011. Landscape of next-generation sequencing technologies. Anal. Chem. 83 (12), 4327–4341.

Nikolaki, S., Tsiamis, G., 2013. Microbial diversity in the era of omic technologies. Biomed. Res. Int. 2013.

Nossa, C.W., Oberdorf, W.E., Yang, L., Aas, J.A., Paster, B.J., DeSantis, T.Z., et al., 2010. Design of 16S rRNA gene primers for 454 pyrosequencing of the human foregut microbiome. World J. Gastroenterol. 16 (33), 4135.

Oliver, G.R., Hart, S.N., Klee, E.W., 2015. Bioinformatics for clinical next generation sequencing. Clin. Chem. 61 (1), 124–135.

Ozsolak, F., Milos, P.M., 2011. RNA sequencing: advances, challenges and opportunities. Nat. Rev. Genet. 12 (2), 87–98.

Pareek, C.S., Smoczynski, R., Tretyn, A., 2011. Sequencing technologies and genome sequencing. J. Appl. Genet. 52 (4), 413–435.

Pedrós-Alió, C., 2006. Marine microbial diversity: can it be determined? Trends Microbiol. 14 (6), 257–263.

Podar, M., Abulencia, C.B., Walcher, M., Hutchison, D., Zengler, K., Garcia, J.A., et al., 2007. Targeted access to the genomes of low-abundance organisms in complex microbial communities. Appl. Environ. Microbiol. 73 (10), 3205–3214.

Quail, M.A., Smith, M., Coupland, P., Otto, T.D., Harris, S.R., Connor, T.R., et al., 2012. A tale of three next generation sequencing platforms: comparison of Ion Torrent, Pacific Bioscience and Illumina MiSeq Sequencers. BMC Genomics 13 (1), 1–13.

Raghunathan, A., Ferguson, H.R., Bornarth, C.J., Song, W., Driscoll, M., Lasken, R.S., 2005. Genomic DNA amplification from a single bacterium. Appl. Environ. Microbiol. 71 (6), 3342–3347.

Rajendhran, J., Gunasekaran, P., 2011. Microbial phylogeny and diversity: small subunit ribosomal RNA sequence analysis and beyond. Microbiol. Res. 166 (2), 99–110.

Rath, C.M., Janto, B., Earl, J., Ahmed, A., Hu, F.Z., Hiller, L., et al., 2011. Meta-omic characterization of the marine invertebrate microbial consortium that produces the chemotherapeutic natural product ET-743. ACS Chem. Biol. 6 (11), 1244–1256.

Rinke, C., Schwientek, P., Sczyrba, A., Ivanova, N.N., Anderson, I.J., Cheng, J.F., et al., 2013. Insights into the phylogeny and coding potential of microbial dark matter. Nature 499 (7459), 431–437.

Robertson, C.E., Baumgartner, L.K., Harris, J.K., Peterson, K.L., Stevens, M.J., Frank, D.N., et al., 2013. Culture-independent analysis of aerosol microbiology in a metropolitan subway system. Appl. Environ. Microbiol. 79 (11), 3485–3493.

Roesch, L.F., Fulthorpe, R.R., Riva, A., Casella, G., Hadwin, A.K., Kent, A.D., et al., 2007. Pyrosequencing enumerates and contrasts soil microbial diversity. ISME J. 1 (4), 283–290.

Ronaghi, M., Karamohamed, S., Pettersson, B., Uhlén, M., Nyrén, P., 1996. Real-time DNA sequencing using detection of pyrophosphate release. Anal. Biochem. 242 (1), 84–89.

Rothberg, J.M., Hinz, W., Rearick, T.M., Schultz, J., Mileski, W., Davey, M., et al., 2011. An integrated semiconductor device enabling non-optical genome sequencing. Nature 475 (7356), 348–352.

Rusch, D.B., Halpern, A.L., Sutton, G., Heidelberg, K.B., Williamson, S., Yooseph, S., et al., 2007. The Sorcerer II global ocean sampling expedition: northwest Atlantic through eastern tropical Pacific. PLoS Biol. 5 (3).

Sanger, F., Nicklen, S., Coulson, A.R., 1977. DNA sequencing with chain-terminating inhibitors. Proc. Natl. Acad. Sci. USA 74 (12), 5463–5467.

Schadt, E.E., Turner, S., Kasarskis, A., 2010. A window into third-generation sequencing. Hum. Mol. Genet. 19 (R2), R227–R240.

Schloss, P.D., Handelsman, J., 2003. Biotechnological prospects from metagenomics. Curr. Opin. Biotechnol. 14 (3), 303–310.

Schmidt, C., 2017. Living in a microbial world. Nat. Biotechnol. 35 (5), 401–403.

Shalem, O., Sanjana, N.E., Zhang, F., 2015. High-throughput functional genomics using CRISPR–Cas9. Nat. Rev. Genet. 16 (5), 299–311.

Shendure, J., Ji, H., 2008. Next-generation DNA sequencing. Nat. Biotechnol. 26 (10), 1135–1145.

Siegl, A., Kamke, J., Hochmuth, T., Piel, J., Richter, M., Liang, C., et al., 2011. Single-cell genomics reveals the lifestyle of Poribacteria, a candidate phylum symbiotically associated with marine sponges. ISME J. 5 (1), 61–70.

Soon, W.W., Hariharan, M., Snyder, M.P., 2013. High-throughput sequencing for biology and medicine. Mol. Syst. Biol. 9 (1), 640.

Staley, C., Unno, T., Gould, T.J., Jarvis, B., Phillips, J., Cotner, J.B., et al., 2013. Application of Illumina next-generation sequencing to characterize the bacterial community of the Upper Mississippi River. J. Appl. Microbiol. 115 (5), 1147–1158.

Stoeck, T., Bass, D., Nebel, M., Christen, R., Jones, M.D., Breiner, H.W., et al., 2010. Multiple marker parallel tag environmental DNA sequencing reveals a highly complex eukaryotic community in marine anoxic water. Mol. Ecol. 19, 21–31.

Suzuki, M.T., Giovannoni, S.J., 1996. Bias caused by template annealing in the amplification of mixtures of 16S rRNA genes by PCR. Appl. Environ. Microbiol. 62 (2), 625–630.

Tang, F., Barbacioru, C., Wang, Y., Nordman, E., Lee, C., Xu, N., et al., 2009. mRNA-Seq whole-transcriptome analysis of a single cell. Nat. Methods 6 (5), 377–382.

Thomas, T., Moitinho-Silva, L., Lurgi, M., et al., 2016. Diversity, structure and convergent evolution of the global sponge microbiome. Nat. Commun. 7, 11870.

Thompson, J.F., Steinmann, K.E., 2010. Single molecule sequencing with a HeliScope genetic analysis system. Curr. Protoc. Mol. Biol. 92 (1), 7–10.

Tiwari, S., Singh, C., Boudh, S., Rai, P.K., Gupta, V.K., Singh, J.S., 2019. Land use change: a key ecological disturbance declines soil microbial biomass in dry tropical uplands. J. Environ. Manage. 242, 1–10.

Torsvik, V., Øvreås, L., 2002. Microbial diversity and function in soil: from genes to ecosystems. Curr. Opin. Microbiol. 5 (3), 240–245.

Tringe, S.G., Mering, Von, Kobayashi, C., Salamov, A., Chen, A.A., Chang, K., et al., 2005. Comparative metagenomics of microbial communities. Science 308 (5721), 554–557.

Tsiamis, G., Katsaveli, K., Ntougias, S., Kyrpides, N., Andersen, G., Piceno, Y., et al., 2008. Prokaryotic community profiles at different operational stages of a Greek solar saltern. Res. Microbiol. 159 (9–10), 609–627.

Tsiamis, G., Tzagkaraki, G., Chamalaki, A., Xypteras, N., Andersen, G., Vayenas, D., et al., 2012. Olive-mill wastewater bacterial communities display a cultivar specific profile. Curr. Microbiol. 64 (2), 197–203.

Tyson, G.W., Chapman, J., Hugenholtz, P., Allen, E.E., Ram, R.J., Richardson, P.M., et al., 2004. Community structure and metabolism through reconstruction of microbial genomes from the environment. Nature 428 (6978), 37–43.

Vasileiadis, S., Puglisi, E., Arena, M., Cappa, F., Cocconcelli, P.S., Trevisan, M., 2012. Soil bacterial diversity screening using single 16S rRNA gene V regions coupled with multi-million read generating sequencing technologies. PLoS One 7 (8).

Vasileiadis, S., Puglisi, E., Arena, M., Cappa, F., Van Veen, J.A., Cocconcelli, P.S., et al., 2013. Soil microbial diversity patterns of a lowland spring environment. FEMS Microbiol. Ecol. 86 (2), 172–184.

Venter, J.C., Adams, M.D., Myers, E.W., Li, P.W., Mural, R.J., Sutton, G.G., et al., 2001. The sequence of the human genome. Science 291 (5507), 1304–1351.

Venter, J.C., Remington, K., Heidelberg, J.F., Halpern, A.L., Rusch, D., Eisen, J.A., et al., 2004. Environmental genome shotgun sequencing of the Sargasso Sea. Science 304 (5667), 66–74.

Vieites, J.M., Guazzaroni, M.E., Beloqui, A., Golyshin, P.N., Ferrer, M., 2008. Metagenomics approaches in systems microbiology. FEMS Microbiol. Rev. 33 (1), 236–255.

Wang, Y., Qian, P.Y., 2009. Conservative fragments in bacterial 16S rRNA genes and primer design for 16S ribosomal DNA amplicons in metagenomic studies. PLoS One 4 (10).

Wang, Z., Gerstein, M., Snyder, M., 2009. RNA-Seq: a revolutionary tool for transcriptomics. Nat. Rev. Genet. 10 (1), 57–63.

Wang, B., Wang, J., Zhang, W., Meldrum, D.R., 2012. Application of synthetic biology in cyanobacteria and algae. Front. Microbiol. 3, 344.

Wang, Q., Lu, Q., Zhao, H., 2015. A review of study designs and statistical methods for genomic epidemiology studies using next generation sequencing. Front. Genet. 6, 149.

Weinstock, G.M., 2012. Genomic approaches to studying the human microbiota. Nature 489 (7415), 250–256.

Whiteford, N., Skelly, T., Curtis, C., Ritchie, M.E., Löhr, A., Zaranek, A.W., et al., 2009. Swift: primary data analysis for the Illumina Solexa sequencing platform. Bioinformatics 25 (17), 2194–2199.

Woyke, T., Tighe, D., Mavromatis, K., Clum, A., Copeland, A., Schackwitz, W., et al., 2010. One bacterial cell, one complete genome. PLoS One 5 (4).

Wu, L., Thompson, D.K., Li, G., Hurt, R.A., Tiedje, J.M., Zhou, J., 2001. Development and evaluation of functional gene arrays for detection of selected genes in the environment. Appl. Environ. Microbiol. 67 (12), 5780–5790.

Wu, D., Hugenholtz, P., Mavromatis, K., Pukall, R., Dalin, E., Ivanova, N.N., et al., 2009. A phylogeny-driven genomic encyclopaedia of Bacteria and Archaea. Nature 462 (7276), 1056–1060.

Xie, J., He, Z., Liu, X., Liu, X., Van Nostrand, J.D., Deng, Y., et al., 2011. GeoChip-based analysis of the functional gene diversity and metabolic potential of microbial communities in acid mine drainage. Appl. Environ. Microbiol. 77 (3), 991–999.

Yoon, H.S., Price, D.C., Stepanauskas, R., Rajah, V.D., Sieracki, M.E., Wilson, W.H., et al., 2011. Single-cell genomics reveals organismal interactions in uncultivated marine protists. Science 332 (6030), 714–717.

Zhang, X., Xia, P., Wang, P., Yang, J., Baird, D.J., 2018. Omics advances in ecotoxicology. Environ. Sci. Technol. 52, 3842–3851.

Zhou, J., Kang, S., Schadt, C.W., Garten, C.T., 2008. Spatial scaling of functional gene diversity across various microbial taxa. Proc. Natl. Acad. Sci. USA 105 (22), 7768–7773.

Zhou, J., Wu, L., Deng, Y., Zhi, X., Jiang, Y.H., Tu, Q., et al., 2011. Reproducibility and quantitation of amplicon sequencing-based detection. ISME J. 5 (8), 1303–1313.

Zhu, T., 2003. Global analysis of gene expression using GeneChip microarrays. Curr. Opin. Plant Biol. 6 (5), 418–425.

Index

Note: Page numbers followed by "*f*" and "*t*" refer to figures and tables, respectively.

F

S

FSC
www.fsc.org
MIX
Papier | Fördert
gute Waldnutzung
FSC® C083411

Kolibri 360 GmbH
Ferdinand-Jühlke-Straße 7
99095 Erfurt, Deutschland
produktsicherheit@kolibri360.de